T0331263

COMPUTATIONAL
NUMBER THEORY

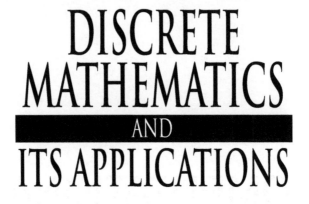

DISCRETE MATHEMATICS AND ITS APPLICATIONS

Series Editor

Kenneth H. Rosen, Ph.D.

Titles (continued)

Richard A. Mollin, Algebraic Number Theory, Second Edition

Richard A. Mollin, Codes: The Guide to Secrecy from Ancient to Modern Times

Richard A. Mollin, Fundamental Number Theory with Applications, Second Edition

Richard A. Mollin, An Introduction to Cryptography, Second Edition

Richard A. Mollin, Quadratics

Richard A. Mollin, RSA and Public-Key Cryptography

Carlos J. Moreno and Samuel S. Wagstaff, Jr., Sums of Squares of Integers

Goutam Paul and Subhamoy Maitra, RC4 Stream Cipher and Its Variants

Dingyi Pei, Authentication Codes and Combinatorial Designs

Kenneth H. Rosen, Handbook of Discrete and Combinatorial Mathematics

Douglas R. Shier and K.T. Wallenius, Applied Mathematical Modeling: A Multidisciplinary Approach

Alexander Stanoyevitch, Introduction to Cryptography with Mathematical Foundations and Computer Implementations

Jörn Steuding, Diophantine Analysis

Douglas R. Stinson, Cryptography: Theory and Practice, Third Edition

Roberto Togneri and Christopher J. deSilva, Fundamentals of Information Theory and Coding Design

W. D. Wallis, Introduction to Combinatorial Designs, Second Edition

W. D. Wallis and J. C. George, Introduction to Combinatorics

Jiacun Wang, Handbook of Finite State Based Models and Applications

Lawrence C. Washington, Elliptic Curves: Number Theory and Cryptography, Second Edition

DISCRETE MATHEMATICS AND ITS APPLICATIONS

Series Editor KENNETH H. ROSEN

COMPUTATIONAL NUMBER THEORY

ABHIJIT DAS

CRC Press
Taylor & Francis Group
Boca Raton London New York

CRC Press is an imprint of the
Taylor & Francis Group, an **informa** business

A CHAPMAN & HALL BOOK

CRC Press
Taylor & Francis Group
6000 Broken Sound Parkway NW, Suite 300
Boca Raton, FL 33487-2742

© 2013 by Taylor & Francis Group, LLC
CRC Press is an imprint of Taylor & Francis Group, an Informa business

No claim to original U.S. Government works

ISBN-13: 978-1-4398-6615-3 (hbk)

Visit the Taylor & Francis Web site at
http://www.taylorandfrancis.com

and the CRC Press Web site at
http://www.crcpress.com

Dedicated to

C. E. Veni Madhavan

Contents

Preface

This book is a result of my teaching a masters-level course with the same name for five years at the Indian Institute of Technology Kharagpur. The course was attended mostly by MTech and final-year BTech students from the Department of Computer Science and Engineering. Students from the Department of Mathematics and other engineering departments (mostly Electronics and Electrical Engineering, and Information Technology) also attended the course. Some research students enrolled in the MS and PhD programs constituted the third section of the student population. Historically, therefore, the material presented in this book is designed to cater to the need and taste of engineering students in advanced undergraduate and beginning graduate levels. However, several topics that could not be covered in a one-semester course have also been included in order to make this book a comprehensive and complete treatment of number-theoretic algorithms.

A justification is perhaps needed to explain why *another* textbook on computational number theory is necessary. Some (perhaps not many) textbooks on this subject are already available to international students. These books vary widely with respect to their coverage and technical sophistication. I believe that a textbook specifically targeting the engineering population is missing. This book should be accessible (but is not restricted) to students who have not attended any course on number theory. My teaching experience shows that heavy use of algebra (particularly, advanced topics like commutative algebra or algebraic number theory) often demotivates students. While I have no intention to underestimate the role played by algebra in number theory, I believe that a large part of number theory can still reach students not conversant with sophisticated algebraic tools. It is, of course, meaningless to avoid algebra altogether. For example, when one talks about finite fields or elliptic curves, one expects the audience to be familiar with the notion of basic algebraic structures like groups, rings and fields (and some linear algebra, too). But that is all that I assume on the part of the reader. Although I made an attempt to cover this basic algebra in an appendix, the concise appendix is perhaps more suited as a reference than as a tool to learn the subject. Likewise, students who have not attended a course on algorithms may pick up the basic terminology (asymptotic notations, types of algorithms, complexity classes) from another appendix. Any sophisticated topic has been treated in a self-contained manner in the main body of the text. For example, some basic algebraic geometry (projective curves, rational functions, divisors) needed for understanding pairing has been developed from scratch. It is not a book on only computational aspects of elementary number theory. Instead its treatment is kept as elementary as possible.

This book is not above the allegation that it is meant for cryptographers. As a practitioner of cryptography, I do not fully deny this allegation. Having said that, I would also add that cryptography and cryptanalysis make heavy use of many important number-theoretic tools. Prime numbers and integer factorization are as important in the fundamental theorem of (computational) number theory as they are in the RSA cryptosystem. It is difficult to locate cryptography-free corners in computational number theory. Some fun stuff in number theory is omitted, like calculation of the digits of π, generalized Mersenne and Fermat numbers, and numerous simply stated but hard-to-prove conjectures of older origins. If one equates dubbing these issues as nonserious with my affinity to cryptology, I am not going to debate. It appears too subjective to tell what is important and fundamental from what is not. However, a dedicated chapter on applications of number theory in public-key cryptography is a product of my background. I think that this book without this chapter still makes sense, and there is no harm in highlighting practically important and interesting engineering applications of the material with which the remaining chapters deal. Arguably, privacy is as ancient as number theory, and when they go hand in hand, I should not be blamed for pointing that out. On the contrary, inclusion of recent developments in pairing-based cryptography is expected to be a value added to this book.

Emphasis on implementation issues is another distinctive feature of this book. Arithmetic of integers and polynomials is covered at the very basic level. The rest of computational number theory is built on top of that. It, however, appears too annoying to invoke low-level subroutines for complicated algorithms. Consequently, the freely available number-theory calculator GP/PARI has been taken up as the medium to demonstrate arithmetic computations. The reader may wonder why GP/PARI and not sage has been promoted as the demonstration environment. A partial justification is this: Work on this book started in 2006. It was a time when sage existed but had not gained its present popularity. Although sage, being a collection of all freely available mathematical utilities including GP/PARI itself, is more efficient and versatile than each of its individual components, GP/PARI is still a convenient and low-footprint package sufficient for most of this book.

Many examples and exercises accompany the technical presentation. While the examples are mostly illustrative in nature, the exercises can be broadly classified in two categories. Some of them are meant for deepening the understanding of the material presented in the chapters and for filling out certain missing details. The rest are used to develop additional theory which I could not cover in the main text of the book. No attempts have been made to classify exercises as easy or difficult.

Proofs of theorems, propositions, and correctness of algorithms are in many places omitted in the book, particularly when these proofs are long and/or involved and/or too sophisticated. Although this omission may alienate readers from mathematical intricacies, I believe that the risk is not beyond control. After all, every author of a book has to make a compromise among the bulk, the coverage, and the details. I achieved this in a way I found most suitable.

I have not made an attempt to formally cite every contribution discussed in the text. Some key references are presented as on-line comments and/or footnotes. I personally find citations like [561] or [ABD2c] rather distracting, suited to technical papers and research monographs, not to a textbook.

I am not going to describe the technical organization of this book. The table of contents already accomplishes this task. I instead underline the impossibility of covering the entire material of this book in a standard three-to-four hour per week course in a semester (or quarter). Chapters 1 and 2 form the backbone of computational number theory, and may be covered in the first half of a course. In the second half, the instructor may choose from a variety of topics. The most reasonable coverage is that of Chapters 5 and 6, followed, if time permits, by excerpts from Chapters 3 and/or 7. A second course might deal with the rest of the book. A beginners' course on elliptic curves may concentrate on Chapters 1, 2 and 4. Suitable portions from Chapters 1, 2, 5, 6 and 9 make a course on introductory public-key cryptology. The entire book is expected to be suitable for self study, for students starting a research career in this area, and for practitioners of cryptography in industry.

While efforts have been made to keep this book as error-free as possible, complete elimination of errors is a dream for which any author can hope. The onus lies on the readers, too, to detect errors and omissions at any level, typographical to conceptual to philosophical. Any suggestion will improve future editions of this book. I can be reached at abhij@cse.iitkgp.ernet.in and also at SadTijihba@gmail.com.

No project like authoring this book can be complete without the active help and participation of others. No amount of words suffice to describe the contribution of my PhD supervisor C. E. Veni Madhavan. It is he who introduced me to the wonderful world of computational number theory and thereby changed the course of my life forever. I will never forget the days of working with him as his student on finite-field arithmetic and the discrete-logarithm problem. Those were, without any shred of doubt, the sweetest days in my academic life. Among my other teachers, A. K. Nandakumaran, Basudeb Datta, Dilipkumar Premchand Patil, Sathya S. Keerthi and Vijay Chandru, all from the Indian Institute of Science, Bangalore, deserve specific mention for teaching me various aspects of pure and applied mathematics. I also gratefully acknowledge Tarun Kumar Mukherjee from Jadavpur University, Calcutta, who inculcated in me a strong affinity for mathematics in my undergraduate days. My one-year stay with Uwe Storch and Hartmut Wiebe at the Ruhr-Universität in Bochum, Germany, was a mathematically invigorating experience.

In the early 2000s, some of my colleagues in IIT Kharagpur developed a taste for cryptography, and I joined this research group with an eye on public-key algorithms. I have been gladly supplementing their areas of interest in symmetric cryptology and hardware-based implementations of crypto protocols. Their constructive suggestions for this book have always been an asset to me. To Debdeep Mukhopadhyay, Dipanwita Roy Choudhury, Indranil Sengupta and Rajat Subhra Chakraborty, thanks a lot to all of you. I must also thank another colleague, Goutam Biswas (he is not a cryptologist though), who has taken up the responsibility of continuing to teach the course after me. He referred to this book quite often, and has pointed out many errors and suggestions for its improvement. Jayanta Mukherjee, neither a cryptologist nor any type of number theorist, is also gratefully acknowledged for setting up my initial contact with CRC Press. The remaining faculty members of my department, too, must be thanked for always extending to me the requisite help and moral support toward the completion of this book.

I express my gratitude to Bimal Kumar Roy, Kishan Chand Gupta, Palash Sarkar, Rana Barua and Subhomoy Maitra of the Indian Statistical Institute, Calcutta, for our frequent exchange of ideas. I acknowledge various forms of number-theoretic and cryptographic interactions with Debasis Giri from the Haldia Institute of Technology; Chandan Mazumdar, Goutam Paul and Indranath Sengupta from Jadavpur University; Gagan Garg from IIT Mandi; Sugata Gangopadhyay from ISI Chennai; Sanjay Burman from DRDO; and Aravind Iyer and Bhargav Bellur from General Motors, ISL, Bangalore.

No course or book is complete without the audience—the students. It is a group of them whose active persuasion let me introduce *Computational Number Theory* to our list of masters-level elective courses. In my five years of teaching, this course gained significant popularity, the semester registration count rocketing from 10 to over 100. I collectively thank this entire student population here. Special thanks are due to Angshuman Karmakar, Aniket Nayak, Anup Kumar Bhattacharya, Binanda Sengupta, Debajit Dey, Debojyoti Bhattacharya, Mathav Kishore, Pratyay Mukherjee, Rishiraj Bhattacharyya, Sabyasachi Karati, Satrajit Ghosh, Somnath Ghosh, Souvik Bhattacherjee, Sourav Basu, Sourav Sen Gupta, Utsab Bose and Yatendra Dalal, whose association with the book at all stages of its conception has been a dream to cherish forever.

Anonymous referees arranged by CRC Press provided invaluable suggestions for improving the content and presentation of this book. The publishing team at CRC Press deserves their share of thanks for always being nice and friendly to me. Last, but not the least, I must acknowledge the support and encouragement of my parents and my brother.

Abhijit Das
Kharagpur

Chapter 1

Arithmetic of Integers

Loosely speaking, number theory deals with the study of integers, also called whole numbers. It is an ancient branch of mathematics, that has enjoyed study for millenniums and attracted a variety of researchers ranging from professional to amateur. Recently, in particular, after the invention of public-key cryptology, number theory has found concrete engineering applications and has turned out to be an interesting and important area of study for computer scientists and engineers and also for security experts. Several outstanding computational challenges pertaining to the theory of numbers continue to remain unsolved and are expected to boost fascinating research in near future.

Let me reserve some special symbols in the blackboard-bold font to denote the following important sets.

$$\begin{aligned}
\mathbb{N} &= \{1, 2, 3, \ldots\} = \text{the set of natural numbers,} \\
\mathbb{N}_0 &= \{0, 1, 2, 3, \ldots\} = \text{the set of non-negative integers,} \\
\mathbb{Z} &= \{\ldots, -3, -2, -1, 0, 1, 2, 3, \ldots\} = \text{the set of all integers,} \\
\mathbb{Q} &= \left\{\frac{a}{b} \mid a \in \mathbb{Z}, b \in \mathbb{N}\right\} = \text{the set of rational numbers,} \\
\mathbb{R} &= \text{the set of real numbers,} \\
\mathbb{C} &= \text{the set of complex numbers,} \\
\mathbb{P} &= \{2, 3, 5, 7, 11, 13, \ldots\} = \text{the set of (positive) prime numbers.}^{[1]}
\end{aligned}$$

1.1 Basic Arithmetic Operations

All common programming languages provide built-in data types for doing arithmetic on integers. Usually, each such data type is of a fixed limited size and so can store only finitely many possible integer values. For example, the data type `int` provided in C is typically a 32-bit signed integer value capable of representing integers in the range $-2^{31}, \ldots, 2^{31} - 1$. If we want to store bigger integers, we may use floating-point data types, but doing so leads to loss of precision. The integer 100! requires 158 decimal digits or 525 bits for an exact representation. A truncated floating-point number like $9.332621544394 \times 10^{157}$ is only an approximate representation of 100! (for that matter, also of 100!+1, and even of 100! + 50!). In number theory, we need to store the exact values of big integers, that is, 100! should be stored as

$$\begin{aligned}
100! \ = \ & 93326215443944152681699238856266700490715968264381621 \\
& 46859296389521759999322991560894146397615651828625369 \\
& 7920827223758251185210916864000000000000000000000000.
\end{aligned}$$

User-defined data types are needed to this effect. Moreover, it is necessary to write specific subroutines to implement the standard arithmetic operations $(+,-,*,/,\%)$ suitable for these data types.[2]

---

[1] The negative primes $-2, -3, -5, -7, -11, -13, \ldots$ are primes in an algebraic sense.

[2] Donald Ervin Knuth (1938–) provides a comprehensive treatment of integers and floating-point numbers of arbitrary precisions in his second volume (*Seminumerical Algorithms*, Chapter 4) of *The Art of Computer Programming*.

1.1.1 Representation of Big Integers

It is customary to express a big integer n in some predetermined base B, and store the B-ary digits of n in an array of integers. The base B should be so chosen that each B-ary digit can fit in a single built-in integer data type. In order to maximize efficiency, it is worthwhile to take B as a power of 2 and as large as possible. In 32-bit machines, the typical choice is $B = 2^{32}$, for which a B-ary digit (a value among $0, 1, 2, \ldots, 2^{32} - 1$) fits in a 32-bit unsigned integer variable. In 64-bit machines, it is preferable to choose the base $B = 2^{64}$. We refer to an integer in such a representation as a *multiple-precision integer*.

Example 1.1 For the purpose of illustration, I choose a small base $B = 2^8 = 256$ in this and many subsequent examples. For this choice, a B-ary digit is a value between 0 and 255 (both inclusive), and can be stored in an eight-bit unsigned integer variable (like `unsigned char` in C).

Let $n = 12345678987654321$. The base-256 expansion of n is

$$n = 12345678987654321$$
$$= 43 \times B^6 + 220 \times B^5 + 84 \times B^4 + 98 \times B^3 + 145 \times B^2 + 244 \times B + 177$$
$$= (43, 220, 84, 98, 145, 244, 177)_B$$

Here, 43 is the most significant (or leftmost) digit, whereas 177 is the least significant (or rightmost) digit. (In general, an integer n having the base-B representation $(n_{s-1}, n_{s-2}, \ldots, n_1, n_0)_B$ with $n_{s-1} \neq 0$ needs s B-ary digits with n_{s-1} and n_0 being respectively the most and least significant digits.) An array of size seven suffices to store n of this example. While expressing an integer in some base, it is conventional to write the most significant digit first. On the other hand, if one stores n in an array with zero-based indexing, it is customary to store n_0 in the zeroth index, n_1 in the first index, and so on. For example, the above seven-digit number has the following storage in an array.

Array index	0	1	2	3	4	5	6
Digit	177	244	145	98	84	220	43

The sign of an integer can be stored using an additional bit with 0 meaning positive and 1 meaning negative. Negative integers may be represented in the standard 1's-complement or 2's-complement format. In what follows, I will stick to the signed magnitude representation only. □

1.1.1.1 Input and Output

Users prefer to input/output integers in the standard decimal notation. A long integer is typically typed as a string of decimal digits. It is not difficult to convert a string of decimal digits to any base-B representation. The converse transformation too is not difficult. I am illustrating these conversion procedures using a couple of examples.

Suppose that the user reads the input string $d_0 d_1 d_2 \ldots d_{t-1}$, where each d_i is a decimal digit (0–9). This is converted to the base-B representation $(n_{s-1}, n_{s-2}, \ldots, n_1, n_0)_B$. Call this integer n. Now, let another digit d_t come in the input string, that is, we now need to represent the integer n' whose decimal representation is $d_0 d_1 d_2 \ldots d_{t-1} d_t$. We have $n' = 10n + d_t$. This means that we multiply the representation $(n_{s-1}, n_{s-2}, \ldots, n_1, n_0)_B$ by ten and add d_t to the least significant digit. Multiplication and addition algorithms are described in the next section. For the time being, it suffices to understand that the string to base-B conversion boils down to a sequence of elementary arithmetic operations on multiple-precision integers.

Example 1.2 We convert the string "123454321" to an integer in base 256.

Input digit	Operation	B-ary representation
	Initialization	$()_B$
1	Multiply by 10	$()_B$
	Add 1	$(1)_B$
2	Multiply by 10	$(10)_B$
	Add 2	$(12)_B$
3	Multiply by 10	$(120)_B$
	Add 3	$(123)_B$
4	Multiply by 10	$(4, 206)_B$
	Add 4	$(4, 210)_B$
5	Multiply by 10	$(48, 52)_B$
	Add 5	$(48, 57)_B$
4	Multiply by 10	$(1, 226, 58)_B$
	Add 4	$(1, 226, 62)_B$
3	Multiply by 10	$(18, 214, 108)_B$
	Add 3	$(18, 214, 111)_B$
2	Multiply by 10	$(188, 96, 86)_B$
	Add 2	$(188, 96, 88)_B$
1	Multiply by 10	$(7, 91, 195, 112)_B$
	Add 1	$(7, 91, 195, 113)_B$

The base-256 representation of 123454321 is, therefore, $(7, 91, 195, 113)_{256}$. □

For the sake of efficiency, one may process, in each iteration, multiple decimal digits from the input. For example, if n'' has the decimal representation $d_0 d_1 d_2 \ldots d_{t-1} d_t d_{t+1}$, we have $n'' = 100n + (d_t d_{t+1})_{10} = 100n + (10 d_t + d_{t+1})$. In fact, the above procedure can easily handle chunks of k digits simultaneously from the input so long as $10^k < B$.

The converse transformation is similar, albeit a little more involved. Now, we have to carry out arithmetic in a base B' which is a suitable integral power of ten. For example, we may choose $B' = 10^l$ with $10^l < B < 10^{l+1}$. First, we express the representation base B in the base B' used for output: $B = HB' + L$ with $H, B \in \{0, 1, \ldots, B' - 1\}$. Let $n = (n_{s-1}, n_{s-2}, \ldots, n_1, n_0)_B$ be available in the base-B representation. Denote $N_i = (n_{s-1}, n_{s-2}, \ldots, n_i)_B$. We have $N_s = 0$ and $N_0 = n$. We iteratively compute the base-B' representations of $N_{s-1}, N_{s-2}, \ldots, N_0$ with the initialization $N_s = (\,)_{B'}$. We have $N_i = N_{i+1}B + n_i = N_{i+1}(HB'+L)+n_i$. Since n_i is a B-ary digit, it may consist of two B'-ary digits, so we write $n_i = h_iB'+l_i$. This gives $N_i = N_{i+1}(HB'+L)+(h_iB'+l_i)$. Therefore, from the base-B' representation of N_{i+1}, we can compute the base-B' representation of N_i using multiple-precision arithmetic under base B'.

Example 1.3 Like the previous two examples, we choose the base $B = 256$. The base for decimal output can be chosen as $B' = 100$. We, therefore, need to convert integers from the base-256 representation to the base-100 representation. We write $B = 2 \times B' + 56$, that is, $H = 2$ and $L = 56$. For the input $n = (19, 34, 230, 113)_B$, the conversion procedure works as follows.

i	B-ary digit n_i	h_i	l_i	Operation	B'-ary representation
4				Initialization	$(\,)_{B'}$
3	19	0	19	Multiply by 256	$(\,)_{B'}$
				Add 19	$(19)_{B'}$
2	34	0	34	Multiply by 256	$(48, 64)_{B'}$
				Add 34	$(48, 98)_{B'}$
1	230	2	30	Multiply by 256	$(1, 25, 38, 88)_{B'}$
				Add 230	$(1, 25, 41, 18)_{B'}$
0	113	1	13	Multiply by 256	$(3, 21, 5, 42, 8)_{B'}$
				Add 113	$(3, 21, 5, 43, 21)_{B'}$

If we concatenate the B'-ary digits from the most significant end to the least significant end, we obtain the decimal representation of n. There is only a small catch here. The digit 5 should be printed as 05, that is, each digit (except the most significant one) in the base $B' = 10^l$ must be printed as an l-digit integer after padding with the requisite number of leading zeros (whenever necessary). In the example above, l is 2, so the 100-ary digits 21, 43 and 21 do not require this padding, whereas 5 requires this. The most significant digit 3 too does not require a leading zero (although there is no harm if one prints it). To sum up, we have $n = (19, 34, 230, 113)_B = 321054321$. □

1.1.2 Schoolbook Arithmetic

We are used to doing arithmetic operations (addition, subtraction, multiplication and Euclidean division) on integers in the standard decimal representation. When we change the base of representation from ten to any other base B (a power of two or a power of ten), the basic procedures remain the same.

1.1.2.1 Addition

Let two multiple-precision integers $a = (a_{s-1}, a_{s-2}, \ldots, a_1, a_0)_B$ and $b = (b_{t-1}, b_{t-2}, \ldots, b_1, b_0)_B$ be available in the base-B representation. Without loss of generality, we may assume that $s = t$ (if not, we pad the smaller operand with leading zero digits). We keep on adding $a_i + b_i$ with carry adjustments in the sequence $i = 0, 1, 2, \ldots$.

A small implementation-related issue needs to be addressed in this context. Suppose that we choose a base $B = 2^{32}$ in a 32-bit machine. When we add a_i and b_i (possibly also an input carry), the sum may be larger than $2^{32} - 1$ and can no longer fit in a 32-bit integer. If such a situation happens, we know that the output carry is 1 (otherwise, it is 0). But then, how can we detect whether an overflow has occurred, particularly if we work only with high-level languages (without access to assembly-level instructions)? A typical behavior of modern computers is that whenever the result of some unsigned arithmetic operation is larger than $2^{32} - 1$ (or $2^{64} - 1$ for a 64-bit machine), the least significant 32 (or 64) bits are returned. In other words, addition, subtraction and multiplication of 32-bit (or 64-bit) unsigned integers are actually computed modulo 2^{32} (or 2^{64}). Based upon this assumption about our computer, we can detect overflows without any assembly-level support.

First, suppose that the input carry is zero, that is, we add two B-ary digits a_i and b_i only. If there is no overflow, the modulo-B sum $a_i + b_i \pmod{2^{32}}$ is not smaller than a_i and b_i. On the other hand, if an overflow occurs, the sum $a_i + b_i \pmod{2^{32}}$ is smaller than both a_i and b_i. Therefore, by inspecting the return value of the high-level sum, one can deduce the output carry. If the input carry is one, $a_i + b_i + 1 \pmod{2^{32}}$ is at most as large as a_i if and only if an overflow occurs. Thus, if there is no overflow, this sum has to be larger than a_i. This observation lets us detect the presence of the output carry.

Example 1.4 Consider the two operands in base $B = 256 = 2^8$:

$$a = 9876543210 = (2, 76, 176, 22, 234)_B,$$
$$b = 1357902468 = (80, 239, 242, 132)_B.$$

Word-by-word addition proceeds as follows. We use c_{in} and c_{out} to denote the input and the output carries, respectively.

i	a_i	b_i	c_{in}	$a_i + b_i + c_{in} \pmod{256}$	Carry?	c_{out}
0	234	132	0	110	$(110 < 234)$?	1
1	22	242	1	9	$(9 \leqslant 22)$?	1
2	176	239	1	160	$(160 \leqslant 176)$?	1
3	76	80	1	157	$(157 \leqslant 76)$?	0
4	2	0	0	2	$(2 < 2)$?	0

It therefore follows that $a + b = (2, 157, 160, 9, 110)_B = 11234445678$.   □

1.1.2.2   Subtraction

The procedure for the subtraction of multiple-precision integers is equally straightforward. In this case, we need to handle the borrows. The input and output borrows are denoted, as before, by c_{in} and c_{out}. The check whether computing $a_i - b_i - c_{in}$ results in an output borrow can be performed before the operation is carried out. If $c_{in} = 0$, the output borrow c_{out} is one if and only if $a_i < b_i$. On the other hand, for $c_{in} = 1$, the output borrow is one if and only if $a_i \leqslant b_i$. Even in the case that there is an output borrow, one may blindly compute the mod B operation $a_i - b_i - c_{in}$ and keep the returned value as the output word, provided that the CPU supports 2's-complement arithmetic. If not, $a_i - b_i - c_{in}$ may be computed as $a_i + (B - b_i - c_{in})$ with $B - b_i - c_{in}$ computed using appropriate bit operations. More precisely, $B - b_i - 1$ is the bit-wise complement of b_i, and $B - b_i$ is one more than that.

Example 1.5 Let us compute $a - b$ for the same operands as in Example 1.4.

i	a_i	b_i	c_{in}	Borrow?	c_{out}	$a_i - b_i - c_{in} \pmod{256}$
0	234	132	0	$(234 < 132)$?	0	102
1	22	242	0	$(22 < 242)$?	1	36
2	176	239	1	$(176 \leqslant 239)$?	1	192
3	76	80	1	$(76 \leqslant 80)$?	1	251
4	2	0	1	$(2 \leqslant 0)$?	0	1

We obtain $a - b = (1, 251, 192, 36, 102)_B = 8518640742$. In this example, $a > b$. If $a < b$, then $a - b$ can be computed as $-(b - a)$. While performing addition and subtraction of signed multiple-precision integers, a subroutine for comparing the absolute values of two operands turns out to be handy. □

1.1.2.3 Multiplication

Multiplication of two multiple-precision integers is somewhat problematic. Like decimal (or polynomial) multiplication, one may multiply every word of the first operand with every word of the second. But a word-by-word multiplication may lead to a product as large as $(B-1)^2$. For $B = 2^{32}$, the product may be a 64-bit value, whereas for $B = 2^{64}$, the product may be a 128-bit value. Correctly obtaining all these bits is trickier than a simple check for an output carry or borrow as we did for addition and subtraction.

Many compilers support integer data types (perhaps non-standard) of size twice the natural word size of the machine. If so, one may use this facility to compute the double-sized intermediate products. Assembly-level instructions may also allow one to retrieve all the bits of word-by-word products. If neither of these works, one possibility is to break each operand word in two half-sized integers. That is, one writes $a_i = h_i\sqrt{B} + l_i$ and $b_j = h'_j\sqrt{B} + l'_j$, and computes $a_i b_j$ as $h_i h'_j B + (h_i l'_j + l_i h'_j)\sqrt{B} + l_i l'_j$. Here, $h_i h'_j$ contributes only to the more significant word of $a_i b_j$, and $l_i l'_j$ to only the less significant word. But $h_i l'_j$ and $l_i h'_j$ contribute to both the words. With appropriate bit-shift and extraction operations, these contributions can be separated and added to appropriate words of the product. When $a_i b_j$ is computed as $hB + l = (h, l)_B$, the less significant word l is added to the $(i + j)$-th position of the output, whereas h is added to the $(i + j + 1)$-st position. Each such addition may lead to a carry which needs to be propagated to higher positions until the carry is absorbed in some word of the product.

Example 1.6 We compute the product of the following two operands available in the representation to base $B = 2^8 = 256$.

$$a = 1234567 = (18, 214, 135)_B,$$
$$b = 76543210 = (4, 143, 244, 234)_B.$$

The product may be as large as having $3 + 4 = 7$ B-ary words. We initialize the product c as an array of seven eight-bit values, each initialized to zero. In the following table, c is presented in the B-ary representation with the most significant digit written first.

i	a_i	j	b_j	$a_ib_j = (h,l)_B$	Operation			c			
					Initialization	$(0,$	$0,$	$0,$	$0,$	$0,$	$0,$	$0)_B$
0	135	0	234	$(123, 102)_B$	Add 102 at pos 0	$(0,$	$0,$	$0,$	$0,$	$0,$	$0, 102)_B$
					Add 123 at pos 1	$(0,$	$0,$	$0,$	$0,$	$0, 123, 102)_B$
		1	244	$(128, 172)_B$	Add 172 at pos 1	$(0,$	$0,$	$0,$	$0,$	$1,$	$39, 102)_B$
					Add 128 at pos 2	$(0,$	$0,$	$0,$	$0, 129,$	$39, 102)_B$
		2	143	$(75, 105)_B$	Add 105 at pos 2	$(0,$	$0,$	$0,$	$0, 234,$	$39, 102)_B$
					Add 75 at pos 3	$(0,$	$0,$	$0,$	$75, 234,$	$39, 102)_B$
		3	4	$(2, 28)_B$	Add 28 at pos 3	$(0,$	$0,$	$0, 103, 234,$	$39, 102)_B$
					Add 2 at pos 4	$(0,$	$0,$	$2, 103, 234,$	$39, 102)_B$
1	214	0	234	$(195, 156)_B$	Add 156 at pos 1	$(0,$	$0,$	$2, 103, 234, 195, 102)_B$
					Add 195 at pos 2	$(0,$	$0,$	$2, 104, 173, 195, 102)_B$
		1	244	$(203, 248)_B$	Add 248 at pos 2	$(0,$	$0,$	$2, 105, 165, 195, 102)_B$
					Add 203 at pos 3	$(0,$	$0,$	$3,$	$52, 165, 195, 102)_B$
		2	143	$(119, 138)_B$	Add 138 at pos 3	$(0,$	$0,$	$3, 190, 165, 195, 102)_B$
					Add 119 at pos 4	$(0,$	$0, 122, 190, 165, 195, 102)_B$
		3	4	$(3, 88)_B$	Add 88 at pos 4	$(0,$	$0, 210, 190, 165, 195, 102)_B$
					Add 3 at pos 5	$(0,$	$3, 210, 190, 165, 195, 102)_B$
2	18	0	234	$(16, 116)_B$	Add 116 at pos 2	$(0,$	$3, 210, 191,$	$25, 195, 102)_B$
					Add 16 at pos 3	$(0,$	$3, 210, 207,$	$25, 195, 102)_B$
		1	244	$(17, 40)_B$	Add 40 at pos 3	$(0,$	$3, 210, 247,$	$25, 195, 102)_B$
					Add 17 at pos 4	$(0,$	$3, 227, 247,$	$25, 195, 102)_B$
		2	143	$(10, 14)_B$	Add 14 at pos 4	$(0,$	$3, 241, 247,$	$25, 195, 102)_B$
					Add 10 at pos 5	$(0, 13, 241, 247,$	$25, 195, 102)_B$
		3	4	$(0, 72)_B$	Add 72 at pos 5	$(0, 85, 241, 247,$	$25, 195, 102)_B$
					Add 0 at pos 6	$(0, 85, 241, 247,$	$25, 195, 102)_B$

The product of a and b is, therefore, $c = ab = (0, 85, 241, 247, 25, 195, 102)_B = (85, 241, 247, 25, 195, 102)_B = 94497721140070$. □

1.1.2.4   Euclidean Division

Euclidean division turns out to be the most notorious among the basic arithmetic operations. Given any integers a, b with $b \neq 0$, there exist unique integers q, r satisfying $a = bq + r$ and $0 \leqslant r \leqslant |b| - 1$. We call q the *quotient* of Euclidean division of a by b, whereas r is called the *remainder* of Euclidean division of a by b. We denote $q = a$ quot b and $r = a$ rem b. For simplicity, let us assume that a and b are positive multiple-precision integers, and $a \geqslant b$. The quotient q and the remainder r are again multiple-precision integers in general. Computing these essentially amounts to efficiently guessing the B-ary digits of q from the most significant end. The process resembles the long division procedure we are accustomed to for decimal integers.

In order that the guesses of the quotient words quickly converge to the correct values, a certain precaution needs to be taken. We will assume that the most significant word of b is at least as large as $B/2$. If this condition is not satisfied, we multiply both a and b by a suitable power of 2. The remainder computed for these modified operands will also be multiplied by the same power of 2, whereas the quotient will remain unchanged. Preprocessing b in this manner is often referred to as *normalization*.

Suppose that $a = (a_{s-1}, a_{s-2}, \ldots, a_1, a_0)_B$ and $b = (b_{t-1}, b_{t-2}, \ldots, b_1, b_0)_B$ with $b_{t-1} \geqslant B/2$. The quotient may be as large as having $s - t + 1$ B-ary words. So we represent q as an array with $s - t + 1$ cells each initialized to zero. Let us denote $q = (q_{s-t}, q_{s-t-1}, \ldots, q_1, q_0)_B$.

If $a_{s-1} \geqslant b_{t-1}$ and $a \geqslant B^{s-t}b$, we increment q_{s-t} by one, and subtract $B^{s-t}b$ from a. Since b is normalized, this step needs to be executed at most once. Now, assume that $a_{s-1} \leqslant b_{t-1}$ and $a < B^{s-t}b$. We now guess the next word q_{s-t-1} of the quotient. The initial guess is based upon only two most significant words of a and one most significant word of b, that is, if $a_{s-1} = b_{t-1}$, we set $q_{s-t-1} = B - 1$, otherwise we set $q_{s-t-1} = \left\lfloor \frac{a_{s-1}B + a_{s-2}}{b_{t-1}} \right\rfloor$. Computing this division is easy if arithmetic routines for double-word-sized integer operands are provided by the compiler.

The initial guess for q_{s-t-1} may be slightly larger than the correct value. If b is normalized, the error cannot be more than three. In order to refine the guess, we first consider one more word from the most significant end of each of a and b. This time we do not require division, but we will keep on decrementing q_{s-t-1} so long as $a_{s-1}B^2 + a_{s-2}B + a_{s-3} < (b_{t-1}B + b_{t-2})q_{s-t-1}$ (this computation involves multiplications only). For a normalized b, this check needs to be carried out at most twice. This means that the guessed value of the word q_{s-t-1} is, at this stage, either the correct value or just one more than the correct value. We compute $c = q_{s-t-1}B^{s-t-1}b$. If $c > a$, the guess is still incorrect, so we decrement q_{s-t-1} for the last time and subtract bB^{s-t-1} from c. Finally, we replace a by $a - c$.

Suppose that after the operations described in the last two paragraphs, we have reduced a to $(a'_{s-2}, a'_{s-3}, \ldots, a'_1, a'_0)_B$. We repeat the above operations on this updated a with s replaced by $s - 1$. The division loop is broken when all the B-ary digits of q are computed, that is, when a is reduced to an integer with at most t B-ary digits. At this stage, the reduced a may still be larger than b. If so, q_0 is incremented by 1, and a is replaced by $a - b$. The value stored in a now is the remainder $r = a$ rem b.

Example 1.7 Let me explain the working of the above division algorithm on the following two operands available in the base-256 representation.

$$a = 369246812345567890 = (5, 31, 212, 4, 252, 77, 138, 146)_B,$$
$$b = 19283746550 = (4, 125, 102, 158, 246)_B.$$

The most significant word of b is too small. Multiplying both a and b by $2^5 = 32$ completes the normalization procedure, and the operands change to

$$a = 1463015397982818880 = (163, 250, 128, 159, 137, 177, 82, 64)_B,$$
$$b = 617079889600 = (143, 172, 211, 222, 192)_B.$$

We initially have $s = 8$ and $t = 5$, that is, the quotient can have at most $s - t + 1 = 4$ digits to the base $B = 256$. The steps of the division procedure are now illustrated in the following table.

s	Condition	Operation	Intermediate values
		Initialization	$q = (0, \ 0, \ 0, \ 0)_B$
			$a = (163, 250, 128, 159, 137, 177, \ 82, \ 64)_B$
8	$(a_7 \geqslant b_4)$?	Yes. Increment q_3	$q = (1, \ 0, \ 0, \ 0)_B$
		Subtract $B^3 b$ from a	$a = (\ 20, \ 77, 172, 192, 201, 177, \ 82, \ 64)_B$
	$(a_7 = b_4)$?	No. Set $q_2 = \left\lfloor \frac{a_7 B + a_6}{b_4} \right\rfloor$	$q = (1, \ 36, \ 0, \ 0)_B$
	$(a_7 B^2 + a_6 B + a_5 < q_2(b_4 B + b_3))$?	No. Do nothing.	
		Compute $c = q_2 B^2 b$	$c = (\ 20, \ 52, \ 77, 203, \ 83, \ 0, \ 0, \ 0)_B$
	$(c > a)$?	No. Do nothing.	
		Set $a := a - c$	$a = (25, \ 94, 245, 118, 177, \ 82, \ 64)_B$
7	$(a_6 \geqslant b_4)$?	No. Do nothing.	
	$(a_6 = b_4)$?	No. Set $q_1 = \left\lfloor \frac{a_6 B + a_5}{b_4} \right\rfloor$	$q = (1, \ 36, \ 45, \ 0)_B$
	$(a_6 B^2 + a_5 B + a_4 < q_1(b_4 B + b_3))$?	No. Do nothing.	
		Compute $c = q_1 B b$	$c = (25, \ 65, \ 97, \ 62, \ 39, 192, \ 0)_B$
	$(c > a)$?	No. Do nothing.	
		Set $a := a - c$	$a = (29, 148, \ 56, 137, 146, \ 64)_B$
6	$(a_5 \geqslant b_4)$?	No. Do nothing.	
	$(a_5 = a_4)$?	No. Set $q_0 = \left\lfloor \frac{a_5 B + a_4}{b_4} \right\rfloor$	$q = (1, \ 36, \ 45, \ 52)_B$
	$(a_5 B^2 + a_4 B + a_3 < q_0(b_4 B + b_3))$?	No. Do nothing.	
		Compute $c = q_0 b$	$c = (29, \ 47, \ 27, \ 9, \ 63, \ 0)_B$
	$(c > a)$?	No. Do nothing.	
		Set $a := a - c$	$a = (101, \ 29, 128, \ 83, \ 64)_B$
5	$(a_4 \geqslant b_4)$?	No. Do nothing.	

Let us again use the letters a, b to stand for the original operands (before normalization). We have computed $(32a)$ quot $(32b) = (1, 36, 45, 52)_B$ and

$(32a) \text{ rem } (32b) = (101, 29, 128, 83, 64)_B$. For the original operands, we then have $a \text{ quot } b = (32a) \text{ quot } (32b) = (1, 36, 45, 52)_B = 19148084$, and $a \text{ rem } b = [(32a) \text{ rem } (32b)]/32 = (3, 40, 236, 2, 154)_B = 13571457690$. $\qquad\square$

1.1.3 Fast Arithmetic

Multiple-precision arithmetic has been an important field of research since the advent of computers. The schoolbook methods for addition and subtraction of two n-digit integers take $O(n)$ time, and cannot be improved further (at least in the big-Oh notation). On the contrary, the $O(n^2)$-time schoolbook algorithms for multiplying two n-digit integers and for dividing a $2n$-digit integer by an n-digit integer are far from optimal. In this section, I explain some multiplication algorithms that run faster than in $O(n^2)$ time. When we study modular arithmetic, some efficient algorithms for modular multiplication (multiplication followed by Euclidean division) will be discussed.

1.1.3.1 Karatsuba–Ofman Multiplication

Apparently, the first fast integer-multiplication algorithm is proposed by Karatsuba and Ofman.[3] Let a, b be two multiple-precision integers each with n B-ary words. For the sake of simplicity, assume that n is even. Let $m = n/2$. We can write $a = A_1 B^m + A_0$ and $b = B_1 B^m + B_0$, where A_0, A_1, B_0, B_1 are multiple-precision integers each having $n/2$ B-ary digits. We have $ab = (A_1 B_1)B^2 + (A_1 B_0 + A_0 B_1)B + (A_0 B_0)$. Therefore, if we compute the four products $A_1 B_1$, $A_1 B_0$, $A_0 B_1$ and $A_0 B_0$ of $n/2$-digit integers, we can compute ab using a few additions of n-digit integers. This immediately does not lead to any improvement in the running time of schoolbook multiplication.

The Karatsuba–Ofman trick is to compute ab using only *three* products of $n/2$-digit integers, namely $A_1 B_1$, $A_0 B_0$ and $(A_1 + A_0)(B_1 + B_0)$. These products give $A_1 B_0 + A_0 B_1 = (A_1 + A_0)(B_1 + B_0) - A_1 B_1 - A_0 B_0$. This decrease in the number of multiplications is achieved at the cost of an increased number of additions and subtractions. But since this is only an additional $O(n)$-time overhead, we can effectively speed up the computation of ab (unless the size n of the operands is too small).

A small trouble with the above strategy is that $A_1 + A_0$ and/or $B_1 + B_0$ may be too large to fit in $m = n/2$ digits. Consequently, the subproduct $(A_1 + A_0)(B_1 + B_0)$ may be of $(m+1)$-digit integers. It is, therefore, preferable to compute the quantities $A_1 - A_0$ and $B_1 - B_0$ which may be negative but must fit in m digits. Subsequently, the product $(A_1 - A_0)(B_1 - B_0)$ is computed, and we obtain $A_1 B_0 + A_0 B_1 = A_1 B_1 + A_0 B_0 - (A_1 - A_0)(B_1 - B_0)$.

---

[3] A. Karatsuba and Yu. Ofman, Multiplication of many-digital numbers by automatic computers, *Doklady Akad. Nauk. SSSR*, Vol. 145, 293–294, 1962. The paper gives the full credit of the multiplication algorithm to Karatsuba only.

Example 1.8 Take $B = 256$, $a = 123456789 = (7, 91, 205, 21)_B$, and $b = 987654321 = (58, 222, 104, 177)_B$. We have $A_1 = (7, 91)_B$, $A_0 = (205, 21)_B$, $B_1 = (58, 222)_B$ and $B_0 = (104, 177)_B$. The subproducts are computed as

$$
\begin{aligned}
A_1 B_1 &= (1, 176, 254, 234)_B, \\
A_0 B_0 &= (83, 222, 83, 133)_B, \\
A_1 - A_0 &= -(197, 186)_B, \\
B_1 - B_0 &= -(45, 211)_B, \\
(A_1 - A_0)(B_1 - B_0) &= (35, 100, 170, 78)_B.
\end{aligned}
$$

It follows that

$$A_1 B_0 + A_0 B_1 = A_1 B_1 + A_0 B_0 - (A_1 - A_0)(B_1 - B_0) = (50, 42, 168, 33)_B.$$

The three subproducts are added with appropriate shifts to obtain ab.

1	176	254	234				
	50	42	168	33			
			83	222	83	133	
1	177	49	20	251	255	83	133

Therefore, $ab = (1, 177, 49, 20, 251, 255, 83, 133)_B = 121932631112635269$. ☐

So far, we have used the Karatsuba–Ofman trick only once. If $m = n/2$ is large enough, we can recursively apply the same trick to compute the three subproducts $A_1 B_1$, $A_0 B_0$ and $(A_1 - B_1)(A_0 - B_0)$. Each such subproduct leads to three subsubproducts of $n/4$-digit integers. If $n/4$ too is large, we can again compute these subsubproducts using the Karatsuba–Ofman algorithm. Recursion stops when either the operands become too small or the level of recursion reaches a prescribed limit. Under this recursive invocation, Karatsuba–Ofman multiplication achieves a running time of $O(n^{\log_2 3})$, that is, about $O(n^{1.585})$. This is much better than the $O(n^2)$ time of schoolbook multiplication. In practice, the advantages of using the Karatsuba–Ofman algorithm show up for integer operands of size at least a few hundred bits.

Example 1.9 Let me recursively compute $A_1 B_1$ of Example 1.8. Since $A_1 = 7B + 91$ and $B_1 = 58B + 222$, we compute $7 \times 58 = (1, 150)_B$, $91 \times 222 = (78, 234)_B$, and $(7 - 91)(58 - 222) = (53, 208)_B$. Finally, $(1, 150)_B + (78, 234)_B - (53, 208)_B = (26, 176)_B$, so $A_1 B_1$ is computed as:

1	150		
	26	176	
		78	234
1	176	254	234

This gives $A_1 B_1 = (1, 176, 254, 234)_B$. The other two subproducts $A_0 B_0$ and $(A_1 - A_0)(B_1 - B_0)$ can be similarly computed. ☐

1.1.3.2 Toom–Cook Multiplication

Before we go to asymptotically faster multiplication algorithms, it is worthwhile to view Karatsuba–Ofman multiplication from a slightly different angle. Let us write $a = A_1 R + A_0$ and $b = B_1 R + B_0$, where $R = B^m = B^{n/2}$. For a moment, treat R as an indeterminate, so a and b behave as linear polynomials in one variable. The product $c = ab$ can be expressed as the quadratic polynomial $c = C_2 R^2 + C_1 R + C_0$ whose coefficients are $C_2 = A_1 B_1$, $C_1 = A_1 B_0 + A_0 B_1$ and $C_0 = A_0 B_0$. In the Karatsuba–Ofman algorithm, we have computed these three coefficients. Indeed, these three coefficients can be fully determined from the values $c(k)$ at three distinct points. Moreover, since $c(k) = a(k)b(k)$, we need to evaluate the polynomials a and b at three points and compute the three subproducts of the respective values. Once C_2, C_1, C_0 are computed, the polynomial c is *evaluated* at R to obtain an integer value.

The three evaluation points k chosen as $\infty, 0, 1$ yield the following three linear equations in C_2, C_1, C_0.

$$
\begin{aligned}
c(\infty) &= & C_2 &= a(\infty)b(\infty) &= & A_1 B_1, \\
c(0) &= & C_0 &= a(0)b(0) &= & A_0 B_0, \\
c(1) &= C_2 + C_1 + C_0 &&= a(1)b(1) &= (A_1 + A_0)(B_1 + B_0).
\end{aligned}
$$

Solving the system for C_2, C_1, C_0 gives the first version of the Karatsuba–Ofman algorithm. If we choose the evaluation point $k = -1$ instead of $k = 1$, we obtain the equation

$$
\begin{aligned}
c(-1) &= & C_2 - C_1 + C_0 &= a(-1)b(-1) \\
&= & (A_0 - A_1)(B_0 - B_1) &= (A_1 - A_0)(B_1 - B_0).
\end{aligned}
$$

This equation along with the equations for $c(\infty)$ and $c(0)$ yield the second version of the Karatsuba–Ofman algorithm (as illustrated in Example 1.8).

This gives us a way to generalize the Karatsuba–Ofman algorithm. Toom[4] and Cook[5] propose representing a and b as polynomials of degrees higher than one. Writing them as quadratic polynomials gives an algorithm popularly known as *Toom-3 multiplication*.

Let a and b be n-digit integers. Take $m = \lceil n/3 \rceil$, and write

$$
\begin{aligned}
a &= A_2 R^2 + A_1 R + A_0, \\
b &= B_2 R^2 + B_1 R + B_0,
\end{aligned}
$$

where $R = B^m$. The product $c = ab$ can be expressed as the polynomial

$$
c = C_4 R^4 + C_3 R^3 + C_2 R^2 + C_1 R + C_0
$$

---

[4]Andrei L. Toom, The complexity of a scheme of functional elements realizing the multiplication of integers, *Doklady Akad. Nauk. SSSR*, Vol. 4, No. 3, 714–716, 1963.

[5]Stephen A. Cook, On the minimum computation time of functions, PhD thesis, Department of Mathematics, Harvard University, 1966.

with the coefficients

$$
\begin{aligned}
C_4 &= A_2 B_2, \\
C_3 &= A_2 B_1 + A_1 B_2, \\
C_2 &= A_2 B_0 + A_1 B_1 + A_0 B_2, \\
C_1 &= A_1 B_0 + A_0 B_1, \\
C_0 &= A_0 B_0.
\end{aligned}
$$

A straightforward computation of these coefficients involves computing nine subproducts $A_i B_j$ for $i, j = 0, 1, 2$, and fails to improve upon the schoolbook method. However, since we now have only five coefficients to compute, it suffices to compute only *five* subproducts of $n/3$-digit integers. We choose five suitable evaluation points k, and obtain $c(k) = a(k)b(k)$ at these points. The choices $k = \infty, 0, 1, -1, -2$ lead to the following equations.

$$
\begin{aligned}
c(\infty) &= & C_4 & & = & & A_2 B_2, \\
c(0) &= & C_0 & & = & & A_0 B_0, \\
c(1) &= & C_4 + C_3 + C_2 + C_1 + C_0 &=& (A_2 + A_1 + A_0)(B_2 + B_1 + B_0), \\
c(-1) &= & C_4 - C_3 + C_2 - C_1 + C_0 &=& (A_2 - A_1 + A_0)(B_2 - B_1 + B_0), \\
c(-2) &= 16C_4 - 8C_3 + 4C_2 - 2C_1 + C_0 &=& (4A_2 - 2A_1 + A_0)(4B_2 - 2B_1 + B_0).
\end{aligned}
$$

This system can be written in the matrix notation as

$$
\begin{pmatrix} c(\infty) \\ c(0) \\ c(1) \\ c(-1) \\ c(-2) \end{pmatrix} =
\begin{pmatrix}
1 & 0 & 0 & 0 & 0 \\
0 & 0 & 0 & 0 & 1 \\
1 & 1 & 1 & 1 & 1 \\
1 & -1 & 1 & -1 & 1 \\
16 & -8 & 4 & -2 & 1
\end{pmatrix}
\begin{pmatrix} C_4 \\ C_3 \\ C_2 \\ C_1 \\ C_0 \end{pmatrix}.
$$

We can invert the coefficient matrix in order to express C_4, C_3, C_2, C_1, C_0 in terms of the five subproducts $c(k)$. The coefficient matrix is independent of the inputs a and b, so the formulas work for all input integers.

$$
\begin{pmatrix} C_4 \\ C_3 \\ C_2 \\ C_1 \\ C_0 \end{pmatrix} =
\begin{pmatrix}
1 & 0 & 0 & 0 & 0 \\
2 & -\frac{1}{2} & \frac{1}{6} & \frac{1}{2} & -\frac{1}{6} \\
-1 & -1 & \frac{1}{2} & \frac{1}{2} & 0 \\
-2 & \frac{1}{2} & \frac{1}{3} & -1 & \frac{1}{6} \\
0 & 1 & 0 & 0 & 0
\end{pmatrix}
\begin{pmatrix} c(\infty) \\ c(0) \\ c(1) \\ c(-1) \\ c(-2) \end{pmatrix}.
$$

This can be rewritten as

$$
\begin{aligned}
C_4 &= c(\infty), \\
C_3 &= (12c(\infty) - 3c(0) + c(1) + 3c(-1) - c(-2))/6, \\
C_2 &= (-2c(\infty) - 2c(0) + c(1) + c(-1))/2, \\
C_1 &= (-12c(\infty) + 3c(0) + 2c(1) - 6c(-1) + c(-2))/6, \\
C_0 &= c(0).
\end{aligned}
$$

These formulas involve multiplications and divisions by small integers (like $2, 3, 6, 12$). Multiplying or dividing an m-digit multiple-precision integer by a single-precision integer can be completed in $O(m)$ time, so this is no trouble. Although some of these expressions involve denominators larger than 1, all the coefficients C_i evaluate to integral values. If the subproducts $a(k)b(k)$ too are computed recursively using Toom-3 multiplication, we get a running time of $O(n^{\log_3 5})$, that is, about $O(n^{1.465})$ which is better than the running time of Karatsuba–Ofman multiplication. However, this theoretical improvement shows up only when the bit sizes of the input integers are sufficiently large. On the one hand, the operands of the subproducts (like $4A_2 - 2A_1 + A_0$) may now fail to fit in m words. On the other hand, the formulas for C_i are more cumbersome than in the Karatsuba–Ofman method. Still, as reported in the literature, there is a range of bit sizes (several hundreds and more), for which Toom-3 multiplication is practically the fastest known multiplication algorithm.

Example 1.10 We take $B = 256$ and multiply $a = 1234567 = (18, 214, 135)_B$ and $b = 7654321 = (116, 203, 177)_B$ by the Toom-3 multiplication algorithm. Here, $n = 3$, that is, $m = 1$, that is, the coefficients A_i and B_j are single-precision integers. In fact, we have $A_2 = 18$, $A_1 = 214$, $A_0 = 135$, $B_2 = 116$, $B_1 = 203$, and $B_0 = 177$. The five subproducts are computed as follows.

$$c(\infty) = A_2 B_2 = (8, 40)_B,$$
$$c(0) = A_0 B_0 = (93, 87)_B,$$
$$c(1) = (A_2 + A_1 + A_0)(B_2 + B_1 + B_0) = (1, 111)_B \times (1, 240)_B = (2, 199, 16)_B,$$
$$c(-1) = (A_2 - A_1 + A_0)(B_2 - B_1 + B_0) = -61 \times 90 = -(21, 114)_B,$$
$$c(-2) = (4A_2 - 2A_1 + A_0)(4B_2 - 2B_1 + B_0) = -221 \times 235 = -(202, 223)_B.$$

The formulas for C_i as linear combinations of these five subproducts give us $C_4 = (8, 40)_B$, $C_3 = (111, 62)_B$, $C_2 = (243, 80)_B$, $C_1 = (255, 3)_B$, and $C_0 = (93, 87)_B$. Adding appropriate shifts of these values yields c as follows.

8	40				
	111	62			
		243	80		
			255	3	
				93	87
8	152	50	79	96	87

Thus, we have computed $ab = (8, 152, 50, 79, 96, 87)_B = 9449772114007$. □

Toom's 3-way multiplication can be readily generalized to any k-way multiplication. For $k = 1$ we have the schoolbook method, for $k = 2$ we get the Karatsuba–Ofman method, whereas $k = 3$ gives the Toom-3 method. For $k = 4$, we get the Toom-4 method that calls for only seven multiplications of one-fourth-sized operands. The schoolbook method invokes sixteen such multiplications, whereas two levels of recursion of the Karatsuba–Ofman method generate nine such multiplications. In general, Toom-k runs in

$O(n^{\log(2k-1)/\log k})$ time in which the exponent can be made arbitrarily close to one by choosing large values of k. Toom and Cook suggest taking k adaptively based on the size of the input. The optimal choice is shown as $k = 2^{\lceil \log r \rceil}$, where each input is broken in k parts each of size r digits. This gives an asymptotic running time of $O(n2^{5\sqrt{\log n}})$ for the optimal Toom–Cook method. Unfortunately, practical implementations do not behave well for $k \geqslant 4$.

1.1.3.3   FFT-Based Multiplication

An alternative method based again on polynomial evaluations and interpolations turns out to be practically significant. Proposed by Schönhage and Strassen,[6] this method achieves a running time of $O(n \log n \log \log n)$, and is practically the fastest known integer-multiplication algorithm for operands of bit sizes starting from a few thousands. The Schönhage–Strassen algorithm is a bit too involved to be discussed in this introductory section. So I present only a conceptually simpler version of the algorithm.

Suppose that each of the input operands a and b consists of n digits in base $B = 2^r$. Let $2^{t-1} < n \leqslant 2^t$. Define $N = 2^{t+1}$, and pad a and b with leading zero words so that each is now represented as an N-digit integer in base B. Denote $a = (a_{N-1}, a_{N-2}, \ldots, a_1, a_0)_B$ and $b = (b_{N-1}, b_{N-2}, \ldots, b_1, b_0)_B$, where each a_i or b_j is an r-bit integer. We have

$$
\begin{aligned}
a &= a_{N-1}B^{N-1} + a_{N-2}B^{N-2} + \cdots + a_1 B + a_0, \\
b &= b_{N-1}B^{N-1} + b_{N-2}B^{N-2} + \cdots + b_1 B + b_0.
\end{aligned}
$$

Because of the padding of a and b in an N-digit space, we have $a_i = b_j = 0$ for $N/2 \leqslant i, j < N$.

The *cyclic convolution* of the two sequences $(a_{N-1}, a_{N-2}, \ldots, a_1, a_0)$ and $(b_{N-1}, b_{N-2}, \ldots, b_1, b_0)$ is the sequence $(c_{N-1}, c_{N-2}, \ldots, c_1, c_0)$, where

$$
c_k = \sum_{\substack{0 \leqslant i,j \leqslant N-1 \\ i+j=k \text{ or } i+j=k+N}} a_i b_j
$$

for all $k \in \{0, 1, 2, \ldots, N-1\}$. Since $a_i = b_j = 0$ for $N/2 \leqslant i, j < N$, we have

$$
\begin{aligned}
c_k &= \sum_{\substack{0 \leqslant i,j \leqslant N-1 \\ i+j=k \text{ or } i+j=k+N}} a_i b_j \\
&= a_k b_0 + a_{k-1}b_1 + \cdots + a_0 b_k + (a_{N-1}b_{k+1} + a_{N-2}b_{k+2} + \cdots + a_{k+1}b_{N-1}) \\
&= a_k b_0 + a_{k-1}b_1 + \cdots + a_0 b_k = \sum_{\substack{0 \leqslant i,j \leqslant k \\ i+j=k}} a_i b_j,
\end{aligned}
$$

that is, the product of a and b can be expressed as

$$
ab = c_{N-1}B^{N-1} + c_{N-2}B^{N-2} + \cdots + c_1 B + c_0.
$$

---

[6]Arnold Schönhage and Volker Strassen, Schnelle Multiplikation großer Zahlen, *Computing*, 7, 281–292, 1971.

Now, let ω_N be a primitive N-th root of unity. Depending upon the field in which we are working, this root ω_N can be appropriately defined. For the time being, let us plan to work in the field of complex numbers so that we can take $\omega_N = e^{i\frac{2\pi}{N}}$ (where $i = \sqrt{-1}$). The *discrete Fourier transform* (DFT) of the sequence $(a_{N-1}, a_{N-2}, \ldots, a_1, a_0)$ is defined as the sequence $(A_{N-1}, A_{N-2}, \ldots, A_1, A_0)$, where for all $k \in \{0, 1, 2, \ldots, N-1\}$, we have

$$A_k = \sum_{0 \leqslant i < N} \omega_N^{ki} a_i.$$

A_k is the value of the *polynomial a* evaluated at ω_N^k (replace B by ω_N^k). Likewise, let $(B_{N-1}, B_{N-2}, \ldots, B_1, B_0)$ be the DFT of $(b_{N-1}, b_{N-2}, \ldots, b_1, b_0)$, and $(C_{N-1}, C_{N-2}, \ldots, C_1, C_0)$ the DFT of $(c_{N-1}, c_{N-2}, \ldots, c_1, c_0)$. Since B_k is b evaluated at ω_N^k, and c_k is c evaluated at ω_N^k, we have

$$C_k = A_k B_k$$

for all $k = 0, 1, 2, \ldots, N-1$. Therefore, if we can efficiently compute the DFTs of the polynomials a and b, we can compute, using only N additional multiplications, the DFT of the product $c = ab$. Computing the sequence $(c_{N-1}, c_{N-2}, \ldots, c_1, c_0)$ from its DFT $(C_{N-1}, C_{N-2}, \ldots, C_1, C_0)$ is called the *inverse discrete Fourier transform* (IDFT). Let $(\hat{C}_{N-1}, \hat{C}_{N-2}, \ldots, \hat{C}_1, \hat{C}_0)$ be the DFT of $(C_{N-1}, C_{N-2}, \ldots, C_1, C_0)$. One can check (Exercise 1.9) that

$$(c_{N-1}, c_{N-2}, \ldots, c_1, c_0) = \frac{1}{N}(\hat{C}_1, \hat{C}_2, \ldots, \hat{C}_{N-1}, \hat{C}_0), \tag{1.1}$$

that is, the IDFT of a sequence can be easily computed from its DFT. So it suffices to compute the DFT as efficiently as possible. A naïve application of the DFT formula leads to $\mathrm{O}(N^2)$ running time. A divide-and-conquer procedure for computing the DFT $(A_{N-1}, A_{N-2}, \ldots, A_1, A_0)$ of $(a_{N-1}, a_{N-2}, \ldots, a_1, a_0)$ is now presented. This procedure uses only $\mathrm{O}(n \log n)$ operations in the underlying field ($\mathbb{C}$ for the time being), and is called the *fast Fourier transform* (FFT) of the input sequence. Let us write the *polynomial a* as

$$a = a^{(e)}(B^2) + B \times a^{(o)}(B^2),$$

where

$$\begin{aligned}
a^{(e)}(B) &= a_{N-2}B^{N/2-1} + a_{N-4}B^{N/2-2} + \cdots + a_2 B + a_0, \text{ and} \\
a^{(o)}(B) &= a_{N-1}B^{N/2-1} + a_{N-3}B^{N/2-2} + \cdots + a_3 B + a_1
\end{aligned}$$

are polynomials obtained from a by taking the terms at even and odd positions, respectively. But $\omega_{\frac{N}{2}} = \omega_N^2$ is a primitive $\frac{N}{2}$-th root of unity. Moreover, $a^{(e)}$ and $a^{(o)}$ are polynomials with $\frac{N}{2}$ terms. We recursively compute the DFT (actually, FFT) $(A_{\frac{N}{2}-1}^{(e)}, A_{\frac{N}{2}-2}^{(e)}, \ldots, A_1^{(e)}, A_0^{(e)})$ of $a^{(e)}$, and the DFT $(A_{\frac{N}{2}-1}^{(o)}, A_{\frac{N}{2}-2}^{(o)}, \ldots, A_1^{(o)}, A_0^{(o)})$ of $a^{(o)}$. Finally, $\omega_N^{\frac{N}{2}} = -1$, so for all $k = 0, 1, 2, \ldots, \frac{N}{2} - 1$ we have

$$A_k = A_k^{(e)} + \omega_N^k A_k^{(o)} \quad \text{and} \quad A_{\frac{N}{2}+k} = A_k^{(e)} - \omega_N^k A_k^{(o)}.$$

Example 1.11 Let me illustrate the FFT-based multiplication algorithm on the following two integers represented in base $B = 2^8 = 256$.

$$a = 1234567890 = (73, 150, 2, 210)_B,$$
$$b = 1357924680 = (80, 240, 73, 72)_B.$$

First, we need to pad a and b with leading zero digits so that each is of length $N = 8$. The product $c = ab$ is given as

$$c = \text{IDFT}(\text{DFT}(0, 0, 0, 0, 73, 150, 2, 210) \cdot \text{DFT}(0, 0, 0, 0, 80, 240, 73, 72)).$$

A primitive eighth roots of unity is $\omega_8 = e^{2\pi i/8} = (1 + i)/\sqrt{2}$. In the recursive calls, we also need a primitive fourth root of unity $\omega_4 = e^{2\pi i/4} = i$, and a primitive second root of unity $\omega_2 = -1$. For a sequence (x_1, x_0) of length two,

$$\text{DFT}(x_1, x_0) = (x_0 - x_1, x_0 + x_1),$$

whereas for a sequence (x_3, x_2, x_1, x_0) of length four,

$$\begin{aligned}
\text{DFT}(x_3, x_2, x_1, x_0) &= \text{COMBINE}(\text{DFT}(x_2, x_0), \text{DFT}(x_3, x_1)) \\
&= \text{COMBINE}((x_0 - x_2, x_0 + x_2), (x_1 - x_3, x_1 + x_3)) \\
&= \Big((x_0 - x_2) - i(x_1 - x_3), (x_0 + x_2) - (x_1 + x_3), \\
&\quad (x_0 - x_2) + i(x_1 - x_3), (x_0 + x_2) + (x_1 + x_3)\Big).
\end{aligned}$$

Here, COMBINE stands for the combination of the DFTs of the two recursive calls. For computing the DFT $(X_7, X_6, X_5, X_4, X_3, X_2, X_1, X_0)$ of the sequence $(x_7, x_6, x_5, x_4, x_3, x_2, x_1, x_0)$ of length eight, recursive calls are made on (x_6, x_4, x_2, x_0) and (x_7, x_5, x_3, x_1) to get the two sub-DFTs

$$\begin{aligned}
(Y_3, Y_2, Y_1, Y_0) &= \Big((x_0 - x_4) - i(x_2 - x_6), (x_0 + x_4) - (x_2 + x_6), \\
&\quad (x_0 - x_4) + i(x_2 - x_6), (x_0 + x_4) + (x_2 + x_6)\Big), \\
(Z_3, Z_2, Z_1, Z_0) &= \Big((x_1 - x_5) - i(x_3 - x_7), (x_1 + x_5) - (x_3 + x_7), \\
&\quad (x_1 - x_5) + i(x_3 - x_7), (x_1 + x_5) + (x_3 + x_7)\Big),
\end{aligned}$$

respectively. Finally, the combine step gives

$$\begin{array}{ll}
X_0 = Y_0 + Z_0, & X_4 = Y_0 - Z_0, \\
X_1 = Y_1 + \left(\frac{1+i}{\sqrt{2}}\right) Z_1, & X_5 = Y_1 - \left(\frac{1+i}{\sqrt{2}}\right) Z_1, \\
X_2 = Y_2 + i Z_2, & X_6 = Y_2 - i Z_2, \\
X_3 = Y_3 + \left(\frac{-1+i}{\sqrt{2}}\right) Z_3, & X_7 = Y_3 - \left(\frac{-1+i}{\sqrt{2}}\right) Z_3.
\end{array}$$

For the first operand a, we have

$$\begin{aligned}
(Y_3, Y_2, Y_1, Y_0) &= \text{DFT}(0, 0, 150, 210) = (210 - 150i, 60, 210 + 150i, 360), \\
(Z_3, Z_2, Z_1, Z_0) &= \text{DFT}(0, 0, 73, 2) = (2 - 73i, -71, 2 + 73i, 75).
\end{aligned}$$

Therefore, the combining formulas give the DFT of a as

$$\left(\left(\frac{-71-75i}{\sqrt{2}} \right) + (210 - 150i), 60 + 71i, \left(\frac{71-75i}{\sqrt{2}} \right) + (210 + 150i), 285, \right.$$
$$\left. \left(\frac{71+75i}{\sqrt{2}} \right) + (210 - 150i), 60 - 71i, \left(\frac{-71+75i}{\sqrt{2}} \right) + (210 + 150i), 435 \right).$$

Likewise, we compute the DFT of b as

$$\left(\left(\frac{-7-153i}{\sqrt{2}} \right) + (72 - 240i), -168 + 7i, \left(\frac{7-153i}{\sqrt{2}} \right) + (72 + 240i), 159, \right.$$
$$\left. \left(\frac{7+153i}{\sqrt{2}} \right) + (72 - 240i), -168 - 7i, \left(\frac{-7+153i}{\sqrt{2}} \right) + (72 + 240i), 465 \right).$$

Now, we make a point-by-point multiplication of the two DFTs to obtain the DFT C of the product $c = ab$:

$$\begin{aligned} C = \Big(&(-23766 - 9720i)\sqrt{2} + (-26369 - 55506i), -10577 - 11508i, \\ &(23766 - 9720i)\sqrt{2} + (-26369 + 55506i), 45315, \\ &(23766 + 9720i)\sqrt{2} + (-26369 - 55506i), -10577 + 11508i, \\ &(-23766 + 9720i)\sqrt{2} + (-26369 + 55506i), 202275 \Big). \end{aligned}$$

In order to recover $c = \text{IDFT}(C)$ from C, we first take the DFT of C (the steps are not shown here):

$$\text{DFT}(C) = (123792, 490768, 267888, 331912, 236160, 46720, 0, 120960).$$

We then obtain c as

$$\begin{aligned} c &= \text{IDFT}(C) \\ &= \frac{1}{8}(0, 46720, 236160, 331912, 267888, 490768, 123792, 120960) \\ &= (0, 5840, 29520, 41489, 33486, 61346, 15474, 15120). \end{aligned}$$

This implies that

$$\begin{aligned} c &= 0 \times B^7 + 5840 \times B^6 + 29520 \times B^5 + 41489 \times B^4 + 33486 \times B^3 + \\ &\quad 61346 \times B^2 + 15474 \times B + 15120 \\ &= 1676450206966525200. \end{aligned}$$

One can check that this is indeed the value of $1234567890 \times 1357924680$.

Throughout this example, I used hybrid arithmetic, that is, integer arithmetic in conjunction with arithmetic associated with the algebraic numbers i and $\sqrt{2}$. Moreover, I have not shown the integer arithmetic in base 256. So long as this example is meant for illustrating FFT-based multiplication, this abstraction is fine. In practice, one may resort to floating-point arithmetic. □

In the FFT procedure, computing the DFT of a sequence of length N is recursively replaced by the computations of the DFTs of two sequences of length $N/2$ each. The additional effort in this process is $O(N)$ field operations. So the running time of FFT can be expressed as $O(N \log N)$ field operations. Since $N = \Theta(n)$, this quantity is $O(n \log n)$ too.

We now review the issues associated with complex arithmetic involved in the process. Knuth (Footnote 2) shows that a floating-point precision of $6(t+1)$ bits (where $N = 2^{t+1}$) suffices for these computations, leading to a running time of $O(n \log n \log \log n \log \log \log n \cdots)$ for FFT-based multiplication.

Schönhage and Strassen point out that one can use an integer-only arithmetic in this algorithm. They suggest working in the ring $\mathbb{Z}_{2^s+1}$ with 2 used as a primitive $2s$-th root of unity (for a suitable s). In this way, they achieve an $O(n \log n \log \log n)$ running time. The details are omitted here. Let me conclude this topic with a relevant advice to potential implementers. For very small input operands, the schoolbook method is the best choice. Beyond that, the Karatsuba–Ofman method takes over, followed by the Toom-3 method. Eventually, for large enough operands, the Schönhage–Strassen method is the fastest alternative. The crossover points are to be located experimentally.

1.1.4 An Introduction to GP/PARI

Does everybody interested in computational number theory need to write all the above basic functions? Fortunately, no. There exist good computational libraries that implement multiple-precision integer arithmetic. Recently, the GMP (the GNU Multi-Precision) library has gained popularity. Some other public-domain libraries are LiDIA, LIP, NTL, SIMATH, and ZEN. Each such library can be freely downloaded from the Internet. The package sage combines many open-source mathematical packages and provides a python-based interface (visit http://www.sagemath.org/). One can read the accompanying usage instructions to learn how to work with these libraries/packages.

In this book, I use the GP/PARI calculator for illustrating multiple-precision arithmetic. It is free and efficient, with a simple text-based interface. When run from a command prompt (shell), it displays a welcome note and then runs an interpreter that waits for the user to enter instructions, parses each instruction, executes it (if error-free), and displays the output of the instruction. In my machine, running GP/PARI invokes the interpreter as shown below. The prompt issued by this interpreter is shown as gp > (in some versions, the prompt is ?).

```
bash$ gp

            GP/PARI CALCULATOR Version 2.1.7 (released)
            i686 running linux (ix86 kernel) 32-bit version
              compiled: Feb 24 2011, gcc-4.4.3 (GCC)
         (readline v6.1 enabled, extended help available)

                 Copyright (C) 2002 The PARI Group
```

```
PARI/GP is free software, covered by the GNU General Public License, and
comes WITHOUT ANY WARRANTY WHATSOEVER.

Type ? for help, \q to quit.
Type ?12 for how to get moral (and possibly technical) support.

   realprecision = 28 significant digits
   seriesprecision = 16 significant terms
   format = g0.28

parisize = 4000000, primelimit = 500000
gp >
```

One can enter an arithmetic expression against the prompt. GP/PARI evaluates the expression and displays the result. This result is actually stored in a variable for future references. These variables are to be accessed as %1,%2,%3,.... The last returned result is stored in the variable %%.

Here follows a simple conversation between me and GP/PARI. I ask GP/PARI to calculate the expressions $2^{2^3} + 3^{2^2}$ and $\binom{100}{25} = \frac{100!}{25!75!}$. GP/PARI uses conventional precedence and associativity rules for arithmetic operators. For example, the exponentiation operator ^ is right-associative and has a higher precedence than the addition operator +. Thus, 2^2^3+3^2^2 is interpreted as (2^(2^3))+(3^(2^2)). One can use explicit disambiguating parentheses.

```
gp > 2^2^3+3^2^2
%1 = 337
gp > 100!/(25!*75!)
%2 = 242519269720337121015504
gp >
```

GP/PARI supports many built-in arithmetic and algebraic functions. For example, the binomial coefficient $\binom{n}{r}$ can be computed by invoking binomial().

```
gp > binomial(100,25)
%3 = 242519269720337121015504
gp >
```

One can also define functions at the GP/PARI prompt. For example, one may choose to redefine the binomial() function as follows.

```
gp > choose1(n,r) = n!/(r!*(n-r)!)
gp > choose1(100,25)
%4 = 242519269720337121015504
gp >
```

Here is an alternative implementation of the `binomial()` function, based on the formula $\binom{n}{r} = \frac{n(n-1)\cdots(n-r+1)}{r!}$. It employs sophisticated programming styles (like `for` loops). The interpreter of GP/PARI reads instructions from the user line by line. If one instruction is too big to fit in a single line, one may let the instruction span over multiple lines. In that case, one has to end each line (except the last) by the special character \.

```
gp > choose2(n,r) = \
       num=1; den=1; \
       for(k=1, r, \
           num*=n; den*=r; \
           n=n-1; r=r-1 \
       ); \
       num/den
gp > choose2(100,25)
%5 = 242519269720337121015504
gp >
```

All variables in GP/PARI are global by default. In the function `choose2()`, the variables `num` and `den` accumulate the numerator and the denominator. When the `for` loop terminates, `num` stores $n(n-1)\cdots(n-r+1)$, and `den` stores $r!$. These values can be printed subsequently. If a second call of `choose2()` is made with different arguments, the values stored in `num` and `den` are overwritten.

```
gp > num
%6 = 37617673321873894319687391903177156670695936000000
gp > den
%7 = 15511210043330985984000000
gp > choose2(55,34)
%8 = 841728816603675
gp > num
%9 = 248505954558196590544596278440992435848871936000000000
gp > den
%10 = 295232799039604140847618609643520000000
gp > 34!
%11 = 295232799039604140847618609643520000000
gp >
```

All local variables to be used in a function must be specified by the `local()` declaration. Here is a function that, upon input x, computes $2x^2 + 3x^3$. In this function, the variables y,z,u,v are local, whereas the variable w is global.

```
gp > f(x) = local(y=x*x,z=x*x*x,u=2,v); \
           v=u-1; w=u+v; \
           u*eval(y) + w*eval(z)
gp > f(43)
```

```
%12 = 242219
gp > x
%13 = x
gp > y
%14 = y
gp > z
%15 = z
gp > u
%16 = u
gp > v
%17 = v
gp > w
%18 = 3
gp >
```

Now, I present a more complicated example in order to illustrate how we can use GP/PARI as a programmable calculator. Consider the expression $f(a, b) = \frac{a^2 + b^2}{ab - 1}$ for all positive integers a, b (except $a = b = 1$). It is known that $f(a, b)$ assumes integer values for infinitely many pairs (a, b). Moreover, whenever $f(a, b)$ evaluates to an integer value, $f(a, b) = 5$. Here is a GP/PARI function that accepts an argument L and locates all pairs (a, b) with $a, b \leqslant L$, for which $f(a, b)$ is an integer. Because of symmetry, we concentrate only on those pairs with $a \leqslant b$. Since $f(a, b)$ cannot be an integer if $a = b$, our search may be restricted only to the pairs (a, b) satisfying $1 \leqslant a < b \leqslant L$.

```
gp > #
   timer = 1 (on)
gp > \
searchPair(L) = \
   for (a=1, L, \
      for (b=a+1, L, \
         x=(a^2+b^2)/(a*b-1); \
         if (x == floor(x), \
            print("   a = ", a, ", b = ", b, ", x = ", x, ".") \
         ) \
      ) \
   )
gp > searchPair(10)
   a = 1, b = 2, x = 5.
   a = 1, b = 3, x = 5.
   a = 2, b = 9, x = 5.
time = 1 ms.
gp > searchPair(100)
   a = 1, b = 2, x = 5.
   a = 1, b = 3, x = 5.
   a = 2, b = 9, x = 5.
   a = 3, b = 14, x = 5.
   a = 9, b = 43, x = 5.
   a = 14, b = 67, x = 5.
time = 34 ms.
gp > searchPair(1000)
   a = 1, b = 2, x = 5.
```

```
      a = 1, b = 3, x = 5.
      a = 2, b = 9, x = 5.
      a = 3, b = 14, x = 5.
      a = 9, b = 43, x = 5.
      a = 14, b = 67, x = 5.
      a = 43, b = 206, x = 5.
      a = 67, b = 321, x = 5.
      a = 206, b = 987, x = 5.
time = 3,423 ms.
gp >
```

In the above illustration, I turned the timer on by using the special directive #. Subsequently, GP/PARI displays the time taken for executing each instruction. The timer can be turned off by typing the directive # once again.

GP/PARI provides text-based plotting facilities also.

```
gp > plot(X=0,2*Pi,sin(X))
```

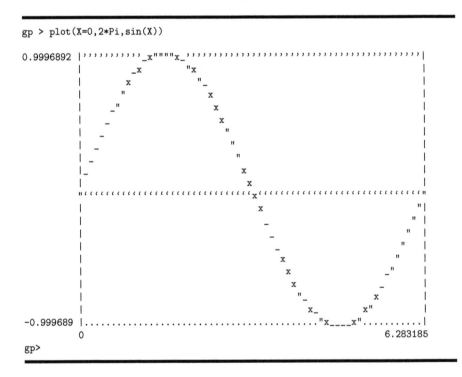

```
gp>
```

This book is not meant to be a tutorial on GP/PARI. One can read the manual supplied in the GP/PARI distribution. One can also use the extensive on-line help facility bundled with the calculator. Entering ? at the GP/PARI prompt yields an overview of help topics. One may follow these instructions in order to obtain more detailed help. Some examples are given below.

```
gp > ?
```

```
Help topics:
   0: list of user-defined identifiers (variable, alias, function)
   1: Standard monadic or dyadic OPERATORS
   2: CONVERSIONS and similar elementary functions
   3: TRANSCENDENTAL functions
   4: NUMBER THEORETICAL functions
   5: Functions related to ELLIPTIC CURVES
   6: Functions related to general NUMBER FIELDS
   7: POLYNOMIALS and power series
   8: Vectors, matrices, LINEAR ALGEBRA and sets
   9: SUMS, products, integrals and similar functions
   10: GRAPHIC functions
   11: PROGRAMMING under GP
   12: The PARI community

Further help (list of relevant functions): ?n (1<=n<=11).
Also:
   ? functionname (short on-line help)
   ?\             (keyboard shortcuts)
   ?.             (member functions)
Extended help looks available:
   ??             (opens the full user's manual in a dvi previewer)
   ??  tutorial   (same with the GP tutorial)
   ??  refcard    (same with the GP reference card)

   ??  keyword    (long help text about "keyword" from the user's manual)
   ??? keyword    (a propos: list of related functions).
gp > ?4

addprimes       bestappr        bezout          bezoutres       bigomega
binomial        chinese         content         contfrac        contfracpnqn
core            coredisc        dirdiv          direuler        dirmul
divisors        eulerphi        factor          factorback      factorcantor
factorff        factorial       factorint       factormod       ffinit
fibonacci       gcd             hilbert         isfundamental   isprime
ispseudoprime   issquare        issquarefree    kronecker       lcm
moebius         nextprime       numdiv          omega           precprime
prime           primes          qfbclassno      qfbcompraw      qfbhclassno
qfbnucomp       qfbnupow        qfbpowraw       qfbprimeform    qfbred
quadclassunit   quaddisc        quadgen         quadhilbert     quadpoly
quadray         quadregulator   quadunit        removeprimes    sigma
sqrtint         znlog           znorder         znprimroot      znstar


gp > ?znorder
znorder(x): order of the integermod x in (Z/nZ)*.


gp > znorder(Mod(19,101))
%19 = 25
gp > ?11

addhelp     alias       allocatemem break       default     error
extern      for         fordiv      forprime    forstep     forsubgroup
forvec      getheap     getrand     getstack    gettime     global
if          input       install     kill        next        print
print1      printp      printp1     printtex    quit        read
reorder     return      setrand     system      trap        type
until       whatnow     while       write       write1      writetex
```

```
gp > ?while
while(a,seq): while a is nonzero evaluate the expression sequence seq.
Otherwise 0.

gp > n=50; t=1; while(n, t*=n; n--)
gp > t
%20 = 30414093201713378043612608166064768844377641568960512000000000000
gp > ?if
if(a,seq1,seq2): if a is nonzero, seq1 is evaluated, otherwise seq2. seq1 and
seq2 are optional, and if seq2 is omitted, the preceding comma can be omitted
also.

gp > MAX(a,b,c) = \
     if (a>b, \
        if(a>c, return(a), return(c)), \
        if(b>c, return(b), return(c)) \
     )
gp > MAX(3,7,5)
%21 = 7
gp > Fib(n) = \
     if(n==0, return(0)); \
     if(n==1, return(1)); \
     return(Fib(n-1)+Fib(n-2))
? Fib(10)
%22 = 55
gp > ?printp
printp(a): outputs a (in beautified format) ending with newline.

gp > ?printtex
printtex(a): outputs a in TeX format.

gp > print(x^13+2*x^5-5*x+4)
x^13 + 2*x^5 - 5*x + 4
gp > printp(x^13+2*x^5-5*x+4)
(x^13 + 2 x^5 - 5 x + 4)
gp > printtex(x^13+2*x^5-5*x+4)
x^{13}
  + 2 x^5
  - 5 x
  + 4
gp >
```

One can close the GP/PARI interpreter by typing \q or by hitting control-d.

```
gp > \q
Good bye!
bash$
```

I will not explain the syntax of GP/PARI further in this book, but will use
the calculator for demonstrating arithmetic (and algebraic) calculations.

1.2 GCD

Divisibility is an important property of integers. Let $a, b \in \mathbb{Z}$. We say that a *divides* b and denote this as $a|b$ if there exists an integer c for which $b = ca$. For example, $31|1023$ (since $1023 = 33 \times 31$), $-31|1023$ (since $1023 = (-33) \times (-31)$), every integer a (including 0) divides 0 (since $0 = 0 \times a$). Let $a|b$ with $a \neq 0$. The (unique) integer c with $b = ca$ is called the *cofactor* of a in b. If $a|b$ with both a, b non-zero, then $|a| \leqslant |b|$. By the notation $a \nmid b$, we mean that a does not divide b.

Definition 1.12 Let $a, b \in \mathbb{Z}$ be not both zero. The largest positive integer d that divides both a and b is called the *greatest common divisor* or the *gcd* of a and b. It is denoted as $\gcd(a, b)$. Clearly, $\gcd(a, b) = \gcd(b, a)$. For $a \neq 0$, we have $\gcd(a, 0) = |a|$. The value $\gcd(0, 0)$ is left undefined. Two integers a, b are called *coprime* or *relatively prime* if $\gcd(a, b) = 1$. ◁

In addition to this usual divisibility, $\mathbb{Z}$ supports another notion of division, that turns out to be very useful for the computation of gcd's.

Theorem 1.13 [*Euclidean division*][7] *For $a, b \in \mathbb{Z}$ with $b \neq 0$, there exist unique integers q, r such that $a = qb + r$ and $0 \leqslant r \leqslant |b| - 1$. We call q the quotient and r the remainder of Euclidean division of a by b. We denote $q = a$ quot b and $r = a$ rem b.* ◁

Example 1.14 We have $1023 = 11 \times 89 + 44$, so 1023 quot $89 = 11$, and 1023 rem $89 = 44$. Similarly, $1023 = (-11) \times (-89) + 44$, that is, 1023 quot $(-89) = -11$, and 1023 rem $(-89) = 44$. Also $-1023 = (-11) \times 89 - 44 = (-12) \times 89 + (89 - 44) = (-12) \times 89 + 45$, that is, (-1023) quot $89 = -12$, and (-1023) rem $89 = 45$. Finally, $-1023 = 12 \times (-89) + 45$, so that (-1023) quot $(-89) = 12$, and (-1023) rem $(-89) = 45$. The last two examples demonstrate that while computing $r = a$ rem b with $a < 0$, we force r to a value $0 \leqslant r \leqslant |b| - 1$. This convention sometimes contradicts the remainder operators in programming languages (like % in C), which may return negative values. □

1.2.1 Euclidean GCD Algorithm

At an early age, we have learned the repeated Euclidean division procedure for computing the gcd of two integers. The correctness of this procedure is based on Euclid's gcd theorem.

---

[7]The Greek mathematician and philosopher Euclid (ca. 325–265 BC) is especially famous for his contributions to geometry. His book *Elements* influences various branches of mathematics even today.

Theorem 1.15 [*Euclidean gcd*]   *Let $a, b \in \mathbb{Z}$ with $b \neq 0$, and $r = a$ rem b. Then, $\gcd(a, b) = \gcd(b, r)$.*   ◁

Suppose that we want to compute $\gcd(a, b)$ with both a and b positive. We compute a sequence of remainders as follows. We first let $r_0 = a$ and $r_1 = b$.

$$
\begin{aligned}
r_0 &= q_2 r_1 + r_2, & 0 < r_2 &\leqslant r_1 - 1, \\
r_1 &= q_3 r_2 + r_3, & 0 < r_3 &\leqslant r_2 - 1, \\
r_2 &= q_4 r_3 + r_4, & 0 < r_4 &\leqslant r_3 - 1, \\
&\cdots \\
r_{i-2} &= q_i r_{i-1} + r_i, & 0 < r_i &\leqslant r_{i-1} - 1, \\
&\cdots \\
r_{k-2} &= q_k r_{k-1} + r_k, & 0 < r_k &\leqslant r_{k-1} - 1, \\
r_{k-1} &= q_{k+1} r_k.
\end{aligned}
$$

Euclid's gcd theorem guarantees that $\gcd(a, b) = \gcd(r_0, r_1) = \gcd(r_1, r_2) = \gcd(r_2, r_3) = \cdots = \gcd(r_{k-2}, r_{k-1}) = \gcd(r_{k-1}, r_k) = r_k$, that is, $\gcd(a, b)$ is the last non-zero remainder in the sequence $r_0, r_1, r_2, r_3, \ldots$.

Example 1.16 Let us compute $\gcd(252, 91)$ by Euclid's procedure.

$$
\begin{aligned}
252 &= 2 \times 91 + 70, \\
91 &= 1 \times 70 + 21, \\
70 &= 3 \times 21 + 7, \\
21 &= 3 \times 7.
\end{aligned}
$$

Therefore, $\gcd(252, 91) = \gcd(91, 70) = \gcd(70, 21) = \gcd(21, 7) = 7$.   □

Some important points need to be noted in connection with the Euclidean gcd procedure. First, the remainder sequence follows the chain of inequalities

$$ r_1 > r_2 > r_3 > \cdots > r_{k-1} > r_k , $$

that is, the procedure terminates after finitely many steps. More specifically, it continues for at most $r_1 = b$ iterations.

In practice, the number of iterations is much less than b. In order to see why, let us look at the computation of the i-th remainder r_i.

$$ r_{i-2} = q_i r_{i-1} + r_i. $$

Without loss of generality, we may assume that $a \geqslant b$ (if not, a single Euclidean division swaps the two operands). Under this assumption, $r_{i-1} \geqslant r_i$ for all $i \geqslant 1$. It then follows that $q_i \geqslant 1$ for all $i \geqslant 2$. Therefore, $r_{i-2} = q_i r_{i-1} + r_i \geqslant r_{i-1} + r_i \geqslant r_i + r_i = 2r_i$, that is, $r_i \leqslant \frac{r_{i-2}}{2}$, that is, after two iterations the remainder reduces by at least a factor of two. Therefore, after $O(\lg(\min(a, b)))$ iterations, the remainder reduces to zero, and the algorithm terminates.

A C function for computing the gcd of two integers a, b is presented now. Here, we assume signed single-precision integer operands only. The basic algorithm can be readily extended to multiple-precision integers by replacing the standard operators -, %, =, etc. by appropriate function calls. If one uses a language that supports operator overloading (like C++), this gcd implementation continues to make sense; however, the default data type `int` is to be replaced by a user-defined name. The following function needs to remember only the last two remainders in the sequence, and these remainders are always stored in the formal arguments `a` and `b`.

Algorithm 1.1: EUCLIDEAN GCD

```
int gcd ( int a, int b )
{
    int r;

    if ((a == 0) && (b == 0)) return 0;
    if (a < 0) a = -a;
    if (b < 0) b = -b;
    while (b != 0) {
        r = a % b;
        a = b;
        b = r;
    }
    return a;
}
```

1.2.2 Extended GCD Algorithm

An extremely important byproduct available during the computation of gcd's is the *Bézout relation* defined by the following theorem.[8]

Theorem 1.17 [*Bézout relation*] *For $a, b \in \mathbb{Z}$, not both zero, there exist integers u, v satisfying $\gcd(a, b) = ua + vb$.* ◁

The computation of the multipliers u, v along with $\gcd(a, b)$ is referred to as the *extended gcd computation*. I first establish that the multipliers do exist. The remainder sequence obtained during the computation of $\gcd(a, b)$ gives $\gcd(a, b) = r_k = r_{k-2} - q_k r_{k-1} = r_{k-2} - q_k(r_{k-3} - q_{k-1}r_{k-2}) = -q_k r_{k-3} + (1 + q_k q_{k-1})r_{k-2} = \cdots$, that is, r_k is expressed first as a linear combination of r_{k-2} and r_{k-1}, then as a linear combination of r_{k-3} and r_{k-2}, and so on until r_k is available as a linear combination of r_0 and r_1 (that is, of a and b). The quotients $q_k, q_{k-1}, \ldots, q_2$ are used in that order for computing the desired Bézout relation and so must be remembered until the values of the multipliers u, v for r_0, r_1 are computed.

---

[8]Étienne Bézout (1730–1783) was a French mathematician renowned for his work on solutions of algebraic equations (see Theorem 4.27 for example).

There is an alternative way to compute the multipliers u, v without remembering the intermediate quotients. Algorithm 1.2 implements this new idea. The proof of its correctness is based on a delicate invariant of the extended Euclidean gcd loop.

For computing $\gcd(a, b)$, we obtain the remainder sequence $r_0, r_1, r_2, \ldots$. In addition, we also compute two sequences $u_0, u_1, u_2, \ldots$ and $v_0, v_1, v_2, \ldots$ such that for all $i \geqslant 0$ we maintain the relation $r_i = u_i a + v_i b$. We initialize the three sequences as $r_0 = a, r_1 = b, u_0 = 1, u_1 = 0, v_0 = 0, v_1 = 1$, so that the relation $r_i = u_i a + v_i b$ is satisfied for $i = 0, 1$.

Assume that during the computation of r_i, we have $r_j = u_j a + v_j b$ for all $j = 0, 1, \ldots, i - 1$. Euclidean division gives $r_i = r_{i-2} - q_i r_{i-1}$. We analogously assign $u_i = u_{i-2} - q_i u_{i-1}$ and $v_i = v_{i-2} - q_i v_{i-1}$. By the induction hypothesis,

$$r_{i-2} = u_{i-2}a + v_{i-2}b, \quad \text{and}$$
$$r_{i-1} = u_{i-1}a + v_{i-1}b, \quad \text{so that}$$
$$r_i = r_{i-2} - q_i r_{i-1} = (u_{i-2} - q_i u_{i-1})a + (v_{i-2} - q_i v_{i-1}) = u_i a + v_i b.$$

Therefore, the invariance continues to hold for $j = i$ too.

Example 1.18 The following table illustrates the extended gcd computation for $252, 91$ (Also see Example 1.16).

i	q_i	r_i	u_i	v_i	$u_i a + v_i b$
Initialization					
0	–	252	1	0	252
1	–	91	0	1	91
Iterations					
2	2	70	1	−2	70
3	1	21	−1	3	21
4	3	7	4	−11	7
5	3	0	−13	36	0

Therefore, $\gcd(252, 91) = 7 = 4 \times 252 + (-11) \times 91$. □

Before the algorithm implementing this idea is presented, some comments are in order. The computation of r_i, u_i, v_i in an iteration requires the values $r_{i-1}, u_{i-1}, v_{i-1}$ and $r_{i-2}, u_{i-2}, v_{i-2}$ only from two previous iterations. We, therefore, need to store only the values from two previous iterations, that is, we store r_i, u_i, v_i as r_0, u_0, v_0 (refers to the current iteration), $r_{i-1}, u_{i-1}, v_{i-1}$ as r_1, u_1, v_1 (the previous iteration), and $r_{i-2}, u_{i-2}, v_{i-2}$ as r_2, u_2, v_2 (the second previous iteration). Moreover, the quotient q_i calculated by the Euclidean division of r_{i-2} by r_{i-1} is never needed after r_i, u_i, v_i are calculated. This eliminates the need for storing the quotient sequence. Finally, observe that there is no need to compute both the u and v sequences inside the loop. This is because the relation $r_i = u_i a + v_i b$ is always maintained, so we can compute

$v_i = (r_i - u_i a)/b$ from the values r_i, u_i, a, b only. Furthermore, the computation of $u_i = u_{i-2} - q_i u_{i-1}$ does not require the values from the v sequence.

Algorithm 1.2 incorporates all these intricate details. Each iteration of the gcd loop performs only a constant number of integer operations. Moreover, like the basic gcd calculation (Algorithm 1.1), the loop is executed $O(\lg(\min(a, b)))$ times. To sum up, the extended gcd algorithm is slower by the basic gcd algorithm by only a small constant factor.

Algorithm 1.2: EXTENDED EUCLIDEAN GCD

```
int egcd ( int a, int b, int *u, int *v )
{
    int q0, r0, r1, r2, u0, u1, u2;

    if ((a == 0) && (b == 0)) { *u = *v = 0;  return 0; }
    r2 = (a < 0) ? -a : a;   r1 = (b < 0) ? -b : b;
    u2 = 1;  u1 = 0;
    while (r1 != 0) {
        q0 = r2 / r1;
        r0 = r2 - q0 * r1;   r2 = r1;   r1 = r0;
        u0 = u2 - q0 * u1;   u2 = u1;   u1 = u0;
    }
    *u = u2;  if (a < 0) *u = -(*u);
    *v = (b == 0) ? 0 : (r2 - (*u) * a) / b;
    return r2;
}
```

GP/PARI provides the built-in function gcd for computing the gcd of two integers. An extended gcd can be computed by the built-in function bezout which returns the multipliers and the gcd in the form of a 3-dimensional vector.

```
gp > gcd(2^77+1,2^91+1)
%1 = 129
gp > bezout(2^77+1,2^91+1)
%2 = [-151124951386849816870911, 9223935021170032768, 129]
gp > (-151124951386849816870911) * (2^77+1) + 9223935021170032768 * (2^91+1)
%3 = 129
```

1.2.3   Binary GCD Algorithm

Since computing gcd's is a very basic operation in number theory, several alternatives to the Euclidean algorithm have been studied extensively by the research community. The Euclidean algorithm involves a division in each step. This turns out to be a costly operation, particularly for multiple-precision integers. Here, I briefly describe the binary gcd algorithm[9] (Algorithm 1.3) which performs better than the Euclidean gcd algorithm. This improvement in performance is achieved by judiciously replacing Euclidean division by considerably faster subtraction and bit-shift operations. Although the number of iterations is typically larger in the binary gcd loop than in the Euclidean gcd loop, the reduction of running time per iteration achieved by the binary gcd algorithm usually leads to a more efficient algorithm for gcd computation.

Let $\gcd(a, b)$ be computed. We first write $a = 2^s a'$ and $b = 2^t b'$ with a', b' odd. Since $\gcd(a, b) = 2^{\min(s,t)} \gcd(a', b')$, we may assume that we are going to compute the gcd of two odd integers, that is, a, b themselves are odd. First assume that $a > b$. We have $\gcd(a, b) = \gcd(a - b, b)$. But $a - b$ is even and so can be written as $a - b = 2^r \alpha$. Since b is odd, it turns out that $\gcd(a, b) = \gcd(\alpha, b)$. Since $\alpha = (a - b)/2^r \leqslant (a - b)/2 \leqslant a/2$, replacing the computation of $\gcd(a, b)$ by $\gcd(\alpha, b)$ implies reduction of the bit-size of the first operand by at least one. The case $a < b$ can be symmetrically treated. Finally, if $a = b$, then $\gcd(a, b) = a$.

Algorithm 1.3: BINARY GCD

```
int bgcd ( int a, int b )
{
    int s,t;

    if ((a == 0) && (b == 0)) return 0;
    if (a < 0) a = -a;   if (b < 0) b = -b;
    if (a == 0) return b;   if (b == 0) return a;
    s = 0; while ((a & 1) == 0) { ++s;   a = a >> 1; }
    t = 0; while ((b & 1) == 0) { ++t;   b = b >> 1; }
    while (b > 0) {
        if (a > b) {
            a = a - b; while ((a & 1) == 0) a = a >> 1;
        } else if (a < b) {
            b = b - a; while ((b & 1) == 0) b = b >> 1;
        } else b = 0;
    }
    return a << ((s <= t) ? s : t);
}
```

It is evident that the binary gcd algorithm performs at most $\lg a + \lg b$ iterations of the gcd loop. For improving the efficiency of the algorithm, bit operations (like >> for shifting, & for checking the least significant bit) are used instead of equivalent arithmetic operations. If a and b vary widely in

---

[9]Josef Stein, Computational problems associated with Racah algebra, *Journal of Computational Physics*, 1(3), 397–405, 1967. This algorithm seems to have been known in ancient China.

size, replacing the first iteration by a Euclidean division often improves the performance of the algorithm considerably.

Like the extended Euclidean gcd algorithm, one can formulate the extended binary gcd algorithm. The details are left to the reader as Exercise 1.13.

Sorenson[10] extends the concept of binary gcd to k-ary gcd for $k \geqslant 2$. Another gcd algorithm tailored to multiple-precision integers is the *Lehmer gcd algorithm.*[11]

1.3 Congruences and Modular Arithmetic

The notion of divisibility of integers leads to the concept of congruences which have far-reaching consequences in number theory.[12]

Definition 1.19 Let $m \in \mathbb{N}$. Two integers $a, b \in \mathbb{Z}$ are called *congruent modulo m*, denoted $a \equiv b \pmod{m}$, if $m \mid (a - b)$ or, equivalently, if a rem $m = b$ rem m. In this case, m is called the *modulus* of the congruence. ◁

Example 1.20 Consider the modulus $m = 5$. The integers congruent to 0 modulo 5 are $0, \pm 5, \pm 10, \pm 15, \ldots$. The integers congruent to 1 modulo 5 are $\ldots, -9, -4, 1, 6, 11, \ldots$, and those congruent to 3 modulo 5 are $\ldots, -7, -2, 3, 8, 13, \ldots$.

In general, the integers congruent to a modulo m are $a + km$ for all $k \in \mathbb{Z}$. □

Some basic properties of the congruence relation are listed now.

Proposition 1.21 *Let $m \in \mathbb{N}$ and $a, b, c, d \in \mathbb{Z}$ be arbitrary.*
(a) $a \equiv a \pmod{m}$.
(b) If $a \equiv b \pmod{m}$, then $b \equiv a \pmod{m}$.
(c) If $a \equiv b \pmod{m}$ and $b \equiv c \pmod{m}$, then $a \equiv c \pmod{m}$.
(d) If $a \equiv c \pmod{m}$ and $b \equiv d \pmod{m}$, then $a + b \equiv c + d \pmod{m}$, $a - b \equiv c - d \pmod{m}$, and $ab \equiv cd \pmod{m}$.
(e) If $a \equiv b \pmod{m}$, and $f(x)$ is a polynomial with integer coefficients, then $f(a) \equiv f(b) \pmod{m}$.
(f) If $a \equiv b \pmod{m}$ and d is a positive integer divisor of m, then $a \equiv b \pmod{d}$.
(g) $ab \equiv ac \pmod{m}$ if and only if $b \equiv c \pmod{\frac{m}{\gcd(a,m)}}$. ◁

---

[10]Jonathan Sorenson, Two fast GCD algorithms, *Journal of Algorithms*, 16(1), 110–144, 1994.

[11]Jonathan Sorenson, An analysis of Lehmer's Euclidean gcd algorithm, *ISSAC*, 254–258, 1995.

[12]The concept of congruences was formalized by the Swiss mathematician Leonhard Euler (1707–1783) renowned for his contributions to several branches of mathematics. Many basic notations we use nowadays (including congruences and functions) were introduced by Euler.

All the parts of the proposition can be easily verified using the definition of congruence. Part (g) in the proposition indicates that one should be careful while canceling a common factor from the two sides of a congruence relation. Such a cancellation should be accompanied by dividing the modulus by the gcd of the modulus with the factor being canceled.

Definition 1.22 Let $m \in \mathbb{N}$. A set of m integers $a_0, a_1, a_2, \ldots, a_{m-1}$ is said to constitute a *complete residue system* modulo m if every integer $a \in \mathbb{Z}$ is congruent modulo m to one and only one of the integers a_i for $0 \leqslant i \leqslant m-1$. Evidently, no two distinct integers a_i, a_j in a complete residue system can be congruent to one another. ◁

Example 1.23 The integers $0, 1, 2, \ldots, m-1$ constitute a complete residue system modulo m. So do the integers $1, 2, 3, \ldots, m$. If $a_0, a_1, \ldots, a_{m-1}$ constitute a complete residue system modulo m, then so also do $aa_0, aa_1, \ldots, aa_{m-1}$ for an integer a if and only if a is coprime to m. For example, if m is odd, then $0, 2, 4, \ldots, 2m-2$ is a complete residue system modulo m. □

A complete residue system modulo m is denoted by $\mathbb{Z}_m$. Take arbitrary elements $a, b \in \mathbb{Z}_m$. Consider the integer sum $a + b$. Since $\mathbb{Z}_m$ is a complete residue system modulo m, there exists a unique integer $c \in \mathbb{Z}_m$ such that $c \equiv a + b \pmod{m}$. We define the sum of a and b in $\mathbb{Z}_m$ to be this integer c. The difference and product of a and b in $\mathbb{Z}_m$ can be analogously defined. These operations make $\mathbb{Z}_m$ a commutative ring with identity. The additive identity of $\mathbb{Z}_m$ is its member congruent to 0 modulo m, whereas the multiplicative identity of $\mathbb{Z}_m$ is that member of $\mathbb{Z}_m$, which is congruent to 1 modulo m.

$\mathbb{Z}$ is a commutative ring (in fact, an integral domain) under addition and multiplication of integers. Congruence modulo a fixed modulus m is an equivalence relation on $\mathbb{Z}$. Indeed the set $m\mathbb{Z} = \{km \mid k \in \mathbb{Z}\}$ is an ideal of $\mathbb{Z}$. The quotient ring $\mathbb{Z}/m\mathbb{Z}$ (that is, the set of all equivalence classes under congruence modulo m) is the algebraic description of the set $\mathbb{Z}_m$.

Example 1.24 The most common representation of $\mathbb{Z}_m$ is the set $\{0, 1, 2, \ldots, m-1\}$. For $a, b \in \mathbb{Z}_m$, we can define the arithmetic operations as follows.

$$a + b \;(\text{mod } m) \;=\; \begin{cases} a + b & \text{if } a + b < m, \\ a + b - m & \text{if } a + b \geqslant m. \end{cases}$$

$$a - b \;(\text{mod } m) \;=\; \begin{cases} a - b & \text{if } a \geqslant b, \\ a - b + m & \text{if } a < b. \end{cases}$$

$$ab \;(\text{mod } m) \;=\; (ab) \text{ rem } m.$$

Here, the operations and relations on the right sides pertain to standard integer arithmetic.

As a specific example, take $m = 17$, $a = 5$, and $b = 8$. Then $a + b$ as an integer is 13 which is less than the modulus. So $a + b \pmod{m}$ is 13. Since $a < b$ as integers, the difference $a - b \pmod{m}$ is equal to $5 - 8 + 17 = 14$. Also $5 \times 8 \pmod{m} = (5 \times 8) \text{ rem } 17 = 6$.

Unless otherwise mentioned, I will let $\mathbb{Z}_m$ stand for the standard residue system $\{0, 1, 2, \ldots, m-1\}$ rather than for any arbitrary residue system. □

GP/PARI supports the standard representation of $\mathbb{Z}_m$ as $\{0, 1, 2, \ldots, m-1\}$. An element $a \in \mathbb{Z}_m$ is represented as Mod(a,m). This notation is meant to differentiate between a as an integer and a considered as an element of $\mathbb{Z}_m$. A sample conversation with the GP/PARI interpreter follows.

```
gp > m = 17
%1 = 17
gp > a = Mod(5,m)
%2 = Mod(5, 17)
gp > b = Mod(25,m)
%3 = Mod(8, 17)
gp > a + b
%4 = Mod(13, 17)
gp > a - b
%5 = Mod(14, 17)
gp > a * b
%6 = Mod(6, 17)
gp > a / b
%7 = Mod(7, 17)
gp > 7 * a
%8 = Mod(1, 17)
gp > a^7
%9 = Mod(10, 17)
```

Here, $7a$ stands for $a + a + \cdots + a$ (7 times), whereas a^7 stands for $a \times a \times \cdots \times a$ (7 times). But what is $a/b \pmod{m}$? Like other situations, a/b is defined as the product of a and the multiplicative inverse of b. Let us, therefore, look at what is meant by modular multiplicative inverses and how to compute them.

Definition 1.25 An element $a \in \mathbb{Z}_m$ is said to be *invertible* modulo m (or a *unit* in $\mathbb{Z}_m$) if there exists an integer u (in $\mathbb{Z}_m$) such that $ua \equiv 1 \pmod{m}$. ◁

Example 1.26 Let $m = 15$. The element 7 is invertible modulo 15, since $13 \times 7 \equiv 1 \pmod{15}$. On the other hand, the element 6 of $\mathbb{Z}_{15}$ is not invertible. I prove this fact by contradiction, that is, I assume that u is an inverse of 6 modulo 15. This means that $6u \equiv 1 \pmod{15}$, that is, $15|(6u - 1)$, that is, $6u - 1 = 15k$ for some $k \in \mathbb{Z}$, that is, $3(2u - 5k) = 1$. This is impossible, since the left side is a multiple of 3, whereas the right side is not. □

Theorem 1.27 *An element $a \in \mathbb{Z}_m$ is invertible if and only if* $\gcd(a, m) = 1$. *Proof* First, suppose that $\gcd(a, m) = 1$. By Bézout's theorem, there exist integers u, v for which $1 = ua + vm$. But then $ua \equiv 1 \pmod{m}$. Conversely, suppose that a is invertible modulo m, that is, $ua \equiv 1 \pmod{m}$ for some integer u. But then $ua - km = 1$ for some $k \in \mathbb{Z}$. The left side is divisible by $\gcd(a, m)$, that is, $\gcd(a, m)|1$, that is, $\gcd(a, m) = 1$. ◁

The proof of Theorem 1.27 indicates that in order to compute the inverse of $a \in \mathbb{Z}_m$, one can compute the extended gcd $d = ua + vm$. If $d > 1$, then a is not invertible modulo m. If $d = 1$, (the integer in $\mathbb{Z}_m$ congruent modulo m to) u is the (unique) inverse of a modulo m.

Example 1.28 Let us compute the inverse of 11 modulo 15. Extended gcd calculations give $\gcd(11, 15) = 1 = (-4) \times 11 + 3 \times 15$, that is, $11^{-1} \equiv -4 \equiv 11 \pmod{15}$, that is, 11 is its own inverse modulo 15.

On the other hand, if we try to invert 12 modulo 15, we obtain the Bézout relation $\gcd(12, 15) = 3 = (-1) \times 12 + 1 \times 15$. Since 12 and 15 are not coprime, 12 does not have a multiplicative inverse modulo 15. $\qquad\square$

Definition 1.29 Let $m \in \mathbb{N}$. A set of integers $a_1, a_2, \ldots, a_l$ is called a *reduced residue system* modulo m if every integer a coprime to m is congruent to one and only one of the integers a_i, $1 \leqslant i \leqslant l$. Elements of a reduced residue system are themselves coprime to m and are not congruent modulo m to one another. Every complete residue system modulo m contains a reduced residue system modulo m.

Every reduced residue system modulo m has the same size. This size is denoted by $\phi(m)$. The function $\phi(m)$ of m is called *Euler's phi function* or *Euler's totient function*. $\qquad\triangleleft$

Example 1.30 The complete residue system $\{0, 1, 2, \ldots, m - 1\}$ modulo m contains the reduced residue system $\{a \mid 0 \leqslant a \leqslant m - 1, \gcd(a, m) = 1\}$. In view of this, $\phi(m)$ is often defined as the number of integers between 0 and $m - 1$ (or between 1 and m), that are coprime to m.

If $a_1, a_2, \ldots, a_l$ constitute a reduced residue system modulo m and if $\gcd(a, m) = 1$, then $aa_1, aa_2, \ldots, aa_l$ again constitute a reduced residue system modulo m.

As a specific example, take $m = 15$. The integers between 0 and 14 that are coprime to 15 are $1, 2, 4, 7, 8, 11, 13, 14$. Thus $\phi(15) = 8$. Since 15 is odd, $\{2, 4, 8, 14, 16, 22, 26, 28\}$ is another reduced residue system modulo 15. $\qquad\square$

A reduced residue system modulo m is denoted by the symbol $\mathbb{Z}_m^*$. The standard representation of $\mathbb{Z}_m^*$ is the set

$$\mathbb{Z}_m^* = \{a \mid 0 \leqslant a \leqslant m - 1, \ \gcd(a, m) = 1\}.$$

Algebraically, $\mathbb{Z}_m^*$ is the set of all units of $\mathbb{Z}_m$. It is a commutative group under multiplication modulo m.

Two important theorems involving the phi function follow.

Theorem 1.31 [*Euler's theorem*] *Let $m \in \mathbb{N}$ and $\gcd(a, m) = 1$. Then $a^{\phi(m)} \equiv 1 \pmod{m}$.*

Proof Let $a_1, a_2, \ldots, a_{\phi(m)}$ constitute a reduced residue system modulo m. Since $\gcd(a, m) = 1$, the integers $aa_1, aa_2, \ldots, aa_{\phi(m)}$ too constitute a reduced residue system modulo m. In particular, we have $(aa_1)(aa_2)\cdots(aa_{\phi(m)}) \equiv$

$a_1 a_2 \cdots a_{\phi(m)}$ (mod m). Each a_i is invertible modulo m, that is, $\gcd(a_i, m) = 1$. Thus we can cancel the factors a_i for $i = 1, 2, \ldots, \phi(m)$ from the congruence, and obtain the desired result. $\triangleleft$

The special case of Euler's theorem applied to prime moduli follows.

Theorem 1.32 [*Fermat's little theorem*][13] *Let $p \in \mathbb{P}$, and a an integer not divisible by p. Then, $a^{p-1} \equiv 1$ (mod p). For any integer b, we have $b^p \equiv b$ (mod p).*
Proof Since p is a prime, every integer between 1 and $p - 1$ is coprime to p and so $\phi(p) = p - 1$. Euler's theorem implies $a^{p-1} \equiv 1$ (mod p). For proving the second part, first consider the case $p \nmid b$. In that case, $\gcd(b, p) = 1$, so $b^{p-1} \equiv 1$ (mod p). Multiplying by b gives $b^p \equiv b$ (mod p). On the other hand, if $p | b$, both b^p and b are congruent to 0 modulo p. $\triangleleft$

An explicit formula for $\phi(m)$ can be derived.

Theorem 1.33 *Let $m = p_1^{e_1} \ldots p_r^{e_r}$ be the prime factorization of m with pairwise distinct primes $p_1, \ldots, p_r$ and with each of $e_1, \ldots, e_r$ positive. Then,*

$$\phi(m) = (p_1^{e_1} - p_1^{e_1-1}) \cdots (p_r^{e_r} - p_r^{e_r-1}) = m \prod_{p|m} \left(1 - \frac{1}{p}\right),$$

where the last product is over the set of all (distinct) prime divisors of m.
Proof Consider the standard residue system modulo m. An integer (between 0 and $m - 1$) is coprime to m if and only if it is divisible by neither of the primes $p_1, \ldots, p_r$. We can use a combinatorial argument based on the principle of inclusion and exclusion in order to derive the given formula for $\phi(m)$. $\triangleleft$

Example 1.34 $98 = 2 \times 7^2$, so $\phi(98) = (2 - 1) \times (7^2 - 7) = 42$. As another example, $100 = 2^2 \times 5^2$, so $\phi(100) = (2^2 - 2) \times (5^2 - 5) = 40$. Finally, 101 is a prime, so $\phi(101) = 100$. $\square$

The call `eulerphi(m)` returns $\phi(m)$ in the `GP/PARI` calculator.

```
gp > eulerphi(98)
%1 = 42
gp > eulerphi(99)
%2 = 60
gp > eulerphi(100)
%3 = 40
gp > eulerphi(101)
%4 = 100
gp > factor(2^101-1)
```

---

[13]Pierre de Fermat (1601–1665), a French lawyer, was an amateur mathematician having significant contributions in number theory. Fermat is famous for his last theorem which states that the equation $x^n + y^n = z^n$ does not have integer solutions with $xyz \neq 0$ for all integers $n \geqslant 3$. See Footnote 1 in Chapter 4 for a historical sketch on Fermat's last theorem.

```
%5 =
[7432339208719 1]

[341117531003194129 1]

gp > (7432339208719-1)*(341117531003194129-1)
%6 = 2535301200456117678030064007904
gp > eulerphi(2^101-1)
%7 = 2535301200456117678030064007904
```

1.3.1 Modular Exponentiation

Modular exponentiation is a very important primitive useful in a variety of computational problems. Given $m \in \mathbb{N}$ and integers a, e, our task is to compute $a^e \pmod{m}$. If $e = -d$ is negative (where $d > 0$), then $a^e \equiv a^{-d} \equiv (a^{-1})^d \pmod{m}$, that is, we first compute the modular inverse of a and then raise this inverse to the positive exponent d. In view of this, it suffices to restrict our attention to positive exponents only. We have $a^e \equiv a \times a \times \cdots \times a$ (e times) $\pmod{m}$, that is, $e - 1$ modular multiplications yield $a^e \pmod{m}$. If e is large, this is a very inefficient algorithm for computing the modular exponentiation. A technique called the *square-and-multiply algorithm* computes $a^e \pmod{m}$ using only $\mathrm{O}(\log e)$ modular multiplications.

Let $e = (e_{s-1}e_{s-2}\ldots e_1 e_0)_2$ be the binary representation of the exponent e, where each $e_i \in \{0, 1\}$. Define the partial exponents $x_i = (e_{s-1}e_{s-2}\ldots e_i)_2$ for $i = s, s-1, \ldots, 1, 0$, where x_s is to be interpreted as 0. Also, denote $b_i \equiv a^{x_i} \pmod{m}$. We initially have $b_s \equiv 1 \pmod{m}$, and then keep on computing $b_{s-1}, b_{s-2}, \ldots$ in that sequence until $b_0 \equiv a^e \pmod{m}$ is computed. Since $x_i = 2x_{i+1} + e_i$, we have $b_i \equiv a^{x_i} \equiv (a^{x_{i+1}})^2 \times a^{e_i} \equiv b_{i+1}^2 \times (a^{e_i}) \pmod{m}$. These observations lead to the following modular exponentiation algorithm.

Algorithm 1.4: SQUARE-AND-MULTIPLY MODULAR EXPONENTIATION

```
Initialize t = 1 (mod m).
for (i = s - 1; i >= 0; --i) {
    Set t = t² (mod m).
    If (eᵢ equals 1), set t = ta (mod m).
}
Return t.
```

Algorithm 1.4 requires the computation of s modular squares and $\leqslant s$ modular multiplications. The bit-size s of the exponent e satisfies $s = \mathrm{O}(\lg e)$, that is, the square-and-multiply loop is executed at most $\mathrm{O}(\lg e)$ times.

Example 1.35 Let us compute 7^{13} (mod 31). Here $m = 31$, $a = 7$, and $e = 13 = (1101)_2$. The following table summarizes the steps of the square-and-multiply algorithm on these parameters.

i	e_i	$x_i = (e_3 \ldots e_i)_2$	t (after sqr)	t (after mul)	b_i
4	–	0	–	–	1
3	1	$(1)_2 = 1$	$1^2 \equiv 1$ (mod 31)	$1 \times 7 \equiv 7$ (mod 31)	7
2	1	$(11)_2 = 3$	$7^2 \equiv 18$ (mod 31)	$18 \times 7 \equiv 2$ (mod 31)	2
1	0	$(110)_2 = 6$	$2^2 \equiv 4$ (mod 31)	(multiplication skipped)	4
0	1	$(1101)_3 = 13$	$4^2 \equiv 16$ (mod 31)	$16 \times 7 \equiv 19$ (mod 31)	19

Thus, $7^{13} \equiv 19$ (mod 31). □

1.3.2 Fast Modular Exponentiation

A modular multiplication (or squaring) is an integer multiplication followed by computing remainder with respect to the modulus. The division operation is costly. There are many situations where the overhead associated with these division operations can be substantially reduced. In a modular exponentiation algorithm (like Algorithm 1.4), the division is always carried out by the same modulus m. This fact can be exploited to speed up the exponentiation loop at the cost of some moderate precomputation before the loop starts. I will now explain one such technique called *Barrett reduction*.[14]

Let B be the base (like 2^{32} or 2^{64}) for representing multiple-precision integers. Barrett reduction works for any $B > 3$. Let the modulus m have the base-B representation $(m_{l-1}, m_{l-2}, \ldots, m_1, m_0)_B$. The quantity $T = \lfloor B^{2l}/m \rfloor$ is precomputed. Given a $2l$-digit integer product $x = (x_{2l-1}, x_{2l-2}, \ldots, x_1, x_0)_B$ of two elements of $\mathbb{Z}_m$ and the precomputed value T, Algorithm 1.5 computes x (mod m) without invoking the division algorithm.

Algorithm 1.5: BARRETT REDUCTION

```
Compute Q = ⌊x/B^(l-1)⌋, Q = QT, and Q = ⌊Q/B^(l+1)⌋.
Compute R = (x − Qm) (mod B^(l+1)).
While R ⩾ m, set R = R − m.
Return R.
```

Barrett reduction works as follows. One can express $x = qm + r$ with $0 \leqslant r \leqslant m-1$. We are interested in computing the remainder r. Algorithm 1.5 starts by computing an approximate value Q for the quotient q. Computing $q = \lfloor x/m \rfloor$ requires a division procedure. We instead compute

$$Q = \left\lfloor \frac{\left\lfloor \frac{x}{B^{l-1}} \right\rfloor \left\lfloor \frac{B^{2l}}{m} \right\rfloor}{B^{l+1}} \right\rfloor.$$

---

[14]Paul D. Barrett, Implementing the Rivest Shamir and Adleman public key encryption algorithm on a standard digital signal processor, *CRYPTO'86*, 311–332, 1987.

Easy calculations show that $q - 2 \leqslant Q \leqslant q$, that is, the approximate quotient Q may be slightly smaller than the actual quotient q. The computation of Q involves an integer multiplication QT and two divisions by powers of B. Under the base-B representation, $\lfloor y/B^i \rfloor$ is obtained by throwing away the i least significant B-ary digits from y. Therefore, Q can be efficiently computed.

In the next step, an approximate remainder R is computed. The actual remainder is $r = x - qm$. Algorithm 1.5 uses the approximate quotient Q to compute $R = x - Qm$. But $q - 2 \leqslant Q \leqslant q$, so R computed in this way satisfies $0 \leqslant R < 3m$. Moreover, since $m < B^l$ and $B > 3$, we have $R < B^{l+1}$. While performing the subtraction $x - Qm$, we may, therefore, pretend that x and Qm are $(l+1)$-digit integers. Even if the last subtraction results in a borrow, we are certain that this borrow will be adjusted in the higher digits. Thus, the computation of R involves one integer multiplication and one subtraction of $(l+1)$-digit integers, that is, R too can be computed efficiently.

From the approximate R, the actual r is computed by subtracting m as often as is necessary. Since $R < 3m$, at most two subtractions are required. Barrett remarks that in 90% of the cases, we have $R = r$, in 9% of the cases, we have $R = r + m$, and in only 1% of the cases, we have $R = r + 2m$.

Example 1.36 Let us represent multiple-precision integers in base $B = 2^8 = 256$. I now illustrate one modular multiplication step using Barrett reduction. The modulus is chosen as $m = 12345678 = (188, 97, 78)_B$ (so $l = 3$). The elements $a = 10585391 = (161, 133, 47)_B$ and $b = 8512056 = (129, 226, 56)_B$ are multiplied. Integer multiplication gives the product $x = ab = 90103440973896 = (81, 242, 215, 151, 160, 72)_B$. The quantity to be pre-computed is $T = \lfloor B^6/m \rfloor = 22799474 = (1, 91, 228, 114)_B$. In this example, T is an $(l+1)$-digit integer. We call Algorithm 1.5 with x, m and T as input.

Integer division of x by B^2 gives $Q = 1374869399 = (81, 242, 215, 151)_B$ (discard two least significant digits of x). Multiplication of this value with the precomputed T gives $31346299115896126 = (111, 93, 74, 255, 211, 125, 62)_B$. Finally, we divide this by B^4 (that is, discard four least significant digits) to obtain the approximate quotient as $Q = 7298378 = (111, 93, 74)_B$. It turns out that this Q is one smaller than the actual quotient $q = \lfloor x/m \rfloor$, but we do not need to worry about (nor even detect) this at this moment.

The approximate remainder R is computed as follows:

$$
\begin{array}{rrl}
x = & 90103440973896 = & (81, 242, 215, 151, 160, 72)_B \\
Qm = & 90103424710284 = & (81, 242, 214, 159, 118, 140)_B \\
\hline
R = x - Qm = & 16263612 \quad = \;(& 0, 248, \;\; 41, 188)_B
\end{array}
$$

This R happens to be larger than m (but smaller than $2m$), so the correct value of $r = x$ rem m is $r = R - m = 3917934 = (59, 200, 110)_B$. $\qquad\square$

1.4   Linear Congruences

Solving congruences modulo m is the same as solving equations in the ring $\mathbb{Z}_m$. A linear congruence is of the form

$$ax \equiv b \;(\text{mod } m),$$

where $m \in \mathbb{N}$ is a given modulus, and a, b are integers. In order to find all integers x satisfying the congruence, we use the following theorem.

Theorem 1.37 *Let $d = \gcd(a, m)$. The congruence $ax \equiv b \;(\text{mod } m)$ is solvable (for x) if and only if $d|b$. If $d|b$, then all solutions are congruent to each other modulo m/d, that is, there is a unique solution modulo m/d. In particular, if $\gcd(a, m) = 1$, then the congruence $ax \equiv b \;(\text{mod } m)$ has a unique solution modulo m.*

Proof   First, suppose that $ax \equiv b \;(\text{mod } m)$ is solvable. For each such solution x we have $ax - b = km$ for some $k \in \mathbb{Z}$. Since $d|a$ and $d|m$, we have $d|(ax - km)$, that is, $d|b$. Conversely, suppose $d|b$, that is, $b = db'$ for some $b' \in \mathbb{Z}$. Also let $a = da'$ and $m = dm'$ for some integers a', m'. The congruence $ax \equiv b \;(\text{mod } m)$ can be rewritten as $da'x \equiv db' \;(\text{mod } dm')$. Canceling the factor d gives an equivalent congruence $a'x \equiv b' \;(\text{mod } m')$. Since $\gcd(a', m') = 1$, a' is invertible modulo m'. Therefore, any integer $x \equiv (a')^{-1}b' \;(\text{mod } m')$ satisfies the congruence $a'x \equiv b' \;(\text{mod } m')$.

Let x_1, x_2 be two solutions of $ax \equiv b \;(\text{mod } m)$. We have $ax_1 \equiv b \equiv ax_2 \;(\text{mod } m)$. Canceling a yields $x_1 \equiv x_2 \;(\text{mod } m/d)$. $\triangleleft$

Since the original congruence is provided modulo m, it is desirable that we supply the solutions of the congruence modulo m (instead of modulo m/d). Let x_0 be a solution (unique) modulo m/d. All the elements in $\mathbb{Z}_m$ that are congruent to x_0 modulo m/d are $x_i = x_0 + i(m/d)$ for $i = 0, 1, 2, \ldots, d - 1$. Thus, $x_0, x_1, \ldots, x_{d-1}$ are all the solutions modulo m. In particular, there are exactly $d = \gcd(a, m)$ solutions modulo m.

Example 1.38 Take the congruence $21x \equiv 9 \;(\text{mod } 15)$. Here, $a = 21$, $b = 9$, and $m = 15$. Since $d = \gcd(a, m) = 3$ divides b, the congruence is solvable. Canceling 3 gives $7x \equiv 3 \;(\text{mod } 5)$, that is, $x \equiv 7^{-1} \times 3 \equiv 3 \times 3 \equiv 4 \;(\text{mod } 5)$. The solutions modulo 15 are $4, 9, 14$.

The congruence $21x \equiv 8 \;(\text{mod } 15)$ is not solvable, since $3 = \gcd(21, 15)$ does not divide 8. $\square$

1.4.1   Chinese Remainder Theorem

We now investigate how congruences modulo several moduli can be solved.

Theorem 1.39 [*Chinese Remainder Theorem (CRT)*]   *Let $m_1, m_2, \ldots, m_t$ be pairwise coprime moduli (that is, $\gcd(m_i, m_j) = 1$ whenever $i \neq j$). Moreover, let $a_1, a_2, \ldots, a_t$ be arbitrary integers. Then, the congruences*

$$x \equiv a_1 \pmod{m_1}, \quad x \equiv a_2 \pmod{m_2}, \quad \ldots, \quad x \equiv a_t \pmod{m_t},$$

are simultaneously solvable. The solution is unique modulo $m_1 m_2 \cdots m_t$.

Proof   A constructive proof of the Chinese remainder theorem is given here. Call $M = m_1 m_2 \cdots m_t$, and $n_i = M/m_i$ for all $i = 1, 2, \ldots, t$. For each i, compute the inverse of n_i modulo m_i, that is, we find an integer u_i for which $u_i n_i \equiv 1 \pmod{m_i}$. Since the moduli $m_1, m_2, \ldots, m_t$ are pairwise coprime,

$$n_i = \prod_{\substack{j=1 \\ j \neq i}}^{t} m_j \text{ is coprime to } m_i, \text{ and so the inverse } u_i \text{ exists. Finally, we take}$$

$x = \sum_{i=1}^{t} u_i n_i a_i$. It is an easy check that this x is a simultaneous solution of the given congruences.

In order to prove the desired uniqueness, let x, y be two integers satisfying the given congruences. We then have $x \equiv y \pmod{m_i}$, that is, $m_i | (x - y)$ for all i. Since m_i are pairwise coprime, it follows that $M | (x - y)$.   ◁

Example 1.40 Legend has it that Chinese generals counted soldiers using CRT. Suppose that there are $x \leqslant 1000$ soldiers in a group. The general asks the soldiers to stand in rows of seven. The number of soldiers left over is counted by the general. Let this count be five. The soldiers are subsequently asked to stand in rows of eleven and in rows of thirteen. In both cases, the numbers of leftover soldiers are counted (say, three and two). The general then uses a magic formula to derive the exact number of soldiers in the group.[15]

Basically, the general is finding the common solution of the congruences:

$$x \equiv 5 \pmod 7,$$
$$x \equiv 3 \pmod{11},$$
$$x \equiv 2 \pmod{13}.$$

We have $m_1 = 7$, $m_2 = 11$, $m_3 = 13$, $a_1 = 5$, $a_2 = 3$, and $a_3 = 2$. $M = m_1 m_2 m_3 = 1001$, and so $n_1 = M/m_1 = m_2 m_3 = 143$, $n_2 = M/m_2 = m_1 m_3 = 91$, and $n_3 = M/m_3 = m_1 m_2 = 77$. The inverses are $u_1 \equiv 143^{-1} \equiv 3^{-1} \equiv 5 \pmod 7$, $u_2 \equiv 91^{-1} \equiv 3^{-1} \equiv 4 \pmod{11}$, and $u_3 \equiv 77^{-1} \equiv 12^{-1} \equiv 12 \pmod{13}$. Therefore, the simultaneous solution is $x \equiv u_1 n_1 a_1 + u_2 n_2 a_2 + u_3 n_3 a_3 \equiv 5 \times 143 \times 5 + 4 \times 91 \times 3 + 12 \times 77 \times 2 \equiv 6515 \equiv 509 \pmod{1001}$. One can verify that this solution is correct: $509 = 72 \times 7 + 5 = 46 \times 11 + 3 = 39 \times 13 + 2$.

If it is known that $0 \leqslant x \leqslant 1000$ (as in the case of the Chinese general), the unique solution is $x = 509$.   □

---

[15]Irrespective of the veracity of this story, there is something Chinese about CRT. The oldest reference to the theorem appears in a third-century book by the Chinese mathematician Sun Tzu. In the sixth and seventh centuries, Indian mathematicians Aryabhatta and Brahmagupta studied the theorem more rigorously.

The proof of the CRT can be readily converted to Algorithm 1.6.

Algorithm 1.6: CHINESE REMAINDER THEOREM

```
int CRT ( int a[], int m[], int t, int *Mptr ) {
    int i, M = 1, n, x = 0, u;

    for (i=0; i<t; ++i) M *= m[i];
    for (i=0; i<t; ++i) {
        n = M / m[i];
        u = invMod(n%m[i],m[i]);
        x += n * u * a[i];
    }
    x %= M;
    if (x < 0) x += M;
    if (Mptr != NULL) *Mptr = M;
    return x;
}
```

GP/PARI supports the call chinese() for CRT-based combination. The function takes only two modular elements as arguments. The function combines the two elements, and returns an element modulo the product of the input moduli. In order to run CRT on more than two moduli, we need to make nested calls of chinese(). The function chinese() can handle non-coprime moduli also. An integer modulo the lcm of the input moduli is returned in this case. However, the input congruences may now fail to have a simultaneous solution. For example, there cannot exist an integer x satisfying both $x \equiv 5 \pmod{12}$ and $x \equiv 6 \pmod{18}$, since such an integer is of the form $18k + 6$ (a multiple of 6) and at the same time of the form $12k + 5$ (a non-multiple of 6).

```
gp > chinese(Mod(5,7),Mod(3,11))
%1 = Mod(47, 77)
gp > chinese(Mod(5,7),Mod(-3,11))
%2 = Mod(19, 77)
gp > chinese(chinese(Mod(5,7),Mod(3,11)),Mod(2,13))
%3 = Mod(509, 1001)
gp > chinese(Mod(47,77),Mod(2,13))
%4 = Mod(509, 1001)
gp > chinese(Mod(5,12),Mod(11,18))
%5 = Mod(29, 36)
gp > chinese(Mod(5,12),Mod(6,18))
  ***   incompatible arguments in chinois.
```

The incremental way of combining congruences for more than two moduli, as illustrated above for GP/PARI, may be a bit faster (practically, but not in terms of the order notation) than Algorithm 1.6 (see Exercise 1.44).

1.5   Polynomial Congruences

Linear congruences are easy to solve. We now look at polynomial congruences of higher degrees. This means that we plan to locate all the roots of a polynomial $f(x)$ with integer coefficients modulo a given $m \in \mathbb{N}$. We can substitute x in $f(x)$ by the elements of $\mathbb{Z}_m$ one by one, and find out for which values of x we get $f(x) \equiv 0 \pmod{m}$. This method is practical only if m is small. However, if the complete factorization of the modulus m is available, a technique known as *Hensel lifting* efficiently solves polynomial congruences.[16]

1.5.1   Hensel Lifting

Let $m = p_1^{e_1} \cdots p_r^{e_r}$ with pairwise distinct primes $p_1, \ldots, p_r$ and with $e_i \in \mathbb{N}$. If we know the roots of $f(x)$ modulo each $p_i^{e_i}$, we can combine these roots by the CRT in order to obtain all the roots of $f(x)$ modulo m. So it suffices to look at polynomial congruences of the form

$$f(x) \equiv 0 \pmod{p^e},$$

where p is a prime and $e \in \mathbb{N}$. Hensel lifting is used to obtain all the solutions of $f(x) \equiv 0 \pmod{p^{\epsilon+1}}$ from the solutions of $f(x) \equiv 0 \pmod{p^\epsilon}$. This means that the roots of $f(x)$ are first computed modulo p. These roots are then lifted to the roots of $f(x)$ modulo p^2, and the roots modulo p^2 are lifted to the roots modulo p^3, and so on.

Let ξ be a solution of $f(x) \equiv 0 \pmod{p^\epsilon}$. All integers that satisfy this congruence and that are congruent to this solution modulo p^ϵ are $\xi + kp^\epsilon$ for $k \in \mathbb{Z}$. We investigate which of these solutions continue to remain the solutions of $f(x) \equiv 0 \pmod{p^{\epsilon+1}}$ also. Let us write the polynomial $f(x)$ as

$$f(x) = a_d x^d + a_{d-1} x^{d-1} + \cdots + a_1 x + a_0.$$

Substituting $x = \xi + kp^\epsilon$ yields

$$\begin{aligned}
&f(\xi + kp^\epsilon) \\
=~ &a_d(\xi + kp^\epsilon)^d + a_{d-1}(\xi + kp^\epsilon)^{d-1} + \cdots + a_1(\xi + kp^\epsilon) + a_0 \\
=~ &(a_d \xi^d + a_{d-1}\xi^{d-1} + \cdots + a_1\xi + a_0) \\
&+ kp^\epsilon(da_d\xi^{d-1} + (d-1)a_{d-1}\xi^{d-2} + \cdots + a_1) + p^{2\epsilon}\alpha \\
=~ &f(\xi) + kp^\epsilon f'(\xi) + p^{2\epsilon}\alpha,
\end{aligned}$$

where α is some polynomial expression in ξ, and $f'(x)$ is the (formal) derivative of $f(x)$. Since $\epsilon \geqslant 1$, we have $2\epsilon \geqslant \epsilon + 1$, so modulo $p^{\epsilon+1}$ the above equation

---

[16]Kurt Wilhelm Sebastian Hensel (1861–1941) is a German mathematician whose major contribution is the introduction of p-adic numbers that find many uses in analysis, algebra and number theory. If the Hensel lifting procedure is carried out for all $e \in \mathbb{N}$, in the limit we get the p-adic solutions of $f(x) = 0$.

reduces to $f(\xi + kp^\epsilon) \equiv f(\xi) + kp^\epsilon f'(\xi) \pmod{p^{\epsilon+1}}$. We need to identify all values of k for which $f(\xi + kp^\epsilon) \equiv 0 \pmod{p^{\epsilon+1}}$. These are given by $kp^\epsilon f'(\xi) \equiv -f(\xi) \pmod{p^{\epsilon+1}}$. Since ξ is a solution for $f(x) \equiv 0 \pmod{p^\epsilon}$, we have $p^\epsilon | f(\xi)$. Canceling p^ϵ yields the linear congruence

$$f'(\xi)k \equiv -\frac{f(\xi)}{p^\epsilon} \pmod{p},$$

which has 0, 1, or p solutions for k depending on the values of $f'(\xi)$ and $\frac{f(\xi)}{p^\epsilon}$.

Each lifting step involves solving a linear congruence only. The problem of solving a polynomial congruence then reduces to solving the congruence modulo each prime divisor of the modulus. We will study root-finding algorithms for polynomials later in a more general setting.

Example 1.41 We find out all the solutions of the congruence

$$2x^3 - 7x^2 + 189 \equiv 0 \pmod{675}.$$

We have $f(x) = 2x^3 - 7x^2 + 189$, and so $f'(x) = 6x^2 - 14x$. The modulus admits the prime factorization $m = 3^3 \times 5^2$. We proceed step by step in order to obtain all the roots.

Solutions of $f(x) \equiv 0 \pmod 3$

We have $f(0) \equiv 189 \equiv 0 \pmod 3$, $f(1) \equiv 2 - 7 + 189 \equiv 1 \pmod 3$ and $f(2) \equiv 16 - 28 + 189 \equiv 0 \pmod 3$. Thus, the roots modulo 3 are $0, 2$.

Solutions of $f(x) \equiv 0 \pmod{3^2}$

Let us first lift the root $x \equiv 0 \pmod 3$. Since $f(0)/3 = 189/3 = 63$ and $f'(0) = 0$, the congruence $f'(0)k \equiv -\frac{f(0)}{3} \pmod 3$ is satisfied by $k = 0, 1, 2 \pmod 3$, and the lifted roots are $0, 3, 6$ modulo 9.

For lifting the root $x \equiv 2 \pmod 3$, we calculate $f(2)/3 = 177/3 = 59$ and $f'(2) = 24 - 28 = -4$. So the congruence $f'(2)k \equiv -\frac{f(2)}{3} \pmod 3$, that is, $-4k \equiv -59 \pmod 3$, has a unique solution $k \equiv 2 \pmod 3$. So there is a unique lifted root $2 + 2 \times 3 = 8$.

Therefore, all the roots of $f(x) \equiv 0 \pmod{3^2}$ are $0, 3, 6, 8$.

Solutions of $f(x) \equiv 0 \pmod{3^3}$

Let us first lift the root $x \equiv 0 \pmod{3^2}$. We have $f(0)/3^2 = 189/9 = 21$ and $f'(0) = 0$. The congruence $f'(0)k \equiv -\frac{f(0)}{3^2} \pmod 3$ is satisfied by $k = 0, 1, 2$, that is, there are three lifted roots $0, 9, 18$.

Next, we lift the root $x \equiv 3 \pmod{3^2}$. We have $f(3)/3^2 = (2 \times 27 - 7 \times 9 + 189)/9 = 6 - 7 + 21 = 20$, whereas $f'(3) = 6 \times 9 - 14 \times 3 = 12$. Thus, the congruence $f'(3)k \equiv -\frac{f(3)}{3^2} \pmod 3$ has no solutions, that is, the root $3 \pmod{3^2}$ does not lift to a root modulo 3^3.

For the root $x \equiv 6 \pmod{3^2}$, we have $f(6)/3^2 = (2 \times 216 - 7 \times 36 + 189)/9 = 48 - 28 + 21 = 41$ and $f'(6) = 216 - 84 = 132$, so there is no solution for k in the congruence $f'(6)k \equiv -\frac{f(6)}{3^2} \pmod 3$, that is, the root 6 does not lift to a root modulo 3^3.

Finally, consider the lifting of the last root $x \equiv 8 \pmod{3^2}$. We have $f(8)/3^2 = (2 \times 512 - 7 \times 64 + 189)/9 = 765/9 = 85$ and $f'(8) = 6 \times 64 - 14 \times 8 = 272$. Therefore, the congruence $f'(8)k \equiv -\frac{f(8)}{3^2} \pmod 3$, that is, $272k \equiv -85 \pmod 3$, that is, $2k \equiv 2 \pmod 3$ has a unique solution $k \equiv 1 \pmod 3$. That is, the unique lifted root is $8 + 1 \times 9 = 17$.

To sum up, all solutions modulo 3^3 are $0, 9, 17, 18$.

Solutions of $f(x) \equiv 0 \pmod 5$

We evaluate $f(x)$ at $x = 0, 1, 2, 3, 4$ and discover that $f(x) \equiv 0 \pmod 5$ only for $x = 3, 4$.

Solutions of $f(x) \equiv 0 \pmod{5^2}$

First, we try to lift the root $x \equiv 3 \pmod 5$. We have $f(3)/5 = (2 \times 27 - 7 \times 9 + 189)/5 = 180/5 = 36$ and $f'(3) = 6 \times 9 - 14 \times 3 = 12$. The congruence $f'(3)k \equiv -\frac{f(3)}{5} \pmod 5$, that is, $2k \equiv -1 \pmod 5$, has a unique solution $k \equiv 2 \pmod 5$, that is, there is a unique lifted root $3 + 2 \times 5 = 13$.

Next, we investigate the lifting of the root $x \equiv 4 \pmod 5$. We have $f(4)/5 = (2 \times 64 - 7 \times 16 + 189)/5 = 205/5 = 41$ and $f'(4) = 6 \times 16 - 14 \times 4 = 40$. The congruence $f'(4)k \equiv -\frac{f(4)}{5} \pmod 5$, that is, the congruence $40k \equiv -41 \pmod 5$ does not have a solution, that is, the root $4 \pmod 5$ does not lift to a root modulo 5^2.

To sum up, the only solution of $f(x) \equiv 0 \pmod{5^2}$ is $x \equiv 13 \pmod{25}$.

Solutions of $f(x) \equiv 0 \pmod{675}$

We finally combine the solutions modulo 27 and 25 using the CRT. Since $25^{-1} \equiv 13 \pmod{27}$ and $27^{-1} \equiv 13 \pmod{25}$, all the solutions of $f(x) \equiv 0 \pmod{675}$ are $25 \times 13 \times a + 27 \times 13 \times b \pmod{675}$ for $a \in \{0, 9, 17, 18\}$ and $b \in \{13\}$. The solutions turn out to be $x \equiv 63, 288, 513, 638 \pmod{675}$.   □

If the factorization of the modulus m is not available, we do not know any easy way to solve polynomial congruences in general. All we can do is first factoring m and then applying the above procedure. Even solving quadratic congruences modulo a composite m is known to be probabilistic polynomial-time equivalent to the problem of factoring m.

1.6   Quadratic Congruences

A special case of polynomial congruences is of the form $ax^2 + bx + c \equiv 0 \pmod m$. In view of the theory described in the previous section, it suffices to concentrate only on prime moduli p. The case $p = 2$ is easy to handle. So we assume that the modulus is an odd prime p. Since we are going to investigate (truly) quadratic polynomial congruences, we would assume $p \nmid a$. In that case, $2a$ is invertible modulo p, and the congruence $ax^2 + bx + c \equiv 0 \pmod p$ can be rewritten as $y^2 \equiv \alpha \pmod p$, where $y \equiv x + b(2a)^{-1} \pmod p$ and

$\alpha \equiv (b^2 - 4ac)(4a^2)^{-1} \pmod{p}$. This implies that it suffices to concentrate only on quadratic congruences of the special form

$$x^2 \equiv a \pmod{p}.$$

Gauss was the first mathematician to study quadratic congruences formally.[17]

1.6.1 Quadratic Residues and Non-Residues

In order to solve the quadratic congruence $x^2 \equiv a \pmod{p}$, we first need to know whether it is at all solvable. This motivates us to define the following.

Definition 1.42 Let p be an odd prime. An integer a with $\gcd(a, p) = 1$ is called a *quadratic residue* modulo p if the congruence $x^2 \equiv a \pmod{p}$ has a solution. An integer a with $\gcd(a, p) = 1$ is called a *quadratic non-residue* modulo p if the congruence $x^2 \equiv a \pmod{p}$ does not have a solution. An integer divisible by p is treated neither as a quadratic residue nor as a quadratic non-residue modulo p. ◁

Example 1.43 Consider the prime $p = 19$. Squaring individual elements of $\mathbb{Z}_{19}$ gives $0^2 \equiv 0 \pmod{19}$, $1^2 \equiv 18^2 \equiv 1 \pmod{19}$, $2^2 \equiv 17^2 \equiv 4 \pmod{19}$, $3^2 \equiv 16^2 \equiv 9 \pmod{19}$, $4^2 \equiv 15^2 \equiv 16 \pmod{19}$, $5^2 \equiv 14^2 \equiv 6 \pmod{19}$, $6^2 \equiv 13^2 \equiv 17 \pmod{19}$, $7^2 \equiv 12^2 \equiv 11 \pmod{19}$, $8^2 \equiv 11^2 \equiv 7 \pmod{19}$, and $9^2 \equiv 10^2 \equiv 5 \pmod{19}$. Thus, the quadratic residues in $\mathbb{Z}_{19}$ are $1, 4, 5, 6, 7, 9, 11, 16, 17$. Therefore, the quadratic non-residues modulo 19 are $2, 3, 8, 10, 12, 13, 14, 15, 18$. □

The above example can be easily generalized to conclude that modulo an odd prime p, there are exactly $(p-1)/2$ quadratic residues and exactly $(p-1)/2$ quadratic non-residues.

1.6.2 Legendre Symbol

Definition 1.44 Let p be an odd prime and a an integer not divisible by p. The *Legendre symbol* $\left(\frac{a}{p}\right)$ is defined as[18]

$$\left(\frac{a}{p}\right) = \begin{cases} 1 & \text{if } a \text{ is a quadratic residue modulo } p, \\ -1 & \text{if } a \text{ is a quadratic non-residue modulo } p. \end{cases}$$

It is sometimes convenient to take $\left(\frac{a}{p}\right) = 0$ for an integer a divisible by p. ◁

---

[17] Johann Carl Friedrich Gauss (1777–1855) was a German mathematician celebrated as one of the most gifted mathematicians of all ages. Gauss is often referred to as the *prince of mathematics* and also as the *last complete mathematician* (in the sense that he was the last mathematician who was conversant with all branches of contemporary mathematics). In his famous book *Disquisitiones Arithmeticae* (written in 1798 and published in 1801), Gauss introduced the terms *quadratic residues* and *non-residues*.

[18] Adrien-Marie Legendre (1752–1833) was a French mathematician famous for pioneering research in several important branches of mathematics.

Some elementary properties of the Legendre symbol are listed now.

Proposition 1.45 *Let p be an odd prime, and a, b integers. Then, we have:*

(a) *If $a \equiv b \pmod{p}$, then $\left(\frac{a}{p}\right) = \left(\frac{b}{p}\right)$.*

(b) *$\left(\frac{a}{p}\right) = \left(\frac{a \operatorname{rem} p}{p}\right)$.*

(c) *$\left(\frac{ab}{p}\right) = \left(\frac{a}{p}\right)\left(\frac{b}{p}\right)$.*

(d) *$\left(\frac{0}{p}\right) = 0$, $\left(\frac{1}{p}\right) = 1$, and $\left(\frac{a^2}{p}\right) = 1$ if $\gcd(a, p) = 1$.*

(e) *$\left(\frac{-1}{p}\right) = (-1)^{(p-1)/2}$, that is, $\left(\frac{-1}{p}\right) = 1$ if and only if $p \equiv 1 \pmod{4}$.*

(f) *$\left(\frac{2}{p}\right) = (-1)^{(p^2-1)/8}$, that is, $\left(\frac{2}{p}\right) = 1$ if and only if $p \equiv \pm 1 \pmod{8}$.* ◁

The following theorem leads to an algorithmic solution to the problem of computing the Legendre symbol $\left(\frac{a}{p}\right)$.

Theorem 1.46 *[Euler's criterion]* *Let p be an odd prime, and a any integer. Then, $\left(\frac{a}{p}\right) \equiv a^{(p-1)/2} \pmod{p}$.*

Proof The result is evident for $a \equiv 0 \pmod{p}$. So consider the case that $\gcd(a, p) = 1$. Let $\left(\frac{a}{p}\right) = 1$, that is, $x^2 \equiv a \pmod{p}$ has a solution, say b. But then $a^{(p-1)/2} \equiv b^{p-1} \equiv 1 \pmod{p}$ by Fermat's little theorem.

$\mathbb{Z}_p$ is a field, since every non-zero element in it is invertible. Every $a \in \mathbb{Z}_p^*$ satisfies $a^{p-1} - 1 \equiv (a^{(p-1)/2} - 1)(a^{(p-1)/2} + 1) \equiv 0 \pmod{p}$, that is, $a^{(p-1)/2} \equiv \pm 1 \pmod{p}$. All quadratic residues in $\mathbb{Z}_p^*$ satisfy $a^{(p-1)/2} \equiv 1 \pmod{p}$. But the congruence $a^{(p-1)/2} \equiv 1 \pmod{p}$ cannot have more than $(p-1)/2$ roots, so all the quadratic non-residues in $\mathbb{Z}_p^*$ must satisfy $a^{(p-1)/2} \equiv -1 \pmod{p}$. ◁

Example 1.47 Let $p = 541$. We have $\left(\frac{41}{541}\right) \equiv 41^{(541-1)/2} \equiv 1 \pmod{541}$, that is, 41 is a quadratic residue modulo 541. Also, $\left(\frac{51}{541}\right) \equiv 51^{(541-1)/2} \equiv -1 \pmod{541}$, that is, 51 is a quadratic non-residue modulo 541. □

A more efficient algorithm for the computation of $\left(\frac{a}{p}\right)$ follows from the *quadratic reciprocity law* which is stated without proof here.[19]

Theorem 1.48 *[The law of quadratic reciprocity]* *Let p, q be odd primes. Then, $\left(\frac{p}{q}\right) = (-1)^{(p-1)(q-1)/4} \left(\frac{q}{p}\right)$.* ◁

Example 1.49 Using the quadratic reciprocity law, we compute $\left(\frac{51}{541}\right)$ as

$$\left(\frac{51}{541}\right) = \left(\frac{3}{541}\right)\left(\frac{17}{541}\right)$$

---

[19]Conjectured by Legendre, the quadratic reciprocity law was first proved by Gauss. Indeed, Gauss himself published eight proofs of this law. At present, hundreds of proofs of this law are available in the mathematics literature.

$$= (-1)^{(3-1)(541-1)/4} \left(\frac{541}{3}\right) (-1)^{(17-1)(541-1)/4} \left(\frac{541}{17}\right)$$

$$= \left(\frac{541}{3}\right) \left(\frac{541}{17}\right) = \left(\frac{1}{3}\right) \left(\frac{14}{17}\right) = \left(\frac{14}{17}\right) = \left(\frac{2}{17}\right) \left(\frac{7}{17}\right)$$

$$= (-1)^{(17^2-1)/8} \left(\frac{7}{17}\right) = \left(\frac{7}{17}\right) = (-1)^{(7-1)(17-1)/4} \left(\frac{17}{7}\right) = \left(\frac{17}{7}\right)$$

$$= \left(\frac{3}{7}\right) = (-1)^{(3-1)(7-1)/4} \left(\frac{7}{3}\right) = -\left(\frac{7}{3}\right) = -\left(\frac{1}{3}\right) = -1.$$

Thus, 51 is a quadratic non-residue modulo 541. □

1.6.3 Jacobi Symbol

Calculating $\left(\frac{a}{p}\right)$ as in the last example has a drawback. We have to factor several integers during the process. For example, in the very first step, we need to factor 51. It would be useful if we can directly apply the quadratic reciprocity law to $\left(\frac{51}{541}\right)$, that is, if we can write $\left(\frac{51}{541}\right) = (-1)^{(51-1)(541-1)/4} \left(\frac{541}{51}\right) = \left(\frac{541}{51}\right) = \left(\frac{31}{51}\right)$. However, $\left(\frac{a}{p}\right)$ has so far been defined only for odd primes p, and so we require an extension of the Legendre symbol to work for non-prime denominators also.[20]

Definition 1.50 Let b be an odd positive integer having prime factorization $b = p_1 p_2 \cdots p_r$, where the (odd) primes $p_1, p_2, \ldots, p_r$ are not necessarily all distinct. For an integer a, we define the *Jacobi symbol* $\left(\frac{a}{b}\right)$ as

$$\left(\frac{a}{b}\right) = \prod_{i=1}^{r} \left(\frac{a}{p_i}\right).$$

Here, each $\left(\frac{a}{p_i}\right)$ is the Legendre symbol (extended to include the case $p_i | a$). ◁

If b is prime, the Jacobi symbol $\left(\frac{a}{b}\right)$ is the same as the Legendre symbol $\left(\frac{a}{b}\right)$. However, for composite b, the Jacobi symbol $\left(\frac{a}{b}\right)$ has no direct relationship with the solvability of the congruence $x^2 \equiv a \pmod{b}$. If $\left(\frac{a}{b}\right) = -1$, the above congruence is not solvable modulo at least one prime divisor of b, and consequently modulo b too. However, the value $\left(\frac{a}{b}\right) = 1$ does not immediately imply that the congruence $x^2 \equiv a \pmod{b}$ is solvable. For example, the congruences $x^2 \equiv 2 \pmod 3$ and $x^2 \equiv 2 \pmod 5$ are both unsolvable, so that $\left(\frac{2}{3}\right) = \left(\frac{2}{5}\right) = -1$. By definition, we then have $\left(\frac{2}{15}\right) = \left(\frac{2}{3}\right) \left(\frac{2}{5}\right) = 1$, whereas the congruence $x^2 \equiv 2 \pmod{15}$ is clearly unsolvable.

A loss of connection with the solvability of quadratic congruences is not a heavy penalty to pay. We instead gain something precious, namely the law

---

[20]The Jacobi symbol was introduced in 1837 by the German mathematician Carl Gustav Jacob Jacobi (1804–1851).

of quadratic reciprocity continues to hold for the generalized Jacobi symbol. That was precisely the motivation for the generalization. Indeed, the Jacobi symbol possesses properties identical to the Legendre symbol.

Proposition 1.51 *For odd positive integers b, b', and for any integers a, a', we have:*

(a) $\left(\frac{aa'}{b}\right) = \left(\frac{a}{b}\right)\left(\frac{a'}{b}\right)$.

(b) $\left(\frac{a}{bb'}\right) = \left(\frac{a}{b}\right)\left(\frac{a}{b'}\right)$.

(c) *If $a \equiv a' \pmod{b}$, then $\left(\frac{a}{b}\right) = \left(\frac{a'}{b}\right)$.*

(d) $\left(\frac{a}{b}\right) = \left(\frac{a \text{ rem } b}{b}\right)$.

(e) $\left(\frac{-1}{b}\right) = (-1)^{(b-1)/2}$.

(f) $\left(\frac{2}{b}\right) = (-1)^{(b^2-1)/8}$.

(g) *[Law of quadratic reciprocity]* $\left(\frac{b}{b'}\right) = (-1)^{(b-1)(b'-1)/4}\left(\frac{b'}{b}\right)$. ◁

Example 1.52 Let us compute $\left(\frac{51}{541}\right)$ without making any factoring attempts. At some steps, we may have to extract powers of 2, but that is doable efficiently by bit operations only. $\left(\frac{51}{541}\right) = (-1)^{(51-1)(541-1)/4}\left(\frac{541}{51}\right) = \left(\frac{31}{51}\right) = (-1)^{(31-1)(51-1)/4}\left(\frac{51}{31}\right) = -\left(\frac{51}{31}\right) = -\left(\frac{20}{31}\right) = -\left(\frac{2}{31}\right)^2\left(\frac{5}{31}\right) = -\left(\frac{5}{31}\right) = -(-1)^{(5-1)(31-1)/4}\left(\frac{31}{5}\right) = -\left(\frac{31}{5}\right) = -\left(\frac{1}{5}\right) = -1$. □

The **GP/PARI** interpreter computes the Jacobi symbol $\left(\frac{a}{b}\right)$, when the call `kronecker(a,b)` is made.[21] Here are some examples.

```
gp > kronecker(41,541)
%1 = 1
gp > kronecker(51,541)
%2 = -1
gp > kronecker(2,15)
%3 = 1
gp > kronecker(2,45)
%4 = -1
gp > kronecker(21,45)
%5 = 0
```

For an odd prime p, the congruence $x^2 \equiv a \pmod{p}$ has exactly $1 + \left(\frac{a}{p}\right)$ solutions. We first compute the Legendre symbol $\left(\frac{a}{p}\right)$, and if the congruence is found to be solvable, the next task is to compute the roots of the congruence. We postpone the study of root finding until Chapter 3. Also see Exercises 1.58 and 1.59.

---

[21]Kronecker extended the Jacobi symbol to all non-zero integers b, including even and negative integers (see Exercise 1.65). Leopold Kronecker (1823–1891) was a German mathematician who made significant contributions to number theory and algebra. A famous quote from him is: *Die ganzen Zahlen hat der liebe Gott gemacht, alles andere ist Menschenwerk.* (The dear God has created the whole numbers, everything else is man's work.)

1.7 Multiplicative Orders

Let $a \in \mathbb{Z}_m^*$. Since $\mathbb{Z}_m^*$ is finite, the elements $a, a^2, a^3, \ldots$ modulo m cannot be all distinct, that is, there exist $i, j \in \mathbb{N}$ with $i > j$ such that $a^i \equiv a^j \pmod{m}$. Since a is invertible modulo m, we have $a^{i-j} \equiv 1 \pmod{m}$, that is, there exist positive exponents e for which $a^e \equiv 1 \pmod{m}$. This observation leads to the following important concept.

Definition 1.53 Let $a \in \mathbb{Z}_m^*$. The smallest positive integer e for which $a^e \equiv 1 \pmod{m}$ is called the *multiplicative order* (or simply the *order*) of a modulo m and is denoted by $\operatorname{ord}_m a$. If $e = \operatorname{ord}_m a$, we also often say that a *belongs to the exponent* e modulo m. ◁

Example 1.54 (1) Take $m = 35$. We have $8^1 \equiv 8 \pmod{35}$, $8^2 \equiv 64 \equiv 29 \pmod{35}$, $8^3 \equiv 29 \times 8 \equiv 22 \pmod{35}$, and $8^4 \equiv 22 \times 8 \equiv 1 \pmod{35}$. Therefore, $\operatorname{ord}_{35} 8 = 4$.

(2) For every $m \in \mathbb{N}$, we have $\operatorname{ord}_m 1 = 1$.

(3) If p is an odd prime, then $\operatorname{ord}_p a = 2$ if and only if $a \equiv -1 \pmod{p}$. For a composite m, there may exist more than one elements of order 2. For example, $6, 29, 34$ are all the elements of order 2 modulo 35.

(4) The order $\operatorname{ord}_m a$ is not defined for integers a not coprime to m. Indeed, if $d = \gcd(a, m) > 1$, then $a, a^2, a^3, \ldots$ modulo m are all multiples of d, and so none of them equals 1 modulo m. □

The order $\operatorname{ord}_m a$ is returned by the call `znorder(Mod(a,m))` in `GP/PARI`.

```
gp > znorder(Mod(1,35))
%1 = 1
gp > znorder(Mod(2,35))
%2 = 12
gp > znorder(Mod(4,35))
%3 = 6
gp > znorder(Mod(6,35))
%4 = 2
gp > znorder(Mod(7,35))
  ***    not an element of (Z/nZ)* in order.
```

1.7.1 Primitive Roots

The order of an integer modulo m satisfies a very important property.

Theorem 1.55 Let $a \in \mathbb{Z}_m^*$, $e = \operatorname{ord}_m a$, and $h \in \mathbb{Z}$. Then, $a^h \equiv 1 \pmod{m}$ if and only if $e \mid h$. In particular, $e \mid \phi(m)$, where $\phi(m)$ is the Euler phi function.

Proof　Let $e|h$, that is, $h = ke$ for some $k \in \mathbb{Z}$. But then $a^h \equiv (a^e)^k \equiv 1^k \equiv 1 \pmod{m}$. Conversely, let $a^h \equiv 1 \pmod{m}$. Euclidean division of h by e yields $h = ke + r$ with $0 \leqslant r < e$. Since $a^e \equiv 1 \pmod{m}$, we have $a^r \equiv 1 \pmod{m}$. By definition, e is the smallest positive integer with $a^e \equiv 1 \pmod{m}$, that is, we must have $r = 0$, that is, $e|h$. The fact $e|\phi(m)$ follows directly from Euler's theorem: $a^{\phi(m)} \equiv 1 \pmod{m}$.　　　　　　　　　　◁

Definition 1.56 If $\mathrm{ord}_m\, a = \phi(m)$ for some $a \in \mathbb{Z}_m^*$, we call a a *primitive root* modulo m.[22]　　　　　　　　　　◁

Primitive roots do not exist for all moduli m. We prove an important fact about primes in this context.

Theorem 1.57 *Every prime p has a primitive root.*
Proof　Since $\mathbb{Z}_p$ is a field, a non-zero polynomial of degree d over $\mathbb{Z}_p$ can have at most d roots. By Fermat's little theorem, the polynomial $x^{p-1} - 1$ has $p - 1$ roots (all elements of $\mathbb{Z}_p^*$). Let d be a divisor of $p - 1$. We have $x^{p-1} - 1 = (x^d - 1)f(x)$ for a polynomial $f(x)$ of degree $p - 1 - d$. Since $f(x)$ cannot have more than $p - 1 - d$ roots, it follows that there are exactly d roots of $x^d - 1$ (and exactly $p - 1 - d$ roots of $f(x)$) modulo p.

Let $p - 1 = p_1^{e_1} \cdots p_r^{e_r}$ with pairwise distinct primes $p_1, \ldots, p_r$ and with each $e_i \in \mathbb{N}$. As argued in the last paragraph, $\mathbb{Z}_p^*$ contains exactly $p_i^{e_i}$ elements of orders dividing $p_i^{e_i}$, and exactly $p_i^{e_i - 1}$ elements of orders dividing $p_i^{e_i - 1}$. This implies that $\mathbb{Z}_p^*$ contains at least one element (in fact, $p_i^{e_i} - p_i^{e_i - 1}$ elements) of order equal to $p_i^{e_i}$. Let a_i be any such element. By Exercise 1.66(b), the element $a = a_1 \cdots a_r$ is of order $p_1^{e_1} \cdots p_r^{e_r} = p - 1$.　　　　　◁

Primes are not the only moduli to have primitive roots. The following theorem characterizes all moduli that have primitive roots.

Theorem 1.58 *The only positive integers > 1 that have primitive roots are $2, 4, p^e, 2p^e$, where p is any odd prime, and e is any positive integer.*　　◁

Example 1.59 (1)　Take $p = 17$. Since $\phi(p) = p - 1 = 16 = 2^4$, the order of every element of $\mathbb{Z}_p^*$ is of the form 2^i for some $i \in \{0, 1, 2, 3, 4\}$. We have $\mathrm{ord}_{17}\, 1 = 1$, so 1 is not a primitive root modulo 17. Also $2^1 \equiv 2 \pmod{16}$, $2^2 \equiv 4 \pmod{17}$, $2^4 \equiv 16 \pmod{17}$, and $2^8 \equiv 1 \pmod{17}$, that is, $\mathrm{ord}_{17}\, 2 = 8$, that is, 2 too is not a primitive root of 17. We now investigate powers of 3 modulo 17. Since $3^1 \equiv 3 \pmod{17}$, $3^2 \equiv 9 \pmod{17}$, $3^4 \equiv 81 \equiv 13 \pmod{17}$, $3^8 \equiv 169 \equiv 16 \pmod{17}$, and $3^{16} \equiv 1 \pmod{17}$, 3 is a primitive root of 17.

(2)　The modulus $m = 18 = 2 \times 3^2$ is of the form $2p^e$ for an odd prime p and so has a primitive root. We have $\phi(18) = 18 \times \left(1 - \frac{1}{2}\right)\left(1 - \frac{1}{3}\right) = 6$. So every element of $\mathbb{Z}_{18}^*$ is of order $1, 2, 3$, or 6. We have $5^1 \equiv 1 \pmod{18}$,

---

[22]The term *primitive root* was coined by Euler. Gauss studied primitive roots in his book *Disquisitiones Arithmeticae*. In particular, Gauss was the first to prove Theorem 1.58.

$5^2 \equiv 7 \pmod{18}$, $5^3 \equiv 17 \pmod{18}$, and $5^6 \equiv 1 \pmod{18}$. Thus, 5 is a primitive root modulo 18.

(3) The modulus $m = 16$ does not have a primitive root, that is, an element of order $\phi(m) = 8$. One can check that $\mathrm{ord}_{16}\, 1 = 1$, $\mathrm{ord}_{16}\, 7 = \mathrm{ord}_{16}\, 9 = \mathrm{ord}_{16}\, 15 = 2$, and $\mathrm{ord}_{16}\, 3 = \mathrm{ord}_{16}\, 5 = \mathrm{ord}_{16}\, 11 = \mathrm{ord}_{16}\, 13 = 4$. □

The call `znprimroot(m)` in the `GP/PARI` calculator returns a primitive root modulo its argument m, provided that such a root exists.

```
gp > znprimroot(47)
%1 = Mod(5, 47)
gp > znprimroot(49)
%2 = Mod(3, 49)
gp > znprimroot(50)
%3 = Mod(27, 50)
gp > znprimroot(51)
 ***   primitive root does not exist in gener
```

1.7.2 Computing Orders

We do not know any efficient algorithm for computing $\mathrm{ord}_m\, a$ unless the complete prime factorization of $\phi(m)$ is provided. Let $\phi(m) = p_1^{e_1} \cdots p_r^{e_r}$ with pairwise distinct primes $p_1, \ldots, p_r$ and with each $e_i \in \mathbb{N}$. Since $\mathrm{ord}_m\, a \mid \phi(m)$, we must have $\mathrm{ord}_m\, a = p_1^{h_1} \cdots p_r^{h_r}$ for some h_i with $0 \leqslant h_i \leqslant e_i$ (for all i). It suffices to compute only the exponents $h_1, \ldots, h_r$. For computing h_i, we raise other prime divisors p_j, $j \neq i$, to the highest possible exponents e_j. Then, we try h_i in the range $0, 1, \ldots, e_i$ in order to detect its exact value. The following algorithm elaborates this idea. We do not use explicit variables storing h_i, but accumulate the product $p_i^{h_i}$ in a variable e to be returned.

Algorithm 1.7: Computing the order $\mathrm{ord}_m\, a$ of $a \in \mathbb{Z}_m^*$

Let $\phi(m) = p_1^{e_1} \cdots p_r^{e_r}$ be the prime factorization of $\phi(m)$.
Initialize $e = 1$.
For $i = 1, 2, \ldots, r$ {
    Compute $b = a^{[\phi(m)/p_i^{e_i}]} \pmod{m}$.
    While ($b \not\equiv 1 \pmod{m}$) {
        Set $b = b^{p_i} \pmod{m}$, and $e = e p_i$.
    }
}
Return e.

1.8 Continued Fractions

John Wallis and Lord William Brouncker, in an attempt to solve a problem posed by Pierre de Fermat, developed the theory of continued fractions. However, there is evidence that this theory in some form was known to the Indian mathematician Bhaskaracharya in the twelfth century A.D.

1.8.1 Finite Continued Fractions

An expression of the form

$$
x_0 + \cfrac{1}{x_1 + \cfrac{1}{x_2 + \cfrac{\ddots}{\quad + \cfrac{1}{x_{k-1} + \cfrac{1}{x_k}}}}}
$$

with $x_0, x_1, \ldots, x_k \in \mathbb{R}$, all positive except perhaps x_0, is called a (finite) *continued fraction*. If all x_i are integers, this continued fraction is called *simple*. In that case, the integers x_i are called the *partial quotients* of the continued fraction. By definition, x_0 may be positive, negative, or zero, whereas all subsequent x_i must be positive. Let us agree to denote the above continued fraction by the compact notation $\langle x_0, x_1, x_2, \ldots, x_{k-1}, x_k \rangle$.

If we start folding the stair from bottom, we see that a finite simple continued fraction $\langle a_0, \ldots, a_k \rangle$ evaluates to a rational number. For example, $\langle 9, 1, 10, 4, 2 \rangle$ represents $1001/101$. The converse holds too, that is, given any rational h/k with $k > 0$, one can develop a finite simple continued fraction expansion of h/k. This can be done using the procedure for computing $\gcd(h, k)$.

Example 1.60 We plan to compute the simple continued fraction expansion of $1001/101$. Repeated Euclidean divisions yield

$$
\begin{aligned}
1001 &= 9 \times 101 + 92, \\
101 &= 1 \times 92 + 9, \\
92 &= 10 \times 9 + 2, \\
9 &= 4 \times 2 + 1, \\
2 &= 2 \times 1.
\end{aligned}
$$

Thus, $1001/101 = \langle 1001/101 \rangle = \langle 9, 101/92 \rangle = \langle 9, 1, 92/9 \rangle = \langle 9, 1, 10, 9/2 \rangle = \langle 9, 1, 10, 4, 2 \rangle$. $\square$

Let $r = \langle a_0, \ldots, a_k \rangle$ be a simple continued fraction. If $a_k > 1$, the continued fraction $\langle a_0, \ldots, a_k - 1, 1 \rangle$ also evaluates to r. On the other hand, if $a_k = 1$, the continued fraction $\langle a_0, \dots, a_{k-1} + 1\rangle$ also evaluates to r. In either case, these are the only two simple continued fractions for r. For example, $\langle 9, 1, 10, 4, 2\rangle$ and $\langle 9, 1, 10, 4, 1, 1\rangle$ are the only simple continued fractions for $1001/101$.

1.8.2 Infinite Continued Fractions

Finite continued fractions appear not to be so interesting. Assume that an infinite sequence $a_0, a_1, a_2, \dots$ of integers, all positive except perhaps a_0, is given. We want to assign a meaning (and value) to the *infinite simple continued fraction* $\langle a_0, a_1, a_2, \dots\rangle$. To this end, we inductively define two infinite sequences h_n and k_n as follows.

$$h_{-2} = 0, \quad h_{-1} = 1, \quad h_n = a_n h_{n-1} + h_{n-2} \quad \text{for } n \geqslant 0.$$
$$k_{-2} = 1, \quad k_{-1} = 0, \quad k_n = a_n k_{n-1} + k_{n-2} \quad \text{for } n \geqslant 0.$$

We also define the rational numbers $r_n = h_n/k_n$ for $n \geqslant 0$. This is allowed, since we have $1 = k_0 \leqslant k_1 < k_2 < k_3 < \cdots < k_n < \cdots$.

Theorem 1.61 *With the notations just introduced, we have* $r_n = \langle a_0, \dots, a_n\rangle$ *for every* $n \in \mathbb{N}_0$. *Furthermore, the rational numbers* r_n *satisfy the inequalities:* $r_0 < r_2 < r_4 < \cdots < r_5 < r_3 < r_1$. *The limit* $\xi = \lim_{n \to \infty} r_n$ *exists. For every* $m, n \in \mathbb{N}_0$, *we have* $r_{2m} < \xi < r_{2n+1}$. ◁

This theorem allows us to let the unique real number ξ stand for the infinite continued fraction $\langle a_0, a_1, a_2, \dots\rangle$, that is,

$$\langle a_0, a_1, a_2, \dots\rangle = \lim_{n \to \infty} \langle a_0, a_1, \dots, a_n\rangle = \lim_{n \to \infty} r_n = \lim_{n \to \infty} \frac{h_n}{k_n} = \xi.$$

Theorem 1.62 *The value of an infinite simple continued fraction is irrational. Moreover, two different infinite simple continued fractions evaluate to different values, that is, if* $\langle a_0, a_1, a_2, \dots\rangle = \langle b_0, b_1, b_2, \dots\rangle$, *then* $a_n = b_n$ *for all* $n \in \mathbb{N}_0$. ◁

Definition 1.63 *For each* $n \in \mathbb{N}_0$, *the rational number* r_n *is called the n-th convergent to the irrational number* $\xi = \langle a_0, a_1, a_2, \dots\rangle$. ◁

Example 1.64 Let us compute the irrational number ξ with continued fraction expansion $\langle 1, 2, 2, 2, \dots\rangle$. We have $\xi = 1 + \frac{1}{\lambda}$, where $\lambda = \langle 2, 2, 2, \dots\rangle$. Observe that $\lambda = 2 + \frac{1}{\lambda}$, so $\lambda^2 - 2\lambda - 1 = 0$, that is, $\lambda = 1 \pm \sqrt{2}$. Since λ is positive, we have $\lambda = 1 + \sqrt{2}$. Therefore, $\xi = 1 + \frac{1}{1+\sqrt{2}} = 1 + (\sqrt{2} - 1) = \sqrt{2}$. □

The converse question is: Does any irrational ξ expand to an infinite simple continued fraction? The answer is: *yes*. We inductively generate $a_0, a_1, a_2, \dots$ with $\xi = \langle a_0, a_1, a_2, \dots\rangle$ as follows. We start by setting $\xi_0 = \xi$ and $a_0 = \lfloor \xi_0 \rfloor$. When $\xi_0, \dots, \xi_n$ and $a_0, \dots, a_n$ are known for some $n \geqslant 0$, we calculate $\xi_{n+1} = 1/(\xi_n - a_n)$ and $a_{n+1} = \lfloor \xi_{n+1} \rfloor$. Since ξ is irrational, it follows that each ξ_n is also irrational. In addition, the integers $a_1, a_2, a_3, \dots$ are all positive. Only a_0 may be positive, negative or zero.

Example 1.65 (1) Let us first obtain the infinite simple continued fraction expansion of $\sqrt{2}$.

$$\xi_0 = \sqrt{2} = 1.4142135623\ldots, \quad a_0 = \lfloor \xi_0 \rfloor = 1,$$

$$\xi_1 = \frac{1}{\xi_0 - a_0} = \frac{1}{\sqrt{2} - 1} = 1 + \sqrt{2} = 2.4142135623\ldots, \quad a_1 = \lfloor \xi_1 \rfloor = 2,$$

$$\xi_2 = \frac{1}{\xi_1 - a_1} = \frac{1}{\sqrt{2} - 1} = 1 + \sqrt{2} = 2.4142135623\ldots, \quad a_2 = \lfloor \xi_2 \rfloor = 2,$$

and so on. Therefore, $\sqrt{2} = \langle 1, 2, 2, 2, \ldots \rangle$. The first few convergents to $\sqrt{2}$ are $r_0 = \langle 1 \rangle = 1$, $r_1 = \langle 1, 2 \rangle = \frac{3}{2} = 1.5$, $r_2 = \langle 1, 2, 2 \rangle = \frac{7}{5} = 1.4$, $r_3 = \langle 1, 2, 2, 2 \rangle = \frac{17}{12} = 1.4166666666\ldots$, $r_4 = \langle 1, 2, 2, 2, 2 \rangle = \frac{41}{29} = 1.4137931034\ldots$. It is apparent that the convergents $r_0, r_1, r_2, r_3, r_4, \ldots$ go successively closer to $\sqrt{2}$.

(2) Let us now develop the infinite simple continued fraction expansion of $\pi = 3.1415926535\ldots$.

$$\xi_0 = \pi = 3.1415926535\ldots, \quad a_0 = \lfloor \xi_0 \rfloor = 3,$$

$$\xi_1 = \frac{1}{\xi_0 - a_0} = 7.0625133059\ldots, \quad a_1 = \lfloor \xi_1 \rfloor = 7,$$

$$\xi_2 = \frac{1}{\xi_1 - a_1} = 15.996594406\ldots, \quad a_2 = \lfloor \xi_2 \rfloor = 15,$$

$$\xi_3 = \frac{1}{\xi_2 - a_2} = 1.0034172310\ldots, \quad a_3 = \lfloor \xi_3 \rfloor = 1,$$

and so on. Thus, the first few convergents to π are $r_0 = \langle 3 \rangle = 3$, $r_1 = \langle 3, 7 \rangle = \frac{22}{7} = 3.1428571428\ldots$, $r_2 = \langle 3, 7, 15 \rangle = \frac{333}{106} = 3.1415094339\ldots$, $r_3 = \langle 3, 7, 15, 1 \rangle = \frac{355}{113} = 3.1415929203\ldots$. Here too, the convergents $r_0, r_1, r_2, r_3, \ldots$ go successively closer to π. This is indeed true, in general. $\square$

Lemma 1.66 *Let h_n/k_n, $n \in \mathbb{N}_0$, be the convergents to the irrational number ξ. Then, $|\xi - \frac{h_n}{k_n}| < \frac{1}{k_n k_{n+1}}$ or equivalently, $|\xi k_n - h_n| < \frac{1}{k_{n+1}}$ for all $n \geqslant 0$.* $\triangleleft$

Theorem 1.67 *Let h_n/k_n, $n \in \mathbb{N}_0$, be the convergents to the irrational number ξ. Then, for all $n \geqslant 1$, we have $|\xi k_n - h_n| < |\xi k_{n-1} - h_{n-1}|$. In particular, $|\xi - \frac{h_n}{k_n}| < |\xi - \frac{h_{n-1}}{k_{n-1}}|$ for all $n \geqslant 1$.* $\triangleleft$

The convergents h_n/k_n to the irrational number ξ are called *best possible approximations* of ξ in the sense that if a rational a/b is closer to ξ than h_n/k_n, the denominator b has to be larger than k_n. More precisely, we have:

Theorem 1.68 *Let $a \in \mathbb{Z}$ and $b \in \mathbb{N}$ with $|\xi - \frac{a}{b}| < |\xi - \frac{h_n}{k_n}|$ for some $n \geqslant 1$. Then, $b > k_n$.* $\triangleleft$

The continued fraction of a real number x is returned by the call `contfrac(x)` in the GP/PARI calculator. The number of terms returned in the output depends on the precision of the calculator. The expansion is truncated after these terms. The user may optionally specify the number of terms that (s)he desires in the expansion.

```
gp > contfrac(1001/101)
%1 = [9, 1, 10, 4, 2]
gp > contfrac(sqrt(11))
%2 = [3, 3, 6, 3, 6, 3, 6, 3, 6, 3, 6, 3, 6, 3, 6, 3, 6, 3, 6, 3]
gp > Pi
%3 = 3.1415926535897932384626433383
gp > contfrac(Pi)
%4 = [3, 7, 15, 1, 292, 1, 1, 1, 2, 1, 3, 1, 14, 2, 1, 1, 2, 2, 2, 2, 1, 84, 2,
1, 1, 15, 3]
gp > contfrac(Pi,10)
%5 = [3, 7, 15, 1, 292, 1, 1, 1, 3]
gp > contfrac(Pi,100)
%6 = [3, 7, 15, 1, 292, 1, 1, 1, 2, 1, 3, 1, 14, 2, 1, 1, 2, 2, 2, 2, 1, 84, 2,
1, 1, 15, 3]
```

The last two convergents h_n/k_n and h_{n-1}/k_{n-1} are returned in the form of the 2×2 matrix $\begin{pmatrix} h_n & h_{n-1} \\ k_n & k_{n-1} \end{pmatrix}$ by the call `contfracpnqn()` which accepts a continued fraction expansion as its only input.

```
gp > contfrac(Pi)
%1 = [3, 7, 15, 1, 292, 1, 1, 1, 2, 1, 3, 1, 14, 2, 1, 1, 2, 2, 2, 2, 1, 84, 2,
1, 1, 15, 3]
gp > contfracpnqn(contfrac(Pi))
%2 =
[428224593349304 139755218526789]

[136308121570117 44485467702853]

gp > contfrac(Pi,3)
%3 = [3, 7, 15]
gp > contfracpnqn(contfrac(Pi,3))
%4 =
[333 22]

[106 7]

gp > contfrac(Pi,5)
%5 = [3, 7, 15, 1, 292]
gp > contfracpnqn(contfrac(Pi,5))
%6 =
[103993 355]

[33102 113]

gp > contfracpnqn(contfrac(1001/101))
%7 =
[1001 446]

[101 45]
```

1.9 Prime Number Theorem and Riemann Hypothesis

Euclid (ca. 300 BC) was seemingly the first to prove that there are infinitely many primes. Euclid's proof, given below (in modern terminology), is still an inspiring and influential piece of reasoning.

Theorem 1.69 *There are infinitely many primes.*
Proof The assertion is proved by contradiction. Suppose that there are only finitely many primes $p_1, p_2, \ldots, p_r$. The integer $n = p_1 p_2 \cdots p_r + 1$ is evidently divisible by neither of the primes $p_1, p_2, \ldots, p_r$. So all prime factors of n (n itself may be prime) are not present in our initial exhaustive list $p_1, p_2, \ldots, p_r$ of primes, a contradiction. ◁

At the first sight, prime numbers are distributed somewhat erratically. There is no formula involving simple functions like polynomials, that can generate only prime numbers. There are arbitrarily long gaps in the sequence of primes (for example, $(n+1)! + i$ is composite for all $i = 2, 3, \ldots, n+1$). Still, mathematicians tried to harness some patterns in the distribution of primes.

For a positive real number x, we denote by $\pi(x)$ the number of primes between 1 and x. There exists no simple formula to describe $\pi(x)$ for all (or almost all) values of x. For about a century, mathematicians tried to establish the following assertion, first conjectured by Legendre in 1797 or 1798.

Theorem 1.70 [*Prime number theorem (PNT)*] *$\pi(x)$ approaches the quantity $x/\ln x$ as $x \to \infty$. Here, the term "approaches" means that the limit*
$$\lim_{x \to \infty} \frac{\pi(x)}{x/\ln x} \text{ is equal to 1.} \qquad ◁$$

Several branches of mathematics (most notably, the study of analytic functions in complex analysis) got enriched during attempts to prove the prime number theorem. The first complete proof of the PNT (based mostly on the ideas of Riemann and Chebyshev) was given independently by the French mathematician Hadamard and by the Belgian mathematician de la Vallée Poussin in 1896. Their proof is regarded as one of the most important achievements of modern mathematics. An elementary proof (that is, a proof not based on results from analysis or algebra) of the theorem was found by Paul Erdös (1949) and Atle Selberg (1950).

Although the formula $x/\ln x$ for $\pi(x)$ is asymptotic, it is a good approximation for $\pi(x)$ for all values of x. In fact, it can be proved much more easily than the PNT that for all sufficiently large values of x, we have $0.922 < \dfrac{\pi(x)}{x/\ln x} < 1.105$. These inequalities indicate that $\pi(x) = \Theta(x/\ln x)$—a result as useful to a computational number theorist as the PNT. Proving the PNT, however, has been a landmark in the history of mathematics.

A better approximation of $\pi(x)$ is provided by Gauss's Li function defined by the logarithmic integral $\mathrm{Li}(x) = \int_2^x \dfrac{\mathrm{d}x}{\ln x}$. The quantity $\mathrm{Li}(x)$ approaches $\pi(x)$ as $x \to \infty$. (This was first conjectured by Dirichlet in 1838.)

Although the ratio $\dfrac{\pi(x)}{x/\ln x}$ (or the ratio $\dfrac{\pi(x)}{\mathrm{Li}(x)}$) approaches 1 asymptotically, the difference $\pi(x) - \dfrac{x}{\ln x}$ (or the difference $\pi(x) - \mathrm{Li}(x)$) is not necessarily zero, nor even convergent to a finite value. However, mathematicians treat the distribution of primes as *well-behaved* if this difference is *not too large*. Riemann proposed the following conjecture in 1859.[23]

Conjecture 1.71 [*Riemann hypothesis (RH)*] $\pi(x) - \mathrm{Li}(x) = \mathrm{O}(\sqrt{x}\ln x)$. ◁

It has been proved (for example, by Vallée Poussin) that $\pi(x) - \mathrm{Li}(x) = \mathrm{O}\left(\dfrac{x}{\ln x}e^{-\alpha\sqrt{\ln x}}\right)$ for some constant α. However, the tighter bound indicated by the Riemann hypothesis stands unproven even in the twenty-first century. A generalization of Theorem 1.69 is proved by Dirichlet.[24]

Theorem 1.72 [*Dirichlet's theorem on primes in arithmetic progression*] *Let $a, b \in \mathbb{N}$ with $\gcd(a, b) = 1$. There exist infinitely many primes in the sequence $a, a+b, a+2b, a+3b, \ldots$.* ◁

Let a, b be as above. We denote by $\pi_{a,b}(x)$ the number of primes of the form $a + kb$, that are $\leqslant x$. The prime number theorem can be generalized as follows: $\pi_{a,b}(x)$ approaches $\dfrac{x}{\phi(b)\ln x}$ as $x \to \infty$. Moreover, the Riemann hypothesis can be extended as follows.

Conjecture 1.73 [*Extended Riemann hypothesis (ERH)*]
$$\pi_{a,b}(x) = \frac{1}{\phi(b)}\mathrm{Li}(x) + \mathrm{O}(\sqrt{x}\ln x).$$
◁

The ERH has significant implications in computational number theory. Certain algorithms are known to run in polynomial time only under the assumption that the ERH is true. However, like the RH, this extended version continues to remain unproven. The RH is indeed a special case (corresponding to $a = b = 1$) of the ERH.

As an example of the usefulness of the ERH, let us look at the following implication of the ERH.

Conjecture 1.74 *The smallest positive quadratic non-residue modulo a prime p is $< 2\ln^2 p$.* ◁

---

[23] Georg Friedrich Bernhard Riemann (1826–1866) was a German mathematician whose works have deep impacts in several branches of mathematics including complex analysis, analytic number theory and geometry.

[24] Johann Peter Gustav Lejeune Dirichlet (1805–1859) was a German mathematician having important contributions to analytic and algebraic number theory.

In Exercise 1.59, a probabilistic polynomial-time algorithm (by Tonelli and Shanks) is described. If Conjecture 1.74 is true, this algorithm can be readily converted to a deterministic polynomial-time algorithm. However, if we do not assume the ERH, the best provable bound on the smallest positive quadratic non-residue modulo p turns out to be $O(p^\alpha)$ for some positive constant α. Consequently, we fail to arrive at a deterministic algorithm that runs in provably polynomial time.

1.10   Running Times of Arithmetic Algorithms

In introductory courses on algorithms, it is often conventional to treat the size of an integer as a constant. For example, when analyzing the time (and space) complexities of algorithms that sort arrays of integers, only the size of the array is taken into consideration.

In number theory, we work with multiple-precision integers. Treating each integer operand as having a constant size is a serious loss of generality. The amount of time taken by arithmetic operations on multiple-precision integers and even by simple copy and comparison operations depends heavily on the size of the operands. For example, multiplying two million-bit integers is expected to take sufficiently longer than multiplying two thousand-bit integers.

In view of this, the input size for an arithmetic algorithm is measured by the total number of bits needed to encode its operands. Of course, the encoding involved must be reasonable. The binary representation of an integer is a reasonable encoding of the integer, whereas its unary representation which is exponential in the size of its binary representation is not reasonable. Henceforth, we take the size of an integer n as $\log_2 n = \lg n$, or ignoring constant factors as $\log_e n = \ln n$, or even as $\log n$ to an unspecified (but fixed) base.

When arithmetic modulo an integer m is considered, it is assumed that $\mathbb{Z}_m$ has the standard representation $\{0, 1, 2, \ldots, m-1\}$. Thus, each element of $\mathbb{Z}_m$ is at most as big as $m-1$, that is, has a bit size which is no larger than $\lg m$. Moreover, since $a^{\phi(m)} \equiv 1 \pmod{m}$ for any $a \in \mathbb{Z}_m^*$, an exponentiation of the form $a^e \pmod{m}$ is carried out under the assumption that e is available modulo $\phi(m)$. But $\phi(m) < m$, that is, e can also be encoded using $\leqslant \lg m$ bits. To sum up, an algorithm involving only a constant number of inputs from $\mathbb{Z}_m$ is considered to have an input size of $\lg m$ (or $\ln m$ or $\log m$).

A polynomial $f(x)$ of degree d has a size needed to encode its $d+1$ coefficients. If each coefficient is known to have an upper bound t on its bit length, the size of the polynomial is $\leqslant (d+1)t$. For example, if $f(x) \in \mathbb{Z}_m[x]$, the size of $f(x)$ is $\leqslant (d+1)\lg m$, or more simplistically $d\lg m$. Analogously, the size of a $k \times l$ matrix with entries from $\mathbb{Z}_m$ is $kl\lg m$.

Having defined the notion of the size of arithmetic operands, we are now ready to concentrate on the running times of arithmetic algorithms. First, we look at the basic arithmetic operations. Under the assumption of standard schoolbook arithmetic, we have the following complexity figures. Running times of faster arithmetic operations (like Karatsuba or Toom-3 or FFT multiplication) are already elaborated in the text.

Operation	Running time
Copy $x = a$	$O(\lg a)$
Comparison of a with b	$O(\max(\lg a, \lg b))$
Addition $a + b$	$O(\max(\lg a, \lg b))$
Subtraction $a - b$	$O(\max(\lg a, \lg b))$
Multiplication ab	$O(\lg a \lg b)$
Square a^2	$O(\lg^2 a)$
Euclidean division a quot b and/or a rem b with $a \geqslant b$	$O(\lg^2 a)$
Euclidean gcd $\gcd(a, b)$ with $a \geqslant b$	$O(\lg^2 a)$
Extended Euclidean gcd $\gcd(a, b) = ua + vb$ with $a \geqslant b$	$O(\lg^2 a)$
Binary gcd $\gcd(a, b)$ with $a \geqslant b$	$O(\lg^2 a)$
Extended binary gcd $\gcd(a, b) = ua + vb$ with $a \geqslant b$	$O(\lg^2 a)$

The running times of modular arithmetic operations in $\mathbb{Z}_m$ are as follows.

Operation	Running time
Addition $a + b \,(\mathrm{mod}\ m)$	$O(\lg m)$
Subtraction $a - b \,(\mathrm{mod}\ m)$	$O(\lg m)$
Multiplication $ab \,(\mathrm{mod}\ m)$	$O(\lg^2 m)$
Inverse $a^{-1} \,(\mathrm{mod}\ m)$	$O(\lg^2 m)$
Exponentiation $a^e \,(\mathrm{mod}\ m)$	$O(\lg^3 m)$

Finally, the running times of some other important algorithms discussed in this chapter are compiled.

Operation	Running time
Chinese remainder theorem (Algorithm 1.6)	$O(t \lg^2 M)$
Legendre-symbol $\left(\frac{a}{p}\right)$ with $a \in \mathbb{Z}_p$	$O(\lg^3 p)$
Order $\mathrm{ord}_m\, a$ (Algorithm 1.7)	$O(\lg^4 m)$

A derivation of these running times is left to the reader as an easy exercise.

Exercises

1. Describe an algorithm to compare the absolute values of two multiple-precision integers.

2. Describe an algorithm to compute the product of a multiple-precision integer with a single-precision integer.

3. Squaring is a form of multiplication where both the operands are the same. Describe how this fact can be exploited to speed up the schoolbook multiplication algorithm for multiple-precision integers. What about Karatsuba multiplication?

4. Describe an efficient algorithm to compute the Euclidean division of a multiple-precision integer by a non-zero single-precision integer.

5. Describe how multiple-precision division by an integer of the form $B^l \pm m$ (B is the base, and m is a small integer) can be efficiently implemented.

6. Explain how multiplication and division of multiple-precision integers by powers of 2 can be implemented efficiently using bit operations.

7. Describe the details of the Toom-4 multiplication method. Choose the evaluation points as $k = \infty, 0, \pm 1, -2, \pm \frac{1}{2}$.

8. Toom's multiplication can be adapted to work for unbalanced operands, that is, when the sizes of the operands vary considerably. Suppose that the number of digits of a is about two-thirds the number of digits of b. Write a as a polynomial of degree two, and b as a polynomial of degree three. Describe how you can compute the product ab in this case using a Toom-like algorithm.

9. Derive Equation (1.1).

10. Verify the following assertions. Here, a, b, c, x, y are arbitrary integers.
    (a) $a|a$.
    (b) If $a|b$ and $b|c$, then $a|c$.
    (c) If $a|b$ and $b|a$, then $a = \pm b$.
    (d) If $a|b$ and $a|c$, then $a|(bx + cy)$.
    (e) If $a|(bc)$ and $\gcd(a, b) = 1$, then $a|c$.

11. Let p be a prime. If $p|(ab)$, show that $p|a$ or $p|b$. More generally, show that if $p|(a_1 a_2 \cdots a_n)$, then $p|a_i$ for some $i \in \{1, 2, \dots, n\}$.

12. Suppose that $\gcd(r_0, r_1)$ is computed by the repeated Euclidean division algorithm. Suppose also that $r_0 > r_1 > 0$. Let r_{i+1} denote the remainder obtained by the i-th division (that is, in the i-th iteration of the Euclidean loop). So the computation proceeds as $\gcd(r_0, r_1) = \gcd(r_1, r_2) = \gcd(r_2, r_3) = \cdots$ with $r_0 > r_1 > r_2 > \cdots > r_k > r_{k+1} = 0$ for some $k \geqslant 1$.
    (a) If the computation of $\gcd(r_0, r_1)$ requires exactly k Euclidean divisions, show that $r_0 \geqslant F_{k+2}$ and $r_1 \geqslant F_{k+1}$. Here, F_n is the n-th Fibonacci number: $F_0 = 0$, $F_1 = 1$, and $F_n = F_{n-1} + F_{n-2}$ for $n \geqslant 2$.

(b) Modify the Euclidean gcd algorithm slightly so as to ensure that $r_i \leqslant \frac{1}{2}r_{i-1}$ for $i \geqslant 2$. Here, r_i need not be the remainder r_{i-2} rem r_{i-1}.
(c) Explain the speedup produced by the modified algorithm. Assume that $F_n \approx \frac{1}{\sqrt{5}}\rho^n$, where $\rho = \frac{1+\sqrt{5}}{2} = 1.6180339887\ldots$ is the golden ratio.

13. Modify the binary gcd algorithm (Algorithm 1.3) so that two integers u, v satisfying $\gcd(a, b) = ua + vb$ are computed along with $\gcd(a, b)$. Your algorithm should run in quadratic time as the original binary gcd algorithm.

14. Let $a, b \in \mathbb{N}$ with $d = \gcd(a, b) = ua + vb$ for some $u, v \in \mathbb{Z}$. Demonstrate that u, v are not unique. Now, assume that $(a, b) \neq (1, 1)$. Prove that:
    (a) If $d = 1$, then u, v can be chosen to satisfy $|u| < b$ and $|v| < a$.
    (b) In general, u, v can be chosen to satisfy $|u| < b/d$ and $|v| < a/d$.

15. Define the n-th *continuant polynomial* $K_n(x_1, x_2, \ldots, x_n)$ recursively as:

$$K_0() = 1,$$
$$K_1(x_1) = x_1,$$
$$K_n(x_1, x_2, \ldots, x_n) = x_n K_{n-1}(x_1, x_2, \ldots, x_{n-1}) + K_{n-2}(x_1, x_2, \ldots, x_{n-2}), \ n \geqslant 2.$$

    (a) Find $K_2(x_1, x_2)$, $K_3(x_1, x_2, x_3)$ and $K_4(x_1, x_2, x_3, x_4)$.
    (b) Prove that $K_n(x_1, x_2, \ldots, x_n)$ is the sum of all subproducts of $x_1 x_2 \cdots x_n$ with zero or more non-overlapping contiguous pairs $x_i x_{i+1}$ removed.
    (c) Conclude that $K_n(x_1, x_2, \ldots, x_n) = K_n(x_n, x_{n-1}, \ldots, x_1)$.
    (d) Prove that the number of terms in $K_n(x_1, x_2, \ldots, x_n)$ is F_{n+1}.
    (e) Deduce that for all $n \geqslant 1$, the continuant polynomials satisfy the identity

$$K_n(x_1, \ldots, x_n)K_n(x_2, \ldots, x_{n+1}) - K_{n+1}(x_1, \ldots, x_{n+1})K_{n-1}(x_2, \ldots, x_n) = (-1)^n.$$

16. Consider the extended Euclidean gcd algorithm described in Section 1.2.2. Suppose that the algorithm terminates after computing $r_j = 0$ (so that r_{j-1} is the gcd of $r_0 = a$ and $r_1 = b$). Assume that $a \geqslant b$ and let $d = \gcd(a, b)$. Finally, let $q_2, q_3, \ldots, q_j$ be the quotients obtained during the Euclidean divisions.
    (a) Show that

$$|u_1| < |u_2| \leqslant |u_3| < |u_4| < |u_5| < \cdots < |u_j|, \text{ and}$$
$$|v_0| < |v_1| \leqslant |v_2| < |v_3| < |v_4| < \cdots < |v_j|.$$

    (b) Prove that

$$|u_i| = K_{i-2}(q_3, \ldots, q_i) \text{ for all } i = 2, 3, \ldots, j, \text{ and}$$
$$|v_i| = K_{i-1}(q_2, \ldots, q_i) \text{ for all } i = 1, 2, \ldots, j.$$

    (c) Prove that $\gcd(u_i, v_i) = 1$ for all $i = 0, 1, 2, \ldots, j$.
    (d) Prove that $|u_j| = b/d$ and $|v_j| = a/d$.
    (e) Prove that the extended gcd algorithm returns the multipliers u, v with $|u| \leqslant b/d$ and $|v| \leqslant a/d$ with strict inequalities holding if $b \nmid a$.

17. Let $a_1, a_2, \ldots, a_n$ be non-zero integers with $d = \gcd(a_1, a_2, \ldots, a_n)$.

(a) Prove that there exist integers $u_1, u_2, \ldots, u_n$ satisfying $u_1 a_1 + u_2 a_2 + \cdots + u_n a_n = d$.

(b) How can you compute $u_1, u_2, \ldots, u_n$ along with d?

18. Assume that a randomly chosen non-zero even integer is divisible by 2^t but not by 2^{t+1} with probability $1/2^t$ for all $t \in \mathbb{N}$. Prove that the average number of iterations of the outer loop of the binary gcd algorithm (Algorithm 1.3) is at most $\max(\lg a, \lg b)$.

19. Prove that the maximum number of iterations of the outer loop of the binary gcd algorithm (Algorithm 1.3) is at most $1 + \max(\lg a, \lg b)$. Prove that this bound is quite tight. (**Hint:** Establish the tighter (in fact, achievable) upper bound $\lg(a + b)$ on the number of iterations. Use induction.)

(**Remark:** The inner `while` loops of the binary gcd algorithm may call for an effort proportional to $\lg a + \lg b$ in the entire execution of Algorithm 1.3.)

20. Argue that for two odd integers a, b, either $a + b$ or $a - b$ is a multiple of four. How can you exploit this observation to modify the binary gcd algorithm? What is the expected benefit of this modification?

21. Propose how, in an iteration of the binary gcd algorithm with $a > b$, you can force the least significant word of a to become zero by subtracting a suitable multiple of b from a. What (and when) do you gain from this modification?

22. Consider the following variant of the binary gcd algorithm which attempts to remove one or more *most* significant bits of one operand in each iteration.[25]

Algorithm 1.8: LEFT-SHIFT BINARY GCD

Assume that the inputs a, b are positive and odd, and that $a \geqslant b$.
While $(b \neq 0)$ {
    Determine $e \in \mathbb{N}_0$ such that $2^e b \leqslant a < 2^{e+1} b$.
    Compute $t = \min(a - 2^e b, 2^{e+1} b - a)$.
    Set $a = b$ and $b = t$.
    If $(a < b)$, swap a with b.
}
Return a.

(a) Prove that Algorithm 1.8 terminates and correctly computes $\gcd(a, b)$.

(b) How can you efficiently implement Algorithm 1.8 using bit operations?

(c) Prove that the number of iterations of the `while` loop is $O(\lg a + \lg b)$.

(d) Argue that Algorithm 1.8 can be so implemented to run in $O(\lg^2 a)$ time (where a is the larger input operand).

23. Let $a, b \in \mathbb{N}$ with $\gcd(a, b) = 1$. Assume that $a \neq 1$ and $b \neq 1$.

(a) Prove that any integer $n \geqslant ab$ can be expressed as $n = sa + tb$ with integers $s, t \geqslant 0$.

(b) Devise a polynomial-time (in $\log n$) algorithm to compute s, t of Part (a).

---

[25] Jeffrey Shallit and Jonathan Sorenson, Analysis of a left-shift binary gcd algorithm, *Journal of Symbolic Computation*, 17, 473–486, 1994.

(c) Determine the running time of your algorithm.

(**Remark:** The *Frobenius coin change problem* deals with the determination of the largest positive integer that cannot be represented as a linear non-negative integer combination of some given positive integers $a_1, a_2, \ldots, a_k$ with $\gcd(a_1, a_2, \ldots, a_k) = 1$. For $k = 2$, this integer is $a_1 a_2 - a_1 - a_2$.)

24. Let $n \in \mathbb{N}$, and p a prime. The *multiplicity* of p in n, denoted $v_p(n)$, is the largest (non-negative) exponent e for which $p^e | n$. Prove that $v_p(n!) = \sum_{k \in \mathbb{N}} \lfloor n/p^k \rfloor$. Conclude that the number of trailing zeros in $n!$ is $v_5(n!)$. Propose an algorithm to compute the number of trailing zeros in $n!$.

25. Prove that for any $r \in \mathbb{N}$, the product of r consecutive integers is divisible by $r!$. (Please avoid the argument that the binomial coefficient $\binom{n}{r}$ is an integer.)

26. Algorithm 1.4 is called left-to-right exponentiation, since the bits in the exponent are considered from left to right. Rewrite the square-and-multiply exponentiation algorithm in such a way that the exponent bits are considered from right to left. In other words, if $e = (e_{l-1} e_{l-2} \ldots e_1 e_0)_2$, then for $i = 0, 1, 2, \ldots, l - 1$ in that order, the i-th iteration should compute $a^{(e_i e_{i-1} \ldots e_1 e_0)_2} \pmod{m}$.

27. Algorithm 1.4 can be speeded up by constant factors using several tricks one of which is explained here. Let w be a small integer (typically, 2, 4 or 8). One precomputes $a^i \pmod{m}$ for $i = 0, 1, \ldots, 2^w - 1$. The exponent is broken into chunks, each of size w bits. Inside the square-and-multiply loop, w successive square operations are performed. Subsequently, depending on the current w-bit chunk in the exponent, a single multiplication is carried out. The precomputed table is looked up for this multiplication. Work out the details of this *windowed exponentiation* algorithm. Argue how this variant speeds up the basic exponentiation algorithm.

28. Suppose that we want to compute $x^r y^s \pmod{m}$, where r and s are positive integers of the same bit size. By the repeated square-and-multiply algorithm, one can compute $x^r \pmod{m}$ and $y^s \pmod{m}$ independently, and then multiply these two values. Alternatively, one may rewrite the square-and-multiply algorithm using only one loop in which the bits of both the exponents r and s are simultaneously considered. After each square operation, one multiplies by 1, x, y or xy.

(a) Elaborate the algorithm outlined above. What speedup is this modification expected to produce?

(b) Generalize the concept to the computation of $x^r y^s z^t \pmod{m}$, and analyze the speedup.

29. Show that the quotient Q and the remainder R computed by the Barrett reduction algorithm (Algorithm 1.5) satisfy $q - 2 \leqslant Q \leqslant q$ and $0 \leqslant R < 3m$.

30. For $n \in \mathbb{N}$, the integer $x = \lfloor \sqrt{n} \rfloor$ is called the *integer square-root* of n. Newton's iteration can be used to obtain x from n. Assume that we want to find a zero of the function $f(x)$. In the specific case of computing integer square-roots, we have $f(x) = x^2 - n$. We start with an initial approximation x_0 of the zero.

Next, we enter a loop in which the approximation x_i is modified to a better approximation x_{i+1}. We generate the new approximation as $x_{i+1} = x_i - \frac{f(x_i)}{f'(x_i)}$. For the special case $f(x) = x^2 - n$, we have $x_{i+1} = \frac{x_i+(n/x_i)}{2}$. Since we plan to perform integer operations only, we generate $x_{i+1} = \left\lfloor \frac{x_i+\lfloor n/x_i \rfloor}{2} \right\rfloor$. Propose a polynomial-time algorithm for computing the integer square-root x of n using this idea. Suggest how an initial approximation x_0 can be obtained in the range $\lfloor \sqrt{n} \rfloor \leqslant x_0 \leqslant 2\lfloor \sqrt{n} \rfloor$ using bit operations only. Determine a termination criterion for the Newton loop.

31. Let $k \geqslant 2$ be a constant integer. Design a polynomial-time algorithm that uses Newton's iteration for computing the integer k-th root $\lfloor \sqrt[k]{n} \rfloor$ of $n \in \mathbb{N}$.

32. Let $n \in \mathbb{N}$. Design a polynomial-time algorithm for checking whether n is a perfect power, that is, $n = m^k$ for some $m, k \in \mathbb{N}$ with $k \geqslant 2$.

33. Assume that you are given a polynomial-time algorithm that, given positive integers n and k, determines whether n is a perfect k-th power (that is, whether $n = m^k$ for some positive integer m). Devise a polynomial-time algorithm that, given two positive integers n_1, n_2, determines whether there exist positive integers e_1, e_2 such that $n_1^{e_1} = n_2^{e_2}$ and if so, computes a pair of such positive integers e_1, e_2.

34. Let p be a prime and $a, b \in \mathbb{Z}$. Prove the following assertions.
(a) $\binom{p}{k} \equiv 0 \pmod{p}$ for all $k = 1, \ldots, p - 1$, where $\binom{p}{k} = \frac{p!}{k!(p-k)!}$ is the binomial coefficient.
(b) $(a+b)^p \equiv a^p + b^p \pmod{p}$ or, more generally, $(a+b)^{p^r} \equiv a^{p^r} + b^{p^r} \pmod{p}$ for every $r \in \mathbb{N}$.
(c) If $a^p \equiv b^p \pmod{p}$, then $a^p \equiv b^p \pmod{p^2}$.

35. Let a, b, c be non-zero integers, and $d = \gcd(a, b)$.
(a) Prove that the equation $ax + by = c$ is solvable in *integer values* of x, y if and only if $d \mid c$.
(b) Suppose that $d \mid c$, and (s, t) is one solution of the congruence of Part (a). Prove that all the solutions of this congruence can be given as $(s + k(b/d), t - k(a/d))$ for all $k \in \mathbb{Z}$. Describe how one solution (s, t) can be efficiently computed.
(c) Compute all the (integer) solutions of the equation $21x + 15y = 60$.

36. Prove that the multivariate linear congruence $a_1 x_1 + a_2 x_2 + \cdots + a_n x_n \equiv b \pmod{m}$ is solvable for integer-valued variables $x_1, x_2, \ldots, x_n$ if and only if $\gcd(a_1, a_2, \ldots, a_n, m) \mid b$.

37. Find the simultaneous solution of the congruences: $7x \equiv 8 \pmod 9$, $x \equiv 9 \pmod{10}$, and $2x \equiv 3 \pmod{11}$.

38. Compute all the solutions of the following congruences:
(a) $x^2 + x + 1 \equiv 0 \pmod{91}$.
(b) $x^2 + x - 1 \equiv 0 \pmod{121}$.
(c) $x^2 + 5x + 24 \equiv 0 \pmod{36}$.
(d) $x^{50} \equiv 10 \pmod{101}$.

39. Compute all the simultaneous solutions of the congruences: $5x \equiv 3 \pmod{47}$, and $3x^2 \equiv 5 \pmod{49}$.

40. Let p be a prime.
(a) Show that

$$x^{p-1} - 1 \equiv (x-1)(x-2)\cdots(x-(p-1)) \pmod{p}, \quad \text{and}$$
$$x^p - x \equiv x(x-1)(x-2)\cdots(x-(p-1)) \pmod{p},$$

where $f(x) \equiv g(x) \pmod{p}$ means that the coefficient of x^i in the polynomial $f(x)$ is congruent modulo p to the coefficient of x^i in $g(x)$ for all $i \in \mathbb{N}_0$.
(b) [*Wilson's theorem*]   Prove that $(p-1)! \equiv -1 \pmod{p}$.
(c) If $m \in \mathbb{N}$ is composite and > 4, prove that $(m-1)! \equiv 0 \pmod{m}$.

41. [*Generalized Euler's theorem*]   Let $m \in \mathbb{N}$, and a any integer (not necessarily coprime to m). Prove that $a^m \equiv a^{m-\phi(m)} \pmod{m}$.

42. Let $\sigma(n)$ denote the sum of positive integral divisors of $n \in \mathbb{N}$. Let $n = pq$ with two distinct primes p, q. Devise a polynomial-time algorithm to compute p, q from the knowledge of n and $\sigma(n)$.

43. (a) Let $n = p^2 q$ with p, q distinct odd primes, $p \nmid (q-1)$ and $q \nmid (p-1)$. Prove that factoring n is polynomial-time equivalent to computing $\phi(n)$.
(b) Let $n = p^2 q$ with p, q odd primes satisfying $q = 2p + 1$. Argue that one can factor n in polynomial time.

44. (a) Let m_1, m_2 be coprime moduli, and let $a_1, a_2 \in \mathbb{Z}$. By the extended gcd algorithm, one can compute integers u, v with $um_1 + vm_2 = 1$. Prove that $x \equiv um_1a_2 + vm_2a_1 \pmod{m_1m_2}$ is the simultaneous solution of the congruences $x \equiv a_i \pmod{m_i}$ for $i = 1, 2$.
(b) Let $m_1, m_2, \ldots, m_t$ be pairwise coprime moduli, and $a_1, a_2, \ldots, a_t \in \mathbb{Z}$. Write an incremental procedure for the Chinese remainder theorem that starts with the solution $x \equiv a_1 \pmod{m_1}$ and then runs a loop, the i-th iteration of which (for $i = 2, 3, \ldots, t$ in that order) computes the simultaneous solution of $x \equiv a_j \pmod{m_j}$ for $j = 1, 2, \ldots, i$.

45. [*Generalized Chinese remainder theorem*]   Let $m_1, m_2, \ldots, m_t$ be t moduli (not necessarily coprime to one another). Prove that the congruences $x \equiv a_i \pmod{m_i}$ for $i = 1, 2, \ldots, t$ are simultaneously solvable if and only if $\gcd(m_i, m_j) | (a_i - a_j)$ for every pair (i, j) with $i \neq j$. Show also that in this case the solution is unique modulo $\text{lcm}(m_1, m_2, \ldots, m_t)$.

46. (a) Design an algorithm that, given moduli m_1, m_2 and integers a_1, a_2 with $\gcd(m_1, m_2) | (a_1 - a_2)$, computes a simultaneous solution of the congruences $x \equiv a_i \pmod{m_i}$ for $i = 1, 2$.
(b) Design an algorithm to implement the generalized CRT on $t \geqslant 2$ moduli.

47. [*Theoretical foundation of the RSA cryptosystem*]   Let $m = p_1 p_2 \cdots p_k$ be a product of $k \geqslant 2$ distinct primes. Prove that the map $\mathbb{Z}_m \to \mathbb{Z}_m$ that takes a to $a^e \pmod{m}$ is a bijection if and only if $\gcd(e, \phi(m)) = 1$. Describe the inverse of this exponentiation map.

48. Let $m = pq$ be a product of two distinct known primes p, q. Assume that q^{-1} (mod p) is available. Suppose that we want to compute $b \equiv a^e$ (mod m) for $a \in \mathbb{Z}_m^*$ and $0 \leqslant e < \phi(m)$. To that effect, we first compute $e_p = e$ rem $(p-1)$ and $e_q = e$ rem $(q-1)$, and then the modular exponentiations $b_p \equiv a^{e_p}$ (mod p) and $b_q \equiv a^{e_q}$ (mod q). Finally, we compute $t \equiv q^{-1}(b_p - b_q)$ (mod p).

(a) Prove that $b \equiv b_q + tq$ (mod m).

(b) Suppose that p, q are both of bit sizes roughly half of that of m. Explain how computing b in this method speeds up the exponentiation process. You may assume classical (that is, schoolbook) arithmetic for the implementation of products and Euclidean division.

49. Let $m \in \mathbb{N}$ be odd and composite. Show that if there exists $b \in \mathbb{Z}_m^*$ for which $b^{m-1} \not\equiv 1$ (mod m), then at least half of the elements $a \in \mathbb{Z}_m^*$ satisfy $a^{m-1} \not\equiv 1$ (mod m).

50. Let $m \in \mathbb{N}$, $m > 1$. Prove that the number of solutions of $x^{m-1} \equiv 1$ (mod m) is $\prod\limits_{p \mid m} \gcd(p - 1, m - 1)$, where the product is over the set of distinct prime divisors p of m.

51. A composite number m is called a *Carmichael number* if $a^{m-1} \equiv 1$ (mod m) for every a coprime to m. Show that m is a Carmichael number if and only if m is square-free and $(p - 1) \mid (m - 1)$ for every prime $p \mid m$.

(a) Let $k \in \mathbb{N}$ be such that $p_1 = 6k + 1$, $p_2 = 12k + 1$ and $p_3 = 18k + 1$ are all primes. Show that $p_1 p_2 p_3$ is a Carmichael number.

(b) Show that there are no even Carmichael numbers.

(c) Show that a Carmichael number must be the product of at least three distinct odd primes.

(d) Verify that $561 = 3 \times 11 \times 17$, $41041 = 7 \times 11 \times 13 \times 41$, $825265 = 5 \times 7 \times 17 \times 19 \times 73$, and $321197185 = 5 \times 19 \times 23 \times 29 \times 37 \times 137$ are Carmichael numbers. (**Remark:** These are the smallest Carmichael numbers having three, four, five and six prime factors, respectively. R. D. Carmichael first found the existence of Carmichael numbers in 1910, studied their properties, and conjectured that there are infinitely many of them. This conjecture has been proved by Alford, Granville and Pomerance in 1994. All the Carmichael numbers $< 10^5$ are 561, 1105, 1729, 2465, 2821, 6601, 8911, 10585, 15841, 29341, 41041, 46657, 52633, 62745, 63973, and 75361.)

52. (a) Find all the solutions of the congruence $2x^3 + x + 9 \equiv 0$ (mod 11).

(b) Find all the solutions of the congruence $2x^3 + x + 9 \equiv 0$ (mod 121) using Hensel's lifting procedure.

53. In Section 1.5.1, we lifted solutions of polynomial congruences of the form $f(x) \equiv 0$ (mod p^e) to the solutions of $f(x) \equiv 0$ (mod p^{e+1}). In this exercise, we investigate lifting the solutions of $f(x) \equiv 0$ (mod p^e) to solutions of $f(x) \equiv 0$ (mod p^{2e}), that is, the exponent in the modulus doubles every time (instead of getting incremented by only 1).

(a) Let $f(x) \in \mathbb{Z}[x]$, $e \in \mathbb{N}$, and ξ a solution of $f(x) \equiv 0 \pmod{p^e}$. Write $\xi' = \xi + kp^e$. Show how we can compute all values of k for which ξ' satisfies $f(\xi') \equiv 0 \pmod{p^{2e}}$.

(b) It is given that the only solution of $2x^3 + 4x^2 + 3 \equiv 0 \pmod{25}$ is $14 \pmod{25}$. Using the lifting procedure of Part (a), compute all the solutions of $2x^3 + 4x^2 + 3 \equiv 0 \pmod{625}$.

54. Find all the solutions of the following congruences:
 (a) $x^2 \equiv 71 \pmod{713}$. (**Remark:** $713 = 23 \times 31$).
 (b) $x^2 + x + 1 \equiv 0 \pmod{437}$. (**Remark:** $437 = 19 \times 23$).

55. Let $m = p_1^{e_1} \cdots p_r^{e_r}$ be the prime factorization of an odd modulus m. Also, let $a \in \mathbb{Z}$. Prove that the quadratic congruence $x^2 \equiv a \pmod{m}$ has exactly
$$\prod_{i=1}^{r}\left(1 + \left(\frac{a}{p_i}\right)\right) \text{ solutions modulo } m.$$ In particular, if $\gcd(a, m) = 1$, there
are either 0 or 2^r solutions.

56. Imitate the binary gcd algorithm for computing the Jacobi symbol $\left(\frac{a}{b}\right)$.

57. Let p be a prime > 3. Prove that 3 is a quadratic residue modulo p if and only if $p \equiv \pm 1 \pmod{12}$.

58. Let p be an odd prime and $a \in \mathbb{Z}_p^*$ be a quadratic residue modulo p. Prove the following assertions.
 (a) If $p \equiv 3 \pmod 4$, then a modular square-root of a is $a^{(p+1)/4} \pmod p$.
 (b) If $p \equiv 5 \pmod 8$, then a modular square-root of a is $a^{(p+3)/8} \pmod p$ if $a^{(p-1)/4} \equiv 1 \pmod p$, or $2a \cdot (4a)^{(p-5)/8} \pmod p$ if $a^{(p-1)/4} \equiv -1 \pmod p$.

59. Let p be an odd prime, and $\left(\frac{a}{p}\right) = 1$. If $p \equiv 3 \pmod 4$ or $p \equiv 5 \pmod 8$, a modular square-root of a can be obtained by performing some modular exponentiation(s) as described in Exercise 1.58. If $p \equiv 1 \pmod 8$, Algorithm 1.9 can be used to compute a square-root of a modulo p. The algorithm is, however, valid for any odd prime p.

Algorithm 1.9: TONELLI AND SHANKS ALGORITHM FOR COMPUTING A SQUARE-ROOT OF $a \in \mathbb{Z}_p^*$ MODULO A PRIME p

---

If $\left(\frac{a}{p}\right) = -1$, return "failure."
Write $p - 1 = 2^v q$ with q odd.
Find any quadratic non-residue b modulo p.
Compute $g \equiv b^q \pmod p$, and $x \equiv a^{(q+1)/2} \pmod p$.
While (1) {
    Find the smallest $i \in \{0, 1, \ldots, v - 1\}$ with $(x^2 a^{-1})^{2^i} \equiv 1 \pmod p$.
    If i is 0, return x.
    Set $x \equiv x \cdot g^{2^{v-i-1}} \pmod p$.
}

(a) Prove the correctness of Algorithm 1.9.

(b) Under the assumption that a quadratic non-residue b modulo p can be located by randomly trying $O(1)$ elements of $\mathbb{Z}_p^*$, prove that Algorithm 1.9 runs in polynomial time. Justify this assumption.

(c) Convert Algorithm 1.9 to a deterministic polynomial-time algorithm assuming that Conjecture 1.74 is true.

60. Let p be an odd prime. Prove that the congruence $x^2 \equiv -1 \,(\mathrm{mod}\ p)$ is solvable if and only if $p \equiv 1 \,(\mathrm{mod}\ 4)$.

61. Let p be a prime.
   (a) Let x be an integer not divisible by p. Show that there exist integers a, b satisfying $0 < |a| < \sqrt{p}$, $0 < |b| < \sqrt{p}$, and $a \equiv bx \,(\mathrm{mod}\ p)$.
   (b) Show that p can be expressed in the form $a^2 + b^2$ with $a, b \in \mathbb{Z}$ if and only if $p = 2$ or $p \equiv 1 \,(\mathrm{mod}\ 4)$.
   (c) Let m be a positive integer. Prove that m can be expressed as the sum of two positive integers if and only if every prime divisor of m, that is of the form $4k + 3$, is of even multiplicity in m.

62. Let p be a prime of the form $4k + 1$, and let a and b be positive integers with a odd and $a^2 + b^2 = p$. Show that $\left(\frac{a}{p}\right) = 1$. (**Hint:** Use quadratic reciprocity.)

63. Let p be a prime.
   (a) Prove that the congruence $x^2 \equiv -2 \,(\mathrm{mod}\ p)$ is solvable if and only if $p = 2$ or $p \equiv 1$ or $3 \,(\mathrm{mod}\ 8)$.
   (b) Show that p can be expressed as $a^2 + 2b^2$ with $a, b \in \mathbb{Z}$ if and only if $p = 2$ or $p \equiv 1$ or $3 \,(\mathrm{mod}\ 8)$.

64. Let p be a prime.
   (a) Prove that the congruence $x^2 \equiv -3 \,(\mathrm{mod}\ p)$ is solvable if and only if $p = 3$ or $p \equiv 1 \,(\mathrm{mod}\ 3)$.
   (b) Show that p can be expressed as $a^2 + 3b^2$ with $a, b \in \mathbb{Z}$ if and only if $p = 3$ or $p \equiv 1 \,(\mathrm{mod}\ 3)$.
   (**Remark:** For a prime p and an integer constant d, $1 \leqslant d < p$, a decomposition $p = a^2 + db^2$ can be computed using the *Cornacchia algorithm* (1908). One first obtains a square root of $-d$ modulo p. If a square root does not exist, the equation $p = a^2 + db^2$ does not have a solution. Otherwise, we run the Euclidean gcd algorithm on p and this square root, and stop the gcd loop as soon as a remainder a less than $\sqrt{p}$ is obtained. If $(p - a^2)/d$ is an integer square b^2, then (a, b) is the desired solution (see Algorithm 1.10).[26])

65. Kronecker extended Jacobi's symbol to $\left(\frac{a}{b}\right)$ for all integers a, b with $b \neq 0$. Write $b = up_1 p_2 \cdots p_t$, where $u \in \{1, -1\}$, and p_i are primes (not necessarily all distinct). Define

$$\left(\frac{a}{b}\right) = \left(\frac{a}{u}\right) \prod_{i=1}^{t} \left(\frac{a}{p_i}\right).$$

---

[26] A proof for the correctness of the Cornacchia algorithm is not very easy and can be found, for example, in the paper J. M. Basilla, On the solution of $x^2 + dy^2 = m$, *Proc. Japan Acad.*, 80, Series A, 40–41, 2004.

Algorithm 1.10: CORNACCHIA ALGORITHM FOR SOLVING $a^2 + db^2 = p$

Compute the Legendre symbol $\left(\frac{-d}{p}\right)$.

If $\left(\frac{-d}{p}\right) = -1$, return "failure."

Compute s with $s^2 \equiv -d \pmod{p}$.

If $s \leqslant p/2$, replace s by $p - s$.

Set $x = p$ and $y = s$.

while $(y > \lfloor \sqrt{p} \rfloor)$ {

    $r = x$ rem y, $x = y$, $y = r$.

}

$a = y$.

Set $b = \sqrt{(p - a^2)/d}$.

If b is an integer, return (a, b), else return "failure."

This extension requires the following special cases:

$$\left(\frac{a}{1}\right) = 1$$

$$\left(\frac{a}{-1}\right) = \begin{cases} 1 & \text{if } a > 0 \\ -1 & \text{if } a < 0 \end{cases}$$

$$\left(\frac{a}{2}\right) = \begin{cases} 0 & \text{if } a \text{ is even} \\ \left(\frac{2}{|a|}\right) & \text{if } a \text{ is odd} \end{cases}$$

Describe an efficient algorithm to compute the Kronecker symbol.

66. Let $h = \text{ord}_m a$, $k = \text{ord}_m b$, and $l \in \mathbb{Z}$. Prove the following assertions.
  (a) $\text{ord}_m a^l = h/\gcd(h, l)$.
  (b) If $\gcd(h, k) = 1$, then $\text{ord}_m(ab) = hk$.
  (c) In general, $\text{ord}_m(ab) | \text{lcm}(h, k)$.
  (d) There exist m, a, b for which $\text{ord}_m(ab) < \text{lcm}(h, k)$.

67. Let m be a modulus having a primitive root, and $a \in \mathbb{Z}_m^*$. Prove that a is a primitive root modulo m if and only if $a^{\phi(m)/q} \not\equiv 1 \pmod{m}$ for every prime divisor q of $\phi(m)$. Design an algorithm that, given $a \in \mathbb{Z}_m^*$ and the prime factorization of $\phi(m)$, determines whether a is a primitive root modulo m.

68. Suppose that m is a modulus having a primitive root. Prove that $\mathbb{Z}_m^*$ contains exactly $\phi(\phi(m))$ primitive roots modulo m. In particular, a prime p has exactly $\phi(p - 1)$ primitive roots.

69. Let g, g' be two primitive roots modulo an odd prime p. Prove that:
  (a) gg' is not a primitive root modulo p.
  (b) $g^e \pmod{p}$ is a quadratic residue modulo p if and only if e is even.

70. Let p be an odd prime, $a \in \mathbb{Z}_p^*$, and $e \in \mathbb{N}$. Prove that the multiplicative order of $1 + ap$ modulo p^e is p^{e-1}. (**Remark:** This result can be used to obtain primitive roots modulo p^e.)

71. Expand the following irrational numbers as infinite simple continued fractions:
$\sqrt{2} - 1$, $1/\sqrt{3}$ and $\sqrt{15}$.

72. Let h_n/k_n be the convergents to an irrational ξ, and let F_n, $n \in \mathbb{N}_0$, denote the Fibonacci numbers.

(a) Show that $k_n \geqslant F_{n+1}$ for all $n \in \mathbb{N}_0$.

(b) Deduce that $k_n \geqslant \left\lfloor \dfrac{1}{\sqrt{5}} \left(\dfrac{1 + \sqrt{5}}{2} \right)^{n+1} \right\rfloor$ for all $n \in \mathbb{N}_0$. (**Remark:** This shows that the denominators in the convergents to an irrational number grow quite rapidly (at least exponentially) in n.)

(c) Does there exist an irrational ξ for which $k_n = F_{n+1}$ for all $n \in \mathbb{N}_0$?

73. (a) Prove that the continued fraction $\langle a_0, a_1, \ldots, a_n \rangle$ equals $\frac{K_{n+1}(a_0, a_1, \ldots, a_n)}{K_n(a_1, a_2, \ldots, a_n)}$, where K_n is the n-th continuant polynomial (Exercise 1.15).

(b) Let h_n/k_n be the n-th convergent to an irrational number ξ with h_n, k_n defined as in Section 1.8. Prove that $h_n = K_{n+1}(a_0, a_1, \ldots, a_n)$ and $k_n = K_n(a_1, a_2, \ldots, a_n)$ for all $n \geqslant 0$.

(c) Argue that $\gcd(h_n, k_n) = 1$, that is, the fraction h_n/k_n is in lowest terms.

74. A real number of the form $\frac{a + \sqrt{b}}{c}$ with $a, b, c \in \mathbb{Z}$, $c \neq 0$, and $b \geqslant 2$ *not* a perfect square, is called a *quadratic irrational*. An infinite simple continued fraction $\langle a_0, a_1, a_2, \ldots \rangle$ is called *periodic* if there exist $s \in \mathbb{N}_0$ and $t \in \mathbb{N}$ such that $a_{n+t} = a_n$ for all $n \geqslant s$. One can rewrite a periodic continued fractions as $\langle a_0, \ldots, a_{s-1}, \overline{b_0, \ldots, b_{t-1}} \rangle$, where the bar over the block of terms $b_0, \ldots, b_{t-1}$ indicates that this block is repeated *ad infinitum*. If $s = 0$, this continued fraction can be written as $\langle \overline{b_0, \ldots, b_{t-1}} \rangle$ and is called *purely periodic*. Show that a periodic simple continued fraction represents a quadratic irrational. (**Hint:** First consider the case of purely periodic continued fractions, and then adapt to the general case.)

75. Evaluate the periodic continued fractions $\langle 1, 2, 3, \overline{4} \rangle$ and $\langle 1, 2, \overline{3, 4} \rangle$.

76. Prove that there are infinitely many solutions in positive integers of both the equations $x^2 - 2y^2 = 1$ and $x^2 - 2y^2 = -1$. (**Hint:** Compute $h_n^2 - 2k_n^2$, where h_n/k_n is the n-th convergent to $\sqrt{2}$.)

77. (a) Compute the infinite simple continued fraction expansion of $\sqrt{3}$.

(b) For all $k \geqslant 1$, write $a_k + b_k\sqrt{3} = (2 + \sqrt{3})^k$ with a_k, b_k integers. Prove that for all $n \geqslant 0$, the $(2n + 1)$-th convergent of $\sqrt{3}$ is $r_{2n+1} = a_{n+1}/b_{n+1}$.

(**Remark:** a_k, b_k for $k \geqslant 1$ constitute all the non-zero solutions of the Pell equation $a^2 - 3b^2 = 1$. Proving this needs tools of algebraic number theory.)

78. (a) Compute the continued fraction expansion of $\sqrt{5}$.

(b) It is known that all the solutions of the Pell equation $x^2 - 5y^2 = 1$ with $x, y > 0$ are of the form $x = h_n$ and $y = k_n$, where h_n/k_n is a convergent to $\sqrt{5}$. Find the solution of the Pell equation $x^2 - 5y^2 = 1$ with the smallest possible $y > 0$.

(c) Let (a, b) denote the smallest solution obtained in Part (b). Define the sequence of pairs (x_n, y_n) of positive integers recursively as follows.

$$(x_0, y_0) = (a, b) \text{ and}$$
$$(x_n, y_n) = (ax_{n-1} + 5by_{n-1}, bx_{n-1} + ay_{n-1}) \text{ for } n \geqslant 1.$$

Prove that each (x_n, y_n) is a solution of $x^2 - 5y^2 = 1$. (In particular, there are infinitely many solutions in positive integers of the equation $x^2 - 5y^2 = 1$.)

79. Propose a polynomial-time algorithm that, given $t \in \mathbb{N}$, returns a prime of bit-length t. You may assume that you have at your disposal a polynomial-time algorithm for proving the primality or otherwise of an integer.

80. By Exercise 1.12(a), the number of iterations in the computation of the Euclidean gcd of a and b is $O(\lg \max(a, b))$. Since each Euclidean division done in the algorithm can take as much as $\Theta(\lg^2 \max(a, b))$ time, the running time of Euclidean gcd is $O(\lg^3 \max(a, b))$. This, however, turns out to be a gross overestimate. Prove that Euclidean gcd runs in $O(\lg^2 \max(a, b))$ time.

Programming Exercises

Use the GP/PARI calculator to solve the following problems.

81. For $n \in \mathbb{N}$ denote by $S_7(n)$ the sum of the digits of n expanded in base 7. We investigate those primes p for which $S_7(p)$ is composite. It turns out that for small values of p, most of the values $S_7(p)$ are also prime. Write a GP/PARI program that determines all primes $\leqslant 10^6$, for which $S_7(p)$ is composite. Provide a theoretical argument justifying the scarcity of small primes p for which $S_7(p)$ is composite.

82. Let B be a positive integral bound. Write a GP/PARI program that locates all pairs a, b of positive integers with $1 \leqslant a \leqslant b \leqslant B$, for which $\frac{a^2+b^2}{ab+1}$ is an integer. Can you detect a pattern in these integer values of the expression $\frac{a^2+b^2}{ab+1}$? Try to prove your guess.

83. Let B be a positive integral bound. Write a GP/PARI program that locates all pairs a, b of positive integers with $1 \leqslant a \leqslant b \leqslant B$ and $ab > 1$, for which $\frac{a^2+b^2}{ab-1}$ is an integer. Can you detect a pattern in these integer values of the expression $\frac{a^2+b^2}{ab-1}$? Try to prove your guess.

84. It can be proved that given any $a \in \mathbb{N}$, there exists an exponent $e \in \mathbb{N}$ for which the decimal expansion of 2^e starts with a (at the most significant end). For example, if $a = 7$, the smallest exponent e with this property is $e = 46$. Indeed, $2^{46} = 70368744177664$. Write a GP/PARI program that, given a, finds the smallest exponent e with the above property. Using the program, compute the value of this exponent e for $a = 2013$.

Chapter 2

Arithmetic of Finite Fields

A *field* is a commutative ring with identity, in which every non-zero element is invertible. More informally, a field F is a set with two commutative operations (addition and multiplication), in which one can add, subtract, and multiply any two elements, divide any element by any non-zero element, and the multiplication operation distributes over addition. It is a convention to disregard the zero ring (the ring with $0 = 1$) as a field.

Common examples of fields are $\mathbb{Q}$ (the field of rational numbers), $\mathbb{R}$ (the field of real numbers), and $\mathbb{C}$ (the field of complex numbers). On the other hand, $\mathbb{Z}$ (the ring of integers) is not a field, because the only elements of $\mathbb{Z}$ that have inverses (in $\mathbb{Z}$ itself) are ± 1. The fields $\mathbb{Q}, \mathbb{R}, \mathbb{C}$ are all infinite, since they contain infinitely many elements.

Definition 2.1 A field containing only finitely many elements is called a *finite field* or a *Galois field*.[1] ◁

Finite fields possess many properties (occasionally counter-intuitive) that the common infinite fields mentioned above do not. It is important in number theory and algebra to study these (and other) properties of finite fields. Lidl and Niederreiter[2] provide a comprehensive mathematical treatment of finite fields. In this chapter, we study some of these mathematical results along with computational issues associated with them.

2.1 Existence and Uniqueness of Finite Fields

The first counter-intuitive notion about finite fields is that they have positive characteristics as explained below.

If R is a ring with identity $1_R = 1$, then to every $m \in \mathbb{Z}$ we associate a unique element m_R of R. If $m = 0$, we take $0_R = 0$ (the additive identity). If $m > 0$, we take $m_R = 1 + 1 + \cdots + 1$ (m times). Finally, if $m < 0$, we take $m_R = -(-m)_R$. It is customary to denote the element m_R simply by m.

Definition 2.2 Let R be a ring with multiplicative identity 1. The smallest positive integer m for which $m = m_R = 1 + 1 + \cdots + 1$ (m times) $= 0$ is called the *characteristic* of R, denoted char R. If no such positive integer m exists, we take char $R = 0$. ◁

The elements $1, 2, 3, \ldots$ of a finite field F cannot be all distinct, that is, there exist positive integers m_1, m_2 with $m_1 < m_2$ such that $m_1 = m_2$. But then, $m = m_2 - m_1 = 0$ implying that finite fields have positive characteristics.

Proposition 2.3 *The characteristic of a finite field is prime.*
Proof Suppose not, that is, char $F = m = uv$ with $1 < u, v < m$ for some finite field F. The integers m, u, v are identified with the elements m_F, u_F, v_F of F. By the distributivity property, we have $0 = m_F = u_F v_F$. By definition,

---

[1]Finite fields are called *Galois fields* after the French mathematician Évariste Galois (1811–1832). Galois's seminal work at the age of twenty solved a contemporary open mathematical problem that states that univariate polynomial equations cannot, in general, be solved by radicals unless the degree of the polynomial is less than five. Galois's work has multiple ramifications in modern mathematics. In addition to the theory of finite fields, Galois introduced *Galois theory* and a formal treatment of *group theory*. Galois died at the age of twenty of a bullet injury sustained in a duel. Although Galois's paper did not receive immediate acceptance by mathematicians, Joseph Liouville (1809–1882) eventually understood its importance and was instrumental in publishing the article in the *Journal de Mathématiques Pures et Appliquées* in 1846.
[2]Rudolf Lidl and Harald Niederreiter, *Introduction to finite fields and their applications*, Cambridge University Press, 1994.

u_F and v_F are non-zero (u, v are smaller than m). It is easy to argue that a field is an integral domain, that is, the product of two non-zero elements cannot be 0. Thus, m cannot admit a factorization as assumed above. Moreover, if $m = 1$, then F is the zero ring (not a field by definition). ◁

The simplest examples of finite fields are the rings $\mathbb{Z}_p$ with p prime. (Every element of $\mathbb{Z}_p \setminus \{0\}$ is invertible. Other field properties are trivially valid.)

Let F be a finite field of size q and characteristic p. It is easy to verify that F contains $\mathbb{Z}_p$ as a subfield. (Imagine the way $\mathbb{Z}$ and $\mathbb{Q}$ are embedded in fields like $\mathbb{R}$ and $\mathbb{C}$.) Thus, F is an *extension* (see below) of $\mathbb{Z}_p$. From algebra it follows that F is a finite-dimensional vector space over $\mathbb{Z}_p$, that is, $q = p^n$, where n is the dimension of F over $\mathbb{Z}_p$.

Proposition 2.4 *Every finite field is of size p^n for some $p \in \mathbb{P}$ and $n \in \mathbb{N}$.* ◁

The converse of this is also true (although I will not prove it here).

Proposition 2.5 *For every $p \in \mathbb{P}$ and $n \in \mathbb{N}$, there exists a finite field with exactly p^n elements.* ◁

Let F, F' be two finite fields of the same size $q = p^n$. Both F and F' are extensions of $\mathbb{Z}_p$. It can be proved (not very easily) that there exists an isomorphism $\varphi : F \to F'$ of fields, that fixes the subfield $\mathbb{Z}_p$ element-wise. This result implies that any two finite fields of the same size follow the same arithmetic. In view of this, it is customary to talk about *the* finite field of size q (instead of *a* finite field of size q).

Definition 2.6 The finite field of size $q = p^n$ is denoted by $\mathbb{F}_q = \mathbb{F}_{p^n}$. If q itself is prime (corresponding to $n = 1$), the field $\mathbb{F}_q = \mathbb{F}_p$ is called a *prime field*. If $n > 1$, we call $\mathbb{F}_q$ an *extension field*. An alternative notation for $\mathbb{F}_q$ is $GF(q)$ (Galois field of size q). ◁

For a prime p, the two notations $\mathbb{F}_p$ and $\mathbb{Z}_p$ stand for the same algebraic object. However, when $q = p^n$ with $n > 1$, the notations $\mathbb{F}_q$ and $\mathbb{Z}_q$ refer to two different rings. They exhibit different arithmetic. $\mathbb{F}_q$ is a field, and so every non-zero element of it is invertible. On the other hand, $\mathbb{Z}_q$ is not a field (nor even an integral domain). Indeed $\phi(p^n) = p^{n-1}(p - 1)$, that is, $\mathbb{Z}_q$ contains $p^{n-1} - 1 > 0$ non-zero non-invertible elements, namely $p, 2p, \ldots, (p^{n-1} - 1)p$.

Throughout the rest of this chapter, we take p to be a prime and $q = p^n$ for some $n \in \mathbb{N}$.

2.2 Representation of Finite Fields

The prime field $\mathbb{F}_p$ is the same as $\mathbb{Z}_p$, that is, the arithmetic of $\mathbb{F}_p$ can be carried out as the integer arithmetic modulo the prime p. Chapter 1 already deals with this modular arithmetic. Here. we concentrate on extension fields.

2.2.1   Polynomial-Basis Representation

Let us recall the process of defining $\mathbb{C}$ as an extension of $\mathbb{R}$. We know that the polynomial $x^2 + 1$ with real coefficients has no real roots, and so is an irreducible polynomial over $\mathbb{R}$. Let us imagine that i is a root of $x^2 + 1$. Since i $\notin \mathbb{R}$, we need to introduce a structure $\mathbb{C}$ strictly bigger than $\mathbb{R}$. We want this bigger structure $\mathbb{C}$ to be a field again and contain both $\mathbb{R}$ and i.

$\mathbb{C}$ must be closed under arithmetic operations. To start with, we restrict to addition, subtraction and multiplication. A general element of a set containing $\mathbb{R}$ and i, and closed under these three operations must be of the form $t(i)$ for some polynomial $t(x) \in \mathbb{R}[x]$. Since i is a root of $x^2 + 1$, we have $i^2 = -1$, $i^3 = -i$, $i^4 = 1$, $i^5 = i$, and so on. This implies that the polynomial $t(x)$ may have degree $\geqslant 2$, but the element $t(i)$ can be simplified as $u + iv$ for some real numbers u, v. In short, $\mathbb{C}$ must contain all elements of the form $u + iv$.

However, $\mathbb{C}$ would be a field, and so every non-zero element of it must be invertible. Take an element of the form $x + iy$ with real numbers x, y not both zero. We have $(x + iy)^{-1} = \left(\frac{x}{x^2 + y^2} \right) + i \left(\frac{-y}{x^2 + y^2} \right)$, that is, the inverse of $x + iy$ can again be represented in the form $u + iv$. That is, polynomial expressions $t(i)$ of degrees < 2 suffice for making $\mathbb{C}$ a field.

We say that $\mathbb{C}$ is obtained by *adjoining* to $\mathbb{R}$ a root i of the irreducible polynomial $x^2 + 1$, and denote this as $\mathbb{C} = \mathbb{R}(i)$. An analogous construction applies to any field. Let $F = \mathbb{F}_p$, and let $f(x) \in F[x]$ be an irreducible polynomial of degree $n \geqslant 2$. In order that the construction works, there must exist such an irreducible polynomial $f(x)$ of degree n. We fail to extend $\mathbb{C}$ using the above construction, since any non-constant polynomial in $\mathbb{C}[x]$ has a root in $\mathbb{C}$ (*the fundamental theorem of algebra*). However, for every $p \in \mathbb{P}$ and $n \in \mathbb{N}$, there exists (at least) one irreducible polynomial of degree n in $\mathbb{F}_p[x]$.

Imagine that θ is a root of $f(x)$ in a (smallest) field K containing $F = \mathbb{F}_p$. Then, every polynomial expression $t(\theta)$ with $t(x) \in \mathbb{F}_p[x]$ must also reside in K. We have $f(\theta) = 0$ and $\deg f = n$. Thus, θ^n can be expressed as an F-linear combination of $1, \theta, \theta^2, \ldots, \theta^{n-1}$. But then, $\theta^{n+1} = \theta \times \theta^n$ can also be so expressed. More generally, θ^k for all $k \geqslant n$ can be expressed as F-linear combinations of $1, \theta, \theta^2, \ldots, \theta^{n-1}$. Consequently, even if $\deg t(x) \geqslant n$, we can express $t(\theta)$ as a polynomial expression (in θ) of degree $< n$.

Now, let us take any non-zero element $t(\theta)$ of degree $< n$. Since $f(x)$ is irreducible of degree n, we have $\gcd(t(x), f(x)) = 1$. Therefore, by Bézout's theorem for polynomials, there exist $u(x), v(x) \in F[x]$ such that $u(x)t(x) + v(x)f(x) = 1$. Since $f(\theta) = 0$, we have $u(\theta)t(\theta) = 1$, that is, $t(\theta)^{-1} = u(\theta)$. If $u(x)$ is of degree $\geqslant n$, we reduce $u(\theta)$ to a polynomial in θ of degree $< n$. It therefore follows that K needs to contain only the polynomial expressions of the form $t(\theta)$ with $t(x) \in F[x]$ and with $\deg t(x) < n$.

Let us finally take two polynomials $s(\theta), t(\theta)$ each of degree $< n$. The polynomial $r(x) = s(x) - t(x)$ is also of degree $< n$. Assume that $r(x) \neq 0$ but $r(\theta) = 0$. But then, θ is a root of both $r(x)$ and $f(x)$. Since $\deg r(x) < n$ and $f(x)$ is irreducible, we have $\gcd(r(x), f(x)) = 1$, that is, θ is a root of 1, an absurdity. It follows that $r(\theta) = 0$ if and only if $r(x) = 0$, that is, $s(x) = t(x)$. This implies that different polynomials $s(x), t(x)$ of degrees $< n$ correspond to different elements $s(\theta), t(\theta)$ of K. Thus, K can be represented by the set

$$K = \{t(\theta) \mid t(x) \in F[x], \ \deg t(x) < n\}.$$

A polynomial $t(x)$ of this form has n coefficients (those of $1, x, x^2, \ldots, x^{n-1}$), and each of these coefficients can assume any of the p values from $F = \mathbb{F}_p$. Consequently, the size of K is p^n, that is, K is a concrete realization of the field $\mathbb{F}_q = \mathbb{F}_{p^n}$. This representation is called the *polynomial-basis representation* of $\mathbb{F}_q$ over $\mathbb{F}_p$, because each element of K is an $\mathbb{F}_p$-linear combination of the *polynomial basis* $1, \theta, \theta^2, \ldots, \theta^{n-1}$. We denote this as $K = F(\theta)$.

$\mathbb{F}_q$ is an n-dimensional vector space over $\mathbb{F}_p$. Any set of n elements $\theta_0, \theta_1, \ldots, \theta_{n-1}$ constitute an $\mathbb{F}_p$-basis of $\mathbb{F}_q$ if and only if these elements are linearly independent over $\mathbb{F}_p$. The elements $1, \theta, \theta^2, \ldots, \theta^{n-1}$ form such a basis.

To sum up, an irreducible polynomial $f(x)$ of degree n in $\mathbb{F}_p[x]$ is needed to represent the extension $\mathbb{F}_q = \mathbb{F}_{p^n}$. Let $s(\theta), t(\theta)$ be two elements of $\mathbb{F}_q$, where

$$
\begin{aligned}
s(x) &= a_0 + a_1 x + a_2 x^2 + \cdots + a_{n-1} x^{n-1}, \\
t(x) &= b_0 + b_1 x + b_2 x^2 + \cdots + b_{n-1} x^{n-1}.
\end{aligned}
$$

Arithmetic operations on these elements are defined as follows.

$$
\begin{aligned}
s(\theta) + t(\theta) &= (a_0 + b_0) + (a_1 + b_1)\theta + (a_2 + b_2)\theta^2 + \cdots + (a_{n-1} + b_{n-1})\theta^{n-1}, \\
s(\theta) - t(\theta) &= (a_0 - b_0) + (a_1 - b_1)\theta + (a_2 - b_2)\theta^2 + \cdots + (a_{n-1} - b_{n-1})\theta^{n-1}, \\
s(\theta)t(\theta) &= r(\theta), \text{ where } r(x) = (s(x)t(x)) \text{ rem } f(x), \\
s(\theta)^{-1} &= u(\theta), \text{ where } u(x)s(x) + v(x)f(x) = 1 \text{ (provided that } s(\theta) \neq 0).
\end{aligned}
$$

Addition and subtraction in this representation of $\mathbb{F}_q$ do not require the irreducible polynomial $f(x)$, but multiplication and division do. A more detailed implementation-level description of these operations follows in Section 2.3.

Example 2.7 (1) Let us look at the polynomial-basis representation of $\mathbb{F}_4 = \mathbb{F}_{2^2}$. The polynomials of degree two in $\mathbb{F}_2[x]$ are $x^2, x^2 + x, x^2 + 1, x^2 + x + 1$. The first two in this list are clearly reducible. Also, $x^2 + 1 \equiv (x+1)^2 \pmod 2$ is reducible. The polynomial $x^2 + x + 1$ is irreducible. So we take $f(x) = x^2 + x + 1$ as the defining polynomial, and represent

$$\mathbb{F}_4 = \mathbb{F}_2(\theta) = \{a_1\theta + a_0 \mid a_1, a_0 \in \{0, 1\}\}, \quad \text{where } \theta^2 + \theta + 1 = 0.$$

The elements of $\mathbb{F}_4$ are, therefore, $0, 1, \theta, \theta + 1$. The addition and multiplication tables for $\mathbb{F}_4$ are given below.

	0	1	θ	$\theta + 1$
0	0	1	θ	$\theta + 1$
1	1	0	$\theta + 1$	θ
θ	θ	$\theta + 1$	0	1
$\theta + 1$	$\theta + 1$	θ	1	0

Addition in $\mathbb{F}_4$

	0	1	θ	$\theta + 1$
0	0	0	0	0
1	0	1	θ	$\theta + 1$
θ	0	θ	$\theta + 1$	1
$\theta + 1$	0	$\theta + 1$	1	θ

Multiplication in $\mathbb{F}_4$

Consider the elements θ and $\theta+1$. Their sum is $\theta+\theta+1 = 2\theta+1 = 1$ modulo 2, whereas their product is $\theta(\theta+1) = \theta^2+\theta = (\theta^2+\theta+1)+1 = 1$. In any ring of characteristic two, subtraction is same as addition (since $-1 = 1$).

(2) Let us now represent $\mathbb{F}_8 = \mathbb{F}_{2^3}$. First, we need an irreducible polynomial in $\mathbb{F}_2[x]$ of degree three. The polynomials of degree three that split into linear factors are x^3, $x^2(x+1) = x^3+x^2$, $x(x+1)^2 = x(x^2+1) = x^3+x$, and $(x+1)^3 = x^3+x^2+x+1$. On the other hand, the polynomials of degree three that factor into one linear factor and one quadratic irreducible factor are $x(x^2+x+1) = x^3+x^2+x$, and $(x+1)(x^2+x+1) = x^3+1$. This leaves us with only two irreducible polynomials of degree three: x^3+x+1 and x^3+x^2+1. Let us take the defining polynomial $f(x) = x^3+x^2+1$ so that

$$\mathbb{F}_8 = \mathbb{F}_2(\theta) = \{a_2\theta^2 + a_1\theta + a_0 \mid a_2, a_1, a_0 \in \{0,1\}\},$$

where $\theta^3 + \theta^2 + 1 = 0$. The addition table for $\mathbb{F}_8$ follows now.

	0	1	θ	$\theta+1$	θ^2	θ^2+1	$\theta^2+\theta$	$\theta^2+\theta+1$
0	0	1	θ	$\theta+1$	θ^2	θ^2+1	$\theta^2+\theta$	$\theta^2+\theta+1$
1	1	0	$\theta+1$	θ	θ^2+1	θ^2	$\theta^2+\theta+1$	$\theta^2+\theta$
θ	θ	$\theta+1$	0	1	$\theta^2+\theta$	$\theta^2+\theta+1$	θ^2	θ^2+1
$\theta+1$	$\theta+1$	θ	1	0	$\theta^2+\theta+1$	$\theta^2+\theta$	θ^2+1	θ^2
θ^2+1	θ^2	θ^2+1	$\theta^2+\theta$	$\theta^2+\theta+1$	0	1	θ	$\theta+1$
$\theta^2+\theta$	θ^2+1	θ^2	$\theta^2+\theta+1$	$\theta^2+\theta$	1	0	$\theta+1$	θ
$\theta^2+\theta+1$	$\theta^2+\theta$	$\theta^2+\theta+1$	θ^2	θ^2+1	θ	$\theta+1$	0	1
	$\theta^2+\theta+1$	$\theta^2+\theta$	θ^2+1	θ^2	$\theta+1$	θ	1	0

Multiplication in $\mathbb{F}_8$ involves multiplying two elements of $\mathbb{F}_8$ as polynomials over $\mathbb{F}_2$. If the product has degree three or more, one uses the equation $\theta^3 = \theta^2 + 1$ repeatedly, in order to reduce the product to a polynomial of degree less than three. This leads to the following multiplication table.

	0	1	θ	$\theta+1$	θ^2	θ^2+1	$\theta^2+\theta$	$\theta^2+\theta+1$
0	0	0	0	0	0	0	0	0
1	0	1	θ	$\theta+1$	θ^2	θ^2+1	$\theta^2+\theta$	$\theta^2+\theta+1$
θ	0	θ	θ^2	$\theta^2+\theta$	θ^2+1	$\theta^2+\theta+1$	1	$\theta+1$
$\theta+1$	0	$\theta+1$	$\theta^2+\theta$	θ^2+1	1	θ	$\theta^2+\theta+1$	θ^2
θ^2	0	θ^2	θ^2+1	1	$\theta^2+\theta+1$	$\theta+1$	θ	$\theta^2+\theta$
θ^2+1	0	θ^2+1	$\theta^2+\theta+1$	θ	$\theta+1$	$\theta^2+\theta$	θ^2	1
$\theta^2+\theta$	0	$\theta^2+\theta$	1	$\theta^2+\theta+1$	θ	θ^2	$\theta+1$	θ^2+1
$\theta^2+\theta+1$	0	$\theta^2+\theta+1$	$\theta+1$	θ^2	$\theta^2+\theta$	1	θ^2+1	θ

As a specific example, let us multiply $\theta^2+\theta$ with $\theta^2+\theta+1$. The product of the polynomials over $\mathbb{F}_2$ is $(\theta^2+\theta)(\theta^2+\theta+1) = \theta^4+2\theta^3+2\theta^2+\theta = \theta^4+\theta$. Now, we use the fact that $\theta^3+\theta^2+1 = 0$ in order to obtain $\theta^4+\theta = (\theta^4+\theta^3+\theta)+\theta^3 = \theta(\theta^3+\theta^2+1)+\theta^3 = \theta^3 = (\theta^3+\theta^2+1)+\theta^2+1 = \theta^2+1$.

Let us finally compute $(\theta^2+\theta)^{-1}$ in this representation of $\mathbb{F}_8$. Obviously, we can locate the desired inverse by looking at the above multiplication table. However, if q is large, the entire multiplication table for $\mathbb{F}_q$ cannot be computed or stored, and one should resort to computing a Bézout relation involving the polynomial to be inverted and the defining polynomial. In this example, we have $x(x^2+x)+1(x^3+x^2+1) = 1$. Substituting $x = \theta$ gives $(\theta^2+\theta)^{-1} = \theta$.

(3) Rijndael (accepted as the Advanced Encryption Standard (AES)), is a cryptographic cipher whose working is based on the arithmetic of the field $\mathbb{F}_{256} = \mathbb{F}_{2^8}$ represented by the irreducible polynomial $x^8 + x^4 + x^3 + x + 1$.

(4) Let us now look at a finite field of characteristic larger than two, namely at the field $\mathbb{F}_9 = \mathbb{F}_{3^2}$ of characteristic three. Since 2 is a quadratic non-residue modulo 3, the polynomial $x^2 - 2$ is irreducible in $\mathbb{F}_3[x]$. But $-2 \equiv 1 \pmod 3$, that is, we take $f(x) = x^2 + 1$ as the defining polynomial. (Like complex numbers) we then have the following representation of $\mathbb{F}_9$.

$$\mathbb{F}_9 = \mathbb{F}_3(\theta) = \{a_1\theta + a_0 \mid a_1, a_0 \in \{0, 1, 2\}\}, \text{ where } \theta^2 + 1 = 0.$$

We could have also taken $2(x^2 + 1)$ as the defining polynomial. However, it often turns out to be convenient to take a monic[3] irreducible polynomial as the defining polynomial. The addition table for $\mathbb{F}_9$ is as follows.

	0	1	2	θ	$\theta+1$	$\theta+2$	2θ	$2\theta+1$	$2\theta+2$
0	0	1	2	θ	$\theta+1$	$\theta+2$	2θ	$2\theta+1$	$2\theta+2$
1	1	2	0	$\theta+1$	$\theta+2$	θ	$2\theta+1$	$2\theta+2$	2θ
2	2	0	1	$\theta+2$	θ	$\theta+1$	$2\theta+2$	2θ	$2\theta+1$
θ	θ	$\theta+1$	$\theta+2$	2θ	$2\theta+1$	$2\theta+2$	0	1	2
$\theta+1$	$\theta+1$	$\theta+2$	θ	$2\theta+1$	$2\theta+2$	2θ	1	2	0
$\theta+2$	$\theta+2$	θ	$\theta+1$	$2\theta+2$	2θ	$2\theta+1$	2	0	1
2θ	2θ	$2\theta+1$	$2\theta+2$	0	1	2	θ	$\theta+1$	$\theta+2$
$2\theta+1$	$2\theta+1$	$2\theta+2$	2θ	1	2	0	$\theta+1$	$\theta+2$	θ
$2\theta+2$	$2\theta+2$	2θ	$2\theta+1$	2	0	1	$\theta+2$	θ	$\theta+1$

The multiplication table for $\mathbb{F}_9$ is as follows.

	0	1	2	θ	$\theta+1$	$\theta+2$	2θ	$2\theta+1$	$2\theta+2$
0	0	0	0	0	0	0	0	0	0
1	0	1	2	θ	$\theta+1$	$\theta+2$	2θ	$2\theta+1$	$2\theta+2$
2	0	2	1	2θ	$2\theta+2$	$2\theta+1$	θ	$\theta+2$	$\theta+1$
θ	0	θ	2θ	2	$\theta+2$	$2\theta+2$	1	$\theta+1$	$2\theta+1$
$\theta+1$	0	$\theta+1$	$2\theta+2$	$\theta+2$	2θ	1	$2\theta+1$	2	θ
$\theta+2$	0	$\theta+2$	$2\theta+1$	$2\theta+2$	1	θ	$\theta+1$	2θ	2
2θ	0	2θ	θ	1	$2\theta+1$	$\theta+1$	2	$2\theta+2$	$\theta+2$
$2\theta+1$	0	$2\theta+1$	$\theta+2$	$\theta+1$	2	2θ	$2\theta+2$	θ	1
$2\theta+2$	0	$2\theta+2$	$\theta+1$	$2\theta+1$	θ	2	$\theta+2$	1	2θ

As a specific example, consider the product $(\theta+2)(2\theta+1) = 2\theta^2 + 5\theta + 2 = 2\theta^2 + 2\theta + 2 = 2(\theta^2 + 1) + 2\theta = 2\theta$. □

2.2.2 Working with Finite Fields in GP/PARI

GP/PARI supports arithmetic over finite fields. The arithmetic of the prime field $\mathbb{F}_p$ is the modular arithmetic of $\mathbb{Z}_p$ and is discussed earlier. Let us focus

---

[3]The coefficient of the non-zero term of the highest degree in a non-zero polynomial $f(x)$ is called the *leading coefficient* of the polynomial, denoted lc $f(x)$. If lc $f(x) = 1$, we call $f(x)$ monic. If $f(x)$ is any non-zero polynomial over a field F with $a = $ lc $f(x)$, multiplying $f(x)$ by $a^{-1} \in F$ gives a monic polynomial.

on an extension field $\mathbb{F}_q = \mathbb{F}_{p^n}$. This involves modular arithmetic of two types. First, all polynomial coefficients are reduced modulo p, that is, the coefficient arithmetic is the modular arithmetic of $\mathbb{Z}_p$. Second, the arithmetic of $\mathbb{F}_q$ is the polynomial arithmetic of $\mathbb{Z}_p[x]$ modulo the defining polynomial $f(x)$.

As an example, let us represent $\mathbb{F}_8$ as in Example 2.7(2). First, we fix a defining polynomial.

```
gp > f = Mod(1,2)*x^3+Mod(1,2)*x^2+Mod(1,2)
%1 = Mod(1, 2)*x^3 + Mod(1, 2)*x^2 + Mod(1, 2)
```

Next, we take two elements of $\mathbb{F}_8$ as two polynomials in $\mathbb{F}_2[x]$ modulo $f(x)$.

```
gp > a = Mod(Mod(1,2)*x^2+Mod(1,2)*x, f)
%2 = Mod(Mod(1, 2)*x^2 + Mod(1, 2)*x, Mod(1, 2)*x^3 + Mod(1, 2)*x^2 + Mod(1, 2))
gp > b = Mod(Mod(1,2)*x^2+Mod(1,2)*x+Mod(1,2), f)
%3 = Mod(Mod(1, 2)*x^2 + Mod(1, 2)*x + Mod(1, 2), Mod(1, 2)*x^3 + Mod(1, 2)*x^2
+ Mod(1, 2))
```

Now, we can carry out arithmetic operations on a and b.

```
gp > a + b
%4 = Mod(Mod(1, 2), Mod(1, 2)*x^3 + Mod(1, 2)*x^2 + Mod(1, 2))
gp > a * b
%5 = Mod(Mod(1, 2)*x^2 + Mod(1, 2), Mod(1, 2)*x^3 + Mod(1, 2)*x^2 + Mod(1, 2))
gp > a^(-1)
%6 = Mod(Mod(1, 2)*x, Mod(1, 2)*x^3 + Mod(1, 2)*x^2 + Mod(1, 2))
gp > a / b
%7 = Mod(Mod(1, 2)*x^2, Mod(1, 2)*x^3 + Mod(1, 2)*x^2 + Mod(1, 2))
gp > a^4
%8 = Mod(Mod(1, 2)*x^2 + Mod(1, 2), Mod(1, 2)*x^3 + Mod(1, 2)*x^2 + Mod(1, 2))
```

The fact that the inverse a^{-1} is correctly computed can be verified by invoking the extended gcd function on polynomials.

```
gp > bezout(Mod(1,2)*x^2+Mod(1,2)*x,f)
%9 = [Mod(1, 2)*x, Mod(1, 2), Mod(1, 2)]
```

The expressions handled by GP/PARI may appear a bit clumsy. But if one looks closely at these expressions, the exact structure of the elements becomes absolutely clear. Our simpler (and more compact) mathematical notations are meaningful only under the assumption that certain symbols are implicitly understood from the context (like θ is a root of the defining polynomial $f(x)$). Given that GP/PARI provides only a text-based interface and supports a variety of objects, this explicit and verbose representation seems unavoidable (perhaps undesirable too). However, if you insist that you do not want to see the defining polynomial, you can `lift()` a field element. If, in addition, you do not want the Mod's in the coefficients, make another `lift()`. But `lift()` destroys information, and should be used with proper precaution.

```
gp > c = lift(a * b)
%10 = Mod(1, 2)*x^2 + Mod(1, 2)
gp > d = lift(c)
%11 = x^2 + 1
gp > c^4
%12 = Mod(1, 2)*x^8 + Mod(1, 2)
gp > d^4
%13 = x^8 + 4*x^6 + 6*x^4 + 4*x^2 + 1
gp > e = a * b
%14 = Mod(Mod(1, 2)*x^2 + Mod(1, 2), Mod(1, 2)*x^3 + Mod(1, 2)*x^2 + Mod(1, 2))
gp > e^4
%15 = Mod(Mod(1, 2)*x + Mod(1, 2), Mod(1, 2)*x^3 + Mod(1, 2)*x^2 + Mod(1, 2))
gp > lift(e^4)
%16 = Mod(1, 2)*x + Mod(1, 2)
gp > lift(lift(e^4))
%17 = x + 1
```

2.2.3   Choice of the Defining Polynomial

The irreducible polynomial $f(x) \in \mathbb{F}_p[x]$ used to define the extension $\mathbb{F}_{p^n}$ has a bearing on the running time of the arithmetic routines of $\mathbb{F}_{p^n}$. A random irreducible polynomial in $\mathbb{F}_p[x]$ of a given degree n can be obtained using the procedure described in Section 3.1. This procedure is expected to produce dense irreducible polynomials (polynomials with $\Theta(n)$ non-zero coefficients).

Multiplication in $\mathbb{F}_{p^n}$ is a multiplication in $\mathbb{F}_p[x]$ of two polynomials of degrees $< n$, followed by reduction modulo $f(x)$. The reduction step involves a long division of a polynomial of degree $\leqslant 2n - 2$ by a polynomial of degree n. A straightforward implementation of this division may take $\Theta(n^2)$ time.

If $f(x)$ is sparse (that is, has only a few non-zero coefficients), this reduction step can be significantly more efficient than the case of a dense $f(x)$, because for a sparse $f(x)$, only a few coefficients need to be adjusted in each step of polynomial division. In view of this, irreducible binomials, trinomials, quadrinomials and pentanomials (that is, polynomials with exactly two, three, four and five non-zero terms) in $\mathbb{F}_p[x]$ are often of importance to us. They lead to a running time of the order $O(n)$ for the reduction step.

An irreducible *binomial* in $\mathbb{F}_p[x]$ must be of the form $x^n + a$ with $a \in \mathbb{F}_p^*$. We can characterize all values of a for which this polynomial is irreducible.

Theorem 2.8 *The binomial $x^n + a \in \mathbb{F}_p[x]$ is irreducible if and only if both the following conditions are satisfied:*

(1) Every prime factor of n must divide $\mathrm{ord}_p(-a)$, but not $(p-1)/\mathrm{ord}_p(-a)$.
(2) If $n \equiv 0 \pmod{4}$, then $p \equiv 1 \pmod{4}$. ◁

These two conditions are somewhat too restrictive. For example, we do not have any irreducible binomials of degree $4k$ over a prime field of size $4l + 3$. It is, therefore, advisable to study irreducible *trinomials* $x^n + ax^k + b$ with $1 \leqslant k \leqslant n - 1$ and $a, b \in \mathbb{F}_p^*$. Complete characterizations of irreducible trinomials over all prime fields $\mathbb{F}_p$ are not known. Some partial results are, however, available. For example, the following result is useful when $p \gg n$.

Theorem 2.9 *The number of irreducible trinomials in $\mathbb{F}_p[x]$ of the form $x^n + x + b$ (with $b \in \mathbb{F}_p^*$) is asymptotically equal to p/n.* ◁

This result indicates that after trying $O(n)$ random values of b, we expect to obtain an irreducible trinomial of the form $x^n + x + b$. The choice $k = 1$ in Theorem 2.9 is particularly conducive to efficient implementations. However, if p is small, there are not many choices for b for the random search to succeed with high probability. In that case, we need to try with other values of k.

For every finite field $\mathbb{F}_p$ and every degree n, an irreducible binomial or trinomial or quadrinomial may fail to exist. An example ($p = 2$ and $n = 8$) is covered in Exercise 2.5. However, the following conjecture[4] is interesting.

Conjecture 2.10 *For any finite field $\mathbb{F}_q$ with $q \geqslant 3$ and for any $n \in \mathbb{N}$, there exists an irreducible polynomial in $\mathbb{F}_q[x]$ with degree n and with at most four non-zero terms.* ◁

2.3   Implementation of Finite Field Arithmetic

Given `GP/PARI` functions implementing all elementary operations in finite fields, writing these functions ourselves may sound like a waste of time. Still, as we have done with multiple-precision integers, let us devote some time here to these implementation issues. Impatient readers may skip this section.

2.3.1   Representation of Elements

The standard polynomial-basis representation of $\mathbb{F}_q = \mathbb{F}_{p^n}$ calls for a defining irreducible polynomial $f(x) \in \mathbb{F}_p[x]$ of degree n. This polynomial acts as the modulus for all field operations. An element $\alpha \in \mathbb{F}_q$ is a polynomial of degree $< n$, and can be represented by its n coefficients, each of which is an

---

[4]Joachim von zur Gathen, Irreducible trinomials over finite fields, *Mathematics of Computation*, 72, 1987–2000, 2003.

integer available modulo p. One then needs to write polynomial-manipulation routines to implement modular arithmetic on these polynomials.

The case $p = 2$ turns out to be practically the most important. In view of this, I focus mostly on fields of characteristic two (also called *binary fields*) in this section. An element of $\mathbb{F}_2$ is essentially a bit, so an element of $\mathbb{F}_{2^n}$ is an array of n bits. It is preferable to pack multiple bits in a single word. This promotes compact representation and efficient arithmetic routines. The word size (in bits) is denoted by w. Two natural choices for w are 32 and 64.

Example 2.11 For the sake of illustration, I take an artificially small $w = 8$, that is, eight bits of a field element are packed in a word. Let us represent $\mathbb{F}_{2^{19}}$ (with extension degree $n = 19$) as $\mathbb{F}_2(\theta)$, where θ is a root of the irreducible pentanomial $f(x) = x^{19} + x^5 + x^2 + x + 1$. (Incidentally, there is no irreducible trinomial of degree 19 in $\mathbb{F}_2[x]$.)

An element of $\mathbb{F}_{2^{19}}$ is a polynomial of the form

$$a_{18}\theta^{18} + a_{17}\theta^{17} + \cdots + a_1\theta + a_0,$$

where each $a_i \in \{0, 1\}$. The bit string $a_{18}a_{17}\ldots a_1a_0$ represents this element. All these 19 bits do not fit in a word. We need three words of size $w = 8$ bits to store this bit array. This representation is denoted as

$$a_{18}a_{17}a_{16} \quad a_{15}a_{14}a_{13}a_{12}a_{11}a_{10}a_9a_8 \quad a_7a_6a_5a_4a_3a_2a_1a_0,$$

with spaces indicating word boundaries. As a concrete example, the element

$$\theta^{17} + \theta^{12} + \theta^{11} + \theta^9 + \theta^7 + \theta^5 + \theta^2 + \theta + 1$$

is represented as

010 00011010 10100111.

The leftmost word is not fully used (since n is not a multiple of w). The unused bits in this word may store any value, but it is safe to pad the leftmost word with zero bits (equivalently, the polynomial with leading zero coefficients). □

Extension fields of characteristic three have found recent applications (for example, in pairing calculations). An element of $\mathbb{F}_{3^n}$ is represented by a polynomial of degree $\leqslant n - 1$ with coefficients from $\{0, 1, 2\}$. The coefficients are now three-valued, whereas bit-level representations are more convenient in digital computers. One possibility is to represent each coefficient by a pair of bits. Although the natural choices for the coefficients $0, 1, 2$ are respectively $00, 01$ and 10 (with the bit pattern 11 left undefined), it is not mandatory to be natural. Kawahara et al.[5] show that using the encoding $11, 01$ and 10 for $0, 1, 2$ respectively is more profitable from the angle of efficient implementation of the arithmetic of $\mathbb{F}_{3^n}$.

---

[5] Yuto Kawahara, Kazumaro Aoki and Tsuyoshi Takagi, Faster implementation of η_T pairing over GF(3^m) using minimum number of logical instructions for GF(3)-addition, *Pairing*, 282–296, 2008.

Therefore, an element

$$a_{n-1}\theta^{n-1} + a_{n-2}\theta^{n-2} + \cdots + a_1\theta + a_0$$

with each $a_i \in \{0, 1, 2\}$ being encoded by Kawahara et al.'s scheme can be represented by the bit string $h_{n-1}l_{n-1}h_{n-2}l_{n-2} \ldots h_1 l_1 h_0 l_0$ of length $2n$, where $h_i l_i$ is the two-bit encoding of a_i. It is advisable to separately store the high-order bits and the low-order bits. That means that the above element is to be stored as two bit arrays $h_{n-1}h_{n-2} \ldots h_1 h_0$ and $l_{n-1}l_{n-2} \ldots l_1 l_0$ each of size n. Each of these bit arrays can be packed individually in an array of w-bit words, as done for binary fields.

Example 2.12 Let us represent $\mathbb{F}_{3^{19}}$ as $\mathbb{F}_3(\theta)$, where θ is a root of $f(x) = x^{19} + x^2 + 2$. Consider the element

$$2\theta^{18} + \theta^{16} + \theta^{13} + 2\theta^{10} + \theta^6 + \theta^5 + 2\theta^2$$

of $\mathbb{F}_{3^{19}}$. As a sequence of ternary digits, this polynomial can be represented as 2010010020001100200. Under Kawahara et al.'s encoding, the bit representation of this element is as follows. We take words of size $w = 8$ bits.

High-order bit array   110   11011111   10011111
Low-order bit array   011   11111011   11111011

Different words are separated by spaces.   □

2.3.2   Polynomial Arithmetic

Arithmetic in an extension field $\mathbb{F}_{p^n}$ under the polynomial-basis representation is the modular polynomial arithmetic of $\mathbb{F}_p[x]$. The irreducible polynomial that defines the extension $\mathbb{F}_{p^n}$ is used as the modulus in all these operations.

2.3.2.1   Addition and Subtraction

Addition in binary fields is the bit-by-bit XOR operation. Since multiple coefficients are packed per word, word-level XOR operations add multiple (w) coefficients simultaneously. This makes the addition operation very efficient.

Example 2.13 We use the representation of $\mathbb{F}_{2^{19}}$ as in Example 2.11. Consider the following two operands with the bit-vector representations:

$$\theta^{18} + \theta^{16} + \theta^{14} + \theta^{12} + \theta^{10} + \theta^9 + \theta^8 + \theta^2 + 1 \qquad 101 \;\; 01010111 \;\; 00000101$$
$$\theta^{17} + \theta^{11} + \theta^9 + \theta^8 + \theta^7 + \theta^6 + \theta^5 + \theta^4 + \theta^3 + 1 \qquad 010 \;\; 00001011 \;\; 11111001$$

Applying word-level XOR operations on the two arrays gives the array

$$111 \;\; 01011100 \;\; 11111100$$

which represents the sum of the input operands, that is, the polynomial

$$\theta^{18} + \theta^{17} + \theta^{16} + \theta^{14} + \theta^{12} + \theta^{11} + \theta^{10} + \theta^7 + \theta^6 + \theta^5 + \theta^4 + \theta^3 + \theta^2. \;\; \square$$

Addition in fields of characteristic three is somewhat more involved. We have adopted Kawahara et al.'s encoding scheme so as to minimize the word-level bit-wise operations on the high and low arrays of the operands. Let the input operands be α and β. Denote the high and low bit arrays of α as α_h and α_l. Likewise, use the symbols β_h and β_l for the two arrays of β. The two arrays γ_h and γ_l of the sum $\gamma = \alpha + \beta$ need to be computed. Kawahara et al. show that this is possible with six bit-wise operations only:

$$\gamma_h = (\alpha_l \text{ XOR } \beta_l) \text{ OR } ((\alpha_h \text{ XOR } \beta_h) \text{ XOR } \alpha_l),$$
$$\gamma_l = (\alpha_h \text{ XOR } \beta_h) \text{ OR } ((\alpha_l \text{ XOR } \beta_l) \text{ XOR } \alpha_h).$$

Kawahara et al. also demonstrate that no encoding scheme can achieve this task using less than six bit-wise operations (XOR, OR, AND and NOT only).

Example 2.14 Take the following two elements α, β in the representation of $\mathbb{F}_{3^{19}}$ given in Example 2.12.

α	$=$	$2\theta^{18} + \theta^{16} + \theta^{13} + 2\theta^{10} + \theta^6 + \theta^5 + 2\theta^2$
α_h	:	110 11011111 10011111
α_l	:	011 11111011 11111011
β	$=$	$\theta^{17} + \theta^{14} + 2\theta^{13} + \theta^{10} + \theta^8 + 2\theta^3 + 2\theta^2 + 1$
β_h	:	101 10111010 11111110
β_l	:	111 11011111 11110011

The six operations are shown now. Temporary arrays $\tau_1, \tau_2, \tau_3, \tau_4$ are used.

$$
\begin{aligned}
\tau_1 &= \alpha_h \text{ XOR } \beta_h &&= 011 \ 01100101 \ 01100001 \\
\tau_2 &= \alpha_l \text{ XOR } \beta_l &&= 100 \ 00100100 \ 00001000 \\
\tau_3 &= \tau_1 \text{ XOR } \alpha_l &&= 000 \ 10011110 \ 10011010 \\
\tau_4 &= \tau_2 \text{ XOR } \alpha_h &&= 010 \ 11111011 \ 10010111 \\
\gamma_h &= \tau_2 \text{ OR } \tau_3 &&= 100 \ 10111110 \ 10011010 \\
\gamma_l &= \tau_1 \text{ OR } \tau_4 &&= 011 \ 11111111 \ 11110111
\end{aligned}
$$

Thus, γ corresponds to the sequence 211 01000001 01102101 of ternary digits, that is, to the following polynomial which is clearly $\alpha + \beta$ modulo 3.

$$2\theta^{18} + \theta^{17} + \theta^{16} + \theta^{14} + \theta^8 + \theta^6 + \theta^5 + 2\theta^3 + \theta^2 + 1. \qquad \square$$

Subtraction in $\mathbb{F}_{2^n}$ is same as addition. For subtraction in $\mathbb{F}_{3^n}$, it suffices to note that $\alpha - \beta = \alpha + (-\beta)$, and that the representation of $-\beta$ is obtained from that of β by swapping the high- and low-order bit arrays.

2.3.2.2 Multiplication

Multiplication in $\mathbb{F}_{p^n}$ involves two basic operations. First, the two operands are multiplied as polynomials in $\mathbb{F}_p[x]$. The result is a polynomial of degree $\leqslant 2(n-1)$. Subsequently, this product is divided by the defining polynomial $f(x)$. The remainder (a polynomial of degree $< n$) is the canonical representative of the product in the field. In what follows, I separately discuss these two primitive operations. As examples, I concentrate on binary fields only.

The first approach towards multiplying two polynomials of degrees $< n$ is to initialize the product as the zero polynomial with (formal) degree $2(n-1)$. For each non-zero term $b_i x^i$ in the second operand, the first operand is multiplied by b_i, shifted by i positions, and added to the product. For binary fields, the only non-zero value of b_i is 1, so only shifting and adding (XOR) suffice.

Example 2.15 Let us multiply the two elements α and β of Example 2.13. The exponents i in the non-zero terms θ^i of β and the corresponding shifted versions of α are shown below. When we add (XOR) all these shifted values, we obtain the desired product.

i	$x^i \alpha(x)$
0	101 01010111 00000101
3	101010 10111000 00101
4	1010101 01110000 0101
5	10101010 11100000 101
6	1 01010101 11000001 01
7	10 10101011 10000010 1
8	101 01010111 00000101
9	1010 10101110 0000101
11	101010 10111000 00101
17	1010 10101110 0000101
	01010 10001000 01100101 00011011 00011101

Storing the product, a polynomial of degree $\leqslant 36$, needs five eight-bit words. $\square$

2.3.2.3   Comb Methods

The above multiplication algorithm can be speeded up in a variety of ways. Here, I explain some tricks for binary fields.[6] An important observation regarding the above shift-and-add algorithm is that the shifts $x^i \alpha(x)$ and $x^j \alpha(x)$ differ only by trailing zero words if $i \equiv j \pmod{w}$. Therefore, the shifted polynomials $x^j \alpha(x)$ need to be computed only for $j = 0, 1, 2, \ldots, w-1$. Once a shifted value $x^j \alpha(x)$ is computed, it can be used for all non-zero coefficients $b_i = 1$ with $i = j + kw$ for $k = 0, 1, 2, \ldots$. All we need is to add $x^j \alpha(x)$ starting at the k-th word (from right) of the product. This method is referred to as the *right-to-left comb method*.

Example 2.16 In the multiplication of Example 2.15, we need to compute $x^j \alpha(x)$ for $j = 0, 1, 2, \ldots, 7$ only. We have $x^0 \alpha(x) = \alpha(x)$, and $x^j \alpha(x) = x \times (x^{j-1} \alpha(x))$ for $j = 1, 2, \ldots, 7$, that is, $x^j \alpha(x)$ is obtained from $x^{j-1} \alpha(x)$

---

[6] Julio López and Ricardo Dahab, High-speed software multiplication in $\mathbb{F}_{2^m}$, *IndoCrypt*, 203–212, 2000.

by a left-shift operation. In the current example, $x^0\alpha(x)$ is used for $i = 0, 8$, $x^1\alpha(x)$ for $i = 9, 17$, $x^2\alpha(x)$ is not used, $x^3\alpha(x)$ is used for $i = 3, 11$, $x^4\alpha(x)$ for $i = 4$, $x^5\alpha(x)$ for $i = 5$, $x^6\alpha(x)$ for $i = 6$, and $x^7\alpha(x)$ for $i = 7$. ☐

The *left-to-right comb method* shifts the product polynomial (instead of the first operand). This method processes bit positions $j = w - 1, w - 2, \ldots, 1, 0$ in a word, in that sequence. For all words in β with the j-th bit set, α is added to the product with appropriate word-level shifts. When a particular j is processed, the product is multiplied by x (left-shifted by one bit), so that it is aligned at the next (the $(j-1)$-st) bit position in the word. The left-to-right comb method which shifts the product polynomial is expected to be slower than the right-to-left comb method which shifts the first operand.

Example 2.17 The left-to-right comb method is now illustrated for the multiplication of Example 2.15. The product polynomial γ is always maintained as a polynomial of (formal) degree 36. The k-th word of γ is denoted by γ_k.

$$
\begin{aligned}
\alpha =&& 101\ 01010111\ 00000101 \\
x^8\alpha =&& 101\ 01010111\ 00000101 \\
x^{16}\alpha =& 101\ 01010111\ 00000101 &
\end{aligned}
$$

j	i	Operation	γ				
			γ_4	γ_3	γ_2	γ_1	γ_0
		Initialize γ to 0	00000	00000000	00000000	00000000	00000000
7	7	Add α from word 0	00000	00000000	00000101	01010111	00000101
		Left-shift γ	00000	00000000	00001010	10101110	00001010
6	6	Add α from word 0	00000	00000000	00001111	11111001	00001111
		Left-shift γ	00000	00000000	00011111	11110010	00011110
5	5	Add α from word 0	00000	00000000	00011010	10100101	00011011
		Left-shift γ	00000	00000000	00110101	01001010	00110110
4	4	Add α from word 0	00000	00000000	00110000	00011101	00110011
		Left-shift γ	00000	00000000	01100000	00111010	01100110
3	3	Add α from word 0	00000	00000000	01100101	01101101	01100011
	11	Add α from word 1	00000	00000101	00110010	01101000	01100011
		Left-shift γ	00000	00001010	01100100	11010000	11000110
2		Left-shift γ	00000	00010100	11001001	10100001	10001100
1	9	Add α from word 1	00000	00010001	10011110	10100100	10001100
	17	Add α from word 2	00101	01000110	10011011	10100100	10001100
		Left-shift γ	01010	10001101	00110111	01001001	00011000
0	0	Add α from word 0	01010	10001101	00110010	00011110	00011101
	8	Add α from word 1	01010	10001000	01100101	00011011	00011101

The final value of γ is the same polynomial computed in Example 2.15. For the last bit position $j = 0$, γ is not left-shifted since no further alignment of γ is necessary. The shifted polynomials $x^8\alpha$ and $x^{16}\alpha$ need not be explicitly computed. They are shown above only for the convenience of the reader. ☐

2.3.2.4 Windowed Comb Methods

The comb methods can be made faster using precomputation and table lookup. The basic idea is to use a *window* of some size k (as in the windowed exponentiation algorithm of Exercise 1.27). The products $\alpha\delta$ are precomputed and stored for all of the 2^k binary polynomials δ of degree $< k$. In the multiplication loop, k bits of β are processed at a time. Instead of adding α (or a shifted version of α) for each one-bit of β, we add the precomputed polynomial $\alpha\delta$ (or its suitably shifted version), where δ is the k-bit chunk read from the second operand β. A practically good choice for k is four.

Example 2.18 The working of the *right-to-left windowed comb method* applied to the multiplication of Example 2.16 is demonstrated now. We take the window size $k = 2$. The four products are precomputed as

$$
\begin{aligned}
(00)\alpha(x) &= (0x + 0)\alpha(x) &=& \quad 0000\ 00000000\ 00000000 \\
(01)\alpha(x) &= (0x + 1)\alpha(x) &=& \quad 0101\ 01010111\ 00000101 \\
(10)\alpha(x) &= (1x + 0)\alpha(x) &=& \quad 1010\ 10101110\ 00001010 \\
(11)\alpha(x) &= (1x + 1)\alpha(x) &=& \quad 1111\ 11111001\ 00001111
\end{aligned}
$$

The multiplication loop runs as follows.

i	Bits $b_{i+1}b_i$	$x^i(b_{i+1}x + b_i)\alpha$
0	01	0101 01010111 00000101
2	10	101010 10111000 001010
4	11	11111111 10010000 1111
6	11	11 11111110 01000011 11
8	11	1111 11111001 00001111
10	10	101010 10111000 001010
12	00	
14	00	
16	10	1010 10101110 00001010
18	00	
		01010 10001000 01100101 00011011 00011101

For the bit pattern 00, no addition needs to be made. In Example 2.15, 2.16 or 2.17, ten XOR operations are necessary, whereas the windowed comb method needs only seven. Of course, we now have the overhead of precomputation. In general, two is not a good window size (even for larger extensions than used in these examples). A good tradeoff among the overhead of precomputation (and storage), the number of XOR operations, and programming convenience (the window size k should divide the word size w) is $k = 4$. □

A trouble with the right-to-left windowed comb method is that it requires many bit-level shifts (by amounts which are multiples of k) of many or all of the precomputed polynomials. Effectively handling all these shifts is a nuisance. The solution is to shift the product instead of the precomputed polynomials.

That indicates that we need to convert the left-to-right comb method to the windowed form. Since k bits are simultaneously processed from β, the product γ should now be left-shifted by k bits.

Example 2.19 The *left-to-right windowed comb* method works on the multiplication of Example 2.15 as follows. We take the window size $k = 2$ for our illustration. The four precomputed polynomials are the same as in Example 2.18. The main multiplication loop is now unfolded.

j	i	$b_{i+1}b_i$	Value of variable	Op
			$\gamma = $ 00000 00000000 00000000 00000000 00000000	Init
6	6	11	$(11)\alpha = $ 1111 11111001 00001111	
			$\gamma = $ 00000 00000000 00001111 11111001 00001111	Add
	14	00		
	22	00		
			$\gamma = $ 00000 00000000 00111111 11100100 00111100	Shift
4	4	11	$(11)\alpha = $ 1111 11111001 00001111	
			$\gamma = $ 00000 00000000 00110000 00011101 00110011	Add
	12	00		
	20	00		
			$\gamma = $ 00000 00000000 11000000 01110100 11001100	Shift
2	2	10	$(10)\alpha = $ 1010 10101110 00001010	
			$\gamma = $ 00000 00000000 11001010 11011010 11000110	Add
	10	10	$x^8(10)\alpha = $ 1010 10101110 00001010	
			$\gamma = $ 00000 00001010 01100100 11010000 11000110	Add
	18	00		
			$\gamma = $ 00000 00101001 10010011 01000011 00011000	Shift
0	0	01	$(01)\alpha = $ 0101 01010111 00000101	
			$\gamma = $ 00000 00101001 10010110 00010100 00011101	Add
	8	11	$x^8(11)\alpha = $ 1111 11111001 00001111	
			$\gamma = $ 00000 00100110 01101111 00011011 00011101	Add
	16	10	$x^{16}(10)\alpha = $ 1010 10101110 00001010	
			$\gamma = $ 01010 10001000 01100101 00011011 00011101	Add

In the above table, the operation "Add" stands for adding a *word-level* shift of a precomputed polynomial to γ. These word-level shifts are not computed explicitly, but are shown here for the reader's convenience. The "Shift" operation stands for two-bit left shift of γ. As in Example 2.18, only seven XOR operations suffice. The number of *bit-level* shifts of the product (each by k bits) is always $(w/k) - 1$ (three in this example), independent of the operands. □

Other fast multiplication techniques (like Karatsuba–Ofman multiplication) can be used. I am not going to discuss this further here. Let me instead concentrate on the second part of modular multiplication, that is, reduction modulo the defining polynomial $f(x)$.

2.3.2.5   Modular Reduction

We assume that we have a polynomial $\gamma(x)$ of degree $\leqslant 2(n-1)$. Our task is to compute the remainder $\rho(x) = \gamma(x) \operatorname{rem} f(x)$, where $\deg f(x) = n$.

Euclidean division of polynomials keeps on removing terms of degrees larger than n by subtracting suitable multiples of $f(x)$ from $\gamma(x)$. It is natural to remove non-zero terms one by one from $\gamma(x)$ in the decreasing order of their degrees. Subtraction of a multiple of $f(x)$ may introduce new non-zero terms, so it is, in general, not easy to eliminate multiple non-zero terms simultaneously. For polynomials over $\mathbb{F}_2$, the only non-zero coefficient is 1. In order to remove the non-zero term x^i from $\gamma(x)$, we need to subtract (that is, add or XOR) $x^{i-n}f(x)$ from $\gamma(x)$, where $x^{i-n}f(x)$ can be efficiently computed by left shifting $f(x)$ by $i - n$ bits. Eventually, $\gamma(x)$ reduces to a polynomial of degree $< n$. This is the desired remainder $\rho(x)$.

Example 2.20 Let us reduce the product $\gamma(x)$ computed in Examples 2.15–2.19. Elimination of its non-zero terms of degrees $\geqslant 19$ is illustrated below.

i		Intermediate values					Operation
	$\gamma(x) =$ 01010	10001000	01100101	00011011	00011101		Init
35	$x^{16}f(x) =$ 1000	00000000	00100111				Shift
	$\gamma(x) =$ 00010	10001000	01000010	00011011	00011101		Add
33	$x^{14}f(x) =$ 10	00000000	00001001	11			Shift
	$\gamma(x) =$ 00000	10001000	01001011	11011011	00011101		Add
31	$x^{12}f(x) =$	10000000	00000010	0111			Shift
	$\gamma(x) =$ 00000	00001000	01001001	10101011	00011101		Add
27	$x^{8}f(x) =$	1000	00000000	00100111			Shift
	$\gamma(x) =$ 00000	00000000	01001001	10001100	00011101		Add
22	$x^{3}f(x) =$		1000000	00000001	00111		Shift
	$\gamma(x) =$ 00000	00000000	00001001	10001101	00100101		Add
19	$x^{0}f(x) =$		1000	00000000	00100111		Shift
	$\gamma(x) =$ 00000	00000000	00000001	10001101	00000010		Add

Here, "Shift" is the shifted polynomial $x^{i-n}f(x)$, and "Add" is the addition of this shifted value to $\gamma(x)$. After six iterations of term cancellation, $\gamma(x)$ reduces to a polynomial of degree 16. It follows that the product of α and β (of Example 2.15) in $\mathbb{F}_{2^{19}}$ is $\theta^{16} + \theta^{15} + \theta^{11} + \theta^{10} + \theta^8 + \theta$. $\qquad\square$

Modular reduction can be made efficient if the defining polynomial $f(x)$ is chosen appropriately. The first requirement is that $f(x)$ should have as few non-zero terms as possible. Irreducible binomials, trinomials, quadrinomials and pentanomials are very helpful in this regard. Second, the degrees of the non-zero terms in $f(x)$ (except x^n itself) should be as low as possible. In other words, the largest degree n_1 of these terms should be sufficiently smaller than n. If $n - n_1 \geqslant w$ (where w is the word size), cancellation of a non-zero term ax^i

by subtracting $ax^{i-n}f(x)$ from $\gamma(x)$ does not affect other coefficients residing in the same word of $\gamma(x)$ storing the coefficient of x^i. This means that we can now cancel an entire word together.

To be more precise, let us concentrate on binary fields, and write $f(x) = x^n + f_1(x)$ with $n_1 = \deg f_1(x) \leqslant n - w$. We want to cancel the leftmost non-zero word μ from $\gamma(x)$. Clearly, μ is a polynomial of degree $\leqslant w - 1$. If μ is the r-th word in γ, we need to add (XOR) $x^{rw-n}\mu f(x)$ to $\gamma(x)$. But $x^{rw-n}\mu f(x) = x^{rw}\mu + x^{rw-n}\mu f_1(x)$. The first part $x^{rw}\mu$ is precisely the r-th word of γ, so we can set this word to zero without actually performing the addition. The condition $n_1 \leqslant n - w$ indicates that the second part $x^{rw-n}\mu f_1(x)$ does not have non-zero terms in the r-th word of γ. Since multiplication by x^{rw-n} is a left shift, the only non-trivial computation is that of $\mu f_1(x)$. But μ has a small degree ($\leqslant w - 1$). If $f_1(x)$ too has only a few non-zero terms, this multiplication can be quite efficient. We can use a comb method for this multiplication in order to achieve higher efficiency. Since $f_1(x)$ is a polynomial dependent upon the representation of the field (but not on the operands), the precomputation for a windowed comb method needs to be done only once, for all reduction operations in the field. Even eight-bit windows can be feasible in terms of storage if $f_1(x)$ has only a few non-zero coefficients.

Example 2.21 Let us perform the division of Example 2.20 by word-based operations. In our representation of $\mathbb{F}_{2^{19}}$, we choose the irreducible polynomial $f(x) = x^{19} + f_1(x)$, where $f_1(x) = x^5 + x^2 + x + 1$, for which the degree $n_1 = 5$ is sufficiently smaller than $n = 19$ (our choice for w is eight). Therefore, it is safe to reduce $\gamma(x)$ word by word. The calculations are given below. The words of γ are indexed by the variable r. Since f_1 fits in a word, we treat it as a polynomial of degree seven, so μf_1 is of degree 12 and fits in two words.

r	μ	Intermediate values				
		$\gamma(x) =$ 01010	10001000	01100101	00011011	00011101
4	00001010	$x^{13}\mu f_1(x) =$ 00000	00101110	110		
		$\gamma(x) =$ 00000	10001000	01001011	11011011	00011101
3	10001000	$x^5\mu f_1(x) =$	00010	01010111	000	
		$\gamma(x) =$ 00000	00000000	01001001	10001100	00011101
2	00001001	$\mu f_1(x) =$		00000001	00011111	
		$\gamma(x) =$ 00000	00000000	00000001	10001101	00000010

The last iteration (for $r = 2$) is a bit tricky. This word of γ indicates the non-zero terms x^{22}, x^{19} and x^{16}. We need to remove the first two of these, but we cannot remove x^{16}. Since $n = 19$, we consider only the coefficients of x^{19} to x^{23}. The word is appropriately right shifted to compute μ in this case. □

For f_1 of special forms, further optimizations can be made. For instance, if f_1 contains only a few non-zero terms with degrees sufficiently separated from one another, the computation of μf_1 does not require a true multiplication. Word-level shift and XOR operations can subtract $x^{rw-n}\mu f_1$ from γ.

Example 2.22 NIST recommends[7] the representation of $\mathbb{F}_{2^{233}}$ using the irreducible polynomial $f(x) = x^{233} + x^{74} + 1$. As in a real-life implementation, we now choose $w = 64$. An element of $\mathbb{F}_{2^{233}}$ fits in four words. Moreover, an unreduced product γ of two field elements is a polynomial of degree at most 464, and fits in eight words. Let us denote the words of γ as $\gamma_0, \gamma_1, \ldots, \gamma_7$. We need to eliminate $\gamma_7, \gamma_6, \gamma_5, \gamma_4$ completely and γ_3 partially. For $r = 7, 6, 5, 4$ (in that sequence), we need to compute $x^{rw-n}\mu f_1 = x^{64r-233}\gamma_r(x^{74} + 1) = (x^{64r-159} + x^{64r-233})\gamma_r = (x^{64(r-3)+33} + x^{64(r-4)+23})\gamma_r = x^{64(r-3)}(x^{33}\gamma_r) + x^{64(r-4)}(x^{23}\gamma_r)$. Subtracting (XORing) this quantity from γ is equivalent to the following four word-level XOR operations:

$$\begin{aligned}
\gamma_{r-3} &\quad \text{is XORed with} \quad \text{LEFT-SHIFT}(\gamma_r, 33), \\
\gamma_{r-2} &\quad \text{is XORed with} \quad \text{RIGHT-SHIFT}(\gamma_r, 31), \\
\gamma_{r-4} &\quad \text{is XORed with} \quad \text{LEFT-SHIFT}(\gamma_r, 23), \\
\gamma_{r-3} &\quad \text{is XORed with} \quad \text{RIGHT-SHIFT}(\gamma_r, 41).
\end{aligned}$$

Removal of the coefficients of x^{255} through x^{233} in γ_3 can be similarly handled. The details are left to the reader as Exercise 2.10. □

Modular reduction involves division by the defining polynomial $f(x)$. In other situations (like gcd computations), we may have to divide any $a(x)$ by any non-zero $b(x)$. The standard coefficient-removal procedure continues to work here too. But a general $b(x)$ does not always enjoy the nice structural properties of $f(x)$. Consequently, the optimized division procedures that work for these $f(x)$ values are no longer applicable to a general Euclidean division.

2.3.3 Polynomial GCD and Inverse

Given a non-zero element α in a field, we need to compute the field element u such that $u\alpha = 1$. All algorithms in this section compute the extended gcd of α with the defining polynomial f. Since f is irreducible, and α is non-zero with degree smaller than $n = \deg f$, we have $\gcd(\alpha, f) = 1 = u\alpha + vf$ for some polynomials u, v. Computing u is of concern to us. We pass α and f as the two input parameters to the gcd algorithms. We consider only binary fields $\mathbb{F}_{2^n}$, adaptations to other fields $\mathbb{F}_{p^n}$ being fairly straightforward. I will restart using the polynomial notation for field elements (instead of the bit-vector notation).

2.3.3.1 Euclidean Inverse

Like integers, the Euclidean gcd algorithm generates a remainder sequence initialized as $r_0 = f$ and $r_1 = \alpha$. Subsequently, for $i = 2, 3, \ldots$, one computes $r_i = r_{i-2} \text{ rem } r_{i-1}$. We maintain two other sequences u_i and v_i satisfying $u_i\alpha + v_if = r_i$ for all $i \geqslant 0$. We initialize $u_0 = 0$, $u_1 = 1$, $v_0 = 1$, $v_1 = 0$ so that the invariance is satisfied for $i = 0, 1$. If $q_i = r_{i-2} \text{ quot } r_{i-1}$, then r_i

---

[7]http://csrc.nist.gov/groups/ST/toolkit/documents/dss/NISTReCur.pdf

can be written as $r_i = r_{i-2} - q_i r_{i-1}$. We update both the u and v sequences analogously, that is, $u_i = u_{i-2} - q_i u_{i-1}$ and $v_i = v_{i-2} - q_i v_{i-1}$. One can easily verify that these new values continue to satisfy $u_i \alpha + v_i f = r_i$.

For inverse calculations, it is not necessary to explicitly compute the v sequence. Even if v_i is needed at the end of the gcd loop, we can obtain this as $v_i = (r_i - u_i \alpha)/f$. Since each r_i and each u_i depend on only two previous terms, we need to store data only from two previous iterations.

Example 2.23 Let us define $\mathbb{F}_{2^7}$ by the irreducible polynomial $f(x) = x^7 + x^3 + 1$, and compute the inverse of $\alpha(x) = x^6 + x^3 + x^2 + x$. (Actually, $\alpha = \theta^6 + \theta^3 + \theta^2 + \theta$, where θ is a root of f. In the current context, we prefer to use x instead of θ to highlight that we are working in $\mathbb{F}_2[x]$.) Iterations of the extended Euclidean gcd algorithm are tabulated below. For clarity, we continue to use the indexed notation for the sequences r_i and u_i.

i	q_i	r_i	u_i
0		$x^7 + x^3 + 1$	0
1		$x^6 + x^3 + x^2 + x$	1
2	x	$x^4 + x^2 + 1$	x
3	$x^2 + 1$	$x^3 + x^2 + x + 1$	$x^3 + x + 1$
4	$x + 1$	x^2	$x^4 + x^3 + x^2 + x + 1$
5	$x + 1$	$x + 1$	$x^5 + x^3 + x$
6	$x + 1$	1	$x^6 + x^5 + 1$

It follows that $\alpha^{-1} = x^6 + x^5 + 1$. (Actually, $\alpha^{-1} = \theta^6 + \theta^5 + 1$.) □

2.3.3.2 Binary Inverse

The binary inverse algorithm in $\mathbb{F}_{2^n}$ is a direct adaptation of the extended binary gcd algorithm for integers, Now, x plays the role of 2. Algorithm 2.1 describes the binary inverse algorithm. The two polynomials $\alpha(x)$ and $f(x)$ are fed to this algorithm as input. Algorithm 2.1 maintains the invariance

$$u_1 \alpha + v_1 f = r_1,$$
$$u_2 \alpha + v_2 f = r_2.$$

Here, r_1, r_2 behave like the remainder sequence of Euclidean gcd. The other sequences u and v are subject to the same transformations as the r sequence. We do not maintain the v sequence explicitly. This sequence is necessary only for understanding the correctness of the algorithm.

The trouble now is that when we force r_1 (or r_2) to be divisible by x, the polynomial u_1 (or u_2) need not be divisible by x. Still, we have to extract an x from u_1 (or u_2). This is done by mutually adjusting the u and v values. Suppose $x | r_1$ but $x \nmid u_1$, and we want to cancel x from $u_1 \alpha + v_1 f = r_1$. Since $x \nmid u_1$, the constant term in u_1 must be 1. Moreover, since f is an irreducible polynomial, its constant term too must be 1. But then, the constant term of $u_1 + f$ is zero, that is, $x | (u_1 + f)$. We rewrite the invariance formula as $(u_1 + f)\alpha + (v_1 + \alpha)f = r_1$. Since x divides both r_1 and $u_1 + f$ (but not f), x must divide $v_1 + \alpha$ too. So we can now cancel x throughout the equation.

Algorithm 2.1: BINARY INVERSE ALGORITHM

```
Initialize r₁ = α, r₂ = f, u₁ = 1 and u₂ = 0.
Repeat {
    While (r₁ is divisible by x) {
        Set r₁ = r₁/x.
        If (u₁ is not divisible by x), set u₁ = u₁ + f.
        Set u₁ = u₁/x.
        If (r₁ = 1), return u₁.
    }
    While (r₂ is divisible by x) {
        Set r₂ = r₂/x.
        If (u₂ is not divisible by x), set u₂ = u₂ + f.
        Set u₂ = u₂/x.
        If (r₂ = 1), return u₂.
    }
    If (deg r₁ ⩾ deg r₂) {
        Set r₁ = r₁ + r₂ and u₁ = u₁ + u₂.
    } else {
        Set r₂ = r₂ + r₁ and u₂ = u₂ + u₁.
    }
}
```

Example 2.24 Under the representation of $\mathbb{F}_{2^7}$ by the irreducible polynomial $f(x) = x^7 + x^3 + 1$, we compute α^{-1}, where $\alpha = x^6 + x^3 + x^2 + x$. The computations are listed in the following table (continued on the next page).

r_1	r_2	u_1	u_2
$x^6 + x^3 + x^2 + x$	$x^7 + x^3 + 1$	1	0
Repeatedly remove x from r_1. Adjust u_1.			
$x^5 + x^2 + x + 1$	$x^7 + x^3 + 1$	$x^6 + x^2$	0
Set $r_2 = r_2 + r_1$ and $u_2 = u_2 + u_1$.			
$x^5 + x^2 + x + 1$	$x^7 + x^5 + x^3 + x^2 + x$	$x^6 + x^2$	$x^6 + x^2$
Repeatedly remove x from r_2. Adjust u_2.			
$x^5 + x^2 + x + 1$	$x^6 + x^4 + x^2 + x + 1$	$x^6 + x^2$	$x^5 + x$
Set $r_2 = r_2 + r_1$ and $u_2 = u_2 + u_1$.			
$x^5 + x^2 + x + 1$	$x^6 + x^5 + x^4$	$x^6 + x^2$	$x^6 + x^5 + x^2 + x$
Repeatedly remove x from r_2. Adjust u_2.			
$x^5 + x^2 + x + 1$	$x^5 + x^4 + x^3$	$x^6 + x^2$	$x^5 + x^4 + x + 1$
$x^5 + x^2 + x + 1$	$x^4 + x^3 + x^2$	$x^6 + x^2$	$x^6 + x^4 + x^3 + x^2 + 1$
$x^5 + x^2 + x + 1$	$x^3 + x^2 + x$	$x^6 + x^2$	$x^6 + x^5 + x^3 + x$
$x^5 + x^2 + x + 1$	$x^2 + x + 1$	$x^6 + x^2$	$x^5 + x^4 + x^2 + 1$

r_1	r_2	u_1	u_2
Set $r_1 = r_1 + r_2$ and $u_1 = u_1 + u_2$.			
x^5	$x^2 + x + 1$	$x^6 + x^5 + x^4 + 1$	$x^5 + x^4 + x^2 + 1$
Repeatedly remove x from r_1. Adjust u_1.			
x^4	$x^2 + x + 1$	$x^6 + x^5 + x^4 + x^3 + x^2$	$x^5 + x^4 + x^2 + 1$
x^3	$x^2 + x + 1$	$x^5 + x^4 + x^3 + x^2 + x$	$x^5 + x^4 + x^2 + 1$
x^2	$x^2 + x + 1$	$x^4 + x^3 + x^2 + x + 1$	$x^5 + x^4 + x^2 + 1$
x	$x^2 + x + 1$	$x^6 + x^3 + x + 1$	$x^5 + x^4 + x^2 + 1$
1	$x^2 + x + 1$	$x^6 + x^5 + 1$	$x^5 + x^4 + x^2 + 1$

In this example, r_1 eventually becomes 1, so the inverse of α is the value of u_1 at that time, that is, $x^6 + x^5 + 1$. □

For integers, binary gcd is usually faster than Euclidean gcd, since Euclidean division is significantly more expensive than addition and shifting. For polynomials, binary inverse and Euclidean inverse have comparable performances. Here, Euclidean inverse can be viewed as a sequence of removing the most significant terms from one of the remainders. In binary inverse, a term is removed from the least significant end, and subsequently divisions by x restore the least significant term back to 1. Both these removal processes use roughly the same number and types (shift and XOR) of operations.

2.3.3.3 Almost Inverse

The almost inverse algorithm[8] is a minor variant of the binary inverse algorithm. In the binary inverse algorithm, we cancel x from both r_1 and u_1 (or from r_2 and u_2) inside the gcd loop. The process involves conditional adding of f to u_1 (or u_2). In the almost inverse algorithm, we do not extract the x's from the u (and v) sequences (but remember how many x's need to be extracted). After the loop terminates, we extract all these x's from u_1 or u_2.

More precisely, we now maintain the invariances

$$u_1\alpha + v_1 f = x^k r_1,$$
$$u_2\alpha + v_2 f = x^k r_2,$$

for some integer $k \geqslant 0$. The value of k changes with time (and should be remembered), but must be the same in both the equations at the same point of time. Suppose that both r_1 and r_2 have constant terms 1, and $\deg r_1 \geqslant \deg r_2$. In that case, we add the second equation to the first to get

$$(u_1 + u_2)\alpha + (v_1 + v_2)f = x^k(r_1 + r_2).$$

Renaming $u_1 + u_2$ as u_1, $v_1 + v_2$ as v_1, and $r_1 + r_2$ as r_1 gives $u_1\alpha + v_1 f = x^k r_1$. Now, r_1 is divisible by x. Let t be the largest exponent for which x^t divides r_1. We extract x^t from r_1, and rename r_1/x^t as r_1 to get

$$u_1\alpha + v_1 f = x^{k+t} r_1.$$

---

[8]R. Schroeppel, H. Orman, S. O'Malley and O. Spatscheck, Fast key exchange with elliptic curve systems, *CRYPTO*, 43–56, 1995.

We do not update u_1 and v_1 here. However, since the value of k has changed to $k + t$, the other equation must be updated to agree with this, that is, the second equation is transformed as

$$(x^t u_2)\alpha + (x^t v_2)f = x^{k+t} r_2.$$

Renaming $x^t u_2$ as u_2 and $x^t v_2$ as v_2 restores both the invariances. Algorithm 2.2 implements this idea. We do not need to maintain v_1, v_2 explicitly.

Algorithm 2.2: ALMOST INVERSE ALGORITHM

```
Initialize r₁ = α, r₂ = f, u₁ = 1, u₂ = 0, and k = 0.
Repeat {
    if (r₁ is divisible by x) {
        Let xᵗ|r₁ but xᵗ⁺¹∤r₁.
        Set r₁ = r₁/xᵗ, u₂ = xᵗu₂, and k = k + t.
        If (r₁ = 1), return x⁻ᵏu₁ (mod f).
    }
    if (r₂ is divisible by x) {
        Let xᵗ|r₂ but xᵗ⁺¹∤r₂.
        Set r₂ = r₂/xᵗ, u₁ = xᵗu₁, and k = k + t.
        If (r₂ = 1), return x⁻ᵏu₂ (mod f).
    }
    If (deg r₁ ⩾ deg r₂) {
        Set r₁ = r₁ + r₂ and u₁ = u₁ + u₂.
    } else {
        Set r₂ = r₂ + r₁ and u₂ = u₂ + u₁.
    }
}
```

We have to extract the required number of x's after the termination of the loop. Suppose that the loop terminates because of the condition $r_1 = 1$. In that case, we have $u_1\alpha + v_1 f = x^k$ for some k. But then, $\alpha^{-1} = x^{-k} u_1$ modulo f. Likewise, if r_2 becomes 1, we need to compute $x^{-k} u_2$ modulo f.

Suppose that $x^{-k} u$ needs to be computed modulo f for some u. One possibility is to divide u by x as long as possible. When the constant term in u becomes 1, f is added to it, and the process continues until x is removed k times. This amounts to doing the same computations as in Algorithm 2.1 (albeit at a different location in the algorithm).

If f is of some special form, the removal process can be made somewhat efficient. Suppose that x^l is the non-zero term in f with smallest degree > 0. Let h denote the sum of the non-zero terms of u with degrees $< l$ (terms involving $x^0, x^1, \ldots, x^{l-1}$). Since $u + hf$ is divisible by x^l, l occurrences of x can be simultaneously removed from $u + hf$. For small l ($l = 1$ in the worst case), the removal of x's requires too many iterations. If l is not too small, many x's are removed per iteration, and we expect the almost inverse algorithm to run a bit more efficiently than the binary inverse algorithm.

Example 2.25 As in the previous two examples, take $f = x^7 + x^3 + 1$ and $\alpha = x^6 + x^3 + x^2 + x$. The iterations of the main gcd loop are shown first.

r_1	r_2	u_1	u_2	t
$x^6 + x^3 + x^2 + x$	$x^7 + x^3 + 1$	1	0	0
Remove x^1 from r_1. Adjust u_2.				
$x^5 + x^2 + x + 1$	$x^7 + x^3 + 1$	1	0	1
Set $r_2 = r_2 + r_1$ and $u_2 = u_2 + u_1$.				
$x^5 + x^2 + x + 1$	$x^7 + x^5 + x^3 + x^2 + x$	1	1	1
Remove x^1 from r_2. Adjust u_1.				
$x^5 + x^2 + x + 1$	$x^6 + x^4 + x^2 + x + 1$	x	1	2
Set $r_2 = r_2 + r_1$ and $u_2 = u_2 + u_1$.				
$x^5 + x^2 + x + 1$	$x^6 + x^5 + x^4$	x	$x + 1$	2
Remove x^4 from r_2. Adjust u_1.				
$x^5 + x^2 + x + 1$	$x^2 + x + 1$	x^5	$x + 1$	6
Set $r_1 = r_1 + r_2$ and $u_1 = u_1 + u_2$.				
x^5	$x^2 + x + 1$	$x^5 + x + 1$	$x + 1$	6
Remove x^5 from r_1. Adjust u_2.				
1	$x^2 + x + 1$	$x^5 + x + 1$	$x^6 + x^5$	11

The loop terminates because of $r_1 = 1$. At that instant, $k = 11$, and $u_1 = x^5 + x + 1$. So we compute $x^{-11}(x^5 + x + 1)$ modulo f. For $f(x) = x^7 + x^3 + 1$, we have $l = 3$, that is, we can remove x^3 in one iteration. In the last iteration, only x^2 is removed. The removal procedure is illustrated below.

l	h	$u + hf$	$(u + hf)/x^l$ (renamed as u)
			$x^5 + x + 1$
3	$x + 1$	$x^8 + x^7 + x^5 + x^4 + x^3$	$x^5 + x^4 + x^2 + x + 1$
3	$x^2 + x + 1$	$x^9 + x^8 + x^7 + x^3$	$x^6 + x^5 + x^4 + 1$
3	1	$x^7 + x^6 + x^5 + x^4 + x^3$	$x^4 + x^3 + x^2 + x + 1$
2	$x + 1$	$x^8 + x^7 + x^2$	$x^6 + x^5 + 1$

$\square$

2.4 Some Properties of Finite Fields

Some mathematical properties of finite fields, that have important bearings on the implementations of finite-field arithmetic, are studied in this section.

2.4.1 Fermat's Little Theorem for Finite Fields

Theorem 2.26 [*Fermat's little theorem for $\mathbb{F}_q$*] Let $\alpha \in \mathbb{F}_q$. Then, we have $\alpha^q = \alpha$. Moreover, if $\alpha \neq 0$, then $\alpha^{q-1} = 1$.

Proof First, take $\alpha \neq 0$, and let $\alpha_1, \ldots, \alpha_{q-1}$ be all the elements of $\mathbb{F}_q^* = \mathbb{F}_q \setminus \{0\}$. Then, $\alpha\alpha_1, \ldots, \alpha\alpha_{q-1}$ is a permutation of $\alpha_1, \ldots, \alpha_{q-1}$, so $\prod_{i=1}^{q-1} \alpha_i = \prod_{i=1}^{q-1}(\alpha\alpha_i) = \alpha^{q-1}\prod_{i=1}^{q-1}\alpha_i$. Canceling the product yields $\alpha^{q-1} = 1$ and so $\alpha^q = \alpha$. For $\alpha = 0$, we have $0^q = 0$. $\triangleleft$

A very important consequence of this theorem follows.

Theorem 2.27 *The polynomial $x^q - x \in \mathbb{F}_p[x]$ splits into linear factors over $\mathbb{F}_q$ as $x^q - x = \prod_{\alpha \in \mathbb{F}_q}(x - \alpha)$.* $\triangleleft$

Proposition 2.28 *Let $q = p^n$, and d a positive divisor of n. Then, $\mathbb{F}_q$ contains a unique intermediate field of size $\mathbb{F}_{p^d}$. Moreover, an element $\alpha \in \mathbb{F}_q$ belongs to this intermediate field if and only if $\alpha^{p^d} = \alpha$.*
Proof Consider the set $E = \{\alpha \in \mathbb{F}_q \mid \alpha^{p^d} = \alpha\}$. It is easy to verify that E satisfies all axioms for a field and that E contains exactly p^d elements. $\triangleleft$

Example 2.29 Let us represent $\mathbb{F}_{64} = \mathbb{F}_{2^6}$ as

$$\mathbb{F}_2(\theta) = \{a_5\theta^5 + a_4\theta^4 + a_3\theta^3 + a_2\theta^2 + a_1\theta + a_0 \mid a_i \in \{0,1\}\},$$

where $\theta^6 + \theta + 1 = 0$. The element $a_5\theta^5 + a_4\theta^4 + a_3\theta^3 + a_2\theta^2 + a_1\theta + a_0 \in \mathbb{F}_{64}$ is abbreviated as $a_5a_4a_3a_2a_1a_0$. The square of this element is calculated as

$$
\begin{aligned}
& (a_5\theta^5 + a_4\theta^4 + a_3\theta^3 + a_2\theta^2 + a_1\theta + a_0)^2 \\
=\ & a_5\theta^{10} + a_4\theta^8 + a_3\theta^6 + a_2\theta^4 + a_1\theta^2 + a_0 \\
=\ & a_5\theta^4(\theta+1) + a_4\theta^2(\theta+1) + a_3(\theta+1) + a_2\theta^4 + a_1\theta^2 + a_0 \\
=\ & a_5\theta^5 + (a_5 + a_2)\theta^4 + a_4\theta^3 + (a_4 + a_1)\theta^2 + a_3\theta + (a_3 + a_0).
\end{aligned}
$$

The smallest positive integer t for which $\alpha^{2^t} = \alpha$ is listed for each $\alpha \in \mathbb{F}_{64}$.

α	t	α	t	α	t	α	t
000000	1	010000	6	100000	6	110000	6
000001	1	010001	6	100001	6	110001	6
000010	6	010010	6	100010	6	110010	6
000011	6	010011	6	100011	6	110011	6
000100	6	010100	6	100100	6	110100	6
000101	6	010101	6	100101	6	110101	6
000110	6	010110	3	100110	6	110110	6
000111	6	010111	3	100111	6	110111	6
001000	6	011000	3	101000	6	111000	6
001001	6	011001	3	101001	6	111001	6
001010	6	011010	6	101010	6	111010	2
001011	6	011011	6	101011	6	111011	2
001100	6	011100	6	101100	6	111100	6
001101	6	011101	6	101101	6	111101	6
001110	3	011110	6	101110	6	111110	6
001111	3	011111	6	101111	6	111111	6

The proper divisors of the extension degree 6 are $1, 2, 3$. The unique intermediate field of $\mathbb{F}_{64}$ of size 2^1 is $\{0, 1\}$. The intermediate field of size 2^2 is

$$\{0, 1, \theta^5 + \theta^4 + \theta^3 + \theta, \theta^5 + \theta^4 + \theta^3 + \theta + 1\}.$$

Finally, the intermediate field of size 2^3 is

$$\left\{0, 1, \theta^3 + \theta^2 + \theta, \theta^3 + \theta^2 + \theta + 1, \theta^4 + \theta^2 + \theta, \theta^4 + \theta^2 + \theta + 1, \theta^4 + \theta^3, \theta^4 + \theta^3 + \theta\right\}. \quad \square$$

2.4.2 Multiplicative Orders of Elements in Finite Fields

The concept of multiplicative orders modulo p can be generalized for $\mathbb{F}_q$.

Definition 2.30 Let $\alpha \in \mathbb{F}_q$, $\alpha \neq 0$. The smallest positive integer e for which $\alpha^e = 1$ is called the order of α and is denoted by $\operatorname{ord}\alpha$. By Fermat's little theorem for $\mathbb{F}_q$, we have $\operatorname{ord}\alpha | (q - 1)$. If $\operatorname{ord}\alpha = q - 1$, then α is called a *primitive element* of $\mathbb{F}_q$.

If $\mathbb{F}_q = \mathbb{F}_p(\theta)$, where θ is a root of the irreducible polynomial $f(x) \in \mathbb{F}_p[x]$, and if θ is a primitive element of $\mathbb{F}_q$, we call $f(x)$ a *primitive polynomial*. $\quad \triangleleft$

Theorem 2.31 *Every finite field has a primitive element. (In algebraic terms, the group $\mathbb{F}_q^*$ is cyclic.)*
Proof Follow an argument as in the proof of Theorem 1.57. $\quad \triangleleft$

Example 2.32 As in Example 2.29, we continue to represent $\mathbb{F}_{64}$ as $\mathbb{F}_2(\theta)$, where $\theta^6 + \theta + 1 = 0$. The orders of all the elements of $\mathbb{F}_{64}$ are listed now.

α	$\operatorname{ord}\alpha$	α	$\operatorname{ord}\alpha$	α	$\operatorname{ord}\alpha$	α	$\operatorname{ord}\alpha$
000000	–	010000	63	100000	63	110000	63
000001	1	010001	21	100001	63	110001	63
000010	63	010010	21	100010	63	110010	63
000011	21	010011	63	100011	63	110011	21
000100	63	010100	9	100100	63	110100	63
000101	21	010101	63	100101	63	110101	63
000110	9	010110	7	100110	63	110110	21
000111	63	010111	7	100111	63	110111	63
001000	21	011000	7	101000	21	111000	63
001001	63	011001	7	101001	63	111001	21
001010	63	011010	9	101010	63	111010	3
001011	9	011011	63	101011	21	111011	3
001100	63	011100	9	101100	63	111100	63
001101	21	011101	63	101101	63	111101	63
001110	7	011110	63	101110	63	111110	21
001111	7	011111	9	101111	63	111111	63

The field $\mathbb{F}_q$ has exactly $\phi(q-1)$ primitive elements. For $q = 64$, this number is $\phi(63) = 36$. The above table shows that θ itself is a primitive element. Therefore, the defining polynomial $x^6 + x + 1$ is a primitive polynomial. $\quad \square$

2.4.3 Normal Elements

Let $\mathbb{F}_q = \mathbb{F}_p(\theta)$, where θ is a root of an irreducible polynomial $f(x) \in \mathbb{F}_p[x]$ of degree n. Since $f(x)$ has a root θ in $\mathbb{F}_q$, the polynomial is no longer irreducible in $\mathbb{F}_q[x]$. But what happens to the other $n-1$ roots of $f(x)$?

In order to answer this question, we first review the extension of $\mathbb{R}$ by adjoining a root i of $x^2 + 1$. The other root of this polynomial is $-\mathrm{i}$ which is included in $\mathbb{C}$. Thus, the polynomial $x^2 + 1$ splits into linear factors over $\mathbb{C}$ as $x^2 + 1 = (x - \mathrm{i})(x + \mathrm{i})$. Since the defining polynomial is of degree two, and has one root in the extension, the other root must also reside in that extension.

Let us now extend the field $\mathbb{Q}$ by a root θ of the polynomial $x^3 - 2$. The three roots of this polynomial are $\theta_0 = 2^{1/3}$, $\theta_1 = 2^{1/3} e^{\mathrm{i} 2\pi/3}$, and $\theta_2 = 2^{1/3} e^{\mathrm{i} 4\pi/3}$. Let us take $\theta = \theta_0$. Since this root is a real number, adjoining it to $\mathbb{Q}$ gives a field contained in $\mathbb{R}$. On the other hand, the roots θ_1, θ_2 are properly complex numbers, that is, $\mathbb{Q}(\theta)$ does not contain θ_1, θ_2. Indeed, the defining polynomial factors over the extension as $x^3 - 2 = (x - 2^{1/3})(x^2 + 2^{1/3}x + 2^{2/3})$, the second factor being irreducible in this extension.

The above two examples illustrate that an extension may or may not contain all the roots of the defining polynomial. Let us now concentrate on finite fields. Let us write $f(x)$ explicitly as

$$f(x) = a_0 + a_1 x + a_2 x^2 + \cdots + a_n x^n$$

with each $a_i \in \mathbb{F}_p$. Exercise 1.34 implies $f(x)^p = a_0^p + a_1^p x^p + a_2^p x^{2p} + \cdots + a_n^p x^{np}$. By Fermat's little theorem, $a_i^p = a_i$ in $\mathbb{F}_p$, and so $f(x)^p = a_0 + a_1 x^p + a_2 x^{2p} + \cdots + a_n x^{np} = f(x^p)$. Putting $x = \theta$ yields $f(\theta^p) = f(\theta)^p = 0^p = 0$, that is, θ^p is again a root of $f(x)$. Moreover, $\theta^p \in \mathbb{F}_q$. We can likewise argue that $\theta^{p^2} = (\theta^p)^p$, $\theta^{p^3} = (\theta^{p^2})^p$, ... are roots of $f(x)$ and lie in $\mathbb{F}_q$. One can show that the roots $\theta, \theta^p, \theta^{p^2}, \ldots, \theta^{p^{n-1}}$ of $f(x)$ are pairwise distinct and so must be all the roots of $f(x)$. In other words, $f(x)$ splits into linear factors over $\mathbb{F}_q$:

$$f(x) = a_n (x - \theta)(x - \theta^p)(x - \theta^{p^2}) \cdots (x - \theta^{p^{n-1}}).$$

Definition 2.33 The elements θ^{p^i} for $i = 0, 1, 2, \ldots, n-1$ are called *conjugates* of θ. (More generally, the roots of an irreducible polynomial over any field are called conjugates of one another.)

If $\theta, \theta^p, \theta^{p^2}, \ldots, \theta^{p^{n-1}}$ are linearly independent over $\mathbb{F}_p$, θ is called a *normal element* of $\mathbb{F}_q$, and $\theta, \theta^p, \theta^{p^2}, \ldots, \theta^{p^{n-1}}$ a *normal basis* of $\mathbb{F}_q$ over $\mathbb{F}_p$.

If a normal element θ is also a primitive element of $\mathbb{F}_q$, we call θ a *primitive normal element*, and $\theta, \theta^p, \theta^{p^2}, \ldots, \theta^{p^{n-1}}$ a *primitive normal basis*. ◁

Example 2.34 We represent $\mathbb{F}_{64}$ as in Examples 2.29 and 2.32. The elements θ^{2^i} for $0 \leqslant i \leqslant 5$ are now expressed in the polynomial basis $1, \theta, \ldots, \theta^5$.

$$
\begin{aligned}
\theta &= & \theta \\
\theta^2 &= & \theta^2 \\
\theta^4 &= & & \theta^4 \\
\theta^8 &= & \theta^2 + \theta^3 & \\
\theta^{16} &= 1 + \theta & & + \theta^4 \\
\theta^{32} &= 1 & + \theta^3 &
\end{aligned}
$$

In matrix notation, we have:

$$
\begin{pmatrix} \theta \\ \theta^2 \\ \theta^4 \\ \theta^8 \\ \theta^{16} \\ \theta^{32} \end{pmatrix}
=
\begin{pmatrix}
0 & 1 & 0 & 0 & 0 & 0 \\
0 & 0 & 1 & 0 & 0 & 0 \\
0 & 0 & 0 & 0 & 1 & 0 \\
0 & 0 & 1 & 1 & 0 & 0 \\
1 & 1 & 0 & 0 & 1 & 0 \\
1 & 0 & 0 & 1 & 0 & 0
\end{pmatrix}
\begin{pmatrix} 1 \\ \theta \\ \theta^2 \\ \theta^3 \\ \theta^4 \\ \theta^5 \end{pmatrix}
$$

The last column of the 6×6 matrix consists only of 0, that is, the transformation matrix is singular, that is, the elements θ^{2^i} for $0 \leqslant i \leqslant 5$ are not linearly independent. In fact, $\theta + \theta^2 + \theta^4 + \theta^8 + \theta^{16} + \theta^{32} = 0$. Therefore, θ is not a normal element of $\mathbb{F}_{64}$. By Example 2.32, θ is a primitive element of $\mathbb{F}_q$. Thus, being a primitive element is not sufficient for being a normal element. $\qquad\square$

2.4.4 Minimal Polynomials

The above discussion about conjugates and normal elements extends to any arbitrary element $\alpha \in \mathbb{F}_q$. The conjugates of α are $\alpha, \alpha^p, \alpha^{p^2}, \ldots, \alpha^{p^{t-1}}$, where t is the smallest positive integer for which $\alpha^{p^t} = \alpha$. The polynomial

$$
f_\alpha(x) = (x - \alpha)(x - \alpha^p)(x - \alpha^{p^2}) \cdots (x - \alpha^{p^{t-1}})
$$

is an irreducible polynomial in $\mathbb{F}_p[x]$. We call $f_\alpha(x)$ the *minimal polynomial* of α over $\mathbb{F}_p$, and t is called the *degree* of α. The element α is a root of a polynomial $g(x) \in \mathbb{F}_p[x]$ if and only if $f_\alpha(x) | g(x)$ in $\mathbb{F}_p[x]$.

Example 2.35 Extend $\mathbb{F}_2$ by θ satisfying $\theta^6 + \theta + 1 = 0$ to obtain $\mathbb{F}_{64}$.

(1) Let $\alpha = \theta^5 + \theta^4 + \theta^3 + \theta$. By Example 2.29, α is of degree two. We have $\alpha^2 = \theta^{10} + \theta^8 + \theta^6 + \theta^2 = \theta^4(\theta + 1) + \theta^2(\theta + 1) + (\theta + 1) + \theta^2 = \theta^5 + \theta^4 + \theta^3 + \theta + 1 = \alpha + 1$, and $\alpha^{2^2} = (\alpha + 1)^2 = \alpha^2 + 1 = (\alpha + 1) + 1 = \alpha$. The minimal polynomial of α is $f_\alpha(x) = (x + \alpha)(x + \alpha^2) = (x + \alpha)(x + \alpha + 1) = x^2 + (\alpha + \alpha + 1)x + \alpha(\alpha + 1) = x^2 + x + (\alpha^2 + \alpha) = x^2 + x + (\alpha + 1 + \alpha) = x^2 + x + 1$.

(2) By Example 2.29, $\beta = \theta^4 + \theta^3$ is of degree three. We have $\beta^2 = \theta^8 + \theta^6 = \theta^3 + \theta^2 + \theta + 1$, $\beta^4 = (\beta^2)^2 = \theta^6 + \theta^4 + \theta^2 + 1 = \theta^4 + \theta^2 + \theta$, and $\beta^8 = (\beta^4)^2 = \theta^8 + \theta^4 + \theta^2 = \theta^2(\theta + 1) + \theta^4 + \theta^2 = \theta^4 + \theta^3 = \beta$. So the minimal polynomial of β is $(x + \beta)(x + \beta^2)(x + \beta^4) = x^3 + (\beta + \beta^2 + \beta^4)x^2 + (\beta^3 + \beta^5 + \beta^6)x + \beta^7$. Calculations in $\mathbb{F}_{64}$ show $\beta + \beta^2 + \beta^4 = 1$, $\beta^3 + \beta^5 + \beta^6 = 0$, and $\beta^7 = 1$, that is, $f_\beta(x) = x^3 + x^2 + 1$. $\mathbb{F}_8$ is defined by this polynomial in Example 2.7(2).

(3) The element $\gamma = \theta^5 + 1$ of degree six has the conjugates

$$
\begin{aligned}
\gamma &= \theta^5 + 1, \\
\gamma^2 &= \theta^5 + \theta^4 + 1, \\
\gamma^4 &= \theta^5 + \theta^4 + \theta^3 + \theta^2 + 1, \\
\gamma^8 &= \theta^5 + \theta^3 + \theta^2 + \theta, \\
\gamma^{16} &= \theta^5 + \theta^2 + \theta + 1, \text{ and} \\
\gamma^{32} &= \theta^5 + \theta^2 + 1.
\end{aligned}
$$

($\gamma^{64} = \theta^5 + 1 = \gamma$, as expected.) The minimal polynomial of γ is, therefore,
$f_\gamma(x) = (x + \gamma)(x + \gamma^2)(x + \gamma^4)(x + \gamma^8)(x + \gamma^{16})(x + \gamma^{32}) = x^6 + x^5 + 1$. $\square$

An element $\alpha \in \mathbb{F}_{p^n}$ is called *normal* if $\alpha, \alpha^p, \alpha^{p^2}, \ldots, \alpha^{p^{n-1}}$ are linearly independent over $\mathbb{F}_p$. If the degree of α is a proper divisor of n, these elements cannot be linearly independent. So a necessary condition for α to be normal is that α has degree n. If α is a normal element of $\mathbb{F}_q$, the basis $\alpha, \alpha^p, \alpha^{p^2}, \ldots, \alpha^{p^{n-1}}$ is called a *normal basis* of $\mathbb{F}_{p^n}$ over $\mathbb{F}_p$. If, in addition, α is primitive, it is called a *primitive normal element* of $\mathbb{F}_{p^n}$, and the basis $\alpha, \alpha^p, \alpha^{p^2}, \ldots, \alpha^{p^{n-1}}$ is called a *primitive normal basis*.

Theorem 2.36 *For every $p \in \mathbb{P}$ and $n \in \mathbb{N}$, the extension $\mathbb{F}_{p^n}$ contains a normal element.[9] Moreover, for every $p \in \mathbb{P}$ and $n \in \mathbb{N}$, there exists a primitive normal element in $\mathbb{F}_{p^n}$.[10]*   ◁

Example 2.37 (1) Consider the element $\gamma \in \mathbb{F}_{64}$ of Example 2.35. We have

$$
\begin{pmatrix} \gamma \\ \gamma^2 \\ \gamma^4 \\ \gamma^8 \\ \gamma^{16} \\ \gamma^{32} \end{pmatrix} = \begin{pmatrix} 1 & 0 & 0 & 0 & 0 & 1 \\ 1 & 0 & 0 & 0 & 1 & 1 \\ 1 & 0 & 1 & 1 & 1 & 1 \\ 0 & 1 & 1 & 1 & 0 & 1 \\ 1 & 1 & 1 & 0 & 0 & 1 \\ 1 & 0 & 1 & 0 & 0 & 1 \end{pmatrix} \begin{pmatrix} 1 \\ \theta \\ \theta^2 \\ \theta^3 \\ \theta^4 \\ \theta^5 \end{pmatrix}.
$$

The 6×6 transformation matrix has determinant 1 modulo 2. Therefore, γ is a normal element of $\mathbb{F}_{64}$ and $\gamma, \gamma^2, \gamma^4, \gamma^8, \gamma^{16}, \gamma^{32}$ constitute a normal basis of $\mathbb{F}_{64}$ over $\mathbb{F}_2$. By Example 2.32, $\mathrm{ord}\,\gamma = 63$, that is, γ is also a primitive element of $\mathbb{F}_{64}$. Therefore, γ is a primitive normal element of $\mathbb{F}_{64}$, and the basis $\gamma, \gamma^2, \gamma^4, \gamma^8, \gamma^{16}, \gamma^{32}$ is a primitive normal basis of $\mathbb{F}_{64}$ over $\mathbb{F}_2$.

(2) The conjugates of $\delta = \theta^5 + \theta^4 + \theta^3 + 1$ are

---

[9] Eisenstein (1850) conjectured that normal bases exist for all finite fields. Kurt Hensel (1888) first proved this conjecture. Hensel and Ore counted the number of normal elements in a finite field.

[10] The proof that primitive normal bases exist for all finite fields can be found in the paper: Hendrik W. Lenstra, Jr. and René J. Schoof, Primitive normal bases for finite fields, *Mathematics of Computation*, 48, 217–231, 1986.

$$
\begin{aligned}
\delta &= \theta^5 + \theta^4 + \theta^3 + 1, \\
\delta^2 &= \theta^5 + \theta^4 + \theta^3 + \theta^2 + \theta, \\
\delta^4 &= \theta^5 + \theta^3 + \theta + 1, \\
\delta^8 &= \theta^5 + \theta^4 + \theta^2 + \theta, \\
\delta^{16} &= \theta^5 + \theta^3, \text{ and} \\
\delta^{32} &= \theta^5 + \theta^4 + \theta + 1,
\end{aligned}
$$

so that

$$
\begin{pmatrix} \delta \\ \delta^2 \\ \delta^4 \\ \delta^8 \\ \delta^{16} \\ \delta^{32} \end{pmatrix} = \begin{pmatrix} 1 & 0 & 0 & 1 & 1 & 1 \\ 0 & 1 & 1 & 1 & 1 & 1 \\ 1 & 1 & 0 & 1 & 0 & 1 \\ 0 & 1 & 1 & 0 & 1 & 1 \\ 0 & 0 & 0 & 1 & 0 & 1 \\ 1 & 1 & 0 & 0 & 1 & 1 \end{pmatrix} \begin{pmatrix} 1 \\ \theta \\ \theta^2 \\ \theta^3 \\ \theta^4 \\ \theta^5 \end{pmatrix}.
$$

The transformation matrix has determinant 1 modulo 2, that is, δ is a normal element of $\mathbb{F}_{64}$, and $\delta, \delta^2, \delta^4, \delta^8, \delta^{16}, \delta^{32}$ constitute a normal basis of $\mathbb{F}_{64}$ over $\mathbb{F}_2$. However, by Example 2.32, ord $\delta = 21$, that is, δ is not a primitive element of $\mathbb{F}_{64}$, that is, δ is not a primitive normal element of $\mathbb{F}_{64}$, and $\delta, \delta^2, \delta^4, \delta^8, \delta^{16}, \delta^{32}$ is not a primitive normal basis of $\mathbb{F}_{64}$ over $\mathbb{F}_2$. Combining this observation with Example 2.34, we conclude that being a primitive element is neither necessary nor sufficient for being a normal element. $\square$

The relevant computational question here is how we can efficiently locate normal elements in a field $\mathbb{F}_{p^n}$. The first and obvious strategy is keeping on picking random elements from $\mathbb{F}_{p^n}$ until a normal element is found. Each normality check involves computing the determinant (or rank) of an $n \times n$ matrix with entries from $\mathbb{F}_p$, as demonstrated in Example 2.37. Another possibility is to compute the gcd of two polynomials over $\mathbb{F}_{p^n}$ (see Exercise 3.42). Such a random search is efficient, since the density of normal elements in a finite field is significant. More precisely, a random element of $\mathbb{F}_{p^n}$ is normal over $\mathbb{F}_p$ with probability $\geq 1/34$ if $n \leq p^4$ and with probability $> 1/(16 \log_p n)$ if $n > p^4$. These density estimates, and also a deterministic polynomial-time algorithm based on polynomial root finding, can be found in the paper of Von zur Gathen and Giesbrecht.[11] This paper also proposes a randomized polynomial-time algorithm for finding primitive normal elements.

A more efficient randomized algorithm is based on the following result, proved by Emil Artin. This result is, however, inappropriate if p is small.

Proposition 2.38 *Represent* $\mathbb{F}_{p^n} = \mathbb{F}_p(\theta)$*, where* θ *is a root of the monic irreducible polynomial* $f(x) \in \mathbb{F}_p[x]$ *(of degree* n*). Consider the polynomial*

$$
g(x) = \frac{f(x)}{(x - \alpha) f'(\alpha)} \in \mathbb{F}_{p^n}[x],
$$

---

[11] Joachim Von Zur Gathen and Mark Giesbrecht, Constructing normal bases in finite fields, *Journal of Symbolic Computation*, 10, 547–570, 1990.

where $f'(x)$ is the formal derivative of $f(x)$. Then, there are at least $p - n(n-1)$ elements a in $\mathbb{F}_p$, for which $g(a)$ is a normal element of $\mathbb{F}_{p^n}$ over $\mathbb{F}_p$. ◁

It follows that if $p \geqslant 2n(n-1)$, then for a random $a \in \mathbb{F}_p$, the element $g(a) \in \mathbb{F}_{p^n}$ is normal over $\mathbb{F}_p$ with probability at least $1/2$. Moreover, in this case, a random element in $\mathbb{F}_{p^n}$ is normal with probability at least $1/2$ (as proved by Gudmund Skovbjerg Frandsen from Aarhus University, Denmark).

Deterministic polynomial-time algorithms are known for locating normal elements. For example, see Lenstra's paper cited in Footnote 15 on page 114.

2.4.5 Implementing Some Functions in GP/PARI

Let us now see how we can program the GP/PARI calculator for computing minimal polynomials and for checking normal elements. We first introduce the defining polynomial for representing $\mathbb{F}_{64}$.

```
gp > f = Mod(1,2)*x^6 + Mod(1,2)*x + Mod(1,2)
%1 = Mod(1, 2)*x^6 + Mod(1, 2)*x + Mod(1, 2)
```

Next, we define a function for computing the minimal polynomial of an element of $\mathbb{F}_{64}$. Since the variable x is already used in the representation of $\mathbb{F}_{64}$, we use a separate variable y for the minimal polynomial. The computed polynomial is lifted twice to remove the moduli f and 2 in the output.

```
gp > minimalpoly(a) = \
     p = y - a; \
     b = a * a; \
     while (b-a, p *= (y-b); b = b*b); \
     lift(lift(p))
gp > minimalpoly(Mod(Mod(0,2),f))
%2 = y
gp > minimalpoly(Mod(Mod(1,2),f))
%3 = y + 1
gp > minimalpoly(Mod(Mod(1,2)*x,f))
%4 = y^6 + y + 1
gp > minimalpoly(Mod(Mod(1,2)*x^5+Mod(1,2)*x^4+Mod(1,2)*x^3+Mod(1,2)*x,f))
%5 = y^2 + y + 1
gp > minimalpoly(Mod(Mod(1,2)*x^4+Mod(1,2)*x^3,f))
%6 = y^3 + y^2 + 1
gp > minimalpoly(Mod(Mod(1,2)*x^5+Mod(1,2),f))
%7 = y^6 + y^5 + 1
gp > minimalpoly(Mod(Mod(1,2)*x^5+Mod(1,2)*x^2+Mod(1,2)*x+Mod(1,2),f))
%8 = y^6 + y^5 + 1
```

Now, we define a function for checking whether an element of $\mathbb{F}_{64}$ is normal. We compute the transformation matrix M and subsequently its determinant.

```
gp > pc(p,i) = polcoeff(p,i)
gp > isnormal(a0) = \
     a1 = (a0^2) % f; a2 = (a1^2) % f; a3 = (a2^2) % f; \
     a4 = (a3^2) % f; a5 = (a4^2) % f; \
     M = Mat([ pc(a0,0),pc(a0,1),pc(a0,2),pc(a0,3),pc(a0,4),pc(a0,5); \
               pc(a1,0),pc(a1,1),pc(a1,2),pc(a1,3),pc(a1,4),pc(a1,5); \
               pc(a2,0),pc(a2,1),pc(a2,2),pc(a2,3),pc(a2,4),pc(a2,5); \
               pc(a3,0),pc(a3,1),pc(a3,2),pc(a3,3),pc(a3,4),pc(a3,5); \
               pc(a4,0),pc(a4,1),pc(a4,2),pc(a4,3),pc(a4,4),pc(a4,5); \
               pc(a5,0),pc(a5,1),pc(a5,2),pc(a5,3),pc(a5,4),pc(a5,5) ]); \
     printp("M = ", lift(M)); print("det(M) = ",matdet(M)); \
     if(matdet(M)==Mod(1,2), print("normal");1, print("not normal");0)
```

We pass a polynomial in $\mathbb{F}_2[x]$ of degree less than six as the only argument of isnormal() to check whether this corresponds to a normal element of $\mathbb{F}_{64}$.

```
gp > isnormal(Mod(1,2)*x^5+Mod(1,2)*x^4+Mod(1,2)*x^3+Mod(1,2))
M =
[1 0 0 1 1 1]

[0 1 1 1 1 1]

[1 1 0 1 0 1]

[0 1 1 0 1 1]

[0 0 0 1 0 1]

[1 1 0 0 1 1]

det(M) = Mod(1, 2)
normal
%9 = 1
gp > isnormal(Mod(1,2)*x^5+Mod(1,2)*x)
M =
[0 1 0 0 0 1]

[0 0 1 0 1 1]

[0 0 1 1 0 1]

[1 1 0 0 0 1]

[1 0 1 0 1 1]

[1 0 1 1 0 1]

det(M) = Mod(0, 2)
not normal
%10 = 0
```

2.5   Alternative Representations of Finite Fields

So far, we have used the polynomial-basis representation of extension fields. We are now equipped with sophisticated machinery for investigating several alternative ways of representing extension fields.

2.5.1   Representation with Respect to Arbitrary Bases

$\mathbb{F}_{p^n}$ is an n-dimensional vector space over $\mathbb{F}_p$. For a root θ of an irreducible polynomial in $\mathbb{F}_p[x]$ of degree n, the elements $1, \theta, \theta^2, \ldots, \theta^{n-1}$ form an $\mathbb{F}_p$-basis of $\mathbb{F}_{p^n}$, that is, every element of $\mathbb{F}_{p^n}$ can be written as a unique $\mathbb{F}_p$-linear combination of the basis elements. This representation can be generalized to any arbitrary $\mathbb{F}_p$-basis of $\mathbb{F}_{p^n}$.

Let $\theta_0, \theta_1, \ldots, \theta_{n-1}$ be n linearly independent elements of $\mathbb{F}_{p^n}$. Then, any element $\alpha \in \mathbb{F}_{p^n}$ can be written uniquely as $\alpha = a_0\theta_0 + a_1\theta_1 + \cdots + a_{n-1}\theta_{n-1}$ with each $a_i \in \mathbb{F}_p$. Let $\beta = b_0\theta_0 + b_1\theta_1 + \cdots + b_{n-1}\theta_{n-1}$ be another element of $\mathbb{F}_{p^n}$ in this representation. The sum of these elements can be computed easily as $\alpha + \beta = (a_0 + b_0)\theta_0 + (a_1 + b_1)\theta_1 + \cdots + (a_{n-1} + b_{n-1})\theta_{n-1}$, where each $a_i + b_i$ stands for the addition of $\mathbb{F}_p$.

Multiplication of the elements α and β encounters some difficulty. We have $\alpha\beta = \sum_{i,j} a_ib_j\theta_i\theta_j$. For each pair (i, j), we need to express $\theta_i\theta_j$ in the basis $\theta_0, \theta_1, \ldots, \theta_{n-1}$. Suppose $\theta_i\theta_j = t_{i,j,0}\theta_0 + t_{i,j,1}\theta_1 + \cdots + t_{i,j,n-1}\theta_{n-1}$. Then,

$$\alpha\beta = \left(\sum_{i,j} a_ib_jt_{i,j,0} \right)\theta_0 + \left(\sum_{i,j} a_ib_jt_{i,j,1} \right)\theta_1 + \cdots + \left(\sum_{i,j} a_ib_jt_{i,j,n-1} \right)\theta_{n-1}.$$

It is, therefore, preferable to precompute the values $t_{i,j,k}$ for all indices i, j, k between 0 and $n - 1$. This requires a storage overhead of $O(n^3)$ which is reasonable unless n is large. However, a bigger problem in this regard pertains to the time complexity ($\Theta(n^3)$ operations in the field $\mathbb{F}_p$) of expressing $\alpha\beta$ in the basis $\theta_0, \theta_1, \ldots, \theta_{n-1}$. For the polynomial-basis representation, a product in $\mathbb{F}_{p^n}$ can be computed using only $\Theta(n^2)$ operations in $\mathbb{F}_p$.

Example 2.39 Let $\mathbb{F}_8 = \mathbb{F}_2(\theta)$, where $\theta^3 + \theta + 1 = 0$. The elements $\theta_0 = 1$, $\theta_1 = 1 + \theta$, and $\theta_2 = 1 + \theta + \theta^2$ are evidently linearly independent over $\mathbb{F}_2$. We write $\theta_i\theta_j$ in the basis $\theta_0, \theta_1, \theta_2$.

$$
\begin{aligned}
\theta_0^2 &= \quad 1 \quad = \theta_0, & \theta_0\theta_1 = \theta_1\theta_0 &= \quad 1 + \theta \quad = \theta_1, \\
\theta_1^2 &= 1 + \theta^2 = \theta_0 + \theta_1 + \theta_2, & \theta_0\theta_2 = \theta_2\theta_0 &= 1 + \theta + \theta^2 = \theta_2, \\
\theta_2^2 &= 1 + \theta = \theta_1, & \theta_1\theta_2 = \theta_2\theta_1 &= \quad \theta \quad = \theta_0 + \theta_1.
\end{aligned}
$$

Now, consider the elements $\alpha = \theta_0 + \theta_2$ and $\beta = \theta_1 + \theta_2$ expressed in the basis $\theta_0, \theta_1, \theta_2$. Their sum is $\alpha + \beta = (1 + 0)\theta_0 + (0 + 1)\theta_1 + (1 + 1)\theta_2 = \theta_0 + \theta_1$, whereas their product is $\alpha\beta = (\theta_0 + \theta_2)(\theta_1 + \theta_2) = \theta_0\theta_1 + \theta_0\theta_2 + \theta_2\theta_1 + \theta_2^2 = (\theta_1) + (\theta_2) + (\theta_0 + \theta_1) + (\theta_1) = \theta_0 + \theta_1 + \theta_2.$ □

2.5.2 Normal and Optimal Normal Bases

In general, multiplication is encumbered by a representation of elements of $\mathbb{F}_{p^n}$ in an arbitrary basis. There are some special kinds of bases which yield computational benefit. One such example is a normal basis. Let $\psi \in \mathbb{F}_{p^n}$ be a normal element so that $\psi, \psi^p, \psi^{p^2}, \ldots, \psi^{p^{n-1}}$ constitute an $\mathbb{F}_p$-basis of $\mathbb{F}_{p^n}$, that is, every element $\alpha \in \mathbb{F}_{p^n}$ can be written uniquely as $\alpha = a_0\psi + a_1\psi^p + a_2\psi^{p^2} + \cdots + a_{n-1}\psi^{p^{n-1}}$. Many applications over $\mathbb{F}_{p^n}$ (such as in cryptography) involve exponentiation of elements in $\mathbb{F}_{p^n}$. We expand the exponent e in base p as $e = e_{l-1}p^{l-1} + e_{l-2}p^{l-2} + \cdots + e_1p + e_0$ with each $e_i \in \{0, 1, 2, \ldots, p-1\}$. We then have $\alpha^e = (\alpha^{p^{l-1}})^{e_{l-1}}(\alpha^{p^{l-2}})^{e_{l-2}} \cdots (\alpha^p)^{e_1}(\alpha)^{e_0}$. If α is represented using a normal basis as mentioned above, we have $\alpha^p = a_0^p\psi^p + a_1^p\psi^{p^2} + a_2^p\psi^{p^3} + \cdots + a_{n-1}^p\psi^{p^n} = a_{n-1}\psi + a_0\psi^p + a_1\psi^{p^2} + \cdots + a_{n-2}\psi^{p^{n-1}}$, that is, the representation of α^p in the normal basis can be obtained by cyclically rotating the coefficients $a_0, a_1, a_2, \ldots, a_{n-1}$. Thus, it is extremely easy to compute p-th power exponentiations under the normal-basis representation.

Computing α^e involves some multiplications too. For this, we need to express $\psi^{p^i}\psi^{p^j}$ in the basis $\psi, \psi^p, \psi^{p^2}, \ldots, \psi^{p^{n-1}}$ for all i, j between 0 and $n-1$. For $i \leqslant j$, we have $\psi^{p^i}\psi^{p^j} = \left(\psi\psi^{p^{j-i}}\right)^{p^i}$, that is, the representation of $\psi^{p^i}\psi^{p^j}$ can be obtained by cyclically rotating the representation of $\psi\psi^{p^{j-i}}$ by i positions. So it suffices to store data only for $\psi\psi^{p^i}$ for $i = 0, 1, 2, \ldots, n-1$.

The complexity of the normal basis $\psi, \psi^p, \psi^{p^2}, \ldots, \psi^{p^{n-1}}$ is defined to be the total number of non-zero coefficients in the expansions of $\psi\psi^{p^i}$ for all $i = 0, 1, 2, \ldots, n-1$. Normal bases of small complexities are preferred in software and hardware implementations. The minimum possible value of this complexity is $2n-1$. A normal basis with the minimum possible complexity is called an *optimal normal basis*.[12] Unlike normal and primitive normal bases, optimal normal bases do not exist for all values of p and n.

Example 2.40 Let $\mathbb{F}_8 = \mathbb{F}_2(\theta)$, where $\theta^3 + \theta + 1 = 0$. Take $\psi = \theta + 1$. We have $\psi^2 = \theta^2 + 1$ and $\psi^4 = \theta^2 + \theta + 1$, that is, $\begin{pmatrix} \psi \\ \psi^2 \\ \psi^4 \end{pmatrix} = \begin{pmatrix} 1 & 1 & 0 \\ 1 & 0 & 1 \\ 1 & 1 & 1 \end{pmatrix} \begin{pmatrix} 1 \\ \theta \\ \theta^2 \end{pmatrix}$ with the coefficient matrix having determinant 1 modulo 2. Thus, ψ is a normal element of $\mathbb{F}_8$. We express $\psi_0\psi_i$ in the basis ψ_0, ψ_1, ψ_2, where $\psi_i = \psi^{2^i}$.

$$\psi \cdot \psi = \theta^2 + 1 = \psi_1, \quad \psi \cdot \psi^2 = \theta^2 = \psi_0 + \psi_2, \quad \psi \cdot \psi^4 = \theta = \psi_1 + \psi_2.$$

Consequently, the complexity of the normal basis ψ_0, ψ_1, ψ_2 is $5 = 2 \times 3 - 1$, that is, the basis is optimal. □

---

[12]R. Mullin, I. Onyszchuk, S. Vanstone and R. Wilson, Optimal normal bases in $GF(p^n)$, *Discrete Applied Mathematics*, 22, 149–161, 1988/89.

2.5.3   Discrete-Log Representation

Another interesting representation of a finite field $\mathbb{F}_q$ (q may be prime) results from the observation that $\mathbb{F}_q^*$ contains primitive elements. Take a primitive element γ. Every non-zero $\alpha \in \mathbb{F}_q$ can be represented as $\alpha = \gamma^i$ for some unique i in the range $0 \leqslant i \leqslant q - 2$, that is, we represent α by the index i. Multiplication and exponentiation are trivial in this representation, namely, $\gamma^i \gamma^j = \gamma^k$, where $k \equiv i + j \pmod{q - 1}$, and $(\gamma^i)^e = \gamma^l$, where $l \equiv ie \pmod{q - 1}$. Moreover, the inverse of γ^i is γ^{q-1-i} for $i > 0$.

Carrying out addition and subtraction becomes non-trivial in this representation. For example, given i, j, we need to find out k satisfying $\gamma^i + \gamma^j = \gamma^k$. We do not know an easy way of computing k from i, j. If q is small, one may precompute and store k for each pair (i, j). Addition is then performed by table lookup. The storage requirement is $\Theta(q^2)$ which is prohibitively large except only for small values of q.

A trick can reduce the storage requirement to $\Theta(q)$. We precompute and store only the values k for the pairs $(0, j)$. For each $j \in \{0, 1, 2, \ldots, q - 2\}$, we precompute and store z_j satisfying $\gamma^{z_j} = 1 + \gamma^j$. The quantities z_j are called *Zech's logarithms* or *Jacobi's logarithms*.[13] For $i \leqslant j$, we compute $\gamma^i + \gamma^j = \gamma^i(1 + \gamma^{j-i}) = \gamma^i \gamma^{z_{j-i}} = \gamma^k$, where $k \equiv i + z_{j-i} \pmod{q - 1}$.

Here, we have assumed both the operands γ^i, γ^j to be non-zero and also their sum $\gamma^i + \gamma^j = \gamma^k$ to be non-zero. If one of the operands is zero (or both are zero), one sets $\gamma^i + 0 = 0 + \gamma^i = \gamma^i$ (or $0 + 0 = 0$), whereas if $\gamma^i + \gamma^j = 0$ (equivalently, if $\gamma^{j-i} = -1$), the Zech logarithm z_{j-i} is not defined. An undefined z_{j-i} implies that the sum $\gamma^i + \gamma^j$ would be zero.

Example 2.41 (1) 3 is a primitive element in the prime field $\mathbb{F}_{17}$. The powers of 3 are given in the following table.

i	0	1	2	3	4	5	6	7	8	9	10	11	12	13	14	15	16
$3^i \pmod{17}$	1	3	9	10	13	5	15	11	16	14	8	7	4	12	2	6	1

From this table, the Zech's logarithm table can be computed as follows. For $j \in \{0, 1, 2, \ldots, 15\}$, compute $1 + 3^j \pmod{17}$ and then locate the value z_j for which $3^{z_j} \equiv 1 + 3^j \pmod{17}$.

j	0	1	2	3	4	5	6	7	8	9	10	11	12	13	14	15
z_j	14	12	3	7	9	15	8	13	–	6	2	10	5	4	1	11

The Zech logarithm table can be used as follows. Take $i = 8$ and $j = 13$. Then, $3^i + 3^j \equiv 3^k \pmod{17}$, where $k \equiv i + z_{j-i} \equiv 8 + z_5 \equiv 8 + 15 \equiv 23 \equiv 7 \pmod{16}$.

(2) Let $\mathbb{F}_8 = \mathbb{F}_2(\theta)$, where $\theta^3 + \theta + 1 = 0$. Consider the powers of $\gamma = \theta$.

i	0	1	2	3	4	5	6
γ^i	1	θ	θ^2	$\theta + 1$	$\theta^2 + \theta$	$\theta^2 + \theta + 1$	$\theta^2 + 1$

---

[13]The concept of Zech's logarithms was introduced near the mid-nineteenth century. Reference to Zech's logarithms is found in Jacobi's work.

Zech's logarithms to the base $\gamma = \theta$ are listed next.

j	0	1	2	3	4	5	6
z_j	$-$	3	6	1	5	4	2

Let us compute k with $\gamma^k = \gamma^i + \gamma^j$ for $i = 2$ and $j = 5$. We have $k \equiv i + z_{j-i} \equiv 2 + z_3 \equiv 2 + 1 \equiv 3 \pmod 7$. Indeed, $\gamma^2 + \gamma^5 = \theta^2 + (\theta^2 + \theta + 1) = \theta + 1 = \gamma^3$. $\square$

2.5.4 Representation with Towers of Extensions

In certain situations, we use one or more intermediate fields for representing $\mathbb{F}_{p^n}$. Suppose that $n = st$ with $s, t \in \mathbb{N}$. We first represent $\mathbb{F}_{p^s}$ using one of the methods discussed earlier. Then, we represent $\mathbb{F}_{p^{st}}$ as an extension of $\mathbb{F}_{p^s}$. Let us first concentrate on polynomial-basis representations. Let $f(x)$ be an irreducible polynomial in $\mathbb{F}_p[x]$ of degree s. Adjoining a root θ of $f(x)$ to $\mathbb{F}_p$ gives a representation of the field $\mathbb{F}_{p^s}$. As the next step, we choose an irreducible polynomial $g(y)$ of degree t from $\mathbb{F}_{p^s}[y]$. We adjoin a root ψ of $g(y)$ to $\mathbb{F}_{p^s}$ in order to obtain a representation of $\mathbb{F}_{p^{st}} = \mathbb{F}_{p^n}$.

Example 2.42 Let us represent $\mathbb{F}_{64}$ using the intermediate field $\mathbb{F}_8$. First, represent $\mathbb{F}_8$ as the extension of $\mathbb{F}_2$ obtained by adjoining a root θ of the irreducible polynomial $f(x) = x^3 + x + 1 \in \mathbb{F}_2[x]$. Thus, every element of $\mathbb{F}_8$ is of the form $a_2\theta^2 + a_1\theta + a_0$, and the arithmetic in $\mathbb{F}_8$ is the polynomial arithmetic of $\mathbb{F}_2[x]$ modulo $f(x)$.

Next, consider the polynomial $g(y) = y^2 + (\theta^2 + 1)y + \theta \in \mathbb{F}_8[y]$. One can easily verify that $g(\alpha) \neq 0$ for all $\alpha \in \mathbb{F}_8$, that is, $g(y)$ has no root in $\mathbb{F}_8$. Since the degree of $g(y)$ is 2, it follows that $g(y)$ is irreducible in $\mathbb{F}_8[y]$. Let ψ be a root of $g(y)$, which we adjoin to $\mathbb{F}_8$ in order to obtain the extension $\mathbb{F}_{64}$ of $\mathbb{F}_8$. Thus, every element of $\mathbb{F}_{64}$ is now a polynomial in ψ of degree < 2, that is, of the form $u_1\psi + u_0$, where u_0, u_1 are elements of $\mathbb{F}_8$, that is, polynomials in θ of degrees < 3. The arithmetic of $\mathbb{F}_{64}$ in this representation is the polynomial arithmetic of $\mathbb{F}_8[y]$ modulo the irreducible polynomial $g(y)$. The coefficients of these polynomials follow the arithmetic of $\mathbb{F}_8$, that is, the polynomial arithmetic of $\mathbb{F}_2[x]$ modulo $f(x)$.

As a specific example, consider the elements $\alpha = (\theta + 1)\psi + (\theta^2)$ and $\beta = (\theta^2 + \theta + 1)\psi + (1)$ in $\mathbb{F}_{64}$. Their sum is

$$\alpha + \beta = [(\theta + 1) + (\theta^2 + \theta + 1)]\psi + [\theta^2 + 1] = (\theta^2)\psi + (\theta^2 + 1),$$

and their product is

$$\begin{aligned}
\alpha\beta &= [(\theta+1)(\theta^2+\theta+1)]\psi^2 + [(\theta+1)(1) + (\theta^2)(\theta^2+\theta+1)]\psi + [(\theta^2)(1)]\\
&= (\theta^3 + 1)\psi^2 + (\theta^4 + \theta^3 + \theta^2 + \theta + 1)\psi + (\theta^2)\\
&= (\theta + 1 + 1)\psi^2 + [\theta(\theta+1) + (\theta+1) + \theta^2 + \theta + 1]\psi + (\theta^2)\\
&\qquad\qquad\qquad\qquad\qquad\qquad\qquad [\text{since } \theta^3 + \theta + 1 = 0]
\end{aligned}$$

$$
\begin{aligned}
&= (\theta)\psi^2 + (\theta)\psi + (\theta^2) \\
&= (\theta)[(\theta^2 + 1)\psi + \theta] + (\theta)\psi + (\theta^2) \quad \text{[since } \psi^2 + (\theta^2 + 1)\psi + \theta = 0] \\
&= [\theta(\theta^2 + 1) + \theta]\psi + [(\theta)(\theta) + \theta^2] \\
&= (\theta^3)\psi + 0 \\
&= (\theta + 1)\psi \quad \text{[since } \theta^3 + \theta + 1 = 0].
\end{aligned}
$$

This multiplication involves several modular operations. First, there is reduction modulo 2; second, there is reduction modulo $f(x)$; and finally, there is reduction modulo $g(y)$.

One may view $\mathbb{F}_{64}$ in this representation as an $\mathbb{F}_2$-vector space with a basis consisting of the six elements $\theta^i \psi^j$ for $i = 0, 1, 2$, and $j = 0, 1$. □

The above construction can be readily generalized to other representations of finite fields. In general, if we have a way to perform arithmetic in a field F, we can also perform the arithmetic in the polynomial ring $F[x]$. If $f(x) \in F[x]$ is an irreducible polynomial, we can compute remainders of polynomials in $F[x]$ modulo $f(x)$. This gives us an algorithm to implement the arithmetic of the field obtained by adjoining a root of $f(x)$ to F.

Example 2.43 Let us represent $\mathbb{F}_8$ in the $\mathbb{F}_2$-basis consisting of the elements $\theta_0 = 1$, $\theta_1 = 1 + \theta$, and $\theta_2 = 1 + \theta + \theta^2$, where $\theta^3 + \theta + 1 = 0$ (see Example 2.39). In this basis, the polynomial $g(y)$ of Example 2.42 is written as $g(y) = (\theta_0)y^2 + (\theta_0 + \theta_1 + \theta_2)y + (\theta_0 + \theta_1)$. We represent $\mathbb{F}_{64}$ in the polynomial basis $1, \psi$ over $\mathbb{F}_8$, where $g(\psi) = 0$. The elements α, β of Example 2.42 are expressed as

$$
\begin{aligned}
\alpha &= (\theta_1)\psi + (\theta_1 + \theta_2), \\
\beta &= (\theta_2)\psi + (\theta_0).
\end{aligned}
$$

The sum of these elements is

$$
\alpha + \beta = (\theta_1 + \theta_2)\psi + (\theta_0 + \theta_1 + \theta_2),
$$

whereas their product is

$$
\begin{aligned}
\alpha\beta &= (\theta_1\theta_2)\psi^2 + (\theta_1\theta_0 + \theta_1\theta_2 + \theta_2^2)\psi + (\theta_1\theta_0 + \theta_2\theta_0) \\
&= [(\theta_0 + \theta_1)]\psi^2 + [(\theta_1) + (\theta_0 + \theta_1) + (\theta_1)]\psi + [(\theta_1) + (\theta_2)] \\
&= (\theta_0 + \theta_1)\psi^2 + (\theta_0 + \theta_1)\psi + (\theta_1 + \theta_2) \\
&= (\theta_0 + \theta_1)[(\theta_0 + \theta_1 + \theta_2)\psi + (\theta_0 + \theta_1)] + (\theta_0 + \theta_1)\psi + (\theta_1 + \theta_2) \\
&= [(\theta_0 + \theta_1)(\theta_0 + \theta_1 + \theta_2) + (\theta_0 + \theta_1)]\psi + [(\theta_0 + \theta_1)^2 + (\theta_1 + \theta_2)] \\
&= (\theta_0^2 + \theta_0\theta_1 + \theta_0\theta_2 + \theta_1\theta_0 + \theta_1^2 + \theta_1\theta_2 + \theta_0 + \theta_1)\psi + (\theta_0^2 + \theta_1^2 + \theta_1 + \theta_2) \\
&= (\theta_0^2 + \theta_0\theta_2 + \theta_1^2 + \theta_1\theta_2 + \theta_0 + \theta_1)\psi + (\theta_0^2 + \theta_1^2 + \theta_1 + \theta_2) \\
&= [(\theta_0) + (\theta_2) + (\theta_0 + \theta_1 + \theta_2) + (\theta_0 + \theta_1) + \theta_0 + \theta_1]\psi + \\
&\quad [(\theta_0) + (\theta_0 + \theta_1 + \theta_2) + \theta_1 + \theta_2)] \\
&= (\theta_1)\psi.
\end{aligned}
$$

Since $\theta_1 = \theta + 1$, this result tallies (it should!) with Example 2.42. □

2.6 Computing Isomorphisms among Representations

We have adjoined a root of an irreducible polynomial in order to obtain an extension field. What happens if we adjoin another root of the same irreducible polynomial? Do we get a different field? Let $f(x) \in F[x]$ be an irreducible polynomial, and let θ, ψ be two roots of $f(x)$. Consider the extensions $K = F(\theta)$ and $L = F(\psi)$. The two sets K, L may or may not be the same. However, there exists an isomorphism between the fields K and L. The following theorem implies that K and L are algebraically indistinguishable. So it does not matter which root of $f(x)$ is adjoined to represent the extension.

Theorem 2.44 *There exists an isomorphism* $K \to L$ *of fields that fixes each element of* F *and that maps* θ *to* ψ. ◁

Now, let $f(x), g(x) \in F[x]$ be two *different* irreducible polynomials of the same degree. Let θ be a root of $f(x)$, and ψ a root of $g(x)$. The fields $K = F(\theta)$ and $L = F(\psi)$ may or may not be isomorphic. However, if F is a *finite* field, the fields K and L have the same size, and so are isomorphic. Now, we discuss an algorithm to compute an explicit isomorphism between K and L.

More concretely, let $F = \mathbb{F}_p$, and let both K and L be extensions of $\mathbb{F}_p$ of degree n. Suppose that K is represented in the $\mathbb{F}_p$-basis $\theta_0, \theta_1, \theta_2, \ldots, \theta_{n-1}$, and L is represented in the $\mathbb{F}_p$-basis $\psi_0, \psi_1, \psi_2, \ldots, \psi_{n-1}$. We plan to compute an isomorphism $\mu : K \to L$ that fixes the common subfield $\mathbb{F}_p$ element-wise. It suffices to define μ only for the basis elements $\theta_0, \theta_1, \theta_2, \ldots, \theta_{n-1}$. Write

$$
\begin{aligned}
\mu(\theta_0) &= t_{0,0}\psi_0 + t_{0,1}\psi_1 + t_{0,2}\psi_2 + \cdots + t_{0,n-1}\psi_{n-1}, \\
\mu(\theta_1) &= t_{1,0}\psi_0 + t_{1,1}\psi_1 + t_{1,2}\psi_2 + \cdots + t_{1,n-1}\psi_{n-1}, \\
&\cdots \\
\mu(\theta_{n-1}) &= t_{n-1,0}\psi_0 + t_{n-1,1}\psi_1 + t_{n-1,2}\psi_2 + \cdots + t_{n-1,n-1}\psi_{n-1},
\end{aligned}
$$

where each $t_{i,j} \in \mathbb{F}_p$. The $n \times n$ matrix T whose (i, j)-th element is $t_{i,j}$ is called the *transformation matrix* from the representation K to the representation L of $\mathbb{F}_{p^n}$. In matrix notation, we have

$$
\begin{pmatrix} \mu(\theta_0) \\ \mu(\theta_1) \\ \vdots \\ \mu(\theta_{n-1}) \end{pmatrix} = T \begin{pmatrix} \psi_0 \\ \psi_1 \\ \vdots \\ \psi_{n-1} \end{pmatrix}.
$$

Take an element $\alpha = a_0\theta_0 + a_1\theta_1 + \cdots + a_{n-1}\theta_{n-1} \in K$. By linearity, we have $\mu(\alpha) = a_0\mu(\theta_0) + a_1\mu(\theta_1) + \cdots + a_{n-1}\mu(\theta_{n-1})$, that is,

$$
\mu(\alpha) = \begin{pmatrix} a_0 & a_1 & \cdots & a_{n-1} \end{pmatrix} \begin{pmatrix} \mu(\theta_0) \\ \mu(\theta_1) \\ \vdots \\ \mu(\theta_{n-1}) \end{pmatrix} = \begin{pmatrix} a_0 & a_1 & \cdots & a_{n-1} \end{pmatrix} T \begin{pmatrix} \psi_0 \\ \psi_1 \\ \vdots \\ \psi_{n-1} \end{pmatrix}.
$$

The last expression gives the representation of $\mu(\alpha)$ in the basis $\psi_0, \psi_1, \psi_2,$ $\dots, \psi_{n-1}$. Thus, μ is completely specified by the transformation matrix T.

Let us specialize to polynomial-basis representations of both K and L, that is, $\theta_i = \theta^i$ and $\psi_j = \psi^j$, where $f(\theta) = 0 = g(\psi)$ for some irreducible polynomials $f(x), g(x) \in \mathbb{F}_p[x]$ of degree n. Since μ is an isomorphism of fields, $\mu(\theta^i) = \mu(\theta)^i$ for all i. Therefore, μ is fully specified by the element $\mu(\theta)$ only. Since $0 = \mu(0) = \mu(f(\theta)) = f(\mu(\theta))$, the element $\mu(\theta) \in L$ must be a root of $f(x)$. All roots of an irreducible polynomial being algebraically indistinguishable, the task of computing μ reduces to computing any root θ' of $f(x)$ in L, and setting $\mu(\theta) = \theta'$.

Example 2.45 Consider two representations K, L of $\mathbb{F}_8$, where $K = \mathbb{F}_2(\theta)$ with $\theta^3 + \theta + 1 = 0$, and $L = \mathbb{F}_2(\psi)$ with $\psi^3 + \psi^2 + 1 = 0$. The three roots of $f(x)$ in L are $\psi + 1, \psi^2 + 1, \psi^2 + \psi$. The choice $\theta' = \mu(\theta) = \psi^2 + 1$ gives

$$
\begin{aligned}
\mu(1) &= 1, \\
\mu(\theta) &= \psi^2 + 1, \\
\mu(\theta^2) = \mu(\theta)^2 &= \psi^4 + 1 = \psi(\psi^2 + 1) + 1 = \psi^3 + \psi + 1 \\
&= (\psi^2 + 1) + \psi + 1 = \psi^2 + \psi.
\end{aligned}
$$

Thus, the transformation matrix is $T = \begin{pmatrix} 1 & 0 & 0 \\ 1 & 0 & 1 \\ 0 & 1 & 1 \end{pmatrix}$. Take the elements $\alpha = \theta + 1$ and $\beta = \theta^2 + \theta + 1$ in K. Since $(1 \ \ 1 \ \ 0)T = (0 \ \ 0 \ \ 1)$ and $(1 \ \ 1 \ \ 1)T = (0 \ \ 1 \ \ 0)$, we have $\alpha' = \mu(\alpha) = \psi^2$, and $\beta' = \mu(\beta) = \psi$. We have $\alpha + \beta = \theta^2$, so $\mu(\alpha + \beta) = \mu(\theta^2) = \psi^2 + \psi = \alpha' + \beta' = \mu(\alpha) + \mu(\beta)$, as expected. Moreover, $\alpha\beta = \theta^3 + 1 = \theta$, so $\mu(\alpha\beta) = \mu(\theta) = \psi^2 + 1$, whereas $\alpha'\beta' = \psi^3 = \psi^2 + 1$, that is, $\mu(\alpha\beta) = \mu(\alpha)\mu(\beta)$, as expected. $\square$

In order to complete the description of the algorithm for the computation of μ, we need to describe how the roots of a polynomial can be computed in a finite field. We defer this study until Chapter 3. The root-finding algorithms we study there are randomized, and run in polynomial time. There are deterministic polynomial-time algorithms[14] under the assumption that the extended Riemann hypothesis (ERH) is true. No deterministic polynomial-time algorithm not based on unproven assumptions is known for polynomial root finding (in finite fields). However, the problem of computing isomorphisms between two representations of a finite field can be solved by deterministic polynomial-time algorithms without resorting to root finding.[15] For practical purposes, our randomized algorithm based on root finding suffices.

---

[14]S. A. Evdokimov, Factorization of solvable polynomials over finite fields and the generalized Riemann hypothesis, *Journal of Mathematical Sciences*, 59(3), 842–849, 1992. This is a translation of a Russian article published in 1989.

[15]Hendrik W. Lenstra, Jr., Finding isomorphisms between finite fields, *Mathematics of Computation*, 56(193), 329–347, 1991.

Exercises

1. Let F be a field (not necessarily finite), and $F[x]$ the ring of polynomials in one indeterminate x. Let $f(x), g(x) \in F[x]$ with $g(x) \neq 0$.
   (a) Prove that there exist unique polynomials $q(x), r(x) \in F[x]$ satisfying $f(x) = q(x)g(x) + r(x)$, and either $r(x) = 0$ or $\deg r(x) < \deg g(x)$.
   (b) Prove that $\gcd(f(x), g(x)) = \gcd(g(x), r(x))$. (**Remark:** If $d(x)$ is a gcd of $f(x)$ and $g(x)$, then so also is $ad(x)$ for any non-zero $a \in F$. We can adjust a so that the leading coefficient of $ad(x)$ equals one. This monic gcd is called *the* gcd of $f(x)$ and $g(x)$, and is denoted by $\gcd(f(x), g(x))$.)
   (c) Prove that there exist polynomials $u(x), v(x) \in F[x]$ with the property $\gcd(f(x), g(x)) = u(x)f(x) + v(x)g(x)$.
   (d) Prove that if $f(x)$ and $g(x)$ are non-constant, we may choose $u(x), v(x)$ in such a way that $\deg u(x) < \deg g(x)$ and $\deg v(x) < \deg f(x)$.

2. We have seen that the polynomials $x^2 + x + 1$, $x^3 + x + 1$ and $x^6 + x + 1$ are irreducible in $\mathbb{F}_2[x]$. Prove or disprove: the polynomial $x^n + x + 1$ is irreducible in $\mathbb{F}_2[x]$ for every $n \geq 2$.

3. Consider the extension of $\mathbb{Q}$ obtained by adjoining a root of the irreducible polynomial $x^4 + 1$. Derive how $x^4 + 1$ factors in the extension.

4. (a) List all monic irreducible polynomials of degrees $1, 2, 3, 4$ in $\mathbb{F}_2[x]$.
   (b) List all monic irreducible polynomials of degrees $1, 2, 3$ in $\mathbb{F}_3[x]$.
   (c) List all monic irreducible polynomials of degrees $1, 2, 3$ in $\mathbb{F}_5[x]$.

5. (a) Verify whether $x^8 + x + 1$ and $x^8 + x^3 + 1$ are irreducible in $\mathbb{F}_2[x]$.
   (b) Prove or disprove: There does not exist an irreducible binomial/trinomial/quadrinomial of degree eight in $\mathbb{F}_2[x]$.

6. (a) Prove that the polynomial $f(x) = x^4 + x + 4$ is irreducible in $\mathbb{F}_5[x]$.
   (b) Represent $\mathbb{F}_{625} = \mathbb{F}_{5^4}$ by adjoining a root θ of $f(x)$ to $\mathbb{F}_5$, and let $\alpha = 2\theta^3 + 3\theta + 4$ and $\beta = \theta^2 + 2\theta + 3$. Compute $\alpha + \beta$, $\alpha - \beta$, $\alpha\beta$ and α/β.

7. (a) Which of the polynomials $x^2 \pm 7$ is irreducible modulo 19? Justify.
   (b) Using the irreducible polynomial $f(x)$ of Part (a), represent the field $\mathbb{F}_{361} = \mathbb{F}_{19^2}$ as $\mathbb{F}_{19}(\theta)$, where $f(\theta) = 0$. Compute $(2\theta + 3)^{11}$ in this representation of $\mathbb{F}_{361}$ using left-to-right square-and-multiply exponentiation.

8. Let $\mathbb{F}_{2^n}$ have a polynomial-basis representation. Store each element of $\mathbb{F}_{2^n}$ as an array of w-bit words. Denote the words of $\alpha \in \mathbb{F}_{2^n}$ by $\alpha_0, \alpha_1, \ldots, \alpha_{N-1}$, where $N = \lceil n/w \rceil$. Write pseudocodes for addition, schoolbook multiplication, left-to-right comb multiplication, modular reduction, and inverse in $\mathbb{F}_{2^n}$.

9. Let $\alpha = a_{n-1}\theta^{n-1} + a_{n-2}\theta^{n-2} + \cdots + a_1\theta + a_0 \in \mathbb{F}_{2^n}$ with $a_i \in \mathbb{F}_{2^n}$.
   (a) Prove that $\alpha^2 = a_{n-1}\theta^{2(n-1)} + a_{n-2}\theta^{2(n-2)} + \cdots + a_1\theta^2 + a_0$.
   (b) How can you efficiently square a polynomial in $\mathbb{F}_2[x]$ under the bit-vector representation? Argue that squaring is faster (in general) than multiplication.
   (c) How can precomputation speed up this squaring algorithm?

10. Explain how the coefficients of x^{255} through x^{233} in γ_3 of Example 2.22 can be eliminated using bit-wise shift and XOR operations.

11. Design efficient reduction algorithms (using bit-wise operations) for the following fields recommended by NIST. Assume a packing of 64 bits in a word.
    (a) $\mathbb{F}_{2^{1223}}$ defined by $x^{1223} + x^{255} + 1$.
    (b) $\mathbb{F}_{2^{571}}$ defined by $x^{571} + x^{10} + x^5 + x^2 + 1$.

12. Repeat Exercise 2.11 for a packing of 32 bits per word.

13. An obvious way to compute β/α for $\alpha, \beta \in \mathbb{F}_{2^n}$, $\alpha \neq 0$, is to compute $\beta \times \alpha^{-1}$ which involves one inverse computation and one multiplication. Explain how the multiplication can be avoided altogether by modifying the initialization step of the binary inverse algorithm (Algorithm 2.1).

14. Let $\alpha \in \mathbb{F}_{2^n}^*$.
    (a) Prove that $\alpha^{-1} = \alpha^{2^n - 2}$.
    (b) Use Part (a) and the fact that $2^n - 2 = 2 + 2^2 + 2^3 + \cdots + 2^{n-1}$ to design an algorithm to compute inverses in $\mathbb{F}_{2^n}$.

15. We now investigate another way of computing $\alpha^{-1} = \alpha^{2^n - 2}$ for an $\alpha \in \mathbb{F}_{2^n}^*$.
    (a) Suppose that $\alpha^{2^k - 1}$ has been computed for some $k \geqslant 1$. Explain how $\alpha^{2^{2k} - 1}$ and $\alpha^{2^{2k+1} - 1}$ can be computed.
    (b) Based upon the result of Part (a), devise an algorithm to compute $\alpha^{2^n - 2} = (\alpha^{2^{n-1} - 1})^2$ from the binary representation of $n - 1$.
    (c) Compare the algorithm of Part (b) with that of Exercise 2.14(b).

16. Let $\alpha \in \mathbb{F}_{2^n}$. Prove that the equation $x^2 = \alpha$ has a unique solution in $\mathbb{F}_{2^n}$.

17. Represent $\mathbb{F}_{2^n} = \mathbb{F}_2(\theta)$, and let $\alpha \in \mathbb{F}_{2^n}$.
    (a) Prove that $\sqrt{\theta} = \theta^{2^{n-1}}$.
    (b) How can you express α as $A_0(\theta^2) + \theta \times A_1(\theta^2)$ (for polynomials A_0, A_1)?
    (c) Design an efficient algorithm for computing $\sqrt{\alpha}$.

18. Let $\mathbb{F}_{2^{1223}}$ be defined by the irreducible trinomial $f(x) = x^{1223} + x^{255} + 1$. Show that $\sqrt{x} = x^{612} + x^{128}$ modulo $f(x)$.

19. Let $\mathbb{F}_{2^n} = \mathbb{F}_2(\theta)$, θ being a root of an irreducible trinomial $f(x) = x^n + x^k + 1$.
    (a) Prove that both n and k cannot be even.
    (b) If n and k are both odd, show that $\sqrt{\theta} = \theta^{(n+1)/2} + \theta^{(k+1)/2}$.
    (c) If n is odd and k is even, show that $\sqrt{\theta} = \theta^{-(n-1)/2}(\theta^{k/2} + 1)$.
    (d) Derive a similar formula for $\sqrt{\theta}$ when n is even and k is odd.

20. Write arithmetic routines (addition, subtraction, schoolbook and comb-based multiplication, and modular inverse) for the field $\mathbb{F}_{3^n}$ under Kawahara et al.'s representation scheme. Pack w bits in a word.

21. Supply an efficient reduction algorithm for modular reduction in the NIST-recommended field $\mathbb{F}_{3^{509}}$ defined by the irreducible polynomial $x^{509} - x^{318} - x^{191} + x^{127} + 1$. Represent elements of $\mathbb{F}_{3^{509}}$ as in Exercise 2.20 with $w = 64$.

22. Let the field $\mathbb{F}_{p^n}$ be represented in the polynomial basis $1, \theta, \theta^2, \ldots, \theta^{n-1}$. An element $a_{n-1}\theta^{n-1} + a_{n-2}\theta^{n-2} + \cdots + a_1\theta + a_0$ is identified with the non-negative integer $a_{n-1}p^{n-1} + a_{n-2}p^{n-2} + \cdots + a_1 p + a_0$. Argue that under

this identification, the elements of $\mathbb{F}_{p^n}$ are represented uniquely as integers between 0 and $p^n - 1$. Write pseudocodes that add, subtract, multiply and divide two elements of $\mathbb{F}_{p^n}$ in this representation.

23. Represent an element of $\mathbb{F}_{p^n}$ as an n-tuple of elements of $\mathbb{F}_p$, so an element can be stored using $O(n \lg p)$ bits. Assume that we use schoolbook polynomial arithmetic. Deduce the running times for addition, subtraction and multiplication in $\mathbb{F}_{p^n}$. Also deduce the running time for inverting an element in $\mathbb{F}_{p^n}^*$. (Assume that $\mathbb{F}_{p^n}$ has a polynomial-basis representation.)

24. Show that the p-th power exponentiation in $\mathbb{F}_{p^n}$ can be efficiently computed.

25. [*Itoh–Tsujii inversion* (1988)] Let $\alpha \in \mathbb{F}_{p^n}^*$, and $r = (p^n - 1)/(p - 1) = 1 + p + p^2 + \cdots + p^{n-1}$.
    (a) Prove that $\alpha^r \in \mathbb{F}_p$.
    (b) How can α^{-1} be efficiently computed by the formula $\alpha^{-1} = (\alpha^r)^{-1} \alpha^{r-1}$?

26. Consider a finite field $\mathbb{F}_{q^n} = \mathbb{F}_q(\theta)$ with q large and n small. Elements of $\mathbb{F}_{q^n}$ are polynomials of degree $n - 1$ with coefficients from $\mathbb{F}_q$. Multiplication of two elements of $\mathbb{F}_{q^n}$ first involves a polynomial multiplication over $\mathbb{F}_q$ in order to obtain an intermediate polynomial of degree $2n - 2$. This is followed by reduction modulo the minimal polynomial of θ (over $\mathbb{F}_q$). Schoolbook multiplication requires n^2 $\mathbb{F}_q$-multiplications to compute the intermediate product. Karatsuba–Ofman multiplication can reduce this number. Since $\mathbb{F}_q$ is a large field, this leads to practical improvements. Let ν denote the number of $\mathbb{F}_q$-multiplications used to compute the intermediate product. Prove that:
    (a) If $n = 2$, we can take $\nu = 3$.
    (b) If $n = 3$, we can take $\nu = 6$.
    (c) If $n = 4$, we can take $\nu = 9$.
    (d) If $n = 5$, we can take $\nu = 14$.
    (e) If $n = 6$, we can take $\nu = 18$.

27. Prove that every element $\alpha \in \mathbb{F}_{p^n}$ has a unique p-th root in $\mathbb{F}_{p^n}$. Show that this root is given by $\sqrt[p]{\alpha} = \alpha^{p^{n-1}}$.

28. Assume that p is small (like 3 or 5). Generalize the algorithm of Exercise 2.17 to compute the p-th root of an element $\alpha \in \mathbb{F}_{p^n}$. (**Hint:** Represent $\mathbb{F}_{p^n}$ as $\mathbb{F}_p(\theta)$. Precompute $(\sqrt[p]{\theta})^i$ for $i = 0, 1, 2, \ldots, p - 1$.)

29. Let $\mathbb{F}_{3^{509}}$ be defined by the irreducible pentanomial $g(x) = x^{509} - x^{318} - x^{191} + x^{127} + 1$. Show that $x^{1/3} = x^{467} + x^{361} - x^{276} + x^{255} + x^{170} + x^{85}$, and $x^{2/3} = -x^{234} + x^{128} - x^{43}$ modulo $f(x)$.

30. Find the minimal polynomials of all elements in some representation of $\mathbb{F}_8$.

31. Find the minimal polynomials (over $\mathbb{F}_5$) of α and β of Exercise 2.6.

32. Represent $\mathbb{F}_{16}$ by adjoining to $\mathbb{F}_2$ a root of an irreducible polynomial of degree four in $\mathbb{F}_2[x]$. Find a primitive element in this representation of $\mathbb{F}_{16}$. Also find a normal element in $\mathbb{F}_{16}$. Check whether this normal element is primitive too.

33. Represent $\mathbb{F}_{27}$ by adjoining to $\mathbb{F}_3$ a root of an irreducible polynomial of degree three in $\mathbb{F}_3[x]$. Find a primitive element in this representation of $\mathbb{F}_{27}$. Also find a normal element in $\mathbb{F}_{27}$. Check whether this normal element is primitive too.

34. Represent $\mathbb{F}_{25}$ by adjoining to $\mathbb{F}_5$ a root of an irreducible polynomial of degree two in $\mathbb{F}_5[x]$. Find a primitive element in this representation of $\mathbb{F}_{25}$. Also find a normal element in $\mathbb{F}_{25}$. Check whether this normal element is primitive too.

35. Find a primitive element in the field $\mathbb{F}_{29}$.

36. Represent $\mathbb{F}_4$ by adjoining θ to $\mathbb{F}_2$, where $\theta^2 + \theta + 1 = 0$.
   (a) List all monic irreducible polynomials of degrees $1, 2$ in $\mathbb{F}_4[x]$.
   (b) Let ψ be a root of an irreducible polynomial of $\mathbb{F}_4[x]$ of degree 2. Represent $\mathbb{F}_{16}$ as $\mathbb{F}_4(\psi)$. How can you add and multiply elements in this representation?
   (c) Find a primitive element in this representation of $\mathbb{F}_{16}$.
   (d) Compute the minimal polynomial of the element $(\theta + 1)\psi + 1$ over $\mathbb{F}_2$.
   (e) Compute the minimal polynomial of the element $(\theta + 1)\psi + 1$ over $\mathbb{F}_4$.

37. Represent $\mathbb{F}_{64} = \mathbb{F}_{2^6}$ as $\mathbb{F}_2(\theta)$ with $\theta^6 + \theta^3 + 1 = 0$.
   (a) Find all the conjugates of θ (over $\mathbb{F}_2$ as polynomials in θ of degrees < 6).
   (b) Prove or disprove: θ is a primitive element of $\mathbb{F}_{64}^*$.
   (c) What is the minimal polynomial of θ^3 over $\mathbb{F}_2$?

38. Represent $\mathbb{F}_9$ as $\mathbb{F}_3(\theta)$, where $\theta^2 + \theta + 2 = 0$.
   (a) Find the roots of $x^2 + x + 2$ in $\mathbb{F}_9$.
   (b) Find the roots of $x^2 + x + 2$ in $\mathbb{Z}_9$.
   (c) Prove that θ is a primitive element of $\mathbb{F}_9$.
   (d) Prove that the polynomial $y^2 - \theta$ is irreducible over $\mathbb{F}_9$.
   Represent $\mathbb{F}_{81}$ as $\mathbb{F}_9(\psi)$, where $\psi^2 - \theta = 0$.
   (e) Determine whether ψ is a primitive element of $\mathbb{F}_{81}$.
   (f) Find the minimal polynomial of ψ over $\mathbb{F}_3$.

39. Prove that $\theta + 1$ is a normal element in $\mathbb{F}_{32} = \mathbb{F}_2(\theta)$, where $\theta^5 + \theta^2 + 1 = 0$.

40. Prepare the Zech logarithm tables for the fields $\mathbb{F}_{25}, \mathbb{F}_{27}, \mathbb{F}_{29}$ with respect to some primitive elements.

41. Compute an explicit isomorphism between the two representations of $\mathbb{F}_{16}$ in Exercises 2.32 and 2.36.

42. Prove that the total number of ordered bases of $\mathbb{F}_{p^n}$ over $\mathbb{F}_p$ is

$$(p^n - 1)(p^n - p)(p^n - p^2) \cdots (p^n - p^{n-1}).$$

43. Let $\alpha \in \mathbb{F}_{p^n}$, and $f_\alpha(x)$ the minimal polynomial of α over $\mathbb{F}_p$. Prove that the degree of $f_\alpha(x)$ divides n.

44. Let $f(x), g(x)$ be irreducible polynomials in $\mathbb{F}_p[x]$ of degrees m and n. Let $\mathbb{F}_{p^m}$ be represented by adjoining a root of $f(x)$ to $\mathbb{F}_p$. Prove that:
   (a) If $m = n$, then $g(x)$ splits over $\mathbb{F}_{p^m}$.
   (b) If $\gcd(m, n) = 1$, then $g(x)$ is irreducible in $\mathbb{F}_{p^m}[x]$.

45. Let $q = p^n$, $f(x)$ a polynomial in $\mathbb{F}_q[x]$, and $f'(x)$ the (formal) derivative of $f(x)$. Prove that $f'(x) = 0$ if and only if $f(x) = g(x)^p$ for some $g(x) \in \mathbb{F}_q[x]$.

46. Let p be an odd prime, $n \in \mathbb{N}$, and $q = p^n$. Prove that for every $\alpha \in \mathbb{F}_q$,

$$x^q - x = (x + \alpha)((x + \alpha)^{(q-1)/2} - 1)((x + \alpha)^{(q-1)/2} + 1).$$

47. Let $n \in \mathbb{N}$, and $q = 2^n$. Prove that for every $\alpha \in \mathbb{F}_q$,

$$x^q + x = \left((x + \alpha) + (x + \alpha)^2 + (x + \alpha)^4 + \cdots + (x + \alpha)^{2^{n-1}}\right) \times$$
$$\left(1 + (x + \alpha) + (x + \alpha)^2 + (x + \alpha)^4 + \cdots + (x + \alpha)^{2^{n-1}}\right).$$

48. Modify Algorithm 1.7 in order to compute the order of an element $\alpha \in \mathbb{F}_q^*$. You may assume that the complete prime factorization of $q - 1$ is available.

49. Let $\alpha \in \mathbb{F}_{p^n}^*$. Prove that the orders of $\alpha, \alpha^p, \alpha^{p^2}, \ldots, \alpha^{p^{n-1}}$ are the same. In particular, all conjugates of a primitive element of $\mathbb{F}_{p^n}$ are again primitive.

50. Prove that $n \mid \phi(p^n - 1)$ for every $p \in \mathbb{P}$ and $n \in \mathbb{N}$.

51. Let $q - 1 = p_1^{e_1} \cdots p_r^{e_r}$ be the prime factorization of the size $q - 1$ of $\mathbb{F}_q^*$ with each $e_i \geq 1$. Prove that $\displaystyle\sum_{\alpha \in \mathbb{F}_q^*} \operatorname{ord}\alpha = \prod_{i=1}^{r} \frac{p_i^{2e_i+1} + 1}{p_i + 1}$.

52. [*Euler's criterion for finite fields*] Let $\alpha \in \mathbb{F}_q^*$ with q odd. Prove that the equation $x^2 = \alpha$ has a solution in $\mathbb{F}_q^*$ if and only if $\alpha^{(q-1)/2} = 1$.

53. [*Generalized Euler's criterion*] Let $\alpha \in \mathbb{F}_q^*$, $t \in \mathbb{N}$, and $d = \gcd(t, q-1)$. Prove that the equation $x^t = \alpha$ has a solution in $\mathbb{F}_q^*$ if and only if $\alpha^{(q-1)/d} = 1$.

54. Prove that for any finite field $\mathbb{F}_q$ and for any $\alpha \in \mathbb{F}_q$, the equation $x^2 + y^2 = \alpha$ has at least one solution for (x, y) in $\mathbb{F}_q \times \mathbb{F}_q$.

55. Let γ be a primitive element of $\mathbb{F}_q$, and $r \in \mathbb{N}$. Prove that the polynomial $x^r - \gamma$ has a root in $\mathbb{F}_q$ if and only if $\gcd(r, q - 1) = 1$.

56. Prove that the field $\mathbb{Q}(\sqrt{2})$ is not isomorphic to the field $\mathbb{Q}(\sqrt{3})$.

57. Let θ, ψ be two distinct roots of some non-constant irreducible polynomial $f(x) \in F[x]$, and let $K = F(\theta)$ and $L = F(\psi)$. Give an example where $K = L$ as sets. Give another example where $K \neq L$ as sets.

58. Let $\alpha \in \mathbb{F}_{p^n}$. The *trace* and *norm* of α over $\mathbb{F}_p$ are defined respectively as

$$\operatorname{Tr}(\alpha) = \alpha + \alpha^p + \alpha^{p^2} + \cdots + \alpha^{p^{n-1}},$$
$$\operatorname{N}(\alpha) = \alpha \times \alpha^p \times \alpha^{p^2} \times \cdots \times \alpha^{p^{n-1}}.$$

(a) Prove that $\operatorname{Tr}(\alpha), \operatorname{N}(\alpha) \in \mathbb{F}_p$.
(b) Prove that if $\alpha \in \mathbb{F}_p$, then $\operatorname{Tr}(\alpha) = n\alpha$ and $\operatorname{N}(\alpha) = \alpha^n$.
(c) Prove that $\operatorname{Tr}(\alpha + \beta) = \operatorname{Tr}(\alpha) + \operatorname{Tr}(\beta)$ and $\operatorname{N}(\alpha\beta) = \operatorname{N}(\alpha)\operatorname{N}(\beta)$ for all $\alpha, \beta \in \mathbb{F}_{p^n}$. (Trace is additive, and norm is multiplicative.)
(d) Prove that $\operatorname{Tr}(\alpha) = 0$ if and only if $\alpha = \gamma^p - \gamma$ for some $\gamma \in \mathbb{F}_{p^n}$.

59. Let $\alpha \in \mathbb{F}_{2^n}$.
(a) Prove that $x^2 + x = \alpha$ is solvable for x in $\mathbb{F}_{2^n}$ if and only if $\operatorname{Tr}(\alpha) = 0$.
(b) Let $\operatorname{Tr}(\alpha) = 0$. If n is odd, prove that $\alpha^{2^1} + \alpha^{2^3} + \alpha^{2^5} + \cdots + \alpha^{2^{n-2}}$ is a solution of $x^2 + x = \alpha$. What is the other solution?
(c) Describe a method to solve the general quadratic equation $ax^2 + bx + c = 0$ with $a, b, c \in \mathbb{F}_{2^n}^*$ (assume that n is odd).

60. Let $\mathbb{F}_q$ be a finite field, and let $\gamma \in \mathbb{F}_q^*$ be a primitive element. For every $\alpha \in \mathbb{F}_q^*$, there exists a unique x in the range $0 \leqslant x \leqslant q - 2$ such that $\alpha = \gamma^x$. Denote this x by $\mathrm{ind}_\gamma \, \alpha$ (index of α with respect to γ).
(a) First assume that q is odd. Prove that the equation $x^2 = \alpha$ is solvable in $\mathbb{F}_q$ for $\alpha \in \mathbb{F}_q^*$ if and only if $\mathrm{ind}_\gamma \, \alpha$ is even.
(b) Now, let $q = 2^n$. In this case, for every $\alpha \in \mathbb{F}_q$, there exists a unique $\beta \in \mathbb{F}_q$ such that $\beta^2 = \alpha$. In fact, $\beta = \alpha^{2^{n-1}}$. Suppose that $\alpha, \beta \in \mathbb{F}_q^*$, $k = \mathrm{ind}_\gamma \, \alpha$, and $l = \mathrm{ind}_\gamma \, \beta$. Express l as an efficiently computable formula in k and q.

61. Let $\theta_0, \theta_1, \ldots, \theta_{n-1}$ be elements of $\mathbb{F}_{p^n}$. The *discriminant* $\Delta(\theta_0, \theta_1, \ldots, \theta_{n-1})$ of $\theta_0, \theta_1, \ldots, \theta_{n-1}$ is defined as the determinant of the $n \times n$ matrix

$$
A = \begin{pmatrix}
\mathrm{Tr}(\theta_0\theta_0) & \mathrm{Tr}(\theta_0\theta_1) & \cdots & \mathrm{Tr}(\theta_0\theta_{n-1}) \\
\mathrm{Tr}(\theta_1\theta_0) & \mathrm{Tr}(\theta_1\theta_1) & \cdots & \mathrm{Tr}(\theta_1\theta_{n-1}) \\
\vdots & \vdots & \cdots & \vdots \\
\mathrm{Tr}(\theta_{n-1}\theta_0) & \mathrm{Tr}(\theta_{n-1}\theta_1) & \cdots & \mathrm{Tr}(\theta_{n-1}\theta_{n-1})
\end{pmatrix}.
$$

(a) Prove that $\theta_0, \theta_1, \ldots, \theta_{n-1}$ constitute a basis of $\mathbb{F}_{p^n}$ over $\mathbb{F}_p$ if and only if $\Delta(\theta_0, \theta_1, \ldots, \theta_{n-1}) \neq 0$.
(b) Define the matrix B as

$$
B = \begin{pmatrix}
\theta_0 & \theta_1 & \cdots & \theta_{n-1} \\
\theta_0^p & \theta_1^p & \cdots & \theta_{n-1}^p \\
\vdots & \vdots & \cdots & \vdots \\
\theta_0^{p^{n-1}} & \theta_1^{p^{n-1}} & \cdots & \theta_{n-1}^{p^{n-1}}
\end{pmatrix}
$$

Prove that $B^t B = A$, where B^t denotes the transpose of B. Conclude that $\theta_0, \theta_1, \ldots, \theta_{n-1}$ constitute a basis of $\mathbb{F}_{p^n}$ over $\mathbb{F}_p$ if and only if $\det B \neq 0$.
(c) Let $\theta \in \mathbb{F}_{p^n}$. Prove that $\Delta(1, \theta, \theta^2, \ldots, \theta^{n-1}) = \displaystyle\prod_{0 \leqslant i < j \leqslant n-1} \left(\theta^{p^i} - \theta^{p^j} \right)^2$.

Programming Exercises

62. Write a GP/PARI function for the Euclidean inverse algorithm in $\mathbb{F}_{2^n}$.

63. Write a GP/PARI function for the binary inverse algorithm in $\mathbb{F}_{2^n}$.

64. Write a GP/PARI function for the almost inverse algorithm in $\mathbb{F}_{2^n}$.

65. Write a GP/PARI function for the Euclidean inverse algorithm in $\mathbb{F}_{p^n}$.

66. Write a GP/PARI function for the binary inverse algorithm in $\mathbb{F}_{p^n}$.

67. Write a GP/PARI function for the almost inverse algorithm in $\mathbb{F}_{p^n}$.

68. Generalize the GP/PARI code of Section 2.4 for checking normal elements so as to work for any extension $\mathbb{F}_{2^n}$.

69. Write GP/PARI functions to compute traces and norms of elements in $\mathbb{F}_{p^n}$. Use these functions to compute the traces and norms of all elements in $\mathbb{F}_{64}$.

Chapter 3

Arithmetic of Polynomials

Polynomials are useful in a variety of mathematical and computational contexts. We have already used modular polynomial arithmetic to represent finite fields. In addition to such applications, polynomials themselves constitute an independent area of study. Two most important computational problems pertaining to polynomials are finding roots of polynomials and factoring polynomials. This chapter is an introduction to computations involving polynomials.

The set of all polynomials in one indeterminate (also called variable) x and with coefficients from a ring A is denoted by $A[x]$. If A is an integral domain, then so also is $A[x]$ (and conversely). Polynomial rings over fields enjoy several nice algebraic properties. For example, the polynomial ring $K[x]$ over a field K is a unique factorization domain, that is, every non-constant polynomial in $K[x]$ can be written as a product of irreducible polynomials, and this factorization is unique up to rearrangement of the factors and up to multiplication by non-zero elements of K. This unique factorization property of $K[x]$ is derived from the fact that $K[x]$ is a Euclidean domain, that is, the concept of Euclidean division and Euclidean gcd holds in $K[x]$.

We start our study with polynomials over finite fields, that is, with the ring $\mathbb{F}_q[x]$ for some q. Next, we look at the polynomial ring $\mathbb{Z}[x]$ over integers. In some sense, the study of $\mathbb{Z}[x]$ is the same as the study of $\mathbb{Q}[x]$. Since $\mathbb{Q}$ is a field, the ring $\mathbb{Q}[x]$ is an easier object to study than $\mathbb{Z}[x]$. Of course, $\mathbb{Q} \subseteq \mathbb{R} \subseteq \mathbb{C}$, and so a study of $\mathbb{Q}[x]$ may benefit from a study of $\mathbb{R}[x]$ and $\mathbb{C}[x]$. However, a study of $\mathbb{R}[x]$ and $\mathbb{C}[x]$ may lead us too far away from our focus of interest. Indeed, we cannot represent every real (or complex) number in computers. Every finite representation of real numbers has to be approximate. Algorithmic issues pertaining to such approximate representations (like convergence and numerical stability) are not dealt with in this book.

3.1 Polynomials over Finite Fields

3.1.1 Polynomial Arithmetic

Let $\mathbb{F}_q$ be a finite field for some $q = p^n$ with $p \in \mathbb{P}$ and $n \in \mathbb{N}$. We know how to carry out the arithmetic of the field $\mathbb{F}_q$. Using these computational primitives, we can implement the arithmetic of $\mathbb{F}_q[x]$. For $f(x), g(x) \in \mathbb{F}_q[x]$, we can compute $f(x) + g(x)$, $f(x) - g(x)$, and $f(x)g(x)$ with the coefficient arithmetic being that of $\mathbb{F}_q$. $\mathbb{F}_q[x]$ supports Euclidean division, that is, for $g(x) \neq 0$, we can compute $f(x)$ quot $g(x)$ and $f(x)$ rem $g(x)$. The Euclidean gcd condition can be stated for polynomials as $\gcd(f(x), g(x)) = \gcd(g(x), f(x) \text{ rem } g(x))$ (provided that $g(x) \neq 0$). Finally, a *Bézout relation* for $f(x), g(x) \in \mathbb{F}_q[x]$, not both zero, is of the form $\gcd(f(x), g(x)) = u(x)f(x) + v(x)g(x)$ for some $u(x), v(x) \in \mathbb{F}_q[x]$. Solve Exercise 2.1 for learning the details.

3.1.2 Irreducible Polynomials over Finite Fields

Irreducible polynomials play a crucial role in the arithmetic of $\mathbb{F}_q[x]$. We start with a mathematical study of them. For more details, the reader may consider Lidl and Niederreiter's book (Footnote 2 on page 76).

Let $\mathbb{F}_q = \mathbb{F}_p(\theta)$ with $f(\theta) = 0$, where $f(x) \in \mathbb{F}_p[x]$ is monic and irreducible of degree n. Let $g(x)$ be another monic irreducible polynomial in $\mathbb{F}_p[x]$ of degree $d|n$. Let ψ be any root of $g(x)$. Suppose that $\psi \notin \mathbb{F}_q$. Adjoining ψ to $\mathbb{F}_q$ gives a field K with $|K| > q$. Every element $\alpha \in \mathbb{F}_q$ satisfies $\alpha^q = \alpha$. Moreover, $\psi^q = \psi$, since ψ can be used for representing $\mathbb{F}_{p^d}$ with $d|n$. It follows that the polynomial $x^q - x$ has at least $q+1$ roots in K. This is impossible, since K is a field. Therefore, $\psi \in \mathbb{F}_q$, that is, $g(x)$ has a root ψ and so all the roots $\psi, \psi^p, \psi^{p^2}, \ldots, \psi^{p^{d-1}}$ in $\mathbb{F}_q$. Since $g(x) = (x - \psi)(x - \psi^p)(x - \psi^{p^2}) \cdots (x - \psi^{p^{d-1}})$, we have proved (a part of) the following important result.

Theorem 3.1 *The product of all monic irreducible polynomials in $\mathbb{F}_p[x]$ of degrees dividing n is $x^{p^n} - x$. More generally, for $m \in \mathbb{N}$, the product of all monic irreducible polynomials in $\mathbb{F}_q[x]$ of degrees dividing m is $x^{q^m} - x$.* ◁

Example 3.2 (1) Let $p = 2$ and $n = 6$. All positive integral divisors of 6 are $1, 2, 3, 6$. The polynomial $x^{2^6} - x$ factors over $\mathbb{F}_2$ as

$$
\begin{aligned}
x^{64} + x &= x(x+1)(x^2+x+1)(x^3+x+1)(x^3+x^2+1) \times \\
&\quad (x^6+x+1)(x^6+x^5+1)(x^6+x^3+1)(x^6+x^4+x^2+x+1) \times \\
&\quad (x^6+x^5+x^4+x^2+1)(x^6+x^4+x^3+x+1)(x^6+x^5+x^3+x^2+1) \times \\
&\quad (x^6+x^5+x^2+x+1)(x^6+x^5+x^4+x+1).
\end{aligned}
$$

All the (monic) irreducible polynomials of $\mathbb{F}_2[x]$ of degree one are x and $x+1$. The only irreducible polynomial of $\mathbb{F}_2[x]$ of degree two is $x^2 + x + 1$. The irreducible polynomials of $\mathbb{F}_2[x]$ of degree three are $x^3 + x + 1$ and $x^3 + x^2 + 1$, whereas those of degree six are $x^6 + x + 1$, $x^6 + x^5 + 1$, $x^6 + x^3 + 1$, $x^6 + x^4 + x^2 + x + 1$, $x^6 + x^5 + x^4 + x^2 + 1$, $x^6 + x^4 + x^3 + x + 1$, $x^6 + x^5 + x^3 + x^2 + 1$, $x^6 + x^5 + x^2 + x + 1$, and $x^6 + x^5 + x^4 + x + 1$.

(2) Let us now see how $x^{64} + x = x^{4^3} + x$ factors over $\mathbb{F}_4$, that is, we take $q = 4$ and $m = 3$. We represent $\mathbb{F}_{64} = \mathbb{F}_2(\theta)$, where $\theta^6 + \theta + 1 = 0$. By Example 2.29, the unique copy of $\mathbb{F}_4$ contained in this representation of $\mathbb{F}_{64}$ is $\{0, 1, \xi, \xi+1\}$, where $\xi = \theta^5 + \theta^4 + \theta^3 + \theta$ satisfies $\xi^2 + \xi + 1 = 0$. Since the extension degree of $\mathbb{F}_{64}$ over $\mathbb{F}_4$ is three, $x^{64} + x$ factors over $\mathbb{F}_4$ into irreducible polynomials of degrees one and three only. The four irreducible polynomials in $\mathbb{F}_4[x]$ of degree one are $x, x+1, x+\xi, x+\xi+1$. The irreducible polynomial $x^2 + x + 1$ of $\mathbb{F}_2[x]$ factors in $\mathbb{F}_4[x]$ as $(x+\xi)(x+\xi+1)$. The polynomials $x^3 + x + 1$ and $x^3 + x^2 + 1$ of $\mathbb{F}_2[x]$ continue to remain irreducible in $\mathbb{F}_4[x]$. Each of the nine irreducible polynomials in $\mathbb{F}_2[x]$ of degree six factors into two irreducible polynomials in $\mathbb{F}_4[x]$ of degree three, as shown below.

$$
\begin{aligned}
x^6 + x + 1 &= [x^3 + x^2 + (\xi+1)x + \xi]\,[x^3 + x^2 + \xi x + (\xi+1)], \\
x^6 + x^5 + 1 &= [x^3 + (\xi+1)x^2 + \xi x + \xi]\,[x^3 + \xi x^2 + (\xi+1)x + (\xi+1)], \\
x^6 + x^3 + 1 &= [x^3 + \xi]\,[x^3 + (\xi+1)], \\
x^6 + x^4 + x^2 + x + 1 &= [x^3 + \xi x + 1]\,[x^3 + (\xi+1)x + 1], \\
x^6 + x^5 + x^4 + x^2 + 1 &= [x^3 + \xi x^2 + 1]\,[x^3 + (\xi+1)x^2 + 1], \\
x^6 + x^4 + x^3 + x + 1 &= [x^3 + x^2 + x + \xi]\,[x^3 + x^2 + x + (\xi+1)], \\
x^6 + x^5 + x^3 + x^2 + 1 &= [x^3 + \xi x^2 + \xi x + \xi]\,[x^3 + (\xi+1)x^2 + (\xi+1)x + (\xi+1)], \\
x^6 + x^5 + x^2 + x + 1 &= [x^3 + \xi x^2 + (\xi+1)x + \xi]\,[x^3 + (\xi+1)x^2 + \xi x + (\xi+1)], \\
x^6 + x^5 + x^4 + x + 1 &= [x^3 + (\xi+1)x^2 + x + \xi]\,[x^3 + \xi x^2 + x + (\xi+1)].
\end{aligned}
$$
□

Theorem 3.1 has many consequences. First, it gives us a formula for computing the number $N_{q,m}$ of monic irreducible polynomials in $\mathbb{F}_q[x]$ of degree equal to m. Equating degrees in Theorem 3.1 gives

$$
p^n = \sum_{d \mid n} d N_{p,d}, \quad \text{or, more generally,} \quad q^m = \sum_{d \mid m} d N_{q,d}.
$$

These are still not explicit formulas for $N_{p,n}$ and $N_{q,m}$. In order to derive the explicit formulas, we use an auxiliary result.

Definition 3.3 The *Möbius function* $\mu : \mathbb{N} \to \{0, 1, -1\}$ is defined as[1]

$$\mu(n) = \begin{cases} 1 & \text{if } n = 1, \\ 0 & \text{if } p^2 | n \text{ for some } p \in \mathbb{P}, \\ (-1)^t & \text{if } n \text{ is the product of } t \in \mathbb{N} \text{ pairwise distinct primes.} \end{cases} \lhd$$

Example 3.4 The following table lists $\mu(n)$ for some small values of n.

n	1	2	3	4	5	6	7	8	9	10	11	12	13	14	15	16
$\mu(n)$	1	−1	−1	0	−1	1	−1	0	0	1	−1	0	−1	1	1	0

$\square$

Lemma 3.5 *For all $n \in \mathbb{N}$, the Möbius function satisfies the identity*

$$\sum_{d|n} \mu(d) = \begin{cases} 1 & \text{if } n = 1, \\ 0 & \text{if } n > 1, \end{cases}$$

where the sum $\sum_{d|n} \mu(d)$ extends over all positive integral divisors d of n.

Proof Let $n = p_1^{e_1} \cdots p_t^{e_t}$ be the prime factorization of $n > 1$ with pairwise distinct primes $p_1, \ldots, p_t$ and with each $e_i \in \mathbb{N}$. A divisor d of n is of the form $d = p_1^{r_1} \cdots p_t^{r_t}$ with each r_i in the range $0 \leqslant r_i \leqslant e_i$. If some $r_i > 1$, then $\mu(d) = 0$ by definition. Therefore, $\mu(d)$ is non-zero if and only if each $r_i \in \{0, 1\}$. But then $\sum_{d|n} \mu(d) = \sum_{(r_1, \ldots, r_t) \in \{0,1\}^t} (-1)^{r_1 + \cdots + r_t} = (1 - 1)^t = 0. \lhd$

Proposition 3.6 *[Möbius inversion formula]* Let $f, g : \mathbb{N} \to \mathbb{R}$ satisfy $f(n) = \sum_{d|n} g(d)$ for all $n \in \mathbb{N}$. Then, $g(n) = \sum_{d|n} \mu(d) f(n/d) = \sum_{d|n} \mu(n/d) f(d)$.

Proof We have

$$\sum_{d|n} \mu(d) f(n/d) = \sum_{d|n} \left[\mu(d) \sum_{d'|(n/d)} g(d') \right] = \sum_{(dd')|n} \mu(d) g(d')$$

$$= \sum_{d'|n} \left[g(d') \sum_{d|(n/d')} \mu(d) \right]$$

$$= g(n) \sum_{d|1} \mu(d) + \sum_{\substack{d'|n \\ d' < n}} \left[g(d') \sum_{d|(n/d')} \mu(d) \right] = g(n),$$

where the last equality follows from Lemma 3.5. $\lhd$

---

[1] August Ferdinand Möbius (1790–1868) was a German mathematician who had deep contributions to number theory and geometry. In addition to Möbius function and Möbius inversion formula, Möbius is also well-known for *Möbius transform* and *Möbius strip*.

Corollary 3.7 *The number of monic irreducible polynomials in $\mathbb{F}_p[x]$ of degree n is*

$$N_{p,n} = \frac{1}{n} \sum_{d|n} \mu(d) p^{n/d} = \frac{1}{n} \sum_{d|n} \mu(n/d) p^d.$$

The number of monic irreducible polynomials in $\mathbb{F}_q[x]$ of degree m is

$$N_{q,m} = \frac{1}{m} \sum_{d|m} \mu(d) q^{m/d} = \frac{1}{m} \sum_{d|m} \mu(m/d) q^d.$$ ◁

Example 3.8 (1) First take $p = 2$ and $n = 6$. By the Möbius inversion formula, we have $N_{2,6} = \frac{1}{6}(\mu(1)2^6 + \mu(2)2^3 + \mu(3)2^2 + \mu(6)2^1) = \frac{1}{6}(64 - 8 - 4 + 2) = 9$. These irreducible polynomials are listed in Example 3.2(1).

(2) For $q = 4$ and $m = 3$, we have $N_{4,3} = \frac{1}{3}(\mu(1)4^3 + \mu(3)4^1) = \frac{1}{3}(64-4) = 20$. Example 3.2(2) lists all these irreducible polynomials. □

A close look at the above formulas for $N_{p,n}$ or $N_{q,m}$ indicates that the terms containing p^n or q^m dominate over the other terms, so that $N_{p,n} \approx p^n/n$, and $N_{q,m} \approx q^m/m$. There are p^n monic polynomials of degree n in $\mathbb{F}_p[x]$. Under the assumption that irreducible polynomials are distributed randomly among these monic polynomials, a randomly chosen monic polynomial of degree n in $\mathbb{F}_p[x]$ is irreducible with probability nearly $1/n$. Likewise, a random monic polynomial in $\mathbb{F}_q[x]$ of degree m is irreducible with probability about $1/m$.

3.1.3 Testing Irreducibility of Polynomials

Let $f(x) \in \mathbb{F}_p[x]$ be a monic polynomial of degree $d \geqslant 1$. We want to check whether $f(x)$ is irreducible in $\mathbb{F}_p[x]$. Evidently, $f(x)$ is reducible in $\mathbb{F}_p[x]$ if and only if it has an irreducible factor of some degree $r \leqslant \lfloor d/2 \rfloor$. Moreover, $x^{p^r} - x$ is the product of all monic irreducible polynomials in $\mathbb{F}_p[x]$ of degrees dividing r. Therefore, if $\gcd(f(x), x^{p^r} - x) \neq 1$ for some $r \leqslant \lfloor d/2 \rfloor$, we conclude that $f(x)$ has one or more irreducible factors of degrees dividing r. On the other hand, if $\gcd(f(x), x^{p^r} - x) = 1$ for all $r = 1, 2, \ldots, \lfloor d/2 \rfloor$, we conclude that $f(x)$ is irreducible. Algorithm 3.1 implements this idea with $\mathbb{F}_p$ replaced by a general field $\mathbb{F}_q$. Since the polynomial $x^{q^r} - x$ has a large degree, it is expedient to compute $x^{q^r} - x$ modulo $f(x)$. The correctness of the algorithm follows from the fact that $\gcd(f(x), x^{q^r} - x) = \gcd(f(x), (x^{q^r} - x) \operatorname{rem} f(x))$.

Example 3.9 (1) Let us check whether the polynomial $f(x) = x^8 + x^3 + 1 \in \mathbb{F}_2[x]$ is irreducible. The iterations of Algorithm 3.1 reveal that $f(x)$ has an irreducible factor of degree three, that is, $f(x)$ is reducible.

r	$x^{2^r} \pmod{f(x)}$	$\gcd(x^{2^r} + x, f(x))$
1	x^2	1
2	x^4	1
3	$x^3 + 1$	$x^3 + x + 1$

Algorithm 3.1: CHECKING WHETHER $f(x) \in \mathbb{F}_q[x]$ WITH $d = \deg f(x) \geqslant 1$
IS IRREDUCIBLE

Initialize a temporary polynomial $t(x) = x$.
For $(r = 1;\ r \leqslant \lfloor d/2 \rfloor;\ ++r)$ {
     Set $t(x) = t(x)^q \pmod{f(x)}$.
     If $(\gcd(f(x), t(x) - x) \neq 1)$, return *False* (that is, reducible).
}
Return *True* (that is, irreducible).

---

(2) For $f(x) = x^{12} + x^{11} + x^9 + x^8 + x^6 + x^3 + 1 \in \mathbb{F}_2[x]$, Algorithm 3.1
reveals that $f(x)$ has two irreducible factors of degree three, and is reducible.

r	$x^{2^r} \pmod{f(x)}$	$\gcd(x^{2^r} + x, f(x))$
1	x^2	1
2	x^4	1
3	x^8	$x^6 + x^5 + x^4 + x^3 + x^2 + x + 1$

(3) Now, take $f(x) = x^{15} + x^4 + 1 \in \mathbb{F}_2[x]$. The iterations of Algorithm 3.1
proceed as follows, and indicate that $x^{15} + x^4 + 1$ is irreducible in $\mathbb{F}_2[x]$.

r	$x^{2^r} \pmod{f(x)}$	$\gcd(x^{2^r} + x, f(x))$
1	x^2	1
2	x^4	1
3	x^8	1
4	$x^5 + x$	1
5	$x^{10} + x^2$	1
6	$x^9 + x^5 + x^4$	1
7	$x^{10} + x^8 + x^7 + x^3$	1

(4) Let us represent $\mathbb{F}_4 = \mathbb{F}_2(\theta)$, where $\theta^2 + \theta + 1 = 0$. Take $f(x) = x^6 + \theta x + \theta \in \mathbb{F}_4[x]$. We first compute the Euclidean gcd of $x^4 + x$ with $f(x)$:

$$
\begin{aligned}
x^6 + \theta x + \theta &= x^2 \times (x^4 + x) + (x^3 + \theta x + \theta), \\
x^4 + x &= x \times (x^3 + \theta x + \theta) + (\theta x^2 + (\theta + 1)x) \\
&= x \times (x^3 + \theta x + \theta) + \theta(x^2 + \theta x), \\
x^3 + \theta x + \theta &= (x + \theta) \times (x^2 + \theta x) + (x + \theta), \\
x^2 + \theta x &= x \times (x + \theta).
\end{aligned}
$$

Thus, $\gcd(x^4 + x, f(x)) = x + \theta$, that is, $x^6 + \theta x + \theta$ is reducible. $\qquad\square$

A polynomial of degree d in $\mathbb{F}_q[x]$ is stored as a list of its $d+1$ coefficients. Each such coefficient is a member of $\mathbb{F}_q$, and can be encoded using $O(\log q)$

bits. Thus, the size of the polynomial is $O(d \log q)$. An algorithm involving this polynomial as input is said to run in polynomial time if its running time is a polynomial in both d and $\log q$.

Let us deduce the running time of Algorithm 3.1. The loop continues for a maximum of $\lfloor d/2 \rfloor = O(d)$ times. Each iteration of the loop involves a modular exponentiation followed by a gcd calculation. The exponentiation is done modulo $f(x)$, that is, the degrees of all intermediate products are kept at values $< d$. The exponent is q, that is, square-and-multiply exponentiation makes $O(\log q)$ iterations only. In short, each exponentiation in Algorithm 3.1 requires $O(d^2 \log q)$ field operations. The gcd calculation involves at most d Euclidean divisions with each division requiring $O(d^2)$ operations in $\mathbb{F}_q$. This is actually an overestimate—Euclidean gcd requires $O(d^2)$ field operations only. The arithmetic of $\mathbb{F}_q$ can be implemented to run in $O(\log^2 q)$ time per operation (schoolbook arithmetic). To sum up, Algorithm 3.1 runs in time $O(d^3 \log^3 q)$ which is polynomial in both d and $\log q$.

3.1.4 Handling Irreducible Polynomials in GP/PARI

A GP/PARI function to check the irreducibility of $f(x) \in \mathbb{F}_p[x]$ follows.

```
gp > \
isirr(f,p) = \
   local (t,g); \
   t = Mod(1,p) * x; \
   for (r=1, floor(poldegree(f)/2), \
       t = (t^p)%f; \
       g = gcd(t-Mod(1,p)*x,f); \
       print(lift(g)); \
       if (g-Mod(1,p), print("Not irreducible"); return(0)) \
   ); \
   print("Irreducible"); return(1)
gp > isirr(Mod(1,2)*x^8 + Mod(1,2)*x^3 + Mod(1,2), 2)
1
1
x^3 + x + 1
Not irreducible
%1 = 0
gp > isirr(Mod(1,2)*x^12 + Mod(1,2)*x^11 + Mod(1,2)*x^9 + Mod(1,2)*x^8 + \
     Mod(1,2)*x^6 + Mod(1,2)*x^3 + Mod(1,2), 2)
1
1
x^6 + x^5 + x^4 + x^3 + x^2 + x + 1
Not irreducible
%2 = 0
gp > isirr(Mod(1,2)*x^15 + Mod(1,2)*x^4 + Mod(1,2), 2)
1
1
1
1
1
```

```
1
1
Irreducible
%3 = 1
gp > isirr(Mod(1,7)*x^15 + Mod(1,7)*x^4 + Mod(1,7), 7)
1
1
1
1
1
4*x^6 + 6*x^3 + x^2 + x + 6
Not irreducible
%4 = 0
```

In fact, GP/PARI provides a built-in function polisirreducible() for checking whether its argument is irreducible.

```
gp > polisirreducible(Mod(1,2)*x^8 + Mod(1,2)*x^3 + Mod(1,2))
%5 = 0
gp > polisirreducible(Mod(1,2)*x^12 + Mod(1,2)*x^11 + Mod(1,2)*x^9 + \
    Mod(1,2)*x^8 + Mod(1,2)*x^6 + Mod(1,2)*x^3 + Mod(1,2))
%6 = 0
gp > polisirreducible(Mod(1,2)*x^15 + Mod(1,2)*x^4 + Mod(1,2))
%7 = 1
gp > polisirreducible(Mod(1,7)*x^15 + Mod(1,7)*x^4 + Mod(1,7))
%8 = 0
```

3.2 Finding Roots of Polynomials over Finite Fields

Theorem 3.1 turns out to be instrumental again for designing a probabilistic polynomial-time root-finding algorithm. Let the polynomial $f(x) \in \mathbb{F}_q[x]$
to be factored have the (unknown) factorization

$$f(x) = l_1(x)^{u_1} l_2(x)^{u_2} \cdots l_s(x)^{u_s} g_1(x)^{v_1} g_2(x)^{v_2} \cdots g_t(x)^{v_t},$$

where $l_i(x)$ are linear factors, and $g_j(x)$ are irreducible factors of degrees larger
than one. Assume that $f(x), l_i(x), g_j(x)$ are all monic. Each linear factor of
$f(x)$ corresponds to a root of $f(x)$ in $\mathbb{F}_q$, whereas an irreducible factor of $f(x)$
of degree > 1 does not have a root in $\mathbb{F}_q$. By Theorem 3.1, the polynomial
$x^q - x$ is the product of all monic linear polynomials in $\mathbb{F}_q[x]$, that is, $\bar{f}(x) =
\gcd(x^q - x, f(x)) = l_1(x)l_2(x)\cdots l_s(x)$, that is, $\bar{f}(x)$ has exactly the same
roots as $f(x)$. We may, therefore, assume that $f(x)$ itself is a product of linear
factors, and is square-free (that is, no factor of $f(x)$ appears more than once).

Given such a polynomial $f(x)$, we plan to find all the roots or equivalently
all the linear factors of $f(x)$. If q is small, we can evaluate $f(x)$ at all elements

of $\mathbb{F}_q$, and output all those $\alpha \in \mathbb{F}_q$ for which $f(\alpha) = 0$. This algorithm takes time proportional to q, and is impractical for large q.

We follow an alternative strategy. We try to split $f(x)$ as $f(x) = f_1(x)f_2(x)$ in $\mathbb{F}_q[x]$. If $\deg f_1 = 0$ or $\deg f_2 = 0$, this split is called trivial. A non-trivial split gives two factors of $f(x)$ of strictly smaller degrees. We subsequently try to split $f_1(x)$ and $f_2(x)$ non-trivially. This process is repeated until $f(x)$ is split into linear factors. I now describe a strategy[2] to split $f(x)$.

3.2.1 Algorithm for Fields of Odd Characteristics

First, let q be odd, and write $x^q - x = x \left(x^{(q-1)/2} - 1 \right) \left(x^{(q-1)/2} + 1 \right)$. If x is a factor of $f(x)$ (this can be decided from the constant term), we divide $f(x)$ by x, and assume that zero is not a root of $f(x)$. We then have $f(x) = f_1(x)f_2(x)$, where $f_1(x) = \gcd(x^{(q-1)/2} - 1, f(x))$, and $f_2(x) = \gcd(x^{(q-1)/2} + 1, f(x))$. This gives us a non-trivial split of $f(x)$ unless all linear factors of $f(x)$ divide exactly one of the polynomials $x^{(q-1)/2} \pm 1$. Even if this is not the case, that is, $f_1(x)f_2(x)$ is a non-trivial split of $f(x)$, this method fails to split $f_1(x)$ and $f_2(x)$ further, since $f_1(x)$ divides $x^{(q-1)/2} - 1$, and $f_2(x)$ divides $x^{(q-1)/2} + 1$.

Take any $\alpha \in \mathbb{F}_q$. We have $(x + \alpha)^q = x^q + \alpha^q = x^q + \alpha$, so $(x + \alpha)^q - (x + \alpha) = x^q - x$, that is, irrespective of the choice of α, the polynomial $(x+\alpha)^q - (x+\alpha)$ is the product of all monic linear factors of $\mathbb{F}_q[x]$. We can write $(x+\alpha)^q - (x+\alpha) = (x+\alpha) \left((x + \alpha)^{(q-1)/2} - 1 \right) \left((x + \alpha)^{(q-1)/2} + 1 \right)$. Assuming that $x+\alpha$ does not divide $f(x)$, we can write $f(x) = f_1(x)f_2(x)$, where $f_1(x) = \gcd \left((x + \alpha)^{(q-1)/2} - 1, f(x) \right)$, and $f_2(x) = \gcd \left((x + \alpha)^{(q-1)/2} + 1, f(x) \right)$.

Although $x^q - x$ and $(x+\alpha)^q - (x+\alpha)$ have the same roots (all elements of $\mathbb{F}_q$), the roots of $(x+\alpha)^{(q-1)/2} - 1$ are the roots of $x^{(q-1)/2} - 1$ shifted by α. So we expect to obtain a non-trivial factor $\gcd \left((x + \alpha)^{(q-1)/2} - 1, f(x) \right)$ of $f(x)$ for some α. Indeed, Rabin shows that this gcd is non-trivial with probability at least $1/2$ for a random α. This observation leads to Algorithm 3.2 for finite fields of odd characteristics. The algorithm assumes that the input polynomial $f(x)$ is square-free, and is a product of linear factors only. The polynomial $(x + \alpha)^{(q-1)/2} - 1$ has large degrees for large values of q. However, we need to compute the gcd of this polynomial with $f(x)$. Therefore, $(x + \alpha)^{(q-1)/2} - 1$ may be computed modulo $f(x)$. In that case, all intermediate polynomials are reduced to polynomials of degrees $< \deg f$, and we get a probabilistic algorithm with expected running time polynomial in $\deg f$ and $\log q$.

Example 3.10 (1) Let us find the roots of $f(x) = x^{15} + x^{12} + 2x^6 + 3x^4 + 6x^3 + 3x + 4 \in \mathbb{F}_{73}[x]$. We first compute $t(x) \equiv x^{73} - x \equiv 48x^{14} + 4x^{13} + 23x^{12} + 14x^{11} + 43x^{10} + 6x^9 + 9x^8 + 72x^7 + 65x^6 + 67x^5 + 33x^4 + 39x^3 + 24x^2 + 30 \pmod{f(x)}$, and replace $f(x)$ by $\gcd(f(x), t(x)) = x^6 + 17x^5 + 4x^4 + 67x^3 + 17x^2 + 4x + 66$. Therefore, $f(x)$ has six linear factors and so six roots in $\mathbb{F}_{73}$.

---

[2]This algorithm is first described in: Michael O. Rabin, Probabilistic algorithms in finite fields, *SIAM Journal of Computing*, 9(2), 273–280, 1980.

Algorithm 3.2: FINDING ROOTS OF MONIC $f(x) \in \mathbb{F}_q[x]$ WITH q ODD

Assumption: $f(x)$ *is a product of distinct linear factors.*

```
If (deg f = 0), return the empty list.
If (deg f = 1), return the negative of the constant term of f(x).
Initialize the flag splitfound to zero.
While (splitfound is zero) {
      Choose α ∈ 𝔽_q randomly.
      Compute g(x) = (x + α)^((q−1)/2) − 1 (mod f(x)).
      Compute f₁(x) = gcd(f(x), g(x)).
      if (0 < deg f₁ < deg f) {
            Set splitfound to one.
            Recursively call Algorithm 3.2 on f₁(x).
            Recursively call Algorithm 3.2 on f(x)/f₁(x).
      }
}
```

We try to split $f(x) = x^6 + 17x^5 + 4x^4 + 67x^3 + 17x^2 + 4x + 66$. We pick the element $\alpha = 19 \in \mathbb{F}_{73}$, and compute $g(x) \equiv (x + \alpha)^{(p-1)/2} - 1 \equiv (x + 19)^{36} - 1 \equiv 21x^5 + 29x^4 + 7x^3 + 46x^2 + 4x + 30 \pmod{f(x)}$. We have $f_1(x) = \gcd(g(x), f(x)) = x^2 + 44x + 43$, and $f_2(x) = f(x)/f_1(x) = x^4 + 46x^3 + 54x^2 + 20x + 27$. Thus, we are successful in splitting $f(x)$ non-trivially.

We then try to split $f_1(x) = x^2 + 44x + 43$. For $\alpha = 47$, we compute $g_1(x) \equiv (x + 47)^{36} - 1 \equiv 0 \pmod{f_1(x)}$, so $\gcd(g_1(x), f_1(x)) = f_1(x)$ is a trivial factor of $f_1(x)$. For $\alpha = 57$, we compute $g_1(x) \equiv (x + 57)^{36} - 1 \equiv 71 \pmod{f_1(x)}$, so $\gcd(g_1(x), f_1(x)) = 1$ is again a trivial factor of $f_1(x)$. We then try $\alpha = 67$, and compute $g_1(x) \equiv (x + 67)^{36} - 1 \equiv 66x + 64 \pmod{f_1(x)}$. But then, $\gcd(g_1(x), f_1(x)) = x + 43$ is a non-trivial factor of $f_1(x)$. This factor is linear, and reveals the root $-43 \equiv 30 \pmod{73}$. The cofactor $f_1(x)/(x + 43) = x + 1$ is also linear, and yields the root $-1 \equiv 72 \pmod{73}$.

We now attempt to split $f_2(x) = x^4 + 46x^3 + 54x^2 + 20x + 27$. Take $\alpha = 33$, and compute $g_2(x) \equiv (x + 33)^{36} - 1 \equiv 3x^3 + 19x^2 + 41x + 64 \pmod{f_2(x)}$. We have $f_3(x) = \gcd(g_2(x), f_2(x)) = x^3 + 55x^2 + 38x + 70$. The cofactor $f_2(x)/f_3(x) = x + 64$ is linear, and yields the root $-64 \equiv 9 \pmod{73}$.

For splitting $f_3(x) = x^3 + 55x^2 + 38x + 70$, we take $\alpha = 41$, compute $g_3(x) \equiv (x + 41)^{36} - 1 \equiv 3x^2 + 43x + 4 \pmod{f_3(x)}$ and $\gcd(g_3(x), f_3(x)) = x + 6$. Thus, we get the root $-6 \equiv 67 \pmod{73}$.

It remains to split the cofactor $f_4(x) = f_3(x)/(x + 6) = x^2 + 49x + 36$. For $\alpha = 8$, we have $g_4(x) \equiv (x + 8)^{36} - 1 \equiv 31x + 28 \pmod{f_4(x)}$, and $\gcd(f_4(x), g_4(x)) = 1$, that is, we fail to split $f_4(x)$. For $\alpha = 25$, we have $g_4(x) \equiv (x + 25)^{36} - 1 \equiv 11x + 13$, and $\gcd(g_4(x), f_4(x)) = x + 41$, that is, the root $-41 \equiv 32 \pmod{73}$ is discovered. The last root of $f_4(x)$ is obtained from $f_4(x)/(x + 41) = x + 8$, and equals $-8 \equiv 65 \pmod{73}$.

To sum up, all the roots of $f(x)$ are $9, 30, 32, 65, 67, 72$ modulo 73. Indeed,

$$x^6 + 17x^5 + 4x^4 + 67x^3 + 17x^2 + 4x + 66$$
$$\equiv (x-9)(x-30)(x-32)(x-65)(x-67)(x-72)$$
$$\equiv (x+1)(x+6)(x+8)(x+41)(x+43)(x+64) \pmod{73}.$$

It can be shown that the original polynomial factors in $\mathbb{F}_{73}[x]$ as

$$x^{15} + x^{12} + 2x^6 + 3x^4 + 6x^3 + 3x + 4$$
$$\equiv (x+1)^2(x+6)(x+8)(x+41)(x+43)(x+64) \times$$
$$(x^8 + 55x^7 + 11x^6 + 37x^5 + 5x^4 + 62x^3 + 62x^2 + 46x + 62) \pmod{73},$$

the last factor being irreducible of degree eight.

(2) Let us now consider the extension field $\mathbb{F}_9 = \mathbb{F}_3(\theta)$, where $\theta^2 + 1 = 0$. We plan to find the roots of $f(x) = x^5 + x^4 + \theta x^3 + x^2 + (\theta + 1)x + 2\theta \in \mathbb{F}_9[x]$. We first compute $g(x) \equiv x^9 - x \equiv x^4 + (2\theta + 2)x^3 + (2\theta + 1)x^2 + (\theta + 1)x + (\theta + 1) \pmod{f(x)}$, and replace $f(x)$ by $\gcd(f(x), g(x)) = x^2 + (\theta + 1)x + \theta$.

Our task is now to split $f(x) = x^2 + (\theta + 1)x + \theta$ by computing its gcd with the polynomial $(x + a)^{(9-1)/2} - 1 = (x + a)^4 - 1$ for randomly chosen $a \in \mathbb{F}_9$. Instead of computing $(x + a)^4 - 1$, we compute $g(x) \equiv (x + a)^4 - 1 \pmod{f(x)}$.

We first try $a = 1$. In this case, $g(x) \equiv (x + 1)^4 - 1 \equiv (\theta + 1)x + \theta \pmod{f(x)}$, which gives $\gcd(g(x), f(x)) = 1$, a trivial factor of $f(x)$.

Next, we try $a = \theta + 1$. We obtain $g(x) \equiv (x + \theta + 1)^4 - 1 \equiv 0 \pmod{f(x)}$, so that $\gcd(g(x), f(x)) = x^2 + (\theta + 1)x + \theta$ is again a trivial factor of $f(x)$.

Finally, for $a = 2\theta$, we get $g(x) \equiv (x + 2\theta)^4 - 1 \equiv (\theta + 1)x + (\theta + 2) \pmod{f(x)}$ for which $\gcd(g(x), f(x)) = x + \theta$, that is, $-\theta = 2\theta$ is a root of $f(x)$. The cofactor $f(x)/(x + \theta) = x + 1$ gives the other root as $-1 = 2$.

To sum up, the polynomial $x^5 + x^4 + \theta x^3 + x^2 + (\theta + 1)x + 2\theta \in \mathbb{F}_9[x]$ has the two roots $2, 2\theta$ in $\mathbb{F}_9$. Indeed, we have the factorization

$$x^5 + x^4 + \theta x^3 + x^2 + (\theta + 1)x + 2\theta = (x + 1)(x + \theta)^2(x^2 + \theta x + \theta),$$

the quadratic factor being irreducible in $\mathbb{F}_9[x]$. $\qquad\square$

3.2.2 Algorithm for Fields of Characteristic Two

For fields $\mathbb{F}_q$ of characteristic 2, we cannot factor $x^q - x$ as above, since $(q - 1)/2$ is not an integer. If $q = 2^n$, we instead use the decomposition

$$x^q - x = \left((x + a) + (x + a)^2 + (x + a)^4 + (x + a)^8 + \cdots + (x + a)^{2^{n-1}} \right) \times$$
$$\left(1 + (x + a) + (x + a)^2 + (x + a)^4 + (x + a)^8 + \cdots + (x + a)^{2^{n-1}} \right)$$

for any $a \in \mathbb{F}_q$. We, therefore, compute $g(x) \equiv (x + a) + (x + a)^2 + (x + a)^4 + (x + a)^8 + \cdots + (x + a)^{2^{n-1}} \pmod{f(x)}$ and the factor $f_1(x) = \gcd(g(x), f(x))$

of $f(x)$. If $0 < \deg f_1 < \deg f$, then $f_1(x)$ and the cofactor $f_2(x) = f(x)/f_1(x)$ are split recursively. Algorithm 3.3 elaborates this idea, and assumes that the input polynomial $f(x)$ is square-free and a product of linear factors.

Algorithm 3.3: FINDING ROOTS OF MONIC $f(x) \in \mathbb{F}_{2^n}[x]$

Assumption: $f(x)$ is a product of distinct linear factors.

```
If (deg f = 0), return the empty list.
If (deg f = 1), return the constant term of f(x).
Initialize the flag splitfound to zero.
While (splitfound is zero) {
      Choose α ∈ F_2n randomly.
      Set t(x) = x + α, and s(x) = t(x).
      for i = 1, 2, ..., n − 1 {
            Compute t(x) = t(x)² (mod f(x)), and s(x) = s(x) + t(x).
      }
      Compute f₁(x) = gcd(f(x), s(x)).
      if (0 < deg f₁ < deg f) {
            Set splitfound to one.
            Recursively call Algorithm 3.3 on f₁(x).
            Recursively call Algorithm 3.3 on f(x)/f₁(x).
      }
}
```

Example 3.11 Let us represent $\mathbb{F}_{16} = \mathbb{F}_2(\theta)$, where $\theta^4 + \theta + 1 = 0$. We plan to find the roots of $f(x) = x^{10} + (\theta^3 + 1)x^9 + (\theta + 1)x^5 + (\theta^3 + \theta^2 + 1)x^4 + \theta^3 x^3 + (\theta^3 + \theta^2 + 1)x^2 + \theta^2 x + \theta^3 \in \mathbb{F}_{16}[x]$. We compute $h(x) \equiv x^{16} + x \equiv (\theta^3 + 1)x^9 + (\theta^2 + 1)x^8 + (\theta^3 + \theta^2 + \theta + 1)x^7 + (\theta + 1)x^6 + (\theta + 1)x^5 + \theta x^4 + (\theta^3 + \theta^2 + \theta + 1)x^2 + \theta^3 x + (\theta^3 + \theta^2 + 1)$ (mod $f(x)$), and replace $f(x)$ by $\gcd(f(x), h(x)) = x^2 + (\theta^3 + \theta^2 + 1)x + (\theta^3 + \theta^2 + \theta + 1)$.

To find the two roots of $f(x)$, we try to split $f(x) = x^2 + (\theta^3 + \theta^2 + 1)x + (\theta^3 + \theta^2 + \theta + 1)$. For $\alpha = \theta + 1$, we have $g(x) \equiv (x + \alpha) + (x + \alpha)^2 + (x + \alpha)^4 + (x + \alpha)^8 \equiv \theta^2 x + (\theta^3 + \theta + 1)$ (mod $f(x)$). This produces the split of $f(x)$ into $f_1(x) = \gcd(f(x), g(x)) = x + (\theta^2 + \theta)$ and $f_2(x) = f(x)/f_1(x) = x + (\theta^3 + \theta + 1)$, yielding the roots $\theta^2 + \theta$ and $\theta^3 + \theta + 1$. □

Algorithm 3.3 looks attractive, but has problems and cannot be used without modifications. We discuss this again in Section 3.3.3. See Exercise 3.10 too.

3.2.3   Root Finding with GP/PARI

The GP/PARI interpreter provides the built-in function `polrootsmod()` for computing the roots of polynomials over prime fields. The function takes two arguments: a polynomial $f(x)$ with integer coefficients, and a prime p. The function returns a column vector consisting of the roots of $f(x)$ modulo p. The computation of `polrootsmod()` may fail for a non-prime modulus m. But then, one may ask `GP/PARI` to use a naive root-finding algorithm by providing a third argument with value 1. This is acceptable if m is small, since `polrootsmod()` does not encounter an error in this case, but may fail to produce all the roots. For small m, a better approach is to evaluate the polynomial at all elements of $\mathbb{Z}_m$. For the following snippet, notice that $731 = 17 \times 43$.

```
gp > polrootsmod(x^15 + x^12 + 2*x^6 + 3*x^4 + 6*x^3 + 3*x + 4, 73)
%1 = [Mod(9, 73), Mod(30, 73), Mod(32, 73), Mod(65, 73), Mod(67, 73), Mod(72, 73)]~
gp > polrootsmod(x^15 + x^12 + 2*x^6 + 3*x^4 + 6*x^3 + 3*x + 4, 731)
  ***   impossible inverse modulo: Mod(43, 731).
gp > polrootsmod(x^15 + x^12 + 2*x^6 + 3*x^4 + 6*x^3 + 3*x + 4, 731, 1)
%2 = [Mod(50, 731), Mod(424, 731)]~
gp > polrootsmod(x^2+3*x+2,731)
%3 = [Mod(730, 731)]~
gp > polrootsmod(x^2+3*x+2,731,1)
%4 = [Mod(84, 731), Mod(644, 731)]~
gp > findrootsmod(f,m) = for(x=0, m-1, if((eval(f)%m == 0), print1(x," ")))
gp > findrootsmod(x^2 + 3*x + 2, 731)
84 644 729 730
gp > findrootsmod(x^15 + x^12 + 2*x^6 + 3*x^4 + 6*x^3 + 3*x + 4, 731)
50 424 611 730
```

3.3 Factoring Polynomials over Finite Fields

The root-finding algorithm described above factors a product of linear polynomials. This idea can be extended to arrive at an algorithm for factoring arbitrary polynomials over finite fields. Indeed, the problem of factoring polynomials over finite fields is conceptually akin to root finding.

Berlekamp's Q-matrix method[3] is the first modern factoring algorithm (see Exercise 3.30). Cantor and Zassenhaus[4] propose a probabilistic algorithm of a different type. The algorithm I am going to describe is of the Cantor-Zassenhaus type, and is a simplified version (without the optimizations) of the algorithm of Von zur Gathen and Shoup.[5] Kaltofen and Shoup[6] propose the fastest known variant in the Cantor-Zassenhaus family of algorithms.

---

[3] Elwyn R. Berlekamp, Factoring polynomials over large finite fields, *Mathematics of Computation*, 24(111), 713–735, 1970.

[4] David G. Cantor and Hans Zassenhaus, A new algorithm for factoring polynomials over finite fields, *Mathematics of Computation*, 36(154), 587–592, 1981.

[5] Joachim von zur Gathen and Victor Shoup, Computing Frobenius maps and factoring polynomials, *STOC*, 97–105, 1992.

[6] Eric Kaltofen and Victor Shoup, Subquadratic-time factoring of polynomials over finite fields, *Mathematics of Computation*, 67(223), 1179–1197, 1998.

Our factorization algorithm has three stages described individually below.

3.3.1 Square-Free Factorization

Let $f(x) \in \mathbb{F}_q[x]$ be a non-constant polynomial to be factored. In this stage, one writes $f(x)$ as a product of square-free polynomials. In order to explain this stage, we define the *formal derivative* of

$$f(x) = \alpha_d x^d + \alpha_{d-1} x^{d-1} + \cdots + \alpha_1 x + \alpha_0$$

as the polynomial

$$f'(x) = d\alpha_d x^{d-1} + (d-1)\alpha_{d-1} x^{d-1} + \cdots + 2\alpha_2 x + \alpha_1.$$

Proposition 3.12 *The formal derivative $f'(x)$ of $f(x)$ is 0 if and only if $f(x) = g(x)^p$ for some $g(x) \in \mathbb{F}_q[x]$, where p is the characteristic of $\mathbb{F}_q$. The polynomial $f(x)/\gcd(f(x), f'(x))$ is square-free. In particular, $f(x)$ is square-free if and only if $\gcd(f(x), f'(x)) = 1$.*

Proof For proving the first assertion, let us rewrite $f(x)$ as

$$f(x) = \alpha_1 x^{e_1} + \alpha_2 x^{e_2} + \cdots + \alpha_k x^{e_k}$$

with pairwise distinct exponents $e_1, e_2, \ldots, e_k$, and with each $\alpha_i \neq 0$. Then,

$$f'(x) = e_1 \alpha_1 x^{e_1 - 1} + e_2 \alpha_2 x^{e_2 - 1} + \cdots + e_k \alpha_k x^{e_k - 1}.$$

The condition $f'(x) = 0$ implies that each e_i is a multiple of p (where $q = p^n$). Write $e_i = p\epsilon_i$ for $i = 1, 2, \ldots, k$. Since $\alpha_i \in \mathbb{F}_{p^n}$, we have $\alpha_i^{p^n} = \alpha_i$ for all i. It follows that $f(x) = g(x)^p$, where $g(x) = \alpha_1^{p^{n-1}} x^{\epsilon_1} + \alpha_2^{p^{n-1}} x^{\epsilon_2} + \cdots + \alpha_k^{p^{n-1}} x^{\epsilon_k}$. Conversely, if $f(x) = g(x)^p$, then $f'(x) = pg(x)^{p-1} g'(x) = 0$.

Let $f(x) = \alpha f_1(x)^{s_1} f_2(x)^{s_2} \cdots f_l(x)^{s_l}$ be the factorization of $f(x)$ into monic irreducible factors $f_i(x)$ with $s_i \geqslant 1$ for all i (and with $\alpha \in \mathbb{F}_q^*$). Then,

$$
\begin{aligned}
f'(x) &= \alpha s_1 f_1(x)^{s_1 - 1} f_1'(x) f_2(x)^{s_2} \cdots f_l(x)^{s_l} + \\
&\quad \alpha s_2 f_1(x)^{s_1} f_2(x)^{s_2 - 1} f_2'(x) \cdots f_l(x)^{s_l} + \\
&\quad \cdots + \alpha s_l f_1(x)^{s_1} f_2(x)^{s_2} \cdots f_l(x)^{s_l - 1} f_l'(x) \\
&= \alpha f_1(x)^{s_1 - 1} f_2(x)^{s_2 - 1} \cdots f_l(x)^{s_l - 1} \Big[s_1 f_1'(x) f_2(x) f_3(x) \cdots f_l(x) + \\
&\quad s_2 f_1(x) f_2'(x) f_3(x) \cdots f_l(x) + \cdots + s_l f_1(x) f_2(x) \cdots f_{l-1}(x) f_l'(x) \Big].
\end{aligned}
$$

The factor within square brackets is divisible by $f_i(x)$ if and only if $p|s_i$. Therefore, $\gcd(f(x), f'(x)) = \alpha f_1(x)^{t_1} f_2(x)^{t_2} \cdots f_l(x)^{t_l}$, where each $t_i = s_i$ or $s_i - 1$ according as whether $p|s_i$ or not. That is, $f(x)/\gcd(f(x), f'(x))$ is a divisor of $f_1(x) f_2(x) \cdots f_l(x)$ and is, therefore, square-free. ◁

This proposition shows us a way to compute the square-free factorization of $f(x)$. We first compute $f'(x)$. If $f'(x) = 0$, we compute $g(x) \in \mathbb{F}_q[x]$ satisfying

$f(x) = g(x)^p$, and recursively compute the square-free factorization of $g(x)$. If $f'(x) \neq 0$, we compute $h(x) = \gcd(f(x), f'(x))$. If $h(x) = 1$, then $f(x)$ is itself square-free. Otherwise, we output the square-free factor $f(x)/h(x)$ of $f(x)$, and recursively compute the square-free factorization of $h(x)$.

Example 3.13 Let us compute the square-free factorization of $f(x) = x^{16} + x^8 + x^6 + x^4 + x^2 + 1 \in \mathbb{F}_2[x]$. We have $f'(x) = 0$, so $f(x) = g(x)^2$, where $g(x) = x^8 + x^4 + x^3 + x^2 + x + 1$. Now, $g'(x) = x^2 + 1 \neq 0$, and $h(x) = \gcd(g(x), g'(x)) = x^2 + 1$, that is, $g(x)/h(x) = x^6 + x^4 + x + 1$ is square-free. We compute the square-free factorization of $h(x) = x^2 + 1$. We have $h'(x) = 0$. Indeed, $h(x) = (x+1)^2$, where $x+1$, being coprime to its derivative, is square-free. Thus, the square-free factorization of $f(x)$ is

$$f(x) = (x^6 + x^4 + x + 1)(x^6 + x^4 + x + 1)(x + 1)(x + 1)(x + 1)(x + 1). \quad \square$$

3.3.2 Distinct-Degree Factorization

In view of square-free factorization, we may assume that the polynomial $f(x) \in \mathbb{F}_q[x]$ to be factored is monic and square-free. By Theorem 3.1, the polynomial $x^{q^r} - x$ is the product of all monic irreducible polynomials of $\mathbb{F}_q[x]$ of degrees dividing r. Therefore, $f_r(x) = \gcd(f(x), x^{q^r} - x)$ is the product of all irreducible factors of $f(x)$ of degrees dividing r. If $f(x)$ does not contain factors of degrees smaller than r, then $f_r(x)$ is equal to the product of all irreducible factors of $f(x)$ of degree equal to r. This leads to Algorithm 3.4 for decomposing $f(x)$ into the product $f_1(x)f_2(x) \cdots f_r(x) \cdots$. At the r-th iteration, $f_r(x) = \gcd(f(x), x^{q^r} - x)$ is computed, and $f(x)$ is replaced by $f(x)/f_r(x)$ so that irreducible factors of degree r are eliminated from $f(x)$. It is convenient to compute the polynomial $x^{q^r} - x$ modulo $f(x)$.

Algorithm 3.4: DISTINCT-DEGREE FACTORIZATION OF MONIC SQUARE-FREE $f(x) \in \mathbb{F}_q[x]$

Initialize $r = 0$ and $g(x) = x$.
While $(f(x) \neq 1)$ {
    Increment r.
    Compute $g(x) = g(x)^q \pmod{f(x)}$.
    Compute $f_r(x) = \gcd(f(x), g(x) - x)$.
    Output $(f_r(x), r)$.
    If $(f_r(x) \neq 1)$ { Set $f(x) = f(x)/f_r(x)$ and $g(x) = g(x) \operatorname{rem} f(x)$. }
}

Example 3.14 (1) Let us compute the distinct-degree factorization of $f(x) = x^{20} + x^{17} + x^{15} + x^{11} + x^{10} + x^9 + x^5 + x^3 + 1 \in \mathbb{F}_2[x]$. One can check that $\gcd(f(x), f'(x)) = 1$, that is, $f(x)$ is indeed square-free. We compute $f_1(x) = \gcd(f(x), x^2 + x) = 1$, that is, $f(x)$ has no linear factors. We then compute

$f_2(x) = \gcd(f(x), x^4 + x) = 1$, that is, $f(x)$ does not contain any quadratic factors also. Since $f_3(x) = \gcd(f(x), x^8 + x) = x^6 + x^5 + x^4 + x^3 + x^2 + x + 1$, we discover that $f(x)$ has two irreducible cubic factors. We replace $f(x)$ by $f(x)/f_3(x) = x^{14} + x^{13} + x^{11} + x^{10} + x^9 + x^8 + x^7 + x^6 + x^5 + x^4 + x^3 + x + 1$. Now, we compute $x^{16} + x \equiv x^{11} + x^{10} + x^9 + x^8 + x^7 + x^6 + x^5 + x + 1 \pmod{f(x)}$, where $f(x)$ is the reduced polynomial of degree 14 mentioned above. Since $f_4(x) = \gcd(f(x), (x^{16} + x) \pmod{f(x)}) = x^4 + x^3 + x^2 + x + 1$, $f(x)$ contains a single irreducible factor of degree four, and we replace $f(x)$ by $f(x)/f_4(x) = x^{10} + x^8 + x^7 + x^5 + x^3 + x^2 + 1$. We subsequently compute $x^{32} + x \equiv 0 \pmod{f(x)}$, that is, $f_5(x) = \gcd(f(x), (x^{32} + x) \pmod{f(x)}) = x^{10} + x^8 + x^7 + x^5 + x^3 + x^2 + 1$, that is, $f(x)$ has two irreducible factors of degree five, and we replace $f(x)$ by $f(x)/f_5(x) = 1$, and the distinct-degree factorization loop terminates. We have, therefore, obtained the following factorization of $f(x)$.

$$
\begin{aligned}
f_1(x) &= 1, \\
f_2(x) &= 1, \\
f_3(x) &= x^6 + x^5 + x^4 + x^3 + x^2 + x + 1, \\
f_4(x) &= x^4 + x^3 + x^2 + x + 1, \\
f_5(x) &= x^{10} + x^8 + x^7 + x^5 + x^3 + x^2 + 1.
\end{aligned}
$$

We need to factor $f_3(x)$ and $f_5(x)$ in order to obtain the complete factorization of $f(x)$. This is accomplished in the next stage.

(2) Now, let $f(x) = x^{10} + (\theta + 1)x^9 + (\theta + 1)x^8 + x^7 + \theta x^6 + (\theta + 1)x^4 + \theta x^3 + \theta x^2 + (\theta + 1)x \in \mathbb{F}_4[x]$, where $\mathbb{F}_4 = \mathbb{F}_2(\theta)$ with $\theta^2 + \theta + 1 = 0$. We first compute $f_1(x) = \gcd(f(x), x^4 + x) = x^2 + \theta x$, and replace $f(x)$ by $f(x)/f_1(x) = x^8 + x^7 + x^6 + (\theta + 1)x^5 + (\theta + 1)x^4 + x^3 + x^2 + \theta$. We then compute $g(x) \equiv x^{16} + x \equiv x^7 + x^6 + (\theta + 1)x^5 + \theta x^4 + \theta x^3 + x^2 + \theta x + 1 \pmod{f(x)}$ and subsequently $f_2(x) = \gcd(f(x), g(x)) = x^2 + x + \theta$. We then replace $f(x)$ by $f(x)/f_2(x) = x^6 + (\theta + 1)x^4 + \theta x^2 + (\theta + 1)x + 1$. At the next iteration, we compute $x^{64} + x \equiv 0 \pmod{f(x)}$, that is, $f_3(x) = \gcd(f(x), 0) = x^6 + (\theta + 1)x^4 + \theta x^2 + (\theta + 1)x + 1$, and $f(x)$ is replaced by $f(x)/f_3(x) = 1$. Thus, the distinct-degree factorization of $f(x)$ is as follows.

$$
\begin{aligned}
f_1(x) &= x^2 + \theta x, \\
f_2(x) &= x^2 + x + \theta, \\
f_3(x) &= x^6 + (\theta + 1)x^4 + \theta x^2 + (\theta + 1)x + 1.
\end{aligned}
$$

That is, $f(x)$ has two linear, one quadratic and two cubic factors.   □

3.3.3   Equal-Degree Factorization

In the last stage of factorization, we assume that the polynomial $f(x) \in \mathbb{F}_q[x]$ to be factored is square-free and is a product of monic irreducible factors of the same known degree r. If $\deg f = d$, then $f(x)$ is a product of d/r irreducible factors. Our task is to find all the factors of $f(x)$.

In this stage, we use a strategy similar to the root-finding algorithm. Recall that $x^{q^r} - x$ is the product of all monic irreducible polynomials of $\mathbb{F}_q[x]$ of degrees dividing r. If we can split $x^{q^r} - x$ non-trivially, computing the gcd of $f(x)$ with a non-trivial factor of $x^{q^r} - x$ may yield a non-trivial factor of $f(x)$. We then recursively factor this factor of $f(x)$ and the corresponding cofactor.

First, let q be odd. For any $\alpha \in \mathbb{F}_q$, we have

$$x^{q^r} - x = (x + \alpha)^{q^r} - (x + \alpha)$$
$$= (x + \alpha) \left((x + \alpha)^{(q^r - 1)/2} - 1 \right) \left((x + \alpha)^{(q^r - 1)/2} + 1 \right).$$

So $g(x) \equiv \left((x + \alpha)^{(q^r - 1)/2} - 1 \right) \pmod{f(x)}$ and $f_1(x) = \gcd(f(x), g(x))$ are computed. If $0 < \deg f_1 < \deg f$, we recursively split $f_1(x)$ and $f(x)/f_1(x)$.

Algorithm 3.5: FACTORING $f(x) \in \mathbb{F}_q[x]$, A PRODUCT OF IRREDUCIBLE FACTORS OF DEGREE r, WHERE q IS ODD

```
if (deg f = 0), return.
if (deg f = r) { Output f(x). Return. }
Initialize the flag splitfound to zero.
While (splitfound is zero) {
      Choose α ∈ F_q randomly.
      Compute g(x) = (x + α)^((q^r−1)/2) − 1 (mod f(x)).
      Compute f₁(x) = gcd(f(x), g(x)).
      If (0 < deg f₁ < deg f) {
            Set splitfound to one.
            Recursively call Algorithm 3.5 on f₁(x).
            Recursively call Algorithm 3.5 on f(x)/f₁(x).
      }
}
```

Example 3.15 Let us factor $f(x) = x^9 + 3x^8 + 3x^7 + 2x^6 + 2x + 2 \in \mathbb{F}_5[x]$. It is given that $f(x)$ is the product of three cubic irreducible polynomials of $\mathbb{F}_5[x]$. For $\alpha = 2 \in \mathbb{F}_q$, we have $g(x) \equiv (x+2)^{(5^3-1)/2} - 1 \equiv (x+2)^{62} - 1 \equiv 3x^8 + 3x^7 + x^6 + 2x^5 + 3x^4 + 3x^3 + 3x^2 + x + 4 \pmod{f(x)}$, and $f_1(x) = \gcd(f(x), g(x)) = x^6 + 2x^5 + x^4 + 4x^3 + 2x^2 + x + 1$. The cofactor $f(x)/f_1(x) = x^3 + x^2 + 2$ is an irreducible factor of $f(x)$. The other two factors are obtained from $f_1(x)$.

For $\alpha = 0$, $g_1(x) \equiv x^{62} - 1 \equiv 0 \pmod{f_1(x)}$, that is, $\gcd(f_1(x), g_1(x)) = f_1(x)$ is a trivial factor of $f_1(x)$.

For $\alpha = 3$, we obtain $g_1(x) \equiv (x + 3)^{62} - 1 \equiv 4x^4 + 4x^2 + 4x \pmod{f_1(x)}$, and $\gcd(f_1(x), g_1(x)) = x^3 + x + 1$ is an irreducible factor of $f_1(x)$. The other factor of $f_1(x)$ is $f_1(x)/(x^3 + x + 1) = x^3 + 2x^2 + 1$. Thus, f factors as

$$x^9 + 3x^8 + 3x^7 + 2x^6 + 2x + 2 = (x^3 + x + 1)(x^3 + x^2 + 2)(x^3 + 2x^2 + 1). \quad \square$$

If $q = 2^n$, the polynomial $x^{q^r} - x = x^{2^{nr}} + x$ factors for any $\alpha \in \mathbb{F}_q$ as

$$x^{2^{nr}} + x = (x + \alpha)^{2^{nr}} + (x + \alpha)$$
$$= \left((x + \alpha) + (x + \alpha)^2 + (x + \alpha)^{2^2} + (x + \alpha)^{2^3} + \cdots + (x + \alpha)^{2^{nr-1}} \right) \times$$
$$\left(1 + (x + \alpha) + (x + \alpha)^2 + (x + \alpha)^{2^2} + (x + \alpha)^{2^3} + \cdots + (x + \alpha)^{2^{nr-1}} \right).$$

We compute $g(x) \equiv (x+\alpha) + (x+\alpha)^2 + (x+\alpha)^{2^2} + (x+\alpha)^{2^3} + \cdots + (x+\alpha)^{2^{nr-1}}$ (mod $f(x)$) and $f_1(x) = \gcd(f(x), g(x))$. If $0 < \deg f_1 < \deg f$, we recursively split $f_1(x)$ and $f(x)/f_1(x)$. The steps are explained in Algorithm 3.6.

Algorithm 3.6: FACTORING $f(x) \in \mathbb{F}_{2^n}[x]$, A PRODUCT OF IRREDUCIBLE FACTORS OF DEGREE r

---

```
If (deg f = 0) return.
If (deg f = r) return f(x).
Set the flag splitfound to zero.
While (splitfound is zero) {
    Choose α ∈ F_2n randomly.
    Set t(x) = x + α, and s(x) = t(x).
    For i = 1, 2, ..., nr - 1 {
        Set t(x) = t(x)² (mod f(x)), and s(x) = s(x) + t(x).
    }
    Compute f₁(x) = gcd(f(x), s(x)).
    if (0 < deg f₁ < deg f) {
        Set splitfound to one.
        Recursively call Algorithm 3.6 on f₁(x).
        Recursively call Algorithm 3.6 on f(x)/f₁(x).
    }
}
```

Example 3.16 Let us represent $\mathbb{F}_4 = \mathbb{F}_2(\theta)$ with $\theta^2 + \theta + 1 = 0$, and factor $f(x) = x^6 + \theta x^5 + (\theta+1)x^4 + x^3 + \theta \in \mathbb{F}_4[x]$. It is given that $f(x)$ is the product of two cubic irreducible polynomials. For $\alpha = \theta$, $g(x) \equiv (x + \theta) + (x + \theta)^2 + (x + \theta)^4 + (x + \theta)^8 + (x + \theta)^{16} + (x + \theta)^{32} \equiv \theta x^4 + \theta x^2 + \theta x + 1$ (mod $f(x)$), and $\gcd(f(x), g(x)) = x^3 + \theta x^2 + \theta x + \theta$ is a factor of $f(x)$. The other factor is $f(x)/(x^3 + \theta x^2 + \theta x + \theta) = x^3 + x + 1$. Thus, we get the factorization

$$x^6 + \theta x^5 + (\theta + 1)x^4 + x^3 + \theta = (x^3 + x + 1)(x^3 + \theta x^2 + \theta x + \theta). \qquad \square$$

A big danger is waiting for us. In certain situations, Algorithms 3.5 and 3.6 fail to split $f(x)$ completely into irreducible factors. The most conceivable case is when q is small compared to the number d/r of irreducible factors of $f(x)$. Since there are approximately q^r/r monic irreducible polynomials of degree r in $\mathbb{F}_q[x]$, this situation can indeed arise. That is, if q is small compared to d/r, it may so happen that all elements α of $\mathbb{F}_q$ are tried in the equal-degree factorization algorithm, but a complete split of $f(x)$ is not achieved.

Example 3.17 (1) Let us try to split $f(x) = x^{16} + 2x^{15} + x^{14} + x^{13} + 2x^{12} + x^{11} + x^{10} + 2x^6 + x^5 + 2x^4 + 2x^2 + 2x + 1 \in \mathbb{F}_3[x]$ which is known to be a product of four irreducible factors each of degree four. We try all three values of $\alpha \in \mathbb{F}_3$ in Algorithm 3.5. The splits obtained are listed below.

α	$g_\alpha(x) \equiv x^{40} - 1 \pmod{f(x)}$	$f_{\alpha 1}(x) = \gcd(f(x), g_\alpha(x))$	$f_{\alpha 2}(x) = f(x)/f_{\alpha 1}(x)$
0	0	$f(x)$	1
1	$2x^{13} + x^{12} + x^8 + 2x^7 + 2x^6 + 2x^5 + 2x^4 + 2x^3 + 2x + 1$	$x^4 + x^2 + 2x + 1$	$x^{12} + 2x^{11} + 2x^7 + x^6 + x^5 + x^4 + x^3 + x^2 + 1$
2	$x^{14} + 2x^{13} + 2x^{12} + 2x^{11} + 2x^8 + 2x^7 + 2x^5 + x^3 + x^2 + x$	$x^8 + x^7 + 2x^6 + 2x^5 + x^4 + 2x^3 + 2x^2 + x + 1$	$x^8 + x^7 + x^6 + 2x^5 + x^4 + 2x^2 + x + 1$

Thus, we have discovered the irreducible factor $f_{11}(x) = x^4 + x^2 + 2x + 1$ of $f(x)$. If we plan to split $f_{12}(x)$, we get trivial factorizations for $\alpha = 0, 1$. For $\alpha = 2$, we obtain the non-trivial split $f_{12}(x) = (x^4 + x^3 + 2x + 1)f_{21}(x)$ which reveals yet another irreducible factor of $f(x)$.

We cannot split $f_{21}(x)$ non-trivially for all values of $\alpha \in \{0, 1, 2\}$. On the other hand, $\alpha = 1$ splits $f_{22}(x)$ into known factors $x^4 + x^2 + 2x + 1$ and $x^4 + x^3 + 2x + 1$. Thus, Algorithm 3.5 discovers only the partial factorization

$$
\begin{aligned}
f(x) &= x^{16} + 2x^{15} + x^{14} + x^{13} + 2x^{12} + x^{11} + x^{10} + 2x^6 + x^5 + \\
&\quad 2x^4 + 2x^2 + 2x + 1 \\
&= (x^4 + x^2 + 2x + 1)(x^4 + x^3 + 2x + 1) \times \\
&\quad (x^8 + x^7 + 2x^6 + 2x^5 + x^4 + 2x^3 + 2x^2 + x + 1),
\end{aligned}
$$

since it fails to split the factor of degree eight for all values of $\alpha \in \mathbb{F}_3$.

(2) Let us try to factor $f(x) = x^{15} + x^7 + x^3 + x + 1 \in \mathbb{F}_2[x]$ using Algorithm 3.6. It is given that $f(x)$ is the product of three irreducible polynomials each of degree five. We compute $g(x) \equiv (x + \alpha) + (x + \alpha)^2 + (x + \alpha)^4 + (x + \alpha)^8 + (x + \alpha)^{16} \pmod{f(x)}$ for $\alpha = 0, 1$. For $\alpha = 0$ we get $g(x) = 0$ so that $\gcd(f(x), g(x)) = f(x)$, whereas for $\alpha = 1$ we get $g(x) = 1$ so that $\gcd(f(x), g(x)) = 1$. So Algorithm 3.6 fails to split $f(x)$ at all.

(3) Splitting may fail even in a case where the field can supply more elements than the number of irreducible factors of $f(x)$. By Example 3.14(2), $f(x) = x^6 + (\theta + 1)x^4 + \theta x^2 + (\theta + 1)x + 1 \in \mathbb{F}_4[x]$ is a product of two cubic irreducible factors, where $\mathbb{F}_4 = \mathbb{F}_2(\theta)$ with $\theta^2 + \theta + 1 = 0$. The values of $g(x) \equiv (x+\alpha) + (x+\alpha)^2 + (x+\alpha)^4 + (x+\alpha)^8 + (x+\alpha)^{16} + (x+\alpha)^{32} \pmod{f(x)}$ are listed below for all $\alpha \in \mathbb{F}_4$. No value of $\alpha \in \mathbb{F}_4$ can split $f(x)$ non-trivially.

α	$g(x)$	$\gcd(f(x), g(x))$
0	0	$f(x)$
1	0	$f(x)$
θ	1	1
$\theta + 1$	1	1

□

The root-finding algorithms of Section 3.2 may encounter similar failures, particularly when the underlying field $\mathbb{F}_q$ is small. However, any input polynomial cannot have more than q roots in $\mathbb{F}_q$, so failure cases are relatively uncommon. Moreover, we can evaluate the input polynomial at all elements of $\mathbb{F}_q$. This is feasible (may even be desirable) if q is small.

We can modify Algorithms 3.5 and 3.6 so that this difficulty is removed. (A similar modification applies to root-finding algorithms too.) Let $u(x) = u_0 + u_1 x + u_2 x^2 + \cdots + u_s x^s$ be a non-constant polynomial in $\mathbb{F}_q[x]$. We have $u(x)^{q^r} = u_0 + u_1 x^{q^r} + u_2 x^{2q^r} + \cdots + u_s x^{sq^r}$, so that $u(x)^{q^r} - u(x) = u_1(x^{q^r} - x) + u_2(x^{2q^r} - x^2) + \cdots + u_s(x^{sq^r} - x^s)$. Since $x^{\sigma q^r} - x^\sigma$ is divisible by $x^{q^r} - x$ for all $\sigma \in \mathbb{N}$, the polynomial $u(x)^{q^r} - u(x)$ contains as factors all monic irreducible polynomials of $\mathbb{F}_q[x]$ of degree r (together with other irreducible factors, in general). If q is odd, we have

$$u(x)^{q^r} - u(x) = u(x) \left(u(x)^{(q^r-1)/2} - 1 \right) \left(u(x)^{(q^r-1)/2} + 1 \right).$$

Therefore, $\gcd(f(x), u(x)^{(q^r-1)/2} - 1)$ is potentially a non-trivial factor of $f(x)$. On the other hand, if $q = 2^n$, we have

$$u(x)^{2^{nr}} + u(x) = \left(u(x) + u(x)^2 + u(x)^{2^2} + u(x)^{2^3} + \cdots + u(x)^{2^{nr-1}} \right) \times$$
$$\left(1 + u(x) + u(x)^2 + u(x)^{2^2} + u(x)^{2^3} + \cdots + u(x)^{2^{nr-1}} \right).$$

That is, $\gcd(f(x), u(x) + u(x)^2 + u(x)^{2^2} + u(x)^{2^3} + \cdots + u(x)^{2^{nr-1}})$ is potentially a non-trivial factor of $f(x)$. Algorithms 3.7 and 3.8 incorporate these modifications. Algorithms 3.5 and 3.6 are indeed special cases of these algorithms corresponding to $u(x) = x + \alpha$ for $\alpha \in \mathbb{F}_q$. A convenient way of choosing the polynomial $u(x)$ in Algorithm 3.8 is discussed in Exercise 3.10.

Algorithm 3.7: MODIFICATION OF ALGORITHM 3.5 FOR EQUAL-DEGREE FACTORIZATION OF $f(x) \in \mathbb{F}_q[x]$ WITH q ODD

```
If (deg f = 0), return.
If (deg f = r), return f(x).
Set the flag splitfound to zero.
While (splitfound is zero) {
      Choose random non-constant u(x) ∈ F_q[x] of small degree.
      Compute g(x) = u(x)^(q^r−1)/2 − 1 (mod f(x)).
      Compute f₁(x) = gcd(f(x), g(x)).
      If (0 < deg f₁ < deg f) {
            Set splitfound to one.
            Recursively call Algorithm 3.7 on f₁(x).
            Recursively call Algorithm 3.7 on f(x)/f₁(x).
      }
}
```

Algorithm 3.8: MODIFICATION OF ALGORITHM 3.6 FOR EQUAL-DEGREE FACTORIZATION OF $f(x) \in \mathbb{F}_{2^n}[x]$

```
If (deg f = 0), return.
If (deg f = r), return f(x).
Set the flag splitfound to zero.
While (splitfound is zero) {
        Choose random non-constant u(x) ∈ F₂ⁿ[x] of small degree.
        Set s(x) = u(x).
        for i = 1, 2, ..., nr − 1 {
                Compute u(x) = u(x)² (mod f(x)), and s(x) = s(x) + u(x).
        }
        Compute f₁(x) = gcd(f(x), s(x)).
        If (0 < deg f₁ < deg f) {
                Set splitfound to one.
                Recursively call Algorithm 3.8 on f₁(x).
                Recursively call Algorithm 3.8 on f(x)/f₁(x).
        }
}
```

Example 3.18 Let us now handle the failed attempts of Example 3.17.

(1) We factor $f(x) = x^8 + x^7 + 2x^6 + 2x^5 + x^4 + 2x^3 + 2x^2 + x + 1 \in \mathbb{F}_3[x]$ which is known to be a product of two irreducible factors of degree four. Choose $u(x) = x^2 + x + 2$ for which $g(x) \equiv u(x)^{40} - 1 \equiv 2x^7 + 2x^5 + x^4 + 2x^3 + 2x^2 + 2x + 2 \pmod{f(x)}$. This yields the non-trivial factor $\gcd(f(x), g(x)) = x^4 + x^2 + x + 1$. The other factor of $f(x)$ is $f(x)/(x^4 + x^2 + x + 1) = x^4 + x^3 + x^2 + 1$. Therefore, we have the equal-degree factorization

$$
\begin{aligned}
f(x) &= x^8 + x^7 + 2x^6 + 2x^5 + x^4 + 2x^3 + 2x^2 + x + 1 \\
&= (x^4 + x^2 + x + 1)(x^4 + x^3 + x^2 + 1).
\end{aligned}
$$

(2) Now, we try to factor $f(x) = x^{15} + x^7 + x^3 + x + 1 \in \mathbb{F}_2[x]$ which is known to be the product of three irreducible polynomials each of degree five. Choose $u(x) = x^3 + 1$. This gives $g(x) \equiv u(x) + u(x)^2 + u(x)^4 + u(x)^8 + u(x)^{16} \equiv x^{10} + x^8 + x^6 + x^5 + x^4 + x + 1 \pmod{f(x)}$, and we obtain the split of $f(x)$ into $f_1(x) = \gcd(f(x), g(x)) = x^{10} + x^8 + x^6 + x^5 + x^4 + x + 1$, and the cofactor $f(x)/f_1(x) = x^5 + x^3 + 1$ which is already an irreducible factor of $f(x)$.

It now remains to split $f_1(x)$. Polynomials $u(x)$ of degrees $1, 2, 3, 4$ fail to split $f_1(x)$. However, the choice $u(x) = x^5 + 1$ works. We obtain $g_1(x) \equiv u(x) + u(x)^2 + u(x)^4 + u(x)^8 + u(x)^{16} \equiv x^9 + x^8 + x^7 + x^5 + x^2 + x \pmod{f_1(x)}$, and the factors of $f_1(x)$ are revealed as $\gcd(f_1(x), g_1(x)) = x^5 + x^2 + 1$ and $f_1(x)/(x^5 + x^2 + 1) = x^5 + x^3 + x^2 + x + 1$. To sum up, we have

$$f(x) = x^{15} + x^7 + x^3 + x + 1 = (x^5 + x^2 + 1)(x^5 + x^3 + 1)(x^5 + x^3 + x^2 + x + 1).$$

(3)   We represent $\mathbb{F}_4 = \mathbb{F}_2(\theta)$ with $\theta^2 + \theta + 1 = 0$, and try to factor $f(x) = x^6 + (\theta + 1)x^4 + \theta x^2 + (\theta + 1)x + 1 \in \mathbb{F}_4[x]$ which is a product of two cubic irreducible polynomials. All polynomials $u(x)$ of degrees $1, 2, 3, 4$ fail to split $f(x)$ non-trivially. The choice $u(x) = x^5$ produces a non-trivial split. We have $g(x) \equiv u(x) + u(x)^2 + u(x)^4 + u(x)^8 + u(x)^{16} + u(x)^{32} \equiv \theta x^5 + x^3 + \theta x^2 + (\theta+1)x+(\theta+1) \pmod{f(x)}$ which yields the factors $\gcd(f(x), g(x)) = x^3 + x + 1$ and $f(x)/(x^3 + x + 1) = x^3 + \theta x + 1$ of $f(x)$.   □

The root-finding and factoring algorithms for polynomials over finite fields, as discussed above, are randomized. The best deterministic algorithms known for these problems have running times fully exponential in $\log q$ (the size of the underlying field). The best known deterministic algorithm for factoring a polynomial of degree d in $\mathbb{F}_q[x]$ is from Shoup[7], and is shown by Shparlinski[8] to run in $O(q^{1/2}(\log q)d^{2+\epsilon})$ time, where d^ϵ stands for a polynomial in $\log d$. Computations over finite fields exploit randomization very effectively.

3.3.4   Factoring Polynomials in GP/PARI

The generic factoring function in GP/PARI is factor(). One may supply a polynomial over some prime field as its only argument. The function returns the irreducible factors of the polynomial together with the multiplicity of each factor. One may alternatively use the function factormod() which takes two arguments: a polynomial $f(x) \in \mathbb{Z}[x]$ and a prime modulus p.

```
gp > factor(Mod(1,2)*x^15 + Mod(1,2)*x^7 + Mod(1,2)*x^3 + Mod(1,2)*x + Mod(1,2))
%1 =
[Mod(1, 2)*x^5 + Mod(1, 2)*x^2 + Mod(1, 2) 1]

[Mod(1, 2)*x^5 + Mod(1, 2)*x^3 + Mod(1, 2) 1]

[Mod(1, 2)*x^5 + Mod(1, 2)*x^3 + Mod(1, 2)*x^2 + Mod(1, 2)*x + Mod(1, 2) 1]

gp > factor(Mod(1,2)*x^16 + Mod(1,2)*x^8 + Mod(1,2)*x^6 + Mod(1,2)*x^4 + \
     Mod(1,2)*x^2 + Mod(1,2))
%2 =
[Mod(1, 2)*x + Mod(1, 2) 6]

[Mod(1, 2)*x^2 + Mod(1, 2)*x + Mod(1, 2) 2]

[Mod(1, 2)*x^3 + Mod(1, 2)*x + Mod(1, 2) 2]

gp > factor(Mod(1,3)*x^16 + Mod(2,3)*x^15 + Mod(1,3)*x^14 + Mod(1,3)*x^13 + \
     Mod(2,3)*x^12 + Mod(1,3)*x^11 + Mod(1,3)*x^10 + Mod(2,3)*x^6 + \
     Mod(1,3)*x^5 + Mod(2,3)*x^4 + Mod(2,3)*x^2 + Mod(2,3)*x + Mod(1,3))
%3 =
```

[7]Victor Shoup, On the deterministic complexity of factoring polynomials over finite fields, *Information Processing Letters*, 33, 261–267, 1990.
[8]Igor E. Shparlinski, *Computational problems in finite fields*, Kluwer, 1992.

```
[Mod(1, 3)*x^4 + Mod(1, 3)*x^2 + Mod(1, 3)*x + Mod(1, 3) 1]

[Mod(1, 3)*x^4 + Mod(1, 3)*x^2 + Mod(2, 3)*x + Mod(1, 3) 1]

[Mod(1, 3)*x^4 + Mod(1, 3)*x^3 + Mod(2, 3)*x + Mod(1, 3) 1]

[Mod(1, 3)*x^4 + Mod(1, 3)*x^3 + Mod(1, 3)*x^2 + Mod(1, 3) 1]

gp > factormod(x^20 + x^17 + x^15 + x^11 + x^10 + x^9 + x^5 + x^3 + 1, 2)
%4 =
[Mod(1, 2)*x^3 + Mod(1, 2)*x + Mod(1, 2) 1]

[Mod(1, 2)*x^3 + Mod(1, 2)*x^2 + Mod(1, 2) 1]

[Mod(1, 2)*x^4 + Mod(1, 2)*x^3 + Mod(1, 2)*x^2 + Mod(1, 2)*x + Mod(1, 2) 1]

[Mod(1, 2)*x^5 + Mod(1, 2)*x^2 + Mod(1, 2) 1]

[Mod(1, 2)*x^5 + Mod(1, 2)*x^3 + Mod(1, 2) 1]

gp > factormod(x^9 + 3*x^8 + 3*x^7 + 2*x^6 + 2*x + 2, 5)
%5 =
[Mod(1, 5)*x^3 + Mod(1, 5)*x + Mod(1, 5) 1]

[Mod(1, 5)*x^3 + Mod(1, 5)*x^2 + Mod(2, 5) 1]

[Mod(1, 5)*x^3 + Mod(2, 5)*x^2 + Mod(1, 5) 1]

gp > factormod(x^15 + x^12 + 2*x^6 + 3*x^4 + 6*x^3 + 3*x + 4, 73)
%6 =
[Mod(1, 73)*x + Mod(1, 73) 2]

[Mod(1, 73)*x + Mod(6, 73) 1]

[Mod(1, 73)*x + Mod(8, 73) 1]

[Mod(1, 73)*x + Mod(41, 73) 1]

[Mod(1, 73)*x + Mod(43, 73) 1]

[Mod(1, 73)*x + Mod(64, 73) 1]

[Mod(1, 73)*x^8 + Mod(55, 73)*x^7 + Mod(11, 73)*x^6 + Mod(37, 73)*x^5 + Mod(5, 7
3)*x^4 + Mod(62, 73)*x^3 + Mod(62, 73)*x^2 + Mod(46, 73)*x + Mod(62, 73) 1]
```

Both `factor()` and `factormod()` allow working in $\mathbb{Z}_m[x]$ for a composite m. The factorization attempt of GP/PARI may fail in that case. Although it makes sense to find roots of polynomials in $\mathbb{Z}_m[x]$, factorization in $\mathbb{Z}_m[x]$ does not make good sense, for $\mathbb{Z}_m[x]$ is not a unique factorization domain (not even an integral domain). For example, $x^2 + 7 \equiv (x+1)(x+7) \equiv (x+3)(x+5) \pmod{8}$.

```
gp > factor(Mod(1,99)*x^2 + Mod(20,99)*x + Mod(1,99))
%7 =
```

```
[Mod(1, 99)*x + Mod(10, 99) 2]

gp > factormod(x^2 + 20*x + 1, 99)
%8 =
[Mod(1, 99)*x + Mod(10, 99) 2]

gp > factor(Mod(1,100)*x^2 + Mod(20,100)*x + Mod(1,100))
  ***    impossible inverse modulo: Mod(2, 100).
gp > factormod(x^2 + 20*x + 1,100)
  ***    impossible inverse modulo: Mod(2, 100).
```

GP/PARI also provides facilities for factoring polynomials over extension fields $\mathbb{F}_q = \mathbb{F}_{p^n}$. The relevant function is factorff() which takes three arguments: the first argument is the polynomial $f(x) \in \mathbb{F}_q[x]$ to be factored, the second is the characteristic p of $\mathbb{F}_q$, and the third is the irreducible polynomial $\iota(\theta)$ used to represent $\mathbb{F}_q$ as an extension of $\mathbb{F}_p$. Let us use t in order to denote the element θ (a root of ι) that is adjoined to $\mathbb{F}_p$ for representing the extension $\mathbb{F}_q$. We use the variable x for the polynomial $f(x) \in \mathbb{F}_q[x]$ to be factored.

Here is an example (see Example 3.18(3)). Let us represent $\mathbb{F}_4 = \mathbb{F}_2(\theta)$, where $\theta^2 + \theta + 1 = 0$. The second argument to factorff() will then be 2, and the third argument will be $\iota(t) = t^2 + t + 1$. Suppose that we want to factor $f(x) = x^6 + (\theta + 1)x^4 + \theta x^2 + (\theta + 1)x + 1 \in \mathbb{F}_4[x]$. Thus, we should pass $x^6 + (t+1)x^4 + tx^2 + (t+1)x + 1$ as the first argument to factorff().

```
gp > lift(factorff(x^6 + (t+1)*x^4 + t*x^2 + (t+1)*x + 1, 2, t^2 + t + 1))
%9 =
[Mod(1, 2)*x^3 + Mod(1, 2)*x + Mod(1, 2) 1]

[Mod(1, 2)*x^3 + (Mod(1, 2)*t)*x + Mod(1, 2) 1]
```

The output of GP/PARI is quite elaborate, so we lift the output of factorff to avoid seeing the modulus $\iota(t)$. A second lift suppresses the prime p too. Two other examples follow. The first example corresponds to the factorization

$$x^{10} + (\theta + 1)x^9 + (\theta + 1)x^8 + x^7 + \theta x^6 + (\theta + 1)x^4 + \theta x^3 + \theta x^2 + (\theta + 1)x$$
$$= x(x + \theta)(x^2 + x + \theta)(x^3 + x + 1)(x^3 + \theta x + 1) \in \mathbb{F}_4[x]$$

(see Examples 3.14(2) and 3.18(3)), and the second example to

$$x^{15} + (\theta + 2)x^{12} + (2\theta + 1) = (x + (\theta + 1))^6(x^3 + \theta x + (\theta + 2))^3 \in \mathbb{F}_9[x],$$

where $\mathbb{F}_9 = \mathbb{F}_3(\theta)$ with $\theta^2 + 1 = 0$.

```
gp > lift(lift(factorff( \
     x^10+(t+1)*x^9+(t+1)*x^8+x^7+t*x^6+(t+1)*x^4+t*x^3+t*x^2+(t+1)*x, \
     2, t^2+t+1)))
```

```
%10 =
[x 1]

[x + t 1]

[x^2 + x + t 1]

[x^3 + x + 1 1]

[x^3 + t*x + 1 1]

gp > lift(lift(factorff(x^15 + (t+2)*x^12 + (2*t+1), 3, t^2+1)))
%11 =
[x + (t + 1) 6]

[x^3 + t*x + (t + 2) 3]
```

3.4 Properties of Polynomials with Integer Coefficients

In this and the next sections, we study polynomials with integer coefficients. The basic results of interest are the factorization algorithms of Section 3.5. This section develops some prerequisites for understanding them.

3.4.1 Relation with Polynomials with Rational Coefficients

$\mathbb{Z}$ is a unique factorization domain (UFD) whose field of fractions is $\mathbb{Q}$. This means that given a non-zero $f(x) \in \mathbb{Q}[x]$, we can multiply $f(x)$ by a suitable non-zero integer a in order to obtain $af(x) \in \mathbb{Z}[x]$. Conversely, given a non-zero polynomial $f(x) \in \mathbb{Z}[x]$ with leading coefficient $a \neq 0$, we obtain the monic polynomial $\frac{1}{a}f(x) \in \mathbb{Q}[x]$.

A more profound fact is that factoring in $\mathbb{Z}[x]$ is essentially the same as in $\mathbb{Q}[x]$. Proving this fact requires the following concept.

Definition 3.19 Let $f(x) = a_0 + a_1x + \cdots + a_dx^d \in \mathbb{Z}[x]$ be non-zero. The positive integer $\gcd(a_0, a_1, \ldots, a_d)$ is called the *content* of $f(x)$, and is denoted by cont $f(x)$. If cont $f(x) = 1$, we call $f(x)$ a *primitive polynomial*.[9] ◁

Lemma 3.20 Let $f(x), g(x) \in \mathbb{Z}[x]$ be non-zero. Then, cont$(f(x)g(x)) = $ (cont $f(x)$)(cont $g(x)$). In particular, the product of two primitive polynomials is again primitive.

---

[9]This primitive polynomial has nothing to do with the primitive polynomial of Definition 2.30. The same term used for describing two different objects may create confusion. But we have to conform to conventions.

Proof Let $f(x) = \sum_{i=0}^{m} a_i x^i$ and $g(x) = \sum_{j=0}^{n} b_j x^j$ with $a = \text{cont } f(x)$ and $b = \text{cont } g(x)$. Write $f(x) = a\bar{f}(x)$ and $g(x) = b\bar{g}(x)$, where $\bar{f}(x), \bar{g}(x)$ are primitive polynomials. Since $f(x)g(x) = ab\bar{f}(x)\bar{g}(x)$, it suffices to show that the product of two primitive polynomials is primitive, and we assume without loss of generality that $f(x), g(x)$ are themselves primitive (that is, $a = b = 1$). We proceed by contradiction. Assume that $f(x)g(x)$ is not primitive, that is, there exists a prime $p \mid \text{cont}(f(x)g(x))$, that is, p divides every coefficient of $f(x)g(x)$. Since $f(x)$ is primitive, all coefficients of $f(x)$ are not divisible by p. Let s be the smallest non-negative integer for which $p \nmid a_s$. Analogously, let t be the smallest non-negative integer for which $p \nmid b_t$. The coefficient of x^{s+t} in $f(x)g(x)$ is $a_s b_t + (a_{s-1}b_{t+1} + a_{s-2}b_{t+2} + \cdots) + (a_{s+1}b_{t-1} + a_{s+2}b_{t-2} + \cdots)$. By the choice of s, t, the prime p divides $a_{s-1}, a_{s-2}, \ldots, a_0$ and $b_{t-1}, b_{t-2}, \ldots, b_0$, that is, p divides $a_s b_t$, a contradiction, since $p \nmid a_s$ and $p \nmid b_t$. ◁

Theorem 3.21 *Let $f(x) \in \mathbb{Z}[x]$ be a primitive polynomial. Then, $f(x)$ is irreducible in $\mathbb{Z}[x]$ if and only if $f(x)$ is irreducible in $\mathbb{Q}[x]$.*

Proof The "if" part is obvious. For proving the "only if" part, assume that $f(x) = g(x)h(x)$ is a non-trivial factorization of $f(x)$ in $\mathbb{Q}[x]$. We can write $g(x) = a\bar{g}(x)$ and $h(x) = b\bar{h}(x)$ with $a, b \in \mathbb{Q}^*$ and with primitive polynomials $\bar{g}(x), \bar{h}(x) \in \mathbb{Z}[x]$. We have $f(x) = ab\bar{g}(x)\bar{h}(x)$. Since $f(x)$ and $\bar{g}(x)\bar{h}(x)$ are primitive polynomials in $\mathbb{Z}[x]$, we must have $ab = \pm 1$, that is, $f(x) = (ab\bar{g}(x))(\bar{h}(x))$ is a non-trivial factorization of $f(x)$ in $\mathbb{Z}[x]$. ◁

A standard way to determine the irreducibility or otherwise of a primitive polynomial $f(x)$ in $\mathbb{Z}[x]$ is to factor $f(x)$ in $\mathbb{Z}[x]$ or $\mathbb{Q}[x]$. In Section 3.5, we will study some algorithms for factoring polynomials in $\mathbb{Z}[x]$. There are certain special situations, however, when we can confirm the irreducibility of a polynomial in $\mathbb{Z}[x]$ more easily than factoring the polynomial.

Theorem 3.22 [*Eisenstein's criterion*] *Let $f(x) = a_0 + a_1 x + \cdots + a_d x^d \in \mathbb{Z}[x]$ be a primitive polynomial, and p a prime that divides $a_0, a_1, \ldots, a_{d-1}$, but not a_d. Suppose also that $p^2 \nmid a_0$. Then, $f(x)$ is irreducible.*

Proof Suppose that $f(x) = g(x)h(x)$ is a non-trivial factorization of $f(x)$ in $\mathbb{Z}[x]$, where $g(x) = \sum_{i=0}^{m} b_i x^i$ and $h(x) = \sum_{j=0}^{n} c_j x^j$. Since $f(x)$ is primitive, $g(x)$ and $h(x)$ are primitive too. We have $a_0 = b_0 c_0$. By hypothesis, $p \mid a_0$ but $p^2 \nmid a_0$, that is, p divides exactly one of b_0 and c_0. Let $p \mid c_0$. Since $h(x)$ is primitive, all c_j are not divisible by p. Let t be the smallest positive integer for which $p \nmid c_t$. We have $t \leqslant \deg h(x) = d - \deg g(x) < d$, that is, a_t is divisible by p. But $a_t = b_0 c_t + b_1 c_{t-1} + b_2 c_{t-2} + \cdots$. By the choice of t, all the coefficients $c_{t-1}, c_{t-2}, \ldots, c_0$ are divisible by p. It follows that $p \mid b_0 c_t$, but $p \nmid b_0$ and $p \nmid c_t$, a contradiction. ◁

Example 3.23 I now prove that $f(x) = 1 + x + x^2 + \cdots + x^{p-1} \in \mathbb{Z}[x]$ is irreducible for $p \in \mathbb{P}$. Evidently, $f(x)$ is irreducible if and only if $f(x+1)$ is. But $f(x) = \frac{x^p - 1}{x - 1}$, so $f(x+1) = \frac{(x+1)^p - 1}{(x+1) - 1} = x^{p-1} + \binom{p}{1}x^{p-2} + \binom{p}{2}x^{p-3} + \cdots + \binom{p}{p-1}$ satisfies Eisenstein's criterion. □

3.4.2 Height, Resultant, and Discriminant

In this section, we discuss some auxiliary results pertaining to polynomials. Unless otherwise stated, we deal with polynomials in $\mathbb{C}[x]$. All results for such polynomials are evidently valid for polynomials with integral or rational coefficients. In certain situations, we may generalize the concepts further, and deal with polynomials over an arbitrary field K.

Definition 3.24 Let $f(x) = a_0 + a_1 x + a_2 x^2 + \cdots + a_d x^d \in \mathbb{C}[x]$ with $a_d \neq 0$ (so that $\deg f(x) = d$). Suppose that $z_1, z_2, \ldots, z_d \in \mathbb{C}$ are all the roots of $f(x)$, that is, $f(x) = a_d (x - z_1)(x - z_2) \cdots (x - z_d)$. We define the following quantities associated with $f(x)$.

$$
\begin{aligned}
H(f) &= \text{height of } f(x) &&= \max(|a_0|, |a_1|, \ldots, |a_d|), \\
|f| &= \text{Euclidean norm of } (a_0, a_1, \ldots, a_d) &&= \sqrt{|a_0|^2 + |a_1|^2 + \cdots + |a_d|^2}, \\
M(f) &= \text{measure of } f(x) &&= |a_d| \prod_{i=1}^{d} \max(1, |z_i|). \qquad \triangleleft
\end{aligned}
$$

Now, we specialize to polynomials with integer coefficients. Let $f(x) \in \mathbb{Z}[x]$ be as in Definition 3.24, $g(x) \in \mathbb{Z}[x]$ a factor of $f(x)$ with $\deg g(x) = m$, and $h(x) = f(x)/g(x)$. It may initially appear that the coefficients of $g(x)$ can be arbitrarily *large* (in absolute value). Cancellation of coefficients in the product $g(x)h(x)$ leaves only *small* coefficients in $f(x)$. The following theorem shows that this is not the case. The coefficients of $g(x)$ are instead bounded in absolute value by some function in the degree and coefficients of $f(x)$.

Proposition 3.25 *With the notations introduced in the last paragraph, we have* $H(g) \leqslant 2^m \sqrt{d+1}\, H(f)$.
Proof Renaming the roots $z_1, z_2, \ldots, z_d$ of $f(x)$, we assume that the roots of $g(x)$ are $z_1, z_2, \ldots, z_m$, and write $g(x) = b_0 + b_1 x + b_2 x^2 + \cdots + b_m x^m = b_m (x - z_1)(x - z_2) \cdots (x - z_m)$. For $i \in \{0, 1, 2, \ldots, m\}$, we have $b_i = b_m \sum z_{j_1} z_{j_2} \cdots z_{j_{m-i}}$, where the sum runs over all subsets $\{j_1, j_2, \ldots, j_{m-i}\}$ of size $m - i$ of $\{1, 2, \ldots, m\}$. But $|b_i| = |b_m| \times \left| \sum z_{j_1} z_{j_2} \cdots z_{j_{m-i}} \right| \leqslant |b_m| \sum |z_{j_1} z_{j_2} \cdots z_{j_{m-i}}|$ (by the triangle inequality). Since $g(x)|f(x)$ in $\mathbb{Z}[x]$, we have $b_m | a_d$ in $\mathbb{Z}$, so $|b_m| \leqslant |a_d|$, and each $|z_{j_1} z_{j_2} \cdots z_{j_{m-i}}| \leqslant M(f)$. There are $\binom{m}{m-i} = \binom{m}{i}$ tuples $(j_1, j_2, \ldots, j_{m-i})$, so $|b_i| \leqslant \binom{m}{i} M(f)$. By the inequality of Landau (Exercise 3.37), $M(f) \leqslant |f| = \sqrt{|a_0|^2 + |a_1|^2 + \cdots + |a_d|^2} \leqslant \sqrt{(d+1) \max(|a_0|^2, |a_1|^2, \ldots, |a_d|^2)} = \sqrt{d+1}\, H(f)$. Consequently, each $|b_i| \leqslant \binom{m}{i} \sqrt{d+1}\, H(f) \leqslant 2^m \sqrt{d+1}\, H(f)$, so that $H(g) = \max(|b_0|, |b_1|, \ldots, |b_m|) \leqslant 2^m \sqrt{d+1}\, H(f)$ too. $\triangleleft$

Definition 3.26 Let K be an arbitrary field, and let

$$
\begin{aligned}
f(x) &= a_m x^m + a_{m-1} x^{m-1} + \cdots + a_1 x + a_0, \text{ and} \\
g(x) &= b_n x^n + b_{n-1} x^{n-1} + \cdots + b_1 x + b_0
\end{aligned}
$$

be non-zero polynomials in $K[x]$. The *resultant* $\mathrm{Res}(f(x), g(x))$ of $f(x)$ and $g(x)$ is defined to be the determinant of the $(m+n) \times (m+n)$ matrix

$$\mathrm{Syl}(f(x), g(x)) =$$

$$
\begin{pmatrix}
a_m & a_{m-1} & \cdots & \cdots & \cdots & \cdots & a_1 & a_0 & 0 & 0 & \cdots & 0 & 0 \\
0 & a_m & a_{m-1} & \cdots & \cdots & \cdots & \cdots & a_1 & a_0 & 0 & \cdots & 0 & 0 \\
\cdots & \cdots & \cdots & \cdots & \cdots & \cdots & \cdots & \cdots & \cdots & \cdots & \cdots & \cdots & \cdots \\
0 & \cdots & 0 & a_m & a_{m-1} & \cdots & \cdots & \cdots & \cdots & \cdots & \cdots & a_1 & a_0 \\
b_n & b_{n-1} & \cdots & b_1 & b_0 & 0 & 0 & \cdots & \cdots & \cdots & \cdots & 0 & 0 \\
0 & b_n & b_{n-1} & \cdots & b_1 & b_0 & 0 & 0 & \cdots & \cdots & \cdots & 0 & 0 \\
\cdots & \cdots & \cdots & \cdots & \cdots & \cdots & \cdots & \cdots & \cdots & \cdots & \cdots & \cdots & \cdots \\
0 & 0 & \cdots & \cdots & 0 & 0 & b_n & b_{n-1} & \cdots & \cdots & \cdots & b_1 & b_0
\end{pmatrix}
$$

called the *Sylvester matrix*[10] of $f(x)$ and $g(x)$, that is, $\mathrm{Res}(f(x), g(x)) = \det \mathrm{Syl}(f(x), g(x))$. If $f(x) = 0$ or $g(x) = 0$, we define $\mathrm{Res}(f(x), g(x)) = 0$. ◁

Some elementary properties of resultants are listed now.

Proposition 3.27 *Let $f(x), g(x) \in K[x]$ be as in Definition 3.26.*
(1) $\mathrm{Res}(g(x), f(x)) = (-1)^{mn} \mathrm{Res}(f(x), g(x))$.
(2) Let $m \geqslant n$, and $r(x) = f(x) \operatorname{rem} g(x) \neq 0$. Then, $\mathrm{Res}(f(x), g(x)) = (-1)^{mn} b_n^{m-n} \times \mathrm{Res}(g(x), r(x))$. In particular, resultants can be computed using the Euclidean gcd algorithm for polynomials.
(3) Let $\alpha_1, \alpha_2, \ldots, \alpha_m$ be the roots of $f(x)$, and $\beta_1, \beta_2, \ldots, \beta_n$ the roots of $g(x)$ (in some extension of K). Then, we have

$$\mathrm{Res}(f(x), g(x)) = a_m^n \prod_{i=1}^m g(\alpha_i) = (-1)^{mn} b_n^m \prod_{j=1}^n f(\beta_j) = a_m^n b_n^m \prod_{i=1}^m \prod_{j=1}^n (\alpha_i - \beta_j).$$

In particular, $\mathrm{Res}(f(x), g(x)) = 0$ if and only if $f(x)$ and $g(x)$ have a non-trivial common factor (in $K[x]$). ◁

Example 3.28 (1) For $K = \mathbb{Q}$, $f(x) = 2x^3 + 1$, and $g(x) = x^2 - 2x + 3$,

$$
\mathrm{Res}(f(x), g(x)) = \begin{vmatrix}
2 & 0 & 0 & 1 & 0 \\
0 & 2 & 0 & 0 & 1 \\
1 & -2 & 3 & 0 & 0 \\
0 & 1 & -2 & 3 & 0 \\
0 & 0 & 1 & -2 & 3
\end{vmatrix} = 89.
$$

We can also compute this resultant by Euclidean gcd:

$$
\begin{aligned}
r(x) &= f(x) \operatorname{rem} g(x) &= 2x - 11, \\
s(x) &= g(x) \operatorname{rem} r(x) &= 89/4.
\end{aligned}
$$

Therefore,

$$
\begin{aligned}
\mathrm{Res}(f(x), g(x)) &= (-1)^6 1^3 \mathrm{Res}(g(x), r(x)) = \\
\mathrm{Res}(g(x), r(x)) &= (-1)^2 2^2 \mathrm{Res}(r(x), s(x)) = \\
4 \, \mathrm{Res}(r(x), s(x)) &= 4 \times (89/4) = 89.
\end{aligned}
$$

---

[10]This is named after the English mathematician James Joseph Sylvester (1814–1897).

(2) Take $f(x) = x^4 + x^2 + 1$ and $g(x) = x^3 + 1$ in $\mathbb{Q}[x]$. We have

$$\text{Res}(f(x), g(x)) = \begin{vmatrix} 1 & 0 & 1 & 0 & 1 & 0 & 0 \\ 0 & 1 & 0 & 1 & 0 & 1 & 0 \\ 0 & 0 & 1 & 0 & 1 & 0 & 1 \\ 1 & 0 & 0 & 1 & 0 & 0 & 0 \\ 0 & 1 & 0 & 0 & 1 & 0 & 0 \\ 0 & 0 & 1 & 0 & 0 & 1 & 0 \\ 0 & 0 & 0 & 1 & 0 & 0 & 1 \end{vmatrix} = 0.$$

This is expected, since $f(x) = (x^2 + x + 1)(x^2 - x + 1)$ and $g(x) = (x+1)(x^2 - x + 1)$ have a non-trivial common factor.

(3) Take $K = \mathbb{F}_{89}$, $f(x) = 2x^3 + 1$, and $g(x) = x^2 - 2x + 3$ in $\mathbb{F}_{89}[x]$. Part (1) reveals that $\text{Res}(f(x), g(x)) = 0$ in this case. In $\mathbb{F}_{89}[x]$, we have

$$f(x) = (x + 39)(x^2 + 50x + 8), \quad \text{and} \quad g(x) = (x + 39)(x + 48),$$

that is, $f(x)$ and $g(x)$ share the non-trivial common factor $x + 39$. □

Definition 3.29 Let K be an arbitrary field, and $f(x) \in K[x]$ a polynomial of degree $d \geq 1$ and with leading coefficient a_d. Let $\alpha_1, \alpha_2, \ldots, \alpha_d$ be the roots of $f(x)$ (in some extension of K). The *discriminant* of $f(x)$ is defined as

$$\text{Discr}(f(x)) = a_d^{2d-2} \prod_{i=1}^{d} \prod_{j=i+1}^{d} (\alpha_i - \alpha_j)^2 = (-1)^{d(d-1)/2} a_d^{2d-2} \prod_{i=1}^{d} \prod_{\substack{j=1 \\ j \neq i}}^{d} (\alpha_i - \alpha_j).$$

One can check that $\text{Discr}(f(x)) \in K$. ◁

Discriminants are related to resultants in the following way.

Proposition 3.30 Let $f(x)$ be as in Definition 3.29, $f'(x)$ the formal derivative of $f(x)$, and $k = \deg f'(x)$. Then,

$$\text{Discr}(f(x)) = (-1)^{d(d-1)/2} a_d^{d-k-2} \text{Res}(f(x), f'(x)).$$

If $k = d - 1$ (for example, if the characteristic of K is zero), we have

$$\text{Discr}(f(x)) = (-1)^{d(d-1)/2} a_d^{-1} \text{Res}(f(x), f'(x)).$$

For any field K, $\text{Discr}(f(x)) \neq 0$ if and only if $f(x)$ is square-free.
Proof Use Proposition 3.27(3). ◁

Example 3.31 (1) Discriminants of polynomials of small degrees have the following explicit formulas. A verification of these formulas is left to the reader (Exercise 3.39). In each case, the leading coefficient should be non-zero.

$$\text{Discr}(a_1 x + a_0) = a_1,$$
$$\text{Discr}(a_2 x^2 + a_1 x + a_0) = a_1^2 - 4a_0 a_2, \text{ and}$$
$$\text{Discr}(a_3 x^3 + a_2 x^2 + a_1 x + a_0) = a_1^2 a_2^2 - 4a_0 a_2^3 - 4a_1^3 a_3 + 18 a_0 a_1 a_2 a_3 - 27 a_0^2 a_3^2.$$

(2) Consider the quartic polynomial $f(x) = 4x^4 + 5x - 8 \in \mathbb{Q}[x]$. We have

$$\mathrm{Discr}(f(x)) = (-1)^{4(4-1)/2} 4^{-1} \mathrm{Res}(4x^4 + 5x - 8, 16x^3 + 5)$$

$$= \frac{1}{4} \times \begin{vmatrix} 4 & 0 & 0 & 5 & -8 & 0 & 0 \\ 0 & 4 & 0 & 0 & 5 & -8 & 0 \\ 0 & 0 & 4 & 0 & 0 & 5 & -8 \\ 16 & 0 & 0 & 5 & 0 & 0 & 0 \\ 0 & 16 & 0 & 0 & 5 & 0 & 0 \\ 0 & 0 & 16 & 0 & 0 & 5 & 0 \\ 0 & 0 & 0 & 16 & 0 & 0 & 5 \end{vmatrix}$$

$$= (-34634432)/4 = -8658608.$$

(3) We have the prime factorization $-8658608 = -(2^4 \times 7 \times 97 \times 797)$. Therefore, if $f(x) = 4x^4 + 5x - 8$ is treated as a polynomial in $\mathbb{F}_7[x]$, we have $\mathrm{Discr}(f(x)) = 0$. We have the factorization $f(x) = 4(x + 3)^2(x^2 + x + 6)$ in $\mathbb{F}_7[x]$. The repeated linear factor justifies why $\mathrm{Discr}(f(x)) = 0$ in this case. □

A useful bound on the discriminant of a polynomial $f(x) \in \mathbb{Z}[x]$ follows.

Proposition 3.32 *Let $f(x) \in \mathbb{Z}[x]$ be of degree $d \geqslant 1$. Then, $|\mathrm{Discr}(f(x))| \leqslant (d+1)^{2d-\frac{1}{2}} H(f)^{2d-1}$.*
Proof Let a_d denote the leading coefficient of $f(x)$. We have

$$\begin{aligned} |\mathrm{Discr}(f(x))| &= |a_d|^{-1}|\mathrm{Res}(f(x), f'(x))| \\ &\leqslant |\mathrm{Res}(f(x), f'(x))| && \text{[since } |a_d| \geqslant 1] \\ &\leqslant |f|^{d-1}|f'|^d && \text{[by Exercise 3.43]} \\ &\leqslant d^d |f|^{2d-1} && \text{[since } |f'| \leqslant d|f|] \\ &\leqslant d^d (d+1)^{(2d-1)/2} H(f)^{2d-1} && \text{[since } |f| \leqslant \sqrt{d+1}\, H(f)] \\ &\leqslant (d+1)^{2d-\frac{1}{2}} H(f)^{2d-1}. \end{aligned}$$

$\triangleleft$

In GP/PARI, the function for computing resultants is `polresultant()`, and the function for computing discriminants is `poldisc()`. Some examples follow.

```
gp > polresultant(2*x^3+1,x^2-2*x+3)
%1 = 89
gp > polresultant(x^4+x^2+1,x^3+1)
%2 = 0
gp > polresultant(Mod(2,89)*x^3+Mod(1,89),Mod(1,89)*x^2-Mod(2,89)*x+Mod(3,89))
%3 = 0
gp > poldisc(a*x^2+b*x+c)
%4 = -4*c*a + b^2
gp > poldisc(a*x^3+b*x^2+c*x+d)
%5 = -27*d^2*a^2 + (18*d*c*b - 4*c^3)*a + (-4*d*b^3 + c^2*b^2)
gp > poldisc(4*x^4+5*x-8)
%6 = -8658608
gp > poldisc(Mod(4,7)*x^4+Mod(5,7)*x-Mod(8,7))
%7 = Mod(0, 7)
```

3.4.3 Hensel Lifting

In Section 1.5, we have used the concept of Hensel lifting for solving polynomial congruences modulo p^n. Here, we use a similar technique for lifting factorizations of polynomials modulo p^n to modulo p^{n+1}.

Theorem 3.33 [*Hensel lifting of polynomial factorization*]   *Let p be a prime, and $f(x), g_1(x), h_1(x) \in \mathbb{Z}[x]$ be non-constant polynomials satisfying:*
   (1) $p \nmid \mathrm{Res}(g_1(x), h_1(x))$,
   (2) $g_1(x)$ is monic,
   (3) $\deg g_1(x) + \deg h_1(x) = \deg f(x)$, and
   (4) $f(x) \equiv g_1(x)h_1(x) \pmod{p}$.
Then, for every $n \in \mathbb{N}$, there exist polynomials $g_n(x), h_n(x) \in \mathbb{Z}[x]$ such that:
   (a) $g_n(x) \equiv g_1(x) \pmod{p}$, and $h_n(x) \equiv h_1(x) \pmod{p}$,
   (b) $p \nmid \mathrm{Res}(g_n(x), h_n(x))$,
   (c) $g_n(x)$ is monic, and $\deg g_n(x) = \deg g_1(x)$,
   (d) $\deg g_n(x) + \deg h_n(x) = \deg f(x)$, and
   (e) $f(x) \equiv g_n(x)h_n(x) \pmod{p^n}$.
Proof   We proceed by induction on $n \in \mathbb{N}$. For $n = 1$, the properties (a)–(e) reduce to the properties (1)–(4) and so are valid by hypothesis. In order to prove the inductive step, assume that the polynomials $g_n(x), h_n(x)$ are available for some $n \geqslant 1$. We construct the polynomials $g_{n+1}(x)$ and $h_{n+1}(x)$. Here, $g_n(x)$ and $h_n(x)$ are known modulo p^n, that is, the polynomials

$$g_{n+1}(x) = g_n(x) + p^n u_n(x) \quad \text{and} \quad h_{n+1}(x) = h_n(x) + p^n v_n(x)$$

in $\mathbb{Z}[x]$ satisfy $f(x) \equiv g_{n+1}(x)h_{n+1}(x) \pmod{p^n}$ for any $u_n(x), v_n(x) \in \mathbb{Z}[x]$. Our task is to locate $u_n(x), v_n(x)$ so that $f(x) \equiv g_{n+1}(x)h_{n+1}(x) \pmod{p^{n+1}}$ also. We seek $u_n(x), v_n(x)$ with $\deg u_n(x) < \deg g_1(x)$ and $\deg v_n(x) \leqslant \deg h_1(x)$. We have

$$\begin{aligned}
&f(x) - g_{n+1}(x)h_{n+1}(x) \\
&= (f(x) - g_n(x)h_n(x)) - p^n(v_n(x)g_n(x) + u_n(x)h_n(x)) - p^{2n}u_n(x)v_n(x).
\end{aligned}$$

By the induction hypothesis, $f(x) - g_n(x)h_n(x) = p^n w_n(x)$ for some $w_n(x) \in \mathbb{Z}[x]$ with $\deg w_n(x) \leqslant \deg f(x)$. Since $n \geqslant 1$, we have $2n \geqslant n+1$. Therefore, the condition $f(x) \equiv g_{n+1}(x)h_{n+1}(x) \pmod{p^{n+1}}$ implies

$$v_n(x)g_n(x) + u_n(x)h_n(x) \equiv w_n(x) \pmod{p}.$$

This gives us a linear system modulo p in the unknown coefficients of $u_n(x)$ and $v_n(x)$. The number of variables in the system is $(1 + \deg v_n(x)) + (1 + \deg u_n(x)) = (1 + \deg h_1(x)) + \deg g_1(x) = 1 + \deg f(x)$ which is the same as the number of equations. Moreover, the determinant of the coefficient matrix equals $\mathrm{Res}(g_n(x), h_n(x))$, and is invertible modulo p by the induction hypothesis. Therefore, there exists a unique solution for the polynomials $u_n(x), v_n(x)$ modulo p. That is, there is a unique lift of the factorization $g_n(x)h_n(x)$ modulo p^n to a factorization $g_{n+1}(x)h_{n+1}(x)$ modulo p^{n+1}.

It remains to verify that the lifted polynomials $g_{n+1}(x), h_{n+1}(x)$ continue to satisfy the properties (a)–(d). By construction, $g_{n+1}(x) \equiv g_n(x) \pmod{p^n}$, whereas by the induction hypothesis $g_n(x) \equiv g_1(x) \pmod{p}$, so $g_{n+1}(x) \equiv g_1(x) \pmod{p}$. Analogously, $h_{n+1}(x) \equiv h_1(x) \pmod{p}$. So Property (a) holds.

By construction, $\mathrm{Res}(g_{n+1}(x), h_{n+1}(x)) \equiv \mathrm{Res}(g_n(x), h(x)) \pmod{p}$, and by the induction hypothesis, $p \nmid \mathrm{Res}(g_n(x), h_n(x))$, that is, Property (b) holds.

Now, $\deg u_n(x) < \deg g_1(x)$, $\deg g_n(x) = \deg g_1(x)$, and $g_n(x)$ is monic. It follows that $g_{n+1}(x) = g_n(x) + p^n u_n(x)$ is monic too, with degree equal to $\deg g_1(x)$. Thus, Property (c) holds.

For proving Property (d), assume that $\deg h_{n+1}(x) < \deg h_1(x)$. But Property (a) implies that $f(x) \equiv g_{n+1}(x)h_{n+1}(x) \pmod{p}$ has a degree less than the degree of $g_1(x)h_1(x) \pmod{p}$, contradicting (3) and (4). ◁

Example 3.34 We start with the following values.

$$
\begin{aligned}
f(x) &= 35x^5 - 22x^3 + 10x^2 + 3x - 2 \in \mathbb{Z}[x], \\
p &= 13, \\
g_1(x) &= x^2 + 2x - 2 \in \mathbb{Z}[x], \text{ and} \\
h_1(x) &= -4x^3 - 5x^2 + 6x + 1 \in \mathbb{Z}[x].
\end{aligned}
$$

Let us first verify that the initial conditions (1)–(4) in Theorem 3.33 are satisfied for these choices. We have $\mathrm{Res}(g_1(x), h_1(x)) = 33 = 2 \times 13 + 7$, that is, Condition (1) holds. Clearly, Conditions (2) and (3) hold. Finally, $f(x) - g_1(x)h_1(x) = 39x^5 + 13x^4 - 26x^3 - 13x^2 + 13x = 13 \times (3x^5 + x^4 - 2x^3 - x^2 + x)$, that is, Condition (4) is satisfied.

We now lift the factorization $g_1(x)h_1(x)$ of $f(x)$ modulo p to a factorization $g_2(x)h_2(x)$ of $f(x)$ modulo p^2. First, compute $w_1(x) = (f(x) - g_1(x)h_1(x))/p = 3x^5 + x^4 - 2x^3 - x^2 + x$. Then, we attempt to find the polynomials $u_1(x) = u_{11}x + u_{10}$ and $v_1(x) = v_{13}x^3 + v_{12}x^2 + v_{11}x + v_{10}$ satisfying

$$v_1(x)g_1(x) + u_1(x)h_1(x) \equiv w_1(x) \pmod{p}.$$

Expanding the left side of this congruence and equating the coefficients of x^i, $i = 5, 4, 3, 2, 1, 0$, from both sides give the linear system

$$
\left.
\begin{aligned}
v_{13} &\equiv 3 \\
2v_{13} + v_{12} - 4u_{11} &\equiv 1 \\
-2v_{13} + 2v_{12} + v_{11} - 5u_{11} - 4u_{10} &\equiv -2 \\
-2v_{12} + 2v_{11} + v_{10} + 6u_{11} - 5u_{10} &\equiv -1 \\
-2v_{11} + 2v_{10} + u_{11} + 6u_{10} &\equiv 1 \\
-2v_{10} + u_{10} &\equiv 0
\end{aligned}
\right\} \pmod{p},
$$

that is, the system

$$
\begin{pmatrix}
1 & 0 & 0 & 0 & 0 & 0 \\
2 & 1 & 0 & 0 & -4 & 0 \\
-2 & 2 & 1 & 0 & -5 & -4 \\
0 & -2 & 2 & 1 & 6 & -5 \\
0 & 0 & -2 & 2 & 1 & 6 \\
0 & 0 & 0 & -2 & 0 & 1
\end{pmatrix}
\begin{pmatrix}
v_{13} \\
v_{12} \\
v_{11} \\
v_{10} \\
u_{11} \\
u_{10}
\end{pmatrix}
\equiv
\begin{pmatrix}
3 \\
1 \\
-2 \\
-1 \\
1 \\
0
\end{pmatrix}
\pmod{13}.
$$

The solution of the system is

$$(v_{13} \quad v_{12} \quad v_{11} \quad v_{10} \quad u_{11} \quad u_{10})^{t} \equiv (3 \quad 7 \quad 12 \quad 9 \quad 3 \quad 5)^{t} \pmod{13},$$

that is, $u_1(x) = 3x + 5$, and $v_1(x) = 3x^3 + 7x^2 + 12x + 9$. This gives $g_1(x) + pu_1(x) = x^2 + 41x + 63$, and $h_1(x) + pv_1(x) = 35x^3 + 86x^2 + 162x + 118$. For a reason to be clarified in Section 3.5.1, we plan to leave each coefficient c of $g_n(x), h_n(x)$ in the range $-p^n/2 < c \leqslant p^n/2$, that is, in this case in the range $-84 \leqslant c \leqslant 84$. We, therefore, take

$$g_2(x) = x^2 + 41x + 63, \quad \text{and} \quad h_2(x) = 35x^3 - 83x^2 - 7x - 51.$$

Let us verify that Properties (a)–(e) of Theorem 3.33 are satisfied by the lifted factorization. Properties (a), (c) and (d) hold obviously. $\mathrm{Res}(g_2(x), h_2(x)) = 896640999 = 68972384 \times 13 + 7$, that is, $p \nmid \mathrm{Res}(g_2, h_2)$. Finally, $f(x) - g_2(x)h_2(x) = -1352x^4 + 1183x^3 + 5577x^2 + 2535x + 3211 = 13^2 \times (-8x^4 + 7x^3 + 33x^2 + 15x + 19)$.

Let us now lift this factorization to $g_3(x)h_3(x)$ modulo $p^3 = 2197$. We have $w_2(x) = (f(x) - g_2(x)h_2(x))/p^2 = -8x^4 + 7x^3 + 33x^2 + 15x + 19$. We seek polynomials $u_2(x) = u_{21}x + u_{20}$ and $v_2(x) = v_{23}x^3 + v_{22}x^2 + v_{21}x + v_{20}$ with

$$v_2(x)g_2(x) + u_2(x)h_2(x) \equiv w_2(x) \pmod{p}.$$

The resulting linear system is

$$\left.\begin{array}{rcl}
v_{23} & \equiv & 0 \\
41v_{23} + v_{22} + 35u_{21} & \equiv & -8 \\
63v_{23} + 41v_{22} + v_{21} - 83u_{21} + 35u_{20} & \equiv & 7 \\
63v_{22} + 41v_{21} + v_{20} - 7u_{21} - 83u_{20} & \equiv & 33 \\
63v_{21} + 41v_{20} - 51u_{21} - 7u_{20} & \equiv & 15 \\
63v_{20} - 51u_{20} & \equiv & 19
\end{array}\right\} \pmod{p},$$

that is,

$$\begin{pmatrix}
1 & 0 & 0 & 0 & 0 & 0 \\
41 & 1 & 0 & 0 & 35 & 0 \\
63 & 41 & 1 & 0 & -83 & 35 \\
0 & 63 & 41 & 1 & -7 & -83 \\
0 & 0 & 63 & 41 & -51 & -7 \\
0 & 0 & 0 & 63 & 0 & -51
\end{pmatrix}
\begin{pmatrix}
v_{23} \\
v_{22} \\
v_{21} \\
v_{20} \\
u_{21} \\
u_{20}
\end{pmatrix}
\equiv
\begin{pmatrix}
0 \\
-8 \\
7 \\
33 \\
15 \\
19
\end{pmatrix}
\pmod{13},$$

which has the solution

$$(v_{23} \quad v_{22} \quad v_{21} \quad v_{20} \quad u_{21} \quad u_{20})^{t} \equiv (0 \quad 0 \quad 0 \quad 3 \quad 2 \quad 12)^{t} \pmod{13}.$$

Therefore, $u_2(x) = 2x + 12$ and $v_2(x) = 3$ which yield $g_2(x) + p^2u_2(x) = x^2 + 379x + 2091$ and $h_2(x) + p^2v_2(x) = 35x^3 - 83x^2 - 7x + 456$. We take the coefficients of $g_3(x)$ and $h_3(x)$ between $-p^3/2$ and $p^3/2$, that is, between -1098 and $+1098$, that is,

$$g_3(x) = x^2 + 379x - 106, \quad \text{and} \quad h_3(x) = 35x^3 - 83x^2 - 7x + 456.$$

Since $\mathrm{Res}(g_3(x), h_3(x)) = -861479699866 = (-66267669221) \times 13 + 7$, we have $p \nmid \mathrm{Res}(g_3, h_3)$ (Property (a)). Moreover, $f(x) - g_3(x)h_3(x) = -13182x^4 + 35152x^3 - 6591x^2 - 173563x + 48334 = 13^3 \times (-6x^4 + 16x^3 - 3x^2 - 79x + 22)$ (Property (e)). Other properties in Theorem 3.33 are evidently satisfied. $\square$

3.5 Factoring Polynomials with Integer Coefficients

3.5.1 Berlekamp's Factoring Algorithm

We now possess the requisite background to understand Berlekamp's algorithm for factoring polynomials in $\mathbb{Z}[x]$. We first compute a square-free decomposition of the input polynomial $f(x) \in \mathbb{Z}[x]$. In short, we compute $\gcd(f(x), f'(x))$, where $f'(x)$ is the formal derivative of $f(x)$. Since the characteristic of $\mathbb{Z}$ (or $\mathbb{Q}$) is zero, $f'(x) \neq 0$ for a non-constant $f(x)$. The polynomial $f(x)/\gcd(f(x), f'(x))$ is square-free. The cofactor $\gcd(f(x), f'(x))$, unless equal to 1, is recursively subject to square-free factorization.

We may, therefore, assume that $f(x) \in \mathbb{Z}[x]$ is itself square-free. To start with, let us consider the special case that $f(x)$ is monic. Later, we will remove this restriction. We compute $\Delta = \mathrm{Discr}(f(x))$, and choose a prime $p \nmid \Delta$. Since $f(x)$ is square-free, Δ is non-zero and can have at most $\lg |\Delta|$ prime factors. So the smallest prime that does not divide Δ has to be at most as large as the $(1 + \lg |\Delta|)$-th prime, that is, of the order $O(\log |\Delta| \log \log |\Delta|)$. By Proposition 3.32, $|\Delta| \leqslant (d+1)^{2d-\frac{1}{2}} H(f)^{2d-1}$, where $d = \deg f$, and $H(f)$ is the height of $f(x)$. This implies that we can take $p = O(d \log(dH(f)) \log(d \log(dH(f))))$. There is no need to factor Δ to determine a suitable prime p. One may instead try p from a set of small primes (for example, in the sequence $2, 3, 5, 7, 11, \ldots$), until one not dividing Δ is located.

The motivation behind choosing p with $p \nmid \Delta$ is that the polynomial $f(x)$ continues to remain square-free modulo p. We factor $f(x)$ in $\mathbb{F}_p[x]$ as

$$f(x) = f_1(x)f_2(x) \cdots f_t(x) \in \mathbb{F}_p[x],$$

where $f_1, f_2, \ldots, f_t$ are distinct monic irreducible polynomials in $\mathbb{F}_p[x]$. If $t = 1$, then $f(x)$ is evidently irreducible in $\mathbb{Z}[x]$. So suppose that $t \geqslant 2$.

If $f(x)$ is reducible in $\mathbb{Z}[x]$, it admits a factor $g(x)$ of degree $\leqslant \lfloor d/2 \rfloor$, and we can write $f(x) = g(x)h(x) \in \mathbb{Z}[x]$ with $g(x)$ and $h(x)$ monic. The polynomial $g(x)$, even if irreducible in $\mathbb{Z}[x]$, need not remain so in $\mathbb{F}_p[x]$. However, the factors of $g(x)$ modulo p must come from the set $\{f_1(x), f_2(x), \ldots, f_t(x)\}$. Therefore, we try all possible subsets $\{i_1, i_2, \ldots, i_k\} \subseteq \{1, 2, \ldots, t\}$ for which

$$1 \leqslant \sum_{\substack{j=1 \\ k \leqslant t}}^{k} \deg f_{i_j}(x) \leqslant \lfloor d/2 \rfloor.$$ Such a subset $\{i_1, i_2, \ldots, i_k\}$ corresponds to the

polynomial $g_1(x) = f_{i_1}(x)f_{i_2}(x)\cdots f_{i_k}(x) \in \mathbb{F}_p[x]$ which is a potential reduction of $g(x)$ modulo p. We then compute $h_1(x) = f(x)/g_1(x)$ in $\mathbb{F}_p[x]$. Since $f(x)$ is square-free modulo p, we have $\gcd(g_1(x), h_1(x)) = 1$, that is, $\mathrm{Res}(g_1(x), h_1(x))$ is not divisible by p.

We then lift the factorization $f(x) \equiv g_1(x)h_1(x)$ (mod p) to the (unique) factorization $f(x) \equiv g_n(x)h_n(x)$ (mod p^n) using Theorem 3.33. The polynomial $g_n(x)$ is monic, and is represented so as to have coefficients between $-p^n/2$ and $p^n/2$. We choose n large enough to satisfy $p^n/2 > H(g)$. Then, $g_n(x)$ in this representation can be identified with a polynomial in $\mathbb{Z}[x]$. (Notice that $g(x) \in \mathbb{Z}[x]$ may have negative coefficients. This is why we kept the coefficients of $g_n(x)$ between $-p^n/2$ and $p^n/2$.) Proposition 3.25 gives $H(g) \leqslant 2^{\lfloor d/2 \rfloor}\sqrt{d+1}\,H(f)$, that is, we choose the smallest n satisfying $p^n/2 > 2^{\lfloor d/2 \rfloor}\sqrt{d+1}\,H(f)$ so as to ascertain $p^n/2 > H(g)$.

Once $g_n(x)$ is computed, we divide $f(x)$ by $g_n(x)$ in $\mathbb{Z}[x]$ (actually, $\mathbb{Q}[x]$). If $r(x) = f(x)$ rem $g_n(x) = 0$, we have detected a divisor $g(x) = g_n(x) \in \mathbb{Z}[x]$ of $f(x)$. We then recursively factor $g(x)$ and the cofactor $h(x) = f(x)/g(x)$.

Let us finally remove the restriction that $f(x)$ is monic. Let $a \in \mathbb{Z}$ be the leading coefficient of $f(x)$. We require the reduction of $f(x)$ modulo p to be of degree $d = \deg f$, that is, we require $p \nmid a$. Factoring $f(x)$ in $\mathbb{F}_p[x]$ then gives $f(x) \equiv af_1(x)f_2(x)\cdots f_t(x)$ (mod p) with distinct *monic* irreducible polynomials $f_1, f_2, \ldots, f_t$ in $\mathbb{F}_p[x]$. We start with a divisor $g_1(x) = f_{i_1}(x)f_{i_2}(x)\cdots f_{i_k}(x)$ of $f(x)$ in $\mathbb{F}_p[x]$ with $\deg g_1 \leqslant \lfloor d/2 \rfloor$, and set $h_1(x) = f(x)/g_1(x) \in \mathbb{F}_p[x]$. Here, $g_1(x)$ is monic, and $h_1(x)$ has leading coefficient a (mod p). Using Hensel's lifting, we compute $g_n(x), h_n(x)$ with $g_n(x)$ monic and $f(x) \equiv g_n(x)h_n(x)$ (mod p^n).

A divisor $g(x)$ of $f(x)$ in $\mathbb{Z}[x]$ need not be monic. However, the leading coefficient b of $g(x)$ must divide a. Multiplying $g(x)$ by a/b gives the polynomial $(a/b)g(x)$ with leading coefficient equal to a. Moreover, $(a/b)g(x)$ must divide $f(x)$ in $\mathbb{Q}[x]$. Therefore, instead of checking whether $g_n(x)$ divides $f(x)$, we now check whether $ag_n(x)$ divides $f(x)$ in $\mathbb{Q}[x]$. That is, $ag_n(x)$ is now identified with the polynomial $(a/b)g(x)$. Since $H((a/b)g(x)) \leqslant H(ag(x)) = |a|H(g) \leqslant |a|2^{\lfloor d/2 \rfloor}\sqrt{d+1}\,H(f) \leqslant 2^{\lfloor d/2 \rfloor}\sqrt{d+1}\,H(f)^2$, we now choose n so that $p^n/2 > 2^{\lfloor d/2 \rfloor}\sqrt{d+1}\,H(f)^2$.

Algorithm 3.9 summarizes all these observations in order to arrive at an algorithm for factoring polynomials in $\mathbb{Z}[x]$.

Example 3.35 Let us try to factor the polynomial

$$f(x) = 35x^5 - 22x^3 + 10x^2 + 3x - 2 \in \mathbb{Z}[x]$$

of Example 3.34. We have $a = 35$ and $d = 5$. We compute $\mathrm{Discr}(f(x)) = -17245509120$, and choose the prime $p = 13$ dividing neither this discriminant nor a. The smallest exponent n with $p^n > 2^{\lfloor d/2 \rfloor + 1}(\sqrt{d+1})H(f(x))^2 \approx 24005$ is $n = 4$. (We have $13^4 = 28561 > 24005$.) In $\mathbb{F}_p[x]$, we have the factorization

$$f(x) \equiv 9(x+5)(x+10)(x+11)(x^2+5) \pmod{p}.$$

Algorithm 3.9: BERLEKAMP'S ALGORITHM FOR FACTORING A NON-CONSTANT SQUARE-FREE POLYNOMIAL $f(x) \in \mathbb{Z}[x]$

Let a be the leading coefficient of $f(x)$, and $d = \deg f(x)$.
Choose a prime p such that $p \nmid a$ and $p \nmid \text{Discr}(f(x))$.
Choose the smallest $n \in \mathbb{N}$ such that $p^n > 2^{\lfloor d/2 \rfloor + 1}(\sqrt{d+1})H(f(x))^2$.
Factor $f(x) \equiv af_1(x)f_2(x)\cdots f_t(x) \pmod{p}$,
  where $f_1, f_2, \ldots, f_t \in \mathbb{F}_p[x]$ are distinct, monic and irreducible.
If $(t = 1)$, output the irreducible polynomial $f(x)$, and return.
For each factor $g_1(x) = f_{i_1}(x)\cdots f_{i_k}(x) \in \mathbb{F}_p[x]$ of $f(x)$ with $\deg g_1 \leq \lfloor \frac{d}{2} \rfloor$ {
  Take $h_1(x) = f(x)/g_1(x)$ in $\mathbb{F}_p[x]$.
  Using Hensel's lifting, compute the factorization
    $f(x) \equiv g_n(x)h_n(x) \pmod{p^n}$ with $g_n(x)$ monic,
    $g_n(x) \equiv g_1(x) \pmod{p}$, and $h_n(x) \equiv h_1(x) \pmod{p}$.
  Set $\bar{g}_n(x) = ag_n(x) \in \mathbb{Z}_{p^n}[x]$ with each coefficient of $\bar{g}_n(x)$
    lying between $-p^n/2$ and $p^n/2$.
  Treat $\bar{g}_n(x)$ as a polynomial in $\mathbb{Z}[x]$.
  Compute $g(x) = \bar{g}_n(x)/\text{cont}(\bar{g}_n)$.
  Compute $r(x) = f(x) \text{ rem } g(x)$ in $\mathbb{Q}[x]$.
  if $(r(x) = 0)$ {
    Recursively call Algorithm 3.9 on $g(x)$.
    Recursively call Algorithm 3.9 on $f(x)/g(x)$.
    Return.
  }
}
/* All potential divisors of $f(x)$ tried, but no divisor found */
Output the irreducible polynomial $f(x)$.

The search for potential divisors $g(x)$ of $f(x)$ starts by selecting $g_1(x)$ from

$$\{x+5, x+10, x+11, (x+5)(x+10), (x+5)(x+11), (x+10)(x+11), x^2+5\}.$$

For every i, $1 \leq i \leq n$, we maintain the coefficients of $g_i(x)$ and $h_i(x)$ between $-p^i/2$ and $p^i/2$. For example, $x+10$ is represented as $x-3$, and $(x+5)(x+10) \equiv x^2 + 2x + 11 \pmod{p}$ as $x^2 + 2x - 2$. The computation of $g(x)$ proceeds for different choices of $g_1(x)$ as follows.

Choice 1: $g_1(x) = x + 5 \in \mathbb{Z}_{13}[x]$.
We have $h_1(x) = f(x)/g_1(x) = -4x^4 - 6x^3 - 5x^2 - 4x - 3 \in \mathbb{Z}_p[x]$. Three stages of Hensel lifting produce the following polynomials.

$$\begin{aligned} g_2(x) &= x + 83 \in \mathbb{Z}_{13^2}[x], \\ h_2(x) &= 35x^4 - 32x^3 - 70x^2 + 74x - 55 \in \mathbb{Z}_{13^2}[x], \\ g_3(x) &= x + 421 \in \mathbb{Z}_{13^3}[x], \end{aligned}$$

$$h_3(x) \;=\; 35x^4 + 644x^3 - 915x^2 + 750x + 621 \in \mathbb{Z}_{13^3}[x],$$
$$g_4(x) \;=\; x - 1776 \in \mathbb{Z}_{13^4}[x],$$
$$h_4(x) \;=\; 35x^4 + 5038x^3 + 7873x^2 - 12432x - 1576 \in \mathbb{Z}_{13^4}[x].$$

Subsequently, we compute $\bar{g}_4[x] = ag_4(x) = 35x - 5038 \in \mathbb{Z}_{13^4}[x]$, $g(x) = \bar{g}_4(x)/\operatorname{cont}(\bar{g}_4(x)) = 35x - 5038 \in \mathbb{Z}[x]$, and $r(x) = f(x) \operatorname{rem} g(x) = \frac{3245470620554884228}{1500625} \in \mathbb{Q}[x]$. Since $r(x) \neq 0$, $g(x)$ is not a factor of $f(x)$.

Choice 2: $g_1(x) = x - 3 \in \mathbb{Z}_{13}[x]$.

We have the following sequence of computations.

$$h_1(x) \;=\; f(x)/g_1(x) \;=\; -4x^4 + x^3 - 6x^2 + 5x + 5 \in \mathbb{Z}_{13}[x],$$
$$g_2(x) \;=\; x - 42 \in \mathbb{Z}_{13^2}[x],$$
$$h_2(x) \;=\; 35x^4 - 51x^3 + 33x^2 + 44x - 8 \in \mathbb{Z}_{13^2}[x],$$
$$g_3(x) \;=\; x - 42 \in \mathbb{Z}_{13^3}[x],$$
$$h_3(x) \;=\; 35x^4 - 727x^3 + 202x^2 - 294x + 837 \in \mathbb{Z}_{13^3}[x],$$
$$g_4(x) \;=\; x - 13224 \in \mathbb{Z}_{13^4}[x],$$
$$h_4(x) \;=\; 35x^4 + 5864x^3 + 2399x^2 - 6885x + 5231 \in \mathbb{Z}_{13^4}[x],$$
$$\bar{g}_4(x) \;=\; ag_4(x) \;=\; 35x - 5864 \in \mathbb{Z}_{13^4}[x],$$
$$g(x) \;=\; \bar{g}_4(x)/\operatorname{cont}(\bar{g}_4(x)) \;=\; 35x - 5864 \in \mathbb{Z}[x],$$
$$r(x) \;=\; f(x) \operatorname{rem} g(x) \;=\; \frac{6933621169702778694}{1500625} \in \mathbb{Q}[x].$$

Since $r(x) \neq 0$, this factorization attempt is unsuccessful.

Choice 3: $g_1(x) = x - 2 \in \mathbb{Z}_{13}[x]$.

$$h_1(x) \;=\; f(x)/g_1(x) \;=\; -4x^4 + 5x^3 + x^2 - x + 1 \in \mathbb{Z}_{13}[x],$$
$$g_2(x) \;=\; x - 41 \in \mathbb{Z}_{13^2}[x],$$
$$h_2(x) \;=\; 35x^4 + 83x^3 + x^2 + 51x + 66 \in \mathbb{Z}_{13^2}[x],$$
$$g_3(x) \;=\; x - 379 \in \mathbb{Z}_{13^3}[x],$$
$$h_3(x) \;=\; 35x^4 + 83x^3 + 677x^2 - 456x + 742 \in \mathbb{Z}_{13^3}[x],$$
$$g_4(x) \;=\; x - 13561 \in \mathbb{Z}_{13^4}[x],$$
$$h_4(x) \;=\; 35x^4 - 10902x^3 - 10308x^2 - 9244x - 3652 \in \mathbb{Z}_{13^4}[x],$$
$$\bar{g}_4(x) \;=\; ag_4(x) \;=\; 35x + 10902 \in \mathbb{Z}_{13^4}[x],$$
$$g(x) \;=\; \bar{g}_4(x)/\operatorname{cont}(\bar{g}_4(x)) \;=\; 35x + 10902 \in \mathbb{Z}[x],$$
$$r(x) \;=\; f(x) \operatorname{rem} g(x) \;=\; -\frac{15400260285781371872}{1500625} \in \mathbb{Q}[x].$$

Since $r(x) \neq 0$, this factorization attempt is unsuccessful too.

Choice 4: $g_1(x) = (x + 5)(x + 10) = x^2 + 2x - 2 \in \mathbb{Z}_{13}[x]$.

This choice of $g_1(x)$ is considered in Example 3.34.

$$
\begin{aligned}
h_1(x) &= f(x)/g_1(x) = -4x^3 - 5x^2 + 6x + 1 \in \mathbb{Z}_{13}[x], \\
g_2(x) &= x^2 + 41x + 63 \in \mathbb{Z}_{13^2}[x], \\
h_2(x) &= 35x^3 - 83x^2 - 7x - 51 \in \mathbb{Z}_{13^2}[x], \\
g_3(x) &= x^2 + 379x - 106 \in \mathbb{Z}_{13^3}[x], \\
h_3(x) &= 35x^3 - 83x^2 - 7x + 456 \in \mathbb{Z}_{13^3}[x], \\
g_4(x) &= x^2 + 13561x + 8682 \in \mathbb{Z}_{13^4}[x], \\
h_4(x) &= 35x^3 + 10902x^2 - 7x + 9244 \in \mathbb{Z}_{13^4}[x], \\
\bar{g}_4(x) &= ag_4(x) = 35x^2 - 10902x - 10301 \in \mathbb{Z}_{13^4}[x], \\
g(x) &= \bar{g}_4(x)/\operatorname{cont}(\bar{g}_4(x)) = 35x^2 - 10902x - 10301 \in \mathbb{Z}[x], \\
r(x) &= f(x) \operatorname{rem} g(x) \\
&= \left(\frac{14254770160420556}{42875}\right) x + \frac{13428329145305308}{42875} \in \mathbb{Q}[x].
\end{aligned}
$$

Since $r(x) \neq 0$, this factorization attempt is again unsuccessful.

Choice 5: $g_1(x) = (x + 5)(x + 11) = x^2 + 3x + 3 \in \mathbb{Z}_{13}[x]$.

$$
\begin{aligned}
h_1(x) &= f(x)/g_1(x) = -4x^3 - x^2 + 6x - 5 \in \mathbb{Z}_{13}[x], \\
g_2(x) &= x^2 + 42x - 23 \in \mathbb{Z}_{13^2}[x], \\
h_2(x) &= 35x^3 + 51x^2 - 7x - 44 \in \mathbb{Z}_{13^2}[x], \\
g_3(x) &= x^2 + 42x + 822 \in \mathbb{Z}_{13^3}[x], \\
h_3(x) &= 35x^3 + 727x^2 - 7x + 294 \in \mathbb{Z}_{13^3}[x], \\
g_4(x) &= x^2 + 13224x + 7413 \in \mathbb{Z}_{13^4}[x], \\
h_4(x) &= 35x^3 - 5864x^2 - 7x + 6885 \in \mathbb{Z}_{13^4}[x], \\
\bar{g}_4(x) &= ag_4(x) = 35x^2 + 5864x + 2406 \in \mathbb{Z}_{13^4}[x], \\
g(x) &= \bar{g}_4(x)/\operatorname{cont}(\bar{g}_4(x)) = 35x^2 + 5864x + 2406 \in \mathbb{Z}[x], \\
r(x) &= f(x) \operatorname{rem} g(x) \\
&= \left(\frac{1173724653532041}{42875}\right) x + \frac{482764549856654}{42875} \in \mathbb{Q}[x].
\end{aligned}
$$

Since $r(x) \neq 0$, this factorization attempt is yet again unsuccessful.

Choice 6: $g_1(x) = (x + 10)(x + 11) = x^2 - 5x + 6 \in \mathbb{Z}_{13}[x]$.

$$
\begin{aligned}
h_1(x) &= f(x)/g_1(x) = -4x^3 + 6x^2 + 6x + 4 \in \mathbb{Z}_{13}[x], \\
g_2(x) &= x^2 - 83x + 32 \in \mathbb{Z}_{13^2}[x], \\
h_2(x) &= 35x^3 + 32x^2 - 7x - 74 \in \mathbb{Z}_{13^2}[x], \\
g_3(x) &= x^2 - 421x + 539 \in \mathbb{Z}_{13^3}[x], \\
h_3(x) &= 35x^3 - 644x^2 - 7x - 750 \in \mathbb{Z}_{13^3}[x], \\
g_4(x) &= x^2 + 1776x - 3855 \in \mathbb{Z}_{13^4}[x],
\end{aligned}
$$

$$
\begin{aligned}
h_4(x) &= 35x^3 - 5038x^2 - 7x + 12432 \in \mathbb{Z}_{13^4}[x], \\
\bar{g}_4(x) &= ag_4(x) = 35x^2 + 5038x + 7880 \in \mathbb{Z}_{13^4}[x], \\
g(x) &= \bar{g}_4(x)/\operatorname{cont}(\bar{g}_4(x)) = 35x^2 + 5038x + 7880 \in \mathbb{Z}[x], \\
r(x) &= f(x) \operatorname{rem} g(x) \\
&= \left(\frac{623273765466781}{42875}\right)x + \frac{197140047380562}{8575} \in \mathbb{Q}[x].
\end{aligned}
$$

Since $r(x) \neq 0$, this factorization attempt continues to be unsuccessful.

Choice 7: $g_1(x) = x^2 + 5 \in \mathbb{Z}_{13}[x]$.

$$
\begin{aligned}
h_1(x) &= f(x)/g_1(x) = -4x^3 - 2x - 3 \in \mathbb{Z}_{13}[x], \\
g_2(x) &= x^2 - 34 \in \mathbb{Z}_{13^2}[x], \\
h_2(x) &= 35x^3 - 15x + 10 \in \mathbb{Z}_{13^2}[x], \\
g_3(x) &= x^2 - 879 \in \mathbb{Z}_{13^3}[x], \\
h_3(x) &= 35x^3 - 15x + 10 \in \mathbb{Z}_{13^3}[x], \\
g_4(x) &= x^2 + 5712 \in \mathbb{Z}_{13^4}[x], \\
h_4(x) &= 35x^3 - 15x + 10 \in\in \mathbb{Z}_{13^4}[x], \\
\bar{g}_4(x) &= ag_4(x) = 35x^2 - 7 \in \mathbb{Z}_{13^4}[x], \\
g(x) &= \bar{g}_4(x)/\operatorname{cont}(\bar{g}_4(x)) = 5x^2 - 1 \in \mathbb{Z}[x], \\
r(x) &= f(x) \operatorname{rem} g(x) = 0 \in \mathbb{Q}[x].
\end{aligned}
$$

We at last have $r(x) = 0$, that is, $g(x) = 5x^2 - 1$ is a factor of $f(x)$. The corresponding cofactor is $h(x) = f(x)/g(x) = 7x^3 - 3x^2 + 2$. We then attempt to factor $g(x)$ and $h(x)$ recursively. The steps are not shown here. Instead, we argue logically. Clearly, $g(x)$ is irreducible in $\mathbb{Z}[x]$, since it is irreducible in $\mathbb{Z}_{13}[x]$. The other factor $h(x)$ splits modulo 13 into three linear factors. If $h(x)$ were reducible in $\mathbb{Z}[x]$, it must have at least one linear factor. But we have seen above that neither of the three linear factors of $h(x)$ lifts to a factor of $f(x)$. Consequently, $h(x)$ too is irreducible in $\mathbb{Z}[x]$. □

Let us now look at the running time of Berlekamp's algorithm. The input polynomial $f(x)$ is of degree d and height $H(f)$, and can be encoded using $O(d \log H(f))$ bits. Thus, the input size of the algorithm is taken as $d \log H(f)$. It is easy to argue that under the given choices of p and n, each trial of computing $g(x)$ runs in time polynomial in $d \log H(f)$. However, the algorithm may make many unsuccessful trials. In the worst case, $f(x)$ is irreducible in $\mathbb{Z}[x]$, whereas $f(x)$ splits into d linear factors modulo p (there are examples of this and similar worst-case situations). One then has to attempt each of the $\sum_{k=1}^{\lfloor d/2 \rfloor} \binom{d}{k} \approx 2^{d-1}$ subsets of the factors of $f(x)$ in $\mathbb{F}_p[x]$. All these attempts fail to produce a divisor of $f(x)$ in $\mathbb{Z}[x]$. Since the number of trials is an exponential function of d in this case, Berlekamp's algorithm takes exponential running time in the worst case. On an average, the performance of Berlekamp's algorithm is not always as bad. Still, an algorithm that runs in polynomial time even in the worst case is needed.

3.5.2 Basis Reduction in Lattices

In order to arrive at a polynomial-time algorithm for factoring polynomials with integer coefficients, Lenstra, Lenstra and Lovász[11] involve lattices. In the rest of this chapter, I provide a brief description of this L^3 (or LLL) algorithm. The proof of correctness of the algorithm is not difficult, but quite involved, and so is omitted here. Some auxiliary results are covered as exercises.

Let $\mathbf{b}_1, \mathbf{b}_2, \ldots, \mathbf{b}_n$ be linearly independent vectors in $\mathbb{R}^n$ (written as column vectors). The set L of all integer-linear combinations of these vectors is called a *lattice*, denoted as

$$L = \sum_{i=1}^{n} \mathbb{Z}\mathbf{b}_i = \{r_1\mathbf{b}_1 + r_2\mathbf{b}_2 + \cdots + r_n\mathbf{b}_n \mid r_i \in \mathbb{Z}\}.$$

We say that $\mathbf{b}_1, \mathbf{b}_2, \ldots, \mathbf{b}_n$ constitute a *basis* of L, and also that L is *generated* by $\mathbf{b}_1, \mathbf{b}_2, \ldots, \mathbf{b}_n$. The *determinant* of L is defined as

$$d(L) = |\det(\mathbf{b}_1, \mathbf{b}_2, \ldots, \mathbf{b}_n)|.$$

Example 3.36 Figure 3.1(a) shows some points in the two-dimensional lattice generated by the vectors $\mathbf{b}_1 = \begin{pmatrix} 3 \\ 0 \end{pmatrix}$ and $\mathbf{b}_2 = \begin{pmatrix} 1 \\ 3 \end{pmatrix}$. The area of the shaded region is the determinant of this lattice, which is $\left| \det \begin{pmatrix} 3 & 1 \\ 0 & 3 \end{pmatrix} \right| = 9$.

FIGURE 3.1: A two-dimensional lattice

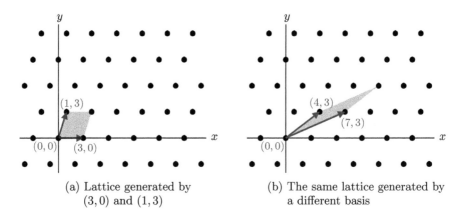

(a) Lattice generated by $(3,0)$ and $(1,3)$	(b) The same lattice generated by a different basis

In Figure 3.1(b), the same lattice is generated by the linearly independent vectors $\mathbf{c}_1 = 2\mathbf{b}_1 + \mathbf{b}_2 = \begin{pmatrix} 7 \\ 3 \end{pmatrix}$ and $\mathbf{c}_2 = \mathbf{b}_1 + \mathbf{b}_2 = \begin{pmatrix} 4 \\ 3 \end{pmatrix}$. Indeed, $\mathbf{b}_1 = \mathbf{c}_1 - \mathbf{c}_2$

---

[11]Arjen K. Lenstra, Hendrik W. Lenstra, Jr. and László Lovász, Factoring polynomials with rational coefficients, *Mathematische Annalen*, 261, 515–534, 1982.

and $\mathbf{b}_2 = -\mathbf{c}_1 + 2\mathbf{c}_2$, so $\mathbf{c}_1, \mathbf{c}_2$ generate the same lattice as $\mathbf{b}_1, \mathbf{b}_2$. The shaded region in Part (b) has the same area $\left| \det \begin{pmatrix} 7 & 4 \\ 3 & 3 \end{pmatrix} \right| = 9$ as in Part (a).

A lattice can have arbitrarily large basis vectors. For example, the two-dimensional lattice of Figure 3.1 is also generated by $\mathbf{d}_1 = 1000\mathbf{b}_1 + 1001\mathbf{b}_2 = \begin{pmatrix} 4001 \\ 3003 \end{pmatrix}$ and $\mathbf{d}_2 = 999\mathbf{b}_1 + 1000\mathbf{b}_2 = \begin{pmatrix} 3997 \\ 3000 \end{pmatrix}$. □

It is evident that a lattice L generated by $\mathbf{b}_1, \mathbf{b}_2, \ldots, \mathbf{b}_n$ is also generated by $\mathbf{c}_1, \mathbf{c}_2, \ldots, \mathbf{c}_n$ if and only if $(\mathbf{c}_1 \ \mathbf{c}_2 \ \cdots \ \mathbf{c}_n) = T(\mathbf{b}_1 \ \mathbf{b}_2 \ \cdots \ \mathbf{b}_n)$ for an $n \times n$ matrix T with integer entries satisfying $\det T = \pm 1$. It follows that the determinant of L is a property of the lattice itself, and not of bases of L.

Given a lattice L generated by $\mathbf{b}_1, \mathbf{b}_2, \ldots, \mathbf{b}_n$, a pertinent question is to find a *shortest vector* in L, where the length of a vector $\mathbf{x} = (x_1 \ x_2 \ \cdots \ x_n)^{\mathrm{t}}$ is its standard Euclidean (or L_2) norm:

$$|\mathbf{x}| = \sqrt{x_1^2 + x_2^2 + \cdots + x_n^2}.$$

This shortest-vector problem turns out to be NP-Complete. Lenstra, Lenstra and Lovász propose an approximation algorithm for the problem. Although the approximation ratio is rather large $(2^{(n-1)/2})$, the algorithm solves many important computational problems in number theory.

The first crucial insight in the problem comes from a simple observation. Example 3.36 demonstrates that the same lattice L can be generated by many bases. The longer the basis vectors are, the more slender is the region corresponding to $\det(L)$ for that basis. Moreover, vectors in a basis have a tendency to be simultaneously long or simultaneously short. For solving the shortest-vector problem, we, therefore, plan to construct a basis of short vectors which are as orthogonal to one another as possible.

For the time being, let us treat $\mathbf{b}_1, \mathbf{b}_2, \ldots, \mathbf{b}_n$ as real vectors. All real-linear combinations of these vectors span the entire $\mathbb{R}^n$. Algorithm 3.10 constructs a basis $\mathbf{b}_1^*, \mathbf{b}_2^*, \ldots, \mathbf{b}_n^*$ of $\mathbb{R}^n$, with the vectors $\mathbf{b}_i^*$ orthogonal to one another. The basis $\mathbf{b}_1^*, \mathbf{b}_2^*, \ldots, \mathbf{b}_n^*$ of $\mathbb{R}^n$ is called the *Gram–Schmidt orthogonalization* of $\mathbf{b}_1, \mathbf{b}_2, \ldots, \mathbf{b}_n$. The inner product (or dot product) of the two vectors $\mathbf{x} = (x_1 \ x_2 \ \cdots \ x_n)^{\mathrm{t}}$ and $\mathbf{y} = (y_1 \ y_2 \ \cdots \ y_n)^{\mathrm{t}}$ is defined as

$$\langle \mathbf{x}, \mathbf{y} \rangle = x_1 y_1 + x_2 y_2 + \cdots + x_n y_n \in \mathbb{R}.$$

Algorithm 3.10: GRAM–SCHMIDT ORTHOGONALIZATION $\mathbf{b}_1^*, \mathbf{b}_2^*, \ldots, \mathbf{b}_n^*$ OF $\mathbf{b}_1, \mathbf{b}_2, \ldots, \mathbf{b}_n$

```
for i = 1, 2, ..., n, compute
```
$$\mathbf{b}_i^* = \mathbf{b}_i - \sum_{j=1}^{i-1} \mu_{i,j} \mathbf{b}_j^*, \text{ where } \mu_{i,j} = \langle \mathbf{b}_i, \mathbf{b}_j^* \rangle / \langle \mathbf{b}_j^*, \mathbf{b}_j^* \rangle.$$

The quantity $\mu_{i,j} \mathbf{b}_j^*$ in Algorithm 3.10 is the component of $\mathbf{b}_i$ in the direction of the vector $\mathbf{b}_j^*$. When all these components are removed from $\mathbf{b}_i$, the vector $\mathbf{b}_i^*$ becomes orthogonal to the vectors $\mathbf{b}_1^*, \mathbf{b}_2^*, \ldots, \mathbf{b}_{i-1}^*$ computed so far.

The multipliers $\mu_{i,j}$ are not necessarily integers, so $\mathbf{b}_i^*$ need not belong to the lattice generated by $\mathbf{b}_1, \mathbf{b}_2, \ldots, \mathbf{b}_n$. Moreover, if the vectors $\mathbf{b}_1, \mathbf{b}_2, \ldots, \mathbf{b}_n$ are already orthogonal to one another, we have $\mathbf{b}_i^* = \mathbf{b}_i$ for all $i = 1, 2, \ldots, n$. The notion of near-orthogonality is captured by the following definition.

Definition 3.37 A basis $\mathbf{b}_1, \mathbf{b}_2, \ldots, \mathbf{b}_n$ of a lattice L (or $\mathbb{R}^n$) is called *reduced* if its Gram–Schmidt orthogonalization satisfies the following two conditions:

$$|\mu_{i,j}| \;\leqslant\; \frac{1}{2} \text{ for all } i,j \text{ with } 1 \leqslant j < i \leqslant n, \tag{3.1}$$

$$|\mathbf{b}_i^* + \mu_{i,i-1}\mathbf{b}_{i-1}^*|^2 \;\geqslant\; \frac{3}{4}|\mathbf{b}_{i-1}^*|^2 \text{ for all } i \text{ with } 2 \leqslant i \leqslant n. \tag{3.2}$$

The constant $\frac{3}{4}$ in Condition (3.2) can be replaced by any real constant in the open interval $\left(\frac{1}{4}, 1\right)$. ◁

Example 3.38 Consider the two-dimensional lattice L of Example 3.36.

(1) First, consider the basis constituted by $\mathbf{b}_1 = \begin{pmatrix} 3 \\ 0 \end{pmatrix}$ and $\mathbf{b}_2 = \begin{pmatrix} 1 \\ 3 \end{pmatrix}$. Its Gram–Schmidt orthogonalization is computed as follows.

$$\mathbf{b}_1^* = \mathbf{b}_1 = \begin{pmatrix} 3 \\ 0 \end{pmatrix},$$

$$\mu_{2,1} = \frac{\langle \mathbf{b}_2, \mathbf{b}_1^* \rangle}{\langle \mathbf{b}_1^*, \mathbf{b}_1^* \rangle} = \frac{1 \times 3 + 3 \times 0}{3 \times 3 + 0 \times 0} = \frac{3}{9} = \frac{1}{3},$$

$$\mathbf{b}_2^* = \mathbf{b}_2 - \mu_{2,1}\mathbf{b}_1^* = \begin{pmatrix} 1 \\ 3 \end{pmatrix} - \frac{1}{3}\begin{pmatrix} 3 \\ 0 \end{pmatrix} = \begin{pmatrix} 0 \\ 3 \end{pmatrix}.$$

We have $|\mathbf{b}_2^* + \mu_{2,1}\mathbf{b}_1^*|^2 = |\mathbf{b}_2|^2 = 1^2 + 3^2 = 10$, and $|\mathbf{b}_1^*|^2 = 3^2 + 0^2 = 9$. Since $10 \geqslant \frac{3}{4} \times 9$, the basis $\mathbf{b}_1, \mathbf{b}_2$ of L is reduced.

(2) Next, consider the basis constituted by $\mathbf{c}_1 = \begin{pmatrix} 7 \\ 3 \end{pmatrix}$ and $\mathbf{c}_2 = \begin{pmatrix} 4 \\ 3 \end{pmatrix}$. For this basis, we have:

$$\mathbf{c}_1^* = \mathbf{c}_1 = \begin{pmatrix} 7 \\ 3 \end{pmatrix},$$

$$\mu_{2,1} = \frac{\langle \mathbf{c}_2, \mathbf{c}_1^* \rangle}{\langle \mathbf{c}_1^*, \mathbf{c}_1^* \rangle} = \frac{4 \times 7 + 3 \times 3}{7 \times 7 + 3 \times 3} = \frac{37}{58},$$

$$\mathbf{c}_2^* = \mathbf{c}_2 - \mu_{2,1}\mathbf{c}_1^* = \begin{pmatrix} 4 \\ 3 \end{pmatrix} - \frac{37}{58}\begin{pmatrix} 7 \\ 3 \end{pmatrix} = \begin{pmatrix} -\frac{27}{58} \\ \frac{63}{58} \end{pmatrix}.$$

Here, $|\mu_{2,1}| > \frac{1}{2}$, so Condition (3.1) is not satisfied. Moreover, $|\mathbf{c}_2^* + \mu_{2,1}\mathbf{c}_1^*|^2 = |\mathbf{c}_2|^2 = 4^2 + 3^2 = 25$, whereas $|\mathbf{c}_1^*|^2 = 7^2 + 3^2 = 58$, that is, Condition (3.2) too is not satisfied. The basis $\mathbf{c}_1, \mathbf{c}_2$ is, therefore, not reduced. □

The attractiveness of reduced bases stems from the following fact.

Proposition 3.39 *Let* $\mathbf{b}_1, \mathbf{b}_2, \ldots, \mathbf{b}_n$ *constitute a reduced basis of a lattice* L. *Then, for any non-zero vector* $\mathbf{x}$ *in* L, *we have*

$$|\mathbf{b}_1|^2 \leqslant 2^{n-1}|\mathbf{x}|^2.$$

Moreover, for any basis $\mathbf{x}_1, \mathbf{x}_2, \ldots, \mathbf{x}_n$ *of* L, *we have*

$$|\mathbf{b}_i|^2 \leqslant 2^{n-1} \max(|\mathbf{x}_1|^2, |\mathbf{x}_2|^2, \ldots, |\mathbf{x}_n|^2)$$

for all $i = 1, 2, \ldots, n$. $\lhd$

Now, let us come to the main question of this section: how we can convert a given basis $\mathbf{b}_1, \mathbf{b}_2, \ldots, \mathbf{b}_n$ of an n-dimensional lattice L to a reduced basis. If the given basis is not reduced, either Condition (3.1) or Condition (3.2) is violated (or both). Algorithm 3.11 repairs the violation of the condition $|\mu_{k,l}| > \frac{1}{2}$ for some k, l satisfying $1 \leqslant l < k \leqslant n$. The correctness of Algorithm 3.11 follows from the formulas for $\mu_{i,j}$ (see Algorithm 3.10). It is important to note that Algorithm 3.11 does not alter any of the orthogonal vectors $\mathbf{b}_i^*$.

Algorithm 3.11: SUBROUTINE FOR HANDLING $|\mu_{k,l}| > \frac{1}{2}$

Let r be the integer nearest to $\mu_{k,l}$.
Replace $\mathbf{b}_k$ by $\mathbf{b}_k - r\mathbf{b}_l$.
Subtract $r\mu_{l,j}$ from $\mu_{k,j}$ for $j = 1, 2, \ldots, l-1$.
Subtract r from $\mu_{k,l}$.

Handling the violation of the second condition is a bit more involved. Let us denote $|\mathbf{b}_i^*|^2$ by B_i. Since the vectors $\mathbf{b}_i^*$ are pairwise orthogonal, the violation of Condition (3.2) for some k in the range $2 \leqslant k \leqslant n$ can be rephrased as:

$$B_k + \mu_{k,k-1}^2 B_{k-1} < \frac{3}{4} B_{k-1}.$$

Now, we swap $\mathbf{b}_{k-1}$ and $\mathbf{b}_k$. This replaces the vector $\mathbf{b}_{k-1}^*$ by (the old vector) $\mathbf{b}_k^* + \mu_{k,k-1}\mathbf{b}_{k-1}^*$. Moreover, $\mathbf{b}_k^*$ (and $\mu_{k,k-1}$) are so updated that the new value of $\mathbf{b}_k^* + \mu_{k,k-1}\mathbf{b}_{k-1}^*$ equals the old vector $\mathbf{b}_{k-1}^*$. Consequently, Condition (3.2) is restored at k. The updating operations are given in Algorithm 3.12.

Algorithm 3.12: SUBROUTINE FOR HANDLING $B_k + \mu_{k,k-1}^2 B_{k-1} < \frac{3}{4} B_{k-1}$

Let $\mu = \mu_{k,k-1}$, and $B = B_k + \mu^2 B_{k-1}$.
Update $\mu_{k,k-1}$ to the value $\mu B_{k-1}/B$.
Set $B_k = B_{k-1}B_k/B$, and $B_{k-1} = B$.
Swap $\mathbf{b}_{k-1}$ and $\mathbf{b}_k$.
Swap $\mu_{k-1,j}$ and $\mu_{k,j}$ for all $j = 1, 2, \ldots, k-2$.
For $i = k+1, k+2, \ldots, n$ {
    Compute $M = \mu_{i,k-1} - \mu\mu_{i,k}$.
    Set $\mu_{i,k-1}$ to $\mu_{i,k} + \mu_{k,k-1}M$.
    Set $\mu_{i,k}$ to M.
}

The Lenstra–Lenstra–Lovász basis-reduction algorithm (Algorithm 3.13) uses the above two subroutines. First, the Gram–Schmidt orthogonalization $\mathbf{b}_1^*, \mathbf{b}_2^*, \ldots, \mathbf{b}_n^*$ is computed. Moreover, the squared norms $B_i = |\mathbf{b}_i^*|^2$ are computed. After this initialization stage, we do not explicitly require the orthogonal vectors $\mathbf{b}_i^*$. Only the values B_i suffice for the rest of the algorithm.

Algorithm 3.13: REDUCTION OF THE BASIS $\mathbf{b}_1, \mathbf{b}_2, \ldots, \mathbf{b}_n$ OF A LATTICE L

```
Compute b₁*, b₂*, ..., bₙ* by Algorithm 3.10.
Compute Bᵢ = |bᵢ*|² for i = 1, 2, ..., n.
Set t = 2.
While (t ⩽ n) {
    If (|μₜ,ₜ₋₁| > ½), call Algorithm 3.11 with k = t and l = t − 1.
    If (Bₜ < (¾ − μ²ₜ,ₜ₋₁)Bₜ₋₁) {
        Call Algorithm 3.12 with k = t.
        If t ⩾ 3, set t = t − 1.
    } else {
        For j = t − 2, t − 3, ..., 1 {
            If (|μₜ,ⱼ| > ½), call Algorithm 3.11 with k = t and l = j.
        }
        Set t = t + 1.
    }
}
Return the reduced basis b₁, b₂, ..., bₙ.
```

The main (while) loop of Algorithm 3.13 maintains the invariance on t that Condition (3.1) is satisfied for $1 \leqslant j < i \leqslant t - 1$, and Condition (3.2) is satisfied for $1 \leqslant i \leqslant t - 1$. Each iteration of the loop attempts to enforce the conditions at the current value of t. First, the condition on $\mu_{t,t-1}$ is checked, and if this condition is violated, it is repaired by invoking Algorithm 3.11. This repairing has no other side effects.

Next, Condition (3.2) is checked for $i = t$. If this condition is satisfied, the other $\mu_{t,j}$ values are handled, t is incremented, and the next iteration of the while loop is started. However, if Condition (3.2) does not hold for $i = t$, the vectors $\mathbf{b}_{t-1}$ and $\mathbf{b}_t$ are swapped, and the relevant updates are carried out by Algorithm 3.12. Although this restores Condition (3.2) for $i = t$, the vector $\mathbf{b}_{t-1}$ and consequently all $\mu_{t-1,j}$ values change in the process. Therefore, t is decremented, and the next iteration of the while loop is started. However, if $t = 2$, there are no valid values of the form $\mu_{t-1,j}$, so t is not decremented.

When the while loop terminates, we have $t = n+1$. By the loop invariance, both the conditions for a reduced basis are satisfied for all relevant values of i and j, that is, $\mathbf{b}_1, \mathbf{b}_2, \ldots, \mathbf{b}_n$ is now a reduced basis of the given lattice.

In order to establish that the while loop terminates after finitely many iterations, Lenstra, Lenstra and Lovász define the quantity

$$d_i = |\det(\,\langle \mathbf{b}_j, \mathbf{b}_k \rangle\,)_{1 \leqslant j,k \leqslant i}| \quad \text{for } i = 0, 1, 2, \ldots, n.$$

It turns out that d_i is the square of the volume of the fundamental region associated with the i-dimensional lattice generated by $\mathbf{b}_1, \mathbf{b}_2, \ldots, \mathbf{b}_i$, that is,

$$d_i = \prod_{j=1}^{i} |\mathbf{b}_j^*|^2 \leqslant \prod_{j=1}^{i} |\mathbf{b}_j|^2 \leqslant B^i,$$

where $B = \max\{|\mathbf{b}_k|^2 \mid 1 \leqslant k \leqslant n\}$. In particular, $d_0 = 1$, and $d_n = d(L)^2$ (where $d(L)$ is the determinant of the lattice L). It also turns out that

$$d_i \geqslant \left(\frac{3}{4}\right)^{i(i-1)/2} M^i$$

for all $i = 0, 1, 2, \ldots, n$, where $M = \min\{|\mathbf{x}|^2 \mid \mathbf{x} \in L \text{ and } \mathbf{x} \neq 0\}$ is a value which depends only on L (and not on any basis for L). Now, let

$$D = d_1 d_2 \cdots d_{n-1}.$$

It follows that D is bounded from below by a positive value determined by the *lattice* L, and from above by a positive value associated with a *basis* of L.

Each adjustment of a $\mu_{k,l}$ by Algorithm 3.11 does not alter any $\mathbf{b}_i^*$, so D remains unaffected. On the other hand, swapping of $\mathbf{b}_{t-1}$ and $\mathbf{b}_t$ by Algorithm 3.12 reduces d_{t-1} (and so D too) by a factor at least as large as $4/3$. Therefore, if B_{init} is the initial value of B, the `while` loop of Algorithm 3.13 goes through at most $O(n^2 \log B_{\text{init}})$ iterations. Since each iteration of the `while` loop can be carried out using $O(n^2)$ integer operations, the running time of the Lenstra–Lenstra–Lovász basis-reduction algorithm is equal to that of $O(n^4 \log B_{\text{init}})$ integer operations. The integers on which these operations are carried out have bit lengths $O(n \log B_{\text{init}})$. With schoolbook integer arithmetic, the running time of Algorithm 3.13 is, therefore, $O(n^6 \log^3 B_{\text{init}})$.

Example 3.40 Let us reduce the basis $\mathbf{c}_1, \mathbf{c}_2$ of Figure 3.1(b). Rename these vectors as $\mathbf{b}_1 = \begin{pmatrix} 7 \\ 3 \end{pmatrix}$ and $\mathbf{b}_2 = \begin{pmatrix} 4 \\ 3 \end{pmatrix}$. In Example 3.38(2), the Gram–Schmidt orthogonalization of this basis is computed as

$$\mathbf{b}_i^* = \begin{pmatrix} 7 \\ 3 \end{pmatrix}, \quad \mu_{2,1} = \frac{37}{58}, \quad \mathbf{b}_2^* = \begin{pmatrix} -\frac{27}{58} \\ \frac{63}{58} \end{pmatrix}.$$

This gives the squared norm values as (we have $B_1 B_2 = d(L)^2$, as expected):

$$B_1 = 7^2 + 3^2 = 58, \quad B_2 = \left(\frac{27}{58}\right)^2 + \left(\frac{63}{58}\right)^2 = \frac{81}{58}.$$

In the first iteration of the `while` loop of Algorithm 3.13, we have $t = 2$, and the condition on $\mu_{2,1}$ is violated (we have $|\mu_{2,1}| > \frac{1}{2}$). The integer closest to $\mu_{2,1} = \frac{37}{58}$ is 1. So we replace $\mathbf{b}_2$ by $\mathbf{b}_2 - \mathbf{b}_1 = \begin{pmatrix} -3 \\ 0 \end{pmatrix}$, and $\mu_{2,1}$ by $\mu_{2,1} - 1 = -\frac{21}{58}$. The values of B_1 and B_2 do not change by this adjustment.

We now have $B_2 = \frac{81}{58}$, and $(\frac{3}{4} - \mu_{2,1}^2)B_1 = \frac{1041}{29}$, that is, Condition (3.2) is violated for $t = 2$. So we invoke Algorithm 3.12. We first compute $B = B_2 + \mu_{2,1}^2 B_1 = 9$, change $\mu_{2,1}$ to $\mu_{2,1}B_1/B = -\frac{7}{3}$, set $B_2 = B_1 B_2 / B = 9$ and $B_1 = B = 9$. Finally, we swap $\mathbf{b}_1$ and $\mathbf{b}_2$, that is, we now have $\mathbf{b}_1 = \begin{pmatrix} -3 \\ 0 \end{pmatrix}$ and $\mathbf{b}_2 = \begin{pmatrix} 7 \\ 3 \end{pmatrix}$. Since $t = 2$, we do not decrement t.

In the second iteration of the `while` loop, we first discover that $|\mu_{2,1}| = \frac{7}{3}$ is again too large. The integer closest to $\mu_{2,1}$ is -2. So we replace $\mathbf{b}_2$ by $\mathbf{b}_2 + 2\mathbf{b}_1 = \begin{pmatrix} 1 \\ 3 \end{pmatrix}$, and $\mu_{2,1}$ by $\mu_{2,1} + 2 = -\frac{1}{3}$.

Since $B_2 = 9 \geqslant \frac{23}{4} = (\frac{3}{4} - \mu_{2,1}^2)B_1$, we do not swap $\mathbf{b}_1$ and $\mathbf{b}_2$. Moreover, there are no $\mu_{t,j}$ values to take care of. So t is increased to three, and the algorithm terminates. The computed reduced basis consists of the vectors $\mathbf{b}_1 = \begin{pmatrix} -3 \\ 0 \end{pmatrix}$ and $\mathbf{b}_2 = \begin{pmatrix} 1 \\ 3 \end{pmatrix}$. Compare this with the basis in Figure 3.1(a). $\square$

3.5.3 Lenstra–Lenstra–Lovász Factoring Algorithm

The Lenstra–Lenstra–Lovász or the L^3 or the LLL algorithm relates polynomial factorization with lattices in a very clever way. Let $f(x) \in \mathbb{Z}[x]$ be a polynomial of degree $d > 0$. For simplicity, assume that f is square-free and monic. I show how the L^3 algorithm is capable of producing a non-trivial split of f, or more precisely, discovering an irreducible factor $g(x)$ of $f(x)$.

Like Berlekamp's algorithm, the L^3 algorithm chooses a prime p not dividing $\mathrm{Discr}(f)$, and also a suitable positive integer k. First, f is factored modulo p, and this factorization is refined to one modulo p^k using Hensel's lifting. Suppose that this gives us a non-constant polynomial $\gamma(x) \in \mathbb{Z}[x]$ satisfying:

(1) $\gamma(x)$ in monic (in $\mathbb{Z}[x]$).
(2) $\gamma(x)$ divides $f(x)$ in $\mathbb{Z}_{p^k}[x]$.
(3) $\gamma(x)$ is irreducible in $\mathbb{F}_p[x]$.
(4) $\gamma(x)^2$ does not divide $f(x)$ in $\mathbb{F}_p[x]$.

Let $l = \deg \gamma(x)$. If $l = d$, the polynomial $f(x)$ is itself irreducible (modulo p and so in $\mathbb{Z}[x]$ too). So we assume that $1 \leqslant l \leqslant d - 1$. The polynomial $\gamma(x)$ needs to be known modulo p^k only. So we may assume that its coefficients are between 0 and $p^k - 1$ (or between $-p^k/2$ and $p^k/2$). In any case, we have

$$|\gamma|^2 \leqslant 1 + lp^{2k}.$$

Lenstra, Lenstra and Lovász prove that the above four conditions uniquely identify a monic irreducible factor $g(x)$ of $f(x)$ (in $\mathbb{Z}[x]$). It is precisely that irreducible factor which is divisible by $\gamma(x)$ in $\mathbb{Z}_{p^k}[x]$. In order to compute this factor $g(x)$, Lenstra et al. use a lattice. An integer m in the range $l \leqslant m \leqslant d$ is chosen to satisfy the following condition (to be justified later):

$$p^{kl} > 2^{md/2} \binom{2m}{m}^{d/2} |f|^{m+d} . \tag{3.3}$$

Let $h(x) \in \mathbb{Z}[x]$ be of degree $\leqslant m$, and treat $\gamma(x)$ as a (monic) polynomial with integer coefficients. Euclidean division of $h(x)$ by $\gamma(x)$ gives

$$h(x) = (q_{m-l}x^{m-l} + q_{m-l-1}x^{m-l-1} + \cdots + q_1 x + q_0)\gamma(x) + (r_{l-1}x^{l-1} + r_{l-2}x^{l-2} + \cdots + r_1 x + r_0)$$

with integers q_i and r_j. The condition $\gamma(x)|h(x)$ modulo p^k is equivalent to having each r_j divisible by p^k. Writing $r_j = p^k s_j$, we get

$$h(x) = (q_{m-l}x^{m-l} + q_{m-l-1}x^{m-l-1} + \cdots + q_1 x + q_0)\gamma(x) + (s_{l-1}p^k x^{l-1} + s_{l-2}p^k x^{l-2} + \cdots + s_1 p^k x + s_0 p^k)$$

We treat a polynomial as a vector of its coefficients. The polynomials $h(x) \in \mathbb{Z}[x]$ of degree $\leqslant m$ and divisible by $\gamma(x)$ modulo p^k, therefore, form an $(m+1)$-dimensional lattice L generated by the vectors $x^i \gamma(x)$ for $i = 0, 1, 2, \ldots, m-l$ and by $p^k x^j$ for $j = 0, 1, 2, \ldots, l-1$. Clearly, the determinant of this lattice is

$$d(L) = p^{kl}.$$

This lattice L is related to the irreducible factor $g(x)$ of $f(x)$ as follows.

Proposition 3.41 *If $b \in L$ satisfies*

$$p^{kl} > |f|^m |b|^d,$$

then $b(x)|g(x)$ in $\mathbb{Z}[x]$. In particular, $\gcd(f(x), b(x)) \neq 1$ in this case. ◁

Proposition 3.41 indicates the importance of short vectors in L. Let $b_1, b_2, \ldots, b_{m+1}$ constitute a reduced basis of L as computed by Algorithm 3.13.[12]

Proposition 3.42 *Suppose that Condition (3.3) is satisfied. Let $g(x) \in \mathbb{Z}[x]$ be the desired irreducible factor of $f(x)$. Then,*

$$\deg g(x) \leqslant m \text{ if and only if } |b_1| < \left(p^{kl}/|f|^m\right)^{1/d} .$$

Let $t \geqslant 1$ be the largest integer for which $|b_t| < \left(p^{kl}/|f|^m\right)^{1/d}$. Then, we have $\deg g(x) = m+1-t$ and, more importantly, $g(x) = \gcd(b_1(x), b_2(x), \ldots, b_t(x))$. Moreover, in this case, $|b_i| < \left(p^{kl}/|f|^m\right)^{1/d}$ for all $i = 1, 2, \ldots, t$. ◁

Example 3.43 Let us factor

$$f(x) = x^5 - 2x^4 + 6x^3 - 5x^2 + 10x - 4$$

by the L^3 algorithm. The discriminant of f is

$$\text{Discr}(f) = 2634468 = 2^2 \times 3 \times 59 \times 61.$$

---

[12]Here, vectors are polynomials too, so I use the notation b_i instead of $\mathbf{b}_i$.

If $f(x)$ is reducible in $\mathbb{Z}[x]$, it must have an irreducible factor of degree $\leqslant 2$. In order to locate this factor, a safe (but optimistic) choice is $m = 2$. Moreover, $d = \deg f = 5$, and $|f| = \sqrt{1^2 + 2^2 + 6^2 + 5^2 + 10^2 + 4^2} \approx 13.491$. This gives us an estimate of the right side of Condition (3.3). We choose the prime $p = 229497501943$ such that Condition (3.3) is satisfied for $k = l = 1$. See below (after this example) how the parameters should actually be chosen. For this sample demonstration, the above choices suffice.

The polynomial $f(x)$ factors modulo p as

$$f(x) \equiv (x + 108272275755)(x + 121225226186)(x + 143510525420) \times$$
$$(x^2 + 85986976523x + 46046039585) \pmod{p}.$$

Let us take $\gamma(x) = x + 108272275755$, so $l = 1$. Let us also take $k = 1$. The values of m for which Condition (3.3) is satisfied are $m = 1$ and $m = 2$. Although $f(x)$ (if reducible in $\mathbb{Z}[x]$) must have an irreducible factor of degree $\leqslant 2$, this factor need not be the desired multiple $g(x)$ of $\gamma(x)$ in $\mathbb{F}_p[x]$. In practice, p and k should be so chosen that Condition (3.3) holds for $m = d - 1$.

We first try with $m = 1$. We consider the two-dimensional lattice generated by $\gamma(x)$ and p, that is, by the vectors $\begin{pmatrix} 1 \\ 108272275755 \end{pmatrix}$ and $\begin{pmatrix} 0 \\ 229497501943 \end{pmatrix}$. Algorithm 3.13 gives us a reduced basis consisting of the polynomials $b_1(x) = 442094x - 122483$ and $b_2(x) = -335557x - 426148$ (equivalently, the vectors $\begin{pmatrix} 442094 \\ -122483 \end{pmatrix}$ and $\begin{pmatrix} -335557 \\ -426148 \end{pmatrix}$). We have

$$\left(p^{kl}/|f|^m \right)^{1/d} \approx 111.21.$$

But b_1 and b_2 have much larger L_2 norms. So the degree of the desired factor $g(x)$ of $f(x)$ is larger than one, and the factoring attempt fails for $m = 1$.

Next, we try with $m = 2$. We consider the three-dimensional lattice generated by $x\gamma(x)$, $\gamma(x)$ and p, that is, by the vectors $\begin{pmatrix} 1 \\ 108272275755 \\ 0 \end{pmatrix}$, $\begin{pmatrix} 0 \\ 1 \\ 108272275755 \end{pmatrix}$ and $\begin{pmatrix} 0 \\ 0 \\ 229497501943 \end{pmatrix}$. A reduced basis for this lattice consists of the three polynomials $b_1(x) = x^2 - 2x + 4$, $b_2(x) = 114648x^2 - 122759x - 90039$ and $b_3(x) = 180082 + 188467 + 49214$, that is, the vectors $\begin{pmatrix} 1 \\ -2 \\ 4 \end{pmatrix}$, $\begin{pmatrix} 114648 \\ -122759 \\ -90039 \end{pmatrix}$ and $\begin{pmatrix} 180082 \\ 188467 \\ 49214 \end{pmatrix}$. For $m = 2$, we have

$$\left(p^{kl}/|f|^m \right)^{1/d} \approx 66.09.$$

Clearly, b_1 has L_2 norm less than this, whereas b_2 and b_3 have L_2 norms larger than this. So $t = 1$ (see Proposition 3.42), that is, $\deg g(x) = m + 1 - t = 2$. Since $t = 1$, no gcd calculation is necessary, and $g(x) = b_1(x) = x^2 - 2x + 4$.

Once the irreducible factor $g(x)$ is discovered, what remains is to factor $f(x)/g(x) = x^3 + 2x - 1$. I do not show this factoring attempt here. Indeed, this cofactor is an irreducible polynomial (in $\mathbb{Z}[x]$). $\qquad\qquad\square$

Some comments on the L^3 factoring algorithm are now in order. First, let me prescribe a way to fix the parameters. Since the irreducible factor $g(x)$ may be of degree as large as $d - 1$, it is preferable to start with $m = d - 1$. For this choice, we compute the right side of Condition (3.3). A prime p and a positive integer k is then chosen to satisfy this condition (may be for the most pessimistic case $l = 1$). The choice $k = 1$ is perfectly allowed. Even if some $k > 1$ is chosen, lifting the factorization of $f(x)$ modulo p to the factorization modulo p^k is an easy effort. Factoring $f(x)$ modulo p can also be efficiently done using the randomized algorithm described in Section 3.3.

If $f(x)$ remains irreducible modulo p, we are done. Otherwise, we choose any irreducible factor of $f(x)$ modulo p^k as $\gamma(x)$. Under the assumption that p does not divide $\mathrm{Discr}(f)$, no factor of f modulo p has multiplicity larger than one. The choice of γ fixes l, and we may again investigate for which values of m, Condition (3.3) holds. We may start with any such value of m. However, the choice $m = d - 1$ is always safe, since a value of m smaller than the degree of $g(x)$ forces us to repeat the basis-reduction process for a larger value of m.

The basic difference between Berlekamp's factoring algorithm and the L^3 algorithm is that in Berlekamp's algorithm, we may have to explore an exponential number of combinations of the irreducible factors of f modulo p. On the contrary, the L^3 algorithm starts with any (and only one) suitable factor of f modulo p in order to discover one irreducible factor of f. Therefore, the L^3 algorithm achieves a polynomial running time even in the worst case. However, both these algorithms are based on factoring f modulo p. Although this can be solved efficiently using randomized algorithms, there is no known polynomial-time deterministic algorithm for this task.

Lenstra et al. estimate that the L^3 algorithm can factor f completely using only $O(d^6 + d^5 \log |f| + d^4 \log p)$ arithmetic operations on integers of bit sizes bounded above by $O(d^3 + d^2 \log |f| + d \log p)$.

3.5.4 Factoring in GP/PARI

The generic factoring function in GP/PARI is `factor()`.

```
gp > factor(35*x^5 - 22*x^3 + 10*x^2 + 3*x - 2)
%1 =
[5*x^2 - 1 1]

[7*x^3 - 3*x + 2 1]

gp > factor(Mod(35,13)*x^5-Mod(22,13)*x^3+Mod(10,13)*x^2+Mod(3,13)*x-Mod(2,13))
%2 =
[Mod(1, 13)*x + Mod(5, 13) 1]
```

```
[Mod(1, 13)*x + Mod(10, 13) 1]

[Mod(1, 13)*x + Mod(11, 13) 1]

[Mod(1, 13)*x^2 + Mod(5, 13) 1]
```

GP/PARI supplies the built-in function qflll for lattice-basis reduction. The initial basis vectors of a lattice should be packed in a matrix. Each column should store one basis vector. The return value is again a matrix which is, however, not the reduced basis vectors packed in a similar format. It is indeed a transformation matrix which, when post-multiplied by the input matrix, gives the reduced basis vectors. This is demonstrated for the two-dimensional lattice of Example 3.40 and the three-dimensional lattice of Example 3.43.

```
gp > M = [ 7, 4; \
           3, 3];
gp > T = qflll(M)
%2 =
[-1 -1]

[1 2]

gp > M * T
%3 =
[-3 1]

[0 3]

gp > M = [1,108272275755,0; 0,1,108272275755; 0,0,229497501943];
gp > M = mattranspose(M);
gp > T = qflll(M)
%6 =
[1 114648 180082]

[-108272275757 -12413199870881999 -19497887962323443]

[51080667973 5856296421718242 9198708849654953]

? M * T
%7 =
[1 114648 180082]

[-2 -122759 188467]

[4 -90039 49214]
```

Exercises

1. [*Multiplicative form of Möbius inversion formula*]   Let f, g be two functions of natural numbers satisfying $f(n) = \prod_{d|n} g(d)$ for all $n \in \mathbb{N}$. Prove that $g(n) = \prod_{d|n} f(d)^{\mu(n/d)} = \prod_{d|n} f(n/d)^{\mu(d)}$ for all $n \in \mathbb{N}$.

2. (a) Find an explicit formula for the product of all monic irreducible polynomials of degree n in $\mathbb{F}_q[x]$.
(b) Find the product of all monic sextic irreducible polynomials of $\mathbb{F}_2[x]$.
(c) Find the product of all monic cubic irreducible polynomials of $\mathbb{F}_4[x]$.

3. Which of the following polynomials is/are irreducible in $\mathbb{F}_2[x]$?
(a) $x^5 + x^4 + 1$.
(b) $x^5 + x^4 + x + 1$.
(c) $x^5 + x^4 + x^2 + x + 1$.

4. Which of the following polynomials is/are irreducible in $\mathbb{F}_3[x]$?
(a) $x^4 + 2x + 1$.
(b) $x^4 + 2x + 2$.
(c) $x^4 + x^2 + 2x + 2$.

5. Prove that a polynomial $f(x) \in \mathbb{F}_q[x]$ of degree two or three is irreducible if and only if $f(x)$ has no roots in $\mathbb{F}_q$.

6. Argue that the termination criterion for the loop in Algorithm 3.4 may be changed to $\deg f(x) \leqslant 2r + 1$. Modify the algorithm accordingly. Explain how this modified algorithm may speed up distinct-degree factorization.

7. Establish that the square-free and the distinct-degree factorization algorithms described in the text run in time polynomial in $\deg f$ and $\log q$.

8. Consider the root-finding Algorithm 3.2. Let $v_\alpha(x) = (x + \alpha)^{(q-1)/2} - 1$, $w_\alpha(x) = (x + \alpha)^{(q-1)/2} + 1$, $v(x) = v_0(x)$, and $w(x) = w_0(x)$.
(a) Prove that the roots of $v(x)$ are all the quadratic residues of $\mathbb{F}_q^*$, and those of $w(x)$ are all the quadratic non-residues of $\mathbb{F}_q^*$.
(b) Let $f(x) \in \mathbb{F}_q[x]$ with $d = \deg f \geqslant 2$ be a product of distinct linear factors. Assume that the roots of $f(x)$ are random elements of $\mathbb{F}_q$. Moreover, assume that the quadratic residues in $\mathbb{F}_q^*$ are randomly distributed in $\mathbb{F}_q^*$. Compute the probability that the polynomial $\gcd(f(x), v_\alpha(x))$ is a non-trivial factor of $f(x)$ for a randomly chosen $\alpha \in \mathbb{F}_q$.
(c) Deduce that the expected running time of Algorithm 3.2 is polynomial in d and $\log q$.

9. (a) Generalize Exercise 3.8 in order to compute the probability that a random $\alpha \in \mathbb{F}_q$ splits $f(x)$ in two non-trivial factors in Algorithm 3.5. Make reasonable assumptions as in Exercise 3.8.
(b) Deduce that the expected running time of Algorithm 3.5 is polynomial in $\deg f$ and $\log q$.

10. Consider the root-finding Algorithm 3.3 over $\mathbb{F}_q$, where $q = 2^n$. Let $v(x) = x + x^2 + x^{2^2} + x^{2^3} + \cdots + x^{2^{n-1}}$, and $w(x) = 1 + v(x)$. Moreover, let $f(x) \in \mathbb{F}_q[x]$ be a product of distinct linear factors, and $d = \deg f(x)$.

(a) Prove that $v(x) = \displaystyle\prod_{\substack{\gamma \in \mathbb{F}_q \\ \mathrm{Tr}(\gamma) = 0}} (x + \gamma)$, and that $w(x) = \displaystyle\prod_{\substack{\gamma \in \mathbb{F}_q \\ \mathrm{Tr}(\gamma) = 1}} (x + \gamma)$, where $\mathrm{Tr}(\gamma) \in \mathbb{F}_2$ is the trace of $\gamma \in \mathbb{F}_q$ as defined in Exercise 2.58.

(b) Prove that $v(x+\alpha)$ is equal to $v(x)$ or $w(x)$ for every $\alpha \in \mathbb{F}_q$. In particular, $\gcd(v(x), f(x))$ is a non-trivial factor of $f(x)$ if and only if $\gcd(v(x+\alpha), f(x))$ is a non-trivial factor of $f(x)$ for each $\alpha \in \mathbb{F}_q$.

This implies that if $v(x)$ fails to split $f(x)$ non-trivially, then so also fails every $v(x + \alpha)$. If so, we need to resort to Algorithm 3.8 with $r = 1$, that is, we choose $u(x) \in \mathbb{F}_q[x]$ and compute $\gcd(v(u(x)), f(x))$ with the hope that this gcd is a non-trivial divisor of $f(x)$. We now propose a way to choose $u(x)$.

(c) Let $i \in \mathbb{N}$, and $\alpha \in \mathbb{F}_q^*$. Prove that $v(\alpha x^{2i}) = v(\alpha^{2^{n-1}} x^i)^2$.

(d) Take any two polynomials $u_1(x), u_2(x) \in \mathbb{F}_q[x]$. Prove that $v(u_1(x) + u_2(x)) = v(u_1(x)) + v(u_2(x)) = w(u_1(x)) + w(u_2(x))$.

(e) Suppose that all (non-constant) polynomials $u(x) \in \mathbb{F}_q[x]$ of degrees $< s$ fail to split $f(x)$. Take $u(x) \in \mathbb{F}_q[x]$ of degree s and leading coefficient α. Prove that $u(x)$ splits $f(x)$ non-trivially if and only if αx^s splits $f(x)$ non-trivially.

(f) Prescribe a strategy to choose the polynomial $u(x)$ in Algorithm 3.8.

11. Generalize the ideas developed in Exercise 3.10 to rewrite Algorithm 3.8 for equal-degree factorization. More precisely, establish that the above sequence of choosing $u(x)$ works for equal-degree factorization too.

12. Let $q = p^n$, $r \in \mathbb{N}$, and $v(x) = x + x^p + x^{p^2} + \cdots + x^{p^{nr-1}}$. Prove that $x^{q^r} - x = \displaystyle\prod_{a \in \mathbb{F}_p}(v(x) - a)$. Prove also that $v(x) - a = \displaystyle\prod_{\substack{\gamma \in \mathbb{F}_{q^r} \\ \mathrm{Tr}(\gamma) = a}} (x - \gamma)$ for each $a \in \mathbb{F}_p$.

13. Find all the roots of $x^6 + x + 5$ in $\mathbb{F}_{17}$.

14. Find all the roots of $x^4 + (\theta + 1)x + \theta$ in $\mathbb{F}_8 = \mathbb{F}_2(\theta)$ with $\theta^3 + \theta + 1 = 0$.

15. Find all the roots of $x^5 + (\theta + 1)x + (2\theta + 1)$ in $\mathbb{F}_9 = \mathbb{F}_3(\theta)$ with $\theta^2 + 1 = 0$.

16. Find the square-free factorization of $x^9 + x^8 + x^7 + x^6 + x^5 + x^4 + x + 1 \in \mathbb{F}_2[x]$.

17. Find the square-free factorization of $x^{10} + 2x^9 + x^8 + 2x^5 + 2x^4 + 1 \in \mathbb{F}_3[x]$.

18. Find the square-free factorization of $x^{20} + (\theta + 1)x^8 + \theta \in \mathbb{F}_8[x]$, where $\mathbb{F}_8 = \mathbb{F}_2(\theta)$ with $\theta^3 + \theta + 1 = 0$.

19. Find the square-free factorization of $x^{15} + (2\theta + 1)x^{12} + (\theta + 2) \in \mathbb{F}_9[x]$, where $\mathbb{F}_9 = \mathbb{F}_3(\theta)$ with $\theta^2 + 1 = 0$.

20. Find the distinct-degree factorization of $x^8 + x^3 + x^2 + 1 \in \mathbb{F}_2[x]$.

21. Find the distinct-degree factorization of $x^8 + x^2 + 1 \in \mathbb{F}_3[x]$.

22. Find the distinct-degree factorization of $x^4 + (\theta + 1)x + \theta \in \mathbb{F}_8[x]$, where $\mathbb{F}_8 = \mathbb{F}_2(\theta)$ with $\theta^3 + \theta + 1 = 0$.

23. Find the distinct-degree factorization of $x^5 + (\theta + 1)x + (2\theta + 1) \in \mathbb{F}_9[x]$, where $\mathbb{F}_9 = \mathbb{F}_3(\theta)$ with $\theta^2 + 1 = 0$.

24. Find the equal-degree factorization of $x^4 + 7x + 2 \in \mathbb{F}_{17}[x]$, which is a product of two quadratic irreducible polynomials.

25. Find the equal-degree factorization of $x^6 + 16x^5 + 3x^4 + 16x^3 + 8x^2 + 8x + 14 \in \mathbb{F}_{17}[x]$, which is a product of three quadratic irreducible polynomials.

26. Represent $\mathbb{F}_8 = \mathbb{F}_2(\theta)$ with $\theta^3 + \theta + 1 = 0$. Find the equal-degree factorization of $x^4 + (\theta+1)x^2 + \theta x + (\theta^2 + \theta + 1) \in \mathbb{F}_8[x]$, which is a product of two quadratic irreducible polynomials.

27. Represent $\mathbb{F}_9 = \mathbb{F}_3(\theta)$ with $\theta^2 + 1 = 0$. Find the equal-degree factorization of $x^4 + \theta x^2 + x + (\theta + 2) \in \mathbb{F}_9[x]$, which is a product of two quadratic irreducible polynomials.

28. [*Chinese remainder theorem for polynomials*] Let K be a field, and $m_1(x)$, $m_2(x), \ldots, m_t(x)$ be pairwise coprime non-constant polynomials in $K[x]$ with $d_i = \deg m_i(x)$. Prove that given polynomials $a_1(x), a_2(x), \ldots, a_t(x) \in K[x]$, there exists a unique polynomial $f(x) \in K[x]$ of degree $< \sum_{i=1}^t d_i$ such that

$$f(x) \equiv a_i(x) \;(\text{mod } m_i(x)) \quad \text{for all } i = 1, 2, \ldots, t.$$

29. Let $f(x) \in \mathbb{F}_q[x]$ be a monic non-constant polynomial, and let $h(x) \in \mathbb{F}_q[x]$ satisfy $h(x)^q \equiv h(x) \;(\text{mod } f(x))$. Prove that $h(x)^q - h(x) = \prod_{\gamma \in \mathbb{F}_q}(h(x) - \gamma)$. Conclude that $f(x) = \prod_{\gamma \in \mathbb{F}_q} \gcd(f(x), h(x) - \gamma)$.

30. [*Berlekamp's Q-matrix factorization*] You are given a monic non-constant square-free polynomial $f(x) \in \mathbb{F}_q[x]$ with t irreducible factors (not necessarily of the same degree). Let $d = \deg f(x)$.
(a) Prove that there are exactly q^t polynomials of degrees less than d satisfying $h(x)^q \equiv h(x) \;(\text{mod } f(x))$.
(b) In order to determine all these polynomials $h(x)$ of Part (a), write $h(x) = \alpha_0 + \alpha_1 x + \alpha_2 x^2 + \cdots + \alpha_{d-1} x^{d-1}$. Derive a $d \times d$ matrix Q such that the unknown coefficients $\alpha_0, \alpha_1, \alpha_2, \ldots, \alpha_{d-1} \in \mathbb{F}_q$ can be obtained by solving the homogeneous linear system $Q \,(\alpha_0 \quad \alpha_1 \quad \alpha_2 \quad \cdots \quad \alpha_{d-1})^t = \mathbf{0}$ in $\mathbb{F}_q$.
(c) Deduce that the matrix Q has rank $d - t$ and nullity t.
(d) Suppose that $t \geq 2$. Let $V \cong \mathbb{F}_q^t$ denote the nullspace of Q. Prove that for every two irreducible factors $f_1(x), f_2(x)$ of $f(x)$ and for every two distinct elements $\gamma_1, \gamma_2 \in \mathbb{F}_q$, there exists an $(\alpha_0, \alpha_1, \ldots, \alpha_{d-1}) \in V$ such that $h(x) \equiv \gamma_1 \;(\text{mod } f_1(x))$ and $h(x) \equiv \gamma_2 \;(\text{mod } f_2(x))$, where $h(x) = \alpha_0 + \alpha_1 x + \cdots + \alpha_{d-1} x^{d-1}$. Moreover, for any basis of V, we can choose distinct $\gamma_1, \gamma_2 \in \mathbb{F}_q$ in such a way that $(\alpha_0, \alpha_1, \ldots, \alpha_{d-1})$ is a vector of the basis.
(e) Assume that q is small. Propose a deterministic polynomial-time algorithm for factoring $f(x)$ based on the ideas developed in this exercise.

31. Factor $x^8 + x^5 + x^4 + x + 1 \in \mathbb{F}_2[x]$ using Berlekamp's Q-matrix algorithm.

32. Let $f(x) \in \mathbb{F}_q[x]$ be as in Exercise 3.30. Evidently, $f(x)$ is irreducible if and only if $t = 1$. Describe a polynomial-time algorithm for checking the irreducibility of $f(x)$, based upon the determination of t. You do not need to assume that q is small. Compare this algorithm with Algorithm 3.1.

33. Using Eisenstein's criterion, prove the irreducibility in $\mathbb{Z}[x]$ of:
(a) $7x^2 - 180$.
(b) $x^{23} - 9x^{12} + 15$.
(c) $x^2 + x + 2$. (**Hint:** Replace x by $x + 3$.)
(d) $x^4 + 2x + 7$.

34. [*Eisenstein's criterion in $K[x, y]$*] Let K be a field. Write $f(x, y) \in K[x, y]$ as $f(x, y) = a_d(y)x^d + a_{d-1}(y)x^{d-1} + \cdots + a_1(y)x + a_0(y)$ with $a_i(y) \in K[y]$. Assume that $d \geqslant 1$, and $\gcd(a_0(y), a_1(y), a_2(y), \ldots, a_d(y)) = 1$. Suppose that there exists an irreducible polynomial $p(y)$ in $K[y]$ with the properties that $p(y) \nmid a_d(y)$, $p(y)|a_i(y)$ for $i = 0, 1, \ldots, d - 1$, and $p(y)^2 \nmid a_0(y)$. Prove that $f(x, y)$ is irreducible in $K[x, y]$.

35. Prove that $x^4y + xy^4 + 4xy^2 - y^3 + 4x - 2y$ is irreducible in $\mathbb{Q}[x, y]$.

36. Let $n \in \mathbb{N}$. A complex number ω satisfying $\omega^n = 1$ is called an *n-th root of unity*. An n-th root ω of unity is called *primitive* if $\omega^m \neq 1$ for $1 \leqslant m < n$. Denote by $P(n)$ the set of all primitive n-th roots of unity. The *n-th cyclotomic polynomial* $\Phi_n(x)$ is defined as $\Phi_n(x) = \prod_{\omega \in P(n)}(x - \omega) \in \mathbb{C}[x]$.
(a) Prove that $\deg \Phi_n(x) = \phi(n)$, where $\phi()$ is Euler's totient function.
(b) Compute $\Phi_n(x)$ for $n = 1, 2, 3, 4, 5, 6$.
(c) Prove that $x^n - 1 = \prod_{d|n} \Phi_d(x)$.
(d) Using Möbius inversion formula, deduce that $\Phi_n(x) = \prod_{d|n}(x^d - 1)^{\mu(n/d)}$.
(e) Prove that $\Phi_p(x) = x^{p-1} + x^{p-2} + \cdots + x + 1$ for any prime p.
(f) Prove that $\Phi_{2^n}(x) = x^{2^{n-1}} + 1$ for any $n \in \mathbb{N}$.
(g) Derive a formula for $\Phi_{p^n}(x)$ for an odd prime p and for $n \in \mathbb{N}$.
(h) Prove that $\Phi_n(x) \in \mathbb{Z}[x]$.
(i) Prove that $\Phi_n(x)$ is irreducible in $\mathbb{Z}[x]$.

37. Let $f(x) = a_0 + a_1x + \cdots + a_dx^d \in \mathbb{C}[x]$ with $d \geqslant 1$, $a_d \neq 0$. Prove that:
(a) $|(x - z)f(x)| = |(\bar{z}x - 1)f(x)|$ for any $z \in \mathbb{C}$ (where $\bar{z}$ is the complex conjugate of z).
(b) [*Inequality of Gonçalves*] $M(f)^2 + |a_0a_d|^2M(f)^{-2} \leqslant |f|^2$.
(c) [*Inequality of Landau*] $M(f) \leqslant |f|$. Moreover, if $f(x)$ is not a monomial, then $M(f) < |f|$.

38. Prove Proposition 3.27.

39. Prove the formulas in Example 3.31(1).

40. Let K be a field, $a \in K$, and $0 \neq f(x), g(x), h(x) \in K[x]$. Prove that:
(a) $\text{Res}((x - a)f(x), g(x)) = g(a)\text{Res}(f(x), g(x))$.
(b) $\text{Res}(f(x), g(x)h(x)) = \text{Res}(f(x), g(x)) \times \text{Res}(f(x), h(x))$.

41. Prove that the non-constant polynomials $f(x), g(x) \in K[x]$ (K a field) have a common non-constant factor if and only if $u(x)f(x) + v(x)g(x) = 0$ for some non-zero $u(x), v(x) \in K[x]$ with $\deg u(x) < \deg g(x)$ and $\deg v(x) < \deg f(x)$.

42. Prove that $\alpha \in \mathbb{F}_{p^n}$ is a normal element if and only if $f(z) = z^n - 1$ and $g(z) = \alpha z^{n-1} + \alpha^p z^{n-2} + \cdots + \alpha^{p^{n-2}}z + \alpha^{p^{n-1}}$ are coprime in $\mathbb{F}_{p^n}[z]$.

43. [*Hadamard's inequality*]   Let M be an $n \times n$ matrix with real entries. Denote the i-th column of M by $\mathbf{b}_i$. Treat $\mathbf{b}_i$ as an n-dimensional vector, and let $|\mathbf{b}_i|$ denote the length of this vector. Prove that $|\det M| \leqslant \prod_{i=1}^{n} |\mathbf{b}_i|$.

44. Let M be an $n \times n$ matrix with integer entries such that each entry of M is ± 1. Prove that $|\det M| \leqslant n^{n/2}$.

45. Find the square-free factorization of $f(x) = x^7 + 6x^6 + 6x^5 - 3x^4 + 27x^3 + 3x^2 - 32x + 12$:
    (a) in $\mathbb{Z}[x]$, and
    (b) in $\mathbb{F}_{17}[x]$.
    (c) Factor the discriminants of the square-free factors of f found in Part (a).

46. Factor $f(x) = 2x^4 + 2x^3 + 7x^2 + x + 3$ in $\mathbb{Z}[x]$ using Berlekamp's algorithm. Note that $\mathrm{Discr}(f) = 64152 = 2^3 \times 3^6 \times 11$. So you may take $p = 5$.

47. Let p be a prime, and $f(x) = x^4 + 1 \in \mathbb{F}_p[x]$.
    (a) Let $p = 2$. Describe how $f(x)$ factors in $\mathbb{F}_2[x]$.
    (b) Let $p \equiv 1 \pmod 4$. Prove that $f(x) = (x^2 + \alpha)(x^2 - \alpha)$ in $\mathbb{F}_p[x]$, where $\alpha^2 \equiv -1 \pmod p$.
    (c) Let $p \equiv 3 \pmod 8$. Prove that $\left(\frac{-2}{p} \right) = 1$. Let $\alpha^2 \equiv -2 \pmod p$. Prove that $f(x) = (x^2 + \alpha x - 1)(x^2 - \alpha x - 1)$ in $\mathbb{F}_p[x]$.
    (d) Let $p \equiv 7 \pmod 8$. Prove that $\left(\frac{2}{p} \right) = 1$. Let $\alpha^2 \equiv 2 \pmod p$. Prove that $f(x) = (x^2 + \alpha x + 1)(x^2 - \alpha x + 1)$ in $\mathbb{F}_p[x]$.

48. (a) Prove that for every $n \in \mathbb{N}$, $n \geqslant 2$, the polynomial $x^{2^n} + 1$ is reducible in $\mathbb{F}_p[x]$ for any prime p.
    (b) Prove that for every $n \in \mathbb{N}$, the polynomial $x^{2^n} + 1$ is irreducible in $\mathbb{Z}[x]$.

49. Prove that the degree of an irreducible polynomial in $\mathbb{R}[x]$ is one or two.

50. Let $\mathbf{b}_1, \mathbf{b}_2, \ldots, \mathbf{b}_n$ constitute a *reduced* basis of a lattice in $\mathbb{R}^n$, and $\mathbf{b}_1^*, \mathbf{b}_2^*, \ldots, \mathbf{b}_n^*$ its Gram–Schmidt orthogonalization. Prove that:
    (a) $|\mathbf{b}_j|^2 \leqslant 2^{i-1} |\mathbf{b}_i^*|^2$ for all i, j with $1 \leqslant j \leqslant i \leqslant n$.
    (b) $d(L) \leqslant \prod_{i=1}^{n} |\mathbf{b}_i| \leqslant 2^{n(n-1)/4} d(L)$.
    (c) $|\mathbf{b}_1| \leqslant 2^{(n-1)/4} d(L)^{1/n}$.

51. Prove Proposition 3.39.

Programming Exercises

52. Write a GP/PARI function that computes the number of monic irreducible polynomials of degree m in $\mathbb{F}_q[x]$. (**Hint:** Use the built-in function `moebius()`.)

53. Write a GP/PARI function that computes the product of all monic irreducible polynomials of degree m in $\mathbb{F}_q[x]$. (**Hint:** Exercise 3.2.)

54. Write a GP/PARI function that checks whether a non-constant polynomial in the prime field $\mathbb{F}_p[x]$ is square-free. (**Hint:** Use the built-in function `deriv()`.)

55. Write a GP/PARI function that computes the square-free factorization of a monic non-constant polynomial in the prime field $\mathbb{F}_p[x]$.

56. Write a `GP/PARI` function that computes the distinct-degree factorization of a monic square-free polynomial in $\mathbb{F}_p[x]$, where p is a prime (Algorithm 3.4).

57. Let $f(x) \in \mathbb{F}_p[x]$ ($p > 2$ is a prime) be monic with each irreducible factor having the same known degree r. Write a `GP/PARI` function that computes the equal-degree factorization of $f(x)$ (Algorithm 3.7).

58. Let $f(x) \in \mathbb{F}_2[x]$ be monic with each irreducible factor having the same known degree r. Write a `GP/PARI` function that computes the equal-degree factorization of $f(x)$ (Algorithm 3.8).

59. Write a `GP/PARI` program implementing Berlekamp's Q-matrix factorization (Exercise 3.30). (**Hint:** Use the built-in functions `polcoeff()`, `matrix()`, `matker()` and `matsize()`.)

60. Write a `GP/PARI` program that implements Hensel lifting of Theorem 3.33.

Chapter 4

Arithmetic of Elliptic Curves

The study of elliptic curves is often called *arithmetic algebraic geometry*. Recent mathematical developments in this area has been motivated to a large extent by attempts to prove *Fermat's last theorem* which states that the equation $x^n + y^n = z^n$ does not have non-trivial solutions in integer values of x, y, z for (integers) $n \geqslant 3$. (The trivial solutions correspond to $xyz = 0$.)[1]

Purely mathematical objects like elliptic curves are used in a variety of engineering applications, most notably in the area of public-key cryptography. Elliptic curves are often preferred to finite fields, because the curves offer a wide range of groups upon which cryptographic protocols can be built, and also because keys pertaining to elliptic curves are shorter (than those pertaining to finite fields), resulting in easier key management. It is, therefore, expedient to look at the arithmetic of elliptic curves from a computational angle. Moreover, we require elliptic curves for an integer-factoring algorithm.

4.1   What Is an Elliptic Curve?

Elliptic curves are plane algebraic curves of genus one.[2] Cubic and quartic equations of special forms in two variables[3] X, Y are elliptic curves. The Greek

---

[1]In 1637, the amateur French mathematician Pierre de Fermat wrote a note in his personal copy of Bachet's Latin translation of the Greek book *Arithmetica* by Diophantus. The note translated in English reads like this: *"It is impossible to separate a cube into two cubes, or a fourth power into two fourth powers, or in general, any power higher than the second into two like powers. I have discovered a truly marvelous proof of this, which this margin is too narrow to contain."* It is uncertain whether Fermat really discovered a proof. However, Fermat himself published a proof for the special case $n = 4$ using a method which is now known as *Fermat's method of infinite descent*.

Given Fermat's proof for $n = 4$, one proves Fermat's last theorem for any $n \geqslant 3$ if one supplies a proof for all primes $n \geqslant 3$. Some special cases were proved by Euler ($n = 3$), by Dirichlet and Legendre ($n = 5$) and by Lamé ($n = 7$). In 1847, Kummer proved Fermat's last theorem for all *regular primes*. However, there exist infinitely many non-regular primes. A general proof for Fermat's last theorem has eluded mathematicians for over three centuries.

In the late 1960s, Hellegouarch discovered a connection between elliptic curves and Fermat's last theorem, which led Gerhard Frey to conclude that if a conjecture known as the *Taniyama–Shimura conjecture* for elliptic curves is true, then Fermat's last theorem holds too. The British mathematician Andrew Wiles, with the help of his student Richard Taylor, finally proved Fermat's last theorem in 1994. Wiles' proof is based on very sophisticated mathematics developed in the 20th century, and only a handful of living mathematicians can claim to have truly understood the entire proof.

It is debatable whether Fermat's last theorem is really a deep theorem that deserved such prolonged attention. Nonetheless, myriads of failed attempts to prove this theorem have, without any shred of doubt, intensely enriched several branches of modern mathematics.

[2]Loosely speaking, the genus of a curve is the number of handles in it. Straight lines and conic sections do not have handles, and have genus zero.

[3]Almost everywhere else in this book, lower-case letters $x, y, z, \ldots$ are used in polynomials. In this chapter, upper-case letters $X, Y, Z, \ldots$ are used as variables in polynomials, whereas the corresponding lower-case letters are reserved for a (slightly?) different purpose (see Section 4.4). I hope this notational anomaly would not be a nuisance to the readers.

mathematician Diophantus seems to be the first to study such curves.[4] Much later, Euler, Gauss, Jacobi, Abel and Weierstrass (among many others) studied these curves in connection with solving *elliptic integrals*. This is the genesis of the name *elliptic curve*. An ellipse is specified by a quadratic equation and has genus zero, so *ellipses are not elliptic curves*. Diophantus observed[5] that a quartic equation of the form

$$Y^2 = (X - a)(X^3 + bX^2 + cX + d) \tag{4.1}$$

can be converted to the cubic equation

$$Y^2 = \alpha X^3 + \beta X^2 + \gamma X + 1 \tag{4.2}$$

by the substitution of X by $(aX + 1)/X$ and of Y by Y/X^2 (where $\alpha = a^3 + a^2b + ac + d$, $\beta = 3a^2 + 2ab + c$, and $\gamma = 3a + b$). Multiplying both sides of Eqn (4.2) by α^2 and renaming αX and αY as X and Y respectively, we obtain an equation of the form

$$Y^2 = X^3 + \mu X^2 + \nu X + \eta. \tag{4.3}$$

Such substitutions allow us to concentrate on cubic curves of the following particular form. It turns out that any plane curve of genus one can be converted to this particular form after suitable substitutions of variables.

Definition 4.1 An elliptic curve E over a field K is defined by the cubic equation

$$E : Y^2 + a_1XY + a_3Y = X^3 + a_2X^2 + a_4X + a_6 \tag{4.4}$$

with $a_1, a_2, a_3, a_4, a_6 \in K$. Eqn (4.4) is called the *Weierstrass equation*[6] for the elliptic curve E. In order that E qualifies as an elliptic curve, we require E to contain no points of singularity in the following sense. ◁

Definition 4.2 Let $C : f(X, Y) = 0$ be the equation of a (plane) curve defined over some field K. A *point of singularity* on the curve C is a point $(h, k) \in K^2$ for which $f(h, k) = 0$, $\frac{\partial f}{\partial X}(h, k) = 0$, and $\frac{\partial f}{\partial Y}(h, k) = 0$. A curve is called *non-singular* or *smooth* if it contains no points of singularity. An elliptic curve is required to be non-singular. ◁

---

[4] *Diophantine equations* are named after Diophantus of Alexandria. These are multivariate polynomial equations with integer coefficients, of which integer or rational solutions are typically investigated.

[5] Diophantus studied special cases, whereas for us it is easy to generalize the results for arbitrary a, b, c, d. Nevertheless, purely algebraic substitutions as illustrated above are apparently not motivated by any geometric or arithmetic significance. Diophantus seems to be the first to use algebraic tools in geometry, an approach popularized by European mathematicians more than one and a half millenniums after Diophantus's time.

[6] Karl Theodor Wilhelm Weierstrass (1815–1897) was a German mathematician famous for his contributions to mathematical analysis. The Weierstrass elliptic (or P) function $\wp$ is named after him.

Example 4.3 (1) Take $K = \mathbb{R}$. Three singular cubic curves are shown in Figure 4.1. In each of these three examples, the point of singularity is the origin $(0, 0)$. If the underlying field is $\mathbb{R}$, we can identify the type of singularity from the *Hessian*[7] of the curve $C : f(X, Y) = 0$, defined as

$$\text{Hessian}(f) = \begin{pmatrix} \frac{\partial^2 f}{\partial X^2} & \frac{\partial^2 f}{\partial X \partial Y} \\ \frac{\partial^2 f}{\partial Y \partial X} & \frac{\partial^2 f}{\partial Y^2} \end{pmatrix}.$$

FIGURE 4.1: Singular cubic curves

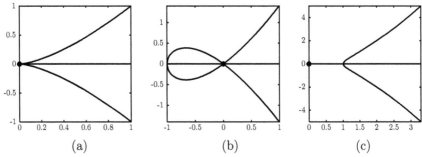

(a)   (b)   (c)

(a) A *cusp* or a *spinode*: $Y^2 = X^3$.
(b) A *loop* or a *double-point* or a *crunode*: $Y^2 = X^3 + X^2$
(c) An *isolated point* or an *acnode*: $Y^2 = X^3 - X^2$

FIGURE 4.2: Elliptic curves

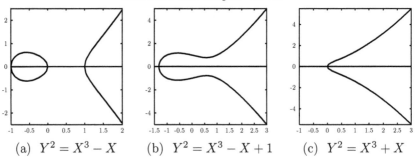

(a) $Y^2 = X^3 - X$ (b) $Y^2 = X^3 - X + 1$ (c) $Y^2 = X^3 + X$

A point of singularity P on C is a cusp if and only if Hessian(f) at P has determinant 0. P is a loop if and only if Hessian(f) at P is negative-definite.[8] Finally, P is an isolated point if and only if Hessian(f) at P is positive-definite. In each case, the tangent is not (uniquely) defined at the point of singularity.

---

[7]The Hessian matrix is named after the German mathematician Ludwig Otto Hesse (1811–1874). Hesse's mathematical contributions are in the area of analytic geometry.

[8]An $n \times n$ matrix A with real entries is called *positive-definite* (resp. *negative-definite*) if for every non-zero column vector $\mathbf{v}$ of size n, we have $\mathbf{v}^t A \mathbf{v} > 0$ (resp. $\mathbf{v}^t A \mathbf{v} < 0$).

(2) Three elliptic curves over $\mathbb{R}$ are shown in Figure 4.2. The partial derivatives $\frac{\partial f}{\partial X}$ and $\frac{\partial f}{\partial Y}$ do not vanish simultaneously at every point on these curves, and so the tangent is defined at every point on these curves. The curve of Part (a) has two disjoint components in the X-Y plane, whereas the curves of Parts (b) and (c) have only single components. The bounded component of the curve in Part (a) is the broken handle. For the other two curves, the handles are not discernible. For real curves, handles may be broken or invisible, since $\mathbb{R}$ is not algebraically closed. Curves over the field $\mathbb{C}$ of complex numbers have discernible handles. But then a plane curve over $\mathbb{C}$ requires four real dimensions, and is impossible to visualize in our three-dimensional world.

(3) Let us now look at elliptic curves over finite fields. Take $K = \mathbb{F}_{17}$. The curve defined by $Y^2 = X^3 + 5X - 1$ is not an elliptic curve, because it contains a singularity at the point $(2, 0)$. Indeed, we have $X^3 + 5X - 1 \equiv (X + 15)^2(X + 13) \pmod{17}$. However, the equation $Y^2 = X^3 + 5X - 1$ defines an elliptic over $\mathbb{R}$, since $X^3 + 5X - 1$ has no multiple roots in $\mathbb{R}$ (or $\mathbb{C}$).

The curve $E_1 : Y^2 = X^3 - 5X + 1$ is non-singular over $\mathbb{F}_{17}$ with 15 points:

$(0, 1), (0, 16), (2, 4), (2, 13), (3, 8), (3, 9), (5, 4), (5, 13), (6, 0), (10, 4),$

$(10, 13), (11, 6), (11, 11), (13, 5), (13, 12).$

There is only one point on E_1 with Y-coordinate equal to 0. We have the factorization $X^3 - 5X + 1 \equiv (X + 11)(X^2 + 6X + 14) \pmod{17}$.

The curve $E_2 : Y^2 = X^3 - 4X + 1$ is non-singular over $\mathbb{F}_{17}$ with 24 points:

$(0, 1), (0, 16), (1, 7), (1, 10), (2, 1), (2, 16), (3, 4), (3, 13), (4, 7), (4, 10), (5, 2),$

$(5, 15), (10, 3), (10, 14), (11, 8), (11, 9), (12, 7), (12, 10), (13, 2), (13, 15),$

$(15, 1), (15, 16), (16, 2), (16, 15).$

E_2 contains no points with Y-coordinate equal to 0, because $X^3 - 4X + 1$ is irreducible in $\mathbb{F}_{17}[X]$.

The non-singular curve $E_3 : Y^2 = X^3 - 3X + 1$ over $\mathbb{F}_{17}$ contains 19 points:

$(0, 1), (0, 16), (1, 4), (1, 13), (3, 6), (3, 11), (4, 6), (4, 11), (5, 3), (5, 14), (7, 0),$

$(8, 8), (8, 9), (10, 6), (10, 11), (13, 0), (14, 0), (15, 4), (15, 13).$

Since $X^3 - 3X + 1 \equiv (X + 3)(X + 4)(X + 10) \pmod{17}$, there are three points on E_3 with Y-coordinate equal to 0.

(4) Take $K = \mathbb{F}_{2^n}$ and a curve $C : Y^2 = X^3 + aX^2 + bX + c$ with $a, b, c \in K$. Write $f(X, Y) = Y^2 - (X^3 + aX^2 + bX + c)$. Then, $\frac{\partial f}{\partial X} = X^2 + b$, and $\frac{\partial f}{\partial Y} = 0$. Every element in $\mathbb{F}_{2^n}$ has a unique square root in $\mathbb{F}_{2^n}$. In particular, $X^2 + b$ has the root $h = b^{2^{n-1}}$. Plugging in this value of X in $Y^2 = X^3 + aX^2 + bX + c$ gives a unique solution for Y, namely $k = (h^3 + ah^2 + bh + c)^{2^{n-1}}$. But then, (h, k) is a point of singularity on C. This means that a curve of the form $Y^2 = X^3 + aX^2 + bX + c$ is never an elliptic curve over $\mathbb{F}_{2^n}$. Therefore, we must have non-zero term(s) involving XY and/or Y on the left side of the Weierstrass equation in order to obtain an elliptic curve over $\mathbb{F}_{2^n}$.

As a specific example, represent $\mathbb{F}_8 = \mathbb{F}_2(\theta)$ with $\theta^3 + \theta + 1 = 0$, and consider the curve $E : Y^2 + XY = X^3 + X^2 + \theta$. Let us first see that the curve is non-singular. Write $f(X,Y) = Y^2 + XY + X^3 + X^2 + \theta$. But then, $\frac{\partial f}{\partial X} = Y + X^2$, and $\frac{\partial f}{\partial Y} = X$. The condition $\frac{\partial f}{\partial X} = \frac{\partial f}{\partial Y} = 0$ implies $X = Y = 0$. But $(0,0)$ is not a point on E. Therefore, E is indeed an elliptic curve over $\mathbb{F}_8$. It contains the following nine points:

$$(0, \theta^2 + \theta), (1, \theta^2), (1, \theta^2 + 1), (\theta, \theta^2), (\theta, \theta^2 + \theta), (\theta + 1, \theta^2 + 1),$$
$$(\theta + 1, \theta^2 + \theta), (\theta^2 + \theta, 1), (\theta^2 + \theta, \theta^2 + \theta + 1).$$

(5) As our final example, represent $K = \mathbb{F}_9 = \mathbb{F}_3(\theta)$, where $\theta^2 + 1 = 0$. Consider the curve $E : Y^2 = X^3 + X^2 + X + \theta$ defined over $\mathbb{F}_9$. Writing $f(X,Y) = Y^2 - (X^3 + X^2 + X + \theta)$ gives the condition for singularity to be $\frac{\partial f}{\partial X} = X - 1 = 0$ and $\frac{\partial f}{\partial Y} = 2Y = 0$, that is, $X = 1$, $Y = 0$. But the point $(1, 0)$ does not lie on the curve, that is, E is an elliptic curve. It contains the following eight points:

$$(0, \theta + 2), (0, 2\theta + 1), (1, \theta + 2), (1, 2\theta + 1), (\theta + 1, \theta), (\theta + 1, 2\theta),$$
$$(2\theta + 2, 1), (2\theta + 2, 2).$$

E contains no points with Y-coordinate equal to 0, since $X^3 + X^2 + X + \theta$ is irreducible in $\mathbb{F}_9[X]$. □

It is convenient to work with simplified forms of the Weierstrass equation (4.4). The simplification depends upon the characteristic of the underlying field K. If char $K \neq 2$, substituting Y by $Y - (a_1 X + a_3)/2$ eliminates the terms involving XY and Y, and one can rewrite Eqn (4.4) as

$$Y^2 = X^3 + b_2 X^2 + b_4 X + b_6. \tag{4.5}$$

If we additionally have char $K \neq 3$, we can replace X by $(X - 3b_2)/36$ and Y by $Y/216$ in order to simplify Eqn (4.5) further as

$$Y^2 = X^3 + aX + b. \tag{4.6}$$

If char $K = 2$, we cannot use the simplified Eqns (4.5) or (4.6). Indeed, we have argued in Example 4.3(d) that curves in these simplified forms are singular over $\mathbb{F}_{2^n}$ and so are not elliptic curves at all. If $a_1 = 0$ in Eqn (4.4), we replace X by $X + a_2$ to eliminate the term involving X^2, and obtain

$$Y^2 + aY = X^3 + bX + c. \tag{4.7}$$

Such a curve is called *supersingular*. If $a_1 \neq 0$, we replace X by $a_1^2 X + \frac{a_3}{a_1}$ and Y by $a_1^3 Y + \frac{a_1^2 a_4 + a_3^2}{a_1^3}$ in order to obtain the equation

$$Y^2 + XY = X^3 + aX^2 + b. \tag{4.8}$$

This curve is called *non-supersingular* or *ordinary*.

All coordinate transformations made above are invertible. Therefore, there is a one-to-one correspondence between the points on the general curve (4.4) and the points on the curves with the simplified equations. That is, if we work with the simplified equations, we do not risk any loss of generality. The following table summarizes the Weierstrass equations.

K	Weierstrass equation
Any field	$Y^2 + (a_1 X + a_3)Y = X^3 + a_2 X^2 + a_4 X + a_6$
char $K \neq 2, 3$	$Y^2 = X^3 + aX + b$
char $K = 2$	$Y^2 + aY = X^3 + bX + c$ (Supersingular curve)
	$Y^2 + XY = X^3 + aX^2 + b$ (Ordinary curve)
char $K = 3$	$Y^2 = X^3 + aX^2 + bX + c$

4.2 Elliptic-Curve Group

Let E be an elliptic curve defined over a field K. A point $(X, Y) \in K^2$ that satisfies the equation of the curve is called a *finite K-rational point* on E. For a reason that will be clear soon, E is assumed to contain a distinguished point $\mathcal{O}$ called the *point at infinity*.[9] This point lies on E even if K itself is a finite field. One has to go to the so-called *projective space* in order to visualize this point. All (finite) K-rational points on E together with $\mathcal{O}$ constitute a set denoted by E_K.[10] We provide E_K with a binary operation, traditionally denoted by addition $(+)$, under which E_K becomes a commutative group.[11]

The basic motivation for defining the group E_K is illustrated in Figure 4.3. For simplicity, we assume that the characteristic of the underlying field K is not two, so we can use the equation $Y^2 = X^3 + aX^2 + bX + c$ for E. The curve in this case is symmetric about the X-axis, that is, if (h, k) is a (finite) K-rational point on E, then so also is its reflection $(h, -k)$ about the X-axis.

Take two points $P = (h_1, k_1)$ and $Q = (h_2, k_2)$ on E. Let $L : Y = \lambda X + \mu$ be the (straight) line passing through these two points. For a moment assume that $P \neq Q$ and that the line L is not vertical. Substituting Y by $\lambda X + \mu$ in the equation for E gives a cubic equation in X. Obviously, $X = h_1$ and $X = h_2$ satisfy this cubic equation, and it follows that the third root $X = h_3$ of the equation must also belong to the underlying field K. In other words,

---

[9]The concept of points at infinity was introduced independently by Johannes Kepler (1571–1630) and Gérard Desargues (1591–1661).

[10]Many authors prefer to use the symbol $E(K)$ instead of E_K. I prefer E_K, because we will soon be introduced to an object denoted as $K(E)$.

[11]In 1901, the French mathematician Jules Henri Poincaré (1854–1912) first proved this group structure.

FIGURE 4.3: Motivating addition of elliptic curve points

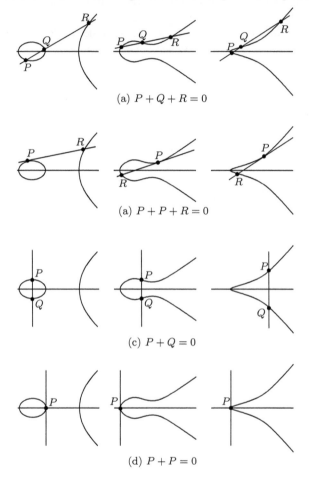

(a) $P + Q + R = 0$

(a) $P + P + R = 0$

(c) $P + Q = 0$

(d) $P + P = 0$

there exists a unique third point R lying on the intersection of E with L. The group operation will satisfy the condition $P + Q + R = 0$ in this case.

Part (b) of Figure 4.3 shows the special case $P = Q = (h, k)$. In this case, the line $L : Y = \lambda X + \mu$ is taken to be the tangent to the curve at P. Substituting $\lambda X + \mu$ for Y in the equation for E gives a cubic equation in X, of which $X = h$ is a double root. The third root identifies a unique intersection point R of the curve E with L. We take $P + P + R = 2P + R = 0$.

Now, take two points $P = (h, k)$ and $Q = (h, -k)$ on the curve (Part (c) of Figure 4.3). The line passing through these two points has the equation $L : X = h$. Substituting X by h in the equation for E gives a quadratic equation in Y, which has the two roots $\pm k$. The line L does not meet the curve E at any other point in K^2. We set $P + Q = 0$ in this case.

A degenerate case of Part (c) is $P = Q = (h, 0)$, and the tangent to E at the point P is vertical (Part (d) of Figure 4.3). Substituting X by h in the equation of E now gives us the equation $Y^2 = 0$ which has a double root at $Y = 0$. We set $P + P = 2P = 0$ in this case.

In view of these observations, we can now state the group law in terms of the so-called *chord-and-tangent rule*. The group operation is illustrated in Figure 4.4. Part (c) illustrates that if P and Q are two distinct points on the curve having the same x-coordinate, then $P + Q = 0$, that is, $Q = -P$, that is, the reflection of P about the X-axis gives the negative of P. We call Q the *opposite* of P (and P the opposite of Q) in this case. In the degenerate case where the vertical line through $P = Q$ is tangential to the curve E, we have a point P which is its own opposite (Part (d)). There are at most three points on the curve, which are opposites of themselves. For $K = \mathbb{R}$, there are either one or three such points. If $K = \mathbb{C}$ (or any algebraically closed field), there are exactly three such points. Finally, if K is a finite field (of characteristic not equal to two), there can be zero, one, or three such points.

Now, consider two points P, Q on E. If Q is the opposite of P, we define $P + Q = 0$. Otherwise, the line passing through P and Q (or the tangent to E at P in the case $P = Q$) is not vertical and meets the curve at a unique third point R. We have $P + Q + R = 0$, that is, $P + Q = -R$, that is, $P + Q$ is defined to be the opposite of the point R (Parts (a) and (b) of Figure 4.4).

An important question now needs to be answered: What is the additive identity 0 in this group? It has to be a point on E that should satisfy $P + 0 = 0 + P = P$ for all points P on E. As can be easily argued from the partial definition of point addition provided so far, no finite point on E satisfies this property. This is where we bring the point $\mathcal{O}$ at infinity in consideration. We assume that $\mathcal{O}$ is an infinitely distant point on E, that lies vertically above every finite point in the X-Y plane. It is so distant from a finite point P that joining P to $\mathcal{O}$ will produce a vertical line. Imagine that you are standing on an infinitely long straight road. Near the horizon, the entire road, no matter how wide, appears to converge to a point.

The question is why does this point lie on E? The direct answer is: for our convenience. The *illuminating* answer is this: Look at the equation of an elliptic curve over $K = \mathbb{R}$. As X becomes large, we can neglect the terms involving X^2, X^1, and X^0, that is, $Y^2 \approx X^3$, that is, $Y \approx X^{3/2}$. But then, for any finite point $P = (h, k) \in \mathbb{R}^2$, we have $\lim_{X \to \infty} \frac{Y - k}{X - h} = \lim_{X \to \infty} \frac{X^{3/2} - k}{X - h} = \infty$, that is, a point on E infinitely distant from P lies vertically above P.

There must be another point $\mathcal{O}'$ on E infinitely distant from and vertically *below* every finite point in the X-Y plane. Why has this point not been taken into consideration? The direct answer is again: for our convenience. The illuminating answer is that if E happened to contain this point also, then some vertical line would pass through four points on E, two finite points and the two points $\mathcal{O}$ and $\mathcal{O}'$. This is a bad situation, where a cubic equation meets a linear equation at more than three points. Moreover, we are going to use $\mathcal{O}$ as

FIGURE 4.4: Addition of elliptic curve points

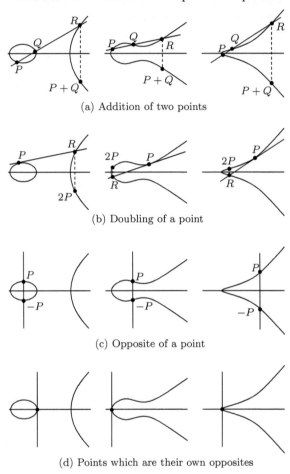

(a) Addition of two points

(b) Doubling of a point

(c) Opposite of a point

(d) Points which are their own opposites

the identity of the elliptic curve group. Thus, we would have $-\mathcal{O} = \mathcal{O}$, that is, $\mathcal{O}'$ and $\mathcal{O}$ are treated as the *same* point on E.

Does it look too imprecise or ad hoc? Perhaps, or perhaps not! The reader needs to understand projective geometry in order to *visualize* the point at infinity. Perfectly rigorous mathematical tools will then establish that there is exactly one point at infinity on any elliptic curve. More importantly, if K is any arbitrary field, even a finite field, the point $\mathcal{O}$ provably exists on the curve. We defer this discussion until Section 4.4.

Let us now logically *deduce* that $P + \mathcal{O} = \mathcal{O} + P = P$ for any finite point P on E_K. The line L passing through P and $\mathcal{O}$ is vertical by the choice of $\mathcal{O}$. Thus, the third point R where L meets E is the opposite of P, that is, $R = -P$. By the chord-and-tangent rule, we then take $P + \mathcal{O} = \mathcal{O} + P = -(-P) = P$.

Finally, what is $\mathcal{O} + \mathcal{O}$? Since $Y^2 \approx X^3$ on the curve for large values of X, the curve becomes vertical as $X \to \infty$. Thus, the tangent L to the curve at $\mathcal{O}$ is indeed a vertical line. The X-coordinate on L is infinity, that is, L does not meet the curve E at any finite point. So the natural choice is $\mathcal{O} + \mathcal{O} = 0$, that is, $\mathcal{O} + \mathcal{O} = \mathcal{O}$. That is how the identity should behave.

Under the addition operation defined in this fashion, E_K becomes a group. This operation evidently satisfies the properties of closure, identity, inverse, and commutativity. However, no simple way of proving the associativity of this addition is known to me. Extremely patient readers may painstakingly furnish a proof using the explicit formulas for point addition given below. A proof naturally follows from the theory of divisors, but a complete exposure to the theory demands serious algebra prerequisites on the part of the readers.

Let us now remove the restriction that the curve is symmetric about the X-axis. This is needed particularly to address fields of characteristic two. Let us work with the unsimplified Weierstrass equation $E : Y^2 + a_1XY + a_3Y = X^3 + a_2X^2 + a_4X + a_6$. First, we define the opposite of a point. We take $-\mathcal{O} = \mathcal{O}$. For any finite point $P = (h, k)$ on E, the vertical line L passing through P has the equation $X = h$. Substituting this in the Weierstrass equation gives $Y^2 + (a_1h + a_3)Y - (h^3 + a_2h^2 + a_4h + a_6) = 0$. This equation already has a root k. Let the other root be k'. We have $k + k' = -(a_1h + a_3)$, that is, $k' = -(k + a_1h + a_3)$. Therefore, we take

$$-P = (h, -(k + a_1h + a_3)).$$

If $a_1 = a_3 = 0$, then $-P$ is indeed the reflection of P about the X-axis.

Now, let us find the point $P+Q$ for $P, Q \in E_K$. We take $P+\mathcal{O} = \mathcal{O}+P = P$ and $\mathcal{O} + \mathcal{O} = \mathcal{O}$. Moreover, if $Q = -P$, then $P + Q = \mathcal{O}$. So let $P = (h_1, k_1)$ and $Q = (h_2, k_2)$ be finite points on E, that are not opposites of one another. In this case, the straight line passing through P and Q (the tangent to E at P if $P = Q$) is not vertical and has the equation $L : Y = \lambda X + \mu$, where

$$\lambda = \begin{cases} \dfrac{k_2 - k_1}{h_2 - h_1}, & \text{if } P \neq Q, \\[2ex] \dfrac{3h_1^2 + 2a_2h_1 + a_4 - a_1k_1}{2k_1 + a_1h_1 + a_3}, & \text{if } P = Q, \text{ and} \end{cases}$$
$$\mu = k_1 - \lambda h_1 = k_2 - \lambda h_2 .$$

Substituting Y by $\lambda X + \mu$ in the equation for E gives a cubic equation in X, of which h_1 and h_2 are already two roots. The third root of this equation is

$$h_3 = \lambda^2 + a_1\lambda - a_2 - h_1 - h_2.$$

That is, the line L meets E at the third point $R = (h_3, \lambda h_3 + \mu)$. We take

$$P + Q = -R = (h_3, -(\lambda + a_1)h_3 - \mu - a_3).$$

We have simpler formulas for the coordinates of $P + Q$ if we work with simplified forms of the Weierstrass equation. We concentrate only on the case that $P = (h_1, k_1)$ and $Q = (h_2, k_2)$ are finite points that are not opposites of one another. Our plan is to compute the coordinates of $P + Q = (h_3, k_3)$.

If char $K \neq 2, 3$, we use the equation $Y^2 = X^3 + aX + b$. In this case,

$$
\begin{aligned}
h_3 &= \lambda^2 - h_1 - h_2, \\
k_3 &= \lambda(h_1 - h_3) - k_1, \quad \text{where} \\
\lambda &= \begin{cases} \dfrac{k_2 - k_1}{h_2 - h_1}, & \text{if } P \neq Q, \\[2mm] \dfrac{3h_1^2 + a}{2k_1}, & \text{if } P = Q. \end{cases}
\end{aligned}
$$

If char $K = 3$, we use the equation $Y^2 = X^3 + aX^2 + bX + c$, and obtain

$$
\begin{aligned}
h_3 &= \lambda^2 - a - h_1 - h_2, \\
k_3 &= \lambda(h_1 - h_3) - k_1, \quad \text{where} \\
\lambda &= \begin{cases} \dfrac{k_2 - k_1}{h_2 - h_1}, & \text{if } P \neq Q, \\[2mm] \dfrac{2ah_1 + b}{2k_1}, & \text{if } P = Q. \end{cases}
\end{aligned}
$$

Finally, let char $K = 2$. For the supersingular curve $Y^2 + aY = X^3 + bX + c$, we have

$$
h_3 = \begin{cases} \left(\dfrac{k_1 + k_2}{h_1 + h_2}\right)^2 + h_1 + h_2, & \text{if } P \neq Q, \\[3mm] \dfrac{h_1^4 + b^2}{a^2}, & \text{if } P = Q, \end{cases}
$$

$$
k_3 = \begin{cases} \left(\dfrac{k_1 + k_2}{h_1 + h_2}\right)(h_1 + h_3) + k_1 + a, & \text{if } P \neq Q, \\[3mm] \left(\dfrac{h_1^2 + b}{a}\right)(h_1 + h_3) + k_1 + a, & \text{if } P = Q, \end{cases}
$$

whereas for the non-supersingular curve $Y^2 + XY = X^3 + aX^2 + b$, we have

$$
h_3 = \begin{cases} \left(\dfrac{k_1 + k_2}{h_1 + h_2}\right)^2 + \dfrac{k_1 + k_2}{h_1 + h_2} + h_1 + h_2 + a, & \text{if } P \neq Q, \\[3mm] h_1^2 + \dfrac{b}{h_1^2}, & \text{if } P = Q, \end{cases}
$$

$$
k_3 = \begin{cases} \left(\dfrac{k_1 + k_2}{h_1 + h_2}\right)(h_1 + h_3) + h_3 + k_1, & \text{if } P \neq Q, \\[3mm] h_1^2 + \left(h_1 + \dfrac{k_1}{h_1} + 1\right)h_3, & \text{if } P = Q. \end{cases}
$$

The opposite of a finite point $P = (h, k)$ on E is

$$
-P = \begin{cases} (h, k + a) & \text{for the supersingular curve,} \\ (h, k + h) & \text{for the non-supersingular curve.} \end{cases}
$$

Example 4.4 (1) Consider the curve $E : Y^2 = X^3 - X + 1$ over $\mathbb{Q}$ and the points $P = (1, -1)$ and $Q = (3, 5)$ on E. First, we compute the point $P + Q$. The straight line L passing through P and Q has the equation $\frac{Y+1}{X-1} = \frac{5+1}{3-1}$, that is, $Y = 3X - 4$. Substituting this value of Y in the equation of E, we get $(3X - 4)^2 = X^3 - X + 1$, that is, $X^3 - 9X^2 + 23X - 15 = 0$, that is, $(X-1)(X-3)(X-5) = 0$. Two of the three roots of this equation are already the X-coordinates of P and Q. The third point of intersection of L with E has the X-coordinate 5, and the corresponding Y-coordinate is $3 \times 5 - 4 = 11$. Therefore, $P + Q = -(5, 11) = (5, -11)$.

We now compute the point $2P$. Since Y^2 equals $X^3 - X + 1$ on the curve, differentiation with respect to X gives $2Y \frac{dY}{dX} = 3X^2 - 1$, that is, $\frac{dY}{dX} = \frac{3X^2-1}{2Y}$. At the point $P = (1, -1)$, we have $\frac{dY}{dX} = \frac{3-1}{-2} = -1$. The tangent T to E at P has the equation $Y = -X + \mu$, where μ is obtained by putting the coordinates of P in the equation: $\mu = -1 + 1 = 0$, so the equation of T is $Y = -X$. Plugging in this value of Y in the equation of E, we get $X^2 = X^3 - X + 1$, that is, $(X - 1)^2(X + 1) = 0$. So the *third* point of intersection of T with E has $X = -1$ and is $(X, -X) = (-1, 1)$, that is, $2P = -(-1, 1) = (-1, -1)$.

Here, for the sake of illustration only, we have computed $P + Q$ and $2P$ using basic principles of elementary coordinate geometry. In practice, we may straightaway apply the formulas presented earlier.

(2) Take $E : Y^2 = X^3 - 5X + 1$ defined over $\mathbb{F}_{17}$, $P = (2, 4)$, and $Q = (13, 5)$. The slope of the line joining P and Q is $\lambda \equiv (5 - 4) \times (13 - 2)^{-1} \equiv 14 \pmod{17}$. Thus, the X-coordinate of $P + Q$ is $\lambda^2 - 2 - 13 \equiv 11 \pmod{17}$, and its Y-coordinate is $\lambda(2 - 11) - 4 \equiv 6 \pmod{17}$, that is, $P + Q = (11, 6)$.

We now compute $2P$. The slope of the tangent to E at P is $\lambda = (3 \times 2^2 - 5) \times (2 \times 4)^{-1} \equiv 3 \pmod{17}$. Thus, the X-coordinate of $2P$ is $\lambda^2 - 2 \times 2 \equiv 5 \pmod{17}$, and its Y-coordinate is $\lambda(2 - 5) - 4 \equiv 4 \pmod{17}$, so $2P = (5, 4)$.

(3) Represent $\mathbb{F}_8 = \mathbb{F}_2(\theta)$ with $\theta^3 + \theta + 1 = 0$. Take the non-supersingular curve $Y^2 + XY = X^3 + X^2 + \theta$ over $\mathbb{F}_8$ and the points $P = (h_1, k_1) = (1, \theta^2)$ and $Q = (h_2, k_2) = (\theta + 1, \theta^2 + 1)$. The line joining P and Q has slope $\lambda = \left(\frac{k_1+k_2}{h_1+h_2}\right) = \left(\frac{1}{\theta}\right) = \theta^2 + 1$. Let $P + Q = (h_3, k_3)$. The explicit formulas presented earlier give $h_3 = \lambda^2 + \lambda + h_1 + h_2 + 1 = (\theta^2 + 1)^2 + (\theta^2 + 1) + 1 + (\theta + 1) + 1 = \theta^4 + \theta^2 + \theta + 1 = \theta(\theta^3 + \theta + 1) + 1 = 1$, and $k_3 = \lambda(h_1 + h_3) + h_3 + k_1 = (\theta^2 + 1)(1 + 1) + 1 + \theta^2 = \theta^2 + 1$, that is, $P + Q = (1, \theta^2 + 1)$.

Finally, we compute $2P = (h_4, k_4)$. Once again we use the explicit formulas given earlier to obtain $h_4 = h_1^2 + \frac{\theta}{h_1^2} = \theta + 1$, and $k_4 = h_1^2 + \left(h_1 + \frac{k_1}{h_1} + 1\right)h_4 = 1 + \left(1 + \frac{\theta^2}{1} + 1\right)(\theta + 1) = 1 + \theta^2(\theta + 1) = \theta^3 + \theta^2 + 1 = (\theta^3 + \theta + 1) + (\theta^2 + \theta) = \theta^2 + \theta$, that is, $2P = (\theta + 1, \theta^2 + \theta)$. $\square$

We now study some useful properties of the elliptic-curve group E_K. The *order* of an element a in an additive group G is defined to be the smallest positive integer m for which $ma = a + a + \cdots + a$ (m times) $= 0$. We denote this as $\operatorname{ord} a = m$. If no such positive integer m exists, we take $\operatorname{ord} a = \infty$.

Proposition 4.5 *Let K be a field with* char $K \neq 2$. *An elliptic curve E defined over K has at most three points of order two. If K is algebraically closed, E has exactly three such points.*

Proof   Suppose that $2P = \mathcal{O}$ with $\mathcal{O} \neq P = (h, k) \in E_K$. Then $P = -P$, that is, P is the opposite of itself. Since char $K \neq 2$, we may assume that the elliptic curve is symmetric about the X-axis and has the equation $Y^2 = X^3 + aX^2 + bX + c$. Since E is non-singular, the right side of this equation has no multiple roots. Now, $P = (h, k) = -P = (h, -k)$ implies $k = 0$, that is, h is a root of $X^3 + aX^2 + bX + c$. Conversely, for any root h of $X^3 + aX^2 + bX + c$, the finite point $(h, 0) \in E_K$ is of order two.   ◁

Clearly, an elliptic curve defined over $\mathbb{R}$ or $\mathbb{C}$ contains infinitely many points. However, an elliptic curve defined over $\mathbb{Q}$ may contain only finitely many points. For example, the curve $Y^2 = X^3 + 6$ contains no points with rational coordinates, that is, the only point on this curve is $\mathcal{O}$. I now state some interesting facts about elliptic curves over $\mathbb{Q}$.

Theorem 4.6 [*Mordell's theorem*][12]   *The group $E_\mathbb{Q}$ of an elliptic curve E defined over $\mathbb{Q}$ is finitely generated.*   ◁

One can write $E_\mathbb{Q} = E_{\text{tors}} \oplus E_{\text{free}}$, where E_{tors} is the subgroup consisting of points of finite orders (the *torsion subgroup* of $E_\mathbb{Q}$), and where $E_{\text{free}} \cong \mathbb{Z}^r$ is the free part of $E_\mathbb{Q}$. Mordell's theorem states that the *rank* r of $E_\mathbb{Q}$ is finite. It is a popular belief that r can be arbitrarily large. At present (March 2012), however, the largest known rank of an elliptic curve is 28. Elkies in 2006 discovered that the following elliptic curve has rank 28.

$Y^2 + XY + Y =$
$X^3 - X^2 - 20067762415575526585033208209338542750930230312178956502X +$
$ 34481611795030556467032985690390720374855944359319180361267 \text{ \textbackslash}$
$ 6008296291939448732243429$

Theorem 4.7 [*Mazur's theorem*][13]   *The torsion subgroup E_{tors} of $E_\mathbb{Q}$ is either a cyclic group of order $m \in \{1, 2, 3, \ldots, 10, 12\}$, or isomorphic to $\mathbb{Z}_2 \oplus \mathbb{Z}_{2m}$ for $m \in \{1, 2, 3, 4\}$. There are infinitely many examples of elliptic curves over $\mathbb{Q}$ for each of these group structures of E_{tors}.*   ◁

Theorem 4.8 [*Nagell–Lutz theorem*][14]   *Let $E : Y^2 = X^3 + aX^2 + bX + c$ be an elliptic curve with integer coefficients a, b, c, and let $\mathcal{O} \neq P \in E_{\text{tors}}$. Then, the coordinates of P are integers.*   ◁

There are two very important quantities associated with elliptic curves.

---

[12]This result was conjectured by Poincaré in 1901, and proved in 1922 by the British mathematician Louis Joel Mordell (1888–1972).

[13]This theorem was proved in 1977 by Barry Charles Mazur (1937–) of Harvard University.

[14]This theorem was proved independently by the Norwegian mathematician Trygve Nagell (1895–1988) and the French mathematician Élisabeth Lutz (1914–2008).

Definition 4.9 For a cubic curve E given by Eqn (4.4), define:

$$
\begin{aligned}
d_2 &= a_1^2 + 4a_2 \\
d_4 &= 2a_4 + a_1 a_3 \\
d_6 &= a_3^2 + 4a_6 \\
d_8 &= a_1^2 a_6 + 4a_2 a_6 - a_1 a_3 a_4 + a_2 a_3^2 - a_4^2 \\
c_4 &= d_2^2 - 24 d_4 \\
\Delta(E) &= -d_2^2 d_8 - 8 d_4^3 - 27 d_6^2 + 9 d_2 d_4 d_6 \qquad (4.9) \\
j(E) &= c_4^3 / \Delta(E), \quad \text{if } \Delta(E) \neq 0. \qquad (4.10)
\end{aligned}
$$

$\Delta(E)$ is called the *discriminant* of E, and $j(E)$ the *j-invariant* of E. $\lhd$

Some important results pertaining to E are now stated.

Theorem 4.10 $\Delta(E) \neq 0$ *if and only if E is smooth. In particular, $j(E)$ is defined for all elliptic curves.* $\lhd$

Theorem 4.11 *Let E and E' be two elliptic curves defined over K. If the groups E_K and E'_K are isomorphic, then $j(E) = j(E')$. Conversely, if $j(E) = j(E')$, then the groups $E_{\bar{K}}$ and $E'_{\bar{K}}$ are isomorphic.* $\lhd$

If char $K \neq 2, 3$, and E is given by the special Weierstrass Eqn (4.6), then $\Delta(E) = -16(4a^3 + 27b^2)$. Up to the constant factor 16, this is the same as the discriminant of the cubic polynomial $X^3 + aX + b$. But $\Delta(X^3 + aX + b) \neq 0$ if and only if $X^3 + aX + b$ does not contain multiple roots—a necessary and sufficient condition for E given by Eqn (4.6) to be smooth.

4.2.1 Handling Elliptic Curves in GP/PARI

GP/PARI provides extensive facilities to work with elliptic curves. An elliptic curve is specified by the coefficients a_1, a_2, a_3, a_4, a_6 in the general Weierstrass Eqn (4.4). The coefficients are to be presented as a row vector. One may call the built-in function ellinit() which accepts this coefficient vector and returns a vector with 19 components storing some additional information about the curve. If the input coefficients define a singular curve, an error message is issued. In the following snippet, the first curve $Y^2 = X^3 + X^2$ is singular over $\mathbb{Q}$, whereas the second curve $Y^2 = X^3 - X + 1$ is an elliptic curve over $\mathbb{Q}$.

```
gp > E1 = ellinit([0,1,0,0,0])
  ***   singular curve in ellinit.
gp > E1 = ellinit([0,0,0,-1,1])
%1 = [0, 0, 0, -1, 1, 0, -2, 4, -1, 48, -864, -368, -6912/23, [-1.32471795724474
6025960908854, 0.6623589786223730129804544272 - 0.56227951206230124389918214494*I
, 0.6623589786223730129804544272 + 0.56227951206230124389918214484*I]~, 4.7070877
61230185561883752116, -2.3535438806150927809418760584 + 1.0982915250610051220258282079*I, -1.2099500630791746535594160804 + 0.E-28*I, 0.60497503153958732677970840420 - 0.94973171956503591227564499833*I, 5.16975459587749284005438911193]
```

A finite point (h, k) on an elliptic curve is represented as the row vector [h,k] without a mention of the specific curve on which this point lies. When one adds or computes multiples of points on an elliptic curve, the curve must be specified, otherwise vector addition and multiplication are performed. The function for point addition is `elladd()`, and that for computing the multiple of a point is `ellpow()`. The opposite of a point can be obtained by computing the (-1)-th multiple of a point. The point at infinity is represented as [0].

```
gp > P1 = [1,-1]
%2 = [1, -1]
gp > Q1 = [3,5]
%3 = [3, 5]
gp > P1 + Q1
%4 = [4, 4]
gp > elladd(E1,P1,Q1)
%5 = [5, -11]
gp > 2*P1
%6 = [2, -2]
gp > ellpow(E1,P1,2)
%7 = [-1, -1]
gp > R1 = ellpow(E1,Q1,-1)
%8 = [3, -5]
gp > elladd(E1,Q1,R1)
%9 = [0]
```

We can work with elliptic curves over finite fields. Here is an example that illustrates the arithmetic of the curve $Y^2 = X^3 - 5X + 1$ defined over $\mathbb{F}_{17}$.

```
gp > E2 = ellinit([Mod(0,17),Mod(0,17),Mod(0,17),Mod(5,17),Mod(-1,17)])
  ***    singular curve in ellinit.
gp > E2 = ellinit([Mod(0,17),Mod(0,17),Mod(0,17),Mod(-5,17),Mod(1,17)])
%10 = [Mod(0, 17), Mod(0, 17), Mod(0, 17), Mod(12, 17), Mod(1, 17), Mod(0, 17),
Mod(7, 17), Mod(4, 17), Mod(9, 17), Mod(2, 17), Mod(3, 17), Mod(3, 17), Mod(14,
17), 0, 0, 0, 0, 0, 0]
gp > P2 = [Mod(2,17),Mod(4,17)];
gp > Q2 = [Mod(13,17),Mod(5,17)];
gp > elladd(E2,P2,Q2)
%13 = [Mod(11, 17), Mod(6, 17)]
gp > ellpow(E2,P2,2)
%14 = [Mod(5, 17), Mod(4, 17)]
```

One can work with curves defined over extension fields. For example, we represent $\mathbb{F}_8 = \mathbb{F}_2(\theta)$ with $\theta^3 + \theta + 1 = 0$, and define the non-supersingular curve $Y^2 + XY = X^3 + X^2 + \theta$ over $\mathbb{F}_8$.

```
gp > f = Mod(1,2)*t^3+Mod(1,2)*t+Mod(1,2)
%15 = Mod(1, 2)*t^3 + Mod(1, 2)*t + Mod(1, 2)
```

```
gp > a1 = Mod(Mod(1,2),f);
gp > a2 = Mod(Mod(1,2),f);
gp > a3 = a4 = 0;
gp > a6 = Mod(Mod(1,2)*t,f);
gp > E3 = ellinit([a1,a2,a3,a4,a6])
%20 = [Mod(Mod(1, 2), Mod(1, 2)*t^3 + Mod(1, 2)*t + Mod(1, 2)), Mod(Mod(1, 2), M
od(1, 2)*t^3 + Mod(1, 2)*t + Mod(1, 2)), 0, 0, Mod(Mod(1, 2)*t, Mod(1, 2)*t^3 +
Mod(1, 2)*t + Mod(1, 2)), Mod(Mod(1, 2), Mod(1, 2)*t^3 + Mod(1, 2)*t + Mod(1, 2)
), Mod(Mod(0, 2), Mod(1, 2)*t^3 + Mod(1, 2)*t + Mod(1, 2)), 0, Mod(Mod(1, 2)*t,
Mod(1, 2)*t^3 + Mod(1, 2)*t + Mod(1, 2)), Mod(Mod(1, 2), Mod(1, 2)*t^3 + Mod(1,
2)*t + Mod(1, 2)), Mod(Mod(1, 2), Mod(1, 2)*t^3 + Mod(1, 2)*t + Mod(1, 2)), Mod(
Mod(1, 2)*t, Mod(1, 2)*t^3 + Mod(1, 2)*t + Mod(1, 2)), Mod(Mod(1, 2)*t^2 + Mod(1
, 2), Mod(1, 2)*t^3 + Mod(1, 2)*t + Mod(1, 2)), 0, 0, 0, 0, 0, 0]
gp > P3 = [Mod(Mod(1,2),f), Mod(Mod(1,2)*t^2,f)];
gp > Q3 = [Mod(Mod(1,2)*t+Mod(1,2),f), Mod(Mod(1,2)*t^2+Mod(1,2),f)];
gp > elladd(E3,P3,Q3)
%22 = [Mod(Mod(1, 2), Mod(1, 2)*t^3 + Mod(1, 2)*t + Mod(1, 2)), Mod(Mod(1, 2)*t^
2 + Mod(1, 2), Mod(1, 2)*t^3 + Mod(1, 2)*t + Mod(1, 2))]
gp > lift(lift(elladd(E3,P3,Q3)))
%23 = [1, t^2 + 1]
gp > lift(lift(ellpow(E3,P3,2)))
%25 = [t + 1, t^2 + t]
```

Given an elliptic curve and an X-coordinate, the function ellordinate()
returns a vector of Y-coordinates of points on the curve having the given
X-coordinate. The vector contains 0, 1, or 2 elements.

```
gp > ellordinate(E1,3)
%26 = [5, -5]
gp > ellordinate(E1,4)
%27 = []
gp > ellordinate(E2,5)
%28 = [Mod(4, 17), Mod(13, 17)]
gp > ellordinate(E2,6)
%29 = [Mod(0, 17)]
```

The function ellorder() computes the additive order of a point on an
elliptic curve defined over $\mathbb{Q}$. If the input point is of infinite order, 0 is returned. This function is applicable to curves defined over $\mathbb{Q}$ only. Curves
defined over finite fields are not supported. Connell and Dujella discovered in 2000 that the curve $Y^2 + XY = X^3 - 15745932530829089880X +
24028219957095969426339278400$ has rank three. Its torsion subgroup is iso-
morphic to $\mathbb{Z}_2 \oplus \mathbb{Z}_8$, and consists of the following 16 points:

$\mathcal{O}, (-4581539664, 2290769832), (-1236230160, 203972501847720),$

$(2132310660, 12167787556920), (2452514160, 12747996298920),$

$(9535415580, 860741285907000), (2132310660, -12169919867580),$

$(-1236230160, -203971265617560), (9535415580, -860750821322580),$

$(2452514160, -12750448813080), (2346026160, -1173013080),$

(1471049760, 63627110794920), (1471049760, −63628581844680),

(3221002560, −82025835631080), (3221002560, 82022614628520),

(8942054015/4, −8942054015/8).

(Argue why the last point in the above list does not contradict the Nagell–Lutz theorem.) The free part of the elliptic curve group is generated by the following three independent points.

(2188064030, −7124272297330),

(396546810000/169, 1222553114825160/2197),

(16652415739760/3481, 49537578975823615480/205379).

```
gp > E4 = ellinit([1,0,0,-15745932530829089880,24028219957095969426339278400])
%30 = [1, 0, 0, -15745932530829089880, 24028219957095969426339278400, 1, -314918
65061658179760, 9611287982838387705357113600, -24793439124139356756716301340277
9136000, 755804761479796314241, -20760382044064624726576831008961, 4358115163151
138213244291572170411842042349568256000000000, 43174654491571340264570089406686688
7232701046835766560206229521/43581151631511382132442915721704118420423495682560
0000000, [2346026160.0000000000000000000, 2235513503.7499999999999999, -4581539
664.00000000000000000000000]~, 0.00008326692542370325455895756925, 0.0000378969084242
9953714081078080*I, 12469.6670944196391654477723, -32053.91346023314093443780493
3*I, 0.0000000031555590475550611737373709609]
gp > P4 = [9535415580, -860750821322580];
gp > Q4 = [-4581539664, 2290769832];
gp > R4 = [0];
gp > S4 = [2188064030, -7124272297330];
gp > ellorder(E4,P4)
%35 = 8
gp > ellorder(E4,Q4)
%36 = 2
gp > ellorder(E4,R4)
%37 = 1
gp > ellorder(E4,S4)
%38 = 0
```

4.3　Elliptic Curves over Finite Fields

Let E be an elliptic curve defined over a finite field $\mathbb{F}_q$. When no confusions are likely, we would use the shorthand notation E_q to stand for the group $E_{\mathbb{F}_q}$. E_q can contain at most $q^2 + 1$ points (all pairs in $\mathbb{F}_q^2$ and the point at infinity, but this is a loose overestimate), that is, E_q is a finite group, and every element of E_q has finite order. The following theorem describes the structure of E_q.

Theorem 4.12 *The group E_q is either cyclic or the direct sum $\mathbb{Z}_{n_1} \oplus \mathbb{Z}_{n_2}$ of two cyclic subgroups with $n_1, n_2 \geqslant 2$, $n_2 | n_1$, and $n_2 | (q-1)$.* ◁

Let m be the size of E_q. If E_q is cyclic, it contains exactly $\phi(m)$ generators. Let P be such a generator. Then, every point Q on E_q can be represented uniquely as $Q = sP$ for some s in the range $0 \leqslant s < m$. If E_q is not cyclic, then $E_q \cong \mathbb{Z}_{n_1} \oplus \mathbb{Z}_{n_2}$ with $m = n_1 n_2$. That is, E_q contains one point P of order n_1 and one point Q of order n_2 such that any point $R \in E_q$ can be uniquely represented as $R = sP + tQ$ with $0 \leqslant s < n_1$ and $0 \leqslant t < n_2$.

Example 4.13 (1) The elliptic curve $E_1 : Y^2 = X^3 - 5X + 1$ defined over $\mathbb{F}_{17}$ consists of the following 16 points.

$$
\begin{aligned}
P_0 &= \mathcal{O}, & P_1 &= (0,1), & P_2 &= (0,16), & P_3 &= (2,4), \\
P_4 &= (2,13), & P_5 &= (3,8), & P_6 &= (3,9), & P_7 &= (5,4), \\
P_8 &= (5,13), & P_9 &= (6,0), & P_{10} &= (10,4), & P_{11} &= (10,13), \\
P_{12} &= (11,6), & P_{13} &= (11,11), & P_{14} &= (13,5), & P_{15} &= (13,12).
\end{aligned}
$$

The multiples of the points in $(E_1)_{\mathbb{F}_{17}}$ and their orders are listed below.

P	$2P$	$3P$	$4P$	$5P$	$6P$	$7P$	$8P$	$9P$	$10P$	$11P$	$12P$	$13P$	$14P$	$15P$	$16P$	ord P
P_0																1
P_1	P_3	P_{14}	P_7	P_{12}	P_{11}	P_6	P_9	P_5	P_{10}	P_{13}	P_8	P_{15}	P_4	P_2	P_0	16
P_2	P_4	P_{15}	P_8	P_{13}	P_{10}	P_5	P_9	P_6	P_{11}	P_{12}	P_7	P_{14}	P_3	P_1	P_0	16
P_3	P_7	P_{11}	P_9	P_{10}	P_8	P_4	P_0									8
P_4	P_8	P_{10}	P_9	P_{11}	P_7	P_3	P_0									8
P_5	P_3	P_{13}	P_7	P_{15}	P_{11}	P_2	P_9	P_1	P_{10}	P_{14}	P_8	P_{12}	P_4	P_6	P_0	16
P_6	P_4	P_{12}	P_8	P_{14}	P_{10}	P_1	P_9	P_2	P_{11}	P_{15}	P_7	P_{13}	P_3	P_5	P_0	16
P_7	P_9	P_8	P_0													4
P_8	P_9	P_7	P_0													4
P_9	P_0															2
P_{10}	P_7	P_4	P_9	P_3	P_8	P_{11}	P_0									8
P_{11}	P_8	P_3	P_9	P_4	P_7	P_{10}	P_0									8
P_{12}	P_{10}	P_2	P_7	P_5	P_4	P_{14}	P_9	P_{15}	P_3	P_6	P_8	P_1	P_{11}	P_{13}	P_0	16
P_{13}	P_{11}	P_1	P_8	P_6	P_3	P_{15}	P_9	P_{14}	P_4	P_5	P_7	P_2	P_{10}	P_{12}	P_0	16
P_{14}	P_{11}	P_5	P_8	P_2	P_3	P_{12}	P_9	P_{13}	P_4	P_1	P_7	P_6	P_{10}	P_{15}	P_0	16
P_{15}	P_{10}	P_6	P_7	P_1	P_4	P_{13}	P_9	P_{12}	P_3	P_2	P_8	P_5	P_{11}	P_{14}	P_0	16

The table demonstrates that the group $(E_1)_{\mathbb{F}_{17}}$ is cyclic. The $\phi(16) = 8$ generators of this group are $P_1, P_2, P_5, P_6, P_{12}, P_{13}, P_{14}, P_{15}$.

(2) The elliptic curve $E_2 : Y^2 = X^3 - 5X + 2$ defined over $\mathbb{F}_{17}$ consists of the following 20 points.

$$
\begin{aligned}
P_0 &= \mathcal{O}, & P_1 &= (0,6), & P_2 &= (0,11), & P_3 &= (1,7), \\
P_4 &= (1,10), & P_5 &= (2,0), & P_6 &= (5,0), & P_7 &= (6,1), \\
P_8 &= (6,16), & P_9 &= (7,2), & P_{10} &= (7,15), & P_{11} &= (8,7), \\
P_{12} &= (8,10), & P_{13} &= (10,0), & P_{14} &= (12,2), & P_{15} &= (12,15), \\
P_{16} &= (13,3), & P_{17} &= (13,14), & P_{18} &= (15,2), & P_{19} &= (15,15).
\end{aligned}
$$

The multiples of these points and their orders are tabulated below.

P	$2P$	$3P$	$4P$	$5P$	$6P$	$7P$	$8P$	$9P$	$10P$	ord P
P_0										1
P_1	P_4	P_{18}	P_8	P_{13}	P_7	P_{19}	P_3	P_2	P_0	10
P_2	P_3	P_{19}	P_7	P_{13}	P_8	P_{18}	P_4	P_1	P_0	10
P_3	P_7	P_8	P_4	P_0						5
P_4	P_8	P_7	P_3	P_0						5
P_5	P_0									2
P_6	P_0									2
P_7	P_4	P_3	P_8	P_0						5
P_8	P_3	P_4	P_7	P_0						5
P_9	P_3	P_{16}	P_7	P_6	P_8	P_{17}	P_4	P_{10}	P_0	10
P_{10}	P_4	P_{17}	P_8	P_6	P_7	P_{16}	P_3	P_9	P_0	10
P_{11}	P_4	P_{14}	P_8	P_5	P_7	P_{15}	P_3	P_{12}	P_0	10
P_{12}	P_3	P_{15}	P_7	P_5	P_8	P_{14}	P_4	P_{11}	P_0	10
P_{13}	P_0									2
P_{14}	P_7	P_{12}	P_4	P_5	P_3	P_{11}	P_8	P_{15}	P_0	10
P_{15}	P_8	P_{11}	P_3	P_5	P_4	P_{12}	P_7	P_{14}	P_0	10
P_{16}	P_8	P_{10}	P_3	P_6	P_4	P_9	P_7	P_{17}	P_0	10
P_{17}	P_7	P_9	P_4	P_6	P_3	P_{10}	P_8	P_{16}	P_0	10
P_{18}	P_7	P_2	P_4	P_{13}	P_3	P_1	P_8	P_{19}	P_0	10
P_{19}	P_8	P_1	P_3	P_{13}	P_4	P_2	P_7	P_{18}	P_0	10

Since $(E_2)_{\mathbb{F}_{17}}$ does not contain a point of order 20, the group $(E_2)_{\mathbb{F}_{17}}$ is not cyclic. The above table shows that $(E_2)_{\mathbb{F}_{17}} \cong \mathbb{Z}_{10} \oplus \mathbb{Z}_2$. We can take the point P_1 as a generator of $\mathbb{Z}_{10}$ and P_5 as a generator of $\mathbb{Z}_2$. Every element of $(E_2)_{\mathbb{F}_{17}}$ can be uniquely expressed as $sP_1 + tP_5$ with $s \in \{0, 1, 2, \ldots, 9\}$ and $t \in \{0, 1\}$. The following table lists this representation of all the points of $E_2(\mathbb{F}_{17})$.

	$s=0$	$s=1$	$s=2$	$s=3$	$s=4$	$s=5$	$s=6$	$s=7$	$s=8$	$s=9$
$t=0$	P_0	P_1	P_4	P_{18}	P_8	P_{13}	P_7	P_{19}	P_3	P_2
$t=1$	P_5	P_{10}	P_{15}	P_{17}	P_{12}	P_6	P_{11}	P_{16}	P_{14}	P_9

(3) The ordinary curve $E_3 : Y^2 + XY = X^3 + \theta X^2 + (\theta^2 + 1)$ defined over $\mathbb{F}_8 = \mathbb{F}_2(\theta)$, where $\theta^3 + \theta + 1 = 0$, contains the following 12 points.

$$P_0 = \mathcal{O}, \qquad P_1 = (0, \theta + 1), \qquad P_2 = (1, \theta),$$
$$P_3 = (1, \theta + 1), \qquad P_4 = (\theta, \theta^2), \qquad P_5 = (\theta, \theta^2 + \theta),$$
$$P_6 = (\theta + 1, 0), \qquad P_7 = (\theta + 1, \theta + 1), \qquad P_8 = (\theta^2, \theta),$$
$$P_9 = (\theta^2, \theta^2 + \theta), \quad P_{10} = (\theta^2 + \theta + 1, \theta), \quad P_{11} = (\theta^2 + \theta + 1, \theta^2 + 1).$$

The multiples of these points and their orders are listed in the table below. The table illustrates that the group $(E_3)_{\mathbb{F}_8}$ is cyclic. The $\phi(12) = 4$ generators of this group are the points P_2, P_3, P_6, P_7.

P	$2P$	$3P$	$4P$	$5P$	$6P$	$7P$	$8P$	$9P$	$10P$	$11P$	$12P$	ord P
P_0												1
P_1	P_0											2
P_2	P_8	P_{11}	P_5	P_6	P_1	P_7	P_4	P_{10}	P_9	P_3	P_0	12
P_3	P_9	P_{10}	P_4	P_7	P_1	P_6	P_5	P_{11}	P_8	P_2	P_0	12
P_4	P_5	P_0										3
P_5	P_4	P_0										3
P_6	P_9	P_{11}	P_4	P_2	P_1	P_3	P_5	P_{10}	P_8	P_7	P_0	12
P_7	P_8	P_{10}	P_5	P_3	P_1	P_2	P_4	P_{11}	P_9	P_6	P_0	12
P_8	P_5	P_1	P_4	P_9	P_0							6
P_9	P_4	P_1	P_5	P_8	P_0							6
P_{10}	P_1	P_{11}	P_0									4
P_{11}	P_1	P_{10}	P_0									4

(4) As a final example, consider the supersingular curve $E_4 : Y^2 + Y = X^3 + X + \theta^2$ defined over $\mathbb{F}_8 = \mathbb{F}_2(\theta)$ with $\theta^3 + \theta + 1 = 0$. The group $(E_4)_{\mathbb{F}_8}$ contains the following five points.

$$P_0 = \mathcal{O}, \qquad P_1 = (0, \theta^2 + \theta), \qquad P_2 = (0, \theta^2 + \theta + 1),$$
$$P_3 = (1, \theta^2 + \theta), \quad P_4 = (1, \theta^2 + \theta + 1).$$

Since the size of $(E_4)_{\mathbb{F}_8}$ is prime, $(E_4)_{\mathbb{F}_8}$ is a cyclic group, and any point on it except $\mathcal{O}$ is a generator of it. $\qquad\square$

The size of the elliptic curve group $E_q = E_{\mathbb{F}_q}$ is trivially upper-bounded by $q^2 + 1$. In practice, this size is much smaller than $q^2 + 1$. The following theorem implies that the size of E_q is $\Theta(q)$.

Theorem 4.14 [*Hasse's theorem*][15] *The size of E_q is $q + 1 - t$, where $-2\sqrt{q} \leqslant t \leqslant 2\sqrt{q}$.* $\qquad\triangleleft$

The integer t in Hasse's theorem is called the *trace of Frobenius*[16] for the elliptic curve E defined over $\mathbb{F}_q$. It is an important quantity associated with the curve. We define several classes of elliptic curves based on the value of t.

Definition 4.15 Let E be an elliptic curve defined over the finite field $\mathbb{F}_q$ of characteristic p, and t the trace of Frobenius for E.
(a) If $t = 1$, that is, if the size of $E_{\mathbb{F}_q}$ is q, we call E an *anomalous curve*.
(b) If $p|t$, we call E a *supersingular* curve, whereas if $p \nmid t$, we call E a *non-supersingular* or an *ordinary* curve. $\qquad\triangleleft$

Recall that for finite fields of characteristic two, we have earlier defined supersingular and non-supersingular curves in a different manner. The earlier definitions turn out to be equivalent to Definition 4.15(b) for the fields $\mathbb{F}_{2^n}$. We have the following important characterization of supersingular curves.

---

[15]This result was conjectured by the Austrian-American mathematician Emil Artin (1898–1962), and proved by the German mathematician Helmut Hasse (1898–1979).
[16]Ferdinand Georg Frobenius (1849–1917) was a German mathematician well-known for his contributions to group theory.

Proposition 4.16 *An elliptic curve E over $\mathbb{F}_q$ and with trace of Frobenius equal to t is supersingular if and only if $t^2 = 0, q, 2q, 3q$, or $4q$. In particular, an elliptic curve defined over $\mathbb{F}_q = \mathbb{F}_{p^n}$ with $p = \operatorname{char} \mathbb{F}_q \neq 2, 3$ and with n odd is supersingular if and only if $t = 0$.* $\lhd$

Example 4.17 (1) The curve $Y^2 = X^3 + X + 3$ defined over $\mathbb{F}_{17}$ contains the following 17 points, and is anomalous.

$$
\begin{aligned}
&P_0 = \mathcal{O}, && P_1 = (2,8), && P_2 = (2,9), && P_3 = (3,4), && P_4 = (3,13), \\
&P_5 = (6,2), && P_6 = (6,15), && P_7 = (7,8), && P_8 = (7,9), && P_9 = (8,8), \\
&P_{10} = (8,9), && P_{11} = (11,6), && P_{12} = (11,11), && P_{13} = (12,3), && P_{14} = (12,14), \\
&P_{15} = (16,1), && P_{16} = (16,16).
\end{aligned}
$$

Evidently, anomalous curves defined over prime fields admit cyclic groups.

(2) The curve $Y^2 = X^3 + X + 1$ defined over $\mathbb{F}_{17}$ contains the following 18 points, and is supersingular.

$$
\begin{aligned}
&P_0 = \mathcal{O}, && P_1 = (0,1), && P_2 = (0,16), && P_3 = (4,1), && P_4 = (4,16), \\
&P_5 = (6,6), && P_6 = (6,11), && P_7 = (9,5), && P_8 = (9,12), && P_9 = (10,5), \\
&P_{10} = (10,12), && P_{11} = (11,0), && P_{12} = (13,1), && P_{13} = (13,16), && P_{14} = (15,5), \\
&P_{15} = (15,12), && P_{16} = (16,4), && P_{17} = (16,13).
\end{aligned}
$$

(3) The curve E_3 of Example 4.13(3), defined over $\mathbb{F}_8$, has trace -3 (not a multiple of two) and is non-supersingular. The curve E_4 of Example 4.13(4), defined again over $\mathbb{F}_8$, has trace four (a multiple of two), and is supersingular.

(4) The non-supersingular curve $Y^2 + XY = X^3 + \theta X^2 + \theta$ over $\mathbb{F}_8 = \mathbb{F}_2(\theta)$, $\theta^3 + \theta + 1 = 0$, contains the following eight points, and is anomalous.

$$
\begin{aligned}
&P_0 = \mathcal{O}, && P_1 = (0, \theta^2 + \theta), && P_2 = (\theta^2, 0), \\
&P_3 = (\theta^2, \theta^2), && P_4 = (\theta^2 + 1, \theta + 1), && P_5 = (\theta^2 + 1, \theta^2 + \theta), \\
&P_6 = (\theta^2 + \theta + 1, 1), && P_7 = (\theta^2 + \theta + 1, \theta^2 + \theta).
\end{aligned}
$$
$\square$

Example 4.76 lists some popular families of supersingular elliptic curves. Also see Exercise 4.66.

Now follows a useful application of the trace of Frobenius. Let E be an elliptic curve over $\mathbb{F}_q$ with trace of Frobenius equal to t. For any $r \in \mathbb{N}$, the curve E continues to remain an elliptic curve over the extension field $\mathbb{F}_{q^r}$. The size of the group $E_{q^r} = E_{\mathbb{F}_{q^r}}$ can be determined from the size of $E_q = E_{\mathbb{F}_q}$.

Proposition 4.18 [*Weil's theorem*][17] *Let the elliptic curve E defined over $\mathbb{F}_q$ have trace t. Let $\alpha, \beta \in \mathbb{C}$ satisfy $W^2 - tW + q = (W - \alpha)(W - \beta)$. Then, for every $r \in \mathbb{N}$, the size of the group E_{q^r} is $q^r + 1 - (\alpha^r + \beta^r)$, that is, the trace of Frobenius for E over $\mathbb{F}_{q^r}$ is $\alpha^r + \beta^r$.* $\lhd$

---

[17]André Abraham Weil (1906–1998) was a French mathematician who made profound contributions in the areas of number theory and algebraic geometry. He was one of the founding members of the mathematicians' (mostly French) group *Nicolas Bourbaki*.

Example 4.19 (1) The group of the curve $E_1 : Y^2 = X^3 - 5X + 1$ over $\mathbb{F}_{17}$ has size 16 (Example 4.13(1)), that is, the trace for this curve is $t = 2$. We have $W^2 - 2W + 17 = (W - \alpha)(W - \beta)$, where $\alpha = 1 + \mathrm{i}4$ and $\beta = 1 - \mathrm{i}4$. Therefore, the size of $(E_1)_{\mathbb{F}_{17^2}}$ is $17^2 + 1 - [(1 + \mathrm{i}4)^2 + (1 - \mathrm{i}4)^2] = 320$.

(2) The anomalous curve of Example 4.17(1) has trace $t = 1$. We have $W^2 - W + 17 = (W - \alpha)(W - \beta)$, where $\alpha = \frac{1 + \mathrm{i}\sqrt{67}}{2}$ and $\beta = \frac{1 - \mathrm{i}\sqrt{67}}{2}$. When the underlying field is $\mathbb{F}_{17^2}$, this curve contains $17^2 + 1 - (\alpha^2 + \beta^2) = 323$ points, and is no longer anomalous.

(3) The supersingular curve E_4 of Example 4.13(4) has trace $t = 4$. We have $W^2 - 4W + 8 = (W - \alpha)(W - \beta)$, where $\alpha = 2 + \mathrm{i}2$ and $\beta = 2 - \mathrm{i}2$. The group $(E_4)_{\mathbb{F}_{512}}$ contains $8^3 + 1 - (\alpha^3 + \beta^3) = 545$ points, and is again supersingular (having trace -32, a multiple of 2). □

4.4 Some Theory of Algebraic Curves

In this section, an elementary introduction to the theory of plane algebraic curves is provided. These topics are essentially parts of algebraic geometry. A treatise on elliptic curves, however, cannot proceed to any reasonable extent without these mathematical tools. Particularly, pairing and point counting on elliptic curves make heavy use of these tools. I have already mentioned that the study of elliptic curves is often referred to as arithmetic algebraic geometry. Studying relevant algebra or geometry is no harm to a number theorist.

The following treatment of algebraic geometry is kept as elementary as possible. In particular, almost all mathematical proofs are omitted. Readers willing to learn the mathematical details may start with the expository reports by Charlap and Robbins,[18] and by Charlap and Coley.[19]

4.4.1 Affine and Projective Curves

For the rest of this chapter, we let K be a field, and $\bar{K}$ its algebraic closure. More often that not, we deal with finite fields, that is, we have $K = \mathbb{F}_q$ with $p = \mathrm{char}\, K$. Mostly for visualization purposes, we would consider $K = \mathbb{R}$.

4.4.1.1 Affine Curves

The two-dimensional plane over K is called the *affine plane* over K:

$$K^2 = \{(h, k) \mid h, k \in K\}.$$

---

[18]Leonard S. Charlap and David P. Robbins, An elementary introduction to elliptic curves, *CRD Expository Report* 31, 1988.

[19]Leonard S. Charlap and Raymond Coley, An elementary introduction to elliptic curves II, *CCR Expository Report* 34, 1990.

We address the two field elements h, k for a point $P = (h, k) \in K^2$ as its X- and Y-coordinates, respectively. These coordinates are called the *affine coordinates* of P, and are unique for any given point. We denote these as

$$h = X(P), \ k = Y(P).$$

Definition 4.20 A plane *affine curve* C over K is defined by a non-zero polynomial as

$$C : f(X, Y) = 0.$$

A K-*rational point* on a curve $C : f(X, Y) = 0$ is a point $P = (h, k) \in K^2$ such that $f(P) = f(h, k) = 0$. A K-rational point is also called a *finite point* on C. The finite points on C are the solutions of $f(X, Y) = 0$ in K^2. ◁

It is customary to consider only *irreducible* polynomials $f(X, Y)$ for defining curves. The reason is that if $f(X, Y)$ admits a non-trivial factorization of the form $u(X, Y)v(X, Y)$ (or $w(X, Y)^a$ for some $a \geqslant 2$), then a study of $f(X, Y)$ is equivalent to a study of curves (u, v or w) of lower degree(s).

Example 4.21 (1) Straight lines are defined by the equation $aX + bY + c = 0$ with at least one of a, b non-zero.

(2) Circles are defined by the equation $(X - a)^2 + (Y - b)^2 = r^2$.

(3) Conic sections are defined by the equation $aX^2 + bXY + cY^2 + dX + eY + f = 0$ with at least one of a, b, c non-zero.

(4) Elliptic curves are smooth curves defined by the Weierstrass equation $Y^2 + (a_1 X + a_3)Y = X^3 + a_2 X^2 + a_4 X + a_6$.

(5) A *hyperelliptic curve* of genus g is a smooth curve defined by an equation of the form $Y^2 + u(X)Y = v(X)$, where $u(X), v(X) \in K[X]$ with $\deg u(X) \leqslant g$, $\deg v(X) = 2g + 1$, and $v(X)$ monic. If char $K \neq 2$, this can be simplified as $Y^2 = w(X)$ for a monic $w(X) \in K[X]$ of degree $2g + 1$.

A parabola is a hyperelliptic curve of genus zero (no handles), whereas an elliptic curve is a hyperelliptic curve of genus one (one handle only). □

4.4.1.2 Projective Curves

By a systematic addition of points at infinity to the affine plane, we obtain the two-dimensional projective plane over K. To that end, we define an equivalence relation $\sim$ on $K^3 \setminus \{(0, 0, 0)\}$ as $(h, k, l) \sim (h', k', l')$ if and only if $h' = \lambda h$, $k' = \lambda k$, and $l' = \lambda l$ for some non-zero $\lambda \in K$. The equivalence class of (h, k, l) is denoted by $[h, k, l]$, and the set of all these equivalence classes is the *projective plane* $\mathbb{P}^2(K)$ over K. For a point $P = [h, k, l] \in \mathbb{P}^2(K)$, the field elements h, k, l are called the *projective coordinates*[20] of P. Projective coordinates are unique up to simultaneous multiplication by non-zero elements of K. The three projective coordinates of a point cannot be simultaneously zero.

---

[20]Möbius (1790–1868) introduced the concept of projective or homogeneous coordinates, thereby pioneering algebraic treatment of projective geometry.

Figure 4.5 explains the relationship between the affine plane K^2 and the projective plane $\mathbb{P}^2(K)$. The equivalence class $P = [h, k, l]$ is identified with the line in K^3 passing through the origin $(0, 0, 0)$ and the point (h, k, l).

FIGURE 4.5: Points in the projective plane

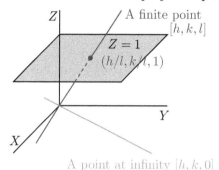

First, let $l \neq 0$. The line in K^3 corresponding to P meets the plane $Z = 1$ at $(h/l, k/l, 1)$. Indeed, $P = [h, k, l] = [h/l, k/l, 1]$. We identify P with the finite point $(h/l, k/l)$. Conversely, given any point $(h, k) \in K^2$, the point $[h, k, 1] \in \mathbb{P}^2(K)$ corresponds to the line passing through $(0, 0, 0)$ and $(h, k, 1)$. Thus, the points in K^2 are in bijection with the points on the plane $Z = 1$.

Next, take $l = 0$. The line of $[h, k, 0]$ lies on the X-Y plane, and does not meet the plane $Z = 1$ at all. Such lines correspond to the points at infinity in $\mathbb{P}^2(K)$. These points are not present in the affine plane. For every slope of straight lines in the X-Y plane, there exists a unique point at infinity.

We assume that a line passes through all the points at infinity. We call this the *line at infinity*. In the projective plane, two lines (parallel or not) meet at exactly one point. This is indeed a case of Bézout's theorem discussed later.

We are now ready to take affine curves to the projective plane.

Definition 4.22 A (multivariate) polynomial is called *homogeneous* if every non-zero term in the polynomial has the same degree. The zero polynomial is considered homogeneous of any degree. ◁

Example 4.23 The polynomial $X^3 + 2XYZ - 3Z^3$ is homogeneous of degree three. The polynomial $X^3 + 2XY - 3Z$ is not homogeneous. □

Definition 4.24 Let $C : f(X, Y) = 0$ be an affine curve of degree d. The *homogenization* of f is defined as $f^{(h)}(X, Y, Z) = Z^d f(X/Z, Y/Z)$. It is a homogeneous polynomial of degree d. The *projective curve* corresponding to C is defined by the equation

$$C^{(h)} : f^{(h)}(X, Y, Z) = 0.$$

Let $[h, k, l] \in \mathbb{P}^2(K)$, and $\lambda \in K^*$. Since $f^{(h)}$ is homogeneous of degree d, we have $f^{(h)}(\lambda h, \lambda k, \lambda l) = \lambda^d f^{(h)}(h, k, l)$. So $f^{(h)}(\lambda h, \lambda k, \lambda l) = 0$ if and only if

$f^{(h)}(h, k, l) = 0$. That is, the zeros of $f^{(h)}$ are not dependent on the choice of the projective coordinates, and a projective curve is a well-defined concept.

A K-*rational point* $[h, k, l]$ on $C^{(h)}$ is a solution of $f^{(h)}(h, k, l) = 0$. The set of all K-rational points on $C^{(h)}$ is denoted by $C_K^{(h)}$. By an abuse of notation, we often describe a curve by its affine equation. But when we talk about the rational points on that curve, we imply all rational points on the corresponding projective curve. In particular, C_K would stand for $C_K^{(h)}$.   ◁

Putting $Z = 1$ gives $f^{(h)}(X, Y, 1) = f(X, Y)$. This gives all the finite points on $C^{(h)}$, that is, all the points on the affine curve C. If, on the other hand, we put $Z = 0$, we get $f^{(h)}(X, Y, 0)$ which is a homogeneous polynomial in X, Y of degree d. The solutions of $f^{(h)}(X, Y, 0) = 0$ give all the points at infinity on $C^{(h)}$. These points are not present on the affine curve C.

Example 4.25 (1) A straight line in the projective plane has the equation $aX + bY + cZ = 0$. Putting $Z = 1$ gives $aX + bY + c = 0$, that is, all points on the corresponding affine line. Putting $Z = 0$ gives $aX + bY = 0$. If $b \neq 0$, then $Y = -(a/b)X$, that is, the line contains only one point at infinity $[1, -(a/b), 0]$. If $b = 0$, we have $X = 0$, that is, $[0, 1, 0]$ is the only point at infinity.

(2) A circle with center at (a, b) and radius r has the projective equation $(X - aZ)^2 + (Y - bZ)^2 = r^2 Z^2$. All finite points on the circle are solutions obtained by putting $Z = 1$, that is, all solutions of $(X - a)^2 + (Y - b)^2 = r^2$. For obtaining the points at infinity on the circle, we put $Z = 0$, and obtain $X^2 + Y^2 = 0$. For $K = \mathbb{R}$, the only solution of this is $X = Y = 0$. But all of the three projective coordinates are not allowed to be zero simultaneously, that is, the circle does not contain any point at infinity. Indeed, a circle does not have a part extending towards infinity in any direction.

However, for $K = \mathbb{C}$, the equation $X^2 + Y^2 = 0$ implies that $Y = \pm iX$, that is, there are two points at infinity: $[1, i, 0]$ and $[1, -i, 0]$.

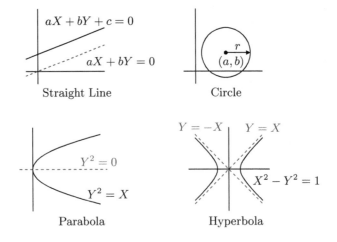

Straight Line   Circle

Parabola   Hyperbola

(3) The parabola $Y^2 = X$ has projective equation $Y^2 = XZ$. Putting $Z = 0$ gives $Y^2 = 0$, that is, $Y = 0$, so $[1, 0, 0]$ is the only point at infinity on this parabola. Since $Y^2 = X$, X grows faster than Y. In the limit $X \to \infty$, the curve becomes horizontal, justifying why its point at infinity satisfies $Y = 0$.

(4) The hyperbola $X^2 - Y^2 = 1$ has projective equation $X^2 - Y^2 = Z^2$. Putting $Z = 0$ gives $X^2 - Y^2 = 0$, that is, $Y = \pm X$, that is, $[1, 1, 0]$ and $[1, -1, 0]$ are the two points at infinity on the hyperbola. From a plot of the hyperbola, we see that the curve asymptotically touches the lines $Y = \pm X$.

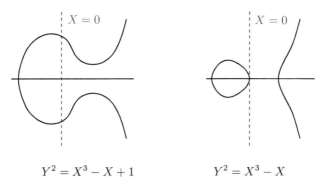

$$Y^2 = X^3 - X + 1 \qquad\qquad Y^2 = X^3 - X$$

Elliptic curves

(5) The homogenization of an elliptic curve given by the Weierstrass equation is $Y^2 Z + a_1 XYZ + a_3 Y Z^2 = X^3 + a_2 X^2 Z + a_4 X Z^2 + a_6 Z^3$. If we put $Z = 0$, we get $X^3 = 0$, that is, $X = 0$, that is, $[0, 1, 0]$ is the only point at infinity on the elliptic curve. In the limit $X \to \infty$, the curve becomes vertical.

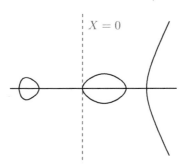

A hyperelliptic curve of genus two: $Y^2 = X(X^2 - 1)(X^2 - 2)$

(6) The homogenization of the hyperelliptic curve (see Example 4.21(5)) is $Y^2 Z^{2g-1} + Z^g u(X/Z)Y Z^g = Z^{2g+1} v(X/Z)$. If $g \geqslant 1$, the only Z-free term in this equation is X^{2g+1}, so the point at infinity on this curve has $X = 0$, and is $[0, 1, 0]$. A hyperelliptic curve of genus $\geqslant 1$ becomes vertical as $X \to \infty$. $\square$

Now that we can algebraically describe points at infinity on a curve, it is necessary to investigate the smoothness of curves at their points at infinity. Definition 4.2 handles only the finite points.

Definition 4.26 Let $f(X,Y) = 0$ be an affine curve, and $f^{(h)}(X,Y,Z) = 0$ the corresponding projective curve. The curve is smooth at the point P (finite or infinite) if the three partial derivatives $\partial f^{(h)}/\partial X$, $\partial f^{(h)}/\partial Y$ and $\partial f^{(h)}/\partial Z$ do not vanish simultaneously at the point P. A (projective) curve is called *smooth* if it is smooth at all points on it (including those at infinity). An elliptic or hyperelliptic curve is required to be smooth by definition.   ◁

In what follows, I often use affine equations of curves, but talk about the corresponding projective curves. The points at infinity on these curves cannot be described by the affine equations and are to be handled separately.

Theorem 4.27 [*Bézout's theorem*]   *An algebraic curve of degree m intersects an algebraic curve of degree n at exactly mn points.*   ◁

As such, the theorem does not appear to be true. A line and a circle must intersect at two points. While this is the case with some lines and circles, there are exceptions. For example, a tangent to a circle meets the circle at exactly one point. But, in this case, the intersection multiplicity is two, that is, we need to count the points of intersection with proper multiplicities. However, there are examples where a line does not meet a circle at all. Eliminating one of the variables X, Y from the equations of a circle and a line gives a quadratic equation in the other variable. If we try to solve this quadratic equation over $\mathbb{R}$, we may fail to get a root. If we solve the same equation over $\mathbb{C}$, we always obtain two roots. These two roots may be the same, implying that this is a case of tangency, that is, a root of multiplicity two. To sum up, it is necessary to work in an algebraically closed field for Bézout's theorem to hold.

But multiplicity and algebraic closure alone do not validate Bézout's theorem. Consider the case of two concentric circles $X^2 + Y^2 = 1$ and $X^2 + Y^2 = 2$. According to Bézout's theorem, they must intersect at four points. Even if we allow X, Y to assume complex values, we end up with an absurd conclusion $1 = 2$, that is, the circles do not intersect at all. The final thing that is necessary for Bézout's theorem to hold is that we must consider projective curves, and take into account the possibilities of intersections of the curves at their common points at infinity. By Example 4.25(2), every circle has two points at infinity over $\mathbb{C}$. These are $[1, i, 0]$ and $[1, -i, 0]$ irrespective of the radius and center of the circle. In other words, any two circles meet at these points at infinity. Two concentric circles touch one another at these points at infinity, so the total number of intersection points is $2 + 2 = 4$.

The equation of the circle $X^2 + Y^2 = a$ can be written as $X^2 - i^2Y^2 = a$, so as complex curves, a circle is a hyperbola too. If we replace i by a real number, we get a hyperbola in the real plane. It now enables us to visualize intersections at the points at infinity. Figure 4.6 illustrates some possibilities.

Parts (a) and (b) demonstrate situations where the two hyperbolas have only two finite points of intersection. Asymptotically, these two hyperbolas become parallel, that is, the two hyperbolas have the same points at infinity, and so the total number of points of intersection is four. Part (c) illustrates the

FIGURE 4.6: Intersection of two hyperbolas

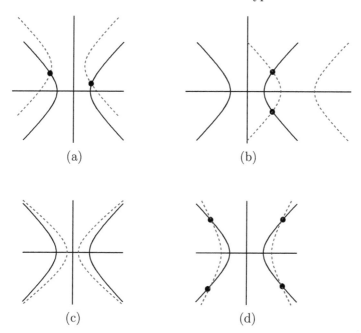

(a)   (b)

(c)   (d)

Equation of the solid hyperbola: $X^2 - Y^2 = 1$ in all the parts.
Equations of the dashed hyperbolas: (a) $(X + \frac{1}{3})^2 - (Y - \frac{4}{3})^2 = 1$,
(b) $(X - 3)^2 - Y^2 = 1$,   (c) $X^2 - Y^2 = \frac{1}{9}$,   (d) $\frac{4}{9}X^2 - Y^2 = 1$.

situation when the two hyperbolas not only have the same points at infinity
but also become tangential to one another at these points at infinity. There
are no finite points of intersection, but each of the two points at infinity is
an intersection point of multiplicity two. This is similar to the case of two
concentric circles. Finally, Part (d) shows a situation where the hyperbolas
have different points at infinity. All the four points of intersection of these
hyperbolas are finite. This is a situation that is not possible for circles. So we
can never see two circles intersecting at four points in the affine plane.

4.4.2   Polynomial and Rational Functions on Curves

Let $C : f(X, Y) = 0$ be defined by an *irreducible polynomial* $f(X, Y)$ in
$K[X, Y]$. For two polynomials $G(X, Y), H(X, Y) \in K[X, Y]$ with $f|(G - H)$,
we have $G(P) = H(P)$ for every rational point P on C (since $f(P) = 0$ on
the curve). Thus, G and H represent the same K-valued function on C. This
motivates us to define the congruence: $G(X, Y) \equiv H(X, Y) \pmod{f(X, Y)}$
if and only if $f|(G - H)$. Congruence modulo f is an equivalence relation
in $K[X, Y]$. Call the equivalence classes of X and Y as x and y. Then, the

equivalence class of a polynomial $G(X,Y) \in K[X,Y]$ is $G(x,y)$. The set of all the equivalence classes of $K[X,Y]$ under congruence modulo f is denoted by

$$K[C] = \{G(x,y) \mid G(X,Y) \in K[X,Y]\} = K[X,Y]/\langle f(X,Y)\rangle.$$

Here, $\langle f \rangle$ is the ideal in the polynomial ring $K[X,Y]$, generated by $f(X,Y)$. $K[C]$ is called the *coordinate ring* of C. Since f is irreducible, the ideal $\langle f \rangle$ is prime. Consequently, $K[C]$ is an integral domain. The set of fractions of elements of $K[C]$ (with non-zero denominators) is a field denoted as

$$K(C) = \{G(x,y)/H(x,y) \mid H(x,y) \neq 0\}.$$

$K(C)$ is called the *function field* of C.

Example 4.28 (1) For the straight line $L : Y = X$, we have $f(X,Y) = Y - X$. Since $Y \equiv X \pmod{f(X,Y)}$, any bivariate polynomial $G(X,Y) \in K[X,Y]$ is congruent to some $g(X) \in K[X]$ modulo f (substitute Y by X). It follows that the coordinate ring $K[L]$ is isomorphic, as a ring, to $K[X]$. Consequently, $K(L) \cong K(X)$, where $K(X)$ is the field of rational functions over K:

$$K(X) = \{g(X)/h(X) \mid g, h \in K[X], \ h \neq 0\}.$$

It is easy to generalize these results to any arbitrary straight line in the plane.

(2) For the circle $C : X^2 + Y^2 = 1$, we have $f(X,Y) = X^2 + Y^2 - 1$. Since $X^2 + Y^2 - 1$ is $(X + Y + 1)^2$ modulo 2, and we require f to be irreducible, we must have char $K \neq 2$. Congruence modulo this f gives the coordinate ring $K[C] = \{G(x,y) \mid G(X,Y) \in K[X,Y]\}$, where $x^2 + y^2 - 1 = 0$.

Since char $K \neq 2$, the elements $x, 1 - y, 1 + y$ are irreducible in $K[C]$, distinct from one another. But then, $x^2 = 1 - y^2 = (1 - y)(1 + y)$ gives two different factorizations of the same element in $K[C]$. Therefore, $K[C]$ is not a unique factorization domain. On the contrary, any polynomial ring over a field is a unique factorization domain. It follows that the coordinate ring of a circle is not isomorphic to a polynomial ring.

However, the *rational map* $K(C) \to K(Z)$ taking $x \mapsto \frac{1-Z^2}{1+Z^2}$ and $y \mapsto \frac{2Z}{1+Z^2}$ can be easily verified to be an isomorphism of fields. Therefore, the function field of a circle is isomorphic to the ring of univariate rational functions. □

Let us now specialize to elliptic (or hyperelliptic) curves. Write the equation of the curve as $C : Y^2 + u(X)Y = v(X)$ (for elliptic curves, $u(X) = a_1 X + a_3$, and $v(X) = X^3 + a_2 X^2 + a_4 X + a_6$). We have $y^2 = -u(x)y + v(x)$. If we take $G(x,y) \in K[C]$, then by repeatedly substituting y^2 by the linear (in y) polynomial $-u(x)y + v(x)$, we can simplify $G(x,y)$ as

$$G(x,y) = a(x) + yb(x)$$

for some $a(X), b(X) \in K[X]$. It turns out that such a representation of a polynomial function on C is unique, that is, every $G(x,y) \in K[C]$ corresponds to unique polynomials $a(X)$ and $b(X)$.

The field $K(C)$ of rational functions on C is a quadratic extension of the field $K(X)$ obtained by adjoining a root y of the irreducible polynomial

$$Y^2 + u(X)Y - v(X) \in K(X)[Y].$$

The other root of this polynomial is $-u(X) - y$. Therefore, we define the *conjugate* of $G(x, y) = a(x) + yb(x)$ as

$$\hat{G}(x, y) = a(x) - (u(x) + y)b(x).$$

The *norm* of G is defined as

$$N(G) = G\hat{G}. \tag{4.11}$$

Easy calculations show that $N(G) = a(x)^2 - a(x)b(x)u(x) - v(x)b(x)^2$. In particular, $N(G)$ is a polynomial in x alone.

Now, take a rational function $R(x, y) = G(x, y)/H(x, y) \in K(C)$. Multiplying both the numerator and the denominator by $\hat{H}$ simplifies R as

$$R(x, y) = s(x) + yt(x),$$

where $s(x), t(x) \in K(x)$ are rational functions in x only.

Let $C : f(X, Y) = 0$ be a plane (irreducible) curve, and $P = (h, k)$ a finite point on C. We plan to *evaluate* a rational function $R(x, y)$ on C at P. The obvious value of R at P should be $R(h, k)$. The following example demonstrates that this evaluation process is not so straightforward.

Example 4.29 Consider the unit circle $C : X^2 + Y^2 - 1 = 0$. Take $R(x, y) = \frac{1-x}{y} \in K(C)$, and $P = (1, 0)$. Plugging in the values $x = 1$ and $y = 0$ in $R(x, y)$ gives an expression of the form $\frac{0}{0}$, and it appears that R is not defined at P. However, $x^2 + y^2 - 1 = 0$, so that $y^2 = (1-x)(1+x)$, and R can also be written as $\frac{y}{1+x}$. Now, if we substitute $x = 1$ and $y = 0$, R evaluates to 0. □

Definition 4.30 The *value* of a polynomial function $G(x, y) \in K[C]$ at $P = (h, k)$ is $G(P) = G(h, k) \in K$. A rational function $R(x, y) \in K(C)$ is *defined* at P if there is a representation $R(x, y) = G(x, y)/H(x, y)$ for polynomials G, H with $H(h, k) \neq 0$. In that case, the *value* of R at P is $R(P) = G(P)/H(P) = G(h, k)/H(h, k) \in K$. If R is not defined at P, we take $R(P) = \infty$. ◁

Definition 4.31 P is called a *zero* of $R(x, y) \in K(C)$ if $R(P) = 0$, whereas P is called a *pole* of R if $R(P) = \infty$. ◁

The notion of the value of a rational function can be extended to the points at infinity on C. We now define this for elliptic curves only. Neglecting lower-degree terms in the Weierstrass equation for an elliptic curve C, we obtain $Y^2 \approx X^3$, that is, Y grows exponentially $3/2$ times as fast as X. In view of this, we give a weight of two to X, and a weight of three to Y. Let $G(x, y) = a(x) + yb(x) \in K[C]$ with $a(x), b(x) \in K[x]$. With x, y assigned the above weights, the degree of $a(x)$ is $2 \deg_x(a)$ (where $\deg_x$ denotes the usual

x-degree), and the degree of $yb(x)$ is $3 + 2 \deg_x(b)$. The larger of these two degrees is taken to be the degree of $G(x, y)$, that is,

$$\deg G = \max\big(2 \deg_x(a), 3 + 2 \deg_x(b)\big).$$

The *leading coefficient* of G, denoted $\mathrm{lc}(G)$, is that of a or b depending upon whether $2 \deg_x(a) > 3 + 2 \deg_x(b)$ or not. The two degrees cannot be equal, since $2 \deg_x(a)$ is even, whereas $3 + 2 \deg_x(b)$ is odd. Now, define the value of $R(x, y) = G(x, y)/H(x, y) \in K(C)$ (with $G, H \in K[C]$) at $\mathcal{O}$ as

$$R(\mathcal{O}) = \begin{cases} 0 & \text{if } \deg G < \deg H, \\ \infty & \text{if } \deg G > \deg H, \\ \mathrm{lc}(G)/\mathrm{lc}(H) & \text{if } \deg G = \deg H. \end{cases}$$

The point $\mathcal{O}$ is a zero (or pole) of R if $R(\mathcal{O}) = 0$ (or $R(\mathcal{O}) = \infty$).

Let C again be any algebraic curve. Although we are now able to uniquely define values of rational functions at points on C, the statement of Definition 4.30 is existential. In particular, nothing in the definition indicates how we can obtain a good representation G/H of R. We use a bit of algebra to settle this issue. The set of rational functions on C defined at P is a local ring with the unique maximal ideal comprising functions that evaluate to zero at P. This leads to the following valuation of rational functions at P. The notion of zeros and poles can be made concrete from this.

Theorem 4.32 *There exists a rational function $U_P(x, y)$ (depending on P) with the following two properties:*
(1) $U_P(P) = 0$.
(2) Any non-zero rational function $R(x, y) \in K(C)$ can be written as $R = U_P^d S$ with $d \in \mathbb{Z}$ and with $S \in K(C)$ having neither a pole nor a zero at P.
    The function U_P is called a uniformizer at the point P. The integer d does not depend upon the choice of the uniformizer. The order of R at P, denoted $\mathrm{ord}_P(R)$, is the integer d. ◁

Theorem 4.33 *If $\mathrm{ord}_P(R) = 0$, then P is neither a pole nor a zero of R. If $\mathrm{ord}_P(R) > 0$, then P is a zero of R. If $\mathrm{ord}_P(R) < 0$, then P is a pole of R.* ◁

Definition 4.34 *The multiplicity of a zero or pole of R at P is the absolute value of the order of R at P, that is, $v_P(R) = |\mathrm{ord}_P(R)|$. If P is neither a zero nor a pole of R, we take $v_P(R) = 0$.* ◁

Example 4.35 Consider the real unit circle $C : X^2 + Y^2 - 1 = 0$, and take a point $P = (h, k)$ on C. The two conditions in Theorem 4.32 indicate that U_P should have a *simple* zero at P. If $k \neq 0$, the vertical line $X = h$ cuts the circle at P, and we take $x - h$ as the uniformizer at P. On the other hand, if $k = 0$, the vertical line $X = h$ touches the circle at P, that is, the multiplicity of the intersection of C with $X = h$ at $(h, 0)$ is two. So we cannot take $x - h$ as the uniformizer at $(h, 0)$. However, the horizontal line $Y = 0$ meets the circle with multiplicity one, so y can be taken as the uniformizer at $(h, 0)$.

As a specific example, consider the rational function

$$R(x,y) = \frac{G(x,y)}{H(x,y)} = \frac{25y^2x + 25yx^2 - 30y^2 - 30yx + 9y - 34x + 30}{5y^2 + 8x - 8}.$$

Case 1: $P = (\frac{3}{5}, \frac{4}{5})$.

In this case, we can take $x - 3/5$ as the uniformizer. Let us instead take $5(x - h) = 5x - 3$ as the uniformizer. Clearly, multiplication of a uniformizer by non-zero field elements does not matter. We have $G(P) = H(P) = 0$, so we need to find an alternative representation of R. We write

$$\begin{aligned}
G(x,y) &= 25y^2x + 25yx^2 - 30y^2 - 30yx + 9y - 34x + 30 \\
&= 5y^2(5x - 3) + y(5x - 3)^2 + (5x - 3)(3x - 5) \\
&= (5x - 3)(5y^2 + y(5x - 3) + 3x - 5).
\end{aligned}$$

The function $5y^2 + y(5x - 3) + 3x - 5$ again evaluates to zero at $P = (\frac{3}{5}, \frac{4}{5})$. So we can factor out $5x - 3$ further from $5y^2 + y(5x - 3) + 3x - 5$.

$$\begin{aligned}
G(x,y) &= (5x - 3)(5y^2 + y(5x - 3) + 3x - 5) \\
&= (5x - 3)(y(5x - 3) + 3x - 5x^2) \\
&= (5x - 3)^2(y - x).
\end{aligned}$$

Since $y - x$ does not evaluate to zero at P, we are done. The denominator is

$$H(x,y) = 5y^2 + 8x - 8 = -5x^2 + 8x - 3 = (5x - 3)(1 - x).$$

It therefore follows that

$$R(x,y) = (5x - 3)\left(\frac{y - x}{1 - x}\right) \tag{4.12}$$

with the rational function $\frac{y-x}{1-x}$ neither zero not ∞ at P. Therefore, $(\frac{3}{5}, \frac{4}{5})$ is a zero of R of multiplicity one (a simple zero).

Case 2: $P = (1, 0)$.

Eqn (4.12) indicates that R has a pole at P. But what is the multiplicity of this pole? Since y (not $x - 1$) is a uniformizer at P, we rewrite $R(x,y)$ as

$$R(x,y) = (5x - 3)\left(\frac{(y - x)(1 + x)}{y^2}\right) = y^{-2}(5x - 3)(y - x)(1 + x).$$

The function $(5x - 3)(y - x)(1 + x)$ is neither zero nor ∞ at P, so $(1,0)$ is a pole of multiplicity two (a double pole) of R.

It is instructive to study uniformizers other than those prescribed above. The line $Y = X - 1$ meets (but not touches) the circle at $P = (1,0)$. So we may take $y - x + 1$ as a uniformizer at $(1,0)$. Since $(y - x + 1)^2 = y^2 + x^2 + 1 - 2yx + 2y - 2x + 1 = 2 - 2x + 2y - 2yx = 2(1 - x)(1 + y)$, Eqn (4.12) gives

$$R(x,y) = (y - x + 1)^{-2}(2(5x - 3)(y - x)(1 + y)).$$

But $2(5x - 3)(y - x)(1 + y)$ has neither a zero nor a pole at $P = (1, 0)$, so $(1, 0)$ is again established as a double pole of R.

It is not necessary to take only linear functions as uniformizers. The circle $(X - 1)^2 + (Y - 1)^2 = 1$ meets the circle $X^2 + Y^2 = 1$ at $P = (1, 0)$ with multiplicity one. So $(x-1)^2 + (y-1)^2 - 1$ may also be taken as a uniformizer at $(1, 0)$. But $(x-1)^2 + (y-1)^2 - 1 = x^2 + y^2 - 1 - 2(x+y-1) = 2(1-x-y)$, and so $[(x-1)^2 + (y-1)^2 - 1]^2 = 4(1 + x^2 + y^2 - 2x - 2y + 2xy) = 4(1 + 1 - 2x - 2y + 2xy) = 8(1 - x)(1 - y)$, and we have the representation

$$R(x, y) = \left((x - 1)^2 + (y - 1)^2 - 1\right)^{-2} \left(8(5x - 3)(y - x)(1 - y)\right),$$

which yet again reveals that the multiplicity of the pole of R at $(1, 0)$ is two.

Let us now pretend to take $x^2 - y^2 - 1$ as a uniformizer at $P = (1, 0)$. We have $x^2 - y^2 - 1 = 2x^2 - 2 - (x^2 + y^2 - 1) = 2x^2 - 2 = 2(x - 1)(x + 1)$, so that

$$R(x, y) = (x^2 - y^2 - 1)^{-1}\left(-2(5x - 3)(y - x)(1 + x)\right).$$

This seems to reveal that $(1, 0)$ is a simple pole of R. This conclusion is wrong, because the hyperbola $X^2 - Y^2 = 1$ *touches* the circle $X^2 + Y^2 = 1$ at $(1, 0)$ (with intersection multiplicity two), that is, $x^2 - y^2 - 1$ is not allowed to be taken as a uniformizer at $(1, 0)$.

If a curve C' meets C at $P = (1, 0)$ with intersection multiplicity larger than two, $R(x, y)$ cannot at all be expressed in terms of the equation of C'. For instance, take C' to be the parabola $Y^2 = 2(1 - X)$ which meets $X^2 + Y^2 = 1$ at $(1, 0)$ with multiplicity four (argue why). We have $y^2 - 2(1 - x) = (x^2 + y^2 - 1) - (x^2 - 2x + 1) = -(1 - x)^2$, that is,

$$[R(x, y)]^2 = \left(y^2 - 2(1 - x)\right)^{-1}\left(-(5x - 3)^2(y - x)^2\right),$$

indicating that $R(x, y)$ itself has a pole at $(1, 0)$ of multiplicity *half*, an absurd conclusion indeed.

At any rate, (non-tangent) linear functions turn out to be the handiest as uniformizers, particularly at all finite points on a curve. □

Some important results pertaining to poles and zeros are now stated.

Theorem 4.36 *Any non-zero rational function has only finitely many zeros and poles.*   ◁

Theorem 4.37 *Suppose that the underlying field K is algebraically closed. If the only poles of a rational function are at the points at infinity, then the rational function is a polynomial function.*   ◁

Theorem 4.38 *For a projective curve defined over an algebraically closed field, the sum of the orders of a non-zero rational function at its zeros and poles is zero.*   ◁

The algebraic closure of K is crucial for Theorem 4.38. If K is not algebraically closed, any non-zero rational function continues to have only finitely many zeros and poles, but the sum of their orders is not necessarily zero.

For elliptic curves $C : Y^2 + u(X)Y = v(X)$, we can supply explicit formulas for the uniformizer. Let $P = (h, k)$ be a finite point on C.

Definition 4.39 The *opposite* of P is defined as $\tilde{P} = -P = (h, -k - u(h))$. P and $\tilde{P}$ are the only points on C with X-coordinate equal to h. Conventionally, the opposite of $\mathcal{O}$ is taken as $\mathcal{O}$ itself.

P is called an *ordinary point* if $\tilde{P} \neq P$, and a *special point* if $\tilde{P} = P$.　◁

Any line passing through a finite point P but not a tangent to C at P can be taken as a uniformizer U_P at P. For example, we may take

$$U_P = \begin{cases} x - h & \text{if } P \text{ is an ordinary point,} \\ y - k & \text{if } P \text{ is a special point.} \end{cases}$$

A uniformizer at $\mathcal{O}$ is x/y.

Theorem 4.32 does not lead to an explicit algorithm for computing orders of rational functions. For elliptic curves, however, orders can be computed without explicitly using uniformizers. Let us first consider polynomial functions. Let $G(x, y) = a(x) + yb(x) \in K[C]$, $P = (h, k)$ a finite point on C, and e the largest exponent for which $(x - h)^e$ divides both $a(x)$ and $b(x)$. Write $G(x, y) = (x - h)^e G_1(x, y)$. If $G_1(h, k) \neq 0$, take $l = 0$, otherwise take l to be the largest exponent for which $(x - h)^l \mid \mathrm{N}(G_1)$ (see Eqn (4.11)). We have

$$\mathrm{ord}_P(G) = \begin{cases} e + l & \text{if } P \text{ is an ordinary point,} \\ 2e + l & \text{if } P \text{ is a special point.} \end{cases} \tag{4.13}$$

The point $\mathcal{O}$ needs special attention:

$$\mathrm{ord}_\mathcal{O}(G) = -\max(2\deg_x a, 3 + 2\deg_x b). \tag{4.14}$$

For a rational function $R(x, y) = G(x, y)/H(x, y) \in K(C)$ with $G, H \in K[C]$,

$$\mathrm{ord}_P(R) = \mathrm{ord}_P(G) - \mathrm{ord}_P(H) \tag{4.15}$$

for any point P on C (including the point $\mathcal{O}$).

Example 4.40 Consider the elliptic curve $C : Y^2 = X^3 - X$ defined over $\mathbb{C}$.

(1) Rational functions involving only x are simpler. The rational function

$$R_1(x, y) = \frac{(x - 1)(x + 1)}{x^3(x - 2)}$$

has simple zeros at $x = \pm 1$, a simple pole at $x = 2$, and a pole of multiplicity three at $x = 0$. The points on C with these X-coordinates are $P_1 = (0, 0)$, $P_2 = (1, 0)$, $P_3 = (-1, 0)$, $P_4 = (2, \sqrt{6})$, and $P_5 = (2, -\sqrt{6})$. P_1, P_2, P_3 are special points, so $\mathrm{ord}_{P_1}(R_1) = -6$, $\mathrm{ord}_{P_2}(R_1) = \mathrm{ord}_{P_3}(R_1) = 2$. P_4 and P_5 are ordinary points, so $\mathrm{ord}_{P_4}(R_1) = \mathrm{ord}_{P_5}(R_1) = -1$. Finally, note that $R_1 \to$

$1/x^2$ as $x \to \infty$. But x has a weight of two, so R_1 has a zero of order four at $\mathcal{O}$. The sum of these orders is $-6 + 2 + 2 - 1 - 1 + 4 = 0$.

(2) Now, consider the rational function $R_2(x,y) = x/y$ involving y. At the point $P_1 = (0,0)$, R_2 appears to be undefined. But $y^2 = x^3 - x$, so $R_2 = y/(x^2 - 1)$ too, and $R_2(P_1) = 0$, that is, R_2 has a zero at P_1. For the numerator y, we have $e = 0$ and $l = 1$, so Eqn (4.13) gives $\mathrm{ord}_{P_1}(y) = 1$. On the other hand, the denominator $x^2 - 1$ has neither a zero nor a pole at P_1. So $\mathrm{ord}_{P_1}(R_2) = 1$ by Eqn (4.15).

Notice that $\mathrm{ord}_{P_1}(x) = 2$ (by Eqn (4.13), since $e = 1$, $l = 0$, and P_1 is a special point), so the representation $R_2 = x/y$ too gives $\mathrm{ord}_{P_1}(R_2) = 2-1 = 1$.

(3) Take the same curve $C : Y^2 = X^3 - X$ defined over $\mathbb{F}_7$. Since 6 is a quadratic non-residue modulo 7, the points $P_4 = (2, \sqrt{6})$ and $P_5 = (2, -\sqrt{6})$ do not lie on the curve over $\mathbb{F}_7$. The only zeros and poles of R_1 are, therefore, P_1, P_2, P_3, and $\mathcal{O}$. The orders of R_1 at these points add up to $-6+2+2+4 = 2 \neq 0$. This example illustrates the necessity of assuming algebraic closure of the field of definition for the elliptic curve. However, C is defined over the algebraic closure $\bar{\mathbb{F}}_7$ of $\mathbb{F}_7$. The field $\bar{\mathbb{F}}_7$ contains the square roots of 6, and the problem associated with $\mathbb{F}_7$ is eliminated. $\square$

Example 4.41 Let us compute all the zeros and poles of the rational function

$$R(x, y) = \frac{G(x, y)}{H(x, y)} = \frac{x + y}{x^2 + y}$$

on the elliptic curve $E : Y^2 = X^3 + X$ defined over the field $\mathbb{F}_5$. We handle the numerator and the denominator of R separately.

Zeros and poles of $G(x, y) = x + y$: A zero of G satisfies $x + y = 0$, that is, $y = -x$. Since x, y also satisfy the equation of E, we have $y^2 = (-x)^2 = x^3 + x$, that is, $x(x^2 - x + 1) = 0$. The polynomial $x^2 - x + 1$ is irreducible over $\mathbb{F}_5$. Let $\theta \in \bar{\mathbb{F}}_5$ be the element satisfying $\theta^2 + 2 = 0$. This element defines the extension $\mathbb{F}_{5^2} = \mathbb{F}_5(\theta)$ in which we have $x(x^2 - x + 1) = x(x + (\theta + 2))(x + (4\theta + 2))$. Therefore, the zeros of $G(x, y)$ correspond to $x = 0, -(\theta + 2), -(4\theta + 2)$, that is, $x = 0, 4\theta + 3, \theta + 3$. Plugging in these values of x in $y = -x$ gives us the three zeros of G as $Q_0 = (0, 0)$, $Q_1 = (4\theta + 3, \theta + 2)$, and $Q_2 = (\theta + 3, 4\theta + 2)$.

In order to compute the multiplicities of these zeros, we write $G(x, y) = a(x) + yb(x)$ with $a(x) = x$ and $b(x) = 1$. We have $\gcd(a(x), b(x)) = 1$, so we compute $N(G) = a(x)^2 - y^2 b(x)^2 = x^2 - (x^3 + x) = -x(x^2 - x + 1)$. This indicates that each of the three zeros of G has $e = 0$ and $l = 1$, and so has multiplicity one (Q_0 is a special point, whereas Q_1 and Q_2 are ordinary).

The degree of $x + y$ is three, so G has a pole of multiplicity three at $\mathcal{O}$.

Zeros and poles of $H(x, y) = x^2 + y$: The zeros of H correspond to $y = -x^2$, that is, $y^2 = (-x^2)^2 = x^3 + x$, that is, $x(x^3 - x^2 - 1) = 0$. The cubic factor being irreducible, all the zeros of H exist on the curve E defined over $\mathbb{F}_{5^3}$. Let $\psi \in \bar{\mathbb{F}}_5$ satisfy $\psi^3 + \psi + 1 = 0$. Then, the zeros of H correspond to $x(x^3 - x^2 - 1) = x(x + (2\psi^2 + 2\psi + 1))(x + (4\psi^2 + 4))(x + (4\psi^2 + 3\psi + 4)) = 0$,

FIGURE 4.7: Zeros and poles of straight lines

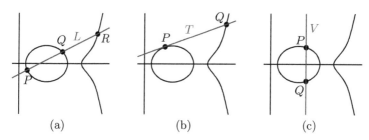

(a) (b) (c)

that is, $x = 0, 3\psi^2 + 3\psi + 4, \psi^2 + 1, \psi^2 + 2\psi + 1$. Since $y = -x^2$ on H, the corresponding y-values are respectively $0, 4\psi^2 + 2\psi + 3, \psi^2 + 4\psi + 1, 4\psi + 2$. Therefore, the zeros of H are $Q_0 = (0,0)$, $Q_3 = (3\psi^2 + 3\psi + 4, 4\psi^2 + 2\psi + 3)$, $Q_4 = (\psi^2 + 1, \psi^2 + 4\psi + 1)$, and $Q_5 = (\psi^2 + 2\psi + 1, 4\psi + 2)$.

The multiplicities of these zeros can be computed from the expression $N(H) = x^4 - y^2 = x^4 - (x^3 - x) = x(x^3 - x^2 - 1)$. It follows that each of the zeros Q_0, Q_3, Q_4, Q_5 has multiplicity one.

The degree of $x^2 + y$ is four, so H has a pole of multiplicity four at $\mathcal{O}$.

Zeros and poles of $R(x, y) = G(x, y)/H(x, y)$: We have $\mathrm{ord}_{Q_0}(R) = 1 - 1 = 0$, $\mathrm{ord}_{Q_1}(R) = 1 - 0 = 1$, $\mathrm{ord}_{Q_2}(R) = 1 - 0 = 1$, $\mathrm{ord}_{Q_3}(R) = 0 - 1 = -1$, $\mathrm{ord}_{Q_4}(R) = 0 - 1 = -1$, $\mathrm{ord}_{Q_5}(R) = 0 - 1 = -1$, and $\mathrm{ord}_{\mathcal{O}}(R) = -3 - (-4) = 1$. To sum up, R has simple zeros at Q_1, Q_2 and $\mathcal{O}$, and simple poles at Q_3, Q_4 and Q_5. The smallest extension of $\mathbb{F}_5$ containing (the coordinates of) all these points is $\mathbb{F}_{5^6} = \mathbb{F}_5(\theta)(\psi)$. Although Q_0 is individually a zero of both G and H, it is neither a zero nor a pole of R. $\square$

Example 4.42 Zeros and poles of straight lines are quite important in the rest of this chapter. Some examples are shown in Figure 4.7.

In Part (a), the non-vertical line L passes through the three points P, Q, R on the elliptic curve. These are the only zeros of L on C. We have $\mathrm{ord}_P(L) = \mathrm{ord}_Q(L) = \mathrm{ord}_R(L) = 1$, and $\mathrm{ord}_{\mathcal{O}}(L) = -3$.

In Part (b), the non-vertical line T is a tangent to the curve at P. The other point of intersection of T with the curve is Q. We have $\mathrm{ord}_P(T) = 2$, $\mathrm{ord}_Q(T) = 1$, and $\mathrm{ord}_{\mathcal{O}}(T) = -3$.

The vertical line V of Part (c) passes through only two points P and Q of the curve. This indicates $\mathrm{ord}_P(V) = \mathrm{ord}_Q(V) = 1$, and $\mathrm{ord}_{\mathcal{O}}(V) = -2$. $\square$

4.4.3 Rational Maps and Endomorphisms on Elliptic Curves

From this section until the end of this chapter, we concentrate on elliptic curves only. Some of the results pertaining to elliptic curves can be generalized to general curves, but I do not plan to be so general. Before starting the discussion, let me once again review our notations.

Let E be an elliptic curve defined over a field K. $\bar{K}$ denotes the algebraic closure of K. Our interests concentrate primarily on finite fields $K = \mathbb{F}_q$ with char $K = p$. This means that we will remove the restriction that K is algebraically closed. For any field extension L of K, we denote by E_L the group of L-rational points on the elliptic curve E (which is defined over L as well). We assume that the point $\mathcal{O}$ belongs to all such groups E_L. When $L = \bar{K}$, we abbreviate $E_{\bar{K}}$ as E. Thus, E would denote[21] both the curve and the set (group) of $\bar{K}$-rational points on it. For $L = \mathbb{F}_{q^k}$, we use the shorthand symbol E_{q^k} to stand for the group $E_{\mathbb{F}_{q^k}}$. A rational function R on E is a member of $\bar{K}(E)$. We say that R is defined over L if R has a representation of the form $R(x, y) = G(x, y)/H(x, y)$ with $G(x, y), H(x, y) \in L[E]$.

Definition 4.43 A *rational map* on E is a function $E \to E$. A rational map α is specified by two rational functions $\alpha_1, \alpha_2 \in \bar{K}(E)$ such that for any $P \in E$, the point $\alpha(P) = \alpha(h, k) = (\alpha_1(h, k), \alpha_2(h, k))$ lies again on E.   ◁

Since $\alpha(P)$ is a point on E, the functions α_1, α_2 satisfy the equation for E, and is a point on the elliptic curve $E_{\bar{K}(E)}$. Denote the point at infinity on this curve by $\mathcal{O}'$. This map stands for the constant function $\mathcal{O}'(P) = \mathcal{O}$ for all $P \in E$. For a non-zero rational map α on $E_{\bar{K}(E)}$ and for a point P on E, either both $\alpha_1(P), \alpha_2(P)$ are defined at P, or both are undefined at P. This is because α_1 and α_2 satisfy the equation of the curve. In the first case, we take $\alpha(P) = (\alpha_1(P), \alpha_2(P))$, whereas in the second case, we take $\alpha(P) = \mathcal{O}$.

Theorem 4.44 *For all rational maps α, β on $E_{\bar{K}(E)}$ and for all points P on E, we have $(\alpha + \beta)(P) = \alpha(P) + \beta(P)$.*   ◁

This is a non-trivial assertion about rational maps. The sum $\alpha + \beta$ is the sum of rational maps on the curve $E_{\bar{K}(E)}$. On the other hand, $\alpha(P) + \beta(P)$ is the sum of points on the curve $E_{\bar{K}}$. Theorem 4.44 states that these two additions are mutually compatible. Another important feature of rational maps (over algebraically closed fields) is the following.

Theorem 4.45 *A rational map is either constant or surjective.*   ◁

Example 4.46 (1) The *zero map* $\mathcal{O}' : E \to E$ taking $P \mapsto \mathcal{O}$ is already discussed as the group identity of $E_{\bar{K}(E)}$.

(2) The *constant map* $\alpha_{h,k} : E \to E$ taking any point P to a fixed point (h, k) on E is a generalization of the zero map. This map corresponds to the two constant rational functions h and k.

(3) The *identity map* id : $E \to E$, $P \mapsto P$, is non-constant (so surjective).

(4) Fix a point $Q \in E$. The *translation map* $\tau_Q : E \to E$ taking $P \mapsto P+Q$ is again surjective.

---

[21] If E is an elliptic curve defined over a field F, we use E_F with dual meaning. It is both a geometric object (the curve, so we can say Q is *on* E_F) and an algebraic object (the group, so we can say Q is *in* E_F, or $Q \in E_F$). In algebraic geometry, they are same anyway.

(5) Take $m \in \mathbb{Z}$. The *multiplication-by-m map* $[m] : E \to E$ takes a point P on E to its m-th multiple mP. For $m \neq 0$, the map is non-constant and, therefore, surjective. That is, given any point Q on E, we can find a point P on E such that $Q = mP$. $\square$

For a moment, suppose that $K = \mathbb{F}_q$ and $(h, k) \in E = E_{\bar{K}}$. Since E is defined over K, the coefficients in the defining equation (like a_i in Eqn (4.4)) are members of K. By Fermat's little theorem for $\mathbb{F}_q$, we conclude that the point (h^q, k^q) lies on the curve E. Of course, if (h, k) is already K-rational, then $(h^q, k^q) = (h, k)$ is K-rational too. But if (h, k) is not K-rational but $\bar{K}$-rational, then (h^q, k^q) is again $\bar{K}$-rational.

Definition 4.47 The *q-th power Frobenius map* is defined as $\varphi_q : E \to E$ taking (h, k) to (h^q, k^q) (where $E = E_{\bar{K}}$). ◁

Definition 4.48 A rational map $\alpha : E \to E$ is called an *endomorphism* or an *isogeny* of E if α is a group homomorphism, that is, $\alpha(P + Q) = \alpha(P) + \alpha(Q)$ for all $P, Q \in E$. The set of all endomorphisms of E is denoted by $\text{End}(E)$. ◁

For $\alpha, \beta \in \text{End}(E)$ and $P \in E$, define $(\alpha + \beta)(P) = \alpha(P) + \beta(P)$. By Theorem 4.44, this addition is the same as the addition in the elliptic curve $E_{\bar{K}(E)}$. Also, define the product of α and β as the composition $(\alpha \circ \beta)(P) = \alpha(\beta(P))$. The set $\text{End}(E)$ is a ring under these operations. Its additive identity is the zero map $\mathcal{O}'$, and its multiplicative identity is the identity map id.

The translation map τ_Q is an endomorphism of E only for $Q = \mathcal{O}$. The multiplication-by-m maps $[m]$ are endomorphisms of E with $[m] \neq [n]$ for $m \neq n$. The set of all the maps $[m]$ is a subring of $\text{End}(E)$, isomorphic to $\mathbb{Z}$.

Definition 4.49 If $\text{End}(E)$ contains an endomorphism other than the maps $[m]$, we call E an elliptic curve with *complex multiplication*. ◁

If E is defined over $K = \mathbb{F}_q$, then the q-th power Frobenius map φ_q taking $(h, k) \in E$ to $(h^q, k^q) \in E$ is an endomorphism of E. We have $\varphi_q \neq [m]$ for any $m \in \mathbb{Z}$. It follows that any elliptic curve defined over any finite field is a curve with complex multiplication.

The notion of rational maps and isogenies can be extended to two different elliptic curves E, E' defined over the same field K.

Definition 4.50 A rational map $\alpha : E \to E'$ is specified by two rational functions $\alpha_1, \alpha_2 \in \bar{K}(E)$ such that for all $P \in E$, the image $(\alpha_1(P), \alpha_2(P))$ is a point on E'. A rational map $E \to E'$, which is also a group homomorphism, is called an *isogeny* of E to E'. An *isomorphism* $E \to E'$ is a bijective isogeny. ◁

Exercise 4.50 provides examples of isogenies between elliptic curves defined over $\mathbb{F}_5$. For elliptic curves, the notion of isomorphism can be characterized in terms of *admissible changes of variables* described in Theorem 4.51.

Theorem 4.51 *Two elliptic curves E, E' defined over K are isomorphic over $\bar{K}$ if and only if there exist $u, r, s, t \in \bar{K}$ with $u \neq 0$ such that substituting X by $u^2 X + r$ and Y by $u^3 Y + s u^2 X + t$ transforms the equation of E to the equation of E'.* ◁

The substitutions made to derive Eqns (4.5), (4.6), (4.7) and (4.8) from the original Weierstrass Eqn (4.4) are examples of admissible changes of variables.

For the rest of this section, we concentrate on the multiplication-by-m endomorphisms. We identify $[m]$ with a pair (g_m, h_m) of rational functions. These rational functions are inductively defined by the chord-and-tangent rule. Consider an elliptic curve defined by Eqn (4.4).

$$g_1 = x, \quad h_1 = y.$$

$$g_2 = -2x + \lambda^2 + a_1 \lambda - a_2, \quad h_2 = -\lambda(g_2 - x) - a_1 g_2 - a_3 - y, \quad (4.16)$$

where $\lambda = \dfrac{3x^2 + 2a_2 x + a_4 - a_1 y}{2y + a_1 x + a_3}$. Finally, for $m \geqslant 3$, we recursively define

$$g_m = -g_{m-1} - x + \lambda^2 + a_1 \lambda - a_2, \quad h_m = -\lambda(g_m - x) - a_1 g_m - a_3 - y, \quad (4.17)$$

where $\lambda = \dfrac{h_{m-1} - y}{g_{m-1} - x}$. The kernel of the map $[m]$ is denoted by $E[m]$, that is,

$$E[m] = \{P \in E = E_{\bar{K}} \mid mP = \mathcal{O}\}.$$

Elements of $E[m]$ are called *m-torsion points* of E. For every $m \in \mathbb{Z}$, $E[m]$ is a subgroup of E.

Theorem 4.52 *Let $p = \operatorname{char} K$. If $p = 0$ or $\gcd(p, m) = 1$, then*

$$E[m] \cong \mathbb{Z}_m \times \mathbb{Z}_m,$$

and so $|E[m]| = m^2$. If $\gcd(m, n) = 1$, then $E[mn] \cong E[m] \times E[n]$. ◁

The rational functions g_m, h_m have poles precisely at the points in $E[m]$. But they have some zeros also. We plan to investigate polynomials having zeros precisely at the points of $E[m]$. Assume that either $p = 0$ or $\gcd(p, m) = 1$. Then, $E[m]$ contains exactly m^2 points. A rational function ψ_m whose only zeros are the m^2 points of $E[m]$ is a polynomial by Theorem 4.37. All these zeros are taken as simple. So ψ_m must have a pole of multiplicity m^2 at $\mathcal{O}$. The polynomial ψ_m is unique up to multiplication by non-zero elements of $\bar{K}$. If we arrange the leading coefficient of ψ_m to be m, then ψ_m becomes unique, and is called the *m-th division polynomial*.

The division polynomials are defined recursively as follows.

$$\begin{aligned}
\psi_0 &= 0 \\
\psi_1 &= 1 \\
\psi_2 &= 2y + a_1 x + a_3
\end{aligned}$$

$$\psi_3 = 3x^4 + d_2x^3 + 3d_4x^2 + 3d_6x + d_8$$

$$\psi_4 = \left[2x^6 + d_2x^5 + 5d_4x^4 + 10d_6x^3 + 10d_8x^2 + \right.$$
$$\left. (d_2d_8 - d_4d_6)x + d_4d_8 - d_6^2\right]\psi_2$$

$$\psi_{2m} = \frac{(\psi_{m+2}\psi_{m-1}^2 - \psi_{m-2}\psi_{m+1}^2)\psi_m}{\psi_2} \quad \text{for } m > 2$$

$$\psi_{2m+1} = \psi_{m+2}\psi_m^3 - \psi_{m-1}\psi_{m+1}^3 \quad \text{for } m \geqslant 2.$$

Here, the coefficients d_i are as in Definition 4.9. The rational functions g_m, h_m can be expressed in terms of the division polynomials as follows.

$$g_m - g_n = -\frac{\psi_{m+n}\psi_{m-n}}{\psi_m^2\psi_n^2}.$$

Putting $n = 1$ gives

$$g_m = x - \frac{\psi_{m+1}\psi_{m-1}}{\psi_m^2}. \tag{4.18}$$

Moreover,

$$h_m = \frac{\psi_{m+2}\psi_{m-1}^2 - \psi_{m-2}\psi_{m+1}^2}{2\psi_2\psi_m^3} - \frac{1}{2}(a_1g_m + a_3) \tag{4.19}$$

$$= y + \frac{\psi_{m+2}\psi_{m-1}^2}{\psi_2\psi_m^3} + (3x^2 + 2a_2x + a_4 - a_1y)\frac{\psi_{m-1}\psi_{m+1}}{\psi_2\psi_m^2}. \tag{4.20}$$

4.4.4 Divisors

Let a_i, $i \in I$, be *symbols* indexed by I. A *finite formal sum* of a_i, $i \in I$, is an expression of the form $\sum_{i \in I} m_i a_i$ with $m_i \in \mathbb{Z}$ such that $m_i = 0$ except for only finitely many $i \in I$. The sum $\sum_{i \in I} m_i a_i$ is formal in the sense that the symbols a_i are not meant to be evaluated. They act as *placeholders*. Define

$$\sum_{i \in I} m_i a_i + \sum_{i \in I} n_i a_i = \sum_{i \in I} (m_i + n_i) a_i, \text{ and } -\sum_{i \in I} m_i a_i = \sum_{i \in I} (-m_i) a_i.$$

Under these definitions, the set of these finite formal sums becomes an Abelian group called the *free Abelian group* generated by the symbols a_i, $i \in I$.

Now, let E be an elliptic curve defined over K. For a moment, let us treat E as a curve defined over the algebraic closure $\bar{K}$ of K.

Definition 4.53 A *divisor* on an elliptic curve E defined over a field K is a formal sum[22] of the rational points on $E = E_{\bar{K}}$. ◁

Let us use the notation $D = \sum_{P \in E} m_P[P]$ to denote a divisor D. Here, the symbol $[P]$ is used to indicate that the sum is formal, that is, the points P must not be evaluated when enclosed within square brackets.

---

[22]In this sense, a divisor is also called a *Weil divisor*.

Definition 4.54 The *support* of a divisor $D = \sum_P m_P[P]$, denoted $\mathrm{Supp}(D)$, is the set of points P for which $m_P \neq 0$. By definition, the support of any divisor is a finite set. The *degree* of D is the sum $\sum_P m_P$. All divisors on E form a group denoted by $\mathrm{Div}_{\bar{K}}(E)$ or $\mathrm{Div}(E)$. The divisors of degree zero form a subgroup denoted by $\mathrm{Div}^0_{\bar{K}}(E)$ or $\mathrm{Div}^0(E)$. ◁

Definition 4.55 The divisor of a non-zero rational function $R \in \bar{K}(E)$ is

$$\mathrm{Div}(R) = \sum_{P \in E} \mathrm{ord}_P(R)[P].$$

Since any non-zero rational function can have only finitely many zeros and poles, $\mathrm{Div}(R)$ is defined (that is, a finite formal sum) for any $R \neq 0$.

A *principal divisor* is the divisor of some rational function. Theorem 4.38 implies that every principal divisor belongs to $\mathrm{Div}^0_{\bar{K}}(E)$. The set of all principal divisors is a subgroup of $\mathrm{Div}^0_{\bar{K}}(E)$, denoted by $\mathrm{Prin}_{\bar{K}}(E)$ or $\mathrm{Prin}(E)$. ◁

Principal divisors satisfy the formulas:

$$\mathrm{Div}(R) + \mathrm{Div}(S) = \mathrm{Div}(RS), \text{ and } \mathrm{Div}(R) - \mathrm{Div}(S) = \mathrm{Div}(R/S). \quad (4.21)$$

Definition 4.56 Two divisors D, D' in $\mathrm{Div}_{\bar{K}}(E)$ are called *equivalent* if they differ by a principal divisor. That is, $D \sim D'$ if and only if $D = D' + \mathrm{Div}(R)$ for some $R(x, y) \in \bar{K}(E)$. ◁

Evidently, equivalence of divisors is an equivalence relation on $\mathrm{Div}(E)$, and also on $\mathrm{Div}^0(E)$.

Definition 4.57 The quotient group $\mathrm{Div}_{\bar{K}}(E)/\mathrm{Prin}_{\bar{K}}(E)$ is called the *divisor class group* or the *Picard group*[23] of E, denoted $\mathrm{Pic}_{\bar{K}}(E)$ or $\mathrm{Pic}(E)$. The quotient group $\mathrm{Div}^0_{\bar{K}}(E)/\mathrm{Prin}_{\bar{K}}(E)$ is called the *Jacobian*[24] of E, denoted $\mathrm{Pic}^0_{\bar{K}}(E)$ or $\mathrm{Pic}^0(E)$ or $\mathbb{J}_{\bar{K}}(E)$ or $\mathbb{J}(E)$. ◁

We have defined divisors, principal divisors, and Jacobians with respect to an algebraically closed field (like $\bar{K}$). If we focus our attention on a field which is not algebraically closed, a principal divisor need not always have degree zero (see Example 4.40(3)). It still makes sense to talk about $\mathrm{Pic}_K(E)$ and $\mathbb{J}_K(E)$ for a field K which is not algebraically closed, but then these groups have to be defined in a different manner (albeit with the help of $\mathrm{Pic}_{\bar{K}}(E)$ and $\mathbb{J}_{\bar{K}}(E)$). The chord-and-tangent rule allows us to bypass this nitty-gritty.

Example 4.58 (1) Consider the lines given in Figure 4.7. We have

$$\mathrm{Div}(L) = [P] + [Q] + [R] - 3[\mathcal{O}] = ([P] - [\mathcal{O}]) + ([Q] - [\mathcal{O}]) + ([R] - [\mathcal{O}]),$$
$$\mathrm{Div}(T) = 2[P] + [Q] - 3[\mathcal{O}] = 2([P] - [\mathcal{O}]) + ([Q] - [\mathcal{O}]), \text{ and}$$
$$\mathrm{Div}(V) = [P] + [Q] - 2[\mathcal{O}] = ([P] - [\mathcal{O}]) + ([Q] - [\mathcal{O}]).$$

---

[23]This is named after the French mathematician Charles Émile Picard (1856–1941).
[24]This is named after Carl Gustav Jacob Jacobi (1804–1851).

(2) The rational function R_1 of Example 4.40(1) on the complex curve $Y^2 = X^3 - X$ has the divisor

$$\mathrm{Div}(R_1) = -6[P_1] + 2[P_2] + 2[P_3] - [P_4] - [P_5] + 4[\mathcal{O}].$$
$$= -6([P_1]-[\mathcal{O}])+2([P_2]-[\mathcal{O}])+2([P_3]-[\mathcal{O}])-([P_4]-[\mathcal{O}])-([P_5]-[\mathcal{O}]).$$

(3) The rational function R of Example 4.41 on the curve $Y^2 = X^3 + X$ defined over $\mathbb{F}_5$ has the divisor

$$\mathrm{Div}(R) = [Q_1] + [Q_2] + [\mathcal{O}] - [Q_3] - [Q_4] - [Q_5].$$
$$= ([Q_1]-[\mathcal{O}]) + ([Q_2]-[\mathcal{O}]) - ([Q_3]-[\mathcal{O}]) - ([Q_4]-[\mathcal{O}]) - ([Q_5]-[\mathcal{O}]).$$

This divisor is not defined over $\mathbb{F}_5$, but over $\mathbb{F}_{5^6}$. $\qquad\square$

For every $D \in \mathrm{Div}^0_{\bar{K}}(E)$, there exist a unique rational point P and a rational function R such that $D = [P]-[\mathcal{O}]+\mathrm{Div}(R)$. But then $D \sim [P]-[\mathcal{O}]$ in $\mathrm{Div}^0_{\bar{K}}(E)$. We identify P with the equivalence class of $[P] - [\mathcal{O}]$ in $\mathbb{J}_{\bar{K}}(E)$. This identification establishes a bijection between the set $E_{\bar{K}}$ of rational points on E and the Jacobian $\mathbb{J}_{\bar{K}}(E)$ of E. As Example 4.58(1) suggests, this bijection also respects the chord-and-tangent rule for addition in E. The motivation for addition of points in an elliptic-curve group, as described in Figures 4.3 and 4.4, is nothing but a manifestation of this bijection. Moreover, it follows that the group $E_{\bar{K}}$ is isomorphic to the Jacobian $\mathbb{J}_{\bar{K}}(E)$.

If K is not algebraically closed, a particular subgroup of $\mathbb{J}_{\bar{K}}(E)$ can be defined to be the Jacobian $\mathbb{J}_K(E)$ of E over K. Thanks to the chord-and-tangent rule, we do not need to worry about the exact definition of $\mathbb{J}_K(E)$. More precisely, if P, Q are K-rational points of E, the explicit formulas for $P+Q$, $2P$, and $-P$ guarantee that these points are defined over K as well. Furthermore, the chord-and-tangent rule provides explicit computational handles on the group $\mathbb{J}_K(E)$. In other words, E_K is the equivalent (and computationally oriented) definition of $\mathbb{J}_K(E)$ (just as $E = E_{\bar{K}}$ was for $\mathbb{J}_{\bar{K}}(E)$). This equivalence proves the following important result.

Theorem 4.59 *A divisor $D = \sum_P m_P[P] \in \mathrm{Div}_K(E)$ is principal if and only if*

(1) $\sum_P m_P = 0$ *(integer sum), and*
(2) $\sum_P m_P P = \mathcal{O}$ *(sum under the chord-and-tangent rule).* $\qquad\triangleleft$

Example 4.60 (1) By Example 4.58(1), we have $P+Q+R = \mathcal{O}$ in Part (a), $2P + Q = \mathcal{O}$ in Part (b), and $P + Q = \mathcal{O}$ in Part (c) of Figure 4.7.

(2) Example 4.58(2) indicates that $-6P_1 + 2P_2 + 2P_3 - P_4 - P_5 = \mathcal{O}$ on the complex elliptic curve $Y^2 = X^3 - X$. This is obvious from the fact that $2P_1 = 2P_2 = 2P_3 = P_4 + P_5 = \mathcal{O}$, and from the expression of the rational function $R_1(x, y)$ in the factored form as presented in Example 4.40(1).

(3) By Example 4.58(3), $Q_1 + Q_2 - Q_3 - Q_4 - Q_5 = \mathcal{O}$ on $Y^2 = X^3 + X$ defined over $\mathbb{F}_{5^6}$. No non-empty proper sub-sum of this sum is equal to $\mathcal{O}$. $\square$

Divisors are instrumental not only for defining elliptic-curve groups but also for proving many results pertaining to elliptic curves. For instance, the concept of pairing depends heavily on divisors. I now highlight some important results associated with divisors, that are needed in the next section.

Let P, Q be points on E_K. By $L_{P,Q}$ we denote the unique (straight) line passing through P and Q. If $P = Q$, then $L_{P,Q}$ is taken to be the tangent to E at the point P. Now, consider the points $P, Q, \pm R$ as shown in Figure 4.8. Here, $P + Q = -R$, that is, $P + Q + R = \mathcal{O}$.

FIGURE 4.8: Divisors of a line and a vertical line

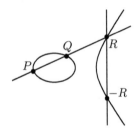

By Example 4.58, we have

$$\mathrm{Div}(L_{P,Q}) = [P] + [Q] + [R] - 3[\mathcal{O}], \text{ and } \mathrm{Div}(L_{R,-R}) = [R] + [-R] - 2[\mathcal{O}].$$

Subtracting the second equation from the first gives

$$\mathrm{Div}(L_{P,Q}/L_{R,-R}) = [P] + [Q] - [-R] - [\mathcal{O}] = [P] + [Q] - [P + Q] - [\mathcal{O}].$$

This implies the following two equivalences.

$$[P] - [\mathcal{O}] \sim [P + Q] - [Q], \text{ and}$$
$$([P] - [\mathcal{O}]) + ([Q] - [\mathcal{O}]) \sim [P + Q] - [\mathcal{O}].$$

In both the cases, the pertinent rational function is $L_{P,Q}/L_{P+Q,-(P+Q)}$ which can be easily computed. We can force this rational function to have leading coefficient one.

Example 4.61 Consider the curve $E : Y^2 = X^3 + X + 5$ defined over $\mathbb{F}_{37}$. Take the points $P = (1,9)$ and $Q = (10,4)$ on E. The equation of the line $L_{P,Q}$ is $y = \left(\frac{4-9}{10-1}\right) x + c = 20x + c$, where $c \equiv 9 - 20 \equiv 26 \pmod{37}$, that is, $L_{P,Q} : y + 17x + 11 = 0$. This line meets the curve at the third point $R = (19, 36)$, and its opposite is the point $-R = (19, -36) = (19, 1)$. The vertical line passing through R and $-R$ is $L_{R,-R} : x - 19 = 0$, that is, $L_{R,-R} : x + 18 = 0$. Therefore, we have

$$[P] - [\mathcal{O}] = [P + Q] - [Q] + \mathrm{Div}\left(\frac{y + 17x + 11}{x + 18}\right), \text{ and}$$

$$([P] - [\mathcal{O}]) + ([Q] - [\mathcal{O}]) = [P + Q] - [\mathcal{O}] + \mathrm{Div}\left(\frac{y + 17x + 11}{x + 18}\right).$$

The leading term of $y + 17x + 11$ is y (recall that y has degree three, and x has degree two), whereas the leading term of $x + 18$ is x. So both the numerator and the denominator of the rational function $\frac{y+17x+11}{x+18}$ are monic. □

Let us now try to evaluate a rational function at a divisor.

Definition 4.62 Let $D = \sum_P n_P[P]$ be a divisor on E, and let $f \in \bar{K}(E)$ be a non-zero rational function, such that the supports of D and $\text{Div}(f)$ are disjoint. Define the value of f at D as

$$f(D) = \prod_{P \in E} f(P)^{n_P} = \prod_{P \in \text{Supp}(D)} f(P)^{n_P}.$$ ◁

Two rational functions f and g have the same divisor if and only if $f = cg$ for a non-zero constant $c \in \bar{K}^*$. In that case, if D has degree zero, then $f(D) = g(D) \prod_P c^{n_P} = g(D)c^{\sum_P n_P} = g(D)c^0 = g(D)$, that is, the value of f at a divisor D of degree zero is dependent upon $\text{Div}(f)$ (rather than on f).

Theorem 4.63 [*Weil's reciprocity theorem*][25] *If f and g are two non-zero rational functions on E such that $\text{Div}(f)$ and $\text{Div}(g)$ have disjoint supports, then $f(\text{Div}(g)) = g(\text{Div}(f))$.* ◁

Example 4.64 For a demonstration of Theorem 4.63, we take the curve of Example 4.61, $f(x,y) = y + 17x + 11$, and $g(x,y) = \dfrac{x+16}{x+4}$. We know that

$$\text{Div}(f) = [P_1] + [P_2] + [P_3] - 3[\mathcal{O}],$$

where $P_1 = (1,9)$, $P_2 = (10,4)$, and $P_3 = (19,36)$. On the other hand, g has a double zero at $P_4 = (-16,0) = (21,0)$ and simple poles at $P_5 = (-4,14) = (33,14)$ and $P_6 = (-4,-14) = (33,23)$. Therefore,

$$\text{Div}(g) = 2[P_4] - [P_5] - [P_6].$$

Therefore, $\text{Div}(f)$ and $\text{Div}(g)$ have disjoint supports. But then,

$$f(\text{Div}(g)) \equiv f(P_4)^2 f(P_5)^{-1} f(P_6)^{-1} \equiv 35^2 \times 31^{-1} \times 3^{-1} \equiv 8 \ (\text{mod } 37), \text{ and}$$
$$g(\text{Div}(f)) \equiv g(P_1)g(P_2)g(P_3)g(\mathcal{O})^{-3} \equiv 33 \times 23 \times 16 \times 1^{-3} \equiv 8 \ (\text{mod } 37).$$

We have $g(\mathcal{O}) = 1$, since both the numerator and the denominator of g are monic, and have the same degree (two). □

---

[25]This theorem was proved by André Weil in 1942.

4.5   Pairing on Elliptic Curves

We are now going to define maps that accept pairs of elliptic-curve points as inputs, and output elements of a finite field. These pairing maps are important from both theoretical and practical angles.

Let $K = \mathbb{F}_q$ with $p = \operatorname{char} K$. We may have $q = p$. Take a positive integer m coprime to p. The set of all m-th roots of unity in $\bar{K}$ is denoted by μ_m. Since all elements of μ_m satisfy the polynomial equation $X^m - 1 = 0$ and $\gcd(m, q) = 1$, there are exactly m elements in μ_m. There are finite extensions of K containing μ_m. The smallest extension $L = \mathbb{F}_{q^k}$ of $K = \mathbb{F}_q$, which contains the set μ_m, has the extension degree $k = \operatorname{ord}_m(q)$ (the multiplicative order of q modulo m). We call this integer k the *embedding degree* (with respect to q and m). In general, the value of k is rather large, namely $|k| \approx |q|$ in terms of bit sizes. It is those particular cases, in which k is small, that are important from a computational perspective.

We denote the set of all points in $E = E_{\bar{K}}$ of orders dividing m by $E[m] = E_{\bar{K}}[m]$. For a field F with $K \subseteq F \subseteq \bar{K}$, let $E_F[m]$ denote those points in $E[m]$, all of whose coordinates are in F. Since $|E[m]| = m^2$, the $2m^2$ coordinates of the elements of $E[m]$ lie in finite extensions of K. Let L' be the smallest extension of K containing the coordinates of all the points of $E[m]$. In general, L' is a field much larger than L. However, think of the following situation which turns out to be computationally the most relevant one: m is a prime divisor of the group size $|E_q|$ with $m \nmid q$ and $m \nmid q - 1$. Then, $L' = L$, where L is the smallest extension of $\mathbb{F}_q$ containing μ_m.

4.5.1   Weil Pairing

Weil pairing is a function

$$e_m : E[m] \times E[m] \to \mu_m$$

defined as follows. Take $P_1, P_2 \in E[m]$. Let D_1 be a divisor equivalent to $[P_1] - [\mathcal{O}]$. Since $mP_1 = \mathcal{O}$, by Theorem 4.59, there exists a rational function f_1 such that $\operatorname{Div}(f_1) = mD_1 \sim m[P_1] - m[\mathcal{O}]$. Similarly, let D_2 be a divisor equivalent to $[P_2] - [\mathcal{O}]$. There exists a rational function f_2 such that $\operatorname{Div}(f_2) = mD_2 \sim m[P_2] - m[\mathcal{O}]$. D_1 and D_2 are chosen to have disjoint supports. Define

$$e_m(P_1, P_2) = \frac{f_1(D_2)}{f_2(D_1)}.$$

We first argue that this definition makes sense. First, note that f_1 and f_2 are defined only up to multiplication by non-zero elements of $\bar{K}$. But we have already established that the values $f_1(D_2)$ and $f_2(D_1)$ are independent of the choices of these constants, since D_1 and D_2 are of degree zero.

Second, we show that the value of $e_m(P_1, P_2)$ is independent of the choices of D_1 and D_2. To that effect, we take a divisor $D_1' = D_1 + \text{Div}(g)$ equivalent to D_1 (where $g \in \bar{K}(E)$) and with support disjoint from that of D_2. Call the corresponding rational function f_1'. We need to look at the ratio $f_1'(D_2)/f_2(D_1')$. Since $mD_1' = mD_1 + m\,\text{Div}(g) = \text{Div}(f_1) + \text{Div}(g^m) = \text{Div}(f_1 g^m)$, we can take $f_1' = f_1 g^m$, and so

$$\frac{f_1'(D_2)}{f_2(D_1')} = \frac{f_1 g^m(D_2)}{f_2(D_1 + \text{Div}(g))} = \frac{f_1(D_2)g^m(D_2)}{f_2(D_1)f_2(\text{Div}(g))} = \frac{f_1(D_2)g(mD_2)}{f_2(D_1)f_2(\text{Div}(g))} =$$
$$= \frac{f_1(D_2)g(\text{Div}(f_2))}{f_2(D_1)f_2(\text{Div}(g))} = \frac{f_1(D_2)g(\text{Div}(f_2))}{f_2(D_1)g(\text{Div}(f_2))} = \frac{f_1(D_2)}{f_2(D_1)},$$

where we have used Weil's reciprocity theorem to conclude that $f_2(\text{Div}(g)) = g(\text{Div}(f_2))$. Analogously, changing D_2 to an equivalent divisor D_2' does not alter the value of $e_m(P_1, P_2)$.

Finally, note that $e_m(P_1, P_2)$ is an m-th root of unity, since

$$e_m(P_1, P_2)^m = f_1(mD_2)/f_2(mD_1)$$
$$= f_1(\text{Div}(f_2))/f_2(\text{Div}(f_1)) = 1 \quad \text{(by Weil reciprocity)}.$$

The divisors $[P_1] - [\mathcal{O}]$ and $[P_2] - [\mathcal{O}]$ do not have disjoint supports. It is customary to choose $D_2 = [P_2] - [\mathcal{O}]$ and $D_1 = [P_1 + T] - [T]$ for any point $T \in E$. The point T need not be in $E[m]$. Indeed, one may choose T randomly from E. However, in order to ensure that D_1 and D_2 have disjoint supports, T must be different from $-P_1$, P_2, $P_2 - P_1$, and $\mathcal{O}$.

Proposition 4.65 *Let P, Q, R be arbitrary points in $E[m]$. Then, we have:*

(1) Bilinearity:

$$e_m(P + Q, R) = e_m(P, R)e_m(Q, R),$$
$$e_m(P, Q + R) = e_m(P, Q)e_m(P, R).$$

(2) Alternating: $e_m(P, P) = 1$.
(3) Skew symmetry: $e_m(Q, P) = e_m(P, Q)^{-1}$.
(4) Non-degeneracy: If $P \neq \mathcal{O}$, then $e_m(P, Q) \neq 1$ for some $Q \in E[m]$.
(5) Compatibility: If $S \in E[mn]$ and $T \in E[n]$, then $e_{mn}(S, T) = e_n(mS, T)$.
(6) Linear dependence: If m is a prime and $P \neq \mathcal{O}$, then $e_m(P, Q) = 1$ if and only if Q lies in the subgroup of $E[m]$ generated by P (that is, $Q = aP$ for some integer a). ◁

4.5.2 Miller's Algorithm

Miller[26] proposes an algorithm similar to the repeated square-and-multiply algorithm for modular exponentiation or the repeated double-and-add algorithm for computing multiples of points on elliptic curves. Miller's algorithm makes use of the following rational functions.

---

[26] Victor Saul Miller, The Weil pairing, and its efficient calculation, *Journal of Cryptology*, 17, 235–261, 2004.

Definition 4.66 Let $P \in E$, and $n \in \mathbb{Z}$. Define the rational function $f_{n,P}$ as having the divisor

$$\text{Div}(f_{n,P}) = n[P] - [nP] - (n-1)[\mathcal{O}].$$

The function $f_{n,P}$ is unique up to multiplication by elements of $\bar{K}^*$. We take unique monic polynomials for the numerator and the denominator of $f_{n,P}$. ◁

The importance of the rational functions $f_{n,P}$ in connection with the Weil pairing lies in the fact that if $P \in E[m]$, then $\text{Div}(f_{m,P}) = m[P] - [mP] - (m-1)[\mathcal{O}] = m[P] - m[\mathcal{O}]$. Therefore, it suffices to compute f_{m,P_1} and f_{m,P_2} in order to compute $e_m(P_1, P_2)$. We can define $f_{n,P}$ inductively as follows.

$$f_{0,P} \;=\; f_{1,P} \;=\; 1, \tag{4.22}$$

$$f_{n+1,P} \;=\; \left(\frac{L_{P,nP}}{L_{(n+1)P,-(n+1)P}} \right) f_{n,P} \text{ for } n \geqslant 1. \tag{4.23}$$

Here, $L_{S,T}$ is the straight line through S and T (or the tangent to E at S if $S = T$), and $L_{S,-S}$ is the vertical line through S (and $-S$). Typically, m in Weil pairing is chosen to be of nearly the same bit size as q. Therefore, it is rather impractical to compute $f_{m,P}$ using Eqn (4.23). A divide-and-conquer approach follows from the following property of $f_{n,P}$.

Proposition 4.67 *For $n, n' \in \mathbb{Z}$, we have*

$$f_{n+n',P} = f_{n,P} \times f_{n',P} \times \left(\frac{L_{nP,n'P}}{L_{(n+n')P,-(n+n')P}} \right). \tag{4.24}$$

In particular, for $n = n'$, we have

$$f_{2n,P} = (f_{n,P})^2 \times \left(\frac{L_{nP,nP}}{L_{2nP,-2nP}} \right). \tag{4.25}$$

Here, $L_{nP,nP}$ is the line tangent to E at the point nP, and $L_{2nP,-2nP}$ is the vertical line passing through $2nP$. ◁

Algorithm 4.1: MILLER'S ALGORITHM FOR COMPUTING $f_{n,P}$

```
Let n = (1n_{s-1}...n_1n_0)_2 be the binary representation of n.
Initialize f = 1 and U = P.
For i = s - 1, s - 2, ..., 1, 0 {
        /* Doubling */
```
Update $f = f^2 \times \left(\frac{L_{U,U}}{L_{2U,-2U}} \right)$ and $U = 2U$.

```
        /* Conditional adding */
```
If $(n_i = 1)$, update $f = f \times \left(\frac{L_{U,P}}{L_{U+P,-(U+P)}} \right)$ and $U = U + P$.
```
}
Return f.
```

Eqn (4.25) in conjunction with Eqn (4.23) give Algorithm 4.1. The function $f_{n,P}$ is usually kept in the factored form. It is often not necessary to compute $f_{n,P}$ explicitly. The value of $f_{n,P}$ at some point Q is only needed. In that case, the functions $L_{U,U}/L_{2U,-2U}$ and $L_{U,P}/L_{U+P,-(U+P)}$ are evaluated at Q before multiplication with f.

We now make the relationship between $e_m(P_1,P_2)$ and $f_{n,P}$ more explicit. We choose a point $T \in E$ not equal to $\pm P_1, -P_2, P_2 - P_1, \mathcal{O}$. We have

$$e_m(P_1,P_2) = \frac{f_{m,P_2}(T)\; f_{m,P_1}(P_2 - T)}{f_{m,P_1}(-T)\; f_{m,P_2}(P_1 + T)}. \tag{4.26}$$

Moreover, if $P_1 \neq P_2$, we also have

$$e_m(P_1,P_2) = (-1)^m \frac{f_{m,P_1}(P_2)}{f_{m,P_2}(P_1)}. \tag{4.27}$$

Eqn (4.27) is typically used when P_1 and P_2 are linearly independent.

It is unnecessary to make four (or two) separate calls of Algorithm 4.1 for computing $e_m(P_1,P_2)$. All these invocations have $n = m$, so a single double-and-add loop suffices. For efficiency, one may avoid the division operations in Miller's loop by separately maintaining the numerator and the denominator. After the loop terminates, a single division is made. Algorithm 4.2 incorporates these ideas, and is based upon Eqn (4.27). The polynomial functions $L_{-,-}$ are first evaluated at appropriate points and then multiplied.

Algorithm 4.2: MILLER'S ALGORITHM FOR COMPUTING $e_m(P_1,P_2)$

If $(P_1 = P_2)$, return 1.
Let $m = (1m_{s-1}\ldots m_1m_0)_2$ be the binary representation of m.
Initialize $f_{num} = f_{den} = 1$, $U_1 = P_1$, and $U_2 = P_2$.
For $i = s - 1, s - 2, \ldots, 1, 0$ {

    /* Doubling */
    Update numerator $f_{num} = f_{num}^2 \times L_{U_1,U_1}(P_2) \times L_{2U_2,-2U_2}(P_1)$.
    Update denominator $f_{den} = f_{den}^2 \times L_{2U_1,-2U_1}(P_2) \times L_{U_2,U_2}(P_1)$.
    Update $U_1 = 2U_1$ and $U_2 = 2U_2$.

    /* Conditional adding */
    If $(m_i = 1)$ {
        Update numerator $f_{num} = f_{num} \times L_{U_1,P_1}(P_2) \times L_{U_2+P_2,-(U_2+P_2)}(P_1)$.
        Update denominator $f_{den} = f_{den} \times L_{U_1+P_1,-(U_1+P_1)}(P_2) \times L_{U_2,P_2}(P_1)$.
        Update $U_1 = U_1 + P_1$ and $U_2 = U_2 + P_2$.
    }
}
Return $(-1)^m f_{num}/f_{den}$.

Example 4.68 Consider the curve $E : Y^2 = X^3 + 3X$ defined over $\mathbb{F}_{43}$. This curve is supersingular with $|E_{\mathbb{F}_{43}}| = 44$. The group $E_{\mathbb{F}_{43}}$ is not cyclic, but of rank two, that is, isomorphic to $\mathbb{Z}_{22} \oplus \mathbb{Z}_2$. We choose $m = 11$. The embedding degree for this choice is $k = 2$. This means that we have to work in the field $\mathbb{F}_{43^2} = \mathbb{F}_{1849}$. Since $p = 43$ is congruent to 3 modulo 4, -1 is a quadratic non-residue modulo p, and we can represent $\mathbb{F}_{43^2}$ as $\mathbb{F}_{43}(\theta) = \{a + b\theta \mid a, b \in \mathbb{F}_{43}\}$, where $\theta^2 + 1 = 0$. The arithmetic of $\mathbb{F}_{43^2}$ resembles that of $\mathbb{C}$. $\mathbb{F}_{43^2}^*$ contains all the 11-th roots of unity. These are 1, $2 + 13\theta$, $2 + 30\theta$, $7 + 9\theta$, $7 + 34\theta$, $11 + 3\theta$, $11 + 40\theta$, $18 + 8\theta$, $18 + 35\theta$, $26 + 20\theta$, and $26 + 23\theta$.

The group $E_{\mathbb{F}_{43^2}}$ contains 44^2 elements, and is isomorphic to $\mathbb{Z}_{44} \oplus \mathbb{Z}_{44}$. Moreover, this group fully contains $E[11]$ which consists of 11^2 elements and is isomorphic to $\mathbb{Z}_{11} \oplus \mathbb{Z}_{11}$. The points $P = (1, 2)$ and $Q = (-1, 2\theta)$ constitute a set of linearly independent elements of $E[11]$. Every element of $E[11]$ can be written as a unique $\mathbb{F}_{11}$-linear combination of P and Q. For example, the element $4P + 5Q = (15 + 22\theta, 5 + 14\theta)$ is again of order 11.

Let us compute $e_m(P_1, P_2)$ by Algorithm 4.2, where $P_1 = P = (1, 2)$, and $P_2 = 4P + 5Q = (15 + 22\theta, 5 + 14\theta)$. The binary representation of 11 is $(1011)_2$. We initialize $f = f_{num}/f_{den} = 1/1$, $U_1 = P_1$, and $U_2 = P_2$. Miller's loop works as shown in the following table. Here, Λ_1 stands for the rational function $L_{U_1,U_1}/L_{2U_1,-2U_1}$ (during doubling) or the function $L_{U_1,P_1}/L_{U_1+P_1,-(U_1+P_1)}$ (during addition), and Λ_2 stands for $L_{2U_2,-2U_2}/L_{U_2,U_2}$ (during doubling) or $L_{U_2+P_2,-(U_2+P_2)}/L_{U_2,P_2}$ (during addition).

i	m_i	Step	Λ_1	Λ_2	f	U_1	U_2
2	0	Dbl	$\dfrac{y+20x+21}{x+32}$	$\dfrac{x+(36+21\theta)}{y+(12+35\theta)x+(26+14\theta)}$	$\dfrac{34+37\theta}{28+\theta}$	$2P_1 =$ $(11, 26)$	$2P_2 =$ $(7+22\theta, 28+7\theta)$
		Add		Skipped			
1	1	Dbl	$\dfrac{y+31x+20}{x+7}$	$\dfrac{x+(2+26\theta)}{y+(18+22\theta)x+(29+2\theta)}$	$\dfrac{12+15\theta}{25+18\theta}$	$4P_1 =$ $(36, 18)$	$4P_2 =$ $(41+17\theta, 6+6\theta)$
		Add	$\dfrac{y+2x+39}{x+33}$	$\dfrac{x+(41+8\theta)}{y+(28+9\theta)x+(31+9\theta)}$	$\dfrac{25+15\theta}{28+20\theta}$	$5P_1 =$ $(10, 16)$	$5P_2 =$ $(2+35\theta, 30+18\theta)$
0	1	Dbl	$\dfrac{y+8x+33}{x+42}$	$\dfrac{x+(28+21\theta)}{y+(19+16\theta)x+(19+16\theta)}$	$\dfrac{10+22\theta}{12+28\theta}$	$10P_1 =$ $(1, 41)$	$10P_2 =$ $(15+22\theta, 38+29\theta)$
		Add	$\dfrac{x+42}{1}$	$\dfrac{1}{x+(28+21\theta)}$	$\dfrac{12\theta}{18+32\theta}$	$11P_1 =$ $\mathcal{O}$	$11P_2 =$ $\mathcal{O}$

From the table, we obtain

$$e_m(P_1, P_2) = (-1)^{11} \left(\frac{12\theta}{18 + 32\theta} \right) = 26 + 20\theta$$

which is indeed an 11-th root of unity in $\mathbb{F}_{43^2}^*$.

Algorithm 4.2 works perfectly when P_1 and P_2 are linearly independent. If they are dependent, the algorithm may encounter an unwelcome situation. Suppose that we want to compute $e_m(P_1, P_2)$ for $P_1 = (1, 2)$ and $P_2 = 3P_1 = (23, 14)$. Now, Miller's loop proceeds as follows.

i	m_i	Step	Λ_1	Λ_2	f	U_1	U_2
2	0	Dbl	$\dfrac{y + 20x + 21}{x + 32}$	$\dfrac{x + 33}{y + 20x + 42}$	$\dfrac{17}{37}$	$2P_1 = (11, 26)$	$2P_2 = (10, 27)$
		Add			Skipped		
1	1	Dbl	$\dfrac{y + 31x + 20}{x + 7}$	$\dfrac{x + 42}{y + 35x + 10}$	$\dfrac{0}{20}$	$4P_1 = (36, 18)$	$4P_2 = (1, 2)$
		Add	$\dfrac{y + 2x + 39}{x + 33}$	$\dfrac{x + 7}{y + 19x + 22}$	$\dfrac{0}{0}$	$5P_1 = (10, 16)$	$5P_2 = (36, 18)$
0	1	Dbl	$\dfrac{y + 8x + 33}{x + 42}$	$\dfrac{x + 20}{y + 3x + 3}$	$\dfrac{0}{0}$	$10P_1 = (1, 41)$	$10P_2 = (23, 29)$
		Add	$\dfrac{x + 42}{1}$	$\dfrac{1}{x + 20}$	$\dfrac{0}{0}$	$11P_1 = \mathcal{O}$	$11P_2 = \mathcal{O}$

During the doubling step in the second iteration, we have $U_2 = 2P_2$. The vertical line ($x + 42 = 0$) passing through $2U_2$ and $-2U_2$ passes through P_1, since $2U_2 = 4P_2 = 12P_1 = P_1$. So the numerator f_{num} becomes 0. During the addition step of the same iteration, we have $U_2 = 4P_2 = P_1$. The line ($y + 19x + 22 = 0$) passing through U_2 and P_2 evaluates to 0 at P_1, and so the denominator f_{den} too becomes 0.

In practice, one works with much larger values of m. If P_2 is a random multiple of P_1, the probability of accidentally hitting upon this linear relation in one of the $\Theta(\log m)$ Miller iterations is rather small, and Algorithm 4.2 successfully terminates with high probability. Nonetheless, if the algorithm fails, we may choose random points T on the curve and use Eqn (4.26) (instead of Eqn (4.27) on which Algorithm 4.2 is based) until $e_m(P_1, P_2)$ is correctly computed. In any case, Proposition 4.65(6) indicates that in this case we are going to get $e_m(P_1, P_2) = 1$ (when m is prime). However, checking whether P_1 and P_2 are linearly dependent is, in general, not an easy computational exercise. Although the situation is somewhat better for supersingular curves, a check for the dependence of P_1 and P_2 should be avoided. ☐

The current versions of GP/PARI do not provide ready supports for Weil (or other) pairings. However, it is not difficult to implement Miller's algorithm using the built-in functions of GP/PARI.

4.5.3 Tate Pairing

Another pairing of elliptic-curve points is called Tate pairing.[27] Let E be an elliptic curve defined over $K = \mathbb{F}_q$ with $p = \operatorname{char} K$. We take $m \in \mathbb{N}$ with $\gcd(m, p) = 1$. Let $k = \operatorname{ord}_m(q)$ be the embedding degree, and $L = \mathbb{F}_{q^k}$. Define

$$E_L[m] = \{P \in E_L \mid mP = \mathcal{O}\}, \quad \text{and} \quad mE_L = \{mP \mid P \in E_L\}.$$

Also, let

$$(L^*)^m = \{a^m \mid a \in L^*\}$$

---

[27] This is named after the American mathematician John Torrence Tate Jr. (1925–).

be the set of m-th powers in L^*. *Tate pairing* is a function

$$\langle -, - \rangle_m : E_L[m] \times (E_L/mE_L) \to L^*/(L^*)^m$$

defined as follows. Let P be a point in $E_L[m]$, and Q a point in E_L, to be treated as a point in E_L/mE_L. Since $mP = \mathcal{O}$, there is a rational function f with $\mathrm{Div}(f) = m[P] - m[\mathcal{O}]$. Let D be any divisor equivalent to $[Q] - [\mathcal{O}]$ with disjoint support from $\mathrm{Div}(f)$. It is customary to choose a point T different from $-P, Q, Q - P, \mathcal{O}$, and take $D = [Q + T] - [T]$. Define

$$\langle P, Q \rangle_m = f(D).$$

Since $\mathrm{Div}(f) = m[P] - m[\mathcal{O}]$ with $P \in E_L$, we can choose f as defined over L. As a result, $f(D) \in L^*$. Although f is unique only up to multiplication by elements of L^*, the value of $f(D)$ is independent of the choice of f, since D is a divisor of degree zero. Still, the value of $f(D)$, as an element of L^*, is not unique because of its dependence on the choice of the divisor D. Let D' be another divisor equivalent to $[Q] - [\mathcal{O}]$. We can write $D' = D + \mathrm{Div}(g)$ for some rational function g. Using Weil reciprocity, we obtain

$$f(D') = f(D + \mathrm{Div}(g)) = f(D)f(\mathrm{Div}(g)) = f(D)g(\mathrm{Div}(f))$$
$$= f(D)g(m[P] - m[\mathcal{O}]) = f(D)\Big(g([P] - [\mathcal{O}])\Big)^m,$$

that is, $f(D')$ and $f(D)$ differ by a multiplicative factor which is an m-th power in L^*. Treating $f(D)$ as an element of $L^*/(L^*)^m$ makes it unique.

Another way of making the Tate pairing unique is based upon the fact that $f(D)^{\frac{q^k-1}{m}} = f(D')^{\frac{q^k-1}{m}}\Big(g([P] - [\mathcal{O}])\Big)^{q^k-1} = f(D')^{\frac{q^k-1}{m}}$, since $a^{q^k-1} = 1$ for all $a \in L^* = \mathbb{F}_{q^k}^*$. The *reduced Tate pairing* of P and Q is defined as

$$\hat{e}_m(P, Q) = (\langle P, Q \rangle_m)^{\frac{q^k-1}{m}} = f(D)^{\frac{q^k-1}{m}}.$$

Raising $\langle P, Q \rangle_m$ to the exponent $(q^k - 1)/m$ is called *final exponentiation*.

Tate pairing is related to Weil pairing as

$$e_m(P, Q) = \frac{\langle P, Q \rangle_m}{\langle Q, P \rangle_m},$$

where the equality is up to multiplication by elements of $(L^*)^m$. Tate pairing shares some (not all) properties of Weil pairing listed in Proposition 4.65.

Proposition 4.69 *For appropriate points P, Q, R on E, we have:*
(1) Bilinearity:

$$\langle P + Q, R \rangle_m = \langle P, R \rangle_m \times \langle Q, R \rangle_m,$$
$$\langle P, Q + R \rangle_m = \langle P, Q \rangle_m \times \langle P, R \rangle_m.$$

(2) Non-degeneracy: For every $P \in E_L[m]$, $P \neq \mathcal{O}$, there exists Q for which $\langle P, Q \rangle_m \neq 1$. For every $Q \notin mE_L$, there exists $P \in E_L[m]$ with $\langle P, Q \rangle_m \neq 1$.

(3) Linear dependence: Let m be a prime divisor of $|E_K|$, and P a generator of a subgroup G of E_K of order m. If $k = 1$ (that is, $L = K$), then $\langle P, P \rangle_m \neq 1$. If $k > 1$, then $\langle P, P \rangle_m = 1$, and so, by bilinearity, $\langle Q, Q' \rangle_m = 1$ for all $Q, Q' \in G$. However, if $k > 1$ and $Q \in E_L$ is linearly independent of P (that is, $Q \notin G$), then $\langle P, Q \rangle_m \neq 1$.

All these properties continue to hold for the reduced Tate pairing. ◁

Miller's algorithm for computing $f_{n,P}$ can be easily adapted to compute the Tate pairing of P and Q. We choose a point $T \neq P, -Q, P - Q, \mathcal{O}$, and take $D = [Q + T] - [T]$. We have

$$\langle P, Q \rangle_m = \frac{f_{m,P}(Q + T)}{f_{m,P}(T)}. \tag{4.28}$$

Moreover, if P and Q are linearly independent, then

$$\langle P, Q \rangle_m = f_{m,P}(Q). \tag{4.29}$$

Algorithm 4.3 describes the computation of the reduced Tate pairing $\hat{e}_m(P, Q)$ using a single Miller loop and using separate variables for the numerator and the denominator. This algorithm is based upon Eqn (4.28).

Algorithm 4.3: MILLER'S ALGORITHM FOR COMPUTING THE REDUCED TATE PAIRING $\hat{e}_m(P, Q)$

Let $m = (1m_{s-1} \ldots m_1 m_0)_2$ be the binary representation of m.
Initialize $f_{num} = f_{den} = 1$, and $U = P$.
Choose a point $T \neq P, -Q, P - Q, \mathcal{O}$.
For $i = s - 1, s - 2, \ldots, 1, 0$ {
    /* Doubling */
    Compute the rational functions $L_{U,U}$ and $L_{2U,-2U}$.
    Update numerator $f_{num} = f_{num}^2 \times L_{U,U}(Q + T) \times L_{2U,-2U}(T)$.
    Update denominator $f_{den} = f_{den}^2 \times L_{2U,-2U}(Q + T) \times L_{U,U}(T)$.
    Update $U = 2U$.

    /* Conditional adding */
    If $(m_i = 1)$ {
        Compute the rational functions $L_{U,P}$ and $L_{U+P,-(U+P)}$.
        Update numerator $f_{num} = f_{num} \times L_{U,P}(Q + T) \times L_{U+P,-(U+P)}(T)$.
        Update denominator $f_{den} = f_{den} \times L_{U+P,-(U+P)}(Q + T) \times L_{U,P}(T)$.
        Update $U = U + P$.
    }
}
Compute $f = f_{num} / f_{den}$.
/* Do the final exponentiation */
Return $f^{(q^k - 1)/m}$.

Algorithm 4.3 is somewhat more efficient than Algorithm 4.2. First, Tate pairing requires only one point U to be maintained and updated in the loop, whereas Weil pairing requires two (U_1 and U_2). Second, in the loop of Algorithm 4.3, only one set of rational functions ($L_{U,U}$ and $L_{2U,-2U}$ during doubling, and $L_{U,P}$ and $L_{U+P,-(U+P)}$ during addition) needs to be computed (but evaluated twice). The loop of Algorithm 4.2 requires the computation of two sets of these functions. To avoid degenerate output, it is a common practice to take the first point P from E_q and the second point Q from E_{q^k}. In this setting, the functions $f_{n,P}$ are defined over $\mathbb{F}_q$, whereas the functions $f_{n,Q}$ are defined over $\mathbb{F}_{q^k}$. This indicates that the Miller's loop for computing $\langle P, Q \rangle_m$ is more efficient than that for computing $\langle Q, P \rangle_m$. Moreover, if P and Q are known to be linearly independent, we use Eqn (4.29) instead of Eqn (4.28). This reduces the number of evaluations of the line functions by a factor of two. As a result, Tate pairing is usually preferred to Weil pairing in practical applications. The reduced Tate pairing, however, calls for an extra final exponentiation. If k is not too small, this added overhead may make Tate pairing less efficient than Weil pairing.

Example 4.70 Let us continue to work with the curve of Example 4.68, and compute the Tate pairing of $P = (1, 2)$ and $Q = (15 + 22\theta, 5 + 14\theta)$. (These points were called P_1 and P_2 in Example 4.68). Miller's loop of Algorithm 4.3 proceeds as in the following table. These computations correspond to the point $T = (36 + 12\theta, 40 + 31\theta)$ for which $Q + T = (19 + 32\theta, 24 + 27\theta)$. Here, we have only one set of points maintained as U (Weil pairing required two: U_1, U_2). The updating rational function Λ is $L_{U,U}/L_{2U,-2U}$ for doubling and $L_{U,P}/L_{U+P,-(U+P)}$ for addition.

i	m_i	Step	Λ	f	U
2	0	Dbl	$\dfrac{y + 20x + 21}{x + 32}$	$\dfrac{41 + 17\theta}{27 + 27\theta}$	$2P = (11, 26)$
		Add		Skipped	
1	1	Dbl	$\dfrac{y + 31x + 20}{x + 32}$	$\dfrac{14 + 31\theta}{15 + 15\theta}$	$4P = (36, 18)$
		Add	$\dfrac{y + 2x + 39}{x + 33}$	$\dfrac{41 + 36\theta}{37 + 16\theta}$	$5P = (10, 16)$
0	1	Dbl	$\dfrac{y + 8x + 33}{x + 42}$	$\dfrac{36 + 24\theta}{11 + 36\theta}$	$10P = (1, 41)$
		Add	$\dfrac{x + 42}{1}$	$\dfrac{9 + 36\theta}{39 + 16\theta}$	$11P = \mathcal{O}$

These computations give

$$\langle P, Q \rangle_m = \frac{9 + 36\theta}{39 + 16\theta} = 14 + 4\theta.$$

The corresponding reduced pairing is

$$\hat{e}_m(P, Q) = (14 + 4\theta)^{(43^2 - 1)/11} = 2 + 13\theta.$$

The value of $\langle P, Q\rangle_m$ depends heavily on the choice of the point T. For example, the choice $T = (34 + 23\theta, 9 + 23\theta)$ (another point of order 44) gives

$$\langle P, Q\rangle_m = 4 + 33\theta.$$

The two values of $\langle P, Q\rangle_m$ differ by a factor which is an m-th power in $\mathbb{F}_{43^2}^*$:

$$\frac{4 + 33\theta}{14 + 4\theta} = 9 + 9\theta = (4 + 23\theta)^m.$$

However, the final exponentiation gives the same value, that is,

$$\hat{e}_m(P, Q) = (4 + 33\theta)^{(43^2 - 1)/11} = 2 + 13\theta.$$

By Example 4.68, $e_m(P, Q) = 26 + 20\theta$. Computing $\langle Q, P\rangle_m$ for the choice $T = (11 + 15\theta, 38 + 25\theta)$ gives $\langle Q, P\rangle_m = 36 + 4\theta$. We have

$$e_m(P, Q) = \left(\frac{\langle P, Q\rangle_m}{\langle Q, P\rangle_m}\right) \times \xi = \left(\frac{14 + 4\theta}{36 + 4\theta}\right) \times \xi = (8 + 4\theta) \times \xi,$$

where $\xi = (26 + 20\theta)/(8 + 4\theta) = 38 + 5\theta = (1 + 37\theta)^m$.

In this example, all multiples of P lie in $E_{\mathbb{F}_{43}}$, that is, P and Q are linearly independent, so we are allowed to use the formula $\langle P, Q\rangle_m = f_{m,P}(Q)$. But then, the line functions are evaluated only at the point Q (instead of at two points $Q + T$ and T). The corresponding calculations in the Miller loop are shown in the following table.

i	m_i	Step	Λ	f	U
2	0	Dbl	$\dfrac{y + 20x + 21}{x + 32}$	$\dfrac{25 + 24\theta}{4 + 22\theta}$	$2P = (11, 26)$
		Add		Skipped	
1	1	Dbl	$\dfrac{y + 31x + 20}{x + 32}$	$\dfrac{5 + 23\theta}{22 + 26\theta}$	$4P = (36, 18)$
		Add	$\dfrac{y + 2x + 39}{x + 33}$	$\dfrac{25 + 14\theta}{11 + 12\theta}$	$5P = (10, 16)$
0	1	Dbl	$\dfrac{y + 8x + 33}{x + 42}$	$\dfrac{13 + 29\theta}{19 + 8\theta}$	$10P = (1, 41)$
		Add	$\dfrac{x + 42}{1}$	$\dfrac{17 + 4\theta}{19 + 8\theta}$	$11P = \mathcal{O}$

We now get $\langle P, Q\rangle_m = \dfrac{17 + 4\theta}{19 + 8\theta} = 15 + 12\theta$. We have seen that Eqn (4.28) with $T = (36 + 12\theta, 40 + 31\theta)$ gives $\langle P, Q\rangle_m = 14 + 4\theta$. The ratio of these two values is $\dfrac{15 + 12\theta}{14 + 4\theta} = 7\theta = (6\theta)^m$ which is an m-th power in $\mathbb{F}_{43^2}^*$. Now, the reduced pairing is $\hat{e}_m(P, Q) = (15 + 12\theta)^{(43^2 - 1)/11} = 2 + 13\theta$ which is again the same as that computed using Eqn (4.28). $\qquad\square$

4.5.4 Non-Rational Homomorphisms

In typical practical applications, one takes a large prime divisor of $|E_q|$ as m. If $k > 1$, there is a unique (cyclic) subgroup of E_q of order m. Let G denote this subgroup. If $k = 1$, there are two copies of $\mathbb{Z}_m$ in E_q, and we take any of these copies as G. The restriction of e_m or $\hat{e}_m$ to the group $G \times G$ is of primary concern to us. The linear dependence property of Weil pairing indicates that $e_m(P, Q) = 1$ for all $P, Q \in G$. The same property for Tate pairing suggests that if $k > 1$, we again get $\hat{e}_m(P, Q) = 1$ for all $P, Q \in G$. Consequently, the maps e_m or $\hat{e}_m$ restricted to $G \times G$ are trivial.

There is a way out of this problem. The pairing of linearly independent points P and Q is a non-trivial value. In order to exploit this property, we first map one of the points $P, Q \in G$, say the second one, to a point Q' linearly independent of P, and then apply the original Weil or Tate pairing on P and Q'. However, some care needs to be adopted so as to maintain the property of bilinearity. Two ways of achieving this are now explained, both of which use group homomorphisms that are not rational (that is, not defined) over $\mathbb{F}_q$.

4.5.4.1 Distortion Maps

Definition 4.71 Let $\phi : E[m] \to E[m]$ be an endomorphism of $E[m]$ with $\phi(P) \notin G$ for some $P \neq \mathcal{O}$ in G. The map ϕ is called a *distortion map* (for G). The *distorted Weil pairing* of $P, Q \in G$ is defined as $e_m(P, \phi(Q))$, whereas the *distorted Tate pairing* of $P, Q \in G$ is defined as $\langle P, \phi(Q) \rangle_m$. ◁

Since $\phi(P)$ is linearly independent of P, we have $e_m(P, \phi(P)) \neq 1$ and $\langle P, \phi(P) \rangle_m \neq 1$. On the other hand, since ϕ is an endomorphism, bilinearity is preserved. Moreover, we achieve an additional property.

Proposition 4.72 *Symmetry: For all* $P, Q \in G$, *we have* $e_m(P, \phi(Q)) = e_m(Q, \phi(P))$, *and* $\langle P, \phi(Q) \rangle_m = \langle Q, \phi(P) \rangle_m$. ◁

Distortion maps, however, exist only for supersingular curves.

Example 4.73 Consider the curve $E : y^2 = X^3 + 3X$ of Examples 4.68 and 4.70. For $m = (p+1)/4 = 11$, the group $E[11]$ contains 11^2 points, and is generated by two linearly independent points $P = (1, 2)$ and $Q = (-1, 2\theta) = (42, 2\theta)$. The subgroup G of $E[11]$ generated by P is a subset of $E_{\mathbb{F}_{43}}$. We want to define a bilinear pairing on $G \times G$.

For $P_1 = P = (1, 2)$ and $P_2 = 3P_1 = (23, 14)$, Algorithm 4.3 with the choice $T = (12 + 7\theta, 5 + 38\theta)$ yields $\langle P_1, P_2 \rangle_m = 22$, and we get the trivial value $\hat{e}_m(P_1, P_2) = 22^{(43^2 - 1)/11} = 1$.

A distortion map $\phi : E[11] \to E[11]$ is fully specified by its images $\phi(P)$ and $\phi(Q)$ only, since P and Q generate $E[11]$. In this example, we may take $\phi(1, 2) = (-1, 2\theta)$, and $\phi(-1, 2\theta) = (1, 2)$. It follows that $\phi(a, b) = (-a, b\theta)$ for all $(a, b) \in E_{\mathbb{F}_{43}}$. In particular, $\phi(P_2) = \phi(23, 14) = (-23, 14\theta) = (20, 14\theta)$.

Tate pairing of $P_1 = (1, 2)$ and $\phi(P_2) = (20, 14\theta)$ is a non-trivial value. Algorithm 4.3 with $T = (37 + 6\theta, 14 + 13\theta)$ gives $\langle P_1, \phi(P_2) \rangle_m = 21 + 2\theta$, and so $\hat{e}_m(P_1, \phi(P_2)) = (21 + 2\theta)^{(43^2-1)/11} = 18 + 8\theta$. Moreover, Algorithm 4.2 for Weil pairing now gives $e_m(P_1, \phi(P_2)) = 11 + 3\theta$, again a non-trivial value.

Let us now compute the pairing of $P_2 = (23, 14)$ and $\phi(P_1) = (-1, 2\theta) = (42, 2\theta)$. Tate pairing with $T = (38 + 21\theta, 19 + 11\theta)$ gives $\langle P_2, \phi(P_1) \rangle_m = 30 + 29\theta = (21 + 2\theta) \times (23\theta) = (21 + 2\theta) \times (30\theta)^m$. The reduced Tate pairing is $\hat{e}_m(P_2, \phi(P_1)) = (30 + 29\theta)^{(43^2-1)/11} = 18 + 8\theta = \hat{e}_m(P_1, \phi(P_2))$. Finally, the Weil pairing of P_2 and $\phi(P_1)$ is $e_m(P_2, \phi(P_1)) = 11 + 3\theta = e_m(P_1, \phi(P_2))$. Symmetry about the two arguments is thereby demonstrated. □

4.5.4.2 Twists

Another way of achieving linear independence of P and Q' is by means of twists which work even for ordinary curves. Suppose that $p \neq 2, 3$, and E is defined by the short Weierstrass equation $E : Y^2 = X^3 + aX + b$. Further, let d be an integer $\geqslant 2$, and $v \in \mathbb{F}_q^*$ a d-th power non-residue. The curve

$$E' : Y^2 = X^3 + v^{4/d}aX + v^{6/d}b$$

is called a *twist of E of degree d*. For $d = 2$ (quadratic twist), E' is defined over $\mathbb{F}_q$ itself. In general, E' is defined over $\mathbb{F}_{q^d}$. E and E' are isomorphic over $\mathbb{F}_{q^d}$ (but need not be over $\mathbb{F}_q$ even when E' is defined over $\mathbb{F}_q$). An explicit isomorphism is given by the map $\phi_d : E' \to E$ taking $(r, s) \mapsto (v^{-2/d}r, v^{-3/d}s)$.

Definition 4.74 Let G be a subgroup of order m in E_{q^k}, and G' a subgroup of order m in E'_{q^k}. For quadratic twists, a natural choice is $G \subseteq E_q$ and $G' \subseteq E'_q$. We now pair points P, Q with $P \in G$ and $Q \in G'$. The *twisted Weil pairing* of P and Q is defined as $e_m(P, \phi_d(Q))$, whereas the *twisted Tate pairing* of P and Q is defined as $\langle P, \phi_d(Q) \rangle_m$. ◁

The domain of definition of twisted pairing is $G \times G'$ (not $G \times G$), but that is not an important issue. Application of ϕ_d takes elements of G' in G, and the original pairing can be applied. The twisted pairing is non-degenerate if $\phi_d(Q)$ is linearly independent of P. Since ϕ_d is a group homomorphism, bilinearity is preserved.

Example 4.75 Let $E : Y^2 = X^3 + 3X$ be the curve of Examples 4.68, 4.70, and 4.73. Since $p \equiv 3 \pmod 8$, two is a quadratic non-residue modulo p. Thus, a quadratic twist of E is $E' : Y^2 = X^3 + 12X$. We now define a pairing on $G \times G'$, where G is the subgroup of $E[11]$ generated by $P = (1, 2)$, and G' is the subgroup of $E'[11]$ generated by $Q = (1, 20)$. Note that G and G' are of order 11, and completely contained in $E_{\mathbb{F}_{43}}$ and $E'_{\mathbb{F}_{43}}$, respectively. Since E and E' are different curves, they have different points. Indeed, the point $(1, 2)$ does not lie on E', and the point $(1, 20)$ does not lie on E.

A square root of two in $\mathbb{F}_{43^2}$ is 16θ (where $\theta^2 + 1 = 0$), so the twist ϕ_2 for E, E' takes $(a, b) \in G'$ to $((16\theta)^{-2}a, (16\theta)^{-3}b) = (22a, 39\theta b)$. Take the points $P_1 = P = (1, 2) \in G$, and $P_2 = 3Q = (11, 42) \in G'$. We have $\phi_2(P_2) = (27, 4\theta)$, which is in $E_{\mathbb{F}_{43^2}}$ but not in $E_{\mathbb{F}_{43}}$. The twisted pairings of P_1 and P_2 are $\langle P_1, \phi_2(P_2) \rangle_m = 18 + 8\theta$ (for $T = (41 + 36\theta, 2 + 38\theta)$), $\hat{e}_m(P_1, \phi_2(P_2)) = (18 + 8\theta)^{(43^2-1)/11} = 11 + 40\theta$, and $e_m(P_1, \phi_2(P_2)) = 23 + \theta$. $\qquad \square$

4.5.5 Pairing-Friendly Curves

Thanks to Miller's algorithm, one can efficiently compute Weil and Tate pairings, provided that the embedding degree k is not too large. For general (like randomly chosen) curves, we have $|k| \approx |m|$. A curve is called *pairing-friendly* if, for a suitable choice of m, the embedding degree k is small, typically $k \leqslant 12$. Only some specific types of curves qualify as pairing-friendly. Nonetheless, many infinite families of curves are known to be pairing-friendly.[28]

By Hasse's theorem, $|E_q| = q+1-t$ for some t with $|t| \leqslant 2\sqrt{q}$. If $p = \operatorname{char} \mathbb{F}_q$ divides t, we call E *supersingular*. A non-supersingular curve is called *ordinary*.

Supersingular curves are known to have small embedding degrees. The only possibilities are $k = 1, 2, 3, 4, 6$. If $\mathbb{F}_q$ is a prime field with $q \geqslant 5$, the only possibility is $k = 2$. Many infinite families of supersingular curves are known.

Example 4.76 (1) Curves of the form $Y^2 + aY = X^3 + bX + c$ (with $a \neq 0$) are supersingular over fields of characteristic two. All supersingular curves over a finite field K of characteristic two have j-invariant equal to 0, and so are isomorphic over $\bar{K}$. For these curves, $k \in \{1, 2, 3, 4\}$.

(2) Curves of the form $Y^2 = X^3 - X \pm 1$ are supersingular over fields of characteristic three. The embedding degree is six for these curves.

(3) Take a prime $p \equiv 2 \pmod 3$, and $a \in \mathbb{F}_p^*$. The curve $Y^2 = X^3 + a$ defined over $\mathbb{F}_p$ is supersingular, and has embedding degree two.

(4) Take a prime $p \equiv 3 \pmod 4$, and $a \in \mathbb{F}_p^*$. The curve $Y^2 = X^3 + aX$ defined over $\mathbb{F}_p$ is supersingular, and has embedding degree two.

Solve Exercise 4.66 for deriving these embedding degrees. $\qquad \square$

Locating ordinary curves (particularly, infinite families) with small embedding degrees is a significantly more difficult task. One method is to fix an embedding degree k and a discriminant Δ, and search for (integer-valued) polynomials $t(x), m(x), q(x) \in \mathbb{Q}[x]$ satisfying the following five conditions:

(1) $q(x) = p(x)^n$ for some $n \in \mathbb{N}$ and $p(x) \in \mathbb{Q}[x]$ representing primes.

(2) $m(x)$ is irreducible with a positive leading coefficient.

(3) $m(x) | (q(x) + 1 - t(x))$.

---

[28] A comprehensive survey on pairing-friendly curves can be found in the article: David Freeman, Michael Scott and Edlyn Teske, A taxonomy of pairing-friendly elliptic curves, *Journal of Cryptology*, 23(2), 224–280, 2010.

(4) $m(x)|\Phi_k(t(x) - 1)$, where Φ_k is the k-th cyclotomic polynomial (see Exercise 3.36).

(5) There are infinitely many integers (x, y) satisfying $\Delta y^2 = 4q(x) - t(x)^2$.

If we are interested in ordinary curves, we additionally require:

(6) $\gcd(q(x), m(x)) = 1$.

For a choice of $t(x), m(x), q(x)$, families of elliptic curves over $\mathbb{F}_q$ of size m, embedding degree k, and discriminant Δ can be constructed using the *complex multiplication method*. If y in Condition (5) can be parametrized by a polynomial $y(x) \in \mathbb{Q}[x]$, the family is called *complete*, otherwise it is called *sparse*. Some sparse families of ordinary pairing-friendly curves are:

- *MNT (Miyaji–Nakabayashi–Takano) curves*[29]: These are ordinary curves of prime orders with embedding degrees three, four, or six. Let $m > 3$ be the order (prime) of an ordinary curve E, $t = q + 1 - m$ the trace of Frobenius, and k the embedding degree of E. The curve E is completely characterized by the following result.

  (1) $k = 3$ if and only if $t = -1 \pm 6x$ and $q = 12x^2 - 1$ for some $x \in \mathbb{Z}$.
  (2) $k = 4$ if and only if $t = -x$ or $t = x + 1$, and $q = x^2 + x + 1$ for some $x \in \mathbb{Z}$.
  (3) $k = 6$ if and only if $t = 1 \pm 2x$ and $q = 4x^2 + 1$ for some $x \in \mathbb{Z}$.

- *Freeman curves*[30]: These curves have embedding degree ten, and correspond to the choices:

$$
\begin{aligned}
t(x) &= 10x^2 + 5x + 3, \\
m(x) &= 25x^4 + 25x^3 + 15x^2 + 5x + 1, \\
q(x) &= 25x^4 + 25x^3 + 25x^2 + 10x + 3.
\end{aligned}
$$

For this family, we have $m(x) = q(x) + 1 - t(x)$. The discriminant Δ of Freeman curves satisfies $\Delta \equiv 43$ or $67 \pmod{120}$.

Some complete families of ordinary pairing-friendly curves are:

- *BN (Barreto–Naehrig) curves*[31]: These curves have embedding degree 12 and discriminant three, and correspond to the following choices.

$$
\begin{aligned}
t(x) &= 6x^2 + 1, \\
m(x) &= 36x^4 + 36x^3 + 18x^2 + 6x + 1, \\
q(x) &= 36x^4 + 36x^3 + 24x^2 + 6x + 1.
\end{aligned}
$$

---

[29] Atsuko Miyaji, Masaki Nakabayashi and Shunzo Takano, New explicit conditions of elliptic curve traces for FR-reductions, *IEICE Transactions on Fundamentals*, E84-A(5), 1234–1243, 2001.

[30] David Freeman, Constructing pairing-friendly elliptic curves with embedding degree 10, *ANTS-VII*, 452–465, 2006.

[31] Paulo S. L. M. Barreto and Michael Naehrig, Pairing-friendly elliptic curves of prime order, *SAC*, 319–331, 2006.

For this family too, we have $m(x) = q(x) + 1 - t(x)$.

- *SB* (*Scott–Barreto*) *curves*[32]: The family of SB curves with embedding degree $k = 6$ corresponds to the choices:

$$
\begin{aligned}
t(x) &= -4x^2 + 4x + 2, \\
q(x) &= 4x^5 - 8x^4 + 3x^3 - 3x^2 + \frac{17}{4}x + 1, \\
m(x) &= 16x^4 - 32x^3 + 12x^2 + 4x + 1.
\end{aligned}
$$

For a square-free positive Δ not dividing $27330 = 2 \times 3 \times 5 \times 911$, the family $(t(\Delta z^2), m(\Delta z^2), q(\Delta z^2))$ parametrized by z is a complete family of ordinary curves of embedding degree six and discriminant Δ.

- *BLS* (*Barreto–Lynn–Scott*)[33] and *BW* (*Brezing–Weng*)[34] *curves*: These families are called *cyclotomic families*, as the construction of these curves is based upon cyclotomic extensions of $\mathbb{Q}$. For example, the choices

$$
\begin{aligned}
t(x) &= -x^2 + 1, \\
m(x) &= \Phi_{4k}(x), \\
q(x) &= \frac{1}{4}\left(x^{2k+4} + 2x^{2k+2} + x^{2k} + x^4 - 2x^2 + 1\right)
\end{aligned}
$$

parametrize a family of BW curves with odd embedding degree $k < 1000$ and with discriminant $\Delta = 1$. Many other families of BLS and BW curves are known.

4.5.6 Efficient Implementation

Although the iterations in Miller's loops resemble repeated square-and-multiply algorithms for exponentiation in finite fields, and repeated double-and-add algorithms for point multiplication in elliptic curves (Exercise 4.26), the computational overhead in an iteration of Miller's loop is somewhat more than that for finite-field exponentiation or elliptic-curve point multiplication. In this section, we study some implementation tricks that can significantly speed up Miller's loop for computing Weil and Tate pairings.

Eisenträger, Lauter and Montgomery[35] describe methods to speed up the computation when the i-th bit m_i in Miller's loop is 1. In this case, computing

---

[32] Michael Scott and Paulo S. L. M. Barreto, Generating more MNT elliptic curves, *Designs, Codes and Cryptography*, 38, 209–217, 2006.

[33] Paulo S. L. M. Barreto, Ben Lynn and Michael Scott, Constructing elliptic curves with prescribed embedding degrees, *SCN*, 263–273, 2002.

[34] Friederike Brezing and Annegret Weng, Elliptic curves suitable for pairing based cryptography, *Designs, Codes and Cryptography*, 37, 133–141, 2005.

[35] Kirsten Eisenträger, Kristin Lauter and Peter L. Montgomery, Improved Weil and Tate pairings for elliptic and hyperelliptic curves, *ANTS*, 169–183, 2004

$2U+P$ as $(U+P)+U$ saves a few field operations (Exercise 4.27). For speeding up the update of f, they propose a second trick of using a parabola.

Another improvement is from Blake, Murty and Xu[36] who replace the computation of four lines by that of only two lines for the case $m_i = 1$ (Exercise 4.61). Although Blake et al. claim that their improvement is useful when most of the bits m_i are 1, the practicality of this improvement is rather evident even for random values of m.

4.5.6.1 Windowed Loop in Miller's Algorithm

By Exercise 1.27, a repeated square-and-multiply or double-and-add algorithm can benefit from consuming chunks of bits in the exponent or multiplier in each iteration, at the cost of some precomputation. Employing a similar strategy for Miller's algorithm is not straightforward, since the updating formulas for f do not immediately yield to efficient generalizations. For example, the formulas given in Exercise 4.62 do not make it obvious how even a two-bit window can be effectively handled.

Blake, Murty and Xu use a separate set of formulas for a two-bit window (see Exercise 4.59). Algorithm 4.4 rewrites Miller's algorithm for the computation of $f_{m,P}(Q)$, and can be readily adapted to handle Miller's algorithms for computing Weil and Tate pairings.

Some further refinements along the line of Blake et al.'s developments are proposed by Liu, Horng and Chen[37] who segment the bitstring for m into patterns of the form $(01)^r$, 01^r, 01^r0, 0^r, and 1^r0, and 1^r. For many such patterns, Liu et al. propose modifications to reduce the number of lines in the updating formulas for f.

4.5.6.2 Final Exponentiation

Although the double-and-add loop for Tate pairing is more efficient than that for Weil pairing, the added overhead of final exponentiation is unpleasant for the reduced Tate pairing. Fortunately, we can choose the curve parameters and tune this stage so as to arrive at an efficient implementation.[38]

Suppose that the basic field of definition of the elliptic curve E is $\mathbb{F}_q$ with $p = \operatorname{char} \mathbb{F}_q$. We take m to be a prime dividing $|E_q|$. If k is the embedding degree for this q and m, the final-exponentiation stage involves an exponent of $(q^k - 1)/m$. In this stage, we do arithmetic in the extension field $\mathbb{F}_{q^k}$.

---

[36] Ian F. Blake, V. Kumar Murty and Guangwu Xu, Refinements of Miller's algorithm for computing the Weil/Tate pairing, *Journal of Algorithms*, 58(2), 134–149, 2006.

[37] Chao-Liang Liu, Gwoboa Horng and Te-yu Chen, Further refinement of pairing computation based on Miller's algorithm, *Applied Mathematics and Computation*, 189(1), 395–409, 2007. This is available also at `http://eprint.iacr.org/2006/106`.

[38] Michael Scott, Naomi Benger, Manuel Charlemagne, Luis J. Dominguez Perez and Ezekiel J. Kachisa, On the final exponentiation for calculating pairings on ordinary elliptic curves, *Pairing*, 78–88, 2009.

Algorithm 4.4: Two-bit windowed loop for computing $f_{m,P}(Q)$

Take the base-4 representation $m = (M_s M_{s-1} \ldots M_1 M_0)_4$, $M_s \neq 0$.
Initialize $f = 1$, and $U = P$.

If $(M_s = 2)$, set $f = \dfrac{L_{P,P}(Q)}{L_{2P,-2P}(Q)}$, and $U = 2P$,

else if $(M_s = 3)$, set $f = -\dfrac{L_{P,P}(Q) \times L_{P,-P}(Q)}{L_{2P,P}(-Q)}$, and $U = 3P$.

For $i = s-1, s-2, \ldots, 1, 0$, do {
  If $(M_i = 0)$ {

    Set $f = -f^4 \times \left(\dfrac{L_{U,U}^2(Q)}{L_{2U,2U}(-Q)} \right)$, and $U = 4U$.

  } else if $(M_i = 1)$ {

    Set $f = -f^4 \times \left(\dfrac{L_{U,U}^2(Q) \times L_{4U,P}(Q)}{L_{4U+P,-(4U+P)}(Q) \times L_{2U,2U}(-Q)} \right)$, and $U = 4U + P$.

  } else if $(M_i = 2)$ {

    Set $f = -f^4 \times \left(\dfrac{L_{U,U}^2(Q) \times L_{2U,P}^2(Q)}{L_{2U,-2U}^2(Q) \times L_{2U+P,2U+P}(-Q)} \right)$, and $U = 4U + 2P$.

  } else if $(M_i = 3)$ {

    Set $f = -f^4 \times \left(\dfrac{L_{U,U}^2(Q) \times L_{2U,P}^2(Q) \times L_{4U+2P,P}(Q)}{L_{2U,-2U}^2(Q) \times L_{2U+P,2U+P}(-Q) \times L_{4U+3P,-(4U+3P)}(Q)} \right)$,

    and $U = 4U + 3P$.

  }
}
Return f.

Let $\beta_0, \beta_1, \ldots, \beta_{k-1}$ be an $\mathbb{F}_q$-basis of $\mathbb{F}_{q^k}$. An element $\alpha = a_0\beta_0 + a_1\beta_1 + \cdots + a_{k-1}\beta_{k-1} \in \mathbb{F}_{q^k}$ (with $a_i \in \mathbb{F}_q$) satisfies $\alpha^q = a_0\beta_0^q + a_1\beta_1^q + \cdots + a_{k-1}\beta_{k-1}^q$, since $a_i^q = a_i$ by Fermat's little theorem for $\mathbb{F}_q$. If the quantities $\beta_0^q, \beta_1^q, \ldots, \beta_{k-1}^q$ are precomputed as linear combinations of the basis elements $\beta_0, \beta_1, \ldots, \beta_{k-1}$, simplifying α^q to its representation in this basis is easy. For example, if $q = p \equiv 3 \pmod 4$, $k = 2$, and we represent $\mathbb{F}_{p^2} = \mathbb{F}_p(\theta)$ with $\theta^2 + 1 = 0$, then $(a_0 + a_1\theta)^p = a_0 - a_1\theta$. Thus, exponentiation to the q-th power is much more efficient than general square-and-multiply exponentiation.

Now, suppose that $k = 2d$ is even, and write $q^k - 1 = (q^d - 1)(q^d + 1)$. Since $k = \text{ord}_q(m)$, we conclude that m does not divide $q^d - 1$. But m is prime, so it must divide $q^d + 1$, and the final exponentiation can be carried out as

$$f^{(q^k-1)/m} = \left(f^{q^d-1} \right)^{(q^d+1)/m}.$$

The inner exponentiation (to the power $q^d - 1$) involves d q-th power exponentiations, followed by multiplication by f^{-1}. Since $|(q^d+1)/m| \approx \frac{1}{2}|(q^k-1)/m|$, this strategy reduces final-exponentiation time by a factor of about two.

Some more optimizations can be performed on the exponent $(q^d + 1)/m$. It follows from Exercise 3.36(c) that the cyclotomic polynomial $\Phi_k(x)$ divides $x^d + 1$ in $\mathbb{Z}[x]$. Moreover, $m | \Phi_k(q)$. Therefore,

$$
f^{(q^k - 1)/m} = \left(\left(f^{q^d - 1} \right)^{(q^d + 1)/\Phi_k(q)} \right)^{\Phi_k(q)/m}.
$$

The intermediate exponent $(q^d + 1)/\Phi_k(q)$ is a polynomial in q, so we can again exploit the efficiency of computing q-th powers in $\mathbb{F}_{q^k}$. Only the outermost exponent $\Phi_k(q)/m$ calls for a general exponentiation algorithm. We have $\deg \Phi_k(x) = \phi(k)$ (Euler totient function), so this exponent is of size $\approx |q^{\phi(k)}/m|$. For $k = 6$, this is two-third the size of $(q^d + 1)/m$.

4.5.6.3 Denominator Elimination

Barreto, Lynn and Scott[39] propose a nice idea in view of which the vertical lines $L_{2U, -2U}$ and $L_{U+P, -(U+P)}$ can be eliminated altogether during the computation of the reduced Tate pairing. Let E be an elliptic curve defined over $\mathbb{F}_q$ of characteristic > 3, m a prime, and $k = \mathrm{ord}_m q$ the embedding degree. Assume that $k = 2d$ is even (we may have $d = 1$), and that E is given by the short Weierstrass equation $Y^2 = X^3 + aX + b$ for some $a, b \in \mathbb{F}_q$.

E is defined over $\mathbb{F}_{q^d}$ as well. Let $v \in \mathbb{F}_{q^d}^*$ be a quadratic non-residue in $\mathbb{F}_{q^d}$. Define the quadratic twist of E over $\mathbb{F}_{q^d}$ in the usual way, that is, $E' : Y^2 = X^3 + v^2 aX + v^3 b$. Choose a non-zero point $P_1 \in E_q[m]$ and a non-zero point $P_2 \in E'_{q^d}[m]$. Let G_1 and G_2 be the groups of order m generated by P_1 and P_2, respectively. We define the reduced Tate pairing on $G_1 \times G_2$ as $\hat{e}_m(P, \phi_2(Q))$ for $P \in G_1$ and $Q \in G_2$. Here, $\phi_2 : E'_{q^d} \to E_{q^k}$ is the twist map $(r, s) \mapsto (v^{-1}r, (v\sqrt{v})^{-1}s)$. Since $v \in \mathbb{F}_{q^d}$ and $G_2 \subseteq E'_{q^d}$, it follows that the X-coordinates of all the points $\phi_2(Q)$ are in $\mathbb{F}_{q^d}$ (although their Y-coordinates may belong to the strictly larger field $\mathbb{F}_{2^k}$).

The vertical lines $L_{2U, -2U}$ and $L_{U+P, -(U+P)}$ are of the form $x + c$ with $c \in \mathbb{F}_q$, and evaluate, at points with X-coordinates in $\mathbb{F}_{q^d}$, to elements of $\mathbb{F}_{q^d}$.

Let $\alpha \in \mathbb{F}_{q^d}^*$. By Fermat's little theorem for $\mathbb{F}_{q^d}$, we have $\alpha^{q^d - 1} = 1$. Since $k = 2d = \mathrm{ord}_m q$, we conclude that $q^d - 1$ is not divisible by m, and so $\alpha^{(q^k - 1)/m} = \left(\alpha^{q^d - 1} \right)^{(q^d + 1)/m} = 1$. This, in turn, implies that the contributions of the values of $L_{2U, -2U}$ and $L_{U+P, -(U+P)}$ at points with X-coordinates in $\mathbb{F}_{q^d}$ are absorbed in the final exponentiation. Consequently, it is not necessary to compute the lines $L_{2U, -2U}$ and $L_{U+P, -(U+P)}$, and to evaluate them. However, in order to make the idea work, we must choose the (random) point T in Algorithm 4.3 from E_{q^d} (or use the formula $f_{m,P}(Q)$ of Eqn (4.29)).

---

[39] Paulo S. L. M. Barreto, Ben Lynn, and Michael Scott, On the selection of pairing-friendly groups, *SAC*, 17–25, 2003.

4.5.6.4 Loop Reduction

Loop reduction pertains to the computation of Tate pairing. In the Miller loop for computing the function $f_{m,P}$, the number of iterations is (about) $\log_2 m$. A loop-reduction technique computes $f_{M,P}$ for some M, and involves $\log_2 M$ iterations. If M is substantially smaller than m, this leads to significant speedup. For example, if $M \approx \sqrt{m}$, the number of iterations in the Miller loop reduces by a factor of nearly two. It is evident that replacing $f_{m,P}$ by $f_{M,P}$ for an arbitrary M is not expected to produce any bilinear pairing at all. We must choose M carefully so as to preserve both bilinearity and non-degeneracy. The idea of loop reduction is introduced by Duursma and Lee.[40]

Eta Pairing

Barreto et al.[41] propose an improvement of the Duursma–Lee construction. Let E be a supersingular elliptic curve defined over $K = \mathbb{F}_q$, m a prime divisor of $|E_q|$, and k the embedding degree. E being supersingular, there exists a distortion map $\phi : G \to G'$ for suitable groups $G \subseteq E_q[m]$ and $G' \subseteq E_{q^k}[m]$ of order m. The distorted Tate pairing is defined as $\langle P, \phi(Q) \rangle_m = f_{m,P}(D)$ for $P, Q \in G$, where D is a divisor equivalent to $[\phi(Q)] - [\mathcal{O}]$. For a suitable choice of M, Barreto et al. define the *eta pairing* of $P, Q \in G$ as

$$\eta_M(P,Q) = f_{M,P}(\phi(Q)).$$

For the original Tate pairing, we have $mP = \mathcal{O}$. Now, we remove this requirement $MP = \mathcal{O}$. We take $M = q - cm$ for some $c \in \mathbb{Z}$ such that for every point $P \in G$ we have $MP = \gamma(P)$ for some automorphism γ of E_q. The automorphism γ and the distortion map ϕ should satisfy the *golden condition*: $\gamma(\phi^q(P)) = \phi(P)$ for all $P \in E_q$. If $M^a + 1 = \lambda m$ for some $a \in \mathbb{N}$ and $\lambda \in \mathbb{Z}$, then $\eta_M(P,Q)$ is related to the Tate pairing $\langle P, \phi(Q) \rangle_m$ as

$$\left(\eta_M(P,Q)^{aM^{a-1}} \right)^{(q^k-1)/m} = \left(\langle P, \phi(Q) \rangle_m^{\lambda} \right)^{(q^k-1)/m} = \hat{e}_m(P, \phi(Q))^{\lambda}.$$

Example 4.77 Eta pairing provides a sizeable speedup for elliptic curves over finite fields of characteristics two and three. For fields of larger characteristics, eta pairing is not very useful. Many families of supersingular curves and distortion maps on them can be found in Example 4.76 and Exercise 4.67.

(1) Let $E : Y^2 + Y = X^3 + X + a$, $a \in \{0, 1\}$, be a supersingular curve defined over $\mathbb{F}_{2^r}$ with odd r. The choice $\gamma = \phi^r$ satisfies the golden condition. In this case, $|E_q| = 2^r \pm 2^{(r+1)/2} + 1$. Suppose that $|E_q|$ is prime, so we take $m = |E_q|$. We choose $M = \mp 2^{(r+1)/2} - 1$, so for $a = 2$ we have $M^2 + 1 = 2M$,

---

[40]Iwan M. Duursma and Hyang-Sook Lee, Tate pairing implementation for hyperelliptic curves $y^2 = x^p - x + d$, *AsiaCrypt*, 111–123, 2003.

[41]Paulo S. L. M. Barreto, Steven Galbraith, Colm Ó hÉigeartaigh and Michael Scott, Efficient pairing computation on supersingular Abelian varieties, *Designs, Codes and Cryptography*, 239–271, 2004.

that is, $\lambda = 2$. Also, $M = q - m$, that is, $c = 1$. Then, $\eta_M(P,Q) = f_{M,P}(\phi(Q))$ involves a Miller loop twice as efficient as the Miller loop for $\langle P, \phi(Q) \rangle_m$. The embedding degree k is four in this case, so

$$\hat{e}_m(P, \phi(Q)) = \langle P, \phi(Q) \rangle_m^{(2^{4r}-1)/m} = \eta_M(P,Q)^{2M(2^{4r}-1)/m}.$$

Thus, the reduced Tate pairing of P and $\phi(Q)$ can be computed by raising $\eta_M(P,Q)$ to an appropriate power.

(2) For the supersingular curve $E : Y^2 = X^3 - X + b$, $b = \pm 1$, defined over $\mathbb{F}_{3^r}$ with $\gcd(r,6) = 1$, the choice $\gamma = \phi^r$ satisfies the golden condition. We have $m = |E_q| = 3^r \pm 3^{(r+1)/2} + 1$. Assume that m is prime. Choose $M = q - m = \mp 3^{(r+1)/2} - 1$. For $a = 3$, we get $M^a + 1 = \lambda m$ with $\lambda = \mp 3^{(r+3)/2}$. The embedding degree is six in this case. This gives

$$\hat{e}_m(P, \phi(Q))^{\lambda} = \left(\langle P, \phi(Q) \rangle_m^{\lambda} \right)^{(3^{6r}-1)/m} = \eta_M(P,Q)^{3M^2(3^{6r}-1)/m}.$$

So a power of the reduced Tate pairing can be obtained from the eta pairing. The speedup is by a factor of about two, since $\log_2 M \approx \frac{1}{2} \log_2 m$. The exponent λ can be removed by another exponentiation, since $\gcd(\lambda, 3^{6r} - 1) = 1$. $\square$

Ate Pairing

Hess et al.[42] extend the idea of eta pairing to ordinary curves. Distortion maps do not exist for ordinary curves. Nonetheless, subgroups $G \subseteq E_q$ and $G' \subseteq E_{q^k}$ of order m can be chosen. Instead of defining a pairing on $G \times G'$, Hess et al. define a pairing on $G' \times G$. If t is the trace of Frobenius for E at q, they take $M = t - 1$, and define the *ate pairing* of $Q \in G'$ and $P \in G$ as

$$a_M(Q,P) = f_{M,Q}(P).$$

The ate pairing is related to the Tate pairing as follows. Let $N = \gcd(M^k - 1, q^k - 1)$, where k is the embedding degree. Write $M^k - 1 = \lambda N$, and $c = \sum_{i=0}^{k-1} M^{k-1-i} q^i \equiv k q^{k-1} \pmod{m}$. Then, we have

$$\left(a_M(Q,P)^c \right)^{(q^k-1)/N} = \left(\langle Q, P \rangle_m^{\lambda} \right)^{(q^k-1)/m} = \hat{e}_m(Q,P)^{\lambda}.$$

It can be shown that a_M is bilinear. Moreover, a_M is non-degenerate if $m \nmid \lambda$.

Let us now see how much saving this is with ate pairing. Since $|t| \leqslant 2\sqrt{q}$ by Hasse's theorem, we have $\log_2 M \leqslant 1 + \frac{1}{2} \log_2 m$, when $\log_2 m \approx \log_2 q$. Therefore, the number of iterations in Miller's loop gets halved. But since a point Q of E_{q^k} has now become the first argument, all line functions and point arithmetic involve working in $\mathbb{F}_{q^k}$. If $k > 1$, each iteration of Miller's loop becomes costlier than that for $\langle P, Q \rangle_m$ (with Q at the second argument). Particularly, for $k \geqslant 6$, we fail to gain any speedup using ate pairing.

---

[42] Florian Hess, Nigel P. Smart and Frederik Vercauteren, The eta pairing revisited, *IEEE Transactions on Information Theory*, 52(10), 4595–4602, 2006.

Twisted Ate Pairing

In order to define an ate-like pairing on $G \times G'$, Hess et al. use quartic (degree-four) and sextic (degree-six) twists. For a twist of degree $d|k$,

$$f_{M^e, P}(Q)$$

is defined as the *twisted ate pairing* of P, Q, where $M = t - 1$ and $e = k/d$. For efficiency, M^e should be small compared to m. If the trace t is large, that is, $|t| \approx \sqrt{q}$, twisted ate pairing fails to provide any speedup. If t is small, that is, $|t| \approx m^{1/\phi(k)}$, twisted ate pairing achieves a speedup of two over basic Tate pairing (here, ϕ is the Euler totient function). Twisted ate pairing in this case is often called an *optimal pairing*, since, under reasonable assumptions, the Miller loop cannot have less than $\Theta\left(\frac{\log m}{\phi(k)}\right)$ iterations.

Example 4.78 I now illustrate how twists can help in speeding up each Miller iteration. The following example is of a modified ate pairing that uses twists, since twisted ate pairing is defined in a slightly different way.

Take a Barreto–Naehrig (BN) curve $E : Y^2 = X^3 + b$ defined over $\mathbb{F}_p$ with a prime $p \equiv 1 \pmod 6$. The embedding degree is $k = 12$. Define a sextic twist of E with respect to a primitive sixth root ζ of unity. The twisted curve can be written as $E' : \mu Y^2 = \nu X^3 + b$, where $\mu \in \mathbb{F}_{p^2}^*$ is a cubic non-residue, and $\nu \in \mathbb{F}_{p^2}^*$ is a quadratic non-residue. E' is defined over $\mathbb{F}_{p^2}$, and we take a subgroup G' of order m in E'_{p^2}. A homomorphism ϕ_6 that maps G' into $E_{p^{12}}$ is given by $(r, s) \mapsto (\nu^{1/3} r, \mu^{1/2} s)$. We use standard ate pairing to define the pairing $a_M(\phi_6(Q), P)$ of $Q \in G'$ and $P \in G$. For $Q \in G$, the point $\phi_6(P)$ is defined over $\mathbb{F}_{p^{12}}$, but not over smaller subfields (in general). Nonetheless, the association of G' with $\phi_6(G')$ allows us to work in $\mathbb{F}_{p^2}$ in some parts of the Miller loop. For example, if $Q_1, Q_2 \in G'$, then $\phi_6(Q_1) + \phi_6(Q_2) = \phi_6(Q_1 + Q_2)$, that is, the point arithmetic in Miller's loop can be carried out in $\mathbb{F}_{p^2}$. $\square$

Ate$_i$ Pairing

Zhao et al.[43] propose another optimization of ate pairing. For an integer i in the range $1 \leqslant i \leqslant k - 1$, they take $M_i \equiv (t - 1)^i \equiv q^i \pmod m$, and define

$$f_{M_i, Q}(P)$$

as the *ate$_i$ pairing* of Q, P. The minimum of $M_1, M_2, \ldots, M_{k-1}$ yields the shortest Miller loop. Some curves offer choices of i for which the number of iterations in the ate$_i$ Miller loop is optimal $(\Theta\left(\frac{\log m}{\phi(k)}\right))$. Ate$_i$ pairings are defined on $G' \times G$, and have the same performance degradation as ate pairing.

[43]Changan Zhao, Fangguo Zhang and Jiwu Huang, A note on the ate pairing, *International Journal of Information Security*, 7(6), 379–382, 2008.

R-ate Pairing

At present, the best loop-reducing pairing is proposed by Lee et al.[44] If e and e' are two pairings, then so also is $e(Q, P)^u e'(Q, P)^v$ for any integers u, v. For $A, B, a, b \in \mathbb{Z}$ with $A = aB + b$, Lee et al. define the *R-ate pairing* as

$$R_{A,B}(Q, P) = f_{a,BQ}(P) f_{b,Q}(P) G_{aBQ,bQ}(P),$$

where $G_{U,V} = \frac{L_{U,V}}{L_{U+V,-(U+V)}}$. Any choice of A, B, a, b does not define a pairing. If $f_{A,Q}(P)$ and $f_{B,Q}(P)$ define non-degenerate bilinear pairings with $\hat{e}_m(Q, P)^{\lambda_1} = f_{A,Q}(P)^{\mu_1}$ and $\hat{e}_m(Q, P)^{\lambda_2} = f_{B,Q}(P)^{\mu_2}$ for $\lambda_1, \lambda_2, \mu_1, \mu_2 \in \mathbb{Z}$, then $R_{A,B}(Q, P)$ is again a non-degenerate bilinear pairing satisfying

$$\hat{e}_m(Q, P)^{\lambda} = R_{A,B}(Q, P)^{\mu},$$

where $\mu = \mathrm{lcm}(\mu_1, \mu_2)$ and $\lambda = (\mu/\mu_1)\lambda_1 - a(\mu/\mu_2)\lambda_2$, provided that $m \nmid \lambda$. There are several choices for A, B, including q, m, and the integers M_i of ate$_i$ pairing. If $A = M_i$ and $B = m$, then $R_{A,B}(Q, P)$ is the ate$_i$ pairing of Q, P.

R-ate pairing makes two invocations of the Miller loop, but a suitable choice of A, B, a, b reduces the total number of Miller iterations compared to the best ate$_i$ pairing. There are examples where loop reduction can be by a factor of six over Tate pairing (for ate and ate$_i$ pairings, the reduction factor can be at most two). Moreover, R-ate pairing is known to be optimal on certain curves for which no ate$_i$ pairing is optimal. Another useful feature of R-ate pairing is that it can handle both supersingular and ordinary curves.

4.6 Elliptic-Curve Point Counting

Let E be an elliptic curve defined over $K = \mathbb{F}_q$. A pertinent question is to find the size of the group $E_K = E_q$. This problem has sophisticated algorithmic solutions. If q is an odd prime or a power of two, there exist polynomial-time (in $\log q$) algorithms to compute $|E_q|$. The first known algorithm of this kind is from Schoof.[45] The SEA algorithm (named after Schoof, Elkies and Atkin)[46] is an efficient modification of Schoof's algorithm.

Another related question is: Given a size s satisfying the Hasse bound (that is, $q + 1 - 2\sqrt{q} \leqslant s \leqslant q + 1 + 2\sqrt{q}$), how can one construct an elliptic curve E over $\mathbb{F}_q$ with $|E_q| = s$? As of now, this question does not have efficient algorithmic solutions, except in some very special cases.

---

[44] Eunjeong Lee, Hyang-Sook Lee and Cheol-Min Park, Efficient and generalized pairing computation on Abelian varieties, *Cryptology ePrint Archive*, Report 2008/040, 2008.

[45] René Schoof, Elliptic curves over finite fields and the computation of square roots mod p, *Mathematics of Computation*, 44(170), 483–494, 1985.

[46] René Schoof, Counting points on elliptic curves over finite fields, *Journal de Théorie des Nombres de Bordeaux*, 7, 219–254, 1995.

4.6.1 A Baby-Step-Giant-Step (BSGS) Method

I first present an exponential-time probabilistic algorithm based on Shank's baby-step-giant-step paradigm,[47] that works for any finite field $\mathbb{F}_q$ (q need not be prime), and runs in $\tilde{O}(q^{1/4})$ time. Let E be an elliptic curve over $\mathbb{F}_q$. By Hasse's theorem, the size of E_q lies between $q+1-2\sqrt{q}$ and $q+1+2\sqrt{q}$. We locate an m in this interval such that $\operatorname{ord} P | m$ for some point $P \in E_q$. If $\operatorname{ord} P$ has a unique multiple m in the interval, then $|E_q| = m$. To promote search, we break the interval into about $2q^{1/4}$ subintervals of about the same size.

Baby Steps: We choose a random non-zero point $P \in E_q$, and compute and store the multiples $\mathcal{O}, \pm P, \pm 2P, \pm 3P, \ldots, \pm sP$, where $s = \lceil q^{1/4} \rceil$. In fact, it suffices to compute only the points $P, 2P, 3P, \ldots, sP$. The opposites of these points can be easily obtained. Some appropriate data structures (like hash tables or sorted arrays or balanced search trees) should be used to store these $2s + 1$ points in order to promote efficient search in the list.

Giant Steps: We compute the points $Q = \lceil q + 1 - 2\sqrt{q} \rceil P$ and $R = (2s+1)P$ by the repeated double-and-add algorithm. Subsequently, for $i = 0, 1, 2, 3, \ldots, 2s+1$, we compute $Q + iR$ (for $i \geqslant 0$, we have $Q + (i+1)R = (Q+iR) + R$). For each i, we check whether $Q + iR$ resides in the list of baby steps, that is, whether $Q + iR = jP$ for some j in the range $-s \leqslant j \leqslant s$. If so, we obtain $mP = \mathcal{O}$, where $m = \lceil q + 1 - 2\sqrt{q} \rceil + (2s+1)i - j$ is a desired multiple of $\operatorname{ord} P$ in the Hasse interval.

Example 4.79 Take $E : Y^2 = X^3 + 3X + 6$ defined over $\mathbb{F}_p = \mathbb{F}_{997}$, and the point $P = (234, 425)$ on E. We have $s = \lceil p^{1/4} \rceil = 6$. The baby steps are computed first. Next, we prepare for the giant steps by computing $Q = \lceil p + 1 - 2\sqrt{p} \rceil P = 935P = (944, 231)$ and $R = (2s+1)P = 13P = (24, 464)$.

	Baby steps			Giant steps	
j	jP	$-jP$	i	$Q + iR$	j
0	$\mathcal{O}$		0	$(944, 231)$	
1	$(234, 425)$	$(234, 572)$	1	$(867, 2)$	
2	$(201, 828)$	$(201, 169)$	2	$(115, 260)$	
3	$(886, 800)$	$(886, 197)$	3	$(527, 565)$	
4	$(159, 57)$	$(159, 940)$	4	$(162, 759)$	
5	$(18, 914)$	$(18, 83)$	5	$(34, 974)$	
6	$(623, 968)$	$(623, 29)$	6	$(549, 677)$	
			7	$(643, 806)$	
			8	$(159, 940)$	-4

We stop making giant steps as soon as we discover i, j with $Q + iR = jP$. This gives $(935 + 8 \times 13 - (-4))P = \mathcal{O}$, that is, $1043P = \mathcal{O}$, that is, $m = 1043$.

---

[47] Daniel Shanks, Class number, a theory of factorization and genera, *Proceedings of Symposia in Pure Mathematics*, 20, 415–440, 1971. The BSGS paradigm is generic enough to be applicable to a variety of computational problems. Its adaptation to point counting and Mestre's improvement are discussed in Schoof's 1995 paper (see Footnote 46). Also see Section 7.1.1 for another adaptation of the BSGS method.

If we complete all the giant steps, we find no other multiple of ord P in the Hasse interval $[935, 1061]$. So the size of $E_{\mathbb{F}_{997}}$ is 1043. Since $1043 = 7 \times 149$ is square-free, this group is cyclic. $\qquad\square$

The BSGS method may fail to supply a unique answer if ord P has more than one multiple in the Hasse interval. For instance, suppose that we start with $149P$ as the base point P in Example 4.79. This point has order 7, so every multiple of 7 in the Hasse interval will be supplied as a possible candidate for $|E_q|$. For this example, the problem can be overcome by repeating the algorithm for different random choices of the base point P. After a few iterations, we expect to find a P with a unique multiple of ord P in the Hasse interval. This is indeed the expected behavior if E_q is a cyclic group.

However, the group E_q need not be cyclic. By Theorem 4.12, we may have $E_q \cong \mathbb{Z}_{n_1} \oplus \mathbb{Z}_{n_2}$ with $n_2 | \gcd(n_1, q-1)$. Every point $P \in E_q$ satisfies $n_1 P = \mathcal{O}$ (indeed, n_1 is the smallest positive integer with this property; we call n_1 the *exponent* of the group E_q). If n_1 is so small that the Hasse interval contains two or more multiples of n_1, the BSGS method fails to supply a unique answer, no matter how many times we run it (with different base points P).

Example 4.80 The curve $E : Y^2 = X^3 + X + 161$ defined over $\mathbb{F}_{1009}$ contains 1024 points and has the group structure $\mathbb{Z}_{64} \oplus \mathbb{Z}_{16}$. The exponent of the group is 64, which has two multiples 960 and 1024 in the Hasse interval $[947, 1073]$. Therefore, both $960P = \mathcal{O}$ and $1024P = \mathcal{O}$ for any point P on E. A point P of order smaller than 64 has other multiples of ord P in the Hasse interval. Trying several random points on E, we can eliminate these extra candidates, but the ambiguity between 960 and 1024 cannot be removed. This is demonstrated below for $P = (6, 49)$. Now, we have $s = 6$, $Q = (947, 339)$, and $R = (947, 670)$.

Baby steps				Giant steps		
j	jP	$-jP$		i	$Q + iR$	j
0	$\mathcal{O}$			0	$(947, 339)$	
1	$(6, 49)$	$(6, 960)$		1	$\mathcal{O}$	0
2	$(3, 47)$	$(3, 962)$		2	$(947, 670)$	
3	$(552, 596)$	$(552, 413)$		3	$(550, 195)$	
4	$(798, 854)$	$(798, 155)$		4	$(588, 583)$	
5	$(455, 510)$	$(455, 499)$		5	$(604, 602)$	
6	$(413, 641)$	$(413, 368)$		6	$(6, 49)$	1
				7	$(717, 583)$	
				8	$(855, 172)$	
				9	$(756, 1000)$	
				10	$(713, 426)$	
				11	$(3, 47)$	2
				12	$(842, 264)$	
				13	$(374, 133)$	

For $(i, j) = (1, 0)$, we get $m = 947 + 13 - 0 = 960$, whereas for $(i, j) = (6, 1)$, we get $m = 947 + 6 \times 13 - 1 = 1024$. The algorithm also outputs $(i, j) = (11, 2)$, for which $m = 947 + 11 \times 13 - 3 = 1088$. This third value of m is outside the Hasse interval $[947, 1073]$, and can be ignored. It is reported because of the approximation of $p^{1/4}$ by the integer s. Indeed, it suffices here to make giant steps for $0 \leqslant i \leqslant 10$ only. $\qquad \square$

4.6.1.1 Mestre's Improvement

J. F. Mestre proposes a way to avoid the problem just mentioned. Let $v \in \mathbb{F}_q^*$ be a quadratic non-residue (this requires q to be odd). Consider the quadratic twist $E' : Y^2 = X^3 + av^2 X + bv^3$ of E, which is again defined over $\mathbb{F}_q$. By Exercise 4.68, $|E_q| + |E'_q| = 2(q + 1)$. Instead of computing $|E_q|$, we may, therefore, calculate $|E'_q|$. Even if E_q is of low exponent, the chance that E'_q too is of low exponent is rather small. Indeed, Mestre proves that for prime fields $\mathbb{F}_p$ with $p > 457$, either E_p or E'_p contains a point of order $\geqslant 4\sqrt{p}$.

Example 4.81 As a continuation of Example 4.80, we take the quadratic non-residue $v = 11 \in \mathbb{F}_{1009}^*$ to define the quadratic twist $E' : Y^2 = X^3 + 121X + 383$ of E. The BSGS method on E' works as follows. We take the base point $P = (859, 587)$, for which $Q = \lceil q + 1 - 2\sqrt{q} \rceil P = 947P = (825, 545)$, and $R = (2s + 1)P = 13P = (606, 90)$.

	Baby steps			Giant steps	
j	jP	$-jP$	i	$Q + iR$	j
0	$\mathcal{O}$		0	$(825, 545)$	
1	$(859, 587)$	$(859, 422)$	1	$(246, 188)$	
2	$(282, 63)$	$(282, 946)$	2	$(746, 845)$	
3	$(677, 631)$	$(677, 378)$	3	$(179, 447)$	
4	$(325, 936)$	$(325, 73)$	4	$(677, 631)$	3
5	$(667, 750)$	$(667, 259)$	5	$(465, 757)$	
6	$(203, 790)$	$(203, 219)$	6	$(80, 735)$	
			7	$(214, 258)$	
			8	$(148, 2)$	
			9	$(557, 197)$	
			10	$(358, 501)$	
			11	$(529, 234)$	
			12	$(144, 55)$	
			13	$(937, 420)$	

The computations show that there is a unique value of i for which $Q + iR$ is found in the list of baby steps. This gives the size of E'_q uniquely as $|E'_q| = 947 + 4 \times 13 - 3 = 996$. Consequently, $|E_q| = 2(1009 + 1) - 996 = 1024$. $\quad \square$

It is easy to verify that the BSGS method makes $\Theta(q^{1/4})$ group operations in E_q, that is, the BSGS method is an exponential-time algorithm, and cannot be used for elliptic-curve point counting, except only when q is small.

4.6.2 Schoof's Algorithm

René Schoof is the first to propose a polynomial-time algorithm to solve the elliptic-curve point counting problem. Schoof's algorithm is particularly suited to prime fields. In this section, we concentrate on elliptic curves defined over a prime field $\mathbb{F}_p$. We also assume that the curve is defined by the short Weierstrass equation $E : Y^2 = X^3 + aX + b$ with $a, b \in \mathbb{F}_p$.

By Hasse's theorem, $|E_p| = p + 1 - t$, where the trace of Frobenius t at p satisfies $-2\sqrt{p} \leqslant t \leqslant 2\sqrt{p}$. Schoof's algorithm computes t modulo many small primes r, and combines the residues by the Chinese remainder theorem to a unique integer in the Hasse interval. By Exercise 4.10, $|E_q| \pmod 2$ can be determined by checking the irreducibility of $X^3 + aX + b$ in $\mathbb{F}_p[X]$. So we assume that r is an odd prime not equal to p.

The Frobenius endomorphism (Definition 4.47) $\varphi : E_{\bar{\mathbb{F}}_p} \to E_{\bar{\mathbb{F}}_p}$ taking (x, y) to (x^p, y^p) fixes $E_{\mathbb{F}_p}$ point by point,[48] and satisfies the identity

$$\varphi^2 - t\varphi + p = 0, \tag{4.30}$$

that is, for any $P = (x, y) \in E_{\bar{\mathbb{F}}_p}$, we have

$$\varphi(\varphi(P)) - t\varphi(P) + pP = (x^{p^2}, y^{p^2}) - t(x^p, y^p) + p(x, y) = \mathcal{O}, \tag{4.31}$$

where the addition, the subtraction and the scalar multiplications correspond to the arithmetic of $E_{\bar{\mathbb{F}}_p}$. Let $t_r \equiv t \pmod r$ and $p_r \equiv p \pmod r$ for a small odd prime r not equal to p. We then have

$$(x^{p^2}, y^{p^2}) - t_r(x^p, y^p) + p_r(x, y) = \mathcal{O} \tag{4.32}$$

for all points (x, y) in the group $E[r]$ of r-torsion points on $E_{\bar{\mathbb{F}}_p}$ (that is, points P with $rP = \mathcal{O}$). By varying t_r in the range $0 \leqslant t_r \leqslant r-1$, we find the correct value of t_r for which Eqn (4.32) holds identically on $E[r]$. For each trial value of t_r, we compute the left side of Eqn (4.32) *symbolically* using the addition formula for the curve. There are, however, two problems with this approach.

The first problem is that the left side of Eqn (4.32) evaluates to a pair of rational functions of degrees as high as $\Theta(p^2)$. Our aim is to arrive at a polynomial-time algorithm (in $\log p$). This problem is solved by using division polynomials (Section 4.4.3). For the reduced Weierstrass equation, we have

$$
\begin{aligned}
\psi_0(x, y) &= 0, \\
\psi_1(x, y) &= 1, \\
\psi_2(x, y) &= 2y, \\
\psi_3(x, y) &= 3x^4 + 6ax^2 + 12bx - a^2, \\
\psi_4(x, y) &= 4y(x^6 + 5ax^4 + 20bx^3 - 5a^2x^2 - 4abx - 8b^2 - a^3),
\end{aligned}
$$

---

[48] Schoof's algorithm performs symbolic manipulation on the coordinates x, y of points on E, which satisfy $y^2 = x^3 + ax + b$. Thus, x, y are actually X, Y modulo $Y^2 - (X^3 + aX + b)$. So the use of lower-case x, y here is consistent with the earlier sections in this chapter.

$$
\begin{aligned}
\psi_5(x,y) \;=\;& 5x^{12} + 62ax^{10} + 380bx^9 - 105a^2x^8 + 240bax^7 + \\
& (-300a^3 - 240b^2)x^6 - 696ba^2x^5 + (-125a^4 - 1920b^2a)x^4 + \\
& (-80ba^3 - 1600b^3)x^3 + (-50a^5 - 240b^2a^2)x^2 + \\
& (-100ba^4 - 640b^3a)x + (a^6 - 32b^2a^3 - 256b^4),
\end{aligned}
$$

$$
\psi_{2m}(x,y) \;=\; \frac{\psi_m(\psi_{m+2}\psi_{m-1}^2 - \psi_{m-2}\psi_{m+1}^2)}{2y},
$$

$$
\psi_{2m+1}(x,y) \;=\; \psi_{m+2}\psi_m^3 - \psi_{m+1}^3\psi_{m-1}.
$$

It is often convenient to define

$$
\psi_{-1}(x,y) \;=\; -1.
$$

Division polynomials possess the following property.

Theorem 4.82 *A point* $(x,y) \in E_{\bar{\mathbb{F}}_p}$ *is in* $E[r]$ *if and only if* $\psi_r(x,y) = 0$. $\triangleleft$

Since we are interested in evaluating Eqn (4.32) for points in $E[r]$, it suffices to do so modulo $\psi_r(x,y)$. But the polynomials $\psi_r(x,y)$ are in two variables x, y. Since $y^2 = x^3 + ax + b$, we can simplify $\psi_m(x,y)$ to either a polynomial in $\mathbb{F}_p[x]$ or y times a polynomial in $\mathbb{F}_p[x]$. In particular, we define

$$
f_m(x) = \begin{cases} \psi_m(x,y) & \text{if } m \text{ is odd,} \\ \psi_m(x,y)/y & \text{if } m \text{ is even.} \end{cases}
$$

The polynomials $f_m(x)$ are in one variable x only, with

$$
\deg f_m(x) = \begin{cases} \frac{1}{2}(m^2 - 1) & \text{if } m \text{ is odd,} \\ \frac{1}{2}(m^2 - 4) & \text{if } m \text{ is even.} \end{cases}
$$

The polynomials $f_m(x)$ are also called division polynomials. We continue to have the following characterization of r-torsion points.

Theorem 4.83 *A point* $(x,y) \in E_{\bar{\mathbb{F}}_p}$ *is in* $E[r]$ *if and only if* $f_r(x) = 0$. $\triangleleft$

To sum up, we compute Eqn (4.32) modulo $y^2 - (x^3 + ax + b)$ and $f_r(x)$. This keeps the degrees of all intermediate polynomials bounded below $\Theta(r^2)$. Since r is a small prime (indeed, $r = O(\log p)$), the polynomial $f_r(x)$ and the left side of Eqn (4.32) can be computed in time polynomial in $\log p$.

The second problem associated with Schoof's algorithm is that during the symbolic computation of Eqn (4.32), we cannot distinguish between the cases of addition and doubling. Since these two cases use different formulas, we need to exercise some care so that Schoof's algorithm works at all. This makes the algorithm somewhat clumsy. First, we rewrite Eqn (4.32) as

$$
(x^{p^2}, y^{p^2}) + \pi(x,y) = \tau(x^p, y^p), \tag{4.33}
$$

where we have used $\pi = p_r$ and $\tau = t_r$ for notational simplicity. Using Eqns (4.18) and (4.19) of Section 4.4.3, we obtain the symbolic equality:

$$(x^{p^2}, y^{p^2}) + \left(x - \frac{\psi_{\pi-1}\psi_{\pi+1}}{\psi_\pi^2}, \frac{\psi_{\pi+2}\psi_{\pi-1}^2 - \psi_{\pi-2}\psi_{\pi+1}^2}{4y\psi_\pi^3}\right)$$

$$= \begin{cases} O & \text{if } \tau = 0, \\[2mm] \left(x^p - \left(\frac{\psi_{\tau-1}\psi_{\tau+1}}{\psi_\tau^2}\right)^p, \left(\frac{\psi_{\tau+2}\psi_{\tau-1}^2 - \psi_{\tau-2}\psi_{\tau+1}^2}{4y\psi_\tau^3}\right)^p\right) & \text{otherwise.} \end{cases} \quad (4.34)$$

The left side of Eqn (4.34) is independent of τ. Name the two summands on this side as $Q = (x^{p^2}, y^{p^2})$ and $R = \pi(x, y)$. The addition formula for the left side depends upon which of the following cases holds: $Q = R$, $Q = -R$, and $Q \neq \pm R$. The first two cases can be treated together using the x-coordinates only, that is, $Q = \pm R$ if and only if the following condition holds:

$$x^{p^2} = x - \frac{\psi_{\pi-1}\psi_{\pi+1}}{\psi_\pi^2}.$$

But ψ_m are polynomials in two variables, so we rewrite this condition as

$$x^{p^2} = \begin{cases} x - \dfrac{f_{\pi-1}(x)f_{\pi+1}(x)(x^3 + ax + b)}{f_\pi^2(x)} & \text{if } \pi \text{ is odd}, \\[4mm] x - \dfrac{f_{\pi-1}(x)f_{\pi+1}(x)}{f_\pi^2(x)(x^3 + ax + b)} & \text{if } \pi \text{ is even}, \end{cases}$$

in terms of the univariate division polynomials $f_m(x)$. This computation is to be done modulo $f_r(x)$. To this effect, we compute the following gcd

$$\gamma(x) = \begin{cases} \gcd\left(f_r(x), (x^{p^2} - x)f_\pi^2(x) + f_{\pi-1}(x)f_{\pi+1}(x)(x^3 + ax + b)\right) \\ \hspace{6cm} \text{if } \pi \text{ is odd}, \\[4mm] \gcd\left(f_r(x), (x^{p^2} - x)f_\pi^2(x)(x^3 + ax + b) + f_{\pi-1}(x)f_{\pi+1}(x)\right) \\ \hspace{6cm} \text{if } \pi \text{ is even}, \end{cases}$$

where the second argument is computed modulo $f_r(x)$.

Case 1: $\gamma(x) = 1$. This is equivalent to the condition that $Q \neq \pm R$ for all $P \in E[r]$. In particular, $\tau \neq 0$. In this case, we use the formula for adding two distinct points not opposite of one another, in order to evaluate the left side of Eqn (4.34). Then, we check for which $\tau \in \{1, 2, \ldots, r-1\}$, equality holds in Eqn (4.34). By multiplying with suitable polynomials, we can clear all the denominators, and check whether equalities of two pairs of polynomial expressions (one for the x-coordinates and the other for the y-coordinates) hold modulo $y^2 - (x^3 + ax + b)$ and $f_r(x)$. I now supply the final y-free formulas without derivations, which use the polynomials $f_m(x)$ instead of $\psi_m(x, y)$.

We first compute two polynomials dependent only on π, modulo $f_r(x)$:

$$\alpha = \begin{cases} (x^3 + ax + b)(f_{\pi+2}f_{\pi-1}^2 - f_{\pi-2}f_{\pi+1}^2) - \\ \hspace{1.5cm} 4(x^3 + ax + b)^{(p^2+1)/2}f_\pi^3 & \text{if } \pi \text{ is odd}, \\[4mm] f_{\pi+2}f_{\pi-1}^2 - f_{\pi-2}f_{\pi+1}^2 - \\ \hspace{1.5cm} 4(x^3 + ax + b)^{(p^2+3)/2}f_\pi^3 & \text{if } \pi \text{ is even}, \end{cases}$$

$$\beta = \begin{cases} 4f_\pi\Big((x - x^{p^2})f_\pi^2 - (x^3 + ax + b)f_{\pi-1}f_{\pi+1}\Big) & \text{if } \pi \text{ is odd,} \\ 4(x^3 + ax + b)f_\pi\Big((x - x^{p^2})(x^3 + ax + b)f_\pi^2 - \\ \hspace{4cm} f_{\pi-1}f_{\pi+1}\Big) & \text{if } \pi \text{ is even.} \end{cases}$$

In order to check the equality of x-coordinates, we compute two other polynomials based upon π only:

$$\delta_1 = \begin{cases} \Big(f_{\pi-1}f_{\pi+1}(x^3 + ax + b) - \\ \hspace{1cm} f_\pi^2(x^{p^2} + x^p + x)\Big)\beta^2(x^3 + ax + b) + f_\pi^2\alpha^2 & \text{if } \pi \text{ is odd,} \\ \Big(f_{\pi-1}f_{\pi+1} - f_\pi^2(x^{p^2} + x^p + x)(x^3 + ax + b)\Big)\beta^2 + \\ \hspace{1cm} f_\pi^2\alpha^2(x^3 + ax + b)^2 & \text{if } \pi \text{ is even,} \end{cases}$$

$$\delta_2 = f_\pi^2\beta^2(x^3 + ax + b).$$

We check whether the following condition holds for the selected value of τ:

$$0 = \begin{cases} \delta_1 f_\tau^{2p} + \delta_2(f_{\tau-1}f_{\tau+1})^p(x^3 + ax + b)^p & \text{if } \tau \text{ is odd,} \\ \delta_1 f_\tau^{2p}(x^3 + ax + b)^p + \delta_2(f_{\tau-1}f_{\tau+1})^p & \text{if } \tau \text{ is even,} \end{cases} \tag{4.35}$$

where the equality is modulo the division polynomial $f_\tau(x)$.

For checking the equality of y-coordinates, we compute (using π only):

$$\delta_3 = \begin{cases} 4(x^3 + ax + b)^{(p-1)/2} \times \\ \quad \left[\Big(f_\pi^2(2x^{p^2} + x) - f_{\pi-1}f_{\pi+1}(x^3 + ax + b)\Big)\alpha\beta^2(x^3 + ax + b) - \right. \\ \qquad \left. f_\pi^2\Big(\alpha^3 + \beta^3(x^3 + ax + b)^{(p^2+3)/2}\Big)\right] & \text{if } \pi \text{ is odd,} \\ 4(x^3 + ax + b)^{(p+1)/2} \times \\ \quad \left[\Big(f_\pi^2(2x^{p^2} + x)(x^3 + ax + b) - f_{\pi-1}f_{\pi+1}\Big)\alpha\beta^2 - \right. \\ \qquad \left. f_\pi^2\Big(\alpha^3 + \beta^3(x^3 + ax + b)^{(p^2-3)/2}\Big)(x^3 + ax + b)^2\right] & \text{if } \pi \text{ is even,} \end{cases}$$

$$\delta_4 = \beta^3 f_\pi^2(x^3 + ax + b).$$

For the selected τ, we check the following equality modulo $f_\tau(x)$:

$$0 = \begin{cases} \delta_3 f_\tau^{3p} - \delta_4\Big(f_{\tau+2}f_{\tau-1}^2 - \\ \hspace{1cm} f_{\tau-2}f_{\tau+1}^2\Big)^p(x^3 + ax + b)^p & \text{if } \tau \text{ is odd,} \\ \delta_3 f_\tau^{3p}(x^3 + ax + b)^{(3p+1)/2} - \delta_4\Big(f_{\tau+2}f_{\tau-1}^2 - \\ \hspace{1cm} f_{\tau-2}f_{\tau+1}^2\Big)^p(x^3 + ax + b)^{(p+1)/2} & \text{if } \tau \text{ is even.} \end{cases} \tag{4.36}$$

There exists a unique τ in the range $1 \leqslant \tau \leqslant r - 1$, for which Eqns (4.35) and (4.36) hold simultaneously. This is the desired value of t_r, that is, the value of t modulo the small prime r.

Case 2: $\gamma(x) \neq 1$, that is, $Q = \pm R$ for some $P \in E[r]$. Now, we need to distinguish between the two possibilities: $Q = R$ and $Q = -R$.

Sub-case 2(a): If $Q = -R$, then $\tau = 0$.

Sub-case 2(b): If $Q = R$, Eqn (4.33) gives $2\pi P = \tau\varphi(P)$, that is, $\varphi(P) = \frac{2\pi}{\tau}P$ (since $\tau \not\equiv 0 \pmod{r}$ in this case) for some non-zero $P \in E[r]$. Plugging in this value of $\varphi(P)$ in Eqn (4.33) eliminates φ as $(\frac{4\pi}{\tau^2} - 1)P = \mathcal{O}$ for some non-zero $P \in E[r]$, that is, $\frac{4\pi}{\tau^2} - 1 \equiv 0 \pmod{r}$ (since r is a prime), that is, $\tau^2 \equiv 4\pi \pmod{r}$. This, in turn, implies that 4π, and so π too, must be quadratic residues modulo r. Let $w \in \mathbb{Z}_r^*$ satisfy $w^2 \equiv \pi \pmod{r}$. Then, $\tau \equiv \pm 2w \pmod{r}$. But $\varphi(P) = \frac{2\pi}{\tau}P = \frac{2w^2}{\tau}P$ (for some P in $E[r]$), so $\varphi(P) = wP$ if $\tau \equiv 2w \pmod{r}$, and $\varphi(P) = -wP$ if $\tau \equiv -2w \pmod{r}$.

In order to identify the correct sub-case, we first compute the Legendre symbol $\left(\frac{\pi}{r}\right)$. If $\left(\frac{\pi}{r}\right) = -1$, then $\tau = 0$. Otherwise, we compute a square root w of π (or p) modulo r. We may use a probabilistic algorithm like the Tonelli and Shanks algorithm (Algorithm 1.9) for computing w. Since $r = O(\log p)$, we can find w by successively squaring $1, 2, 3, \ldots$ until $w^2 \equiv \pi \pmod{r}$ holds.

Now, we check whether $\varphi(P) = wP$ or $\varphi(P) = -wP$ for some $P \in E[r]$. In the first case, $\tau \equiv 2w \pmod{r}$, whereas in the second, $\tau \equiv -2w \pmod{r}$. If no such P exists in $E[r]$, we have $\tau = 0$. If we concentrate only on the x-coordinates, we can detect the existence of such a P. As we have done in detecting whether $Q = \pm R$ (that is, $\varphi^2(P) = \pm \pi P$), checking the validity of the condition $\varphi(P) = \pm wP$ boils down to computing the following gcd:

$$\delta(x) = \begin{cases} \gcd\left(f_r(x), (x^p - x)f_w^2(x) + f_{w-1}(x)f_{w+1}(x)(x^3 + ax + b)\right) \\ \qquad\qquad\qquad\qquad\qquad\qquad\qquad\qquad \text{if } w \text{ is odd,} \\ \gcd\left(f_r(x), (x^p - x)f_w^2(x)(x^3 + ax + b) + f_{w-1}(x)f_{w+1}(x)\right) \\ \qquad\qquad\qquad\qquad\qquad\qquad\qquad\qquad \text{if } w \text{ is even,} \end{cases}$$

where the second argument is computed modulo $f_r(x)$. If $\delta(x) = 1$, then $\tau = 0$. Otherwise, we need to identify which one of the equalities $\varphi(P) = \pm wP$ holds. For deciding this, we consult the y-coordinates, and compute another gcd:

$$\eta(x) = \begin{cases} \gcd\left(f_r(x), 4(x^3 + ax + b)^{(p-1)/2}f_w^3(x) - \right. \\ \qquad\qquad \left. f_{w+2}(x)f_{w-1}^2(x) + f_{w-2}(x)f_{w+1}^2(x)\right) \quad \text{if } w \text{ is odd,} \\ \gcd\left(f_r(x), 4(x^3 + ax + b)^{(p+3)/2}f_w^3(x) - \right. \\ \qquad\qquad \left. f_{w+2}(x)f_{w-1}^2(x) + f_{w-2}(x)f_{w+1}^2(x)\right) \quad \text{if } w \text{ is even,} \end{cases}$$

where again arithmetic modulo $f_r(x)$ is used to evaluate the second argument. If $\eta(x) = 1$, then $\tau \equiv -2w \pmod r$, otherwise $\tau \equiv 2w \pmod r$.

This completes the determination of $\tau \equiv t_r \equiv t \pmod r$. We repeat the above process for $O(\log p)$ small primes r, whose product is larger than the length $4\sqrt{p}+1$ of the Hasse interval. By the prime number theorem, each such small prime is $O(\log p)$. Using CRT, we combine the residues t_r for all small primes r to a unique integer in the Hasse interval.

Example 4.84 Consider the curve $E : Y^2 = X^3 + 3X + 7$ defined over the prime field $\mathbb{F}_p = \mathbb{F}_{997}$. The Hasse interval in this case is $[935, 1061]$ with a width of 127. If we compute the trace t modulo the four small primes $r = 2, 3, 5, 7$, we can uniquely obtain t and consequently $|E_{\mathbb{F}_{997}}|$.

Small prime $r = 2$: $X^3 + 3X + 7$ is irreducible in $\mathbb{F}_{997}[x]$. Therefore, $E_{\mathbb{F}_{997}}$ does not contain a point of order two, that is, $|E_{\mathbb{F}_{997}}|$ and so t are odd.

For working with other small primes, we need the division polynomials. We use only the univariate versions $f_m(x)$. In this example, the polynomials $f_m(x)$ are needed only for $-1 \leqslant m \leqslant 8$, and are listed below. These polynomials are to be treated as members of $\mathbb{F}_{997}[x]$.

$$
\begin{aligned}
f_{-1}(x) &= 996, \\
f_0(x) &= 0, \\
f_1(x) &= 1, \\
f_2(x) &= 2, \\
f_3(x) &= 3x^4 + 18x^2 + 84x + 988, \\
f_4(x) &= 4x^6 + 60x^4 + 560x^3 + 817x^2 + 661x + 318, \\
f_5(x) &= 5x^{12} + 186x^{10} + 666x^9 + 52x^8 + 55x^7 + 80x^6 + 20x^5 + \\
&\quad 753x^4 + 382x^3 + 653x^2 + 586x + 760, \\
f_6(x) &= 6x^{16} + 432x^{14} + 435x^{13} + 427x^{12} + 22x^{10} + 687x^9 + \\
&\quad 986x^8 + 610x^7 + 198x^6 + 994x^5 + 683x^4 + 575x^3 + \\
&\quad 630x^2 + 968x + 704, \\
f_7(x) &= 7x^{24} + 924x^{22} + 689x^{21} + 333x^{20} + 639x^{19} + 154x^{18} + \\
&\quad 666x^{17} + 562x^{16} + 144x^{15} + 365x^{14} + 905x^{13} + 61x^{12} + \\
&\quad 420x^{11} + 552x^{10} + 388x^9 + 457x^8 + 342x^7 + 169x^6 + \\
&\quad 618x^5 + 213x^4 + 507x^3 + 738x^2 + 664x + 223, \\
f_8(x) &= 8x^{30} + 755x^{28} + 322x^{27} + 855x^{26} + 695x^{25} + 229x^{24} + \\
&\quad 103x^{23} + 982x^{22} + 842x^{21} + 842x^{20} + 332x^{19} + 30x^{18} + \\
&\quad 588x^{17} + 864x^{16} + 153x^{15} + 927x^{14} + 834x^{13} + 439x^{12} + \\
&\quad 216x^{11} + 469x^{10} + 109x^9 + 877x^8 + 707x^7 + 879x^6 + \\
&\quad 664x^5 + 490x^4 + 40x^3 + 150x^2 + 333x + 608.
\end{aligned}
$$

Small prime $r = 3$: We first compute the gcd $\gamma(x) = x + 770 \neq 1$, and we go to Case 2 of Schoof's algorithm. We have $\pi = p \operatorname{rem} r = 1$ in this case, that is, p is a quadratic residue modulo r. A square root of p modulo r is $w = 1$. We compute the gcd $\delta(x) = x + 770$ which is not 1 again. In the final step, we compute the gcd $\eta(x) = x + 770$ which is still not 1. We, therefore, conclude that $t \equiv 2w \equiv 2 \pmod{3}$.

Small prime $r = 5$: We get the gcd $\gamma(x) = f_5(x)$ itself. Since 997, that is, 2 is a quadratic non-residue modulo 5, we have $t \equiv 0 \pmod 5$ by Case 2.

Small prime $r = 7$: In this case, we get $\gamma(x) = 1$, that is, we resort to Case 1 of Schoof's algorithm. For this r, $\pi = p \operatorname{rem} r = 3$ is odd. Using the appropriate formulas, we first calculate

$$
\begin{aligned}
\alpha &\equiv (f_5 f_2^2 - f_1 f_4^2)(x^3 + ax + b) - 4(x^3 + ax + b)^{(p^2+1)/2} f_3^3 \\
&\equiv 824x^{23} + 137x^{22} + 532x^{21} + 425x^{20} + 727x^{19} + 378x^{18} + \\
&\quad 669x^{17} + 906x^{16} + 198x^{15} + 305x^{14} + 47x^{13} + 65x^{12} + 968x^{11} + \\
&\quad 985x^{10} + 262x^9 + 867x^8 + 825x^7 + 373x^6 + 771x^5 + 97x^4 + \\
&\quad 921x^3 + 789x^2 + 382x + 296 \pmod{f_7(x)}, \\
\beta &\equiv 4f_3\Big((x - x^{p^2})f_3^2 - f_2 f_4(x^3 + ax + b)\Big) \\
&\equiv 151x^{23} + 501x^{22} + 879x^{21} + 133x^{20} + 288x^{19} + 451x^{18} + 781x^{17} + \\
&\quad 479x^{16} + 809x^{15} + 544x^{14} + 35x^{13} + 923x^{12} + 983x^{11} + 171x^{10} + \\
&\quad 359x^9 + 675x^8 + 633x^7 + 828x^6 + 438x^5 + 955x^4 + 979x^3 + \\
&\quad 367x^2 + 212x + 195 \pmod{f_7(x)}.
\end{aligned}
$$

Subsequently, we compute the four polynomials δ_i as:

$$
\begin{aligned}
\delta_1 &\equiv \Big(f_2 f_4(x^3 + ax + b) - f_3^2(x^{p^2} + x^p + x)\Big)\beta^2(x^3 + ax + b) + f_3^2 \alpha^2 \\
&\equiv 0 \pmod{f_7(x)}, \\
\delta_2 &\equiv \beta^2(x^3 + ax + b)f_3^2 \\
&\equiv 499x^{23} + 124x^{22} + 310x^{21} + 236x^{20} + 74x^{19} + 175x^{18} + 441x^{17} + \\
&\quad 259x^{16} + 557x^{15} + 14x^{14} + 431x^{13} + 339x^{12} + 277x^{11} + 520x^{10} + \\
&\quad 386x^9 + 941x^8 + 281x^7 + 230x^6 + 320x^5 + 201x^4 + 540x^3 + \\
&\quad 89x^2 + 791x + 179 \pmod{f_7(x)}, \\
\delta_3 &\equiv 4(x^3 + ax + b)^{(p-1)/2}\Big[\alpha\beta^2(x^3 + ax + b)\Big(f_3^2(2x^{p^2} + x) - \\
&\quad f_2 f_4(x^3 + ax + b)\Big) - f_3^2\Big(\alpha^3 + \beta^3(x^3 + ax + b)^{(p^2+3)/2}\Big)\Big] \\
&\equiv 148x^{23} + 845x^{22} + 755x^{21} + 370x^{20} + 763x^{19} + 767x^{18} + 338x^{17} + \\
&\quad 938x^{16} + 719x^{15} + 497x^{14} + 885x^{13} + 509x^{12} + 218x^{11} + 467x^{10} + \\
&\quad 586x^9 + 822x^8 + 717x^7 + 680x^6 + 257x^5 + 490x^4 + 488x^3 +
\end{aligned}
$$

$$365x^2 + 16x + 935 \pmod{f_7(x)},$$
$$\delta_4 \equiv \beta^3(x^3 + ax + b)f_3^2$$
$$\equiv 446x^{23} + 533x^{22} + 59x^{21} + 324x^{20} + 148x^{19} + 778x^{18} + 521x^{17} +$$
$$833x^{16} + 640x^{15} + 908x^{14} + 320x^{13} + 608x^{12} + 645x^{11} + 401x^{10} +$$
$$369x^9 + 720x^8 + 629x^7 + 923x^6 + 707x^5 + 427x^4 + 156x^3 +$$
$$688x^2 + 19x + 340 \pmod{f_7(x)}.$$

Finally, we try $\tau = 1, 2, 3, 4, 5, 6$. For each odd τ, we compute the polynomials

$$H_x(x) \equiv \delta_1 f_\tau^{2p} + \delta_2(f_{\tau-1}f_{\tau+1})^p(x^3 + ax + b)^p \pmod{f_7(x)},$$
$$H_y(x) \equiv \delta_3 f_\tau^{3p} - \delta_4(x^3 + ax + b)^p(f_{\tau+2}f_{\tau-1}^2 - f_{\tau-2}f_{\tau+1}^2)^p \pmod{f_7(x)},$$

whereas for each even τ, we compute the same polynomials as

$$H_x(x) \equiv \delta_1 f_\tau^{2p}(x^3 + ax + b)^p + \delta_2(f_{\tau-1}f_{\tau+1})^p \pmod{f_7(x)},$$
$$H_y(x) \equiv \delta_3 f_\tau^{3p}(x^3 + ax + b)^{(3p+1)/2} -$$
$$\delta_4(x^3 + ax + b)^{(p+1)/2}(f_{\tau+2}f_{\tau-1}^2 - f_{\tau-2}f_{\tau+1}^2)^p \pmod{f_7(x)}.$$

It turns out that $H_x(x) \equiv 0 \pmod{f_7(x)}$ for $\tau = 1$ and 6, whereas $H_y(x) \equiv 0 \pmod{f_7(x)}$ only for $\tau = 1$. It follows that $t \equiv 1 \pmod 7$.

To sum up, we have computed the trace t as:

$$t \equiv 1 \pmod 2,$$
$$t \equiv 2 \pmod 3,$$
$$t \equiv 0 \pmod 5,$$
$$t \equiv 1 \pmod 7.$$

By CRT, $t \equiv 155 \pmod{210}$. But t is an integer in the range $[-2\sqrt{p}, 2\sqrt{p}]$, that is, $[-63, 63]$. Thus, $t = -55$, that is, $|E_{\mathbb{F}_{997}}| = 997 + 1 - (-55) = 1053$. $\square$

It is easy to argue that the running time of Schoof's algorithm is polynomial in $\log p$. Indeed, the most time-consuming steps are the exponentiations of polynomials to exponents of values $\Theta(p)$ or $\Theta(p^2)$. These computations are done modulo $f_r(x)$ with $\deg f_r = \Theta(r^2)$, and $r = O(\log p)$. A careful calculation shows that this algorithm runs in $O(\log^8 p)$ time. Although this is a polynomial expression in $\log p$, the large exponent (eight) makes the algorithm somewhat impractical.

Several modifications of this original algorithm are proposed in the literature. The SEA (Schoof–Elkies–Atkin) algorithm reduces the exponent to six by using a suitable divisor of degree $O(r)$ of $f_r(x)$ as the modulus. Adaptations of the SEA algorithm for the fields $\mathbb{F}_{2^n}$ are also proposed. From a practical angle, the algorithms in the SEA family are reasonable, and, in feasible time, can handle fields of sizes as large as 2000 bits. For prime fields, they are the best algorithms known to date. For fields of characteristic two, more efficient algorithms are known.

Exercises

1. Prove that the Weierstrass equation of an elliptic curve is irreducible, that is, the polynomial $Y^2 + (a_1 X + a_3)Y - (X^3 + a_2 X^2 + a_4 X + a_6)$ with $a_i \in K$ is irreducible in $K[X, Y]$.

2. Prove that an elliptic or hyperelliptic curve is smooth at its point at infinity.

3. Let $C : Y^2 = f(X)$ be the equation of a cubic curve C over a field K with char $K \neq 2$, where $f(X) = X^3 + aX^2 + bX + c$ with $a, b, c \in K$. Prove that C is an elliptic curve (that is, smooth or non-singular) if and only if $\mathrm{Discr}(f) \neq 0$ (or, equivalently, if and only if $f(X)$ has no multiple roots).

4. Let K be a finite field of characteristic two, and $a, b, c \in K$. Prove that:
   (a) The curve $Y^2 + aY = X^3 + bX + c$ is smooth if and only if $a \neq 0$.
   (b) The curve $Y^2 + XY = X^3 + aX^2 + b$ is smooth if and only if $b \neq 0$.

5. Determine which of the following curves is/are smooth (that is, elliptic curves).
   (a) $Y^2 = X^3 - X^2 - X + 1$ over $\mathbb{Q}$.
   (b) $Y^2 + 2Y = X^3 + X^2$ over $\mathbb{Q}$.
   (c) $Y^2 + 2XY = X^3 + 1$ over $\mathbb{Q}$.
   (d) $Y^2 + 4XY = X^3 + 4X$ over $\mathbb{Q}$.
   (e) $Y^2 + Y = X^3 + 5$ over $\mathbb{F}_7$.
   (f) $Y^2 + Y = X^3 + 5$ over $\mathbb{F}_{11}$.

6. Let the elliptic curve $E : Y^2 = X^3 + 2X + 3$ be defined over $\mathbb{F}_7$. Take the points $P = (2, 1)$ and $Q = (3, 6)$ on E.
   (a) Compute the points $P + Q$, $2P$ and $3Q$ on the curve.
   (b) Determine the order of P in the elliptic curve group $E(\mathbb{F}_7)$.
   (c) Find the number of points on E treated as an elliptic curve over $\mathbb{F}_{49} = \mathbb{F}_{7^2}$.

7. Let $P = (h, k)$ be a point with $2P = (h', k') \neq \mathcal{O}$ on the elliptic curve $Y^2 = X^3 + aX^2 + bX + c$. Verify that

$$h' = \frac{h^4 - 2bh^2 - 8ch - 4ac + b^2}{4k^2} = \frac{h^4 - 2bh^2 - 8ch - 4ac + b^2}{4h^3 + 4ah^2 + 4bh + 4c}, \text{ and}$$

$$k' = \frac{h^6 + 2ah^5 + 5bh^4 + 20ch^3 + (20ac - 5b^2)h^2 + (8a^2c - 2ab^2 - 4bc)h + (4abc - b^3 - 8c^2)}{8k^3}.$$

8. Let K be a finite field of characteristic two, and $a, b, c \in K$. Prove that:
   (a) The supersingular curve $E_1 : Y^2 + aY = X^3 + bX + c$ contains no points of order two. In particular, the size of $(E_1)_K$ is odd.
   (b) The ordinary curve $E_2 : Y^2 + XY = X^3 + aX^2 + b$ contains exactly one point of order two. In particular, the size of $(E_2)_K$ is even.

9. Let E be an elliptic curve defined over a field K with char $K \neq 2, 3$. Prove that E has at most eight points of order three. If K is algebraically closed, prove that E has exactly eight points of order three.

10. Consider the elliptic curve $E : Y^2 = f(X)$ with $f(X) = X^3 + aX^2 + bX + c \in K[X]$, defined over a finite field K with char $K \neq 2$. Prove that:

(a) The size of the group E_K is odd if and only if $f(X)$ is irreducible in $K[X]$.

(b) If $f(X)$ splits in $K[X]$, then E_K is not cyclic.

11. Take $P = (3, 5)$ on the elliptic curve $E : Y^2 = X^3 - 2$ over $\mathbb{Q}$. Compute the point $2P$ on E. Use the Nagell–Lutz theorem to establish that P has infinite order. Conclude that the converse of the Nagell–Lutz theorem is not true.

12. Consider the elliptic curve $E : Y^2 = X^3 + 17$ over $\mathbb{Q}$. Demonstrate that the points $P_1 = (-1, 4)$, $P_2 = (-2, 3)$, $P_3 = (2, 5)$, $P_4 = (4, 9)$, $P_5 = (8, 23)$, $P_6 = (43, 282)$, $P_7 = (52, 375)$, and $P_8 = (5234, 378661)$ lie on E. Verify that $P_3 = P_1 + 2P_2$, $P_4 = -P_1 - P_2$, $P_5 = -2P_2$, $P_6 = 2P_1 + 3P_2$, $P_7 = -P_1 + P_2$ and $P_8 = 3P_1 + 2P_2$. (**Remark:** It is known that the only points on E with integer coordinates are the points P_1 through P_8, and their opposites.)

13. Consider the ordinary curve $E : Y^2 + XY = X^3 + X^2 + (\theta + 1)$ defined over $\mathbb{F}_8 = \mathbb{F}_2(\theta)$ with $\theta^3 + \theta + 1 = 0$. Find all the points on $E_{\mathbb{F}_8}$. Prepare the group table for $E_{\mathbb{F}_8}$.

14. Let p be an odd prime with $p \equiv 2 \pmod 3$, and let $a \not\equiv 0 \pmod p$. Prove that the elliptic curve $Y^2 = X^3 + a$ defined over $\mathbb{F}_p$ contains exactly $p + 1$ points.

15. Let p be a prime with $p \equiv 3 \pmod 4$, and let $a \not\equiv 0 \pmod p$. Prove that the elliptic curve $Y^2 = X^3 + aX$ defined over $\mathbb{F}_p$ contains exactly $p + 1$ points.

16. Find the order of the torsion point $(3, 8)$ on the curve $E : Y^2 = X^3 - 43X + 166$ defined over $\mathbb{Q}$. Determine the structure of the torsion subgroup of $E_{\mathbb{Q}}$.

17. Let E be an elliptic curve defined over $\mathbb{F}_q$ and so over $\mathbb{F}_{q^n}$ for all $n \in \mathbb{N}$. Let (h, k) be a finite point on $E_{\mathbb{F}_{q^n}}$. Prove that $\varphi(h, k) = (h^q, k^q)$ is again on $E_{\mathbb{F}_{q^n}}$.

18. Prove that a supersingular elliptic curve cannot be anomalous.

19. Let E be an anomalous elliptic curve defined over $\mathbb{F}_q$. Prove that $E_{\mathbb{F}_q}$ is cyclic.

20. Let E denote the ordinary elliptic curve $Y^2 + XY = X^3 + X^2 + 1$ defined over $\mathbb{F}_{2^n}$ for all $n \in \mathbb{N}$. Prove that the size of the group $E_{\mathbb{F}_{2^n}}$ is $2^n + 1 - \frac{1}{2^{n-1}} \left[1 - \binom{n}{2}7 + \binom{n}{4}7^2 - \cdots + (-1)^r \binom{n}{2r}7^r \right]$ for all $n \in \mathbb{N}$, where $r = \lfloor n/2 \rfloor$.

21. Let E denote the elliptic curve $Y^2 = X^3 + X^2 + X + 1$ defined over $\mathbb{F}_{3^n}$ for all $n \in \mathbb{N}$. Prove that the size of the group $E_{\mathbb{F}_{3^n}}$ is $3^n + 1 + (-1)^{n-1} \times 2 \times \left[1 - \binom{n}{2}2 + \binom{n}{4}2^2 - \cdots + (-1)^r \binom{n}{2r}2^r \right]$ for all $n \in \mathbb{N}$, where $r = \lfloor n/2 \rfloor$.

22. Determine the number of points on the curves $Y^2 + Y = X^3 + X$ and $Y^2 + Y = X^3 + X + 1$ over the field $\mathbb{F}_{2^n}$. Conclude that for all $n \in \mathbb{N}$, these curves are supersingular.

23. Determine the number of points on the curves $Y^2 = X^3 \pm X$ and $Y^2 = X^3 + X \pm 1$ over the field $\mathbb{F}_{3^n}$. Conclude that for all $n \in \mathbb{N}$, these curves are supersingular.

24. Determine the number of points on the curves $Y^2 = X^3 - X \pm 1$ over the field $\mathbb{F}_{3^n}$. Conclude that for all $n \in \mathbb{N}$, these curves are supersingular.

25. Let E be a supersingular elliptic curve defined over a prime field $\mathbb{F}_p$ with $p \geqslant 5$. Determine the size of $E_{\mathbb{F}_{p^n}}$, and conclude that E remains supersingular over all extension $\mathbb{F}_{p^n}$, $n \geqslant 1$.

26. Rewrite the square-and-multiply exponentiation algorithm (Algorithm 1.4) for computing the multiple of a point on an elliptic curve. (In the context of elliptic-curve point multiplication, we call this a *double-and-add algorithm.*)

27. [Eisenträger, Lauter and Montgomery] In the double-and-add elliptic-curve point-multiplication algorithm, we need to compute points $2P+Q$ for every 1-bit in the multiplier. Conventionally, this is done as $(P+P)+Q$. Assuming that the curve is given by the short Weierstrass equation, count the field operations used in computing $(P+P)+Q$. Suppose instead that $2P+Q$ is computed as $(P+Q)+P$. Argue that we may avoid computing the Y-coordinate of $P+Q$. What saving does the computation of $(P+Q)+P$ produce (over that of $(P+P)+Q$)?

28. Find the points at infinity (over $\mathbb{R}$ and $\mathbb{C}$) on the following real curves.

(a) Ellipses of the form $\dfrac{X^2}{a^2} + \dfrac{Y^2}{b^2} = 1$.

(b) Hyperbolas of the form $\dfrac{X^2}{a^2} - \dfrac{Y^2}{b^2} = 1$.

(c) Hyperbolas of the form $XY = a$.

29. Let $C : f(X,Y) = 0$ be a curve defined by a non-constant irreducible polynomial $f(X,Y) \in K[X,Y]$. Let d be $\deg f(X,Y)$, and $f_d(X,Y)$ the sum of all non-zero terms of degree d in $f(X,Y)$. Prove that all points at infinity on C are obtained by solving $f_d(X,Y) = 0$. Conclude that all the points at infinity on C can be obtained by solving a *univariate* polynomial equation over K.

30. [*Projective coordinates*] Projective coordinates are often used to speed up elliptic-curve arithmetic. In the projective plane, a finite point (h,k) corresponds to the point $[h', k', l']$ with $l' \neq 0$, $h = h'/l'$, and $k = k'/l'$. Let E be an elliptic curve defined by the special Weierstrass equation $Y^2 = X^3 + aX + b$, and the finite points P_1, P_2 on E have projective coordinates $[h_1, k_1, l_1]$ and $[h_2, k_2, l_2]$. Further, let $P_1 + P_2$ have projective coordinates $[h, k, l]$, and $2P_1$ have projective coordinates $[h', k', l']$.

(a) Express h, k, l as polynomials in $h_1, k_1, l_1, h_2, k_2, l_2$.

(b) Express h', k', l' as polynomials in h_1, k_1, l_1.

(c) Show how the double-and-add point-multiplication algorithm (Exercise 4.26) can benefit from the representation of points in projective coordinates.

31. [*Mixed coordinates*][49] Take the elliptic curve $Y^2 = X^3 + aX + b$. Suppose that the point $P_1 = [h'_1, k'_1, l'_1]$ on the curve is available in projective coordinates, whereas the point $P_2 = (h_2, k_2)$ is available in affine coordinates. Express the projective coordinates of $P_1 + P_2$ as polynomial expressions in $h'_1, k'_1, l'_1, h_2, k_2$. What impact does this have on the point-multiplication algorithm?

---

[49] Henri Cohen, Atsuko Miyaji and Takatoshi Ono, Efficient elliptic curve exponentiation using mixed coordinates, *AsiaCrypt*, 51–65, 1998.

32. [*Generalized projective coordinates*] Let $c, d \in \mathbb{N}$ be constant. Define a relation $\sim$ on $K^3 \setminus \{(0,0,0)\}$ as $(h, k, l) \sim (h', k', l')$ if and only if $h = \lambda^c h'$, $k = \lambda^d k'$, and $l = \lambda l'$ for some non-zero $\lambda \in K$.
(a) Prove that $\sim$ is an equivalence relation.
Denote the equivalence class of (h, k, l) as $[h, k, l]_{c,d}$. (The standard projective coordinates correspond to $c = d = 1$.)
(b) Let $C : f(X, Y) = 0$ be an affine curve defined over K. Replace X by X/Z^c and Y by Y/Z^d in $f(X, Y)$, and clear the denominator by multiplying with the smallest power of Z, in order to obtain the polynomial equation $C^{(c,d)} : f^{(c,d)}(X, Y, Z) = 0$. Demonstrate how the finite points on C can be represented in the generalized projective coordinates $[h, k, l]_{c,d}$.
(c) How can one obtain the points at infinity on $C^{(c,d)}$?

33. [*Jacobian coordinates*] Take $c = 2$, $d = 3$ in Exercise 4.32. Let char $K \neq 2, 3$.
(a) Convert $E : Y^2 = X^3 + aX + b$ to the form $E^{(2,3)} : f(X, Y, Z) = 0$.
(b) What is the point at infinity on $E^{(2,3)}$?
(c) What is the opposite of a finite point on $E^{(2,3)}$?
(d) Write the point-addition formula for finite points on $E^{(2,3)}$.
(e) Write the point-doubling formula for finite points on $E^{(2,3)}$.

34. [*Jacobian-affine coordinates*] Let P_1, P_2 be finite points on $E : Y^2 = X^3 + aX + b$ with P_1 available in Jacobian coordinates, and P_2 in affine coordinates. Derive the Jacobian coordinates of $P_1 + P_2$. (Take char $K \neq 2, 3$.)

35. Repeat Exercise 4.33 for the ordinary curve $E : Y^2 + XY = X^3 + aX^2 + b$ defined over $\mathbb{F}_{2^n}$.

36. [*Chudnovsky coordinates*][50] Let char $K \neq 2, 3$. Take the elliptic curve $E : Y^2 = X^3 + aX + b$ over K. In addition to the usual Jacobian coordinates (X, Y, Z), the Chudnovsky coordinate system stores Z^2 and Z^3, that is, a point is represented as (X, Y, Z, Z^2, Z^3).
(a) Describe how point addition in E becomes slightly faster in the Chudnovsky coordinate system compared to the Jacobian coordinate system.
(b) Compare Chudnovsky and Jacobian coordinates for point doubling in E.

37. [*López–Dahab coordinates*][51] For the ordinary curve $E : Y^2 + XY = X^3 + aX^2 + b$ defined over $\mathbb{F}_{2^n}$, take $c = 1$ and $d = 2$.
(a) What is the point at infinity on $E^{(1,2)}$?
(b) What is the opposite of a finite point on $E^{(1,2)}$?
(c) Write the point-addition formula for finite points on $E^{(1,2)}$.
(d) Write the point-doubling formula for finite points on $E^{(1,2)}$.

38. [*LD-affine coordinates*] Take $K = \mathbb{F}_{2^n}$. Let P_1, P_2 be finite points on $E : Y^2 + XY = X^3 + aX^2 + b$ with P_1 available in López–Dahab coordinates, and P_2 in affine coordinates. Derive the López–Dahab coordinates of $P_1 + P_2$.

---

[50] David V. Chudnovsky and Gregory V. Chudnovsky, Sequences of numbers generated by addition in formal groups and new primality and factorization tests, *Advances in Applied Mathematics*, 7(4), 385–434, 1986.
[51] Julio López and Ricardo Dahab, Improved algorithms for elliptic curve arithmetic in $GF(2^n)$, Technical Report IC-98-39, Relatório Técnico, October 1998.

39. Prove that the norm function defined by Eqn (4.11) is multiplicative, that is, $N(G_1 G_2) = N(G_1) N(G_2)$ for all polynomial functions $G_1, G_2 \in K[C]$.

40. Consider the unit circle $C : X^2 + Y^2 = 1$ as a complex curve. Find all the zeros and poles of the rational function $R(x, y)$ of Example 4.35. Also determine the multiplicities of these zeros and poles. (**Hint:** Use the factored form given in Eqn (4.12). Argue that $1/x$ can be taken as a uniformizer at each of the two points at infinity on C.)

41. Consider the real hyperbola $H : X^2 - Y^2 = 1$. Find all the zeros and poles (and their respective multiplicities) of the following rational function on H:

$$R(x, y) = \frac{2y^4 - 2y^3 x - y^2 + 2yx - 1}{y^2 + yx + y + x + 1}.$$

(**Hint:** Split the numerator and the denominator of R into linear factors.)

42. Repeat Exercise 4.41 treating the hyperbola H as being defined over $\mathbb{F}_5$.

43. Find all the zeros and poles (and their multiplicities) of the rational function x/y on the curve $Y^2 = X^3 - X$ defined over $\mathbb{C}$.

44. Find all the zeros and poles (and their multiplicities) of the function $x^2 + yx$ on the curve $Y^2 = X^3 + X$ defined over $\mathbb{F}_3$.

45. Find all the zeros and poles (and their multiplicities) of the function $1 + yx$ on the curve $Y^2 = X^3 + X - 1$ defined over the algebraic closure $\bar{\mathbb{F}}_7$ of $\mathbb{F}_7$.

46. Prove that the q-th power Frobenius map φ_q (Definition 4.47) is an endomorphism of $E = E_{\bar{\mathbb{F}}_q}$.

47. Prove that an admissible change of variables (Theorem 4.51) does not change the j-invariant (Definition 4.9).

48. Let K be a field of characteristic $\neq 2, 3$. Prove that the elliptic curves $E :$ $Y^2 = X^3 + aX + b$ and $E' : Y^2 = X^3 + a'X + b'$ defined over K are isomorphic over $\bar{K}$ if and only if there exists a non-zero $u \in \bar{K}$ such that replacing X by $u^2 X$ and Y by $u^3 Y$ converts the equation for E to the equation for E'.

49. (a) Find all isomorphism classes of elliptic curves defined over $\mathbb{F}_5$, where isomorphism is over the algebraic closure $\bar{\mathbb{F}}_5$ of $\mathbb{F}_5$.
(b) Argue that the curves $Y^2 = X^3 + 1$ and $Y^2 = X^3 + 2$ are isomorphic over the algebraic closure $\bar{\mathbb{F}}_5$, but not over $\mathbb{F}_5$.
(c) According to Definition 4.50, isomorphism of elliptic curves E and E' is defined by the existence of bijective bilinear maps, not by the isomorphism of the groups E_K and E'_K (where K is a field over which both E and E' are defined). As an example, show that the curves $Y^2 = X^3 + 1$ and $Y^2 = X^3 + 2$ have isomorphic groups over $\mathbb{F}_5$.

50. Consider the elliptic curves $E : Y^2 = X^3 + 4X$ and $E' : Y^2 = X^3 + 4X + 1$ both defined over $\mathbb{F}_5$.
(a) Determine the group structures of $E_{\mathbb{F}_5}$ and $E'_{\mathbb{F}_5}$.
(b) Demonstrate that the rational map

$$\phi(x, y) = \left(\frac{x^2 - x + 2}{x - 1}, \frac{x^2 y - 2xy - y}{x^2 - 2x + 1} \right)$$

is an isogeny from $E_{\mathbb{F}_5}$ to $E'_{\mathbb{F}_5}$.

(c) Demonstrate that the rational map

$$\hat{\phi}(x, y) = \left(\frac{x^2 + 2x + 1}{x + 2}, \frac{x^2 y - xy - 2y}{x^2 - x - 1} \right)$$

is an isogeny from $E'_{\mathbb{F}_5}$ to $E_{\mathbb{F}_5}$.

(d) Verify that $\hat{\phi} \circ \phi$ is the multiplication-by-two map of $E_{\mathbb{F}_5}$, and $\phi \circ \hat{\phi}$ is the multiplication-by-two map of $E'_{\mathbb{F}_5}$.

(**Remark:** It can be proved that a non-trivial isogeny exists between two curves E and E' defined over $\mathbb{F}_q$ if and only if the groups $E_{\mathbb{F}_q}$ and $E'_{\mathbb{F}_q}$ have the same size. Moreover, for every isogeny $\phi : E \to E'$, there exists a unique isogeny $\hat{\phi} : E' \to E$ such that $\hat{\phi} \circ \phi$ is the multiplication-by-m map of $E_{\mathbb{F}_q}$, and $\phi \circ \hat{\phi}$ is the multiplication-by-m map of $E'_{\mathbb{F}_q}$, for some integer m. We call m the *degree* of the isogeny ϕ, and $\hat{\phi}$ the *dual isogeny* for ϕ. For example, the isogeny of Part (b) is of degree two and with dual given by the isogeny of Part (c). Another isogeny $E \to E'$ is defined by the rational map

$$\psi(x, y) = \left(\frac{x^4 - x^3 + x + 1}{x^3 - x^2 - x}, \frac{x^5 y + x^4 y + x^3 y - 2x^2 y + 2xy - 2y}{x^5 + x^4 + 2x^3 - 2x^2} \right).$$

The corresponding dual isogeny $\hat{\psi} : E' \to E$ is

$$\psi'(x, y) = \left(\frac{x^4 - x^3 + 2}{x^3 - x^2 + 2}, \frac{x^5 y + 2x^4 y - x^3 y - x^2 y - xy}{x^5 + 2x^4 - x^3 + x - 1} \right).$$

These isogenies are of degree four.

The curves $Y^2 = X^3 + 1$ and $Y^2 = X^3 + 2$ have isomorphic (and so equinumerous) groups over $\mathbb{F}_5$ (Exercise 4.49(d)), so there exists an isogeny between these curves. But no such isogeny can be bijective. If we allow rational maps over $\bar{\mathbb{F}}_5$, a bijective isogeny can be computed between these curves.)

51. Derive Eqns (4.16) and (4.17).

52. Prove that Weil pairing is bilinear, alternating and skew-symmetric.

53. Prove that Tate pairing is bilinear.

54. Derive Eqns (4.22), (4.23) and (4.24).

55. Rewrite Miller's algorithm using Eqn (4.26), for computing $e_m(P_1, P_2)$.

56. Rewrite Miller's algorithm using Eqn (4.29), for computing $\hat{e}_m(P, Q)$.

57. Write an algorithm that, for two input points P, Q on an elliptic curve E, returns the monic polynomial representing the line $L_{P,Q}$ through P and Q. Your algorithm should handle all possible inputs including $P = \pm Q$, and one or both of P, Q being $\mathcal{O}$.

58. Suppose that $Q = mQ'$ for some $Q' \in E_L$ (see the definition of Tate pairing). Prove that $\langle P, Q \rangle_m = 1$. This observation leads us to define Tate pairing on E_L / mE_L (in the second argument) instead of on E_L itself.

59. [Blake, Murty and Xu]   Let $U \in E[m]$ be non-zero, and $Q \neq \mathcal{O}, U, 2U, \ldots,$ $(m-1)U$. Prove the following assertions.

(a) $\dfrac{L_{U,U}(Q)}{L^2_{U,-U}(Q) L_{2U,-2U}(Q)} = -\dfrac{1}{L_{U,U}(-Q)}$.

(b) $\dfrac{L_{(k+1)U,kU}(Q)}{L_{(k+1)U,-(k+1)U}(Q) L_{(2k+1)U,-(2k+1)U}(Q)} = -\dfrac{L_{kU,-kU}(Q)}{L_{(k+1)U,kU}(-Q)}$ for $k \in \mathbb{Z}$.

(c) $\dfrac{L_{2U,U}(Q)}{L_{2U,-2U}(Q) L_{3U,-3U}(Q)} = -\dfrac{L_{U,-U}(Q)}{L_{2U,U}(-Q)}$.

60. Establish the correctness of Algorithm 4.4.

61. [Blake, Murty and Xu]   Prove that the loop body of Algorithm 4.1 for the computation of $f_{n,P}(Q)$ can be replaced as follows:

If $(n_i = 0)$, then update $f = -f^2 \times \left(\dfrac{L_{2U,-2U}(Q)}{L_{U,U}(-Q)} \right)$ and $U = 2U$,

else update $f = -f^2 \times \left(\dfrac{L_{2U,P}(Q)}{L_{U,U}(-Q)} \right)$ and $U = 2U + P$.

Explain what speedup this modification is expected to produce.

62. Prove that for all $n, n' \in \mathbb{N}$, the functions $f_{n,P}$ in Miller's algorithm satisfy

$$f_{nn',P} = f_{n,P}^{n'} f_{n',nP} = f_{n',P}^{n} f_{n,n'P}.$$

63. Define the functions $f_{n,P,S}$ as rational functions having the divisor

$$\mathrm{Div}(f_{n,P,S}) = n[P+S] - n[S] - [nP] + [\mathcal{O}].$$

Suppose that $mP = \mathcal{O}$, so that $\mathrm{Div}(f_{m,P,S}) = m[P+S] - m[S]$. For Weil and Tate pairing, we are interested in $S = \mathcal{O}$ only. Nonetheless, it is interesting to study the general functions $f_{n,P,S}$. Prove that:

(a) $f_{0,P,S} = 1$, and $f_{1,P,S} = \dfrac{L_{P+S,-(P+S)}}{L_{P,S}}$.

(b) For $n \in \mathbb{N}$, we have $f_{n+1,P,S} = f_{n,P,S} \times f_{1,P,S} \times \dfrac{L_{nP,P}}{L_{(n+1)P,-(n+1)P}}$.

(c) For $n, n' \in \mathbb{N}$, we have $f_{n+n',P,S} = f_{n,P,S} \times f_{n',P,S} \times \dfrac{L_{nP,n'P}}{L_{(n+n')P,-(n+n')P}}$.

64. Rewrite Miller's Algorithm 4.1 to compute the functions $f_{n,P,S}$.

65. Assume that Weil/Tate pairing is restricted to suitable groups of order m. Prove that Weil/Tate pairing under the distortion map is symmetric about its two arguments. Does symmetry hold for twisted pairings too?

66. Deduce that the embedding degrees of the following supersingular curves are as mentioned. Recall that for a supersingular curve, the embedding degree must be one of $1, 2, 3, 4, 6$. Explicit examples for all these cases are given here. In each case, take m to be a suitably large prime divisor of the size of the elliptic-curve group E_q (where E is defined over $\mathbb{F}_q$).
(a) The curve $Y^2 = X^3 + a$ defined over $\mathbb{F}_p$ for an odd prime $p \equiv 2 \pmod 3$ and with $a \not\equiv 0 \pmod p$ has embedding degree two (see Exercise 4.14).

(b) The curve $Y^2 = X^3 + aX$ defined over $\mathbb{F}_p$ for a prime $p \equiv 3 \pmod 4$ and with $a \not\equiv 0 \pmod p$ has embedding degree two (Exercise 4.15).

(c) The curve $Y^2 + Y = X^3 + X + a$ with $a = 0$ or 1, defined over $\mathbb{F}_{2^n}$ with odd n, has embedding degree four (Exercise 4.22).

(d) The curve $Y^2 = X^3 - X + a$ with $a = \pm 1$, defined over $\mathbb{F}_{3^n}$ with n divisible by neither 2 nor 3, has embedding degree six (Exercise 4.24).

(e) Let $p \equiv 5 \pmod 6$ be a prime. Let $a \in \mathbb{F}_{p^2}$ be a square but not a cube. It is known that the curve $Y^2 = X^3 + a$ defined over $\mathbb{F}_{p^2}$ contains exactly $p^2 - p + 1$ points. This curve has embedding degree three.

(f) Let E be a supersingular curve defined over a prime field $\mathbb{F}_p$ with $p \geqslant 5$. E considered as a curve over $\mathbb{F}_{p^n}$ with even n is supersingular (Exercise 4.25), and has embedding degree one.

67. Prove that the following distortion maps are group homomorphisms. (In this exercise, k is not used to denote the embedding degree.)

(a) For the curve $Y^2 = X^3 + a$ defined over $\mathbb{F}_p$ for an odd prime $p \equiv 2 \pmod 3$ and with $a \not\equiv 0 \pmod p$, the map $(h, k) \mapsto (\theta h, k)$, where $\theta^3 = 1$.

(b) For the curve $Y^2 = X^3 + aX$ defined over $\mathbb{F}_p$ for a prime $p \equiv 3 \pmod 4$ and with $a \not\equiv 0 \pmod p$, the map $(h, k) \mapsto (-h, \theta k)$, where $\theta^2 = -1$.

(c) For the curve $Y^2 + Y = X^3 + X + a$ with $a = 0$ or 1, defined over $\mathbb{F}_{2^n}$ with odd n, the map $(h, k) \mapsto (\theta h + \zeta^2, k + \theta \zeta h + \zeta)$, where $\theta \in \mathbb{F}_{2^2}$ satisfies $\theta^2 + \theta + 1 = 0$, and $\zeta \in \mathbb{F}_{2^4}$ satisfies $\zeta^2 + \theta \zeta + 1 = 0$.

(d) For the curve $Y^2 = X^3 - X + a$ with $a = \pm 1$, defined over $\mathbb{F}_{3^n}$ with n divisible by neither 2 nor 3, the map $(h, k) \mapsto (\zeta - h, \theta k)$ where $\theta \in \mathbb{F}_{3^2}$ satisfies $\theta^2 = -1$, and $\zeta \in \mathbb{F}_{3^3}$ satisfies $\zeta^3 - \zeta - a = 0$.

(e) Let $p \equiv 5 \pmod 6$ be a prime, $a \in \mathbb{F}_{p^2}$ a square but not a cube, and let $\gamma \in \mathbb{F}_{p^6}$ satisfy $\gamma^3 = a$. For the curve $Y^2 = X^3 + a$ defined over $\mathbb{F}_{p^2}$, the distortion map is $(h, k) \mapsto (h^p/(\gamma a^{(p-2)/3}), k^p/a^{(p-1)/2})$.

68. Let $E : Y^2 = X^3 + aX + b$ be an elliptic curve defined over a field $\mathbb{F}_q$ of odd characteristic $\geqslant 5$. A quadratic twist of E is defined as $E' : Y^2 = X^3 + v^2 aX + v^3 b$, where $v \in \mathbb{F}_q^*$ is a quadratic non-residue.

(a) Show that the j-invariant of E is $j(E) = 1728 \left(\frac{4a^3}{4a^3 + 27b^2} \right)$.

(b) Conclude that $j(E) = j(E')$. (Thus, E and E' are isomorphic over $\bar{\mathbb{F}}_q$.)

(c) Prove that $|E_q| + |E'_q| = 2(q + 1)$.

69. Let $E : Y^2 = X^3 + aX + b$ be an elliptic curve defined over $\mathbb{F}_p$ with $p \geqslant 5$ a prime. Prove that the trace of Frobenius at p is $- \sum_{x=0}^{p-1} \left(\frac{x^3 + ax + b}{p} \right)$, where $\left(\frac{c}{p} \right)$ is the Legendre symbol.

70. *Edwards curves* were proposed by Harold M. Edwards,[52] and later modified to suit elliptic-curve cryptography by Bernstein and Lange.[53] For finite fields K with char $K \neq 2$, an elliptic curve defined over K is equivalent to an Edwards curve over a suitable extension of K (the extension may be K itself). A unified addition formula (no distinction between addition and squaring, and a uniform treatment of all group elements including the identity) makes Edwards curves attractive and efficient alternatives to elliptic curves. An Edwards curve over a non-binary finite field K is defined by the equation

$$D : X^2 + Y^2 = c^2(1 + dX^2Y^2) \text{ with } 0 \neq c, d \in K \text{ and } dc^4 \neq 1.$$

Suppose also that d is a quadratic non-residue in K. Define an operation on two finite points $P_1 = (h_1, k_1)$ and $P_2 = (h_2, k_2)$ on D as

$$P_1 + P_2 = \left(\frac{h_1 k_2 + k_1 h_2}{c(1 + dh_1 h_2 k_1 k_2)}, \frac{k_1 k_2 - h_1 h_2}{c(1 - dh_1 h_2 k_1 k_2)} \right).$$

Let P, P_1, P_2, P_3 be arbitrary finite points on D. Prove that:
(a) $P_1 + P_2$ is again a *finite* point on D.
(b) $P_1 + P_2 = P_2 + P_1$.
(c) $P + \mathcal{O} = \mathcal{O} + P = P$, where $\mathcal{O}$ is the *finite* point $(0, c)$.
(d) If $P = (h, k)$ on D, we have $P + Q = \mathcal{O}$, where $Q = (-h, k)$.

The *finite* points on D constitute an additive Abelian group. Directly proving the associativity of this addition is very painful.

Now, let $e = 1 - dc^4$, and define the elliptic curve

$$E : \frac{1}{e}Y^2 = X^3 + \left(\frac{4}{e} - 2 \right) X^2 + X.$$

For a finite point $P = (h, k)$ on D, define

$$\phi(P) = \begin{cases} \mathcal{O} & \text{if } P = (0, c), \\ (0, 0) & \text{if } P = (0, -c), \\ \left(\dfrac{c+k}{c-k}, \dfrac{2c(c+k)}{(c-k)h} \right) & \text{if } h \neq 0. \end{cases}$$

(e) Prove that ϕ maps points on D to points on E.

It turns out that for any two points P, Q on D, we have $\phi(P + Q) = \phi(P) + \phi(Q)$, where the additions on the two sides correspond to the curves D and E, respectively. This correspondence can be proved with involved calculations.

---

[52] Harold M. Edwards, A normal form for elliptic curves, *Bulletin of American Mathematical Society*, 44, 393–422, 2007.

[53] Daniel J. Bernstein and Tanja Lange, A complete set of addition laws for incomplete Edwards curves, *Journal of Number Theory*, 131, 858–872, 2011.

Programming Exercises

71. Let p be a small prime. Write a GP/PARI program that, given an elliptic curve E over $\mathbb{F}_p$, finds all the points on $E_p = E_{\mathbb{F}_p}$, calculates the size of E_p, computes the order of each point in E_p, and determines the group structure of E_p.

72. Repeat Exercise 4.71 for elliptic curves over binary fields $\mathbb{F}_{2^n}$ for small n.

73. Write a GP/PARI program that, given a small prime p, an elliptic curve E defined over $\mathbb{F}_p$, and an $n \in \mathbb{N}$, outputs the size of the group $E_{p^n} = E_{\mathbb{F}_{p^n}}$.

74. Write a GP/PARI function that, given an elliptic curve over a finite field (not necessarily small), returns a random point on the curve.

75. Write a GP/PARI function that, given points U, V, Q on a curve, computes the equation of the line passing through U and V, and returns the value of the function at Q. Assume Q to be a finite point, but handle all cases for U, V.

76. Implement the reduced Tate pairing using the function of Exercise 4.75. Consider a supersingular curve $Y^2 = X^3 + aX$ over a prime field $\mathbb{F}_p$ with $p \equiv 3 \pmod 4$ and with $m = (p+1)/4$ a prime.

77. Implement the distorted Tate pairing on the curve of Exercise 4.76.

Chapter 5

Primality Testing

An integer $p > 1$ is called *prime* if its only positive integral divisors are 1 and p. Equivalently, p is prime if and only if $p|ab$ implies $p|a$ or $p|b$. An integer $n > 1$ is called *composite* if it is not prime, that is, if n has an integral divisor u with $1 < u < n$. The integer 1 is treated as neither prime nor composite.

One can extend the notion of primality to the set of all integers. The additive identity 0 and the multiplicative units ± 1 are neither prime nor composite. A non-zero non-unit $p \in \mathbb{Z}$ is called prime if a factorization $p = uv$ necessarily implies that either u or v is a unit. Thus, we now have the *negative primes* $-2, -3, -5, \ldots$. In this book, an unqualified use of the term *prime* indicates positive primes. The set of all (positive) primes is denoted by $\mathbb{P}$.

$\mathbb{P}$ is an infinite set (Theorem 1.69). Given $n \in \mathbb{N}$, there exists a prime (in fact, infinitely many primes) larger than n. The asymptotic density of primes (the prime number theorem) and related results are discussed in Section 1.9.

Theorem 5.1 [*Fundamental theorem of arithmetic*][1] *Any non-zero integer n has a factorization of the form $n = up_1 p_2 \cdots p_t$, where $u = \pm 1$, and $p_1, p_2, \ldots, p_t$ are (positive) primes for some $t \geqslant 0$. Moreover, such a factorization is unique up to rearrangement of the prime factors $p_1, p_2, \ldots, p_t$.* ◁

---

[1]Euclid seems to have been the first to provide a complete proof of this theorem.

Considering the possibility of repeated prime factors, one can rewrite the factorization of n as $n = u q_1^{e_1} q_2^{e_2} \cdots q_r^{e_r}$, where $q_1, q_2, \ldots, q_r$ are pairwise distinct primes, and $e_i \in \mathbb{N}$ is the *multiplicity* of q_i in n, denoted $e_i = \mathrm{v}_{q_i}(n)$.[2]

Problem 5.2 [*Fundamental problem of computational number theory*] *Given a non-zero (usually positive) integer n, compute the decomposition of n into prime factors, that is, compute all the prime divisors p of n together with their respective multiplicities* $\mathrm{v}_p(n)$. ◁

Problem 5.2 is also referred to as the *integer factorization problem* or as *IFP* in short. Solving this demands ability to recognize primes as primes.

Problem 5.3 [*Primality testing*]   *Given a positive integer $n \geqslant 2$, determine whether n is prime or composite.* ◁

The primality testing problem has efficient probabilistic algorithms. The deterministic complexity of primality testing too is polynomial-time. On the contrary, factoring integers appears to be a difficult and challenging computational problem. In this chapter, we discuss algorithms for testing the primality of integers. Integer factorization is studied in Chapter 6.

5.1   Introduction to Primality Testing

The straightforward algorithm to prove whether $n \geqslant 2$ is prime is by *trial division* of n by integers d in the range $2 \leqslant d < n$. We declare n as a prime if and only if no non-trivial divisor d of n can be located. Evidently, n is composite if and only if it admits a factor $d \leqslant \sqrt{n}$. This implies that we can restrict the search for potential divisors d of n to the range $2 \leqslant d \leqslant \lfloor \sqrt{n} \rfloor$.

Trial division is a fully exponential-time algorithm (in the size $\log n$ of n) for primality testing. No straightforward modification of trial division tends to reduce this complexity to something faster than exponential. This is why we do not describe this method further in connection with primality testing. (If integer factorization is of concern, trial division plays an important role.)

5.1.1   Pratt Certificates

The first significant result about the complexity of primality testing comes from Pratt[3] who proves the existence of polynomial-time verifiable certificates for primality. This establishes that primality testing is a problem in the complexity class NP. Pratt certificates are based on the following result.

---

[2]Some authors prefer to state the fundamental theorem only for integers $n \geqslant 2$. But the case $n = 1$ is not too problematic, since 1 factors uniquely into *the empty product* of primes.

[3]Vaughan Pratt, Every prime has a succinct certificate, *SIAM Journal on Computing*, 4, 214–220, 1975.

Theorem 5.4 *A positive integer n is prime if and only if there exists an element $a \in \mathbb{Z}_n^*$ with $\mathrm{ord}_n\, a = n - 1$, that is, with $a^{n-1} \equiv 1 \pmod{n}$, and $a^{(n-1)/p} \not\equiv 1 \pmod{n}$ for every prime divisor p of $n-1$.* ◁

Therefore, the prime factorization of $n-1$ together with an element a of order $n - 1$ seem to furnish a primality certificate for n. However, there is a subtle catch here. We need to certify the primality of the prime divisors of $n - 1$.

Example 5.5 Here is a false certificate about the primality of $n = 17343$. We claim the *prime factorization $n - 1 = 2 \times 8671$* and supply the *primitive root $a \equiv 163 \pmod{n}$*. One verifies that $a^{n-1} \equiv 1 \pmod{n}$, $a^{(n-1)/2} \equiv 3853 \pmod{n}$, and $a^{(n-1)/8671} \equiv 9226 \pmod{n}$, thereby wrongly concluding that $n = 17343$ is prime. However, $n = 3^2 \times 41 \times 47$ is composite. The problem with this certificate is that $8671 = 13 \times 23 \times 29$ is not prime. It is, therefore, necessary to certify every prime divisor of $n - 1$ as prime. □

That looks like a circular requirement. A primality certificate requires other primality certificates which, in turn, require some more primality certificates, and so on. But each prime divisor of $n-1$ is smaller than n, so this inductive (not circular) process stops after finitely many steps. Pratt proves that the total size of a complete certificate for the primality of n is only $O(\log^2 n)$ bits.

Example 5.6 Let us provide a complete primality certificate for $n = 1237$.

(1)	$1237 - 1 = 2^2 \times 3 \times 103$ with primitive root 2
(2)	$2 - 1 = 1$ with primitive root 1
(3)	$3 - 1 = 2$ with primitive root 2
(4)	$2 - 1 = 1$ with primitive root 1
(5)	$103 - 1 = 2 \times 3 \times 17$ with primitive root 5
(6)	$2 - 1 = 1$ with primitive root 1
(7)	$3 - 1 = 2$ with primitive root 2
(8)	$2 - 1 = 1$ with primitive root 1
(9)	$17 - 1 = 2^4$ with primitive root 3
(10)	$2 - 1 = 1$ with primitive root 1

We now verify this certificate. Line (1) supplies the factorization $1237-1 = 2^2 \times 3 \times 103$. Moreover, $2^{1237-1} \equiv 1 \pmod{1237}$, $2^{(1237-1)/2} \equiv 1236 \pmod{1237}$, $2^{(1237-1)/3} \equiv 300 \pmod{1237}$, and $2^{(1237-1)/103} \equiv 385 \pmod{1237}$.

We recursively prove the primality of the divisors 2 (Line (2)), 3 (Lines (3)–(4)) and 103 (Lines (5)–(10)) of $1237-1$. For example, $103-1 = 2 \times 3 \times 17$ with $5^{103-1} \equiv 1 \pmod{103}$, $5^{(103-1)/2} \equiv 102 \pmod{103}$, $5^{(103-1)/3} \equiv 56 \pmod{103}$, and $5^{(103-1)/17} \equiv 72 \pmod{103}$. Thus, 103 is a prime, since 2, 3 and 17 are established as primes in Lines (6), (7)–(8) and (9)–(10), respectively. □

One can verify the Pratt certificate for $n \in \mathbb{P}$ in $O(\log^5 n)$ time. Moreover, a composite number n can be easily *disproved* to be a prime if a non-trivial divisor of n is supplied. Therefore, primality testing is a problem in the class NP ∩ coNP. It is, thus, expected that primality testing is not NP-Complete, since if so, we have P = NP, a fact widely believed to be false.

5.1.2   Complexity of Primality Testing

Is primality testing solvable in polynomial time? This question remained unanswered for quite a long time. Several primality-testing algorithms were devised, that can be classified in two broad categories: deterministic and probabilistic (or randomized). Some deterministic tests happen to have running time very close to polynomial (like $(\log n)^{\log \log \log n}$). For other deterministic tests, rigorous proofs of polynomial worst-case running times were missing.

The probabilistic tests, on the other hand, run in provably polynomial time, but may yield incorrect answers on some occasions. The probability of such incorrect answers can be made arbitrarily small by increasing the number of rounds in the algorithms. Nonetheless, such algorithms cannot be accepted as definite proofs for primality. Under the assumption of certain unproven mathematical facts (like the ERH), one can convert some of these probabilistic algorithms to polynomial-time deterministic algorithms. However, so long as the underlying mathematical conjectures remain unproven, these converted algorithms lack solid theoretical basis.

In August 2002, three Indians (Agarwal, Kayal and Saxena) propose an algorithm that settles the question. Their *AKS primality test* meets three important requirements. First, it is a deterministic algorithm. Second, it has a polynomial worst-case running time. And third, its proof of correctness or complexity does not rely upon any unproven mathematical assumption. Soon after its conception, several improvements of the AKS test were proposed.

These new developments in the area of primality testing, although theoretically deep, do not seem to have significant practical impacts. In practical applications like cryptography, significantly faster randomized algorithms suffice. The failure probability of these randomized algorithms can be made smaller than the probability of hardware failure. Thus, a deterministic primality test is expected to give incorrect answers because of hardware failures more often than a probabilistic algorithm furnishing wrong outputs.

5.1.3   Sieve of Eratosthenes

The Greek mathematician Eratosthenes of Cyrene (circa 276–195 BC) proposed possibly the earliest known sieve algorithm to locate all primes up to a specified positive integer n. We can use this algorithm to quickly generate a set of small primes (like the first million to hundred million primes).

One starts by writing the integers $2, 3, 4, 5, \ldots, n$ in a list. One then enters a loop, each iteration of which discovers a new prime and marks all multiples of that prime (other than the prime itself) as composite.

Initially, no integer in the list is marked prime or composite. The first unmarked integer is 2, which is marked as prime. All even integers in the list larger than 2 are marked as composite. In the second pass, the first unmarked integer 3 is marked as prime. All multiples of 3 (other than 3) are marked as composite. In the third pass, the first unmarked integer 5 is marked as prime, and multiples of 5 (other than itself) are marked composite. This process is repeated until the first unmarked entry exceeds $\sqrt{n}$. All entries in the list, that are marked as prime or that remain unmarked are output as primes.

Example 5.7 The sieve of Eratosthenes finds all primes $\leqslant n = 50$ as follows. Integers discovered as composite are underlined, and appear in bold face in the pass in which they are first marked. The discovered primes are boxed.

Pass 1

[2]	3	**4**	5	**6**	7	**8**	9	**10**	11	**12**	13	**14**	15	**16**	17	
18	19	**20**	21	**22**	23	**24**	25	**26**	27	**28**	29	**30**	31	**32**	33	**34**
35	**36**	37	**38**	39	**40**	41	**42**	43	**44**	45	**46**	47	**48**	49	**50**	

Pass 2

[2]	[3]	4	5	6	7	8	**9**	10	11	12	13	14	**15**	16	17	
18	19	20	**21**	22	23	24	25	26	**27**	28	29	30	31	32	**33**	34
35	36	37	38	**39**	40	41	42	43	44	**45**	46	47	48	49	50	

Pass 3

[2]	[3]	4	[5]	6	7	8	9	10	11	12	13	14	15	16	17	
18	19	20	21	22	23	24	**25**	26	27	28	29	30	31	32	33	34
35	36	37	38	39	40	41	42	43	44	45	46	47	48	49	50	

Pass 4

[2]	[3]	4	[5]	6	[7]	8	9	10	11	12	13	14	15	16	17	
18	19	20	21	22	23	24	25	26	27	28	29	30	31	32	33	34
35	36	37	38	39	40	41	42	43	44	45	46	47	48	**49**	50	

The passes stop here, since the first unmarked entry 11 is larger than $\sqrt{50}$. All primes $\leqslant 50$ are $2, 3, 5, 7, 11, 13, 17, 19, 23, 29, 31, 37, 41, 43, 47$. ☐

5.1.4 Generating Random Primes

In many applications, random primes of given bit sizes are required. One may keep on checking the primality of randomly generated integers of a given size by one or more probabilistic or deterministic tests. After $O(s)$ iterations, one expects to get a random prime of the given bit size, since by the prime number theorem, the fraction of primes among integers of bit size s is $\Theta(1/s)$.

An adaptation of the sieve of Eratosthenes reduces the running time of this search significantly. We start with a random s-bit integer n_0. We concentrate on the integers $n_i = n_0 + i$ in an interval $[n_0, n_{l-1}]$ of length $l = \Theta(s)$. It is of no use to subject the even integers in this interval to a primality test. Likewise, we can discard multiples of 3, the multiples of 5, and so on.

We choose a number t of small primes (typically, the first t primes). We do not perform trial division of all the integers n_i with all these small primes. We instead use an array of size l with each cell initialized to 1. The i-th cell corresponds to the integer n_i for $i = 0, 1, 2, \ldots, l - 1$. For each small prime p, we compute the remainder $r = n_0 \operatorname{rem} p$ which identifies the first multiple of p

in $[n_0, n_{l-1}]$. After this position, every p-th integer in the interval is a multiple of p. We set to zero the array entries at all these positions.

After all the small primes are considered, we look at those array indices i that continue to hold the value 1. These correspond to all those integers n_i in $[n_0, n_{l-1}]$ that are divisible by neither of the small primes. We subject only these integers n_i to one or more primality test(s). This method is an example of *sieving*. Many composite integers are sieved out (eliminated) much more easily than running primality tests individually on all of them. For each small prime p, the predominant cost is that of a division (computation of $n_0 \operatorname{rem} p$). Each other multiple of p is located easily (by adding p to the previous multiple of p). In practice, one may work with 10 to 1000 small primes.

If the length l of the sieving interval is carefully chosen, we expect to locate a prime among the non-multiples of small primes. However, if we are unlucky enough to encounter only composite numbers in the interval, we repeat the process for another random value of n_0. It is necessary to repeat the process also in the case that n_0 is of bit length s, whereas a discovered prime $n_i = n_0 + i$ is of bit length larger than s (to be precise, $s + 1$ for all sufficiently large s). If n_0 is chosen as a random s-bit integer, and if s is not too small, the probability of such an *overflow* is negligibly small.

5.1.5   Handling Primes in the GP/PARI Calculator

Upon start-up, the GP/PARI calculator loads a precalculated list of the first t primes. One can obtain the i-th prime by the command prime(i) for $i = 1, 2, 3, \ldots, t$. The call primes(i) returns a vector of the first i primes. If $i > t$, the calculator issues a warning message. One can add and remove primes to this precalculated list by the directives addprimes and removeprimes.

The primality testing function of GP/PARI is isprime. An optional flag may be supplied to this function in order to indicate the algorithm to use. Consult the GP/PARI manual for more details.

The call nextprime(x) returns the smallest prime $\geqslant x$, whereas the call precprime(x) returns the largest prime $\leqslant x$. Here, x is allowed to be a real number. For $x < 2$, the call precprime(x) returns 0.

A random integer a in the range $0 \leqslant a < n$ can be obtained by the call random(n). Therefore, a random prime of bit length $\leqslant l$ can be obtained by the call precprime(random(2^l)).

```
gp > prime(1000)
%1 = 7919
gp > prime(100000)
  ***    not enough precalculated primes
gp > primes(16)
%2 = [2, 3, 5, 7, 11, 13, 17, 19, 23, 29, 31, 37, 41, 43, 47, 53]
gp > isprime(2^100+277)
%3 = 1
gp > nextprime(2^100)
```

```
%4 = 1267650600228229401496703205653
gp > nextprime(2^100)-2^100
%5 = 277
gp > precprime(2^200)
%6 = 1606938044258990275541962092341162602522202993782792835301301
gp > 2^200-precprime(2^200)
%7 = 75
gp > precprime(random(2^10))
%8 = 811
```

5.2 Probabilistic Primality Testing

In this section, we discuss some probabilistic polynomial-time primality-testing algorithms. These are *No-biased* Monte Carlo algorithms. This means that when the algorithms output *No*, the answer is correct, whereas the output *Yes* comes with some chances of error. In order that these algorithms be useful, the error probabilities should be low. A *Yes-biased* Monte Carlo algorithm produces the answer *Yes* with certainty and the answer *No* with some probability of error. A Yes-biased Monte Carlo algorithm for checking the primality of integers of a particular kind is explored in Exercise 5.21.

5.2.1 Fermat Test

Let n be a prime. By Fermat's little theorem, $a^{n-1} \equiv 1 \pmod{n}$ for every $a \in \mathbb{Z}_n^*$. The converse of this is not true, that is, $a^{n-1} \equiv 1 \pmod{n}$ for some (or many) $a \in \mathbb{Z}_n^*$ does not immediately imply that n is a prime. For example, consider $n = 17343$ and $a = 163$ (see Example 5.5). Nonetheless, this motivates us to define the following concept which leads to Algorithm 5.1.

Definition 5.8 Let $n \in \mathbb{N}$ and $a \in \mathbb{Z}$ with $\gcd(a, n) = 1$. We call n a *pseudoprime* (or a *Fermat pseudoprime*) to the base a if $a^{n-1} \equiv 1 \pmod{n}$.   ◁

Algorithm 5.1: FERMAT TEST FOR TESTING THE PRIMALITY OF n

Fix the number t of iterations.
For $i = 1, 2, \ldots, t$ {
    Choose a random integer $a \in \{2, 3, \ldots, n-1\}$.
    If $(a^{n-1} \not\equiv 1 \pmod{n})$ return "*n is composite*".
}
return "*n is prime*".

A prime n is a pseudoprime to every base a coprime to n. A composite integer may also be a pseudoprime to some bases. If a composite integer n is not a pseudoprime to at least one base $a \in \mathbb{Z}_n^*$, then n is not a pseudoprime to at least half of the bases in $\mathbb{Z}_n^*$.

A base, to which n is not a pseudoprime, is a *witness* to the compositeness of n. If any such witness is found, n is declared as composite. On the other hand, if we do not encounter a witness in t iterations, we declare n to be a prime. If n is indeed prime, then no witnesses for n can be found, and the algorithm correctly declares n as prime. However, if n is composite, and we fail to encounter a witness in t iterations, the decision of the algorithm is incorrect. In this case, suppose that there exists a witness for the compositeness of n. Then, the probability that n is still declared as prime is $1/2^t$. By choosing t appropriately, one can reduce this error probability to a very low value.

Example 5.9 (1) The composite integer $891 = 3^4 \times 11$ is pseudoprime to only the 20 bases 1, 80, 82, 161, 163, 244, 323, 325, 404, 406, 485, 487, 566, 568, 647, 728, 730, 809, 811 and 890. We have $\phi(891) = 540$, that is, the fraction of non-witnesses in $\mathbb{Z}_{891}^*$ is $1/27$.

(2) The composite integer $2891 = 7^2 \times 59$ is pseudoprime only to the four bases 1, 589, 2302, 2890.

(3) The composite integer $1891 = 31 \times 61$ is pseudoprime to 900 bases, that is, to exactly half of the bases in $\mathbb{Z}_{1891}^*$ (we have $\phi(1891) = 30 \times 60 = 1800$).

(4) There are only 22 composite integers $\leqslant 10,000$ that are pseudoprimes to the base 2. These composite integers are 341, 561, 645, 1105, 1387, 1729, 1905, 2047, 2465, 2701, 2821, 3277, 4033, 4369, 4371, 4681, 5461, 6601, 7957, 8321, 8481 and 8911. There are 78 composite pseudoprimes $\leqslant 100,000$ and 245 composite pseudoprimes $\leqslant 1,000,000$ to the base 2. $\square$

There exist composite integers n which are pseudoprimes to *every* base coprime to n. Such composite integers are called *Carmichael numbers*.[4] A Carmichael number n passes the primality test irrespective of how many bases in $\mathbb{Z}_n^*$ are chosen. Here follows a characterization of Carmichael numbers.

Theorem 5.10 *A positive composite integer n is a Carmichael number if and only if n is square-free, and $(p-1)|(n-1)$ for every prime divisor p of n.* ◁

Example 5.11 The smallest Carmichael number is $561 = 3 \times 11 \times 17$. We have $561 - 1 = 280 \times (3-1) = 56 \times (11-1) = 35 \times (17-1)$. All other Carmichael numbers $\leqslant 100,000$ are:

$$
\begin{aligned}
1105 &= 5 \times 13 \times 17, \\
1729 &= 7 \times 13 \times 19, \\
2465 &= 5 \times 17 \times 29,
\end{aligned}
$$

---

[4]These are named after the American mathematician Robert Daniel Carmichael (1879–1967). The German mathematician Alwin Reinhold Korselt (1864–1947) first introduced the concept of Carmichael numbers. Carmichael was the first to discover (in 1910) concrete examples (like 561).

$$2821 = 7 \times 13 \times 31,$$
$$6601 = 7 \times 23 \times 41,$$
$$8911 = 7 \times 19 \times 67,$$
$$10585 = 5 \times 29 \times 73,$$
$$15841 = 7 \times 31 \times 73,$$
$$29341 = 13 \times 37 \times 61,$$
$$41041 = 7 \times 11 \times 13 \times 41,$$
$$46657 = 13 \times 37 \times 97,$$
$$52633 = 7 \times 73 \times 103,$$
$$62745 = 3 \times 5 \times 47 \times 89,$$
$$63973 = 7 \times 13 \times 19 \times 37, \text{ and}$$
$$75361 = 11 \times 13 \times 17 \times 31.$$

The smallest Carmichael numbers with five and six prime factors are $825265 = 5 \times 7 \times 17 \times 19 \times 73$ and $321197185 = 5 \times 19 \times 23 \times 29 \times 37 \times 137$. □

One can show that a Carmichael number must be odd with at least three distinct prime factors. Alford et al.[5] prove that there exist infinitely many Carmichael numbers. That is bad news for the Fermat test. We need to look at its modifications so as to avoid the danger posed by Carmichael numbers.

Here is how the Fermat test can be implemented in GP/PARI. In the function Fpsp(n,t), the first parameter n is the integer whose primality is to be checked, and t is the count of random bases to try.

```
gp > Fpsp(n,t) = \
    for (i=1, t, \
        a = Mod(random(n),n); b = a^(n-1); \
        if (b != Mod(1,n), return(0)) \
    ); \
    return(1);
gp > Fpsp(1001,20)
%1 = 0
gp > Fpsp(1009,20)
%2 = 1
gp > p1 = 601; p2 = 1201; p3 = 1801; n = p1 * p2 * p3
%3 = 1299963601
gp > Fpsp(n,20)
%4 = 1
```

In practical implementations, one may choose single-precision integers (like the first t small primes) as bases a. The error probability does not seem to be affected by small bases, but the modular exponentiation $a^{n-1} \pmod{n}$ becomes somewhat more efficient than in the case of multiple-precision bases.

---

[5] W. R. Alford, Andrew Granville and Carl Pomerance, There are infinitely many Carmichael numbers, *Annals of Mathematics*, 140, 703–722, 1994.

5.2.2 Solovay–Strassen Test

By Euler's criterion, $a^{(n-1)/2} \equiv \left(\frac{a}{n}\right) \pmod{n}$ for every odd prime n and for every base a coprime to n. The converse is again not true, but a probabilistic primality test[6] (Algorithm 5.2) can be based on the following definition.

Definition 5.12 An odd positive integer n is called an *Euler pseudoprime* or a *Solovay–Strassen pseudoprime* to a base a coprime to n if $a^{(n-1)/2} \equiv \left(\frac{a}{n}\right) \pmod{n}$, where $\left(\frac{a}{n}\right)$ is the Jacobi symbol. ◁

Every Euler pseudoprime is a Fermat pseudoprime, but not conversely.

Algorithm 5.2: SOLOVAY–STRASSEN PRIMALITY TEST

```
Fix the number t of iterations.
For i = 1, 2, ..., t {
    Choose a random integer a ∈ {2, 3, ..., n − 1}.
    If ((a/n) = 0), return "n is composite". /* gcd(a, n) > 1 */
    If (a^(n−1)/2 ≢ (a/n) (mod n)), return "n is composite".
}
Return "n is prime".
```

For an odd composite integer n, a base $a \in \mathbb{Z}_n^*$ satisfying $a^{(n-1)/2} \not\equiv \left(\frac{a}{n}\right) \pmod{n}$ is a witness to the compositeness of n. If no such witness is found in t iterations, the number n is declared as composite. Of course, primes do not possess such witnesses. However, a good property of the Solovay–Strassen test is that an odd composite n (even a Carmichael number) has at least $\phi(n)/2$ witnesses to the compositeness of n. Thus, the probability of erroneously declaring a composite n as prime is no more that $1/2^t$.

Example 5.13 (1) The composite integer $891 = 3^4 \times 11$ is a Fermat pseudoprime to 20 bases (Example 5.9(1)), but an Euler pseudoprime to only the following ten of these bases: 1, 82, 161, 163, 404, 487, 728, 730, 809 and 890.

(2) $2891 = 7^2 \times 59$ is an Euler pseudoprime only to the bases 1 and 2890.

(3) $1891 = 31 \times 61$ is an Euler pseudoprime to 450 bases only.

(4) Let $n_F(n)$ denote the count of bases to which n is a Fermat pseudoprime, and $n_E(n)$ the count of bases to which n is an Euler pseudoprime. Clearly, $n_F(n) \geqslant n_E(n)$. These two counts may, however, be equal. For example, for $n = 1681 = 41^2$, there are exactly 40 bases to which n is a Fermat pseudoprime, and n is an Euler pseudoprime to precisely these bases.

(5) The counts $n_F(n), n_E(n)$ together with $\phi(n)$ are listed in the following table for some small Carmichael numbers n. There are no Fermat witnesses to the compositeness of Carmichael numbers, but there do exist Euler witnesses to their compositeness.

---

[6]Robert M. Solovay and Volker Strassen, A fast Monte-Carlo test for primality, *SIAM Journal on Computing*, 6(1), 84–85, 1977.

n	$\phi(n)$	$n_F(n)$	$n_E(n)$
$561 = 3 \times 11 \times 17$	320	320	80
$1105 = 5 \times 13 \times 17$	768	768	192
$1729 = 7 \times 13 \times 19$	1296	1296	648
$2465 = 5 \times 17 \times 29$	1792	1792	896
$2821 = 7 \times 13 \times 31$	2160	2160	540
$6601 = 7 \times 23 \times 41$	5280	5280	1320
$8911 = 7 \times 19 \times 67$	7128	7128	1782
$41041 = 7 \times 11 \times 13 \times 41$	28800	28800	14400
$825265 = 5 \times 7 \times 17 \times 19 \times 73$	497664	497664	124416

□

5.2.3 Miller–Rabin Test

Miller[7] and Rabin[8] propose another variant of the Fermat test, robust against Carmichael numbers. The basic idea behind this test is that there exist non-trivial square roots (that is, square roots other than ± 1) of 1 modulo an odd composite n, provided that n is not a power of a prime (Exercise 1.55). Any Carmichael number has at least six non-trivial square roots of 1.

Suppose that $a^{n-1} \equiv 1 \pmod{n}$ for some odd integer n. Write $n - 1 = 2^s n'$ with $s \in \mathbb{N}$ and with n' odd. Define $b_j \equiv a^{2^j n'} \pmod{n}$ for $j = 0, 1, 2, \ldots, s$. It is given that $b_s \equiv 1 \pmod{n}$. If $b_0 \equiv 1 \pmod{n}$, then $b_j \equiv 1 \pmod{n}$ for all $j = 0, 1, 2, \ldots, s$. On the other hand, if $b_0 \not\equiv 1 \pmod{n}$, there exists (a unique) j in the range $0 \leqslant j < s$ such that $b_j \not\equiv 1 \pmod{n}$ but $b_{j+1} \equiv 1 \pmod{n}$. If $b_j \not\equiv -1 \pmod{n}$ too, then b_j is a non-trivial square root of 1 modulo n. Encountering such a non-trivial square root of 1 establishes the compositeness of n. The following definition uses these notations.

Definition 5.14 An odd composite integer n is called a *strong pseudoprime* or a *Miller–Rabin pseudoprime* to base $a \in \mathbb{Z}_n^*$ if either $b_0 \equiv 1 \pmod{n}$ or $b_j \equiv -1 \pmod{n}$ for some $j \in \{0, 1, 2, \ldots, s - 1\}$. ◁

If n is a strong pseudoprime to base a, then n is evidently a Fermat pseudoprime to base a. It is also true that every strong pseudoprime to base a is an Euler pseudoprime to the same base a.

The attractiveness of the Miller–Rabin test (Algorithm 5.3) lies in the fact that for a composite n, the fraction of bases in $\mathbb{Z}_n^*$ to which n is a strong pseudoprime is no more than $1/4$. Therefore, for t random bases, the probability of missing any witness to the compositeness of n is at most $1/4^t = 1/2^{2t}$.

---

[7]Miller proposed a deterministic primality test which is polynomial-time under the assumption that the ERH is true: Gary L. Miller, Riemann's hypothesis and tests for primality, *Journal of Computer and System Sciences*, 13(3), 300–317, 1976.

[8]Rabin proposed the randomized version: Michael O. Rabin, Probabilistic algorithm for testing primality, *Journal of Number Theory*, 12(1), 128–138, 1980.

Algorithm 5.3: MILLER–RABIN PRIMALITY TEST

```
Fix the number t of iterations.
Write n − 1 = 2ˢn' with s ∈ ℕ and n' odd.
```
Fix the number t of iterations.
Write $n - 1 = 2^s n'$ with $s \in \mathbb{N}$ and n' odd.
For $i = 1, 2, \ldots, t$ {
    Choose a random integer $a \in \{2, 3, \ldots, n-1\}$.
    Compute $b \equiv a^{n'} \pmod{n}$.
    if $(b \not\equiv 1 \pmod{n})$ {
        Set $j = 0$.
        While $((j \leqslant s - 2)$ and $(b \not\equiv -1 \pmod{n}))$ {
            Set $b = b^2 \pmod{n}$. /* We have $b_{j+1} \equiv b_j^2 \pmod{n}$ */
            If $(b \equiv 1 \pmod{n})$ return "*n is composite*".
            ++j.
        }
        If $(b \not\equiv -1 \pmod{n})$, return "*n is composite*".
    }
}
Return "*n is prime*".

Example 5.15 (1) 891 is a strong pseudoprime only to the ten bases 1, 82, 161, 163, 404, 487, 728, 730, 809, 890. These happen to be precisely the bases to which n is an Euler pseudoprime (Example 5.13).

(2) 2891 is a strong pseudoprime only to the bases 1 and 2890.

(3) 1891 is a strong pseudoprime to 450 bases in $\mathbb{Z}_n^*$. These are again all the bases to which n is an Euler pseudoprime.

(4) Let $n_S(n)$ denote the count of bases in $\mathbb{Z}_n^*$ to which n is a strong pseudoprime (also see Example 5.13). We always have $n_S(n) \leqslant n_E(n)$. So far in this example, we had $n_S(n) = n_E(n)$ only. The strict inequality occurs, for example, for the Carmichael number $n = 561$. In that case, $n_F(n) = 320$, $n_E(n) = 80$, and $n_S(n) = 10$. The following table summarizes these counts (along with $\phi(n)$) for some small Carmichael numbers.

n	$\phi(n)$	$n_F(n)$	$n_E(n)$	$n_S(n)$
$561 = 3 \times 11 \times 17$	320	320	80	10
$1105 = 5 \times 13 \times 17$	768	768	192	30
$1729 = 7 \times 13 \times 19$	1296	1296	648	162
$2465 = 5 \times 17 \times 29$	1792	1792	896	70
$2821 = 7 \times 13 \times 31$	2160	2160	540	270
$6601 = 7 \times 23 \times 41$	5280	5280	1320	330
$8911 = 7 \times 19 \times 67$	7128	7128	1782	1782
$41041 = 7 \times 11 \times 13 \times 41$	28800	28800	14400	450
$825265 = 5 \times 7 \times 17 \times 19 \times 73$	497664	497664	124416	486

The table indicates that not only is the Miller–Rabin test immune against Carmichael numbers, but also strong witnesses are often more numerous than

Euler witnesses for the compositeness of an odd composite n. The Miller–Rabin test is arguably the most commonly used primality test of today. □

The most time-consuming step in an iteration of each of the above probabilistic tests is an exponentiation modulo n to an exponent $\leqslant n - 1$. For all these tests, the bases a can be chosen as single-precision integers. Nonetheless, each of these tests runs in $O(\log^3 n)$ time.

5.2.4 Fibonacci Test

The *Fibonacci numbers* $F_0, F_1, F_2, \ldots$ are defined recursively as follows.

$$
\begin{aligned}
F_0 &= 0, \\
F_1 &= 1, \\
F_m &= F_{m-1} + F_{m-2} \text{ for } m \geqslant 2.
\end{aligned}
$$

One can supply an explicit formula for F_m. The characteristic polynomial $x^2 - x - 1$ of the recurrence has two roots $\alpha = \frac{1+\sqrt{5}}{2}$ and $\beta = \frac{1-\sqrt{5}}{2}$. We have:

$$
F_m = \frac{\alpha^m - \beta^m}{\alpha - \beta} = \frac{1}{\sqrt{5}}\left[\left(\frac{1+\sqrt{5}}{2}\right)^m - \left(\frac{1-\sqrt{5}}{2}\right)^m\right] \text{ for all } m \geqslant 0. \quad (5.1)
$$

A relevant property of Fibonacci numbers is given in the next theorem.

Theorem 5.16 *Let $p \in \mathbb{P}$, $p \neq 2, 5$. Then, $F_{p - \left(\frac{5}{p}\right)} \equiv 0 \pmod{p}$, where $\left(\frac{5}{p}\right)$ is the Legendre symbol.*

Proof First, suppose that $\left(\frac{5}{p}\right) = 1$, that is, 5 has (two) square roots modulo p, that is, the two roots α, β of $x^2 - x - 1 \pmod{p}$ belong to $\mathbb{F}_p$. Since $\alpha, \beta \not\equiv 0 \pmod{p}$, we have $\alpha^{p-1} \equiv 1 \pmod{p}$ and $\beta^{p-1} \equiv 1 \pmod{p}$, so $F_{p - \left(\frac{5}{p}\right)} \equiv F_{p-1} \equiv \frac{\alpha^{p-1} - \beta^{p-1}}{\alpha - \beta} \equiv 0 \pmod{p}$.

Now, suppose that $\left(\frac{5}{p}\right) = -1$. The roots α, β of $x^2 - x - 1$ do not belong to $\mathbb{F}_p$ but to $\mathbb{F}_{p^2}$. The p-th power Frobenius map $\mathbb{F}_{p^2} \to \mathbb{F}_{p^2}$ taking $\theta \mapsto \theta^p$ clearly maps a root of $x^2 - x - 1$ to a root of the same polynomial. But $\alpha^p = \alpha$ means $\alpha \in \mathbb{F}_p$, a contradiction. So $\alpha^p = \beta$. Likewise, $\beta^p = \alpha$. Consequently, $\alpha^{p+1} \equiv \beta^{p+1} \equiv \alpha\beta \pmod{p}$, that is, $F_{p - \left(\frac{5}{p}\right)} \equiv F_{p+1} \equiv \frac{\alpha^{p-1} - \beta^{p-1}}{\alpha - \beta} \equiv 0 \pmod{p}$. ◁

This property of Fibonacci numbers leads to the following definition.[9]

Definition 5.17 Let $n \in \mathbb{N}$ with $\gcd(n, 10) = 1$. We call n a *Fibonacci pseudoprime* if $F_{n - \left(\frac{5}{n}\right)} \equiv 0 \pmod{n}$, where $\left(\frac{5}{n}\right)$ is the Jacobi symbol. ◁

---

[9]The Fibonacci and the Lucas tests are introduced by: Robert Baillie and Samuel S. Wagstaff, Jr., Lucas pseudoprimes, *Mathematics of Computation*, 35(152), 1391–1417, 1980.

If $n \neq 2, 5$ is prime, we have $F_{n-\left(\frac{5}{n}\right)} \equiv 0 \pmod{n}$. However, some composite numbers too satisfy this congruence.

Example 5.18 (1) Lehmer proved that there are infinitely many Fibonacci pseudoprimes. Indeed, the Fibonacci number F_{2p} for every prime $p > 5$ is a Fibonacci pseudoprime.

(2) There are only nine composite Fibonacci pseudoprimes $\leqslant 10,000$. These are $323 = 17 \times 19$, $377 = 13 \times 29$, $1891 = 31 \times 61$, $3827 = 43 \times 89$, $4181 = 37 \times 113$, $5777 = 53 \times 109$, $6601 = 7 \times 23 \times 41$, $6721 = 11 \times 13 \times 47$, and $8149 = 29 \times 281$. There are only fifty composite Fibonacci pseudoprimes $\leqslant 100,000$. The smallest composite Fibonacci pseudoprimes with four and five prime factors are $199801 = 7 \times 17 \times 23 \times 73$ and $3348961 = 7 \times 11 \times 23 \times 31 \times 61$.

(3) There is no known composite integer $n \equiv \pm 2 \pmod{5}$ which is simultaneously a Fibonacci pseudoprime and a Fermat pseudoprime to base 2. It is an open question to find such a composite integer or to prove that no such composite integer exists. There, however, exist composite integers $n \equiv \pm 1 \pmod{5}$ which are simultaneously Fibonacci pseudoprimes and Fermat pseudoprimes to base 2. Two examples are $6601 = 7 \times 23 \times 41$ and $30889 = 17 \times 23 \times 79$. □

Algorithm 5.4: FIBONACCI TEST FOR $n > 5$

```
Compute l = (5/n).
If (l = 0) return 0.
Compute F = F_{n-l} modulo n.
If (F = 0), return "n is prime", else return "n is composite".
```

A very important question pertaining to this test is an efficient computation of $F_{n-\left(\frac{5}{n}\right)}$. Using the straightforward iterative method of sequentially computing $F_0, F_1, F_2, \ldots, F_{n-\left(\frac{5}{n}\right)}$ takes time proportional to n, that is, exponential in $\log n$. In order to avoid this difficulty, we look at the following identities satisfied by Fibonacci numbers, that hold for all $k \in \mathbb{N}_0$.

$$
\begin{aligned}
F_{2k} &= F_k(2F_{k+1} - F_k), \\
F_{2k+1} &= F_{k+1}^2 + F_k^2, \\
F_{2k+2} &= F_{k+1}(F_{k+1} + 2F_k).
\end{aligned}
\tag{5.2}
$$

Suppose that we want to compute F_m. Consider the binary representation $m = (m_{s-1} m_{s-2} \ldots m_0)_2$. Denote the subscripts $M_i = (m_{s-1} m_{s-2} \cdots m_i)_2$ for $i = 0, 1, 2, \ldots, s$ (where M_s is to be interpreted as 0). We start with the constant values $(F_0, F_1) = (0, 1)$, a case that corresponds to $i = s$ as explained below. Subsequently, we run a loop for $i = s-1, s-2, \ldots, 0$ (in that sequence) such that at the end of the i-th iteration of the loop, we have computed F_{M_i} and F_{M_i+1}. Computing these two values from the previous values $F_{M_{i+1}}$ and $F_{M_{i+1}+1}$ is done by looking at the i-th bit m_i of m. If $m_i = 0$, then

$M_i = 2M_{i+1}$, so we use first two of the identities (5.2) in order to update $(F_{M_{i+1}}, F_{M_{i+1}+1})$ to $(F_{2M_{i+1}}, F_{2M_{i+1}+1})$. On the other hand, if $m_i = 1$, then $M_i = 2M_{i+1} + 1$, and we use the last two of the identities (5.2) for updating $(F_{M_{i+1}}, F_{M_{i+1}+1})$ to $(F_{2M_{i+1}+1}, F_{2M_{i+1}+2})$.

This clever adaptation of the repeated square-and-multiply exponentiation algorithm is presented as Algorithm 5.5. For $m = O(n)$, the `for` loop in the algorithm runs for $s = O(\log n)$ times, with each iteration involving a constant number of basic arithmetic operations modulo n. Since each such operation can be done in $O(\log^2 n)$ time, Algorithm 5.5 runs in $O(\log^3 n)$ time.

Algorithm 5.5: COMPUTING F_m MODULO n

```
Initialize F = 0 and F_next = 1.
For i = s - 1, s - 2, ..., 1, 0 {
    If (m_i is 0) {
        Compute t ≡ F(2F_next − F) (mod n).
        Compute F_next ≡ F²_next + F² (mod n).
        Assign F = t.
    } else {
        Compute t ≡ F²_next + F² (mod n).
        Compute F_next ≡ F_next(F_next + 2F) (mod n).
        Assign F = t.
    }
}
return F.
```

Example 5.19 We illustrate the working of Algorithm 5.5 in order to establish that 323 is a Fibonacci pseudoprime. Since $\left(\frac{5}{323}\right) = -1$, we compute F_{324} (mod 323). We have $324 = (101000100)_2$. The modular computation of F_{324} is given in the table below.

i	m_i	M_i	F_{M_i} (mod 323)	F_{M_i+1} (mod 323)
9		0	$F_0 = 0$	$F_1 = 1$
8	1	$(1)_2 = 1$	$F_1 \equiv F_1^2 + F_0^2 \equiv 1$	$F_2 \equiv F_1(F_1 + F_0) \equiv 1$
7	0	$(10)_2 = 2$	$F_2 \equiv F_1(2F_2 - F_1) \equiv 1$	$F_3 \equiv F_2^2 + F_1^2 \equiv 2$
6	1	$(101)_2 = 5$	$F_5 \equiv F_3^2 + F_2^2 \equiv 5$	$F_6 \equiv F_3(F_3 + 2F_2) \equiv 8$
5	0	$(1010)_2 = 10$	$F_{10} \equiv F_5(2F_6 - F_5) \equiv 55$	$F_{11} \equiv F_6^2 + F_5^2 \equiv 89$
4	0	$(10100)_2 = 20$	$F_{20} \equiv F_{10}(2F_{11} - F_{10}) \equiv 305$	$F_{21} \equiv F_{11}^2 + F_{10}^2 \equiv 287$
3	0	$(101000)_2 = 40$	$F_{40} \equiv F_{20}(2F_{21} - F_{20}) \equiv 3$	$F_{41} \equiv F_{21}^2 + F_{20}^2 \equiv 5$
2	1	$(1010001)_2 = 81$	$F_{81} \equiv F_{41}^2 + F_{40}^2 \equiv 34$	$F_{82} \equiv F_{41}(F_{41} + 2F_{40}) \equiv 55$
1	0	$(10100010)_2 = 162$	$F_{162} \equiv F_{81}(2F_{82} - F_{81}) \equiv 0$	$F_{163} \equiv F_{82}^2 + F_{81}^2 \equiv 305$
0	0	$(101000100)_2 = 324$	$F_{324} \equiv F_{162}(2F_{163} - F_{162}) \equiv 0$	$F_{325} \equiv F_{163}^2 + F_{162}^2 \equiv 1$

Since $F_{323-\left(\frac{5}{323}\right)} \equiv 0$ (mod 323), 323 is declared as prime. □

Here follows a GP/PARI function implementing Algorithm 5.5.

```
gp > FibMod(m,n) = \
    local(i,s,t,F,Fnext); \
    s = ceil(log(m)/log(2)); \
    F = Mod(0,n); Fnext = Mod(1,n); \
    i = s - 1; \
    while (i>=0, \
       if (bittest(m,i) == 0, \
          t = F * (2 * Fnext - F); \
          Fnext = Fnext^2 + F^2; \
          F = t \
          , \
          t = Fnext^2 + F^2; \
          Fnext = Fnext * (Fnext + 2 * F); \
          F = t \
       ); \
       i--; \
    ); \
    return(F);
gp > FibMod(324,323)
%1 = Mod(0, 323)
gp > F324 = fibonacci(324)
%2 = 23041483585524168262220906489642018075101617466780496790573690289968
gp > F324 % 323
%3 = 0
```

The Fibonacci test is *deterministic*. It recognizes primes as primes, but also certifies certain composite integers as primes, and Fibonacci certificates do not change, no matter how many times we run the test on a fixed input. We now look at generalized versions of the Fibonacci test. These generalized tests have two advantages. First, a concept of variable parameters (like bases) is introduced so as to make the test probabilistic. Second, the tests are made more stringent so that fewer composite numbers are certified as primes.

5.2.5 Lucas Test

An obvious generalization of the Fibonacci sequence is the *Lucas sequence* $U_m = U_m(a,b)$ characterized by two integer parameters a, b.[10]

$$U_0 = 0,$$
$$U_1 = 1,$$
$$U_m = aU_{m-1} - bU_{m-2} \text{ for } m \geqslant 2. \tag{5.3}$$

The characteristic polynomial of this recurrence is $x^2 - ax + b$ with discriminant $\Delta = a^2 - 4b$. We assume that Δ is non-zero and not a perfect square. The two roots α, β of this polynomial are distinct and given by $\alpha = \frac{a+\sqrt{\Delta}}{2}$ and $\beta = \frac{a-\sqrt{\Delta}}{2}$. The sequence U_m can be expressed explicitly in terms of α, β as

---

[10]The Fibonacci sequence corresponds to $a = 1$ and $b = -1$.

$$U_m = U_m(a, b) = \frac{\alpha^m - \beta^m}{\alpha - \beta} \quad \text{for all } m \geqslant 0. \tag{5.4}$$

The generalization of Theorem 5.16 for Lucas sequences is the following.

Theorem 5.20 $U_{p-\left(\frac{\Delta}{p}\right)} \equiv 0 \pmod{p}$ *for a prime p with* $\gcd(p, 2b\Delta) = 1$.

Proof Straightforward modification of the proof of Theorem 5.16. ◁

Definition 5.21 Let $U_m = U_m(a, b)$ be a Lucas sequence, and $\Delta = a^2 - 4b$. An integer n with $\gcd(n, 2b\Delta) = 1$ is called a *Lucas pseudoprime* with parameters (a, b) if $U_{n-\left(\frac{\Delta}{n}\right)} \equiv 0 \pmod{n}$. ◁

The Lucas pseudoprimality test is given as Algorithm 5.6.

Algorithm 5.6: LUCAS TEST FOR n WITH PARAMETERS (a, b)

Compute $\Delta = a^2 - 4b$.
Compute $l = \left(\frac{\Delta}{n}\right)$.
If $(l = 0)$, return "n is composite".
Compute $U = U_{n-l}$ modulo n.
If $(U = 0)$, return "n is prime", else return "n is composite".

We can invoke this test with several parameters (a, b). If any of these invocations indicates that n is composite, then n is certainly composite. On the other hand, if all of these invocations certify n as prime, we accept n as prime. By increasing the number of trials (different parameters a, b), we can reduce the probability that a composite integer is certified as a prime.

We should now supply an algorithm for an efficient computation of the value $U_{n-\left(\frac{\Delta}{n}\right)}$ modulo n. To that effect, we introduce a related sequence $V_m = V_m(a, b)$ as follows.

$$\begin{aligned} V_0 &= 2, \\ V_1 &= a, \\ V_m &= aV_{m-1} - bV_{m-2} \text{ for } m \geqslant 2. \end{aligned} \tag{5.5}$$

As above, let α, β be the roots of the characteristic polynomial $x^2 - ax + b$. An explicit formula for the sequence V_m is as follows.

$$V_m = V_m(a, b) = \alpha^m + \beta^m \quad \text{for all } m \geqslant 0. \tag{5.6}$$

The sequence U_m can be computed from V_m, V_{m+1} by the simple formula:

$$U_m = \Delta^{-1}(2V_{m+1} - aV_m) \quad \text{for all } m \geqslant 0.$$

Therefore, it suffices to compute $V_{n-\left(\frac{\Delta}{n}\right)}$ and $V_{n-\left(\frac{\Delta}{n}\right)+1}$ for the Lucas test. This computation can be efficiently done using the doubling formulas:

$$\begin{aligned} V_{2k} &= V_k^2 - 2b^k, \\ V_{2k+1} &= V_k V_{k+1} - ab^k, \\ V_{2k+2} &= V_{k+1}^2 - 2b^{k+1}. \end{aligned} \tag{5.7}$$

Designing the analog of Algorithm 5.5 for Lucas tests is posed as Exercise 5.17.

Example 5.22 Consider the Lucas sequence $U_m(3,1)$ with $a = 3$ and $b = 1$:

$$\begin{aligned}
U_0 &= 0, \\
U_1 &= 1, \\
U_m &= 3U_{m-1} - U_{m-2} \text{ for } m \geqslant 2.
\end{aligned}$$

Thus, $U_2 = 3U_1 - U_0 = 3$, $U_3 = 3U_2 - U_1 = 8$, $U_4 = 3U_3 - U_2 = 21$, and so on. The discriminant is $a^2 - 4b = 5$, and the roots of the characteristic equation are $\alpha = \frac{3+\sqrt{5}}{2}$ and $\beta = \frac{3-\sqrt{5}}{2}$, that is,

$$U_m = U_m(3,1) = \frac{1}{\sqrt{5}}\left[\left(\frac{3+\sqrt{5}}{2}\right)^m - \left(\frac{3-\sqrt{5}}{2}\right)^m\right] \text{ for all } m \geqslant 0.$$

We show that 21 is a Lucas pseudoprime with parameters 3, 1, that is, $U_{20} \equiv 0 \pmod{21}$ (we have $\left(\frac{5}{21}\right) = 1$). We use the sequence V_m for this computation. Since $b = 1$, we have the simplified formulas:

$$\begin{aligned}
V_{2k} &= V_k^2 - 2, \\
V_{2k+1} &= V_k V_{k+1} - a = V_k V_{k+1} - 3, \\
V_{2k+2} &= V_{k+1}^2 - 2.
\end{aligned}$$

The computation of V_{20} is shown below. Note that $20 = (10100)_2$.

i	m_i	M_i	$V_{M_i} \pmod{21}$	$V_{M_i+1} \pmod{21}$
5		0	$V_0 \equiv 2$	$V_1 \equiv 3$
4	1	$(1)_2 = 1$	$V_1 \equiv V_0 V_1 - 3 \equiv 3$	$V_2 \equiv V_1^2 - 2 \equiv 7$
3	0	$(10)_2 = 2$	$V_2 \equiv V_1^2 - 2 \equiv 7$	$V_3 \equiv V_1 V_2 - 3 \equiv 18$
2	1	$(101)_2 = 5$	$V_5 \equiv V_2 V_3 - 3 \equiv 18$	$V_6 \equiv V_3^2 - 2 \equiv 7$
1	0	$(1010)_2 = 10$	$V_{10} \equiv V_5^2 - 2 \equiv 7$	$V_{11} \equiv V_5 V_6 - 3 \equiv 18$
0	0	$(10100)_2 = 20$	$V_{20} \equiv V_{10}^2 - 2 \equiv 5$	$V_{21} \equiv V_{10} V_{11} - 3 \equiv 18$

Therefore, $U_{20} \equiv \Delta^{-1}(2V_{21} - aV_{20}) \equiv 5^{-1}(2 \times 18 - 3 \times 5) \equiv 0 \pmod{21}$. $\square$

A stronger Lucas test can be developed like the Miller–Rabin test. Let p be an odd prime. Consider the Lucas sequence U_m with parameters a, b. Assume that $\gcd(p, 2b\Delta) = 1$, that is, α, β are distinct in $\mathbb{F}_p$ or $\mathbb{F}_{p^2}$. This implies that $\left(\frac{\Delta}{p}\right) = \pm 1$, that is, $p - \left(\frac{\Delta}{p}\right)$ is even. Write $p - \left(\frac{\Delta}{p}\right) = 2^s t$ with $s, t \in \mathbb{N}$ and with t odd. The condition $U_k \equiv 0 \pmod{p}$ implies $(\alpha/\beta)^k \equiv 1 \pmod{p}$. Since the only square roots of 1 modulo p are ± 1, we have either $(\alpha/\beta)^t \equiv 1 \pmod{p}$ or $(\alpha/\beta)^{2^j t} \equiv -1 \pmod{p}$ for some $j \in \{0, 1, \ldots, s-1\}$. The condition $(\alpha/\beta)^t \equiv 1 \pmod{p}$ implies $U_t \equiv 0 \pmod{p}$, whereas the condition $(\alpha/\beta)^{2^j t} \equiv -1 \pmod{p}$ implies $V_{2^j t} \equiv 0 \pmod{p}$.

Definition 5.23 Let $U_m = U_m(a,b)$ be a Lucas sequence with discriminant $\Delta = a^2 - 4b$. Let $V_m = V_m(a,b)$ be the corresponding sequence as defined by the recurrence and initial conditions (5.5). Let n be a (positive) integer with

$\gcd(n, 2b\Delta) = 1$. We write $n - \left(\frac{\Delta}{n}\right) = 2^s t$ with $s, t \in \mathbb{N}$ and with t odd. We call n a *strong Lucas pseudoprime* with parameters (a, b) if either $U_t \equiv 0 \pmod{n}$ or $V_{2^j t} \equiv 0 \pmod{n}$ for some $j \in \{0, 1, \ldots, s-1\}$. $\lhd$

Obviously, every strong Lucas pseudoprime is also a Lucas pseudoprime (with the same parameters). The converse of this is not true.

Example 5.24 (1) Example 5.22 shows that 21 is a composite Lucas pseudoprime with parameters $3, 1$. In this case, $n - \left(\frac{\Delta}{n}\right) = 20 = 2^2 \times 5$, that is, $s = 2$ and $t = 5$. We have $U_5 \equiv \Delta^{-1}(2V_6 - aV_5) \equiv 5^{-1}(14 - 54) \equiv 13 \not\equiv 0 \pmod{n}$. Moreover, $V_5 \equiv 18 \not\equiv 0 \pmod{21}$, and $V_{10} \equiv 7 \not\equiv 0 \pmod{21}$. That is, 21 is not a strong Lucas pseudoprime with parameters $3, 1$.

There are exactly 21 composite Lucas pseudoprimes $\leqslant 10,000$ with parameter $3, 1$. These are 21, 323, 329, 377, 451, 861, 1081, 1819, 1891, 2033, 2211, 3653, 3827, 4089, 4181, 5671, 5777, 6601, 6721, 8149 and 8557. Only five (323, 377, 1891, 4181 and 5777) of these are strong Lucas pseudoprimes.

(2) There is no composite integer $\leqslant 10^7$, which is a strong Lucas pseudoprime with respect to both the parameters $(3, 1)$ and $(4, 1)$. $\square$

Algorithm 5.7: STRONG LUCAS TEST FOR n WITH PARAMETERS (a, b)

```
Compute Δ = a² − 4b.
Compute l = (Δ/n).
If (l = 0), return "n is composite".
Express n − l = 2ˢt with s, t ∈ ℕ, t odd.
Compute U = Uₜ(a, b) modulo n.
If (U = 0), return "n is prime".
Compute V = Vₜ(a, b) modulo n.
Set j = 0.
while (j < s) {
    If (V = 0), return "n is prime".
    Set V ≡ V² − 2b^(2ʲt) (mod n).
    ++j.
}
Return "n is composite".
```

Algorithm 5.7 presents the strong Lucas primality test. We assume a general parameter (a, b). Evidently, the algorithm becomes somewhat neater and more efficient if we restrict only to parameters of the form $(a, 1)$.

Arnault[11] proves an upper bound on the number of pairs (a, b) to which a composite number n is a strong Lucas pseudoprime with parameters (a, b). More precisely, Arnault takes a discriminant Δ and an odd composite integer n coprime to Δ (but not equal to 9). By $\mathrm{SL}(\Delta, n)$, he denotes the number

---

[11] François Arnault, The Rabin-Monier theorem for Lucas pseudoprimes, *Mathematics of Computation*, 66(218), 869–881, 1997.

of parameters (a, b) with $0 \leqslant a, b < n$, $\gcd(n, b) = 1$, $a^2 - 4b \equiv \Delta \pmod{n}$, and with n being a strong Lucas pseudoprime with parameter (a, b). Arnault proves that $\mathrm{SL}(\Delta, n) \leqslant n/2$ for all values of Δ and n. Moreover, if n is not a product of twin primes of a special form, then $\mathrm{SL}(\Delta, n) \leqslant \frac{4}{15}n$.

5.2.6 Other Probabilistic Tests

An extra strong Lucas test is covered in Exercise 5.18. *Lehmer pseudoprimes* and *strong Lehmer pseudoprimes*[12] are special types of Lucas and strong Lucas pseudoprimes. *Perrin pseudoprimes*[13] are based on (generalized) Perrin sequences[14] (see Exercise 5.20).

Grantham introduces the concept of *Frobenius pseudoprimes.*[15] Instead of working with specific polynomials like $x^2 - ax + b$, he takes a general monic polynomial $f(x) \in \mathbb{Z}[x]$. The Fermat test to base a is a special case with $f(x) = x - a$, whereas the Lucas test with parameter (a, b) is the special case $f(x) = x^2 - ax + b$. Moreover, Perrin pseudoprimes are special cases with $\deg f(x) = 3$. Grantham also defines *strong Frobenius pseudoprimes*, and shows that strong (that is, Miller–Rabin) pseudoprimes and strong and extra strong Lucas pseudoprimes are special cases of strong Frobenius pseudoprimes.

Pomerance, Selfridge and Wagstaff[16] have declared an award of \$620 for solving the open problem reported in Example 5.18(3). Grantham, on the other hand, has declared an award of \$6.20 for locating a composite Frobenius pseudoprime $\equiv \pm 2 \pmod 5$ with respect to the polynomial $x^2 + 5x + 5$ (or for proving that no such pseudoprime exists). Grantham justifies that "the low monetary figure is a reflection of my financial status at the time of the offer, not of any lower confidence level." Indeed, he mentions that "I believe that the two problems are equally challenging."

5.3 Deterministic Primality Testing

The deterministic complexity of primality testing has attracted serious research attention for quite a period of time. Under the assumption that the ERH is true, both the Miller–Rabin test and the Solovay–Strassen test can be derandomized to polynomial-time primality-testing algorithms. Adleman

---

[12] A. Rotkiewicz, On Euler Lehmer pseudoprimes and strong Lehmer pseudoprimes with parameters L, Q in arithmetic progressions, *Mathematics of Computation*, 39(159), 239–247, 1982.

[13] William Adams and Daniel Shanks, Strong primality tests that are not sufficient, *Mathematics of Computation*, 39(159), 255–300, 1982.

[14] R. Perrin introduced this sequence in *L'Intermédiaire des mathématiciens*, Vol. 6, 1899.

[15] Jon Grantham, Frobenius pseudoprimes, *Mathematics of Computation*, 70(234), 873–891, 2001.

[16] Richard Kenneth Guy, *Unsolved problems in number theory* (3rd ed), Springer, 2004.

et al.[17] propose a deterministic primality-testing algorithm with running time $O((\log n)^{\log \log \log n})$. The exponent $\log \log \log n$ grows very slowly with n, but is still not a constant—it grows to infinity as n goes to infinity. The *elliptic-curve primality-proving* algorithm (ECPP), proposed by Goldwasser and Kilian[18] and modified by Atkin and Morain[19], runs in deterministic polynomial time, but its running time analysis is based upon certain heuristic assumptions. Indeed, ECPP has not been proved to run in polynomial time for *all* inputs.

In 2002, Agarwal, Kayal and Saxena[20] proposed the first deterministic polynomial-time algorithm for primality testing. The proof of correctness (and running time) of this AKS algorithm is not based on any unprovable fact or heuristic assumption. Several improvements of the AKS test are proposed. The inventors of the test themselves published a revised version with somewhat reduced running time. Lenstra and Pomerance[21] have proposed some more significant reduction in the running time of the AKS test.

5.3.1 Checking Perfect Powers

I first show that it is computationally easy to determine whether a positive integer is an integral power of a positive integer.

Definition 5.25 Let $k \in \mathbb{N}$, $k \geqslant 2$. A positive integer $n \geqslant 2$ is called a *perfect k-th power* if $n = a^k$ for some positive integer a. For $k = 2$ and $k = 3$, we talk about *perfect squares* and *perfect cubes* in this context. We call n a *perfect power* if n is a perfect k-th power for some $k \geqslant 2$. ◁

Algorithm 5.8 determines whether n is a perfect k-th power. It is based on the *Newton–Raphson method*.[22] We first compute $a = \lfloor \sqrt[k]{n} \rfloor$, and then check whether $n = a^k$. For computing the integer k-th root a of n, we essentially compute a zero of the polynomial $f(x) = x^k - n$. We start with an initial approximation $a_0 \geqslant \sqrt[k]{n}$. A good starting point could be $a_0 = 2^{\lceil l/k \rceil} = 2^{\lfloor (l+k-1)/k \rfloor}$, where l is the bit length of n. Subsequently, we refine the approximation by computing a decreasing sequence $a_1, a_2, \ldots$. For computing a_{i+1} from a_i, we approximate the curve $y = f(x)$ by the tangent to the curve passing through

[17]Leonard M. Adleman, Carl Pomerance and Robert S. Rumely, On distinguishing prime numbers from composite numbers, *Annals of Mathematics*, 117(1), 173–206, 1983.

[18]Shafi Goldwasser and Joe Kilian, Almost all primes can be quickly certified, *STOC*, 316–329, 1986.

[19]A. O. L. Atkin and François Morain, Elliptic curves and primality proving, *Mathematics of Computation*, 61, 29–68, 1993.

[20]Manindra Agrawal, Neeraj Kayal and Nitin Saxena, PRIMES is in P, *Annals of Mathematics*, 160(2), 781–793, 2004. This article can also be downloaded from the Internet site: http://www.cse.iitk.ac.in/users/manindra/algebra/primality.pdf. Their first article on this topic is available at http://www.cse.iitk.ac.in/users/manindra/algebra/primality_original.pdf.

[21]Hendrik W. Lenstra, Jr. and Carl Pomerance, Primality testing with Gaussian periods, available from http://www.math.dartmouth.edu/~carlp/aks041411.pdf, 2011.

[22]This numerical method was developed by Sir Isaac Newton (1642–1727) and Joseph Raphson (1648–1715).

$(a_i, f(a_i))$. This line meets the x-axis at $x = a_{i+1}$. Therefore, $f'(a_i) = \frac{0 - f(a_i)}{a_{i+1} - a_i}$, that is, $a_{i+1} = a_i - \frac{f(a_i)}{f'(a_i)}$. Simple calculations show that $a_{i+1} = \frac{(k-1)a_i^k + n}{ka_i^{k-1}}$. In order to avoid floating-point calculations, we update $a_{i+1} = \left\lfloor \frac{(k-1)a_i^k + n}{ka_i^{k-1}} \right\rfloor$.

Algorithm 5.8: CHECKING WHETHER n IS A PERFECT k-TH POWER

Compute the bit length l of n.
Set the initial approximation $a = 2^{\lceil l/k \rceil} = 2^{\lfloor (l+k-1)/k \rfloor}$.
Repeat until explicitly broken {
    Compute the temporary value $t = a^{k-1}$.

    Compute the next integer approximation $b = \left\lfloor \dfrac{(k-1)at + n}{kt} \right\rfloor$.

    If $(b \geqslant a)$, break the loop, else set $a = b$.
}
If n equals a^k, return "*True*", else return "*False*".

I now prove the correctness of Algorithm 5.8, that is, $a = \lfloor \sqrt[k]{n} \rfloor$ when the loop is broken. Since $k \geqslant 2$, the function $f(x) = x^k - n$ is convex[23] to the right of the real root $\xi = \sqrt[k]{n}$. For resolving ambiguities, let us denote $\alpha_{i+1} = \frac{(k-1)a_i^k + n}{ka_i^{k-1}}$, and $a_{i+1} = \lfloor \alpha_{i+1} \rfloor$. If $a_i > \xi$, the convexity of f implies that $\xi < \alpha_{i+1} < a_i$. Taking floor gives $\lfloor \xi \rfloor \leqslant a_{i+1} < a_i$. This means that the integer approximation stored in a decreases strictly so long as $a > \lfloor \xi \rfloor$, and eventually obtains the value $a_j = \lfloor \xi \rfloor$ for some j. I show that in the next iteration, the loop is broken after the computation of $b = a_{j+1}$ from $a = a_j$.

If n is actually a k-th power, $\lfloor \xi \rfloor = \xi$. In that case, the next integer approximation a_{j+1} also equals $\lfloor \xi \rfloor = \xi$. Thus, $a_{j+1} = a_j$, and the loop is broken. On the other hand, if n is not a perfect k-th power, then $\lfloor \xi \rfloor < \xi$, that is, the current approximation $a_j < \xi$. Since, in this case, we have $f(a_j) < 0$ and $f'(a_j) > 0$, the next real approximation α_{j+1} is larger than a_j, and so its floor a_{j+1} is $\geqslant a_j$. Thus, the condition $b \geqslant a$ is again satisfied.

It is well known from the results of numerical analysis that the Newton–Raphson method converges quadratically. That is, the Newton–Raphson loop is executed at most $O(\log n)$ times. The exponentiation $t = a^{k-1}$ in each iteration can be computed in $O(\log^2 n \log k)$ time. The rest of an iteration runs in $O(\log^2 n)$ time. Therefore, Algorithm 5.8 runs in $O(\log^3 n \log k)$ time.

We now vary k. The maximum possible exponent k for which n can be a perfect k-th power corresponds to the case $n = 2^k$, that is, to $k = \lg n = \frac{\log n}{\log 2}$. Therefore, it suffices to check whether n is a perfect k-th power for $k = 2, 3, \ldots, \left\lfloor \frac{\log n}{\log 2} \right\rfloor$. In particular, we always have $k = O(\log n)$, and checking whether n is a perfect power finishes in $O(\log^4 n \log \log n)$ or $O^{\sim}(\log^4 n)$ time.[24]

---

[23] A real-valued function f is called *convex* in the real interval $[a, b]$ if $f((1 - t)a + tb) \leqslant (1 - t)f(a) + tf(b)$ for every real $t \in [0, 1]$.
[24] The soft-O notation $O^{\sim}(t(n))$ stands for $O(t(n) \log^s(t(n)))$ for some constant s.

In this context, it is worthwhile to mention that there is no need to consider all possible values of k in the range $2 \leqslant k \leqslant \lg n$. One may instead consider only the prime values of k in this range. If k is composite with a prime divisor p, then the condition that n is a perfect k-th power implies that n is a perfect p-th power too. Since $\lg n$ is not a large value in practice, one may use a precomputed table of all primes $\leqslant \lg n$. Skipping Algorithm 5.8 for composite k reduces the running time of the perfect-power-testing algorithm by a factor of $O(\log \log n)$ (by the prime number theorem).

5.3.2 AKS Test

The Agarwal–Kayal–Saxena (AKS) test is based on the following characterization of primes.

Theorem 5.26 *An integer $n \geqslant 2$ is prime if and only if $n \mid \binom{n}{k}$ for all $k = 1, 2, \ldots, n-1$.*

Proof We have $\binom{n}{k} = \frac{n(n-1)\cdots(n-k+1)}{k!}$. If n is prime and $1 \leqslant k \leqslant n-1$, the numerator of this expression for $\binom{n}{k}$ is divisible by n, whereas the denominator is not. On the other hand, suppose that n is composite. Let p be any prime divisor of n, and let $v_p(n) = e$ be the multiplicity of p in n. Take $k = p$. Neither of the factors $n-1, n-2, \ldots, n-p+1$ is divisible by p, whereas $p!$ is divisible by p (but not by p^2). Consequently, $v_p(\binom{n}{p}) = e - 1$, so $n \nmid \binom{n}{p}$. ◁

Corollary 5.27 *Let n be an odd positive integer, and let a be coprime to n. Then, $(x + a)^n \equiv x^n + a \pmod{n}$ if and only if n is a prime.* ◁

A straightforward use of Theorem 5.26 or Corollary 5.27 calls for the computation of $n - 1$ binomial coefficients modulo n, leading to an exponential algorithm for primality testing. This problem can be avoided by taking a polynomial $h(x)$ of small degree and by computing $(x + a)^n$ and $x^n + a$ modulo n and $h(x)$, that is, we now use the arithmetic of the ring $\mathbb{Z}_n[x]/\langle h(x) \rangle$. Let r denote the degree of $h(x)$ modulo n. All intermediate products are maintained as polynomials of degrees $< r$. Consequently, an exponentiation of the form $(x+a)^n$ or x^n can be computed in $O(r^2 \log^3 n)$ time. If $r = O(\log^k n)$ for some constant k, this leads to a polynomial-time test for the primality of n.

Composite integers n too may satisfy $(x + a)^n \equiv x^n + a \pmod{n, h(x)}$, and the AKS test appears to smell like another rotten probabilistic primality test. However, there is a neat way to derandomize this algorithm. In view of the results in Section 5.3.1, we assume that n is not a perfect power.

The AKS algorithm proceeds in two stages. In the first stage, we take $h(x) = x^r - 1$ for some *small* integer r. We call r *suitable* in this context if $\mathrm{ord}_r(n) > \lg^2 n$. A simple argument establishes that for all $n > 5,690,034$, a suitable $r \leqslant \lceil \lg^5 n \rceil$ exists. An efficient computation of $\mathrm{ord}_r(n)$ requires the factorization of r and $\phi(r)$. However, since r is $O(\lg^5 n)$, it is fine to use an exponential (in $\lg r$) algorithm for obtaining these factorizations. Another alternative is to compute $n, n^2, n^3, \ldots$ modulo r until $\mathrm{ord}_r(n)$ is revealed.

Algorithm 5.9: THE AKS PRIMALITY TEST

If n is a perfect power, return "*False*".

```
/* Stage 1 */
```
For $r = \lceil \lg^2 n \rceil + 1, \lceil \lg^2 n \rceil + 2, \lceil \lg^2 n \rceil + 3, \ldots$ {
    Compute the order $t = \mathrm{ord}_r(n)$.
    If $(t > \lg^2 n)$, break.
}
For $a = 2, 3, \ldots, r$ {
    If $(\gcd(a,n) > 1)$, return "*False*".
}

```
/* Stage 2 */
```
For $a = 1, 2, 3, \ldots, \left\lfloor \sqrt{\phi(r)} \lg n \right\rfloor$ {
    If $(x+a)^n \not\equiv x^n + a \pmod{n, x^r - 1}$, return "*False*".
}
return "*True*".

In the second stage, one works with the smallest suitable r available from the first stage. Checking whether $(x+a)^n \equiv x^n + a \pmod{n, x^r - 1}$ for all $a = 1, 2, \ldots, \left\lfloor \sqrt{\phi(r)} \lg n \right\rfloor$ allows one to deterministically conclude about the primality of n. A proof of the fact that only these values of a suffice is omitted here. The AKS test given as Algorithm 5.9 assumes that $n > 5{,}690{,}034$.

Example 5.28 (1) Take $n = 8{,}079{,}493$. The search for a suitable r is shown below. Since $\lg^2 n = 526.511 \ldots$, this search starts from $\lceil \lg^2 n \rceil + 1 = 528$.

$$\mathrm{ord}_{528}(n) = 20, \quad \mathrm{ord}_{529}(n) = 506, \quad \mathrm{ord}_{530}(n) = 52, \quad \mathrm{ord}_{531}(n) = 174,$$
$$\mathrm{ord}_{532}(n) = 9, \quad \mathrm{ord}_{533}(n) = 60, \quad \mathrm{ord}_{534}(n) = 22, \quad \mathrm{ord}_{535}(n) = 212,$$
$$\mathrm{ord}_{536}(n) = 6, \quad \mathrm{ord}_{537}(n) = 89, \quad \mathrm{ord}_{538}(n) = 67, \quad \mathrm{ord}_{539}(n) = 15,$$
$$\mathrm{ord}_{540}(n) = 36, \quad \mathrm{ord}_{541}(n) = 540.$$

Therefore, $r = 541$ (which is a prime), and $\phi(r) = 540$. One then verifies that $\gcd(2, n) = \gcd(3, n) = \cdots = \gcd(541, n) = 1$, that is, n has no small prime factors. One then computes the bound $B = \left\lfloor \sqrt{\phi(r)} \lg n \right\rfloor = 533$, and checks that the congruence $(x+a)^n \equiv x^n + a \pmod{n, x^r - 1}$ holds for all $a = 1, 2, \ldots, B$. For example, $(x+1)^n \equiv x^{199} + 1 \pmod{n, x^{541} - 1}$, and $x^n + 1 \equiv x^{n \bmod 541} + 1 \equiv x^{199} + 1 \pmod{n, x^{541} - 1}$. So $8{,}079{,}493$ is prime.

(2) For $n = 19{,}942{,}739$, we have $\lg^2 n = 588.031 \ldots$. We calculate $\mathrm{ord}_{590}(n) = 58$, $\mathrm{ord}_{591}(n) = 196$, $\mathrm{ord}_{592}(n) = 36$, and $\mathrm{ord}_{593}(n) = 592$. So, $r = 593$ is suitable, and n has no factors $\leqslant 593$. The bound for the second stage is now $B = 590$. However, for $a = 1$, one obtains $(x+1)^n \equiv 9029368x^{592} + 919485x^{591} + 10987436x^{590} + \cdots + 9357097x + 17978236 \pmod{n, x^{593} - 1}$, whereas $x^n + 1 \equiv x^{n \bmod 593} + 1 \equiv x^{149} + 1 \pmod{n, x^{593} - 1}$. We conclude that $19{,}942{,}739$ is not prime. Indeed, $19{,}942{,}739 = 2{,}683 \times 7{,}433$. □

One can easily work out that under schoolbook arithmetic, the AKS algorithm runs in $O(\lg^{16.5} n)$ time. If one uses fast arithmetic (based on FFT), this running time drops to $\tilde{O}(\lg^{10.5} n)$. This exponent is quite high compared to the Miller–Rabin exponent (three). That is, one does not plan to use the AKS test frequently in practical applications. Lenstra and Pomerance's improvement of the AKS test runs in $\tilde{O}(\log^6 n)$ time.

5.4 Primality Tests for Numbers of Special Forms

The primality tests described until now work for general integers. For integers of specific types, more efficient algorithms can be developed.

5.4.1 Pépin Test for Fermat Numbers

A *Fermat number* f_m is of the form $f_m = 2^{2^m} + 1$ for some integer $m \geqslant 0$. In this section, we address the question which of the integers f_m are prime. One easily checks that if $2^a + 1$ is a prime, then a has to be of the form $a = 2^m$ for some $m \geqslant 0$. However, all integers of the form $f_m = 2^{2^m} + 1$ are not prime.

It turns out that $f_0 = 3$, $f_1 = 5$, $f_2 = 17$, $f_3 = 257$, and $f_4 = 65537$ are prime. Fermat conjectured that all numbers of the form $f_m = 2^{2^m} + 1$ are prime. Possibly, $f_5 = 2^{2^5} + 1 = 4294967297$ was too large to Fermat for hand calculations. Indeed, $f_5 = 641 \times 6700417$ is not prime. We know no value of m other than $0, 1, 2, 3, 4$, for which f_m is prime. It is an open question whether the Fermat numbers f_m for $m \geqslant 5$ are all composite.

A deterministic polynomial-time primality test for Fermat numbers can be developed based on the following result.

Theorem 5.29 [*Pépin's test*][25] *The Fermat number f_m for $m \geqslant 1$ is prime if and only if $3^{(f_m-1)/2} \equiv -1 \pmod{f_m}$.*
Proof [if] The condition $3^{(f_m-1)/2} \equiv -1 \pmod{f_m}$ implies $3^{f_m-1} \equiv 1 \pmod{f_m}$, that is, $\text{ord}_{f_m}(3) | f_m - 1 = 2^{2^m}$, that is, $\text{ord}_{f_m}(3) = 2^h$ for some h in the range $1 \leqslant h \leqslant 2^m$. However, if $h < 2^m$, we cannot have $3^{(f_m-1)/2} \equiv -1 \pmod{f_m}$. So $\text{ord}_{f_m}(3) = 2^{2^m} = f_m - 1$, that is, f_m is prime.

[only if] If f_m is prime, we have $3^{(f_m-1)/2} \equiv \left(\frac{3}{f_m}\right) \pmod{f_m}$ by Euler's criterion. By the quadratic reciprocity law, $\left(\frac{3}{f_m}\right) = (-1)^{(f_m-1)(3-1)/4} \left(\frac{f_m}{3}\right) = (-1)^{2^{2^m-1}} \left(\frac{f_m}{3}\right) = \left(\frac{f_m}{3}\right) = \left(\frac{2^{2^m}+1}{3}\right) = \left(\frac{(-1)^{2^m}+1}{3}\right) = \left(\frac{2}{3}\right) = -1.$ ◁

Therefore, Pépin's test involves only a modular exponentiation to the exponent $(f_m - 1)/2 = 2^{2^m-1}$, that can be computed by square operations only.

---

[25] Jean François Théophile Pépin (1826–1904) was a French mathematician.

Example 5.30 (1) We show that $f_3 = 2^{2^3} + 1 = 257$ is prime. We need to compute $3^{2^{2^3-1}}$, that is, 3^{2^7} modulo 257. Repeated squaring gives:

$$3^{2^0} \equiv 3 \pmod{257},$$
$$3^{2^1} \equiv (3^{2^0})^2 \equiv 9 \pmod{257},$$
$$3^{2^2} \equiv (3^{2^1})^2 \equiv 81 \pmod{257},$$
$$3^{2^3} \equiv (3^{2^2})^2 \equiv 136 \pmod{257},$$
$$3^{2^4} \equiv (3^{2^3})^2 \equiv 249 \pmod{257},$$
$$3^{2^5} \equiv (3^{2^4})^2 \equiv 64 \pmod{257},$$
$$3^{2^6} \equiv (3^{2^5})^2 \equiv 241 \pmod{257},$$
$$3^{2^7} \equiv (3^{2^6})^2 \equiv 256 \equiv -1 \pmod{257}.$$

(2) Let us compute $3^{2^{31}}$ modulo $f_5 = 2^{2^5} + 1 = 4294967297$.

$$3^{2^0} \equiv 3 \pmod{4294967297},$$
$$3^{2^1} \equiv (3^{2^0})^2 \equiv 9 \pmod{4294967297},$$
$$3^{2^2} \equiv (3^{2^1})^2 \equiv 81 \pmod{4294967297},$$
$$3^{2^3} \equiv (3^{2^2})^2 \equiv 6561 \pmod{4294967297},$$
$$3^{2^4} \equiv (3^{2^3})^2 \equiv 43046721 \pmod{4294967297},$$
$$3^{2^5} \equiv (3^{2^4})^2 \equiv 3793201458 \pmod{4294967297},$$
$$3^{2^6} \equiv (3^{2^5})^2 \equiv 1461798105 \pmod{4294967297},$$
$$\cdots$$
$$3^{2^{30}} \equiv (3^{2^{29}})^2 \equiv 1676826986 \pmod{4294967297},$$
$$3^{2^{31}} \equiv (3^{2^{30}})^2 \equiv 10324303 \pmod{4294967297}.$$

Since $3^{(f_5-1)/2} \not\equiv -1 \pmod{f_5}$, we conclude that f_5 is not prime. $\square$

5.4.2 Lucas–Lehmer Test for Mersenne Numbers

A *Mersenne number*[26] is of the form $M_n = 2^n - 1$ for $n \in \mathbb{N}$. It is easy to prove that if M_n is prime, then n has to be prime too. The converse of this is not true. For example, $2^{11} - 1 = 2047 = 23 \times 89$ is not prime. The computational question in this context is to find the primes p for which $M_p = 2^p - 1$ are primes. These primes M_p are called *Mersenne primes*.

Mersenne primes are *useful* for a variety of reasons. We do not know an easily computable formula to generate only prime numbers. Mersenne primes turn out to be the largest explicitly *known* primes. The collective Internet effort called the *Great Internet Mersenne Prime Search* (*GIMPS*)[27] bags an

---

[26]The French mathematician Marin Mersenne (1588–1648) studied these numbers.
[27]Look at the Internet site http://www.mersenne.org/.

award of US\$100,000 from the Electronic Frontier Foundation for the first discoverer of a prime with ten million (or more) digits. This prime happens to be the 47-th[28] Mersenne prime $2^{43,112,609} - 1$, a prime with 12,978,189 decimal digits, discovered on August 23, 2008 in the Department of Mathematics, UCLA. Running a deterministic primality test on such huge numbers is out of question. Probabilistic tests, on the other hand, do not furnish iron-clad proofs for primality and are infeasible too for these numbers. A special test known as the *Lucas–Lehmer test*[29] is used for deterministically checking the primality of Mersenne numbers.

A positive integer n is called a *perfect number* if it equals the sum of its proper positive integral divisors. For example, $6 = 1 + 2 + 3$ and $28 = 1 + 2 + 4 + 7 + 14$ are perfect numbers. It is known that n is an *even* perfect number if and only if it is of the form $2^{p-1}(2^p - 1)$ with $M_p = 2^p - 1$ being a (Mersenne) prime. Thus, Mersenne primes have one-to-one correspondence with even perfect numbers. We do not know any odd perfect number. We do not even know whether an odd perfect number exists.

Theorem 5.31 [*Lucas–Lehmer test*]   *The sequence s_i, $i \geqslant 0$, is defined as:*

$$s_0 = 4,$$
$$s_i = s_{i-1}^2 - 2 \text{ for } i \geqslant 1.$$

For $p \in \mathbb{P}$, M_p is prime if and only if $s_{p-2} \equiv 0 \pmod{M_p}$.   ◁

I am not going to prove this theorem here. The theorem implies that we need to compute the *Lucas–Lehmer residue* $s_{p-2} \pmod{M_p}$. The obvious iterative algorithm of computing s_i from s_{i-1} involves a square operation followed by reduction modulo M_p. Since $2^p \equiv 1 \pmod{M_p}$, we write $s_{i-1}^2 - 2 = 2^p n_1 + n_0$, and obtain $s_{i-1}^2 - 2 \equiv n_1 + n_0 \pmod{M_p}$. One can extract n_1, n_0 by bit operations. Thus, reduction modulo M_p can be implemented efficiently.

Example 5.32 (1) We prove that $M_7 = 2^7 - 1 = 127$ is prime. The calculations are shown below.

i	$s_i \pmod{M_7}$
0	4
1	$4^2 - 2 \equiv 14$
2	$14^2 - 2 \equiv 194 \equiv 1 \times 2^7 + 66 \equiv 1 + 66 \equiv 67$
3	$67^2 - 2 \equiv 4487 \equiv 35 \times 2^7 + 7 \equiv 35 + 7 \equiv 42$
4	$42^2 - 2 \equiv 1762 \equiv 13 \times 2^7 + 98 \equiv 13 + 98 \equiv 111$
5	$111^2 - 2 \equiv 12319 \equiv 96 \times 2^7 + 31 \equiv 96 + 31 \equiv 127 \equiv 0$

Since $s_{7-2} \equiv 0 \pmod{M_7}$, M_7 is prime.

---

[28]This is the 45th Mersenne prime to be discovered. Two smaller Mersenne primes were discovered later. It is not yet settled whether there are more undiscovered Mersenne primes smaller than $M_{43,112,609}$.

[29]The French mathematician François Édouard Anatole Lucas (1842–1891) introduced this test in 1856. It was later improved in 1930s by the American mathematician Derrick Henry Lehmer (1905–1991).

(2) We now run the Lucas–Lehmer test on $M_{11} = 2^{11} - 1 = 2047$.

i	$s_i \pmod{M_{11}}$
0	4
1	$4^2 - 2 \equiv 14$
2	$14^2 - 2 \equiv 194$
3	$194^2 - 2 \equiv 37634 \equiv 18 \times 2^{11} + 770 \equiv 18 + 770 \equiv 788$
4	$788^2 - 2 \equiv 620942 \equiv 303 \times 2^{11} + 398 \equiv 303 + 398 \equiv 701$
5	$701^2 - 2 \equiv 491399 \equiv 239 \times 2^{11} + 1927 \equiv 239 + 1927 \equiv 2166$
	$\equiv 1 \times 2048 + 118 \equiv 1 + 118 \equiv 119$
6	$119^2 - 2 \equiv 14159 \equiv 6 \times 2^{11} + 1871 \equiv 6 + 1871 \equiv 1877$
7	$1877^2 - 2 \equiv 3523127 \equiv 1720 \times 2^{11} + 567 \equiv 1720 + 567 \equiv 2287$
	$\equiv 1 \times 2^{11} + 239 \equiv 1 + 239 \equiv 240$
8	$240^2 - 2 \equiv 57598 \equiv 28 \times 2^{11} + 254 \equiv 28 + 254 \equiv 282$
9	$282^2 - 2 \equiv 79522 \equiv 38 \times 2^{11} + 1698 \equiv 38 + 1698 \equiv 1736$

Since $s_{11-2} \not\equiv 0 \pmod{M_{11}}$, M_{11} is not prime.   □

Exercises

1. For a positive integer n, the sum of the reciprocals of all primes $\leqslant n$ asymptotically approaches $\ln \ln n$. Using this fact, derive that the sieve of Eratosthenes can be implemented to run in $O(n \ln \ln n)$ time.

2. Modify the sieve of Eratosthenes so that it runs in $O(n)$ time. (**Hint:** Mark each composite integer only once.)

3. If both p and $2p+1$ are prime, we call p a *Sophie Germain prime*[30], and $2p+1$ a *safe prime*. It is conjectured that there are infinitely many Sophie Germain primes. In this exercise, you are asked to extend the sieve of Section 5.1.4 for locating the smallest Sophie Germain prime $p \geqslant n$ for a given positive integer $n \gg 1$. Sieve over the interval $[n, n + M]$.
   (a) Determine a value of M such that there is (at least) one Sophie Germain prime of the form $n + i$, $0 \leqslant i \leqslant M$, with high probability. The value of M should not be unreasonably large.
   (b) Describe a sieve to throw away the values of $n + i$ for which either $n + i$ or $2(n + i) + 1$ has a prime divisor less than or equal to the t-th prime. Take t as a constant (like 100).
   (c) Describe the gain in the running time that you achieve using the sieve.

4. Let s and t be bit lengths with $s > t$.
   (a) Describe an efficient algorithm to locate a random s-bit prime p such that a random prime of bit length t divides $p - 1$.
   (b) Express the expected running time of your algorithm in terms of s, t.
   (c) How can you adapt the sieve of Section 5.1.4 in this computation?

5. Let p, q be primes, $n = pq$, $a \in \mathbb{Z}_n^*$, and $d = \gcd(p - 1, q - 1)$.
   (a) Prove that n is a pseudoprime to base a if and only if $a^d \equiv 1 \pmod{n}$.
   (b) Prove that n is pseudoprime to exactly d^2 bases in $\mathbb{Z}_n^*$.
   (c) Let $q = 2p - 1$. To how many bases in $\mathbb{Z}_n^*$ is n a pseudoprime?
   (d) Repeat Part (c) for the case $q = 2p + 1$.

6. Let $n \in \mathbb{N}$ be odd and composite. If n is not a pseudoprime to some base in $\mathbb{Z}_n^*$, prove that n is not a pseudoprime to at least half of the bases in $\mathbb{Z}_n^*$.

7. Prove the following properties of any Carmichael number n.
   (a) $(p - 1)|(n - 1)$ for every prime divisor p of n.
   (b) n is odd.
   (c) n is square-free.
   (d) n has at least three distinct prime factors.

8. Suppose that $6k + 1$, $12k + 1$ and $18k + 1$ are all prime for some $k \in \mathbb{N}$. Prove that $(6k + 1)(12k + 1)(18k + 1)$ is a Carmichael number. Find two Carmichael numbers of this form.

---

[30]This is named after the French mathematician Marie-Sophie Germain (1776–1831). The name *safe prime* is attributed to the use of these primes in many cryptographic protocols.

9. Prove that for every odd prime r, there exist only finitely many Carmichael numbers of the form rpq (with p, q primes).

10. Prove that:
   (a) Every Euler pseudoprime to base a is also a pseudoprime to base a.
   (b) Every strong pseudoprime to base a is also a pseudoprime to base a.

11. Let n be an odd composite integer. Prove that:
   (a) There is at least one base $a \in \mathbb{Z}_n^*$, to which n is not an Euler pseudoprime.
   (b) n is not an Euler pseudoprime to at least half of the bases in $\mathbb{Z}_n^*$.

12. Prove that if $n \geqslant 3$ is a pseudoprime to base 2, then $2^n - 1$ is an Euler pseudoprime to base 2 and also a strong pseudoprime to base 2.

13. Let p and $q = 2p - 1$ be primes, and $n = pq$. Prove that:
   (a) n is an Euler pseudoprime to exactly one-fourth of the bases in $\mathbb{Z}_n^*$.
   (b) If $p \equiv 3 \pmod 4$, then n is a strong pseudoprime to exactly one-fourth of the bases in $\mathbb{Z}_n^*$.

14. Deduce the formulas (5.1), (5.4) and (5.6).

15. Prove that for all integers $m \geqslant 1$ and $n \geqslant 0$, the Fibonacci numbers satisfy

$$F_{m+n} = F_m F_{n+1} + F_{m-1} F_n.$$

Deduce the identities (5.2).

16. Prove the doubling formulas (5.7) for V_m defined in Section 5.2.5.

17. Write an analog of Algorithm 5.5 for the computation of $V_m \pmod n$.

18. [*Extra strong Lucas pseudoprime*] Let $U_m = U_m(a, 1)$ be the Lucas sequence with parameters $a, 1$, and $V_m = V_m(a, 1)$ the corresponding V sequence. Take an odd positive integer n with $\gcd(n, 2\Delta a) = 1$, where $\Delta = a^2 - 4$. We write $n - \left(\frac{\Delta}{n}\right) = 2^s t$ with t odd. We call n an extra strong Lucas pseudoprime to base a if either (i) $U_t \equiv 0 \pmod n$ and $V_t \equiv \pm 2 \pmod n$, or (ii) $V_{2^j t} \equiv 0 \pmod n$ for some $j \in \{0, 1, 2, \ldots, s - 1\}$. Prove that:
   (a) If $n \in \mathbb{P}$ does not divide 2Δ, then n is an extra strong Lucas pseudoprime.
   (b) An extra strong Lucas pseudoprime is also a strong Lucas pseudoprime.

19. The *Lehmer sequence* $\bar{U}_m$ with parameters a, b is defined as:

$$
\begin{aligned}
\bar{U}_0 &= 0, \\
\bar{U}_1 &= 1, \\
\bar{U}_m &= \bar{U}_{m-1} - b\bar{U}_{m-2} \text{ if } m \geqslant 2 \text{ is even,} \\
\bar{U}_m &= a\bar{U}_{m-1} - b\bar{U}_{m-2} \text{ if } m \geqslant 3 \text{ is odd.}
\end{aligned}
$$

Let α, β be the roots of $x^2 - \sqrt{a}\, x + b$.
   (a) Prove that $\bar{U}_m = \begin{cases} (\alpha^m - \beta^m)/(\alpha^2 - \beta^2) & \text{if } m \text{ is even,} \\ (\alpha^m - \beta^m)/(\alpha - \beta) & \text{if } m \text{ is odd.} \end{cases}$
   (b) Let $\Delta = a - 4b$, and n a positive integer with $\gcd(n, 2a\Delta) = 1$. We call n is *Lehmer pseudoprime* with parameters a, b if $\bar{U}_{n - \left(\frac{a\Delta}{n}\right)} \equiv 0 \pmod n$. Prove that n is a Lehmer pseudoprime with parameters a, b if and only if n is a Lucas pseudoprime with parameters a, ab.

20. The *Perrin sequence* $P(n)$ is defined recursively as:

$$P(0) = 3,$$
$$P(1) = 0,$$
$$P(2) = 2,$$
$$P(n) = P(n-2) + P(n-3) \text{ for } n \geqslant 3.$$

(a) Let $\omega \in \mathbb{C}$ be a primitive third root of unity. Verify that for $i = 0, 1, 2$,

$$\rho_i = \omega^i \sqrt[3]{\frac{1}{2} + \frac{1}{6}\sqrt{\frac{23}{3}}} + \omega^{2i} \sqrt[3]{\frac{1}{2} - \frac{1}{6}\sqrt{\frac{23}{3}}}$$

satisfy the characteristic equation $x^3 - x - 1 = 0$ of the Perrin sequence.

(b) The real number ρ_0 is called the *plastic number*. Deduce that $P(n)$ is the integer nearest to ρ_0^n for all sufficiently large n. (**Hint:** Show that $P(n) = \rho_0^n + \rho_1^n + \rho_2^n$ for all $n \geqslant 0$.)

(**Remark:** It is proved that if n is prime, then $n|P(n)$. Adams and Shanks (see Footnote 13) discover explicit examples of composite n (like $271441 = 521^2$) satisfying $n|P(n)$; these are (composite) *Perrin pseudoprimes*. Grantham[31] proves that there are infinitely many composite Perrin pseudoprimes.)

21. An odd prime of the form $k2^r + 1$ with $r \geqslant 1$, k odd, and $k < 2^r$, is called a *Proth prime*.[32] The first few Proth primes are $3, 5, 13, 17, 41, 97$.

(a) Describe an efficient way to recognize whether an odd positive integer (not necessarily prime) is of the form $k2^r + 1$ with $r \geqslant 1$, k odd, and $k < 2^r$. Henceforth, we call such an integer a *Proth number*.

(b) Suppose that a Proth number $n = k2^r + 1$ satisfies the condition that $a^{(n-1)/2} \equiv -1 \pmod{n}$ for some integer a. Prove that n is prime.

(c) Devise a yes-biased probabilistic polynomial-time algorithm to test the primality of a Proth number.

(d) Discuss how the algorithm of Part (c) can produce an incorrect answer. Also estimate the probability of this error.

(e) Prove that if the extended Riemann hypothesis (ERH) is true, one can convert the algorithm of Part (c) to a deterministic polynomial-time algorithm to test the primality of a Proth number.

22. [*Pocklington primality test*][33] Let n be a positive odd integer whose primality is to be checked. Write $n - 1 = uv$, where u is a product of small primes, and v has no small prime divisors. Suppose that the complete prime factorization of u is known, whereas no prime factor of v is known. Suppose also that for

---

[31] Jon Grantham, There are infinitely many Perrin pseudoprimes, *Journal of Number Theory*, 130(5), 1117–1128, 2010.

[32] The Proth test was developed by the French farmer François Proth (1852–1879), a self-taught mathematician.

[33] This test was proposed by the English mathematician Henry Cabourn Pocklington (1870–1952) and Lehmer.

some integer a, we have $a^{n-1} \equiv 1 \pmod{n}$, whereas $\gcd(a^{(n-1)/q}, n) = 1$ for all prime divisors q of u.

(a) Prove that every prime factor p of n satisfies $p \equiv 1 \pmod{u}$. (**Hint:** First, show that $u \mid \mathrm{ord}_p(a)$.)

(b) Conclude that if $u \geqslant \sqrt{n}$, then n is prime.

(c) Describe a situation when the criterion of Part (b) leads to an efficient algorithm for determining the primality of n.

23. Suppose that A_y is a yes-biased algorithm for proving the primality of an integer, and A_n a no-biased algorithm for the same purpose. Prove or disprove: By running A_y and A_n alone, we can deterministically conclude about the primality of an integer.

24. [*Binary search algorithm for finding integer k-th roots*]   Suppose that we want to compute $a = \lfloor \sqrt[k]{n} \rfloor$ for positive integers n and k. We start with a lower bound L and an upper bound U on a. We then run a loop, each iteration of which computes $M = (L + U)/2$, and decides whether $a > M$ or $a < M$. Depending on the outcome of this comparison, we refine one of the two bounds L, U. Complete the description of this algorithm for locating a. Determine a suitable condition for terminating the loop. Compare the performance of this method with the Newton–Raphson method discussed in Section 5.3.1, for computing integer k-th roots.

25. Let s_i be the sequence used in the Lucas–Lehmer test for Mersenne numbers. Prove that $s_i = (2 + \sqrt{3})^{2^i} + (2 - \sqrt{3})^{2^i}$ for all $i \geqslant 0$.

26. Prove that for $m \geqslant 2$, the Fermat number $f_m = 2^{2^m} + 1$ is prime if and only if $5^{(f_m - 1)/2} \equiv -1 \pmod{f_m}$.

| **Programming Exercises** |

Write `GP/PARI` functions to implement the following.

27. Obtaining a random prime of a given bit length l.

28. The Solovay–Strassen test.

29. The Miller–Rabin test.

30. The Fibonacci test (you may use the function `FibMod()` of Section 5.2.4).

31. The Lucas test.

32. The strong Lucas test.

33. The AKS test.

34. The Pépin test.

35. The Lucas–Lehmer test.

Chapter 6

Integer Factorization

Now that we are able to *quickly* recognize primes as primes, it remains to compute the prime factorization of (positive) integers. This is the tougher part of the story. Research efforts for decades have miserably failed to produce efficient algorithms for factoring integers. Even randomization does not seem to help here. Today's best integer-factoring algorithms run in subexponential time which, although better than exponential time, makes the factoring problem practically intractable for input integers of size only thousand bits.

This chapter is an introduction to some integer-factoring algorithms. We start with a few fully exponential algorithms. These old algorithms run efficiently in certain specific situations, so we need to study them.

Some subexponential algorithms are discussed next. Assume that n is the (positive) integer to be factored. A subexponential expression in $\log n$ is, in this context, an expression of the form

$$L(n, \omega, c) = \exp \left[(c + \mathrm{o}(1))(\ln n)^{\omega} (\ln \ln n)^{1-\omega} \right],$$

where ω is a real number in the open interval $(0,1)$, and c is a positive real number. Plugging in $\omega = 0$ in $L(n, \omega, c)$ gives a polynomial expression in $\ln n$. On the other hand, for $\omega = 1$, the expression $L(n, \omega, c)$ is fully exponential in $\ln n$. For $0 < \omega < 1$, the expression $L(n, \omega, c)$ is something between polynomial and exponential, and is called a subexponential expression in $\ln n$.

Most modern integer-factoring algorithms have running times of the form $L(n, \omega, c)$. Smaller values of ω lead to expressions closer to polynomial. During 1970s and 1980s, several integer-factoring algorithms are designed with $\omega = 1/2$. It was, for a while, apparent that $\omega = 1/2$ is possibly the best exponent we can achieve. In 1989, the number-field sieve method was proposed. This algorithm corresponds to $\omega = 1/3$, and turns out to be the fastest (both theoretically and practically) known algorithm for factoring integers.

A complete understanding of the number-field sieve method calls for mathematical background well beyond what this book can handle. We will mostly study some $L(n, 1/2, c)$ integer-factoring algorithms. We use the special symbol $L[n, c]$ or $L_n[c]$ to stand for $L(n, 1/2, c)$. If n is understood from the context, we will abbreviate this notation further as $L[c]$.

An integer n we wish to factor is first subjected to a primality test. Only when we are certain that n is composite do we attempt to factor it. In view of the results of Section 5.3.1, we may also assume that n is not a perfect power. An algorithm that computes a non-trivial decomposition $n = uv$ with $1 < u, v < n$ (or that only supplies a non-trivial factor u of n) can be recursively used to factor u and $v = n/u$. So it suffices to compute a non-trivial split of n.

Composite integers of some special forms are often believed to be more difficult to factor than others. For example, the (original) RSA encryption algorithm is based on composite moduli of the form $n = pq$ with p, q being primes of roughly the same bit length. For such composite integers, computing a non-trivial split (or factor) is equivalent to fully factoring n.

The generic factoring function in GP/PARI is `factor()`. The function that specifically handles integers is `factorint()`. One may supply an optional flag to `factorint()` to indicate the user's choice of the factoring algorithm (see the online GP/PARI manual). Here follows a sample conversation with the GP/PARI calculator. The last example illustrates that for about a minute, the calculator tries to use *easy* methods for factoring $2^{301} - 1$. When these easy methods fail, it pulls out a big gun like the MPQS (multiple-polynomial quadratic sieve). A new version of GP/PARI, however, completely factors $2^{301} - 1$ in about two minutes. An attempt to factor $2^{401} - 1$ in this version runs without any warning message, and fails to output the result in fifteen minutes.

```
gp > factor(2^2^5+1)
%1 =
[641 1]

[6700417 1]

gp > factorint(2^2^5+1)
%2 =
[641 1]

[6700417 1]

gp > #
    timer = 1 (on)
```

```
gp > factorint(2^101-1)
time = 68 ms.
%3 =
[7432339208719 1]

[341117531003194129 1]

gp > factorint(2^201-1)
time = 1,300 ms.
%4 =
[7 1]

[1609 1]

[22111 1]

[193707721 1]

[761838257287 1]

[87449423397425857942678833145441 1]

gp > factorint(2^201-1,1)
time = 1,540 ms.
%5 =
[7 1]

[1609 1]

[22111 1]

[193707721 1]

[761838257287 1]

[87449423397425857942678833145441 1]

gp > factorint(2^301-1)
   ***    Warning: MPQS: the factorization of this number will take several hours.
   ***    user interrupt after 1mn, 34,446 ms.
gp >
```

6.1 Trial Division

An obvious way to factor a composite n is to divide n by potential divisors d in the range $2 \leqslant d \leqslant \lfloor \sqrt{n} \rfloor$. A successful trial division by d would also replace n by n/d. The process continues until n reduces to 1 or a prime factor.

It is easy to conceive that trial divisions need to be carried out by only prime numbers. Indeed, if d is a composite divisor of n, then all prime divisors of d divide n and are smaller than d. Therefore, before d is tried, all prime divisors of d are already factored out from n.

However, one then requires a list of primes $\leqslant \sqrt{n}$. It is often not feasible to have such a list. On the other hand, checking every potential divisor d for primality before making a trial division of n by d is a massive investment of time. A practical trade-off can be obtained using the following idea.[1]

After 2 is tried as a potential divisor, there is no point dividing n by even integers. This curtails the space for potential divisors by a factor of 2. Analogously, we should not carry out trial division by multiples of 3 (other than 3 itself) and by multiples of 5 (other than 5 itself). What saving does it produce? Consider $d > 2 \times 3 \times 5 = 30$ with $r = d \operatorname{rem} 30$. If r is not coprime to 30, then d is clearly composite. Moreover, $\phi(30) = (2-1) \times (3-1) \times (5-1) = 8$, that is, only 8 (out of 30) values of r may be prime. Thus, trial division may be skipped by $d > 30$ unless $r = d \operatorname{rem} 30$ is among $1, 7, 11, 13, 17, 19, 23, 29$. This reduces the search space for potential divisors to about one-fourth. One may, if one chooses, use other small primes like $7, 11, 13, \ldots$ in this context. But considering four or more small primes leads to additional bookkeeping, and produces improvements that are not too dramatic. It appears that considering only the first three primes $2, 3, 5$ is a practically optimal choice.

Example 6.1 Let us factor

$$n = 3^{61} + 1 = 127173474825648610542883299604$$

by trial division. We first divide n by 2. Since n is even, 2 is indeed a factor of n. Replacing n by $n/2$ gives

$$63586737412824305271441649802$$

which is again even. So we make another trial division by 2, and reduce n to

$$31793368706412152635720824901.$$

A primality test reveals that this reduced n is composite. So we divide this n by the remaining primes < 30, that is, by $3, 5, 7, 11, 13, 17, 19, 23, 29$. It turns out that n is divisible by neither of these primes.

As potential divisors $d \geqslant 30$ of n, we consider only the values $30k+r$ for $r = 1, 7, 11, 13, 17, 19, 23, 29$. That is, we divide n by $31, 37, 41, 43, 47, 49, 53, 59, 61, 67, 71, 73, 77, 79, 83, 89, 91, 97, \ldots$. Some of these divisors are not prime (like $49, 77, 91$), but that does not matter. Eventually, we detect a divisor $367 = 12 \times 30 + 7$ of n. Clearly, 367 has to be prime. Reducing n by $n/361$ gives

$$86630432442539925437931403.$$

A primality test shows that this reduced n is prime. So there is no need to carry out trial division further, that is, we have the complete factorization

$$n = 3^{61} + 1 = 2^2 \times 367 \times 86630432442539925437931403.$$

---

[1] The trial-division algorithm in this form is presented in: Henri Cohen, *A course in computational algebraic number theory*, Graduate Text in Mathematics, 138, Springer, 1993.

This example illustrates that the method of trial division factors n efficiently if all (except at most one) prime divisors of n are small. $\qquad\square$

A complete factorization of n by trial division calls for a worst-case running time of $O^\sim(\sqrt{n})$. This bound is achievable, for example, for RSA moduli of the form $n = pq$ with bit sizes of the primes p, q being nearly half of that of n. So trial division is impractical except only for small values of n (like $n \leqslant 10^{20}$). For factoring larger integers, more sophisticated ideas are needed.

Before employing these sophisticated algorithms, it is worthwhile to divide n by a set of small primes. That reveals the small factors of n, and may reduce its size considerably so as to make the sophisticated algorithms run somewhat faster. In view of this, it will often be assumed that the number to be factored does not contain small prime divisors.

Example 6.2 Let us use trial division to extract the small prime factors of

$$n = 3^{60} + 1 = 42391158275216203514294433202.$$

Considering all potential divisors $d \leqslant 10^4$ decomposes n as

$$n = 3^{60} + 1 = 2 \times 41 \times 241 \times 6481 \times 330980468807135443441.$$

Primality tests indicate that the last factor is composite. We use sophisticated algorithms for factoring this part of n. $\qquad\square$

In practice, a program for factoring integers would load at startup a pre-calculated list of small primes. This list may be as large as consisting of the first ten million primes. It is useful to make trial divisions by all these primes. If the list is not so large or not available at all, one may divide n by $d \equiv 1, 7, 11, 13, 17, 19, 23, 29 \pmod{30}$, as explained earlier.

6.2 Pollard's Rho Method

Let n be a composite positive integer which is not a perfect power, and p a prime divisor of n. We generate a sequence of integers $x_0, x_1, x_2, \ldots$ modulo n. We start with a random $x_0 \in \mathbb{Z}_n$, and subsequently obtain $x_i = f(x_{i-1})$ for $i \geqslant 1$. Here, f is an easily computable function with the property that $x_0, x_1, x_2, \ldots$ *behaves* like a random sequence in $\mathbb{Z}_n$. The most common choice is $f(x) = x^2 + a \pmod{n}$, where $a \neq 0, -2$ is an element of $\mathbb{Z}_n$.

Since x_i is generated from x_{i-1} using a deterministic formula, and since $\mathbb{Z}_n$ is finite, the sequence $x_0, x_1, x_2, \ldots$ must be eventually periodic. We also consider the sequence $x_0', x_1', x_2', \ldots$, where each x_i' is the reduction of x_i modulo (the unknown prime) p. The reduced sequence $x_0', x_1', x_2', \ldots$ is periodic too, since each x_i' is an element of the finite set $\mathbb{Z}_p$.

Let τ be the (smallest) period of the sequence $x_0, x_1, x_2, \ldots$, and τ' the period of the sequence $x'_0, x'_1, x'_2, \ldots$. It is clear that $\tau' | \tau$. If it so happens that $\tau' < \tau$, then there exist i, j with $i < j$ such that $x_i \not\equiv x_j \pmod{n}$ but $x'_i \equiv x'_j \pmod{p}$, that is, $x_i \equiv x_j \pmod{p}$. In that case, $d = \gcd(x_j - x_i, n)$ is a proper divisor of n. On the other hand, if $\tau' = \tau$, then for all i, j with $i < j$, $\gcd(x_j - x_i, n)$ is either 1 or n.

Pollard's rho method[2] is based on these observations. However, computing $x_j - x_i$ for all i, j with $i < j$ is a massive investment of time. Moreover, we need to store all x_i values until a pair i, j with $\gcd(x_j - x_i, n) > 1$ is located.

6.2.1 Floyd's Variant

The following proposition leads to a significant saving in time and space.

Proposition 6.3 *There exists $k \geqslant 1$ such that $x_k \equiv x_{2k} \pmod{p}$.*
Proof We have $x_i \equiv x_j \pmod{p}$ for some i, j with $i < j$. But then $x_k \equiv x_{k+s(j-i)} \pmod{p}$ for all $k \geqslant i$ and for all $s \in \mathbb{N}$. So take k to be any multiple of $j - i$ larger than or equal to i, and adjust s accordingly. ◁

We compute the sequence $d_k = \gcd(x_{2k} - x_k, n)$ for $k = 1, 2, 3, \ldots$ until a gcd $d_k > 1$ is located. This variant of the Pollard rho method is called *Floyd's variant*,[3] and is supplied as Algorithm 6.1.

Algorithm 6.1: POLLARD'S RHO METHOD (FLOYD'S VARIANT)

```
Initialize x and y to a random element of Z_n.  /* x = y = x_0 */
Repeat until explicitly broken {
    Update x = f(x).      /* x = x_k here */
    Update y = f(f(y)).   /* y = x_2k here */
    Compute d = gcd(y - x, n).
    If (d > 1), break the loop.
}
Return d.
```

Example 6.4 Let us try to factor the Fermat number $n = f_5 = 2^{2^5} + 1 = 4294967297$ by Pollard's rho method. We take the sequence-generating function $f(x) = x^2 + 1 \pmod{n}$. The computations done by Algorithm 6.1 are illustrated in the following table. We start with the initial term $x_0 = 123$. The non-trivial factor 641 of n is discovered by Pollard's rho method. The corresponding cofactor is 6700417. That is, $f_5 = 641 \times 6700417$. Both these factors are prime, that is, we have completely factored f_5. □

---

[2] John M. Pollard, A Monte Carlo method for factorization, *BIT Numerical Mathematics*, 15(3), 331–334, 1975.
[3] Robert W. Floyd, Non-deterministic algorithms, *Journal of ACM*, 14(4), 636–644, 1967. Floyd's paper (1967) presents an algorithn for finding cycles in graphs. Pollard uses this algorithm in his factoring paper (1975). This explains the apparent anachronism.

k	$x_k = f(x_{k-1})$	$x_{2k-1} = f(x_{2(k-1)})$	$x_{2k} = f(x_{2k-1})$	$\gcd(x_{2k} - x_k, n)$
0	123	–	123	–
1	15130	15130	228916901	1
2	228916901	33139238	3137246933	1
3	33139238	2733858014	4285228964	1
4	3137246933	2251701130	1572082836	1
5	2733858014	1467686897	1705858549	1
6	4285228964	1939628362	4277357175	1
7	2251701130	2578142297	3497150839	1
8	1572082836	962013932	1052924614	1
9	1467686897	2363824544	2370126580	1
10	1705858549	2736085524	4145405717	1
11	1939628362	4082917731	786147859	1
12	4277357175	954961871	3660251575	1
13	2578142297	2240785070	2793469517	1
14	3497150839	2846659272	812871200	1
15	962013932	385091241	3158671825	1
16	1052924614	2852659993	4184934804	1
17	2363824544	1777772295	9559945	1
18	2370126580	4234257460	1072318990	1
19	2736085524	1259270631	2648112086	1
20	4145405717	473166199	504356342	1
21	4082917731	1740566372	2252385287	1
22	786147859	2226234309	3261516148	641

The running time of Algorithm 6.1 is based upon the following result.

Proposition 6.5 *Let S be a set of size n. If $k \leqslant n$ elements are selected from S with replacement, the probability p_k of at least one match among the k chosen elements (that is, at least one element is selected more than once) is*

$$p_k = 1 - \prod_{i=1}^{k-1} \left(1 - \frac{i}{n}\right) \approx 1 - e^{\frac{-k^2}{2n}}.$$

We have $p_k \approx 1/2$ for $k \approx 1.177\sqrt{n}$. Also, $p_k \approx 0.99$ for $k \approx 3.035\sqrt{n}$. ◁

Let us now look at the reduced sequence $x_0', x_1', x_2', \ldots$ modulo p. If we assume that this sequence is random, then after $\Theta(\sqrt{p})$ iterations, we expect to obtain a collision, that is, i, j with $i < j$ and $x_i' \equiv x_j' \pmod{p}$. Therefore, we have to make an expected number of $\Theta(\sqrt{p})$ gcd calculations. Since n has a prime divisor $\leqslant \sqrt{n}$, the expected running time of Pollard's rho method is $\tilde{O}(\sqrt[4]{n})$. Although this is significantly better than the running time $\tilde{O}(\sqrt{n})$ of trial division, it is still an exponential function in $\log n$. So Pollard's rho method cannot be used for factoring large integers.

The running time of Pollard's rho method is output-sensitive in the sense that it depends upon the smallest prime factor p of n. If p is small, Pollard's rho method may detect p quite fast, namely in $\tilde{O}(\sqrt{p})$ time. On the other hand, trial division by divisors $\leqslant p$ requires $\tilde{O}(p)$ running time.

6.2.2   Block GCD Calculation

Several modifications of Pollard's rho method reduce the running time by (not-so-small) constant factors. The most expensive step in the loop of Algorithm 6.1 is the computation of the gcd. We can avoid this computation in every iteration. Instead, we accumulate the product $(x_{2k}-x_k)(x_{2k+2}-x_{k+1})$ $(x_{2k+4} - x_{k+2}) \cdots (x_{2k+2r-2} - x_{k+r-1}) \pmod n$ for r iterations. We compute the gcd of this product with n. If all these $x_{2k+2i} - x_{k+i}$ are coprime to n, then the gcd of the product with n is also 1. On the other hand, if some $\gcd(x_{2k+2i} - x_{k+i}, n) > 1$, then the gcd d of the above product with n is larger than 1. If $d < n$, we have found a non-trivial divisor of n. If $d = n$, we compute the individual gcds of $x_{2k+2i} - x_{k+i}$ with n for $i = 0, 1, 2, \ldots, r-1$ until a gcd larger than 1 is located. This strategy is called *block gcd calculation*. In practice, one may use r in the range $20 \leqslant r \leqslant 50$ for optimal performance.

Algorithm 6.2: Pollard's rho method (Brent's variant)

```
Initialize x to a random element of Z_n.         /* Set x = x_0 */
Initialize y to f(x).                            /* Set y = x_1 */
Set t = 1.                          /* t stores the value 2^{r-1} */
Compute d = gcd(y - x, n).
while (d equals 1) {
        Set x = y.                  /* Store in x the element x_{2^r - 1} */
        For s = 1, 2, ..., t, set y = f(y). /* Compute x_{(2^r - 1)+s}, no gcd */
        For s = 1, 2, ..., t {           /* Compute x_{(2^r - 1)+s} with gcd */
                Set y = f(y).
                Compute d = gcd(y - x, n).
                If (d > 1), break the inner (for) loop.
        }
        Set t = 2t.                 /* Prepare for the next iteration */
}
Return d.
```

6.2.3   Brent's Variant

Brent's variant[4] reduces the number of gcd computations considerably. As earlier, τ' stands for the smallest period of the sequence $x'_0, x'_1, x'_2, \ldots$ modulo p. We first compute $d = \gcd(x_1 - x_0, n)$. If this gcd is larger than 1, then $\tau' = 1$, and the algorithm terminates. If $d = 1$, we compute $\gcd(x_k - x_1, n)$ for $k = 3$, that is, we now check whether $\tau' = 2$. Then, we compute $\gcd(x_k - x_3, n)$ for $n = 6, 7$. Here, τ' is searched among $3, 4$. More generally, for $r \in \mathbb{N}$, we compute $\gcd(x_k - x_{2^r - 1}, n)$ for $2^r + 2^{r-1} \leqslant k \leqslant 2^{r+1} - 1$. This means

---

[4]Richard P. Brent, An improved Monte Carlo factorization algorithm, *BIT*, 20, 176–184, 1980.

that we search for values of τ' in the range $2^{r-1} + 1 \leqslant \tau' \leqslant 2^r$. The last value of x_k in the loop is used in the next iteration for the computation of $\gcd(x_k - x_{2^{r+1}-1}, n)$. Since we have already investigated the possibilities for $\tau' \leqslant 2^r$, we start gcd computations from $k = (2^{r+1} - 1) + 2^r + 1 = 2^{r+1} + 2^r$. Algorithm 6.2 formalizes these observations.

Example 6.6 We factor the Fermat number $n = f_5 = 2^{2^5} + 1 = 4294967297$ by Brent's variant of Pollard's rho method. As in Example 6.4, we take $f(x) = x^2 + 1 \pmod{n}$, and start the sequence with $x_0 = 123$. A dash in the last column of the following table indicates that the gcd is not computed.

r	$t = 2^{r-1}$	$2^r - 1$	x_{2^r-1}	k	x_k	$\gcd(x_k - x_{2^r-1}, n)$
0	0	0	123	1	15130	1
1	1	1	15130	2	228916901	—
				3	33139238	1
2	2	3	33139238	4	3137246933	—
				5	2733858014	—
				6	4285228964	1
				7	2251701130	1
3	4	7	2251701130	8	1572082836	—
				9	1467686897	—
				10	1705858549	—
				11	1939628362	—
				12	4277357175	1
				13	2578142297	1
				14	3497150839	1
				15	962013932	1
4	8	15	962013932	16	1052924614	—
				17	2363824544	—
				18	2370126580	—
				19	2736085524	—
				20	4145405717	—
				21	4082917731	—
				22	786147859	—
				23	954961871	—
				24	3660251575	1
				25	2240785070	1
				26	2793469517	641

We get the factorization $f_5 = 641 \times 6700417$. Here, only 11 gcds are computed. Compare this with the number (22) of gcds computed in Example 6.4. □

The working of Algorithm 6.2 is justified if the sequence $x'_0, x'_1, x'_2, \ldots$ is totally periodic. In general, the initial non-periodic part does not turn out to be a big problem, since it suffices to locate a multiple of τ' after the sequence gets periodic. The expected running time of Algorithm 6.2 is $O^{\sim}(\sqrt[4]{n})$. The concept of block gcd calculation can be applied to Brent's variant also.

6.3   Pollard's $p - 1$ Method

Trial division and Pollard's rho method are effective for factoring n if n has small prime factor(s). Pollard's $p-1$ method[5] is effective if $p-1$ has *small* (in the following sense) prime factors for some prime divisor p of n. Note that p itself may be large, but that $p - 1$ has only small prime factors helps.

Definition 6.7 Let $B \in \mathbb{N}$ be a bound. A positive integer n is called B-*power-smooth* if whenever $q^e | n$ for $q \in \mathbb{P}$ and $e \in \mathbb{N}$, we have $q^e \leqslant B$.   ◁

If $p - 1$ is B-power-smooth for a prime divisor p of n for a small bound B, one can extract the prime factor p of n using the following idea. Let a be any integer coprime to n (and so to p too). By Fermat's little theorem, $a^{p-1} \equiv 1 \pmod{p}$, that is, $p | (a^{p-1} - 1)$. More generally, $p | (a^{k(p-1)} - 1)$ for all $k \in \mathbb{N}$. If, on the other hand, $n \nmid (a^{k(p-1)} - 1)$, then $\gcd(a^{k(p-1)} - 1, n)$ is a non-trivial divisor of n. If $p - 1$ is B-power-smooth for a known bound B, a suitable multiple $k(p - 1)$ of $p - 1$ can be obtained easily as follows.

Let $p_1, p_2, \ldots, p_t$ be all the primes $\leqslant B$. For each $i = 1, 2, \ldots, t$, define $e_i = \lfloor \log B / \log p_i \rfloor$. Consider the exponent $E = \prod_{i=1}^{t} p_i^{e_i}$. If $p - 1$ is B-power-smooth, then this exponent E is a multiple of $p - 1$. Note that E may be quite large. However, there is no need to compute E explicitly. We have $a^E \equiv (\cdots ((a^{p_1^{e_1}})^{p_2^{e_2}})^{p_3^{e_3}} \cdots)^{p_t^{e_t}} \pmod{n}$. That is, $a^E \pmod{n}$ can be obtained by a suitable sequence of exponentiations by small primes. It is also worthwhile to consider the fact that instead of computing $a^E \pmod{n}$ at one shot and then computing $\gcd(a^E - 1, n)$, one may sequentially compute $a^{p_1^{e_1}}$, $(a^{p_1^{e_1}})^{p_2^{e_2}}$, $((a^{p_1^{e_1}})^{p_2^{e_2}})^{p_3^{e_3}}$, and so on. After each exponentiation by $p_i^{e_i}$, one computes a gcd. If $p - 1$ is B'-power-smooth for some $B' < B$, this saves exponentiations by prime powers $p_i^{e_i}$ for primes p_i in the range $B' < p_i \leqslant B$.

Pollard's $p - 1$ method is given as Algorithm 6.3. By the prime number theorem, there are $O(B / \log B)$ primes $\leqslant B$. Since each $p_i^{e_i} \leqslant B$, the exponentiation $a^{p_i^{e_i}} \pmod{n}$ can be done in $O(\log B \log^2 n)$ time. Each gcd d can be computed in $O(\log^2 n)$ time. So Algorithm 6.3 runs in $O(B \log^2 n)$ time. Pollard's $p - 1$ method is output-sensitive, and is efficient for small B.

Let us now investigate when Algorithm 6.3 fails to output a non-trivial factor d of n, that is, the cases when the algorithm returns $d = 1$ or $d = n$. The case $d = 1$ indicates that $p-1$ is not B-power-smooth. We can then repeat the algorithm with a larger value of B. Of course, B is unknown *a priori*. So one may repeat Pollard's $p - 1$ method for gradually increasing values of B. However, considering large values of B (like $B > 10^6$) is usually not a good idea. One should employ subexponential algorithms instead.

---

[5] John M. Pollard, Theorems of factorization and primality testing, *Proceedings of the Cambridge Philosophical Society*, 76(3), 521–528, 1974.

Algorithm 6.3: POLLARD'S $p - 1$ METHOD

Choose a base $a \in \mathbb{Z}_n^*$.
For each small prime $q \leqslant B$ {
    Set $e = \lfloor \log B / \log q \rfloor$.
    For $i = 1, 2, \ldots, e$, compute $a = a^q \pmod{n}$.
    Compute $d = \gcd(a - 1, n)$.
    If $(d > 1)$, break the loop.
}
Return d.

The case $d = n$ is trickier. Suppose that $n = pq$ with distinct primes p, q. Suppose that both $p - 1$ and $q - 1$ are B-power-smooth. In that case, both the congruences $a^E \equiv 1 \pmod{p}$ and $a^E \equiv 1 \pmod{q}$ hold. Therefore, $\gcd(a^E - 1, n) = n$, that is, we fail to separate p and q. It is, however, not necessary for $q - 1$ to be B-power-smooth in order to obtain $d = n$. If $\mathrm{ord}_q(a)$ is B-power-smooth, then also we may get $d = n$. For some other base a', we may have $\mathrm{ord}_q(a')$ not B-power-smooth. In that case, $\gcd(a'^E - 1, n)$ is a non-trivial factor (that is, p) of n. This argument can be readily extended to composite integers other than those of the form pq. When we obtain $d = n$ as the output of Algorithm 6.3, it is worthwhile to run the algorithm for a few other bases a. If we always obtain $d = n$, we report failure.

Example 6.8 (1) We factor $n = 1602242212193303$ by Pollard's $p - 1$ method. For the bound $B = 16$ and the base $a = 2$, we have the following sequence of computations. Here, p_i denotes the i-th prime, and $e_i = \lfloor \log B / \log p_i \rfloor$.

i	p_i	e_i	a_{old}	$a_{\text{new}} \equiv a_{\text{old}}^{p_i^{e_i}} \pmod{n}$	$\gcd(a_{\text{new}} - 1, n)$
1	2	4	2	65536	1
2	3	2	65536	50728532231011	1
3	5	1	50728532231011	602671824969697	1
4	7	1	602671824969697	328708173547029	1
5	11	1	328708173547029	265272211830818	1
6	13	1	265272211830818	167535681578625	1

Since the algorithm outputs 1, we try with the increased bound $B = 32$, and obtain the following sequence of computations. The base is again 2.

i	p_i	e_i	a_{old}	$a_{\text{new}} \equiv a_{\text{old}}^{p_i^{e_i}} \pmod{n}$	$\gcd(a_{\text{new}} - 1, n)$
1	2	5	2	4294967296	1
2	3	3	4294967296	98065329688549	1
3	5	2	98065329688549	1142911340053463	1
4	7	1	1142911340053463	1220004434814213	1
5	11	1	1220004434814213	1358948392128938	1
6	13	1	1358948392128938	744812946196424	1
7	17	1	744812946196424	781753012202740	1
8	19	1	781753012202740	1512971883283798	17907121

This gives the factorization $n = 17907121 \times 89475143$ with both the factors prime. Although the bound B is 32, we do not need to raise a to the powers $p_i^{e_i}$ for $p_i = 23, 29, 31$. This is how computing the gcd inside the loop helps.

In order to see why Pollard's $p-1$ method works in this example, we note the factorization of $p-1$ and $q-1$, where $p = 17907121$ and $q = 89475143$.

$$p - 1 = 2^4 \times 3^2 \times 5 \times 7 \times 11 \times 17 \times 19,$$
$$q - 1 = 2 \times 19 \times 2354609.$$

So $p-1$ is 32-power-smooth (indeed, it is 19-power-smooth), but $q-1$ is not.

(2) Let us now try to factor $n = 490428787297681$ using Pollard's $p-1$ method. We work with the bound $B = 24$ and the base 2.

i	p_i	e_i	a_{old}	$a_{\text{new}} \equiv a_{\text{old}}^{p_i^{e_i}} \pmod{n}$	$\gcd(a_{\text{new}} - 1, n)$
1	2	4	2	65536	1
2	3	2	65536	190069571010731	1
3	5	1	190069571010731	185747041897072	1
4	7	1	185747041897072	401041553458073	1
5	11	1	401041553458073	31162570081268	1
6	13	1	31162570081268	248632716971464	1
7	17	1	248632716971464	67661917074372	1
8	19	1	67661917074372	1	490428787297681

We fail to separate p from q in this case. So we try other bases. The base 23 works as given in the next table. The bound remains $B = 24$ as before.

i	p_i	e_i	a_{old}	$a_{\text{new}} \equiv a_{\text{old}}^{p_i^{e_i}} \pmod{n}$	$\gcd(a_{\text{new}} - 1, n)$
1	2	4	23	487183864388533	1
2	3	2	487183864388533	422240241462789	1
3	5	1	422240241462789	64491241974109	1
4	7	1	64491241974109	88891658296507	1
5	11	1	88891658296507	143147690932110	1
6	13	1	143147690932110	244789218562995	1
7	17	1	244789218562995	334411207888980	1
8	19	1	334411207888980	381444508879276	17907121

We obtain $n = pq$ with $p = 17907121$ and $q = 27387361$. We have

$$p - 1 = 2^4 \times 3^2 \times 5 \times 7 \times 11 \times 17 \times 19,$$
$$q - 1 = 2^5 \times 3^2 \times 5 \times 7 \times 11 \times 13 \times 19.$$

This shows that $q-1$ is not 24-power-smooth (it is 32-power-smooth). However, $\text{ord}_q(2) = 1244880 = 2^4 \times 3^2 \times 5 \times 7 \times 13 \times 19$ is 24-smooth, so the base 2 yields the output $d = n$. On the other hand, $\text{ord}_q(23) = 480480 = 2^5 \times 3 \times 5 \times 7 \times 11 \times 13$ is not 24-power-smooth, so the base 23 factors n. $\square$

6.3.1 Large Prime Variation

If Algorithm 6.3 outputs $d = 1$, then $p-1$ is not B-power-smooth. However, it may be the case that $p - 1 = uv$, where u is B-power-smooth, and v is a prime $> B$. In this context, v is called a *large prime* divisor of $p - 1$. If v is not too large, like $v \leqslant B'$ for some bound $B' > B$, adding a second stage to the $p - 1$ factoring algorithm may factor n non-trivially.

Let $p_{t+1}, p_{t+2}, \ldots, p_{t'}$ be all the primes in the range $B < p_i \leqslant B'$. When Algorithm 6.3 terminates, it has already computed $a^E \pmod{n}$ with $u | E$. We then compute $(a^E)^{p_{t+1}}$, $((a^E)^{p_{t+1}})^{p_{t+2}}$, and so on (modulo n) until we get a gcd > 1, or all primes $\leqslant B'$ are exhausted. This method works if v is a square-free product of large primes between B (exclusive) and B' (inclusive).

One could have executed Algorithm 6.3 with the larger bound B' (as B). But then each e_i would be large (compare $\lfloor \log B' / \log p_i \rfloor$ with $\lfloor \log B / \log p_i \rfloor$). Moreover, for all large primes p_i, we take $e_i = 1$. Therefore, using two different bounds B, B' helps. In practice, one may take $B \leqslant 10^6$ and $B' \leqslant 10^8$.

Example 6.9 We attempt to factor $n = 600735950824741$ using Pollard's $p - 1$ algorithm augmented by the second stage. We use the bounds $B = 20$ and $B' = 100$, and select the base $a = 2$. The two stages proceed as follows.

Stage 1

i	p_i	e_i	a_{old}	$a_{\text{new}} \equiv a_{\text{old}}^{p_i^{e_i}} \pmod{n}$	$\gcd(a_{\text{new}} - 1, n)$
1	2	4	2	65536	1
2	3	2	65536	250148431288895	1
3	5	1	250148431288895	404777501817913	1
4	7	1	404777501817913	482691043667836	1
5	11	1	482691043667836	309113846434884	1
6	13	1	309113846434884	297529593613895	1
7	17	1	297529593613895	544042973919022	1
8	19	1	544042973919022	358991192319517	1

Stage 2

i	p_i	e_i	a_{old}	$a_{\text{new}} \equiv a_{\text{old}}^{p_i^{e_i}} \pmod{n}$	$\gcd(a_{\text{new}} - 1, n)$
9	23	1	358991192319517	589515560613570	1
10	29	1	589515560613570	111846253267074	1
11	31	1	111846253267074	593264734044925	1
12	37	1	593264734044925	168270169378399	1
13	41	1	168270169378399	285271807182347	1
14	43	1	285271807182347	538018099945609	15495481

This leads to the factorization $n = pq$ with primes $p = 15495481$ and $q = 38768461$. Let us see why the second stage works. We have the factorizations:

$$p - 1 = 2^3 \times 3^2 \times 5 \times 7 \times 11 \times 13 \times 43,$$
$$q - 1 = 2^2 \times 3 \times 5 \times 79 \times 8179.$$

In particular, $\text{ord}_p(2) = 129129 = 3 \times 7 \times 11 \times 13 \times 43$ is 20-power-smooth apart from the large prime factor 43. On the contrary, $q - 1$ and $\text{ord}_q(2) = 12922820 = 2^2 \times 5 \times 79 \times 8179$ have a prime divisor larger than $B' = 100$. □

6.4   Dixon's Method

Almost all *modern* integer-factoring algorithms rely on arriving at a congruence of the following form (often called a *Fermat congruence*):

$$x^2 \equiv y^2 \pmod{n}.$$

This implies that $n \mid (x - y)(x + y)$, that is, $n = \gcd(x - y, n) \times \gcd(x + y, n)$ (assuming that $\gcd(x, y) = 1$). If $\gcd(x - y, n)$ is a non-trivial factor of n, we obtain a non-trivial split of n.

Example 6.10 We have $899 = 30^2 - 1^2 = (30 - 1)(30 + 1)$, so $\gcd(30 - 1, 899) = 29$ is a non-trivial factor of 899. Moreover, $3 \times 833 = 2499 = 50^2 - 1^2 = (50 - 1)(50 + 1)$, and $\gcd(50 - 1, 833) = 49$ is a non-trivial factor of 833. □

The obvious question now is whether this method always works. Any odd integer n can be expressed as $n = \left(\frac{n+1}{2}\right)^2 - \left(\frac{n-1}{2}\right)^2$. However, this gives us only the trivial factorization $n = \left(\frac{n+1}{2} - \frac{n-1}{2}\right)\left(\frac{n+1}{2} + \frac{n-1}{2}\right) = 1 \times n$.

Since it is easy to verify whether n is a perfect power, we assume that n has $m \geqslant 2$ distinct prime factors. We may also assume that n contains no small prime factors. In particular, n is odd. Then, for any $y \in \mathbb{Z}_n^*$, the congruence $x^2 \equiv y^2 \pmod{n}$ has exactly 2^m solutions for x. The only two trivial solutions are $x \equiv \pm y \pmod{n}$. Each of the remaining $2^m - 2$ solutions yields a non-trivial split of n. If x and y are random elements satisfying $x^2 \equiv y^2 \pmod{n}$, then $\gcd(x - y, n)$ is a non-trivial factor of n with probability $\frac{2^m - 2}{2^m} \geqslant \frac{1}{2}$.

This factoring idea works if we can make available a non-trivial congruence of the form $x^2 \equiv y^2 \pmod{n}$. The modern subexponential algorithms propose different ways of obtaining this congruence. We start with a very simple idea.[6]

We choose a non-zero $x \in \mathbb{Z}_n$ randomly, and compute $a = x^2$ rem n (an integer in $\{0, 1, 2, \ldots, n - 1\}$). If a is a perfect square, say $a = y^2$, then $x^2 \equiv y^2 \pmod{n}$. However, there are only $\lfloor \sqrt{n - 1} \rfloor$ non-zero perfect squares in $\mathbb{Z}_n$. So the probability that a is of the form y^2 is about $1/\sqrt{n}$, that is, after trying $O(\sqrt{n})$ random values of x, we expect to arrive at the desired congruence $x^2 \equiv y^2 \pmod{n}$. This gives an algorithm with exponential running time.

In order to avoid this difficulty, we choose a *factor base* B consisting of the first t primes $p_1, p_2, \ldots, p_t$. We choose a random non-zero $x \in \mathbb{Z}_n^*$, and

[6] John D. Dixon, Asymptotically fast factorization of integers, *Mathematics of Computation*, 36, 255–260, 1981.

compute $a = x^2$ rem n. We now do not check whether a is a perfect square, but we check whether a can be factored completely over B. If so, we have $x^2 \equiv p_1^{\alpha_1} p_2^{\alpha_2} \cdots p_t^{\alpha_t} \pmod{n}$. Such a congruence is called a *relation*. If all α_i are even, the right side of the relation is a perfect square. But, as mentioned in the last paragraph, such a finding is of very low probability.

What we do instead is to collect many such relations. We do not expect to obtain a relation for every random value of x. If x^2 rem n does not factor completely over the factor base B, we discard that value of x. We store only those values of x for which x^2 rem n factors completely over B.

$$
\begin{aligned}
x_1^2 &\equiv p_1^{\alpha_{11}} p_2^{\alpha_{12}} \cdots p_t^{\alpha_{1t}} \pmod{n}, \\
x_2^2 &\equiv p_1^{\alpha_{21}} p_2^{\alpha_{22}} \cdots p_t^{\alpha_{2t}} \pmod{n}, \\
&\cdots \\
x_s^2 &\equiv p_1^{\alpha_{s1}} p_2^{\alpha_{s2}} \cdots p_t^{\alpha_{st}} \pmod{n}.
\end{aligned}
$$

One must not use a factoring algorithm to verify whether x^2 rem n factors completely over B. One should perform trial divisions by the primes $p_1, p_2, \ldots, p_t$. We obtain a relation if and only if trial divisions reduce x^2 rem n to 1.

After many relations are collected, we *combine* the relations so as to obtain a congruence of the desired form $x^2 \equiv y^2 \pmod{n}$. The combination stage involves s variables $\beta_1, \beta_2, \ldots, \beta_s$. The collected relations yield

$$
(x_1^2)^{\beta_1} (x_2^2)^{\beta_2} \cdots (x_s^2)^{\beta_s} \equiv \prod_{j=1}^{t} \left(p_j^{\sum_{i=1}^{s} \alpha_{ij} \beta_i} \right) \pmod{n},
$$

that is,

$$
\left(x_1^{\beta_1} x_2^{\beta_2} \cdots x_s^{\beta_s} \right)^2 \equiv p_1^{\gamma_1} p_2^{\gamma_2} \cdots p_t^{\gamma_t} \pmod{n},
$$

where $\gamma_j = \sum_{i=1}^{s} \alpha_{ij} \beta_i$ for $j = 1, 2, \ldots, t$. The left side of the last congruence is already a square. We adjust the quantities $\beta_1, \beta_2, \ldots, \beta_s$ in such a way that the right side is also a square, that is, all of $\gamma_1, \gamma_2, \ldots, \gamma_t$ are even, that is, $\gamma_j = 2\delta_j$ for all $j = 1, 2, \ldots, t$. We then have the desired congruence

$$
\left(x_1^{\beta_1} x_2^{\beta_2} \cdots x_s^{\beta_s} \right)^2 \equiv \left(p_1^{\delta_1} p_2^{\delta_2} \cdots p_t^{\delta_t} \right)^2 \pmod{n}.
$$

The condition that all γ_j are even can be expressed as follows.

$$
\begin{aligned}
\alpha_{11}\beta_1 + \alpha_{21}\beta_2 + \cdots + \alpha_{s1}\beta_s &\equiv 0 \pmod{2}, \\
\alpha_{12}\beta_1 + \alpha_{22}\beta_2 + \cdots + \alpha_{s2}\beta_s &\equiv 0 \pmod{2}, \\
&\cdots \\
\alpha_{1t}\beta_1 + \alpha_{2t}\beta_2 + \cdots + \alpha_{st}\beta_s &\equiv 0 \pmod{2}.
\end{aligned}
$$

This is a set of t linear equations in s variables $\beta_1, \beta_2, \ldots, \beta_s$. The obvious solution $\beta_1 = \beta_2 = \cdots = \beta_s = 0$ leads to the trivial congruence $1^2 \equiv 1^2 \pmod{n}$. If $s > t$, the system has non-zero solutions. The number of solutions of the system is 2^{s-r}, where r is the rank of the $t \times s$ coefficient matrix (Exercises 6.14

and 8.12). Some of the non-zero solutions are expected to split n. The choices of t and s are explained later.

Example 6.11 We factor $n = 64349$ by Dixon's method. We take the factor base $B = \{2, 3, 5, 7, 11, 13, 17, 19, 23, 29, 31, 37, 41, 43, 47\}$ (the first 15 primes). Relations are obtained for the following random values of x. The values of x, for which x^2 rem n does not split completely over B, are not listed here.

$$
\begin{aligned}
x_1^2 &\equiv 26507^2 \equiv 58667 \equiv 7 \times 17^2 \times 29 && (\bmod\ n) \\
x_2^2 &\equiv 53523^2 \equiv 22747 \equiv 23^2 \times 43 && (\bmod\ n) \\
x_3^2 &\equiv 34795^2 \equiv 29939 \equiv 7^2 \times 13 \times 47 && (\bmod\ n) \\
x_4^2 &\equiv 17688^2 \equiv\ \ 506 \equiv 2 \times 11 \times 23 && (\bmod\ n) \\
x_5^2 &\equiv 58094^2 \equiv\ \ 833 \equiv 7^2 \times 17 && (\bmod\ n) \\
x_6^2 &\equiv 37009^2 \equiv 61965 \equiv 3^6 \times 5 \times 17 && (\bmod\ n) \\
x_7^2 &\equiv 15376^2 \equiv\ \ 3150 \equiv 2 \times 3^2 \times 5^2 \times 7 && (\bmod\ n) \\
x_8^2 &\equiv 31414^2 \equiv 47481 \equiv 3 \times 7^2 \times 19 && (\bmod\ n) \\
x_9^2 &\equiv 62491^2 \equiv 41667 \equiv 3 \times 17 \times 19 \times 43 && (\bmod\ n) \\
x_{10}^2 &\equiv 46770^2 \equiv 17343 \equiv 3^2 \times 41 \times 47 && (\bmod\ n) \\
x_{11}^2 &\equiv 19274^2 \equiv\ \ 299 \equiv 13 \times 23 && (\bmod\ n) \\
x_{12}^2 &\equiv\ \ 4218^2 \equiv 31200 \equiv 2^5 \times 3 \times 5^2 \times 13 && (\bmod\ n) \\
x_{13}^2 &\equiv 23203^2 \equiv 35475 \equiv 3 \times 5^2 \times 11 \times 43 && (\bmod\ n) \\
x_{14}^2 &\equiv 26911^2 \equiv 18275 \equiv 5^2 \times 17 \times 43 && (\bmod\ n) \\
x_{15}^2 &\equiv 58697^2 \equiv 28000 \equiv 2^5 \times 5^3 \times 7 && (\bmod\ n) \\
x_{16}^2 &\equiv 50089^2 \equiv\ \ 4760 \equiv 2^3 \times 5 \times 7 \times 17 && (\bmod\ n) \\
x_{17}^2 &\equiv 25505^2 \equiv\ \ 984 \equiv 2^3 \times 3 \times 41 && (\bmod\ n) \\
x_{18}^2 &\equiv 26820^2 \equiv 19278 \equiv 2 \times 3^4 \times 7 \times 17 && (\bmod\ n) \\
x_{19}^2 &\equiv 18577^2 \equiv\ \ 1242 \equiv 2 \times 3^3 \times 23 && (\bmod\ n) \\
x_{20}^2 &\equiv\ \ 9407^2 \equiv 11774 \equiv 2 \times 7 \times 29^2 && (\bmod\ n)
\end{aligned}
$$

We have collected 20 relations with the hope that at least one non-trivial solution of $\beta_1, \beta_2, \ldots, \beta_{20}$ will lead to a non-trivial decomposition of n. In matrix notation, the above system of linear congruences can be written as

$$
\begin{pmatrix}
0 & 0 & 0 & 1 & 0 & 0 & 1 & 0 & 0 & 0 & 0 & 5 & 0 & 0 & 5 & 3 & 3 & 1 & 1 & 1 \\
0 & 0 & 0 & 0 & 0 & 6 & 2 & 1 & 1 & 2 & 0 & 1 & 1 & 0 & 0 & 0 & 1 & 4 & 3 & 0 \\
0 & 0 & 0 & 0 & 0 & 1 & 2 & 0 & 0 & 0 & 2 & 2 & 2 & 3 & 1 & 0 & 0 & 0 & 0 & 0 \\
1 & 0 & 2 & 0 & 2 & 0 & 1 & 2 & 0 & 0 & 0 & 0 & 0 & 1 & 1 & 0 & 1 & 0 & 1 \\
0 & 0 & 0 & 1 & 0 & 0 & 0 & 0 & 0 & 0 & 0 & 0 & 1 & 0 & 0 & 0 & 0 & 0 & 0 & 0 \\
0 & 0 & 1 & 0 & 0 & 0 & 0 & 0 & 0 & 0 & 1 & 1 & 0 & 0 & 0 & 0 & 0 & 0 & 0 & 0 \\
2 & 0 & 0 & 0 & 1 & 1 & 0 & 1 & 1 & 0 & 0 & 0 & 0 & 1 & 0 & 1 & 0 & 1 & 0 & 0 \\
0 & 0 & 0 & 0 & 0 & 0 & 0 & 1 & 1 & 0 & 0 & 0 & 0 & 0 & 0 & 0 & 0 & 0 & 0 & 0 \\
0 & 2 & 0 & 1 & 0 & 0 & 0 & 0 & 0 & 1 & 0 & 0 & 0 & 0 & 0 & 0 & 0 & 1 & 0 \\
1 & 0 & 0 & 0 & 0 & 0 & 0 & 0 & 0 & 0 & 0 & 0 & 0 & 0 & 0 & 0 & 0 & 0 & 0 & 2 \\
0 & 0 & 0 & 0 & 0 & 0 & 0 & 0 & 0 & 0 & 0 & 0 & 0 & 0 & 0 & 0 & 0 & 0 & 0 & 0 \\
0 & 0 & 0 & 0 & 0 & 0 & 0 & 0 & 0 & 0 & 0 & 0 & 0 & 0 & 0 & 0 & 0 & 0 & 0 & 0 \\
0 & 0 & 0 & 0 & 0 & 0 & 0 & 0 & 0 & 1 & 0 & 0 & 0 & 0 & 0 & 1 & 0 & 0 & 0 \\
0 & 1 & 0 & 0 & 0 & 0 & 0 & 0 & 1 & 0 & 0 & 0 & 1 & 1 & 0 & 0 & 0 & 0 & 0 \\
0 & 0 & 1 & 0 & 0 & 0 & 0 & 0 & 0 & 1 & 0 & 0 & 0 & 0 & 0 & 0 & 0 & 0 & 0 & 0
\end{pmatrix}
\begin{pmatrix}
\beta_1 \\ \beta_2 \\ \beta_3 \\ \beta_4 \\ \beta_5 \\ \beta_6 \\ \beta_7 \\ \beta_8 \\ \beta_9 \\ \beta_{10} \\ \beta_{11} \\ \beta_{12} \\ \beta_{13} \\ \beta_{14} \\ \beta_{15} \\ \beta_{16} \\ \beta_{17} \\ \beta_{18} \\ \beta_{19} \\ \beta_{20}
\end{pmatrix}
\equiv
\begin{pmatrix}
0 \\ 0 \\ 0 \\ 0 \\ 0 \\ 0 \\ 0 \\ 0 \\ 0 \\ 0 \\ 0 \\ 0 \\ 0 \\ 0 \\ 0 \\ 0 \\ 0
\end{pmatrix}
\ (\bmod\ 2).
$$

Call the coefficient matrix A. Since this is a system modulo 2, it suffices to know the exponents α_{ij} modulo 2 only, and the system can be rewritten as

$$
\begin{pmatrix}
0 & 0 & 0 & 1 & 0 & 0 & 1 & 0 & 0 & 0 & 0 & 1 & 0 & 0 & 1 & 1 & 1 & 1 & 1 & 1 \\
0 & 0 & 0 & 0 & 0 & 0 & 0 & 1 & 1 & 0 & 0 & 1 & 1 & 0 & 0 & 0 & 1 & 0 & 1 & 0 \\
0 & 0 & 0 & 0 & 0 & 1 & 0 & 0 & 0 & 0 & 0 & 0 & 0 & 0 & 1 & 1 & 0 & 0 & 0 & 0 \\
1 & 0 & 0 & 0 & 0 & 0 & 1 & 0 & 0 & 0 & 0 & 0 & 0 & 0 & 1 & 1 & 0 & 1 & 0 & 1 \\
0 & 0 & 0 & 1 & 0 & 0 & 0 & 0 & 0 & 0 & 1 & 0 & 0 & 0 & 0 & 0 & 0 & 0 & 0 & 0 \\
0 & 0 & 1 & 0 & 0 & 0 & 0 & 0 & 0 & 1 & 1 & 0 & 0 & 0 & 0 & 0 & 0 & 0 & 0 & 0 \\
0 & 0 & 0 & 0 & 1 & 1 & 0 & 1 & 1 & 0 & 0 & 0 & 1 & 0 & 1 & 0 & 1 & 0 & 0 & 0 \\
0 & 0 & 0 & 0 & 0 & 0 & 0 & 1 & 1 & 0 & 0 & 0 & 0 & 0 & 0 & 0 & 0 & 0 & 0 & 0 \\
0 & 0 & 0 & 1 & 0 & 0 & 0 & 0 & 0 & 1 & 0 & 0 & 0 & 0 & 0 & 0 & 0 & 1 & 0 \\
1 & 0 & 0 & 0 & 0 & 0 & 0 & 0 & 0 & 0 & 0 & 0 & 0 & 0 & 0 & 0 & 0 & 0 & 0 & 0 \\
0 & 0 & 0 & 0 & 0 & 0 & 0 & 0 & 0 & 0 & 0 & 0 & 0 & 0 & 0 & 0 & 0 & 0 & 0 & 0 \\
0 & 0 & 0 & 0 & 0 & 0 & 0 & 0 & 0 & 1 & 0 & 0 & 0 & 0 & 0 & 0 & 1 & 0 & 0 & 0 \\
0 & 1 & 0 & 0 & 0 & 0 & 0 & 0 & 1 & 0 & 0 & 0 & 1 & 1 & 0 & 0 & 0 & 0 & 0 & 0 \\
0 & 0 & 1 & 0 & 0 & 0 & 0 & 0 & 0 & 1 & 0 & 0 & 0 & 0 & 0 & 0 & 0 & 0 & 0 & 0 \\
\end{pmatrix}
\begin{pmatrix}
\beta_1 \\ \beta_2 \\ \beta_3 \\ \beta_4 \\ \beta_5 \\ \beta_6 \\ \beta_7 \\ \beta_8 \\ \beta_9 \\ \beta_{10} \\ \beta_{11} \\ \beta_{12} \\ \beta_{13} \\ \beta_{14} \\ \beta_{15} \\ \beta_{16} \\ \beta_{17} \\ \beta_{18} \\ \beta_{19} \\ \beta_{20}
\end{pmatrix}
\equiv
\begin{pmatrix}
0 \\ 0 \\ 0 \\ 0 \\ 0 \\ 0 \\ 0 \\ 0 \\ 0 \\ 0 \\ 0 \\ 0 \\ 0 \\ 0 \\ 0 \\ 0 \\ 0
\end{pmatrix}
\pmod{2}.
$$

This reduced coefficient matrix (call it $\bar{A}$) has rank 11 (modulo 2), that is, the kernel of the matrix is a 9-dimensional subspace of $\mathbb{Z}_2^{20}$. A basis of this kernel is provided by the following vectors.

$$
\begin{aligned}
\mathbf{v}_1 &= (0\ 1\ 0\ 0\ 0\ 0\ 0\ 1\ 1\ 0\ 0\ 0\ 0\ 0\ 0\ 0\ 0\ 0\ 0\ 0)^t, \\
\mathbf{v}_2 &= (0\ 1\ 0\ 1\ 0\ 0\ 0\ 0\ 0\ 0\ 1\ 1\ 1\ 0\ 0\ 0\ 0\ 0\ 0\ 0)^t, \\
\mathbf{v}_3 &= (0\ 1\ 0\ 0\ 1\ 0\ 0\ 0\ 0\ 0\ 0\ 0\ 0\ 1\ 0\ 0\ 0\ 0\ 0\ 0)^t, \\
\mathbf{v}_4 &= (0\ 0\ 0\ 0\ 1\ 1\ 1\ 0\ 0\ 0\ 0\ 0\ 0\ 0\ 1\ 0\ 0\ 0\ 0\ 0)^t, \\
\mathbf{v}_5 &= (0\ 0\ 0\ 0\ 0\ 1\ 1\ 0\ 0\ 0\ 0\ 0\ 0\ 0\ 0\ 1\ 0\ 0\ 0\ 0)^t, \\
\mathbf{v}_6 &= (0\ 0\ 1\ 0\ 0\ 0\ 0\ 0\ 0\ 1\ 0\ 1\ 0\ 0\ 0\ 0\ 1\ 0\ 0\ 0)^t, \\
\mathbf{v}_7 &= (0\ 0\ 0\ 0\ 1\ 0\ 1\ 0\ 0\ 0\ 0\ 0\ 0\ 0\ 0\ 0\ 0\ 1\ 0\ 0)^t, \\
\mathbf{v}_8 &= (0\ 0\ 0\ 0\ 0\ 0\ 0\ 0\ 0\ 0\ 1\ 1\ 0\ 0\ 0\ 0\ 0\ 0\ 1\ 0)^t, \\
\mathbf{v}_9 &= (0\ 0\ 0\ 0\ 0\ 0\ 1\ 0\ 0\ 0\ 0\ 0\ 0\ 0\ 0\ 0\ 0\ 0\ 0\ 1)^t.
\end{aligned}
$$

There are 2^9 solutions for $\beta_1, \beta_2, \ldots, \beta_{20}$, obtained by all $\mathbb{Z}_2$-linear combinations of $\mathbf{v}_1, \mathbf{v}_2, \ldots, \mathbf{v}_9$. Let us try some random solutions. First, take

$$
\begin{aligned}
&(\beta_1 \quad \beta_2 \quad \cdots \quad \beta_{20})^t \\
={}& \mathbf{v}_2 + \mathbf{v}_3 + \mathbf{v}_5 + \mathbf{v}_6 + \mathbf{v}_9 \\
={}& (0\ 0\ 1\ 1\ 1\ 1\ 0\ 0\ 0\ 1\ 1\ 0\ 1\ 1\ 0\ 1\ 1\ 0\ 0\ 1)^t.
\end{aligned}
$$

For this solution, we have

$$
\mathbf{e} = A\beta = (8\ \ 10\ \ 6\ \ 6\ \ 2\ \ 2\ \ 4\ \ 0\ \ 2\ \ 2\ \ 0\ \ 0\ \ 2\ \ 2\ \ 2)^t.
$$

This gives $x \equiv x_1^{\beta_1} x_2^{\beta_2} \cdots x_{20}^{\beta_{20}} \equiv x_3 x_4 x_5 x_6 x_{10} x_{11} x_{13} x_{14} x_{16} x_{17} x_{20} \equiv 34795 \times 17688 \times 58094 \times 37009 \times 46770 \times 19274 \times 23203 \times 26911 \times 50089 \times 25505 \times 9407 \equiv 53886 \pmod{64349}$. On the other hand, the vector $\mathbf{e}$ gives $y \equiv 2^4 \times 3^5 \times 5^3 \times 7^3 \times 11 \times 13 \times 17^2 \times 23 \times 29 \times 41 \times 43 \times 47 \equiv 53886 \pmod{64349}$. Therefore, $\gcd(x - y, n) = 64349 = n$, that is, the factorization attempt is unsuccessful.

Let us then try

$$(\beta_1 \quad \beta_2 \quad \cdots \quad \beta_{20})^t$$
$$= \mathbf{v}_3 + \mathbf{v}_5 + \mathbf{v}_6 + \mathbf{v}_7 + \mathbf{v}_8$$
$$= (0 \; 1 \; 1 \; 0 \; 0 \; 1 \; 0 \; 0 \; 0 \; 1 \; 1 \; 0 \; 0 \; 1 \; 0 \; 1 \; 1 \; 1 \; 1 \; 0)^t$$

so that

$$\mathbf{e} = A\beta = (8 \quad 16 \quad 4 \quad 4 \quad 0 \quad 2 \quad 4 \quad 0 \quad 4 \quad 0 \quad 0 \quad 0 \quad 2 \quad 2 \quad 2)^t .$$

Therefore, $x \equiv x_2 x_3 x_6 x_{10} x_{11} x_{14} x_{16} x_{17} x_{18} x_{19} \equiv 53523 \times 34795 \times 37009 \times 46770 \times 19274 \times 26911 \times 50089 \times 25505 \times 26820 \times 18577 \equiv 58205 \pmod{64349}$, and $y \equiv 2^4 \times 3^8 \times 5^2 \times 7^2 \times 13 \times 17^2 \times 23^2 \times 41 \times 43 \times 47 \equiv 6144 \pmod{64349}$. In this case, $\gcd(x - y, n) = 1$, and we again fail to split n non-trivially.

As a third attempt, let us try

$$(\beta_1 \quad \beta_2 \quad \cdots \quad \beta_{20})^t$$
$$= \mathbf{v}_3 + \mathbf{v}_6 + \mathbf{v}_7 + \mathbf{v}_9$$
$$= (0 \; 1 \; 1 \; 0 \; 0 \; 0 \; 0 \; 0 \; 0 \; 1 \; 0 \; 1 \; 0 \; 1 \; 0 \; 0 \; 1 \; 1 \; 0 \; 1)^t ,$$

for which

$$\mathbf{e} = A\beta = (10 \quad 8 \quad 4 \quad 4 \quad 0 \quad 2 \quad 2 \quad 0 \quad 2 \quad 2 \quad 0 \quad 0 \quad 2 \quad 2 \quad 2)^t .$$

In this case, $x \equiv x_2 x_3 x_{10} x_{12} x_{14} x_{17} x_{18} x_{20} \equiv 53523 \times 34795 \times 46770 \times 4218 \times 26911 \times 25505 \times 26820 \times 9407 \equiv 10746 \pmod{64349}$. On the other hand, $y \equiv 2^5 \times 3^4 \times 5^2 \times 7^2 \times 13 \times 17 \times 23 \times 29 \times 41 \times 43 \times 47 \equiv 57954 \pmod{64349}$. This gives $\gcd(x - y, n) = 281$, a non-trivial factor of n. The corresponding cofactor is $n/281 = 229$. Since both these factors are prime, we get the complete factorization $64349 = 229 \times 281$. $\qquad \square$

We now derive the (optimal) running time of Dixon's method. If the number t of primes in the factor base is too large, we have to collect too many relations to obtain a system with non-zero solutions (we should have $s \geqslant t$). Moreover, solving the system in that case would be costly. On the contrary, if t is too small, most of the random values of $x \in \mathbb{Z}_n^*$ will fail to generate relations, and we have to iterate too many times before we find a desired number of values of x, for which $x^2 \text{ rem } n$ factors completely over B. For estimating the best trade-off, we need some results from analytic number theory.

Definition 6.12 Let $m \in \mathbb{N}$. An integer x is called *m-smooth* (or *smooth* if m is understood from the context) if all prime factors of x are $\leqslant m$. $\qquad \triangleleft$

Theorem 6.13 supplies the formula for the density of smooth integers.

Theorem 6.13 *Let $m, n \in \mathbb{N}$, and $u = \frac{\ln n}{\ln m}$. For $u \to \infty$ and $u \geqslant \ln^2 n$, the number of m-smooth integers x in the range $1 \leqslant x \leqslant n$ asymptotically approaches $nu^{-u+o(u)}$. That is, the density of m-smooth integers between 1 and n is asymptotically $u^{-u+o(u)}$.* ◁

Corollary 6.14 *The density of $L[\beta]$-smooth positive integers with values of the order $O(n^\alpha)$ is $L[-\frac{\alpha}{2\beta}]$. Here, $L[c]$ stands for $L(n, \frac{1}{2}, c)$, as explained near the beginning of this chapter.*

Proof Integers with values $O(n^\alpha)$ can be taken as $\leqslant kn^\alpha$ for some constant k. The natural logarithm of this is $\alpha \ln n + \ln k \approx \alpha \ln n$. On the other hand, $\ln L[\beta] = (\beta + o(1))\sqrt{\ln n \ln \ln n}$. That is, $u \approx \frac{\alpha}{\beta}\sqrt{\frac{\ln n}{\ln \ln n}}$. Consequently,

$$
u^{-u+o(u)} = \exp[(-u+o(u))\ln u] = \exp[(-1+o(1))u \ln u]
$$

$$
= \exp\left((-1+o(1))\left(\frac{\alpha}{\beta}\sqrt{\frac{\ln n}{\ln \ln n}}\right)\left(\ln \frac{\alpha}{\beta} + \frac{1}{2}\ln \ln n - \frac{1}{2}\ln \ln \ln n\right)\right)
$$

$$
= \exp\left((-1+o(1))(1+o(1))\frac{\alpha}{2\beta}\sqrt{\ln n \ln \ln n}\right)
$$

$$
= \exp\left(\left(-\frac{\alpha}{2\beta}+o(1)\right)\sqrt{\ln n \ln \ln n}\right) = L\left[-\frac{\alpha}{2\beta}\right].
$$
◁

We now deduce the best running time of Dixon's method. Let us choose the factor base to consist of all primes $\leqslant L[\beta]$. By the prime number theorem, $t \approx L[\beta]/\ln L[\beta]$, that is, $t = L[\beta]$ again. The random elements x^2 rem n are $O(n)$, that is, we put $\alpha = 1$ in Corollary 6.14. The probability that such a random element factors completely over B is then $L[-\frac{1}{2\beta}]$, that is, after an expected number $L[\frac{1}{2\beta}]$ of iterations, we obtain one relation (that is, one $L[\beta]$-smooth value). Each iteration calls for trial divisions by $L[\beta]$ primes in the factor base. Finally, we need to generate $s \geqslant t$ relations. Therefore, the total expected time taken by the relation collection stage is $L[\beta]L[\beta]L[\frac{1}{2\beta}] = L[2\beta + \frac{1}{2\beta}]$. The quantity $2\beta + \frac{1}{2\beta}$ is minimized for $\beta = \frac{1}{2}$. For this choice, the running time of the relation-collection stage is $L[2]$.

The next stage of the algorithm solves a $t \times s$ system modulo 2. Since both s and t are expressions of the form $L[1/2]$, using standard Gaussian elimination gives a running time of $L[3/2]$. However, the relations collected lead to equations that are necessarily *sparse*, since each smooth value of x^2 rem n can have only $O(\log n)$ prime factors. Such a sparse $t \times s$ system can be solved in $\tilde{O}(st)$ time using some special algorithms (see Chapter 8). Thus, the system-solving stage can be completed in $L[2/2] = L[1]$ time. To sum up, Dixon's method runs in subexponential time $L[2] = L(n, 1/2, 2)$.

The constant 2 in $L[2]$ makes Dixon's method rather slow. The problem with Dixon's method is that it generates smoothness candidates as large as $O(n)$. By using other algorithms (like CFRAC or QSM), we can generate candidates having values $O(\sqrt{n})$.

6.5 CFRAC Method

The continued fraction (CFRAC) method[7] for factoring n is based upon the continued-fraction expansion of $\sqrt{n}$. Let h_r/k_r denote the r-th convergent to $\sqrt{n}$ for $r = 0, 1, 2, \ldots$. We have

$$
\begin{aligned}
h_r^2 - nk_r^2 &= (h_r^2 - 2\sqrt{n}h_r k_r + nk_r^2) - (2nk_r^2 - 2\sqrt{n}h_r k_r) \\
&= (\sqrt{n}k_r - h_r)^2 - 2\sqrt{n}k_r(\sqrt{n}k_r - h_r),
\end{aligned}
$$

so that

$$
\begin{aligned}
|h_r^2 - nk_r^2| &\leqslant |\sqrt{n}k_r - h_r|^2 + 2\sqrt{n}k_r|\sqrt{n}k_r - h_r| \\
&< \frac{1}{k_{r+1}^2} + \frac{2\sqrt{n}k_r}{k_{r+1}} \quad \text{[by Theorem 1.67]} \\
&\leqslant \frac{2\sqrt{n}(k_{r+1} - 1)}{k_{r+1}} + \frac{1}{k_{r+1}^2} = 2\sqrt{n} - \frac{2\sqrt{n}k_{r+1} - 1}{k_{r+1}^2} \\
&< 2\sqrt{n}.
\end{aligned}
$$

Under the assumption that n is not a perfect square, $\sqrt{n}$ is irrational, that is, $\frac{h_r}{k_r} \neq \sqrt{n}$ for all r, that is, $0 < |h_r^2 - nk_r^2| < 2\sqrt{n}$ for all $r \geqslant 1$.

The CFRAC method uses a factor base $B = \{-1, p_1, p_2, \ldots, p_t\}$, where $p_1, p_2, \ldots, p_t$ are all the small primes $\leqslant L[\beta]$ (the choice of β will be discussed later). One then computes the convergents h_r/k_r for $r = 1, 2, 3, \ldots$. Let $y_r = h_r^2 - nk_r^2$. We have $h_r^2 \equiv y_r \pmod{n}$. We check the smoothness of y_r by trial division of y_r by primes in B. Since some values of y_r are negative, we also include -1 as an element in the factor base. Every smooth y_r gives a relation. As deduced above, we have $|y_r| < 2\sqrt{n}$, that is, the integers tested for smoothness are much smaller in this case than in Dixon's method.

After sufficiently many relations are obtained, they are processed exactly as in Dixon's method. I will not discuss the linear-algebra phase further in connection with the CFRAC method (or other factoring methods).

Let p be a small prime in the factor base. The condition $p|y_r$ implies that $h_r^2 - nk_r^2 \equiv 0 \pmod{p}$, that is, $(h_r k_r^{-1})^2 \equiv n \pmod{p}$, that is, n is a quadratic residue modulo p. Therefore, we need not include those small primes p in the factor base, modulo which n is a quadratic non-residue.

Example 6.15 Let us factor $n = 596333$ by the CFRAC method. We take

$$B = \{-1, 2, 3, 5, 7, 11, 13, 17, 19, 23, 29, 31, 37, 41, 43, 47\}.$$

However, there is no need to consider those primes modulo which n is a quadratic non-residue. This gives the reduced factor base

$$B = \{-1, 2, 11, 13, 23, 29, 37, 43\}.$$

[7]Michael A. Morrison and John Brillhart, A method of factoring and the factorization of F_7, *Mathematics of Computation*, 29(129), 183–205, 1975.

Before listing the relations, I highlight some implementation issues. The denominators k_r are not used in the relation $h_r^2 \equiv y_r \pmod{n}$. We compute only the numerator sequence h_r. We start with $h_{-2} = 0$ and $h_{-1} = 1$. Subsequently, for $r \geqslant 0$, we first obtain the r-th coefficient a_r in the continued-fraction expansion of $\sqrt{n}$. This gives us $h_r = a_r h_{r-1} + h_{r-2}$. The sequence h_r grows quite rapidly with r. However, we need to know the value of h_r modulo n only. We compute $y_r \equiv h_r^2 \pmod{n}$. For even r, we take y_r as a value between $-(n-1)$ and -1. For odd r, we take y_r in the range 1 to $n-1$.

For computing the coefficients a_r, we avoid floating-point arithmetic as follows. We initialize $\xi_0 = \sqrt{n}$, compute $a_r = \lfloor \xi_r \rfloor$ and subsequently $\xi_{r+1} = 1/(\xi_r - a_r)$. Here, each ξ_r is maintained as an expression of the form $t_1 + t_2 \sqrt{n}$ with rational numbers t_1, t_2, that is, we work in the ring $\mathbb{Q}[x]/\langle x^2 - n \rangle$. Only for computing a_r as $\lfloor \xi_r \rfloor$, we need a floating-point value for $\sqrt{n}$.

The following table demonstrates the CFRAC iterations for $0 \leqslant r \leqslant 15$, which give four B-smooth values of y_r. Running the loop further yields B-smooth values of y_r for $r = 20, 23, 24, 27, 28, 39$. So we get ten relations, and the size of the factor base is eight. We then solve the resulting system of linear congruences modulo 2. This step is not shown here.

r	ξ_r	a_r	$h_r(\mathrm{mod}\ n)$	y_r	B-smooth
0	$\sqrt{n}$	772	772	$-349 = (-1) \times 349$	No
1	$(772 + \sqrt{n})/349$	4	3089	$593 = 593$	No
2	$(624 + \sqrt{n})/593$	2	6950	$-473 = (-1) \times 11 \times 43$	Yes
3	$(562 + \sqrt{n})/473$	2	16989	$949 = 13 \times 73$	No
4	$(384 + \sqrt{n})/949$	1	23939	$-292 = (-1) \times 2^2 \times 73$	No
5	$(565 + \sqrt{n})/292$	4	112745	$797 = 797$	No
6	$(603 + \sqrt{n})/797$	1	136684	$-701 = (-1) \times 701$	No
7	$(194 + \sqrt{n})/701$	1	249429	$484 = 2^2 \times 11^2$	Yes
8	$(507 + \sqrt{n})/484$	2	39209	$-793 = (-1) \times 13 \times 61$	No
9	$(461 + \sqrt{n})/793$	1	288638	$613 = 613$	No
10	$(332 + \sqrt{n})/613$	1	327847	$-844 = (-1) \times 2^2 \times 211$	No
11	$(281 + \sqrt{n})/844$	1	20152	$331 = 331$	No
12	$(563 + \sqrt{n})/331$	4	408455	$-52 = (-1) \times 2^2 \times 13$	Yes
13	$(761 + \sqrt{n})/52$	29	535020	$737 = 11 \times 67$	No
14	$(747 + \sqrt{n})/737$	2	285829	$-92 = (-1) \times 2^2 \times 23$	Yes
15	$(727 + \sqrt{n})/92$	16	337620	$449 = 449$	No

$\square$

Let us now deduce the optimal running time of the CFRAC method. The smoothness candidates y_r are $\mathrm{O}(\sqrt{n})$, and have the probability $L[-\frac{1/2}{2\beta}] = L[-\frac{1}{4\beta}]$ for being $L[\beta]$-smooth. We expect to get one $L[\beta]$-smooth value of y_r after $L[\frac{1}{4\beta}]$ iterations. Each iteration involves trial divisions by $L[\beta]$ primes in the factor base. Finally, we need to collect $L[\beta]$ relations, so the running time of the CFRAC method is $L[2\beta + \frac{1}{4\beta}]$. Since $2\beta + \frac{1}{4\beta}$ is minimized for $\beta = \frac{1}{2\sqrt{2}}$, the optimal running time of the relation-collection stage is $L[\sqrt{2}]$. The sparse system involving $L[\beta]$ variables can be solved in $L[2\beta] = L[\frac{1}{\sqrt{2}}]$ time. To sum up, the running time of the CFRAC method is $L[\sqrt{2}]$.

The CFRAC method can run in parallel with each instance handling the continued-fraction expansion of $\sqrt{sn}$ for some $s \in \mathbb{N}$. For a given s, the quantity y_r satisfies the inequality $0 < |y_r| < 2\sqrt{sn}$. If s grows, the probability of y_r being smooth decreases, so only small values of s should be used.

6.6 Quadratic Sieve Method

The Quadratic Sieve method (QSM)[8] is another subexponential-time integer-factoring algorithm that generates smoothness candidates of values $O(\sqrt{n})$. As a result, its performance is similar to the CFRAC method. However, trial divisions in the QSM can be replaced by an efficient sieving procedure. This leads to an $L[1]$ running time of the QSM.

Let $H = \lceil \sqrt{n} \rceil$ and $J = H^2 - n$. If n is not a perfect square, $J \neq 0$. Also, both H and J are of values $O(\sqrt{n})$. For a small integer c (positive or negative), $(H+c)^2 \equiv H^2 + 2cH + c^2 \equiv J + 2cH + c^2 \pmod{n}$. Call $T(c) = J + 2cH + c^2$. Since H, J are $O(\sqrt{n})$ and c is small, we have $T(c) = O(\sqrt{n})$. We check the smoothness of $T(c)$ over a set B of small primes (the factor base). We vary c in the range $-M \leqslant c \leqslant M$. The choices of B and M are explained later. Since c can take negative values, $T(c)$ may be negative. So we add -1 to the factor base. After all smooth values of $T(c)$ are found for $-M \leqslant c \leqslant M$, the resulting linear system of congruences modulo 2 is solved to split n.

The condition $p|T(c)$ implies $(H+c)^2 \equiv n \pmod{p}$. Thus, the factor base B need not contain any small prime modulo which n is a quadratic non-residue.

Example 6.16 We use QSM to factor $n = 713057$, for which $H = \lceil \sqrt{n} \rceil = 845$, and $J = H^2 - n = 968$. All primes < 50, modulo which n is a quadratic residue, constitute the factor base, that is, $B = \{-1, 2, 7, 11, 17, 19, 29, 37, 43\}$. We choose $M = 50$. All the values of c in the range $-50 \leqslant c \leqslant 50$, for which $T(c)$ is smooth over B, are listed below.

c	$T(c)$	c	$T(c)$
-44	$-71456 = (-1) \times 2^5 \times 7 \times 11 \times 29$	-2	$-2408 = (-1) \times 2^3 \times 7 \times 43$
-22	$-35728 = (-1) \times 2^4 \times 7 \times 11 \times 29$	0	$968 = 2^3 \times 11^2$
-15	$-24157 = (-1) \times 7^2 \times 17 \times 29$	2	$4352 = 2^8 \times 17$
-14	$-22496 = (-1) \times 2^5 \times 19 \times 37$	4	$7744 = 2^6 \times 11^2$
-11	$-17501 = (-1) \times 11 \times 37 \times 43$	26	$45584 = 2^4 \times 7 \times 11 \times 37$
-9	$-14161 = (-1) \times 7^2 \times 17^2$	34	$59584 = 2^6 \times 7^2 \times 19$
-4	$-5776 = (-1) \times 2^4 \times 19^2$	36	$63104 = 2^7 \times 17 \times 29$

---

[8] Carl Pomerance, The quadratic sieve factoring algorithm, *Eurocrypt'84*, 169–182, 1985.

There are 14 smooth values of $T(c)$, and the size of the factor base is 9. Solving the resulting system splits n as 761×937 (with both the factors prime). □

Let us now look at the optimal running time of the QSM. Let the factor base B consist of (-1 and) primes $\leqslant L[\beta]$. Since the integers $T(c)$ checked for smoothness over B have values $O(\sqrt{n})$, the probability of each being smooth is $L[-\frac{1}{4\beta}]$. In order that we get $L[\beta]$ relations, the size of the sieving interval (or equivalently M) should be $L[\beta + \frac{1}{4\beta}]$. If we use trial division of each $T(c)$ by all of the $L[\beta]$ primes in the factor base, we obtain a running time of $L[2\beta + \frac{1}{4\beta}]$. As we will see shortly, a sieving procedure reduces this running time by a factor of $L[\beta]$, that is, the running time of the relation-collection stage of QSM with sieving is only $L[\beta + \frac{1}{4\beta}]$. This quantity is minimized for $\beta = \frac{1}{2}$, and we obtain a running time of $L[1]$ for the relation-collection stage of QSM. The resulting sparse system having $L[\frac{1}{2}]$ variables and $L[\frac{1}{2}]$ equations can also be solved in the same time.

6.6.1 Sieving

Both the CFRAC method and the QSM generate smoothness candidates of value $O(\sqrt{n})$. In the CFRAC method, these values are bounded by $2\sqrt{n}$, whereas for the QSM, we have a bound of nearly $2M\sqrt{n}$. The CFRAC method is, therefore, expected to obtain smooth candidates more frequently than the QSM. On the other hand, the QSM offers the possibility of sieving, a process that replaces trial divisions by single-precision subtractions. As a result, the QSM achieves a better running time than the CFRAC method.

In the QSM, the smoothness of the integers $T(c) = J + 2cH + c^2$ is checked for $-M \leqslant c \leqslant M$. To that effect, we use an array A indexed by c in the range $-M \leqslant c \leqslant M$. We initialize the array location A_c to an approximate value of $\log |T(c)|$. We can use only one or two most significant words of $T(c)$ for this initial value. Indeed, it suffices to know $\log |T(c)|$ rounded or truncated after three places of decimal. If so, we can instead store the integer value $\lfloor 1000 \log |T(c)| \rfloor$, and perform only integer operations on the elements of A.

Now, we choose small primes p one by one from the factor base. For each small positive exponent h, we try to find out all the values of c for which $p^h | T(c)$. Since $J = H^2 - n$, this translates to solving $(H + c)^2 \equiv n \pmod{p^h}$. For $h = 1$, we use a root-finding algorithm, whereas for $h > 1$, the solutions can be obtained by lifting the solutions modulo p^{h-1}. In short, all the solutions of $(H + c)^2 \equiv n \pmod{p^h}$ can be obtained in (expected) polynomial time (in $\log p$ and $\log n$).

Let χ be a solution of $(H + c)^2 \equiv n \pmod{p^h}$. For all c in the range $-M \leqslant c \leqslant M$, we subtract $\log p$ from the array element A_c if and only if $c \equiv \chi \pmod{p^h}$. In other words, we first obtain one solution χ, and then update A_c for all $c = \chi \pm kp^h$ with $k = 0, 1, 2, \ldots$. If, for a given c, we have the multiplicity $v = v_p(T(c))$, then $\log p$ is subtracted from A_c exactly v times (once for each of $h = 1, 2, \ldots, v$).

After all primes $p \in B$ and all suitable small exponents h are considered, we look at the array locations A_c for $-M \leqslant c \leqslant M$. If some $T(c)$ is smooth (over B), then all its prime divisors are eliminated during the subtractions of $\log p$ from the initial value of $\log |T(c)|$. Thus, we should have $A_c = 0$. However, since we use only approximate log values, we would get $A_c \approx 0$. On the other hand, a non-smooth $T(c)$ contains a prime factor $\geqslant p_{t+1}$. Therefore, a quantity at least as large as $\log p_{t+1}$ remains in the array location A_c, that is, $A_c \gg 0$ in this case. In short, the post-sieving values of A_c readily identify the smooth values of $T(c)$. Once we know that some $T(c)$ is smooth, we use trial division of that $T(c)$ by the primes in the factor base.

Example 6.17 We factor $n = 713057$ by the QSM. As in Example 6.16, take $B = \{-1, 2, 7, 11, 17, 19, 29, 37, 43\}$, and $M = 50$. Initialize the array entry A_c to $\lfloor 1000 \log |T(c)| \rfloor$ (e is the base of logarithms). Since $T(0) = J = 968$, set $A_0 = \lfloor 1000 \log 968 \rfloor = 6875$. Similarly, $T(20) = J + 40H + 400 = 35168$, so A_{20} is set to $\lfloor 1000 \log 35168 \rfloor = 10467$, and $T(-20) = J - 40H + 400 = -32432$, so A_{-20} is set to $\lfloor 1000 \log 32432 \rfloor = 10386$. The approximate logarithms of the primes in B are $\lfloor 1000 \log 2 \rfloor = 693$, $\lfloor 1000 \log 7 \rfloor = 1945$, $\lfloor 1000 \log 11 \rfloor = 2397$, $\lfloor 1000 \log 17 \rfloor = 2833$, $\lfloor 1000 \log 19 \rfloor = 2944$, $\lfloor 1000 \log 29 \rfloor = 3367$, $\lfloor 1000 \log 37 \rfloor = 3610$, and $\lfloor 1000 \log 43 \rfloor = 3761$.

The following table considers all small primes p and all small exponents h for solving $T(c) \equiv 0 \pmod{p^h}$. For each solution χ, we consider all values of $c \equiv \chi \pmod{p^h}$ with $-M \leqslant c \leqslant M$, and subtract $\log p$ from A_c. For each prime p, we consider $h = 1, 2, 3, \ldots$ in that sequence until a value of h is found, for which there is no solution of $T(c) \equiv 0 \pmod{p^h}$ with $-M \leqslant c \leqslant M$.

p	p^h	χ	$c \equiv \chi \pmod{p^h}, \ -M \leqslant c \leqslant M$
2	2	0	$-50, -48, -46, -44, \ldots, -2, 0, 2, \ldots, \boxed{24}, \mathbf{26}, \ldots, 44, 46, 48, 50$
	4	0	$-48, -44, -40, -36, \ldots, -4, 0, 4, \ldots, \boxed{24}, \ldots, 36, 40, 44, 48$
		2	$-50, -46, -42, -38, \ldots, -2, 2, \ldots, \mathbf{26}, \ldots, 38, 42, 46, 50$
	8	0	$-48, -40, -32, -24, -16, -8, 0, 8, 16, \boxed{24}, 32, 40, 48$
		2	$-46, -38, -30, -22, -14, -6, 2, 10, 18, \mathbf{26}, 34, 42, 50$
		4	$-44, -36, -28, -20, -12, -4, 4, 12, 20, 28, 36, 44$
		6	$-50, -42, -34, -26, -18, -10, -2, 6, 14, 22, 30, 38, 46$
	16	2	$-46, -30, -14, 2, 18, 34, 50$
		4	$-44, -28, -12, 4, 20, 36$
		10	$-38, -22, -6, 10, \mathbf{26}, 42$
		12	$-36, -20, -4, 12, 28, 44$
	32	2	$-30, 2, 34$
		4	$-28, 4, 36$
		18	$-46, -14, 18, 50$
		20	$-44, -12, 20$
	64	2	2
		4	4
		34	$-30, 34$
		36	$-28, 36$

p	p^h	χ	$c \equiv \chi \pmod{p^h}, \quad -M \leqslant c \leqslant M$
2	128	2	2
		36	36
		66	
		100	-28
	256	2	2
		100	
		130	
		228	-28
	512	130	
		228	
		386	
		484	-28
	1024	130	
		228	
		642	
		740	
7	7	5	$-44, -37, -30, -23, -16, -9, -2, 5, 12, 19, \mathbf{26}, 33, 40, 47$
		6	$-50, -43, -36, -29, -22, -15, -8, -1, 6, 13, 20, 27, 34, 41, 48$
	49	34	$-15, 34$
		40	$-9, 40$
	343	132	
		236	
11	11	0	$-44, -33, -22, -11, 0, 11, 22, 33, 44$
		4	$-40, -29, -18, -7, 4, 15, \mathbf{26}, 37, 48$
	121	0	0
		4	4
	1331	242	
		730	
17	17	2	$-49, -32, -15, 2, 19, 36$
		8	$-43, -26, -9, 8, 25, 42$
	289	53	
		280	-9
	4913	53	
		3170	
19	19	5	$-33, -14, 5, \boxed{24}, 43$
		15	$-42, -23, -4, 15, 34$
	361	119	
		357	-4
	6859	480	
		4689	
29	29	7	$-22, 7, 36$
		14	$-44, -15, 14, 43$
	841	72	
		761	

p	p^h	χ	$c \equiv \chi \pmod{p^h},\ -M \leqslant c \leqslant M$
37	37	23	$-14, 23$
		26	$-48, -11, \mathbf{26}$
	1369	245	
		803	
43	43	32	$-11, 32$
		41	$-45, -2, 41$
	1849	471	
		1537	

Let us track the array locations A_{24} and A_{26} in the above table (the bold and the boxed entries). We have $T(24) = 42104$, and $T(26) = 45584$. So A_{24} is initialized to $\lfloor 1000 \log 42104 \rfloor = 10647$, and A_{26} to $\lfloor 1000 \log 45584 \rfloor = 10727$. We subtract $\lfloor 1000 \log 2 \rfloor = 693$ thrice from A_{24}, and $\lfloor 1000 \log 19 \rfloor = 2944$ once from A_{24}. After the end of the sieving process, A_{24} stores the value $10647 - 3 \times 693 - 2944 = 5624$, that is, $T(24)$ is not smooth. In fact, $T(24) = 42104 = 2^3 \times 19 \times 277$. The smallest prime larger than those in B is 53 for which $\lfloor 1000 \log 53 \rfloor = 3970$. Thus, if $T(c)$ is not smooth over B, then after the completion of sieving, A_c would store a value not (much) smaller than 3970.

From A_{26}, $\lfloor 1000 \log 2 \rfloor = 693$ is subtracted four times, $\lfloor 1000 \log 7 \rfloor = 1945$ once, $\lfloor 1000 \log 11 \rfloor = 2397$ once, and $\lfloor 1000 \log 37 \rfloor = 3610$ once, leaving the final value $10727 - 4 \times 693 - 1945 - 2397 - 3610 = 3$ at that array location. So $T(26)$ is smooth. Indeed, $T(26) = 45584 = 2^4 \times 7 \times 11 \times 37$.

This example demonstrates that the final values of A_c for smooth $T(c)$ are clearly separated from those of A_c for non-smooth $T(c)$. As a result, it is easy to locate the smooth values after the sieving process even if somewhat crude approximations are used for representing the logarithms. □

Let us now argue that the sieving process runs in $L[1]$ time for the choice $\beta = 1/2$. The primes in the factor base are $\leqslant L[\beta] = L[1/2]$. On the other hand, $M = L[\beta + \frac{1}{4\beta}] = L[1]$, and so $2M + 1$ is also of the form $L[1]$. First, we have to initialize all of the array locations A_c. Each location demands the computation of $T(c)$ (and its approximate logarithm), a task that can be performed in $O(\ln^k n)$ time for some constant k. Since there are $L[1]$ array locations, the total time for the initialization of A is of the order of

$$(\lg^k n) L[1] = \exp\left[(1 + o(1))\sqrt{\ln n \ln \ln n} + k \ln \ln n\right]$$

$$= \exp\left[\left(1 + o(1) + k\sqrt{\frac{\ln \ln n}{\ln n}}\right)\sqrt{\ln n \ln \ln n}\right],$$

which is again an expression of the form $L[1]$.

Fix a prime $p \in B$. Since $T(c) = O(\sqrt{n})$, one needs to consider only $O(\log n)$ values of h. Finding the solutions χ for each value of h takes $O(\log^l n)$ time for a constant l. If we vary h, we spend a total of $O(\log^{l+1} n)$ time for each prime p. Finally, we vary p, and conclude that the total time taken for computing all the solutions χ is of the order of $(\log^{l+1} n)L[1/2]$, which is again an expression of the form $L[1/2]$.

Now, we derive the total cost of subtracting log values from all array locations. First, take $p = 2$. We have to subtract $\log 2$ from each array location A_c at most $O(\log n)$ times. Since M and so $2M+1$ are expressions of the form $L[1]$, the total effort of all subtractions for $p = 2$ is of the order of $(\log n)L[1]$ which is again of the form $L[1]$. Then, take an odd prime p. Assume that $p \nmid n$, and n is a quadratic residue modulo p. In this case, the congruence $T(c) \equiv 0 \pmod{p^h}$ has exactly two solutions for each value of h. Moreover, the values p^h are $O(\sqrt{n})$. Therefore, the total cost of subtractions for all odd small primes p and for all small exponents h is of the order of $\sum_{p,h} \frac{(2M+1)\times 2}{p^h} <$ $2(2M+1)\sum_{r=1}^n \frac{1}{r} \approx 2(2M+1)\ln n$, which is an expression of the form $L[1]$. Finally, one has to factor the $L[1/2]$ smooth values of $T(c)$ by trial divisions by the $L[1/2]$ primes in the factor base. This process too can be completed in $L[1]$ time. To sum up, the entire relation-collection stage runs in $L[1]$ time.

6.6.2 Incomplete Sieving

There are many practical ways of speeding up the sieving process in the QSM. Here is one such technique. In the sieving process, we solve $T(c) \equiv 0 \pmod{p^h}$ for all $p \in B$ and for all small exponents h. The final value of A_c equals (a number close to) the logarithm of the unfactored part of $T(c)$.

Suppose that now we work with only the value $h = 1$. That is, we solve only $T(c) \equiv 0 \pmod{p}$ and not modulo higher powers of p. This saves the time for lifting and also some time for updating A_c. But then, after the sieving process is over, A_c does not store the correct value of the logarithm of the unfactored part of $T(c)$. We failed to subtract $\log p$ the requisite number of times, and so even for a smooth $T(c)$ (unless square-free), the array location A_c stores a significantly positive quantity (in addition to approximation errors).

But is this situation too bad? If p is small (like 2), its logarithm is small too. So we can tolerate the lack of subtraction of $\log p$ certain number of times. On the other hand, $\log p$ is large for a large member p of B. However, for a large p, we do not expect many values of $T(c)$ to be divisible by p^h with $h \geqslant 2$.

To sum up, considering only the exponent $h = 1$ does not tend to leave a huge value in A_c for a smooth $T(c)$. If $T(c)$ is not smooth, A_c would anyway store a value at least as large as $\log p_{t+1}$. Therefore, if we relax the criterion of smallness of the final values of A_c, we expect to recognize smooth values of $T(c)$ as so. In other words, we suspect $T(c)$ as smooth if and only if the residual value stored in A_c is $\leqslant \xi \log p_{t+1}$ for some constant ξ. In practical situations, the values $1.0 \leqslant \xi \leqslant 2.5$ work quite well.

Since we choose $\xi \geqslant 1$, some non-smooth values of $T(c)$ also pass the selection criterion. Since we do trial division anyway for selected values of $T(c)$, we would discard the non-smooth values of $T(c)$ that pass the liberal selection criterion. An optimized choice of ξ more than compensates for the waste of time on these false candidates by the saving done during the sieving process.

Example 6.18 Let us continue with Examples 6.16 and 6.17. Now, we carry out incomplete sieving. To start with, let us look at what happens to the array locations A_{24} and A_{26}. After the incomplete sieving process terminates, A_{24} stores the value $10647 - 693 - 2944 = 7010$. Since $1000 \log p_{t+1} = 3970$, $T(24)$ is selected as smooth for $\xi \geqslant 7010/3970 \approx 1.766$. On the other hand, A_{26} ends up with the value $10727 - 693 - 1945 - 2397 - 3610 = 2082$ which passes the smoothness test for $\xi \geqslant 2082/3970 \approx 0.524$.

The following table lists, for several values of ξ, all values of c for which $T(c)$ passes the liberal smoothness test.

ξ	c selected with smooth $T(c)$	c selected with non-smooth $T(c)$
0.5	$-15, -11, -2$	None
1.0	$-44, -22, -15, -14, -11, -2, 0, 26$	None
1.5	$-44, -22, -15, -14, -11, -9,$ $-4, -2, 0, 2, 4, 26, 34, 36$	$-33, -23, -1, 5, 15, 19, 41, 43$
2.0	$-44, -22, -15, -14, -11, -9,$ $-4, -2, 0, 2, 4, 26, 34, 36$	$-48, -45, -43, -42, -33, -32, -29,$ $-26, -23, -18, -16, -8, -7, -1, 1, 5,$ $6, 7, 8, 11, 12, 14, 15, 19, 20, 22, 23, 24,$ $25, 32, 33, 41, 42, 43, 48$

Evidently, for small values of ξ, only some (not all) smooth values of $T(c)$ pass the selection criterion. As we increase ξ, more smooth values of $T(c)$ pass the criterion, and more non-smooth values too pass the criterion. □

6.6.3 Large Prime Variation

Suppose that in the relation collection stage of the QSM, we are unable to obtain sufficiently many relations to force non-zero solutions for $\beta_1, \beta_2, \ldots, \beta_s$ needed to split n. This may have happened because t and/or M were chosen to be smaller than the optimal values (possibly to speed up the sieving process). Instead of repeating the entire sieving process with increased values of t and/or M, it is often useful to look at some specific non-smooth values of $T(c)$.

Write $T(c) = uv$ with u smooth over B and with v having no prime factors in B. If v is composite, it admits a prime factor $\leqslant \sqrt{v}$, that is, if $p_t < v < p_t^2$ (where p_t is the largest prime in B), then v is prime. In terms of log values, if the final value stored in A_c satisfies $\log p_t < A_c < 2 \log p_t$, the non-smooth part of $T(c)$ is a prime. Conversely, if $T(c)$ is smooth except for a prime factor q in the range $p_t < q < p_t^2$, the residue left in A_c satisfies $\log p_t < A_c < 2 \log p_t$. Such a prime q is called a *large prime* in the context of the QSM.

Suppose that a large prime q appears as the non-smooth part in two or more values of $T(c)$. In that case, we add q to the factor base, and add all these relations involving q. This results in an increase in the factor-base size by one, whereas the number of relations increases by at least two. After sufficiently many relations involving large primes are collected, we may have more relations than the factor-base size, a situation that is needed to split n.

Example 6.19 Let me illustrate the concept of large primes in connection with Examples 6.16 and 6.17. Suppose we want to keep track of all large primes $\leqslant 430$ (ten times the largest prime in the factor base B). The following table lists all values of c for which $T(c)$ contains such a large prime factor. The final values stored in A_c are also shown. Note that $\lfloor 1000 \log 430 \rfloor = 6063$.

c	A_c	$T(c)$
-48	5573	$-77848 = (-1) \times 2^3 \times 37 \times 263$
-33	5550	$-53713 = (-1) \times 11 \times 19 \times 257$
-32	5948	$-52088 = (-1) \times 2^3 \times 17 \times 383$
-30	4693	$-48832 = (-1) \times 2^6 \times 7 \times 109$
-28	4489	$-45568 = (-1) \times 2^9 \times 89$
-26	5740	$-42296 = (-1)2^3 \times 17 \times 311$
-23	5639	$-37373 = (-1) \times 7 \times 19 \times \boxed{281}$
-18	5803	$-29128 = (-1) \times 2^3 \times 11 \times 331$
-8	5408	$-12488 = (-1) \times 2^3 \times 7 \times 223$
-1	4635	$-721 = (-1) \times 7 \times 103$
5	4264	$9443 = 7 \times 19 \times 71$
6	5294	$11144 = 2^3 \times 7 \times 199$
8	4673	$14552 = 2^3 \times 17 \times \boxed{107}$
12	5253	$21392 = 2^4 \times 7 \times 191$
14	4673	$24824 = 2^3 \times 29 \times \boxed{107}$
15	4845	$26543 = 11 \times 19 \times 127$
19	5639	$33439 = 7 \times 17 \times \boxed{281}$
20	5057	$35168 = 2^5 \times 7 \times 157$
24	5624	$42104 = 2^3 \times 19 \times 277$
32	5094	$56072 = 2^3 \times 43 \times 163$
40	5189	$70168 = 2^3 \times 7^2 \times 179$
41	5477	$71939 = 7 \times 43 \times 239$
42	5602	$73712 = 2^4 \times 17 \times 271$
43	4920	$75487 = 19 \times 29 \times \boxed{137}$
48	4922	$84392 = 2^3 \times 7 \times 11 \times \boxed{137}$

Three large primes have repeated occurrences (see the boxed entries). If we add these three primes to the factor base of Example 6.16, we obtain $B = \{-1, 2, 7, 11, 17, 19, 29, 37, 43, 107, 137, 281\}$ (12 elements) and $14+6 = 20$ relations, that is, we are now guaranteed to have at least $2^{20-12} = 2^8$ solutions. Compare this with the original situation of 14 relations involving 9 elements of B, where the guaranteed number of solutions was $\geqslant 2^{14-9} = 2^5$. □

Relations involving two or more large primes can be considered. If the non-smooth part v in $T(c)$ satisfies $p_t^2 < v < p_t^3$ (equivalently, $2 \log p_t < A_c < 3 \log p_t$ after the sieving process), then v is either a prime or a product of two large primes. In the second case, we identify the two prime factors by trial divisions by large primes. This gives us a relation involving two large primes p, q. If each of these large primes occurs at least once more in other relations (alone or with other repeated large primes), we can add p, q to B.

6.6.4 Multiple-Polynomial Quadratic Sieve Method

The multiple-polynomial quadratic sieve method (MPQSM)[9] is practically considered to be the second fastest among the known integer-factoring algorithms. In the original QSM, we consider smooth values of $T(c) = J + 2cH + c^2$ for $-M \leqslant c \leqslant M$. When $c = \pm M$, the quantity $|T(c)|$ reaches the maximum possible value approximately equal to $2MH \approx 2M\sqrt{n}$. The MPQSM works with a more general quadratic polynomial of the form

$$T(c) = U + 2Vc + Wc^2$$

with $V^2 - UW = n$. The coefficients U, V, W are so adjusted that the maximum value of $T(c)$ is somewhat smaller compared to $2M\sqrt{n}$. We first choose W as a prime close to $\frac{\sqrt{2n}}{M}$, modulo which n is a quadratic residue. We take V to be the smaller square root of n modulo W, so $V \leqslant \frac{1}{2}\frac{\sqrt{2n}}{M} = \frac{\sqrt{\frac{n}{2}}}{M}$. Finally, we take $U = \frac{V^2-n}{W} \approx \frac{-n}{\frac{\sqrt{2n}}{M}} = -\sqrt{\frac{n}{2}}M$. For these choices, the maximum value of $|T(c)|$ becomes $|T(M)| = |U + 2VM + WM^2| \leqslant U + WM + WM^2 \approx U + WM^2 \approx -\sqrt{\frac{n}{2}}M + \frac{\sqrt{2n}}{M} \times M^2 = \frac{1}{\sqrt{2}}M\sqrt{n}$. This value is $2\sqrt{2} \approx 2.828$ times smaller than the maximum value $2M\sqrt{n}$ for the original QSM. Since smaller candidates are now tested for smoothness, we expect to obtain a larger fraction of smooth values of $T(c)$ than in the original QSM.

It remains to establish how we can use the generalized polynomial $T(c) = U + 2Vc + Wc^2$ in order to obtain a relation. Multiplying the expression for $T(c)$ by W gives $WT(c) = (Wc+V)^2 + (UW - V^2) = (Wc+V)^2 - n$, that is,

$$(Wc + V)^2 \equiv WT(c) \pmod{n}.$$

Since the prime W occurs on the right side of every relation, we include W in the factor base B. (For a sufficiently large n, the prime $W \approx \frac{\sqrt{2n}}{M}$ was not originally included in the set B of small prime.)

Example 6.20 We factor $n = 713057$ by the MPQSM. In the original QSM (Example 6.16), we took $B = \{-1, 2, 7, 11, 17, 19, 29, 37, 43\}$, and $M = 50$. Let us continue to work with these choices.

In order to apply the MPQSM, we first choose a suitable polynomial, that is, the coefficients U, V, W. We have $\frac{\sqrt{2n}}{M} \approx 23.884$. The smallest prime larger than this quantity, modulo which n is a quadratic residue, is 29. Since n is an artificially small composite integer chosen for the sake of illustration only, the prime $W = 29$ happens to be already included in the factor base B, that is, the inclusion of W in B does not enlarge B. The two square roots of n modulo W are 11 and 18. We choose the smaller one as V, that is, $V = 11$. Finally, we take $U = \frac{V^2-n}{W} = -24584$. Therefore,

$$T(c) = -24584 + 22c + 29c^2.$$

[9]Robert D. Silverman, The multiple polynomial quadratic sieve, *Mathematics of Computation*, 48, 329–339, 1987. The author, however, acknowledges personal communication with Peter L. Montgomery for the idea.

The following table lists the smooth values of $T(c)$ as c ranges over the interval $-50 \leqslant c \leqslant 50$. The MPQSM yields 19 relations, whereas the original QSM yields only 14 relations (see Example 6.16).

c	$T(c)$	c	$T(c)$
-50	$46816 = 2^5 \times 7 \times 11 \times 19$	6	$-23408 = (-1) \times 2^4 \times 7 \times 11 \times 19$
-38	$16456 = 2^3 \times 11^2 \times 17$	18	$-14792 = (-1) \times 2^3 \times 43^2$
-34	$8192 = 2^{13}$	20	$-12544 = (-1) \times 2^8 \times 7^2$
-32	$4408 = 2^3 \times 19 \times 29$	22	$-10064 = (-1) \times 2^4 \times 17 \times 37$
-29	$-833 = (-1) \times 7^2 \times 17$	26	$-4408 = (-1) \times 2^3 \times 19 \times 29$
-28	$-2464 = (-1) \times 2^5 \times 7 \times 11$	27	$-2849 = (-1) \times 7 \times 11 \times 37$
-14	$-19208 = (-1) \times 2^3 \times 7^4$	28	$-1232 = (-1) \times 2^4 \times 7 \times 11$
-12	$-20672 = (-1) \times 2^6 \times 17 \times 19$	30	$2176 = 2^7 \times 17$
-10	$-21904 = (-1) \times 2^4 \times 37^2$	45	$35131 = 19 \times 43^2$
-3	$-24389 = (-1) \times 29^3$		

In order to exploit the reduction of the values of $T(c)$ in the MPQSM, we could have started with a smaller sieving interval, like $M = 35$. In this case, we have the parameters $U = -19264$, $V = 17$, and $W = 37$, that is, $T(c) = -19264 + 34c + 37c^2$. Smoothness tests of the values of $T(c)$ for $-35 \leqslant c \leqslant 35$ yield 15 smooth values (for $c = -35, -26, -24, -23, -18, -16, -8, -7, 0, 16, 18, 20, 21, 22, 32$).

On the other hand, if we kept $M = 50$ but eliminated the primes 37 and 43 from B, we would have a factor base of size 7. The relations in the above table, that we can now no longer use, correspond to $c = -10, 18, 22, 27, 45$. But that still leaves us with 14 other relations. □

Example 6.20 illustrates that in the MPQSM, we can start with values of t and/or M smaller than optimal. We may still hope to obtain sufficiently many relations to split n. Moreover, we can use different polynomials (for different choices of W), and run different instances of the MPQSM in parallel.

6.7 Cubic Sieve Method

The cubic sieve method (CSM) proposed by Reyneri[10] achieves a running time of $L[\sqrt{2/3}]$, that is, nearly $L[0.8165]$, and so is asymptotically faster than the QSM. Suppose that we know a solution of the congruence

$$x^3 \equiv y^2 z \pmod{n}$$

with $x^3 \neq y^2 z$ as integers, and with x, y, z of absolute values $O(n^\xi)$ for $\xi < 1/2$. Heuristic estimates indicate that we expect to have solutions with $\xi \approx 1/3$.

---

[10]Unpublished manuscript, first reported in: D. Coppersmith, A. M. Odlyzko, R. Schroeppel, Discrete logarithms in $GF(p)$, *Algorithmica*, 1(1), 1–15, 1986.

For small integers a, b, c with $a + b + c = 0$, we have

$$
\begin{aligned}
(x + ay)(x + by)(x + cy) &\equiv x^3 + (a + b + c)x^2 y + (ab + ac + bc)xy^2 + (abc)y^3 \\
&\equiv y^2 z + (ab + ac + bc)xy^2 + (abc)y^3 \\
&\equiv y^2 T(a, b, c) \pmod{n},
\end{aligned}
$$

where

$$
T(a, b, c) = z + (ab + ac + bc)x + (abc)y.
$$

Since each of x, y, z is $O(n^\xi)$, the value of $T(a, b, c)$ is again $O(n^\xi)$. If $\xi \approx 1/3$, then $T(a, b, c)$ is $O(n^{1/3})$, that is, asymptotically smaller than $O(n^{1/2})$, the values of $T(c)$ in the QSM (or MPQSM).

We start with a factor base B, and a bound $M = L[\sqrt{\xi/2}]$. The factor base consists of all of the t primes $\leqslant L[\sqrt{\xi/2}]$ along with the $2M + 1$ integers $x + ay$ for $-M \leqslant a \leqslant M$. The size of the factor base is then $L[\sqrt{\xi/2}]$. If some $T(a, b, c)$ factors completely over the t small primes in B, we get a relation

$$
(x + ay)(x + by)(x + cy) \equiv y^2 p_1^{\alpha_1} p_2^{\alpha_2} \cdots p_t^{\alpha_t} \pmod{n}.
$$

The probability of a $T(a, b, c)$ to be smooth is $L[-\frac{\xi}{2\sqrt{\xi/2}}] = L[-\sqrt{\xi/2}]$. The number of triples a, b, c in the range $-M$ to M with $a + b + c = 0$ is $O(M^2)$, that is, $L[2\sqrt{\xi/2}]$. Therefore, all these $T(a, b, c)$ values are expected to produce $L[\sqrt{\xi/2}]$ relations, which is of the same size as the factor base B.

The CSM supports possibilities of sieving. To avoid duplicate generations of the same relations for different permutations of a, b, c, we force the triples (a, b, c) to satisfy $-M \leqslant a \leqslant b \leqslant c \leqslant M$. In that case, a is always negative, c is always positive, and b varies from $\max(a, -(M + a))$ to $-a/2$. We perform a sieving procedure for each value of $a \in \{-M, -M + 1, \ldots, 0\}$. For a fixed a, the sieving interval is $\max(a, -(M + a)) \leqslant b \leqslant -a/2$. The details of the sieving process are left to the reader (Exercise 6.21).

The CSM with sieving can be shown to run in time $L[2\sqrt{\xi/2}]$. A sparse system involving $L[\sqrt{\xi/2}]$ variables and $L[\sqrt{\xi/2}]$ equations can also be solved in time $L[2\sqrt{\xi/2}]$. That is, the running time of the CSM is $L[2\sqrt{\xi/2}] = L[\sqrt{2\xi}]$. For $\xi = 1/3$, this running time is $L[\sqrt{2/3}]$, as claimed earlier.

Example 6.21 Let us factor $n = 6998891$ using the CSM. We need a solution of $x^3 \equiv y^2 z \pmod{n}$. For $x = 241$, $y = 3$ and $z = -29$, we have $x^3 - y^2 z = 2n$. We take $M = 50$. The factor base B consists of -1, all primes < 100 (there are 25 of them), and all integers of the form $x + ay = 241 + 3a$ for $-50 \leqslant a \leqslant 50$. The size of the factor base is, therefore, $1 + 25 + 101 = 127$. In this case, we have $T(a, b, c) = -29 + 241(ab + ac + bc) + 3abc$. If we vary a, b, c with $-50 \leqslant a \leqslant b \leqslant c \leqslant 50$ and with $a + b + c = 0$, we obtain 162 smooth values of $T(a, b, c)$. Some of these smooth values are listed in the following table.

a	b	c	$T(a,b,c)$
-50	-49	99	$-1043970 = (-1) \times 2 \times 3 \times 5 \times 17 \times 23 \times 89$
-50	-39	89	$-918390 = (-1) \times 2 \times 3 \times 5 \times 11^3 \times 23$
-50	-16	66	$-698625 = (-1) \times 3^5 \times 5^3 \times 23$
-50	14	36	$-556665 = (-1) \times 3 \times 5 \times 17 \times 37 \times 59$
-49	-48	97	$-1016334 = (-1) \times 2 \times 3^3 \times 11 \times 29 \times 59$
-49	-21	70	$-716850 = (-1) \times 2 \times 3^5 \times 5^2 \times 59$
-49	-4	53	$-598598 = (-1) \times 2 \times 7 \times 11 \times 13^2 \times 23$
-49	7	42	$-551034 = (-1) \times 2 \times 3^2 \times 11^3 \times 23$
-49	10	39	$-542010 = (-1) \times 2 \times 3 \times 5 \times 7 \times 29 \times 89$
-49	18	31	$-526218 = (-1) \times 2 \times 3 \times 7 \times 11 \times 17 \times 67$
-48	-38	86	$-872289 = (-1) \times 3^4 \times 11^2 \times 89$
-48	-29	77	$-771894 = (-1) \times 2 \times 3^2 \times 19 \times 37 \times 61$
-48	-19	67	$-678774 = (-1) \times 2 \times 3 \times 29 \times 47 \times 83$
-48	9	39	$-521246 = (-1) \times 2 \times 11 \times 19 \times 29 \times 43$
-48	20	28	$-500973 = (-1) \times 3 \times 11 \times 17 \times 19 \times 47$
			$\cdots$
-5	-4	9	$-14190 = (-1) \times 2 \times 3 \times 5 \times 11 \times 43$
-5	-1	6	$-7410 = (-1) \times 2 \times 3 \times 5 \times 13 \times 19$
-5	2	3	$-4698 = (-1) \times 2 \times 3^4 \times 29$
-4	-3	7	$-8694 = (-1) \times 2 \times 3^3 \times 7 \times 23$
-4	-2	6	$-6633 = (-1) \times 3^2 \times 11 \times 67$
-4	0	4	$-3885 = (-1) \times 3 \times 5 \times 7 \times 37$
-4	1	3	$-3198 = (-1) \times 2 \times 3 \times 13 \times 41$
-3	1	2	$-1734 = (-1) \times 2 \times 3 \times 17^2$
-2	-2	4	$-2873 = (-1) \times 13^2 \times 17$
-1	0	1	$-270 = (-1) \times 2 \times 3^3 \times 5$
0	0	0	$-29 = (-1) \times 29$

We solve for $\beta_1, \beta_2, \ldots, \beta_{162}$ from 127 linear congruences modulo 2. These linear-algebra calculations are not shown here. Since the number of variables is significantly larger than the number of equations, we expect to find a non-zero solution for $\beta_1, \beta_2, \ldots, \beta_{162}$ to split n. Indeed, we get $n = 293 \times 23887$. ☐

The CSM has several problems which restrict the use the CSM in a general situation. The biggest problem is that we need a solution of the congruence $x^3 \equiv y^2 z \pmod{n}$ with $x^3 \neq y^2 z$ and with x, y, z as small as possible. No polynomial-time (nor even subexponential-time) method is known to obtain such a solution. Only when n is of certain special forms (like n or a multiple of n is close to a perfect cube), a solution for x, y, z is available naturally.

A second problem of the CSM is that because of the quadratic and cubic coefficients (in a, b, c), the values of $T(a, b, c)$ are, in practice, rather large. To be precise, $T(a, b, c) = O(L[3\sqrt{\xi/2}]n^\xi)$. Although this quantity is asymptotically $O(n^{\xi+o(1)})$, the expected benefits of the CSM do not show up unless n is quite large. My practical experience with the CSM shows that for integers of bit sizes $\geqslant 200$, the CSM offers some speedup over the QSM. However, for these bit sizes, one would possibly prefer to apply the number-field sieve method. In the special situations where the CSM is readily applicable (as mentioned in the last paragraph), the special number-field sieve method too is applicable, and appears to be a strong contender of the CSM.

6.8 Elliptic Curve Method

Lenstra's elliptic curve method (ECM)[11] is a clever adaptation of Pollard's $p-1$ method. Proposed almost at the same time as the QSM, the ECM has an output-sensitive running time of $L_p(c) = L(p, 1/2, c)$, where p is the smallest prime divisor of n. The ECM can, therefore, be efficient if p is small (that is, much smaller than $\sqrt{n}$), whereas the QSM is not designed to take direct advantage of such a situation.

Let n be an odd composite integer which is not a prime power. Let p be the smallest prime divisor of n. We consider an elliptic curve

$$E : y^2 = x^3 + ax + b$$

defined over $\mathbb{F}_p$. Moreover, let $P = (h, k)$ be a non-zero point on $E_p = E_{\mathbb{F}_p}$. Since p is not known beforehand, we keep the parameters of E and the coordinates of points in E_p reduced modulo n. We carry out curve arithmetic modulo n. The canonical projection $\mathbb{Z}_n \to \mathbb{Z}_p$ lets the arithmetic proceed naturally modulo p as well, albeit behind the curtain. Eventually, we expect the arithmetic to fail in an attempt to invert some integer r which is non-zero modulo n, but zero modulo p. The gcd of r and n divulges a non-trivial factor of n. The ECM is based upon an idea of forcing this failure of computation in a reasonable (subexponential, to be precise) amount of time.

By Hasse's theorem, the group E_p contains between $p + 1 - 2\sqrt{p}$ and $p + 1 + 2\sqrt{p}$ points, that is, $|E_p| \approx p$. We assume that for randomly chosen curve parameters a, b, the size of E_p is a random integer in the Hasse interval. Let the factor base B consist of all small primes $\leqslant L_p[1/\sqrt{2}]$. The size $|E_p|$ is smooth over B with a probability of $L_p[-1/\sqrt{2}]$ (by Corollary 6.14). This means that after trying $L_p[1/\sqrt{2}]$ random curves, we expect to obtain one curve with $|E_p|$ being B-smooth. Such a curve E can split n with high probability as follows.

Let $B = \{p_1, p_2, \ldots, p_t\}$ (where p_i is the i-th prime), $e_i = \lfloor \log p / \log p_i \rfloor$, and $m = \prod_{i=1}^{t} p_i^{e_i}$. Since $|E_p|$ is B-smooth, and the order of the point $P \in E_p$ is a divisor of $|E_p|$, we must have $mP = \mathcal{O}$. This means that at some stage during the computation of mP, we must attempt to add two finite points $Q_1 = (h_1, k_1)$ and $Q_2 = (h_2, k_2)$ satisfying $Q_1 = -Q_2$ in the group E_p, that is, $h_1 \equiv h_2 \pmod{p}$ and $k_1 \equiv -k_2 \pmod{p}$. But p is unknown, so the

---

[11] Hendrik W. Lenstra Jr., Factoring integers with elliptic curves, *Annals of Mathematics*, 126(2), 649–673, 1987.

coefficients h_1, k_1, h_2, k_2 are kept available modulo n. This, in turn, indicates that we expect with high probability (see below why) that the modulo-n representatives of h_1 and h_2 are different (although they are same modulo p). First, assume that Q_1 and Q_2 are ordinary points, that is, different in E_p. Then, the computation of $Q_1 + Q_2$ modulo n attempts to invert $h_2 - h_1$ (see the addition formulas in Section 4.2). We have $p|(h_2 - h_1)$. If $n \nmid (h_2 - h_1)$, then $\gcd(h_2 - h_1, n)$ is a non-trivial divisor (a multiple of p) of n. If Q_1 and Q_2 are special points, $Q_1 + Q_2 = \mathcal{O}$ implies $Q_1 = Q_2$. Such a situation arises when the computation of mP performs a doubling step. Since Q_1 is a special point, we have $k_1 \equiv 0 \pmod{p}$. The doubling procedure involves inverting $2k_1$ modulo n. If $k_1 \not\equiv 0 \pmod{n}$, then $\gcd(k_1, n)$ is a non-trivial factor of n.

Let q be a prime divisor of n, other than p. If p is the smallest prime divisor of n, we have $q > p$. The size of the group E_q is nearly q. It is expected that $|E_p|$ and $|E_q|$ are not B-smooth simultaneously. But then, although $mP = \mathcal{O}$ in E_p, we have $mP \neq \mathcal{O}$ in E_q. Therefore, when $Q_1 + Q_2$ equals $\mathcal{O}$ in the computation of mP in E_p, we have $Q_1 + Q_2 \neq \mathcal{O}$ in E_q, that is, $h_1 \equiv h_2 \pmod{p}$, but $h_1 \not\equiv h_2 \pmod{q}$. By the CRT, $h_1 \not\equiv h_2 \pmod{n}$, that is, $\gcd(h_2 - h_1, n)$ is a multiple of p, but not of q, and so is a proper divisor of n, larger than 1. The case $2Q_1 = \mathcal{O}$ analogously exhibits a high potential of splitting n.

There are some small problems to solve now in order to make the ECM work. First, p itself is unknown until it (or a multiple of it) is revealed. But the determination of the exponents e_i requires the knowledge of p. If an upper bound M on p is known, we take $e_i = \lfloor \log M / \log p_i \rfloor$. These exponents work perfectly for the algorithm, since this new $m = \prod_{i=1}^{t} p_i^{e_i}$ continues to remain a multiple of the order of P in E_p (of course, if $|E_p|$ is B-smooth). If no non-trivial upper bound on p is known, we can anyway take $M = \sqrt{n}$.

A second problem is the choice of the curve parameters (a, b) and the point $P = (h, k)$ on E. If a and b are chosen as random residues modulo n, obtaining a suitable point P on E becomes problematic. The usual way to locate a point on an elliptic curve is to choose the x-coordinate h randomly, and subsequently obtain the y-coordinate k by taking a square root of $h^3 + ah + b$. In our case, n is composite, and computing square roots modulo n is, in general, a difficult computational problem—as difficult as factoring n itself.

This problem can be solved in two ways. The first possibility is to freely choose a random point $P = (h, k)$ and the parameter a. Subsequently, the parameter b is computed as $b \equiv k^2 - (h^3 + ah) \pmod{n}$. The compositeness of n imposes no specific problems in these computations. A second way to avoid the modular square-root problem is to take a randomly, and set $b \equiv c^2 \pmod{n}$ for some random c. We now choose the point $P = (0, c)$ on the curve E. It is often suggested to take $b = c = 1$ (and $P = (0, 1)$) always, and vary a randomly to generate a random sequence of curves needed by the ECM.

The scalar multiplier $m = \prod_{i=1}^{t} p_i^{e_i}$ can be quite large, since t is already a subexponential function of p (or n). The explicit computation of m can, however, be avoided. We instead start with $P_0 = P$, and then compute $P_i = p_i^{e_i} P_{i-1}$ for $i = 1, 2, \ldots, t$. Eventually, we obtain $P_t = mP$.

The curve $E : y^2 = x^3 + ax + b$ is indeed an elliptic curve modulo p if and only if E is non-singular, that is, $p \nmid (4a^3 + 27b^2)$. If this condition is not satisfied, addition in E_p makes no sense. It may, therefore, be advisable to check whether $\gcd(4a^3 + 27b^2, n) > 1$. If so, we have already located a divisor of n. If that divisor is less than n, we return it. If it is equal to n, we choose another random curve. If, on the other hand, $\gcd(4a^3 + 27b^2, n) = 1$, the curve E is non-singular modulo every prime divisor of n, and we proceed to compute the desired multiple mP of P. If a and b are randomly chosen, the probability of obtaining $\gcd(4a^3 + 27b^2, n) > 1$ is extremely low, and one can altogether avoid computing this gcd. But then, since this gcd computation adds only an insignificant overhead to the computation of mP, it does not really matter whether the non-singularity condition is checked or not.

The above observations are summarized as Algorithm 6.4.

Algorithm 6.4: ELLIPTIC CURVE METHOD

```
Repeat until a non-trivial divisor of n is located:
    Choose a, h, k ∈ Zₙ randomly, and take b ≡ k² − (h³ + ah) (mod n).
    E denotes the curve y² = x³ + ax + b, and P the point (h, k).
    For i = 1, 2, ..., t, set P = pᵢᵉⁱP, where eᵢ = ⌊log M/log pᵢ⌋.
    If the for loop encounters an integer r not invertible mod n {
        Compute d = gcd(r, n).
        If d ≠ n, return d.
        If d = n, abort the computation of mP for the current curve.
    }
```

Example 6.22 Let us factor $n = 1074967$ by the ECM. I first demonstrate an unsuccessful attempt. The choices $P = (h, k) = (26, 83)$ and $a = 8$ give $b \equiv k^2 - (h^3 + ah) \equiv 1064072 \pmod{n}$. We have no non-trivial information about a bound on the smallest prime divisor p of n. So we take the trivial bound $M = \lfloor \sqrt{n} \rfloor = 1036$. We choose the factor base $B = \{2, 3, 5, 7, 11\}$. (Since this is an artificially small example meant for demonstration only, the choice of t is not dictated by any specific formula.) The following table illustrates the computation of mP. For each scalar multiplication of a point on the curve, we use a standard repeated double-and-add algorithm.

i	p_i	$e_i = \lfloor \log M/\log p_i \rfloor$	$p_i^{e_i}$	P_i
0				$P_0 = (26, 83)$
1	2	10	1024	$P_1 = 1024P_0 = (330772, 1003428)$
2	3	6	729	$P_2 = 729P_1 = (804084, 683260)$
3	5	4	625	$P_3 = 625P_2 = (742854, 1008597)$
4	7	3	343	$P_4 = 343P_3 = (926695, 354471)$
5	11	2	121	$P_5 = 121P_4 = (730198, 880012)$

The table demonstrates that the computation of $P_5 = mP$ proceeds without any problem, so the attempt to split n for the above choices of a, b, h, k fails.

In order to see what happened behind the curtain, let me reveal that n factors as pq with $p = 541$ and $q = 1987$. The curve E_p is cyclic with prime order 571, and the curve E_q is cyclic of square-free order $1934 = 2 \times 967$. Thus, neither $|E_p|$ nor $|E_q|$ is smooth over the chosen factor base B, and so $mP \neq \mathcal{O}$ in both E_p and E_q, that is, an accidental discovery of a non-invertible element modulo n did not happen during the five scalar multiplications. Indeed, the points P_i on E_p and E_q are listed in the following table.

i	$P_i \pmod n$	$P_i \pmod p$	$P_i \pmod q$
0	$(26, 83)$	$(26, 83)$	$(26, 83)$
1	$(330772, 1003428)$	$(221, 414)$	$(930, 1980)$
2	$(804084, 683260)$	$(158, 518)$	$(1336, 1719)$
3	$(742854, 1008597)$	$(61, 173)$	$(1703, 1188)$
4	$(926695, 354471)$	$(503, 116)$	$(753, 785)$
5	$(730198, 880012)$	$(389, 346)$	$(969, 1758)$

I now illustrate a successful iteration of the ECM. For the choices $P = (h, k) = (81, 82)$ and $a = 3$, we have $b = 550007$. We continue to take $M = 1036$ and $B = \{2, 3, 5, 7, 11\}$. The computation of mP now proceeds as follows.

i	p_i	$e_i = \lfloor \log M / \log p_i \rfloor$	$p_i^{e_i}$	P_i
0				$P_0 = (81, 82)$
1	2	10	1024	$P_1 = 1024 P_0 = (843635, 293492)$
2	3	6	729	$P_2 = 729 P_1 = (630520, 992223)$
3	5	4	625	$P_3 = 625 P_2 = (519291, 923811)$
4	7	3	343	$P_4 = 343 P_3 = (988490, 846127)$
5	11	2	121	$P_5 = 121 P_4 = ?$

The following table lists all intermediate points lP_4 in the left-to-right double-and-add point-multiplication algorithm for computing $P_5 = 121 P_4$.

Step	l	lP_4
Init	1	$(988490, 846127)$
Dbl	2	$(519843, 375378)$
Add	3	$(579901, 1068102)$
Dbl	6	$(113035, 131528)$
Add	7	$(816990, 616888)$
Dbl	14	$(137904, 295554)$
Add	15	$(517276, 110757)$
Dbl	30	$(683232, 158345)$
Dbl	60	$(890993, 947226)$
Dbl	120	$(815911, 801218)$
Add	121	Failure

In the last step, an attempt is made to add $120P_4 = (815911, 801218)$ and $P_4 = (988490, 846127)$. These two points are different modulo n. So we try to invert $r = 988490 - 815911 = 172579$. To that effect, we compute the extended gcd of r and n, and discover that $\gcd(r, n) = 541$ is a non-trivial factor of n.

Let us see what happened behind the curtain to make this attempt success-ful. The choices a, b in this attempt give a curve E_p (where $p = 541$) of order $539 = 7^2 \times 11$, whereas E_q (where $q = 1987$) now has order $1959 = 3 \times 653$. It follows that $mP = \mathcal{O}$ in E_p, but $mP \neq \mathcal{O}$ in E_q. This is the reason why the computation of mP is bound to reveal p at some stage. The evolution of the points P_i is listed below modulo n, p and q.

i	$P_i \pmod{n}$	$P_i \pmod{p}$	$P_i \pmod{q}$
0	$(81, 82)$	$(81, 82)$	$(81, 82)$
1	$(843635, 293492)$	$(216, 270)$	$(1147, 1403)$
2	$(630520, 992223)$	$(255, 29)$	$(641, 710)$
3	$(519291, 923811)$	$(472, 324)$	$(684, 1843)$
4	$(988490, 846127)$	$(83, 3)$	$(951, 1652)$
5	?	$\mathcal{O}$	$(1861, 796)$

It is worthwhile to mention here that the above case of having a B-smooth value of $|E_p|$ and a B-non-smooth value of $|E_q|$ is not the only situation in which the ECM divulges a non-trivial factor of n. It may so happen that for a choice of a, b, the group E_q has B-smooth order, whereas the other group E_p has B-non-smooth order. In this case, the ECM divulges the larger factor q instead of the smaller one p. For example, for the choices $P = (h, k) = (50, 43)$, $a = 6$ and $b = 951516$, we have $|E_p| = 571$, whereas $|E_q| = 1944 = 2^3 \times 3^5$. So the computation of $P_2 = 729P_1 = (729 \times 1024)P$ fails, and reveals the factor $q = 1987$. The details are not shown here.

There are other situations for the ECM to divulge a non-trivial factor of n. Most surprisingly, this may happen even when both $|E_p|$ and $|E_q|$ are non-smooth over B. For instance, the choices $P = (h, k) = (8, 44)$, $a = 1$ and $b = 1076383$ give $|E_p| = 544 = 2^5 \times 17$ and $|E_q| = 1926 = 2 \times 3^2 \times 107$, both non-smooth over B. We compute $P_1 = 1024P_0 = (855840, 602652)$ and $P_2 = 729P_1 = (810601, 360117)$, and then the computation of $P_3 = 625P_2$ involves an addition chain in which we attempt to add $18P_2 = (75923, 578140)$ and $P_2 = (810601, 360117)$. P_2 has order 17 in E_p, so $18P_2 = P_2$, that is, we accidentally try to add P_2 to itself. This results in a wrong choice of the addition formula and a revelation of the factor $p = 541$ of n.

An addition chain in a single double-and-add scalar multiplication involves the computation of only $\Theta(\log n)$ multiples of the base point. Consequently, a situation like the awkward addition just mentioned is rather unlikely in a single scalar multiplication. But then, we make $t = L_p[1/\sqrt{2}]$ scalar multipli-cations in each factoring attempt of the ECM, and we make $L_p[1/\sqrt{2}]$ such attempts before we expect to discover a non-trivial factor of n. Given so many addition chains, *stray* lucky incidents like the one illustrated in the last para-graph need not remain too improbable. This is good news for the ECM, but creates difficulty in analyzing its realistic performance. In fact, the running-time analysis given below takes into account only the *standard* situation: $|E_p|$ is smooth only for the smallest prime factor p of n. $\qquad\square$

For deriving the running time of the ECM, let us first assume that a fairly accurate bound M on the smallest prime factor p of n is available to us. The choices of the factor base and of the exponents e_i depend upon that. More precisely, we let B consist of all primes $\leqslant L_M[1/\sqrt{2}]$. An integer (the size of the group E_p) of value $O(M)$ is smooth over B with probability $L_M[-1/\sqrt{2}]$, that is, about $L_M[1/\sqrt{2}]$ random curves E need to be tried to obtain one B-smooth value of $|E_p|$. For each choice of a curve, we make $t = |B| = L_M[1/\sqrt{2}]$ scalar multiplications. Since each such scalar multiplication can be completed in $O(\log^3 n)$ time (a polynomial in $\log n$), the running time of the ECM is $L_M[\sqrt{2}]$. In the worst case, M is as large as $\sqrt{n}$, and this running time becomes $L_n[1]$ which is the same as that of the QSM. However, if M is significantly smaller than $\sqrt{n}$, the ECM is capable of demonstrating superior performance compared to the QSM. For example, if n is known to be the product of three distinct primes of roughly the same bit size, we can take $M = \sqrt[3]{n}$. In that case, the running time of the ECM is $L_n[\sqrt{2/3}] \approx L_n[0.816]$—the same as the best possible running time of the CSM.

Given our current knowledge of factoring, the hardest nuts to crack are the products of two primes of nearly the same bit sizes. For such integers, the MPQSM has been experimentally found by the researchers to be slightly more efficient than the ECM. The most plausible reason for this is that the ECM has no natural way of quickly sieving out the bad choices. The ECM is still an important algorithm, because it can effectively exploit the presence of small prime divisors—a capability not present at all in the QSM or the MPQSM.

6.9   Number-Field Sieve Method

The most sophisticated and efficient integer-factoring algorithm known to date is the number-field sieve method (NFSM) proposed for special numbers by Lenstra, Lenstra, Manasse and Pollard,[12] and extended later to work for all integers by Buhler, Lenstra and Pomerance.[13] Understanding the NFSM calls for exposure to number fields and rings—topics well beyond the scope of this book. I provide here a very intuitive description of the NFSM.

In order to motivate the development of the NFSM, let us recapitulate the working of the QSM. We take the polynomial $f(x) = x^2 - J$ satisfying $f(H) = n$. Here, n is the integer to be factored, $H = \lceil \sqrt{n} \rceil$, and $J = H^2 - n$. If J is a perfect square y^2, we already have a congruence $H^2 \equiv y^2 \pmod{n}$ capable of splitting n. So assume that J is not a perfect square, that is, $f(x)$ is an irreducible polynomial in $\mathbb{Q}[x]$. Let θ be a root of f in $\mathbb{C}$. By adjoining

[12] Arjen K. Lenstra, Hendrik W. Lenstra, Jr., Mark S. Manasse and John M. Pollard, The number field sieve, *STOC*, 564–572, 1990.

[13] Joe P. Buhler, Hendrik W. Lenstra, Jr. and Carl Pomerance, Factoring integers with the number field sieve, *Lecture Notes in Mathematics*, 1554, Springer, 50–94, 1994.

θ to $\mathbb{Q}$, we get a field $K = \mathbb{Q}(\theta) = \{a + b\theta \mid a, b \in \mathbb{Q}\}$. This is an example of a *quadratic number field*.

Let $\mathfrak{O}_K$ be all the elements of K, which satisfy monic polynomials with integer coefficients.[14] $\mathfrak{O}_K$ turns out to be a ring and so an integral domain. For simplicity, assume that $\mathfrak{O}_K = \{a + b\theta \mid a, b \in \mathbb{Z}\}$ (this is indeed the case if J is a square-free integer congruent to 2 or 3 modulo 4). The product of two elements $a + b\theta$, $c + d\theta \in \mathfrak{O}_K$ is

$$(a + b\theta)(c + d\theta) = ac + (ad + bc)\theta + bd\theta^2 = (bdJ + ac) + (ad + bc)\theta.$$

Since $f(H) \equiv 0 \pmod{n}$, the map $\mathfrak{O}_K \to \mathbb{Z}_n$ taking $a + b\theta$ to $a + bH$ is a ring homomorphism. Applying this homomorphism on the above product gives

$$(a + bH)(c + dH) \equiv (bdJ + ac) + (ad + bc)H \pmod{n}.$$

If the right side of this congruence is smooth over a set of small primes, we obtain a relation of the form

$$(a + bH)(c + dH) \equiv \prod_{i=1}^{t} p_i^{e_i} \pmod{n}.$$

The original QSM takes $a = c$ and $b = d = 1$, so we have

$$(a + bH)(c + dH) \equiv (H + c)^2$$
$$\equiv (bdJ + ac) + (ad + bc)H \equiv (J + c^2) + 2cH \equiv T(c) \pmod{n}.$$

By varying c, we obtain many relations. Also see Exercise 6.19.

With a bit of imagination, let us generalize this idea. Let $f(x) \in \mathbb{Z}[x]$ be an irreducible polynomial such that $f(H)$ is equal to n (or some multiple of n) for some $H \in \mathbb{Z}$. One possibility is to take an H of bit size about $(\lg n)^{2/3}$, and obtain $f(H)$ as the expansion of n to the base H, so $f(H)$ is a polynomial of degree $d \sim (\lg n)^{1/3}$. Adjoining a root $\theta \in \mathbb{C}$ of f to $\mathbb{Q}$ gives us the field

$$K = \mathbb{Q}(\theta) = \{a_0 + a_1\theta + a_2\theta^2 + \cdots + a_{d-1}\theta^{d-1} \mid a_0, a_1, \ldots, a_{d-1} \in \mathbb{Q}\}.$$

This is a number field of degree d. Let $\mathfrak{O}_K$ denote the set of elements of K, that satisfy *monic* irreducible polynomials in $\mathbb{Z}[x]$. $\mathfrak{O}_K$ is an integral domain, called the order of K or the ring of integers of K. For simplicity, assume that $\mathfrak{O}_K$ supports unique factorization of elements into products of primes and units (in $\mathfrak{O}_K$). If we multiply elements $\beta_1, \beta_2, \ldots, \beta_k \in \mathfrak{O}_K$, the product is again a polynomial in θ of degree $\leqslant d - 1$, that is,

$$\beta_1\beta_2 \cdots \beta_k = \alpha = a_0 + a_1\theta + a_2\theta^2 + \cdots + a_{d-1}\theta^{d-1}.$$

Since $f(H) \equiv 0 \pmod{n}$, the map taking $\theta \mapsto H$ naturally extends to a ring homomorphism $\Phi : \mathfrak{O}_K \to \mathbb{Z}_n$. Applying this map to the last equation gives

$$\Phi(\beta_1)\Phi(\beta_2) \cdots \Phi(\beta_k) \equiv \Phi(\alpha) \equiv a_0 + a_1H + a_2H^2 + \cdots + a_{d-1}H^{d-1} \pmod{n}.$$

---

[14]$\mathfrak{O}$ is the Gothic O. The letter O is an acronym for (maximal) *order* of K.

If the right side of this congruence is smooth over a set of small primes, we get a relation of the form

$$\Phi(\beta_1)\Phi(\beta_2)\cdots\Phi(\beta_k) \equiv \prod_{i=1}^{t} p_i^{e_i} \pmod{n}.$$

In the QSM, we choose $H \approx \sqrt{n}$. In order to arrive at a smaller running time, the NFSM chooses a subexponential expression in $\log n$ as H. Moreover, the product α should be a polynomial of small degree in θ so that substituting θ by H in α gives a value much smaller compared to n. A good choice for α is

$$\alpha = a\theta + b$$

for small coprime integers a, b. But then, how can we express such an element $a\theta + b$ as a product of $\beta_1, \beta_2, \ldots, \beta_k$? This is precisely where the algebraic properties of $\mathfrak{O}_K$ come to the forefront. We can identify a set of *small* elements in $\mathfrak{O}_K$. More technically, we choose a set $\mathcal{P}$ of elements of $\mathfrak{O}_K$ of small prime norms, and a generating set $\mathcal{U}$ of units of $\mathfrak{O}_K$. For a choice of a, b, we need an algorithm to factor $a + b\theta$ completely (if possible) into a product of elements from $\mathcal{P} \cup \mathcal{U}$. Indeed, checking the smoothness of $a + b\theta$ reduces to checking the smoothness of the *integer* $(-b)^d f(-a/b)$. Factorization in number rings $\mathfrak{O}_K$ is too difficult a topic to be explained here. Example 6.23 supplies a flavor.

To worsen matters, the ring $\mathfrak{O}_K$ may fail to support unique factorization of elements into products of primes and units. However, all number rings are so-called *Dedekind domains* where unique factorization holds at the level of ideals, and each ideal can be generated by at most two elements. The general number-field sieve method takes all these issues into account, and yields an integer-factoring algorithm that runs in $L(n, 1/3, (64/9)^{1/3})$ time.

Example 6.23 Let us attempt to factor $n = 89478491$ using the NFSM. In this example, I demonstrate how a single relation can be generated. The first task is to choose a suitable polynomial $f(x) \in \mathbb{Z}[x]$ and a positive integer H such that $f(H)$ is a small multiple of n. For the given n, the choices $f(x) = x^7 + x + 1$ and $H = 16$ work. In fact, we have $f(H) = 3n$.

The polynomial $f(x) = x^7 + x + 1$ is irreducible in $\mathbb{Z}[x]$. Let $\theta \in \mathbb{C}$ be a root of $f(x)$. All the roots of an irreducible polynomial are algebraically indistinguishable from one another. So we do not have to identify a specific complex root of f. Instead, we define the number field $K = \mathbb{Q}(\theta)$ as the set

$$K = \mathbb{Q}(\theta) = \{a_0 + a_1\theta + \cdots + a_6\theta^6 \mid a_i \in \mathbb{Q}\},$$

and implement the arithmetic of K as the polynomial arithmetic of $\mathbb{Q}[x]$ modulo the defining irreducible polynomial $f(x)$. This also saves us from addressing the issues associated with floating-point approximations.

The number field K has degree $d = 7$. The number of real roots of the polynomial $f(x)$ is $r_1 = 1$ (in order to see why, plot $f(x)$, or use the fact that the derivative $f'(x) = 7x^6 + 1$ is positive for all real x). This real root is an irrational value $-0.79654\ldots$, but we do not need to know this value explicitly or approximately, because we planned for an algebraic representation of K. The remaining six roots of $f(x)$ are (properly) complex. Since complex roots of a real polynomial occur in complex-conjugate pairs, the number of pairwise non-conjugate complex roots of $f(x)$ is $r_2 = 3$. The pair $(r_1, r_2) = (1, 3)$ is called the *signature* of K.

Having defined the number field K, we now need to concentrate on its ring of integers $\mathfrak{O}_K$, that is, elements of K whose minimal polynomials are in $\mathbb{Z}[x]$ and are monic. It turns out that all elements of $\mathfrak{O}_K$ can be expressed uniquely as $\mathbb{Z}$-linear combinations of $1, \theta, \theta^2, \ldots, \theta^6$, that is,

$$\mathfrak{O}_K = \mathbb{Z}[\theta] = \{a_0 + a_1\theta + \cdots + a_6\theta^6 \mid a_i \in \mathbb{Z}\}.$$

Furthermore, this $\mathfrak{O}_K$ supports unique factorization at the level of elements, so we do not have to worry about factorization at the level of ideals.

The second task is to choose a factor base. The factor base now consists of two parts: small integer primes $p_1, p_2, \ldots, p_t$ for checking the smoothness of the (rational) integers $aH + b$, and some small elements of $\mathfrak{O}_K$ for checking the smoothness of the algebraic integers $a\theta + b$. The so-called small elements of $\mathfrak{O}_K$ are some small primes in $\mathfrak{O}_K$ and some units in $\mathfrak{O}_K$.

In order to understand how small prime elements are chosen from $\mathfrak{O}_K$, we need the concept of norms. Let $\alpha = \alpha(\theta)$ be an element of K. Let $\theta_1, \theta_2, \ldots, \theta_d$ be all the roots of $f(x)$ in $\mathbb{C}$. The *norm* of α is defined as

$$N(\alpha) = \prod_{i=1}^{d} \alpha(\theta_i).$$

It turns out that $N(\alpha) \in \mathbb{Q}$. Moreover, if $\alpha \in \mathfrak{O}_K$, then $N(\alpha) \in \mathbb{Z}$. If $\alpha \in \mathbb{Q}$, then $N(\alpha) = \alpha^d$. Norm is a multiplicative function, that is, for all $\alpha, \beta \in K$, we have $N(\alpha\beta) = N(\alpha)N(\beta)$. We include an element γ of $\mathfrak{O}_K$ in the factor base if and only if $N(\gamma)$ is a small prime.

Let p be a small integer prime. In order to identify all elements of $\mathfrak{O}_K$ of norm p, we factor $f(x)$ over the finite field $\mathbb{F}_p$. Each linear factor of $f(x)$ in $\mathbb{F}_p[x]$ yields an element of $\mathfrak{O}_K$ of norm p.

Modulo the primes $p = 2$ and $p = 5$, the polynomial $f(x)$ remains irreducible. So there are no elements of $\mathfrak{O}_K$ of norm 2 or 5. We say that the (rational) primes 2 and 5 remain *inert* in $\mathfrak{O}_K$.

For the small prime $p = 3$, we have $x^7 + x + 1 \equiv (x+2)(x^6 + x^5 + x^4 + x^3 + x^2 + x + 2) \pmod{3}$. The linear factor yields an element of norm 3. The other factor gives an element of norm 3^6, which is not considered for inclusion in the factor base. For the linear factor $x + 2$, we need to find out an element $\gamma_{3,2}$ (a polynomial in θ of degree $\leqslant d - 1 = 6$) such that all $\mathfrak{O}_K$-linear combinations of 3 and $\theta + 2$ can be written as multiples of $\gamma_{3,2}$. Computation of such an element is an involved process. In this example, we have $\gamma_{3,2} = \theta^6 - \theta^5$.

For $p = 7$, we have $x^7 + x + 1 \equiv (x + 4)(x^2 + x + 3)(x^2 + x + 4)(x^2 + x + 6) \pmod{7}$. The quadratic factors yield elements of $\mathfrak{O}_K$ of norm 7^2. The linear factor corresponds to the element $\gamma_{7,4} = \theta^6 - \theta^5 + \theta^4$ of norm 7.

The small prime $p = 11$ behaves somewhat differently. We have $x^7 + x + 1 \equiv (x + 3)^2(x^5 + 5x^4 + 5x^3 + 2x^2 + 9x + 5) \pmod{11}$. The linear factor $x + 3$ has multiplicity more than one. We say that the prime 11 *ramifies* in $\mathfrak{O}_K$. Nevertheless, this factor gives the element $\gamma_{11,3} = \theta^5 + 1$ of norm 11.

Now that we have seen some examples of how integer primes behave in the number ring $\mathfrak{O}_K$, it remains to understand the units of $\mathfrak{O}_K$ for completing the construction of the factor base. The group of units of $\mathfrak{O}_K$ is isomorphic to the additive group $\mathbb{Z}_s \times \mathbb{Z}^r$ for some $s \in \mathbb{N}$ and $r \in \mathbb{N}_0$. The (torsion group of) units of finite orders is generated by a primitive s-th root ω_s of unity. In this example, the only units of $\mathfrak{O}_K$ of finite orders are ± 1. Besides these (torsion) elements, we can identify r independent units $\xi_1, \xi_2, \ldots, \xi_r$ of infinite order, called a set of *fundamental units* in $\mathfrak{O}_K$. By *Dirichlet's unit theorem*, the rank r is $r_1 + r_2 - 1$, where (r_1, r_2) is the signature of K. In this example, $r_1 = 1$ and $r_2 = 3$, so we have three fundamental units. These units can be chosen as $\xi_1 = \theta$, $\xi_2 = \theta^4 + \theta^2$, and $\xi_3 = \theta^6 - \theta^3 + 1$. These elements have norms ± 1.

To sum up, the factor base needed to factor the algebraic integers $a\theta + b$ consists of the elements γ_{p,c_p} for small integer primes p, the primitive root ω_s of unity, and the fundamental units $\xi_1, \xi_2, \ldots, \xi_r$.

I now demonstrate the generation of a relation. Take the element $\alpha = 2\theta + 1 \in \mathfrak{O}_K$. The norm of this element is $N(\alpha) = (-2)^7 \left((-1/2)^7 + (-1/2) + 1\right) = 1 + 2^6 - 2^7 = -63 = -3^2 \times 7$. Indeed, α factors in $\mathfrak{O}_K$ as $\alpha = \omega_2 \xi_2 \xi_3 \gamma_{3,2}^2 \gamma_{7,4}$. This can be written more explicitly as

$$2\theta + 1 = (-1)(\theta^4 + \theta^2)(\theta^6 - \theta^3 + 1)(\theta^6 - \theta^5)^2(\theta^6 - \theta^5 + \theta^4),$$

that is,

$$2x + 1 \equiv (-1)(x^4 + x^2)(x^6 - x^3 + 1)(x^6 - x^5)^2(x^6 - x^5 + x^4) \pmod{f(x)}.$$

Since $f(H) \equiv 0 \pmod{n}$, substituting x by H gives

$$33 \equiv (-1)(H^4 + H^2)(H^6 - H^3 + 1)(H^6 - H^5)^2(H^6 - H^5 + H^4) \pmod{n}.$$

In fact, putting $H = 16$ lets the right side evaluate to

$$-43118762328978851833388771627827200 = -4818896904394470826948536\,3n + 33.$$

But $33 = 3 \times 11$ is smooth (over a factor base containing at least the first five rational primes), that is, a relation is generated. □

The NFSM deploys two sieves, one for filtering the smooth integers $aH + b$, and the other for filtering the smooth algebraic integers $a\theta + b$. Both these candidates are subexponential expressions in $\log n$. For QSM, the smoothness candidates are exponential in $\log n$. This results in the superior asymptotic performance of the NFSM over the QSM. In practice, this asymptotic superiority shows up for input integers of size at least several hundred bits.

Exercises

1. Let ω and c be constants with $0 < \omega < 1$ and $c > 0$, and

$$L_n(\omega, c) = \exp\left[(c + \mathrm{o}(1))\,(\ln n)^\omega\,(\ln \ln n)^{1-\omega}\right].$$

For a positive constant k, prove that:
(a) The expression $(\ln n)^k L_n(\omega, c)$ is again of the form $L_n(\omega, c)$.
(b) The expression $n^{k+\mathrm{o}(1)} L_n(\omega, c)$ is again of the form $n^{k+\mathrm{o}(1)}$.

2. Take $n \approx 2^{1024}$. Evaluate the expressions $n^{1/4}$, $L_n(1/2, 1)$ and $L_n(1/3, 2)$ (ignore the $\mathrm{o}(1)$ term in subexponential expressions). What do these values tell about known integer-factoring algorithms?

3. You are given a black box that, given two positive integers n and k, returns in one unit of time the decision whether n has a factor d in the range $2 \leqslant d \leqslant k$. Using this black box, devise an algorithm to factor a positive integer n in polynomial (in $\log n$) time. Deduce the running time of your algorithm.

4. Explain why the functions $f(x) = x^2 \pmod{n}$ and $f(x) = x^2 - 2 \pmod{n}$ are not chosen in Pollard's rho method for factoring n.

5. Write a pseudocode implementing Floyd's variant of Pollard's rho method with block gcd calculations.

6. In Floyd's variant of Pollard's rho method for factoring the integer n, we compute the values of x_k and x_{2k} and then $\gcd(x_k - x_{2k}, n)$ for $k = 1, 2, 3, \ldots$. Suppose that we instead choose some $r, s \in \mathbb{N}$, and compute x_{rk+1} and x_{sk} and subsequently $\gcd(x_{rk+1} - x_{sk}, n)$ for $k = 1, 2, 3, \ldots$.
(a) Deduce a condition relating r, s and the period τ' of the cycle such that this method is guaranteed to detect a cycle of period τ'.
(b) Characterize all the pairs (r, s) such that this method is guaranteed to detect cycles of any period.

7. In this exercise, we describe a variant of the second stage for Pollard's $p - 1$ method. Suppose that $p - 1 = uv$ with a B-power-smooth u and with a large prime v (that is, a prime satisfying $B < v < B'$). In the first stage, we have already computed $b \equiv a^E \pmod{n}$.
(a) If $b \not\equiv 1 \pmod{n}$, show that $\mathrm{ord}_p(b) = v$.
(b) Set $b_0 = b$, and for $i = 1, 2, \ldots, t$, compute $b_i = b_{i-1}^{l_i} \pmod{n}$, where l_i is a random integer between 2 and $B' - 1$. Argue that for $t = \mathrm{O}(\sqrt{B'})$, we expect to have distinct i, j with $b_i \equiv b_j \pmod{p}$.
(c) Describe how such a pair (i, j) is expected to reveal the factor p of n.
(d) Compare this variant of the second stage (in terms of running time and space requirement) with the variant presented in Section 6.3.1.

8. [*Euler's factorization method*] We say that a positive integer n can be written as the sum of two squares if $n = a^2 + b^2$ for some positive integers a, b.

(a) Show that if two odd integers m, n can be written as sums of two squares, then their product mn can also be so written.

(b) Prove that no $n \equiv 3 \pmod 4$ can be written as a sum of two squares.

(c) Let a square-free composite integer n be a product of (distinct) primes each congruent to 1 modulo 4. Show that n can be written as a sum of two squares in (at least) two different ways.

(d) Let n be as in Part (c). Suppose that we know two ways of expressing n as a sum of two squares. Describe how n can factored easily.

9. Prove that an integer of the form $4^e (8k + 7)$ (with $e, k \geqslant 0$) cannot be written as a sum of three squares. (**Remark:** The converse of this statement is also true, but is somewhat difficult to prove.)

10. Prove that Carmichael numbers are (probabilistically) easy to factor.

11. (a) Suppose that in Dixon's method for factoring n, we first choose a non-zero $z \in \mathbb{Z}_n$, and generate relations of the form

$$x_i^2 z \equiv p_1^{\alpha_{i1}} p_2^{\alpha_{i2}} \cdots p_t^{\alpha_{it}} \pmod n.$$

Describe how these relations can be combined to arrive at a congruence of the form $x^2 \equiv y^2 \pmod n$.

(b) Now, choose several small values of z (like $1 \leqslant z \leqslant M$ for a small bound M). Describe how you can still generate a congruence $x^2 \equiv y^2 \pmod n$. What, if anything, do you gain by using this strategy (over Dixon's original method)?

12. Dixon's method for factoring an integer n can be combined with a sieve in order to reduce its running time to $L[3/2]$. Instead of choosing random values of $x_1, x_2, \ldots, x_s$ in the relations, we first choose a random value of x, and for $-M \leqslant c \leqslant M$, we check the smoothness of the integers $(x + c)^2 \pmod n$ over t small primes $p_1, p_2, \ldots, p_t$. As in Dixon's original method, we take $t = L[1/2]$.

(a) Determine M for which one expects to get a system of the desired size.

(b) Describe a sieve over the interval $[-M, M]$ for detecting the smooth values of $(x + c)^2 \pmod n$.

(c) Deduce how you achieve a running time of $L[3/2]$ using this sieve.

13. Dixon's method for factoring an integer n can be combined with another sieving idea. We predetermine a factor base $B = \{p_1, p_2, \ldots, p_t\}$, and a sieving interval $[-M, M]$. Suppose that we compute $u \equiv z^2 \pmod n$ for a randomly chosen non-zero $z \in \mathbb{Z}_n$.

(a) Describe a sieve to identify all B-smooth values of $u + cn$ for $-M \leqslant c \leqslant M$. Each such B-smooth value yields a relation of the form

$$z^2 \equiv p_1^{\alpha_{i1}} p_2^{\alpha_{i2}} \cdots p_t^{\alpha_{it}} \pmod n.$$

(b) How can these relations be combined to get a congruence $x^2 \equiv y^2 \pmod n$?

(c) Supply optimal choices for t and M. What is the running time of this variant of Dixon's method for these choices of t and M?

(d) Compare the sieve of this exercise with that of Exercise 6.12.

14. Let α_{ij} be the exponent of the j-th small prime p_j in the i-th relation collected in Dixon's method. We find vectors in the null space of the $t \times s$ matrix $A =$

(α_{ji}). In the linear-algebra phase, it suffices to know the value of α_{ij} modulo 2. Assume that for a small prime p and a small exponent h, the probability that a random square $x^2 \pmod{n}$ has probability $1/p^h$ of being divisible by p^h (irrespective of whether $x^2 \pmod{n}$ is B-smooth or not). Calculate the probability that $\alpha_{ij} \equiv 1 \pmod{2}$. (This probability would be a function of the prime p_j. This probability calculation applies to other subexponential factoring methods like CFRAC, QSM and CSM.)

15. [*Fermat's factorization method*]   Let n be an odd positive composite integer which is not a perfect square, and $H = \lceil \sqrt{n} \rceil$.
    (a) Prove that there exists $c \geqslant 0$ such that $(H + c)^2 - n$ is a perfect square b^2 with $H + c \not\equiv \pm b \pmod{n}$.
    (b) If we keep on trying $c = 0, 1, 2, \ldots$ until $(H + c)^2 - n$ is a perfect square, we obtain an algorithm to factor n. What is its worst-case complexity?
    (c) Prove that if n has a factor u satisfying $\sqrt{n} - u < n^{1/4}$, then $H^2 - n$ is itself a perfect square.

16. Suppose that we want to factor $n = 3337$ using the quadratic sieve method.
    (a) Determine H and J, and write the expression for $T(c)$.
    (b) Let the factor base B be a suitable subset of $\{-1, 2, 3, 5, 7, 11\}$. Find all B-smooth values of $T(c)$ for $-5 \leqslant c \leqslant 5$. You do not have to use a sieve. Find the smooth values by trial division only.

17. (a) Explain how sieving is carried out in the multiple-polynomial quadratic sieve method, that is, for $T(c) = U + 2Vc + Wc^2$ with $V^2 - UW = n$.
    (b) If the factor base consists of $L[1/2]$ primes and the sieving interval is of size $L[1]$, deduce that the sieving process can be completed in $L[1]$ time.

18. In the original QSM, we sieve around $\sqrt{n}$. Let us instead take $H = \lceil \sqrt{2n} \rceil$, and $J = H^2 - 2n$.
    (a) Describe how we can modify the original QSM to work for these values of H and J. (It suffices to describe how we get a relation in the modified QSM.)
    (b) Explain why the modified QSM is poorer than the original QSM. (**Hint:** Look at the approximate average value of $|T(c)|$.)
    (c) Despite the objection in Part (b) about the modified QSM, we can exploit it to our advantage. Suppose that we run two sieves: one around $\sqrt{n}$ (the original QSM), and the other around $\sqrt{2n}$ (the modified QSM), each on a sieving interval of length half of that for the original QSM. Justify why this reduction in the length of the sieving interval is acceptable. Discuss what we gain by using the dual sieve.

19. In the original QSM, we took $T(c) = (H + c)^2 - n = J + 2cH + c^2$. Instead, one may choose c_1, c_2 satisfying $-M \leqslant c_1 \leqslant c_2 \leqslant M$, and consider $T(c_1, c_2) = (H + c_1)(H + c_2) - n = J + (c_1 + c_2)H + c_1 c_2$.
    (a) Describe how we get a relation in this variant of the QSM.
    (b) Prove that if we choose $t = L[1/2]$ primes in the factor base and $M = L[1/2]$, we expect to obtain the required number of relations.
    (c) Describe a sieving procedure for this variant of the QSM.
    (d) Argue that this variant can be implemented to run in $L[1]$ time.

(e) What are the advantages and disadvantages of this variant of the QSM over the original QSM?

20. [*Special-q variant of QSM*] In the original QSM, we sieve the quantities $T(c) = (H + c)^2 - n$ for $-M \leqslant c \leqslant M$. For small values of $|c|$, the values $|T(c)|$ are small and are likely to be smooth. On the contrary, larger values of c in the sieving interval yield larger values of $|T(c)|$ resulting in poorer yields of smooth candidates. In Exercise 6.18, this problem is tackled by using a dual sieve. The MPQSM is another solution. We now study yet another variant.[15] In this exercise, we study this variant for large primes only. See Exercise 6.29 for a potential speedup.
    (a) Let q be a large prime ($B < q < B^2$) and c_0 a small integer such that $q|T(c_0)$. Describe how we can locate such q and c_0 relatively easily.
    (b) Let $T_q(c) = T(c_0 + cq)/q$. How you can sieve $T_q(c)$ for $-M \leqslant c \leqslant M$?
    (c) What do you gain by using this special-q variant of the QSM?

21. Describe a sieve for locating all the smooth values of $T(a, b, c)$ in the CSM.

22. Show that the total number of solutions of the congruence $x^3 \equiv y^2 z \pmod{n}$ with $x^3 \neq y^2 z$ is $\Theta(n^2)$. You may use the formula that $\sum_{1 \leqslant m \leqslant n} d(m) = \Theta(n \ln n)$, where $d(m)$ denotes the number of positive integral divisors of m.

23. Describe a special-q method for the CSM. What do you gain, if anything, by using this special-q variant of the CSM?

24. Show that in the ECM, we can maintain the multiples of P as pairs of rational numbers. Describe what modifications are necessary in the ECM for this representation. What do you gain from this?

25. [*Montgomery ladder*][16] You want to compute nP for a point P on the curve $Y^2 = X^3 + aX + b$. Let $n = (n_{s-1}n_{s-2} \ldots n_1 n_0)_2$ and $N_i = (n_{s-1}n_{s-2} \ldots n_i)_2$.
    (a) Rewrite the left-to-right double-and-add algorithm so that both $N_i P$ and $(N_i + 1)P$ are computed in the loop.
    (b) Prove that given only the X-coordinates of P_1, P_2 and $P_1 - P_2$, we can compute the X-coordinate of $P_1 + P_2$. Handle the case $P_1 = P_2$ too.
    (c) What implication does this have in the ECM?

26. How can a second stage (as in Pollard's $p-1$ method) be added to the ECM?

27. Investigate how the integer primes $13, 17, 19, 23$ behave in the number ring $\mathfrak{O}_K$ of Example 6.23.

28. [*Lattice sieve*] Pollard[17] introduces the concept of lattice sieves in connection with the NFSM. Let B be a bound of small primes in the factor base. One finds out small coprime pairs a, b such that both $a + bm$ (a rational integer) and $a + b\theta$ (an algebraic integer) are B-smooth. The usual way of sieving fixes

[15] James A. Davis and Diane B. Holdridge, Factorization using the quadratic sieve algorithm, *Report SAND* 83–1346, Sandia National Laboratories, Albuquerque, 1983.
    [16] Peter L. Montgomery, Speeding the Pollard and elliptic curve methods of factorization, *Mathematics of Computation*, 48(177), 243–264, 1987.
    [17] John M. Pollard, The lattice sieve, *Lecture Notes in Mathematics*, 1554, Springer, 43–49, 1993.

a, and lets b vary over an interval. This is denoted by *line sieving*. In the rest of this exercise, we restrict our attention to the rational sieve only.

We use a bound $B' < B$. The value $k = B'/B$ lies in the range $[0.1, 0.5]$. All primes $\leqslant B'$ are called *small primes*. All primes p in the range $B' < p \leqslant B$ are called *medium primes*. Assume that no medium prime divides m. First, fix a medium prime q, and consider only those pairs (a, b) with $a + bm \equiv 0 \pmod{q}$. Sieve using all primes $p < q$. This sieve is repeated for all medium primes q. Let us see the effects of this sieving technique.

(a) Let N be the number of (a, b) pairs for which $a + bm$ is checked for smoothness in the line sieve, and N' the same number for the lattice sieve. Show that $N'/N \approx \log(1/k)/\log B$. What is N'/N for $k = 0.25$ and $B = 10^6$?

(b) What smooth candidates are missed in the lattice sieve? Find their relative percentage in the set of smooth integers located in the line sieve, for the values $k = 0.25$, $B = 10^6$, $m = 10^{30}$, and for b varying in the range $0 \leqslant b \leqslant 10^6$. These real-life figures demonstrate that with significantly reduced efforts, one can obtain most of the relations.

(c) Show that all integer solutions (a, b) of $a + bm \equiv 0 \pmod{q}$ form a two-dimensional lattice. Let $V_1 = (a_1, b_1)$ and $V_2 = (a_2, b_2)$ constitute a reduced basis of this lattice.

(d) A solution (a, b) of $a + bm \equiv 0 \pmod{q}$ can be written as $(a, b) = cV_1 + dV_2 = (ca_1 + da_2, cb_1 + db_2)$. Instead of letting a vary from $-M$ to M and b from 1 to M, Pollard suggests letting c vary from $-C$ to C and d from 1 to D. This is somewhat ad hoc, since rectangular regions in the (a, b) plane do not, in general, correspond to rectangular regions in the (c, d) plane. Nonetheless, this is not a practically bad idea. Describe how sieving can be done in the chosen rectangle for (c, d).

29. Describe how the idea of using small and medium primes, introduced in Exercise 6.28, can be adapted to the case of the QSM. Also highlight the expected benefits. Note that this is the special-q variant of the QSM with medium special primes q instead of large special primes q as discussed in Exercise 6.20.

Programming Exercises

Implement the following in GP/PARI.

30. Floyd's variant of Pollard's rho method.

31. Brent's variant of Pollard's rho method.

32. Pollard's $p - 1$ method.

33. The second stage of Pollard's $p - 1$ method.

34. Fermat's factorization method (Exercise 6.15).

35. Dixon's method.

36. The relation-collection stage of the QSM (use trial division instead of a sieve).

37. The sieve of the QSM.

38. Collecting relations involving large primes from the sieve of Exercise 6.37.

Chapter 7

Discrete Logarithms

Let G be a finite group[1] of size n. To start with, assume that G is cyclic, and g is a generator of G. Any element $a \in G$ can be uniquely expressed as $a = g^x$ for some integer x in the range $0 \leqslant x \leqslant n - 1$. The integer x is called the *discrete logarithm* or *index* of a with respect to g, and is denoted by $\text{ind}_g a$. Computing x from G, g and a is called the *discrete logarithm problem* (*DLP*).

We now remove the assumption that G is cyclic. Let $g \in G$ have $\text{ord}(g) = m$, and let H be the subgroup of G generated by g. H is cyclic of order m. We are given an element $a \in G$. If $a \in H$, then $a = g^x$ for some unique integer x in the range $0 \leqslant x \leqslant m - 1$. On the other hand, if $a \notin H$, then a cannot be

---

[1]Unless otherwise stated, a group is a commutative (Abelian) group under multiplication.

expressed as $a = g^x$. The (generalized) discrete logarithm problem ($GDLP$) refers to the determination of whether a can be expressed as g^x and if so, the computation of x. In general, DLP means either the special DLP or the GDLP. In this book, we mostly study the special variant of the DLP.

Depending upon the group G, the computational complexity of the DLP in G varies from easy (polynomial-time) to very difficult (exponential-time). First, consider G to be the additive group $\mathbb{Z}_n$, and let g be a generator of $\mathbb{Z}_n$. It is evident that g generates $\mathbb{Z}_n$ if and only if $g \in \mathbb{Z}_n^*$. Given $a \in \mathbb{Z}_n$, we need to find the unique integer $x \in \{0, 1, \ldots, n-1\}$ such that $a \equiv xg \pmod{n}$ (this is the DLP in an additive setting). Since $g \in \mathbb{Z}_n^*$, we have $\gcd(g, n) = 1$, that is, $ug + vn = 1$ for some integers u, v. But then, $ug \equiv 1 \pmod{n}$, that is, $(ua)g \equiv a \pmod{n}$, that is, the discrete logarithm of a is $x \equiv ua \equiv g^{-1}a \pmod{n}$. To sum up, the discrete logarithm problem is easy in the additive group $\mathbb{Z}_n$.

Next, consider the multiplicative group $\mathbb{Z}_n^*$. For simplicity, assume that $\mathbb{Z}_n^*$ is cyclic, and let $g \in \mathbb{Z}_n^*$ be a primitive root of n. Any $a \in \mathbb{Z}_n^*$ can be written uniquely as $a \equiv g^x \pmod{n}$ for an integer x in the range $0 \leqslant x \leqslant \phi(n) - 1$. In particular, if $n = p$ is a prime, then $1 \leqslant x \leqslant p - 2$. The determination of x in this case is apparently not an easy computational problem.

If we generalize the case of $\mathbb{Z}_p$ to a finite field $\mathbb{F}_q$ and take $G = \mathbb{F}_q^*$, then we talk about the *finite-field discrete-logarithm problem*. For a primitive element g of $\mathbb{F}_q^*$ and for $a \in \mathbb{F}_q^*$, there exists a unique integer x in the range $0 \leqslant x \leqslant q - 2$ such that $a = g^x$. Like the case of $\mathbb{Z}_p^*$, the computation of x in this case again appears to be a difficult computational problem.

Finally, the DLP in the group of rational points on an elliptic curve over a finite field is called the *elliptic-curve discrete-logarithm problem* ($ECDLP$).

In a general group of size n, the discrete logarithm problem can be solved in $O^\sim(\sqrt{n})$ time by algorithms referred to as *square-root methods*. The arithmetic in the group G may allow us to arrive at faster algorithms. For example, the availability of the extended gcd algorithm lets us solve the DLP in $(\mathbb{Z}_n, +)$ in polynomial time. For $\mathbb{Z}_n^*$, or, more generally, for $\mathbb{F}_q^*$, we know several subexponential algorithms. It is a popular belief that the discrete logarithm problem in the multiplicative group of finite fields is computationally as difficult as the integer-factoring problem. Some partial results are known to corroborate this suspected computational equivalence. In practice, many algorithms that we use to solve the finite-field discrete-logarithm problem are adaptations of the subexponential algorithms for factoring integers. For elliptic curves, on the other hand, no algorithms better than the square-root methods are known. Only when the curve is of some special forms, some better (subexponential and even polynomial-time) algorithms are known.

Like the integer-factoring problem, the computational difficulty of the DLP is only apparent. No significant results are known to corroborate the fact that the DLP cannot be solved faster than that achievable by the known algorithms.

The computational complexity of the DLP in a group G may depend upon the representation of the elements of G. For example, if $G = \mathbb{F}_q^*$, and elements of G are already represented by their indices with respect to a primitive element, then computing indices with respect to that or any other primitive element is rather trivial. However, we have argued in Section 2.5 that this representation is not practical except only for small fields.

A related computational problem is called the *Diffie–Hellman problem* (*DHP*) that came to light after the seminal discovery of public-key cryptography by Diffie and Hellman in 1976. Consider a multiplicative group G with $g \in G$. Suppose that the group elements g^x and g^y are given to us for some unknown indices x and y. Computation of g^{xy} from the knowledge of G, g, g^x and g^y is called the Diffie–Hellman problem in G. Evidently, if the DLP in G is easy to solve, the DHP in G is easy too ($g^{xy} = (g^x)^y$ with $y = \text{ind}_g(g^y)$). The converse implication is not clear. It is again only a popular belief that solving the DHP in G is computationally as difficult as solving the DLP in G.

In most of this chapter, I concentrate on algorithms for solving the discrete-logarithm problem in finite fields. I start with some square-root methods that are applicable to any group (including elliptic-curve groups). Later, I focus on two practically important cases: the prime fields $\mathbb{F}_p$, and the binary fields $\mathbb{F}_{2^n}$. Subexponential algorithms, collectively called *index calculus methods*, are discussed for these two types of fields. The DLP in extension fields $\mathbb{F}_{p^n}$ of odd characteristics p is less studied, and not many significant results are known for these fields, particularly when both p and n are allowed to grow indefinitely. At the end, the elliptic-curve discrete-logarithm problem is briefly addressed.

GP/PARI supports computation of discrete logarithms in prime fields. One should call znlog(a,g), where g is a primitive element of $\mathbb{Z}_p^*$ for some prime p.

```
gp > p = nextprime(10000)
%1 = 10007
gp > g = Mod(5,p)
%2 = Mod(5, 10007)
gp > znorder(g)
%3 = 10006
gp > a = Mod(5678,p)
%4 = Mod(5678, 10007)
gp > znlog(a,g)
%5 = 8620
gp > g^8620
%6 = Mod(5678, 10007)
gp > znlog(Mod(0,p),g)
  ***   impossible inverse modulo: Mod(0, 10007).
```

7.1 Square-Root Methods

The square-root methods are DLP algorithms of the *dark age*, but they apply to all groups. Unlike factoring integers, we do not have *modern* algorithms for computing discrete logarithms in all groups. For example, the fastest known algorithms for solving the ECDLP for general elliptic curves are the square-root methods. It is, therefore, quite important to understand the tales from the dark age. In this section, we assume that G is a finite cyclic multiplicative group of size n, and $g \in G$ is a generator of G. We are interested in computing $\text{ind}_g a$ for some $a \in G$.

7.1.1 Shanks' Baby-Step-Giant-Step (BSGS) Method

The baby-step-giant-step method refers to a class of algorithms, proposed originally by the American mathematician Shanks (1917–1996).[2] Let $m = \lceil \sqrt{n} \rceil$. We compute and store g^i for $i = 0, 1, 2, \ldots, m - 1$ as ordered pairs (i, g^i) sorted with respect to the second element. For $j = 0, 1, 2, \ldots$, we check whether ag^{-jm} is a second element in the table of the pairs (i, g^i). If, for some particular j, we obtain such an i, then $ag^{-jm} = g^i$, that is, $a = g^{jm+i}$, that is, $\text{ind}_g a \equiv jm + i \pmod{n}$. The determination of j is called the *giant step*, whereas the determination of i is called the *baby step*. Since $\text{ind}_g a \in \{0, 1, 2, \ldots, n - 1\}$ and $n \leqslant m^2$, we can always express $\text{ind}_g a$ as $jm + i$ for some $i, j \in \{0, 1, \ldots, m - 1\}$.

Example 7.1 Let me illustrate the BSGS method for the group $G = \mathbb{F}_{97}^*$ with generator $g = 23$. Since $n = |G| = 96$, we have $m = \lceil \sqrt{n} \rceil = 10$. The table of baby steps contains (i, g^i) for $i = 0, 1, 2, \ldots, 9$, and is given below. The table is kept sorted with respect to the second element (g^i).

i	0	5	8	6	1	7	3	2	9	4
g^i	1	5	16	18	23	26	42	44	77	93

Let us determine the index of 11 with respect to 23. We first compute $g^{-m} \equiv 66 \pmod{97}$. For $j = 0$, $ag^{-jm} \equiv a \equiv 11 \pmod{97}$ is not present in the above table. For $j = 1$, we have $ag^{-jm} \equiv 47 \pmod{97}$, again not present in the above table. Likewise, $ag^{-2m} \equiv 95 \pmod{97}$ and $ag^{-3m} \equiv 62 \pmod{97}$ are not present in the table. However, $ag^{-4m} \equiv 18 \pmod{97}$ exists in the table against $i = 6$, and we get $\text{ind}_{97}(11) \equiv 4 \times 10 + 6 \equiv 46 \pmod{96}$. $\square$

Computing the powers g^i for $i = 0, 1, 2, \ldots, m-1$ requires a total of $O(m)$ group operations. Sorting the table can be done using $O(m \log m)$ comparisons (and movements) of group elements. The giant steps are taken for $j = 0, 1, 2, \ldots, m - 1$ only. The element $g^{-m} = g^{n-m}$ is precomputed using $O(\log n)$ multiplication and square operations in G. For $j \geqslant 2$, we obtain $g^{-jm} = g^{-(j-1)m} g^{-m}$ by a single group operation. Since the table of baby steps is kept sorted with respect to g^i, searching whether ag^{-jm} belongs to the table can be accomplished by the binary search algorithm, demanding $O(\log m)$ comparisons of group elements for each j. Finally, for a successful finding of

---

[2]Daniel Shanks, Class number, a theory of factorization and genera, *Proceedings of the Symposia in Pure Mathematics*, 20, 415–440, 1971.

(i, j), computing $jm + i$ involves integer arithmetic of $O(\log^2 n)$ total cost. Thus, the running time of the BSGS method is dominated by $\tilde{O}(m)$, that is, $\tilde{O}(\sqrt{n})$ group operations. The space complexity is dominated by the storage for the table of baby steps, and is again $\tilde{O}(m)$, that is, $\tilde{O}(\sqrt{n})$.

7.1.2 Pollard's Rho Method

This method is an adaptation of the rho method for factoring integers (Section 6.2). We generate a random walk in G. We start at a random element $w_0 = g^{s_0} a^{t_0}$ of G. Subsequently, for $i = 1, 2, 3, \ldots$, we jump to the element $w_i = g^{s_i} a^{t_i}$. The sequence $w_0, w_1, w_2, \ldots$ behaves like a random sequence of elements of G. By the birthday paradox, we expect to arrive at a collision $w_i = w_j$ after $O(\sqrt{n})$ iterations. In that case, we have $g^{s_i} a^{t_i} = g^{s_j} a^{t_j}$, that is, $a^{t_i - t_j} = g^{s_j - s_i}$. Since $\text{ord}(g) = n$, we have $(t_i - t_j) \text{ind}_g a \equiv s_j - s_i \pmod{n}$. If $\gcd(t_i - t_j, n) = 1$, we obtain $\text{ind}_g a \equiv (t_i - t_j)^{-1}(s_j - s_i) \pmod{n}$.

We need to realize a suitable function $f : G \to G$ to map w_{i-1} to w_i for all $i \geqslant 1$. We predetermine a small positive integer r, and map $w \in G$ to an element $u \in \{0, 1, 2, \ldots, r - 1\}$. We also predetermine a set of *multipliers* $M_j = g^{\sigma_j} a^{\tau_j}$ for $j = 0, 1, 2, \ldots, r - 1$. We set $f(w) = w \times M_u = w \times g^{\sigma_u} a^{\tau_u}$. In practice, values of $r \approx 20$ work well.

Example 7.2 Let us take $G = \mathbb{F}_{197}^*$. The element $g = 123 \in G$ is a primitive element. We want to compute the discrete logarithm of $a = 111$ with respect to g. Let us agree to take $r = 5$ and the multipliers

$$M_0 = g^1 a^4 = 71, \quad M_1 = g^7 a^5 = 25, \quad M_2 = g^4 a^2 = 7,$$
$$M_3 = g^8 a^5 = 120, \quad M_4 = g^8 a^2 = 168.$$

Given $w \in G$, we take $u = w \text{ rem } r$ and use the multiplier M_u. We start the random walk with $s_0 = 43$ and $t_0 = 24$, so that $w_0 \equiv g^{s_0} a^{t_0} \equiv 189 \pmod{197}$. Some subsequent steps of the random walk are shown below.

i	w_{i-1}	u	M_u	s_i	t_i	w_i	i	w_{i-1}	u	M_u	s_i	t_i	w_i
1	189	4	168	51	26	35	11	48	3	120	109	67	47
2	35	0	71	52	30	121	12	47	2	7	113	69	132
3	121	1	25	59	35	70	13	132	2	7	117	71	136
4	70	0	71	60	39	45	14	136	1	25	124	76	51
5	45	0	71	61	43	43	15	51	1	25	131	81	93
6	43	3	120	69	48	38	16	93	3	120	139	86	128
7	38	3	120	77	53	29	17	128	3	120	147	91	191
8	29	4	168	85	55	144	18	191	1	25	154	96	47
9	144	4	168	93	57	158	19	47	2	7	158	98	132
10	158	3	120	101	62	48	20	132	2	7	162	100	136

The random walk becomes periodic from $i = 11$ (with a shortest period of 7). We have $g^{s_{11}} a^{t_{11}} \equiv g^{s_{18}} a^{t_{18}} \pmod{197}$, that is, $a^{t_{11} - t_{18}} \equiv g^{s_{18} - s_{11}} \pmod{197}$, that is, $a^{-29} \equiv g^{45} \pmod{197}$, that is, $-29 \, \text{ind}_g a \equiv 45 \pmod{196}$, that is, $\text{ind}_g a \equiv (-29)^{-1} \times 45 \equiv 39 \pmod{196}$. $\qquad \square$

Space-saving variants (Floyd's variant and Brent's variant) of the Pollard rho method work in a similar manner as described in connection with factoring integers (Section 6.2). However, the concept of block gcd calculation does not seem to adapt directly to the case of the DLP.

7.1.3   Pollard's Lambda Method

Pollard's lambda method is a minor variant of Pollard's rho method. The only difference is that now we use two random walks $w_0, w_1, w_2, \ldots$ and $w'_0, w'_1, w'_2, \ldots$ in the group G. We use the same update function f to obtain $w_i = f(w_{i-1})$ and $w'_i = f(w'_{i-1})$ for $i \geqslant 1$. Let $w_i = g^{s_i} a^{t_i}$ and $w'_i = g^{s'_i} a^{t'_i}$. After the two random walks intersect, they proceed identically, that is, the two walks together look like the Greek letter λ. This method is also called the method of *wild and tame kangaroos*. Two kangaroos make the two random walks. The tame kangaroo digs a hole at every point it visits, and when the wild kangaroo reaches the same point (later), it gets trapped in the hole.

If $w_i = w'_j$, then $g^{s_i} a^{t_i} = g^{s'_j} a^{t'_j}$, that is, $(t_i - t'_j)\operatorname{ind}_g a \equiv s'_j - s_i \pmod{n}$. If $t_i - t'_j$ is invertible modulo n, we obtain $\operatorname{ind}_g a \equiv (t_i - t'_j)^{-1}(s'_j - s_i) \pmod{n}$.

Example 7.3 As in Example 7.2, we take $G = \mathbb{F}^*_{197}$, $g = 123$, and $a = 111$. We predetermine $r = 5$, and decide the multipliers

$$M_0 = g^3 a^3 = 88, \quad M_1 = g^4 a^2 = 7, \quad M_2 = g^8 a^8 = 142,$$
$$M_3 = g^4 a^9 = 115, \quad M_4 = g^2 a^0 = 157.$$

The two random walks are shown in the following table.

i	s_i	t_i	w_i	s'_i	t'_i	w'_i	i	s_i	t_i	w_i	s'_i	t'_i	w'_i
0	43	24	189	34	42	172	15	102	91	46	108	114	121
1	45	24	123	42	50	193	16	106	93	125	112	116	59
2	49	33	158	46	59	131	17	109	96	165	114	116	4
3	53	42	46	50	61	129	18	112	99	139	116	116	37
4	57	44	125	52	61	159	19	114	99	153	124	124	132
5	60	47	165	54	61	141	20	118	108	62	132	132	29
6	63	50	139	58	63	2	21	126	116	136	134	132	22
7	65	50	153	66	71	87	22	130	118	164	142	140	169
8	69	59	62	74	79	140	23	132	118	138	144	140	135
9	77	67	136	77	82	106	24	136	127	110	147	143	60
10	81	69	164	81	84	151	25	139	130	27	150	146	158
11	83	69	138	85	86	72	26	147	138	91	154	155	46
12	87	78	110	93	94	177	27	151	140	46	158	157	125
13	90	81	27	101	102	115	28	155	142	125	161	160	165
14	98	89	91	104	105	73	29	158	145	165	164	163	139

We get $w_2 = w'_{25}$, so $g^{49} a^{33} \equiv g^{150} a^{146} \pmod{197}$, that is, $(33 - 146)\operatorname{ind}_g a \equiv 150 - 49 \pmod{196}$, that is, $\operatorname{ind}_g a \equiv (-113)^{-1} \times 101 \equiv 39 \pmod{196}$.   □

7.1.4 Pohlig–Hellman Method

Suppose that the complete prime factorization $n = p_1^{e_1} p_2^{e_2} \cdots p_r^{e_r}$ of $n = |G|$ is known. The Pohlig–Hellman method[3] can compute discrete logarithms in G in $\tilde{O}(\sqrt{\max(p_1, p_2, \ldots, p_r)})$ time. If the largest prime divisor of n is small, then this method turns out to be quite efficient. However, in the worst case, this maximum prime divisor is n, and the Pohlig–Hellman method takes an exponential running time $\tilde{O}(\sqrt{n})$.

If $x = \text{ind}_g a$ is known modulo $p_i^{e_i}$ for all $i = 1, 2, \ldots, r$, then the CRT gives the desired value of x modulo n. In view of this, we take p^e to be a divisor of n with $p \in \mathbb{P}$ and $e \in \mathbb{N}$. We first obtain $x \pmod{p}$ by solving a DLP in the subgroup of G of size p. Subsequently, we *lift* this value to $x \pmod{p^2}$, $x \pmod{p^3}$, and so on. Each lifting involves computing a discrete logarithm in the subgroup of size p. One uses the BSGS method or Pollard's rho or lambda method for computing discrete logarithms in the subgroup of size p.

Let $\gamma = g^{n/p^e}$, and $\alpha = a^{n/p^e}$. Since g is a generator of G, we have ord $\gamma = p^e$. Moreover, $a = g^x$ implies that $\alpha = a^{n/p^e} = (g^{n/p^e})^x = \gamma^x = \gamma^{x \text{ rem } p^e}$. We plan to compute $\xi = x \text{ rem } p^e$. Let us write

$$\xi = x_0 + x_1 p + x_2 p^2 + \cdots + x_{e-1} p^{e-1}.$$

We keep on computing the p-ary digits $x_0, x_1, x_2, \ldots$ of ξ one by one. Assume that $x_0, x_1, \ldots, x_{i-1}$ are already computed, and we want to compute x_i. Initially, no x_i values are computed, and we plan to compute x_0. We treat both the cases $i = 0$ and $i > 0$ identically. We already know

$$\lambda = x_0 + x_1 p + \cdots + x_{i-1} p^{i-1}$$

($\lambda = 0$ if $i = 0$). Now, $\alpha = \gamma^\xi$ gives $\alpha \gamma^{-\lambda} = \gamma^{\xi - \lambda} = \gamma^{x_i p^i + x_{i+1} p^{i+1} + \cdots + x_{e-1} p^{e-1}}$. Exponentiation to the power p^{e-i-1} gives

$$(\alpha \gamma^{-\lambda})^{p^{e-i-1}} = \gamma^{x_i p^{e-1} + x_{i+1} p^e + x_{i+2} p^{e+1} + \cdots + x_{e-1} p^{2e-i-2}} = \gamma^{x_i p^{e-1}} = (\gamma^{p^{e-1}})^{x_i}.$$

The order of $\gamma^{p^{e-1}}$ is p, and so x_i is the discrete logarithm of $(\alpha \gamma^{-\lambda})^{p^{e-i-1}}$ to the base $\gamma^{p^{e-1}}$. If we rephrase this in terms of g and a, we see that x_i is the discrete logarithm of $(ag^{-\lambda})^{n/p^{i+1}}$ with respect to the base $g^{n/p}$, that is,

$$x_i = \text{ind}_{g^{n/p}} \left[\left(ag^{-(x_0 + x_1 p + \cdots + x_{i-1} p^{i-1})} \right)^{n/p^{i+1}} \right] \text{ for } i = 0, 1, 2, \ldots, e - 1.$$

These index calculations are done in the subgroup of size p, generated by $g^{n/p}$.

Example 7.4 Let us compute $x = \text{ind}_{123}(111)$ in $G = \mathbb{F}_{197}^*$ by the Pohlig–Hellman method. The size of the group $\mathbb{F}_{197}^*$ is $n = 196 = 2^2 \times 7^2$. That is,

---

[3]Stephen Pohlig and Martin Hellman, An improved algorithm for computing logarithms over GF(p) and its cryptographic significance, *IEEE Transactions on Information Theory*, 24, 106–110, 1978. This algorithm seems to have been first discovered (but not published) by Roland Silver, and is often referred to also as the *Silver-Pohlig–Hellman method*.

we compute x rem $4 = x_0 + 2x_1$ and x rem $49 = x_0' + 7x_1'$. The following table illustrates these calculations.

	$p = 2$					$p = 7$			
i	$g^{n/2}$	λ	$(ag^{-\lambda})^{n/2^{i+1}}$	x_i	i	$g^{n/7}$	λ	$(ag^{-\lambda})^{n/7^{i+1}}$	x_i'
0	196	0	196	1	0	164	0	178	4
1	196	1	196	1	1	164	4	36	5

These calculations yield $x \equiv 1 + 2 \times 1 \equiv 3 \pmod{4}$ and $x \equiv 4 + 7 \times 5 \equiv 39 \pmod{49}$. Combining using the CRT gives $x \equiv 39 \pmod{196}$. □

7.2 Algorithms for Prime Fields

I now present some subexponential algorithms suited specifically to the group $\mathbb{F}_p^*$ with $p \in \mathbb{P}$. These algorithms are collectively called *index calculus methods* (ICM). Known earlier in the works of Kraitchik[4] and of Western and Miller,[5] this method was rediscovered after the advent of public-key cryptography. For example, see Adleman's paper.[6] Coppersmith et al.[7] present several improved variants (LSM, GIM, RLSM and CSM). Odlyzko[8] surveys index calculus methods for fields of characteristic two.

The group $\mathbb{F}_p^*$ is cyclic. We plan to compute the index of a with respect to a primitive element g of $\mathbb{F}_p^*$. We choose a *factor base* $B = \{b_1, b_2, \ldots, b_t\} \subseteq \mathbb{F}_p^*$ such that a *reasonable* fraction of elements of $\mathbb{F}_p^*$ can be written as products of elements of B. We collect *relations* of the form

$$g^\alpha a^\beta \equiv \prod_{i=1}^{t} b_i^{\gamma_i} \pmod{p}.$$

This yields

$$\alpha + \beta \operatorname{ind}_g a \equiv \sum_{i=1}^{t} \gamma_i \operatorname{ind}_g(b_i) \pmod{p-1}.$$

If β is invertible modulo $p - 1$, this congruence gives us $\operatorname{ind}_g a$, provided that the indices $\operatorname{ind}_g(b_i)$ of the elements b_i of the factor base B are known to us. In view of this, an index calculus method proceeds in two stages.

---

[4]M. Kraitchik, *Théorie des nombres*, Gauthier-Villards, 1922.

[5]A. E. Western and J. C. P. Miller, *Tables of indices and primitive roots*, Cambridge University Press, xxxvii–xlii, 1968.

[6]Leonard Max Adleman, A subexponential algorithm for the discrete logarithm problem with applications to cryptography, *FOCS*, 55–60, 1979.

[7]Don Coppersmith, Andrew Michael Odlyzko and Richard Schroeppel, Discrete logarithms in GF(p), *Algorithmica*, 1(1), 1–15, 1986.

[8]Andrew Michael Odlyzko, Discrete logarithms in finite fields and their cryptographic significance, *EuroCrypt* 1984, 224–314, 1985.

Stage 1: Determination of $\mathrm{ind}_g(b_i)$ for $i = 1, 2, \ldots, t$.

In this stage, we generate relations without involving a (with $\beta = 0$), that is,

$$\sum_{i=1}^{t} \gamma_i \, \mathrm{ind}_g(b_i) \equiv \alpha \pmod{p-1},$$

a linear congruence modulo $p - 1$ in the unknown quantities $\mathrm{ind}_g(b_i)$. We collect s such relations for some $s \geqslant t$. We solve the resulting system of linear congruences, and obtain the desired indices of the elements of the factor base.

Stage 2: Determination of $\mathrm{ind}_g a$.

In this stage, we generate a single relation with β invertible modulo $p - 1$.

Different index calculus methods vary in the way the factor base is chosen and the relations are generated. In the rest of this section, we discuss some variants. All these variants have running times of the form

$$L(p, \omega, c) = \exp\left[(c + o(1))(\ln p)^{\omega} (\ln \ln p)^{1-\omega}\right]$$

for real constant values c and ω with $c > 0$ and $0 < \omega < 1$. If $\omega = 1/2$, we abbreviate $L(p, \omega, c)$ as $L_p[c]$, and even as $L[c]$ if p is clear from the context.

7.2.1 Basic Index Calculus Method

The basic index calculus method can be called an adaptation of Dixon's method for factoring integers (Section 6.4).[9] The factor base consists of t small primes, that is, $B = \{p_1, p_2, \ldots, p_t\}$, where p_i is the i-th prime. I will later specify the value of the parameter t so as to optimize the running time of the algorithm. For random α in the range $1 \leqslant \alpha \leqslant p - 2$, we try to express $g^{\alpha} \pmod{p}$ as a product of the primes in B. If the factorization attempt is successful, we obtain a *relation* of the form

$$g^{\alpha} \equiv p_1^{\gamma_1} p_2^{\gamma_2} \cdots p_t^{\gamma_t} \pmod{p}, \text{ that is,}$$

$$\gamma_1 \, \mathrm{ind}_g(p_1) + \gamma_2 \, \mathrm{ind}_g(p_2) + \cdots + \gamma_t \, \mathrm{ind}_g(p_t) \equiv \alpha \pmod{p-1}.$$

We collect s relations of this form:

$$\gamma_{11} \, \mathrm{ind}_g(p_1) + \gamma_{12} \, \mathrm{ind}_g(p_2) + \cdots + \gamma_{1t} \, \mathrm{ind}_g(p_t) \equiv \alpha_1 \pmod{p-1},$$
$$\gamma_{21} \, \mathrm{ind}_g(p_1) + \gamma_{22} \, \mathrm{ind}_g(p_2) + \cdots + \gamma_{2t} \, \mathrm{ind}_g(p_t) \equiv \alpha_2 \pmod{p-1},$$
$$\cdots$$
$$\gamma_{s1} \, \mathrm{ind}_g(p_1) + \gamma_{s2} \, \mathrm{ind}_g(p_2) + \cdots + \gamma_{st} \, \mathrm{ind}_g(p_t) \equiv \alpha_s \pmod{p-1}.$$

---

[9] Historically, the basic index calculus method came earlier than Dixon's method.

This leads to the linear system of congruences:

$$\begin{pmatrix} \gamma_{11} & \gamma_{12} & \cdots & \gamma_{1t} \\ \gamma_{21} & \gamma_{22} & \cdots & \gamma_{2t} \\ \cdots & \cdots & \vdots & \cdots \\ \gamma_{s1} & \gamma_{s2} & \cdots & \gamma_{st} \end{pmatrix} \begin{pmatrix} \text{ind}_g(p_1) \\ \text{ind}_g(p_2) \\ \vdots \\ \text{ind}_g(p_t) \end{pmatrix} \equiv \begin{pmatrix} \alpha_1 \\ \alpha_2 \\ \vdots \\ \alpha_s \end{pmatrix} \pmod{p-1}.$$

For $s \gg t$ (for example, for $s \geqslant 2t$), we expect the $s \times t$ coefficient matrix (γ_{ij}) to be of full column rank (Exercises 7.10 and 8.12). If so, the indices $\text{ind}_g(p_i)$ are uniquely obtained by solving the system. This completes the first stage.

In the second stage, we choose $\alpha \in \{1, 2, \ldots, p-2\}$ randomly, and try to express $ag^\alpha \pmod p$ as a product of the primes in the factor base. A successful factoring attempt gives

$$ag^\alpha \equiv p_1^{\gamma_1} p_2^{\gamma_2} \cdots p_t^{\gamma_t} \pmod{p}, \text{ that is,}$$

$$\text{ind}_g\, a \equiv -\alpha + \gamma_1 \,\text{ind}_g(p_1) + \gamma_2 \,\text{ind}_g(p_2) + \cdots + \gamma_t \,\text{ind}_g(p_t) \pmod{p-1}.$$

We obtain $\text{ind}_g\, a$ using the values of $\text{ind}_g(p_i)$ computed in the first stage.

Example 7.5 Take $p = 821$ and $g = 21$. We intend to compute the discrete logarithm of $a = 237$ to the base g by the basic index calculus method. We take the factor base $B = \{2, 3, 5, 7, 11\}$ consisting of the first $t = 5$ primes.

In the first stage, we compute $g^j \pmod p$ for randomly chosen values of j. After many choices, we come up with the following ten relations.

$$\begin{aligned}
g^{815} &\equiv 90 \equiv 2 \times 3^2 \times 5 \pmod{821} \\
g^{784} &\equiv 726 \equiv 2 \times 3 \times 11^2 \pmod{821} \\
g^{339} &\equiv 126 \equiv 2 \times 3^2 \times 7 \pmod{821} \\
g^{639} &\equiv 189 \equiv 3^3 \times 7 \pmod{821} \\
g^{280} &\equiv 88 \equiv 2^3 \times 11 \pmod{821} \\
g^{295} &\equiv 135 \equiv 3^3 \times 5 \pmod{821} \\
g^{793} &\equiv 375 \equiv 3 \times 5^3 \pmod{821} \\
g^{478} &\equiv 315 \equiv 3^2 \times 5 \times 7 \pmod{821} \\
g^{159} &\equiv 105 \equiv 3 \times 5 \times 7 \pmod{821} \\
g^{635} &\equiv 75 \equiv 3 \times 5^2 \pmod{821}
\end{aligned}$$

The corresponding system of linear congruences is as follows.

$$\begin{pmatrix} 1 & 2 & 1 & 0 & 0 \\ 1 & 1 & 0 & 0 & 2 \\ 1 & 2 & 0 & 1 & 0 \\ 0 & 3 & 0 & 1 & 0 \\ 3 & 0 & 0 & 0 & 1 \\ 0 & 3 & 1 & 0 & 0 \\ 0 & 1 & 3 & 0 & 0 \\ 0 & 2 & 1 & 1 & 0 \\ 0 & 1 & 1 & 1 & 0 \\ 0 & 1 & 2 & 0 & 0 \end{pmatrix} \begin{pmatrix} \text{ind}_g(2) \\ \text{ind}_g(3) \\ \text{ind}_g(5) \\ \text{ind}_g(7) \\ \text{ind}_g(11) \end{pmatrix} \equiv \begin{pmatrix} 815 \\ 784 \\ 339 \\ 639 \\ 280 \\ 295 \\ 793 \\ 478 \\ 159 \\ 635 \end{pmatrix} \pmod{820}.$$

Solving this system yields the indices of the factor-base elements as

$$\text{ind}_g(2) \equiv 19 \;(\text{mod } 820), \quad \text{ind}_g(3) \equiv 319 \;(\text{mod } 820),$$
$$\text{ind}_g(5) \equiv 158 \;(\text{mod } 820), \quad \text{ind}_g(7) \equiv 502 \;(\text{mod } 820),$$
$$\text{ind}_g(11) \equiv 223 \;(\text{mod } 820).$$

In the second stage, we obtain the relation

$$ag^{226} \equiv 280 \equiv 2^3 \times 5 \times 7 \;(\text{mod } 821), \text{ that is,}$$

$$\text{ind}_g a \equiv -226 + 3 \times 19 + 158 + 502 \equiv 491 \;(\text{mod } 820). \qquad \square$$

To optimize the running time of the basic index calculus method, we resort to the density estimate of smooth integers, given in Section 6.4. In particular, we use Corollary 6.14 with $n = p$.

Let the factor base B consist of all primes $\leqslant L[\eta]$, so t too is of the form $L[\eta]$. For randomly chosen values of α, the elements $g^\alpha \in \mathbb{F}_p^*$ are random integers between 1 and $p - 1$, that is, integers of value $O(p)$. The probability that such a value is smooth with respect to B is $L[-\frac{1}{2\eta}]$. Therefore, $L[\frac{1}{2\eta}]$ random choices of α are expected to yield a single relation. We need $s \geqslant 2t$ relations, that is, s is again of the form $L[\eta]$. Thus, the total number of random values of α, that need to be tried, is $L[\eta + \frac{1}{2\eta}]$. The most significant effort associated with each choice of α is the attempt to factor g^α. This is carried out by trial divisions by $L[\eta]$ primes in the factor base. To sum up, the relation-collection stage runs in $L[2\eta + \frac{1}{2\eta}]$ time. The quantity $2\eta + \frac{1}{2\eta}$ is minimized for $\eta = \frac{1}{2}$, leading to a running time of $L[2]$ for the relation-collection stage.

In the linear-algebra stage, a system of $L[\frac{1}{2}]$ linear congruences in $L[\frac{1}{2}]$ variables is solved. Standard Gaussian elimination requires a time of $L[\frac{1}{2}]^3 = L[\frac{3}{2}]$ for solving the system. However, since each relation obtained by this method is necessarily sparse, special sparse system solvers can be employed to run in only $L[\frac{1}{2}]^2 = L[1]$ time. In any case, the first stage of the basic index calculus method can be arranged to run in a total time of $L[2]$.

The second stage involves finding a single smooth value of ag^α. We need to try $L[\frac{1}{2\eta}] = L[1]$ random values of α, with each value requiring $L[\frac{1}{2}]$ trial divisions. Thus, the total time required for the second stage is $L[1 + \frac{1}{2}] = L[\frac{3}{2}]$.

The running time of the basic index calculus method is dominated by the relation-collection phase and is $L[2]$. The space requirement is $L[\eta]$, that is, $L[\frac{1}{2}]$ (assuming that we use a sparse representation of the coefficient matrix).

7.2.2 Linear Sieve Method (LSM)

The basic method generates candidates as large as $\Theta(p)$ for smoothness over a set of small primes. This leads to a running time of $L[2]$, making the basic method impractical. Schroeppel's linear sieve method is an $L[1]$-time algorithm for the computation of discrete logarithms over prime fields. It is an adaptation of the quadratic sieve method for factoring integers, and generates smoothness candidates of absolute values $O(p^{1/2})$. Moreover, we now have the opportunity to expedite trial division by sieving.

7.2.2.1　First Stage

Let $H = \lceil \sqrt{p} \rceil$, and $J = H^2 - p$. Both H and J are $O(\sqrt{p})$. For small integers c_1, c_2 (positive or negative), consider the expression $(H+c_1)(H+c_2) \equiv H^2 + (c_1 + c_2)H + c_1c_2 \equiv J + (c_1 + c_2)H + c_1c_2 \pmod{p}$. Let us denote by $T(c_1, c_2)$ the integer $J + (c_1 + c_2)H + c_1c_2$. Suppose that for some choice of c_1 and c_2, the integer $T(c_1, c_2)$ factors completely over the first t primes $p_1, p_2, \ldots, p_t$. This gives us a relation

$$(H + c_1)(H + c_2) \equiv p_1^{\alpha_1} p_2^{\alpha_2} \cdots p_t^{\alpha_t} \pmod{p}, \text{ that is,}$$

$$\alpha_1 \operatorname{ind}_g(p_1) + \alpha_2 \operatorname{ind}_g(p_2) + \cdots + \alpha_t \operatorname{ind}_g(p_t)$$
$$- \operatorname{ind}_g(H + c_1) - \operatorname{ind}_g(H + c_2) \equiv 0 \pmod{p-1}.$$

This implies that the small primes $p_1, p_2, \ldots, p_t$ should be in the factor base B. Moreover, the integers $H + c$ for some small values of c should also be included in the factor base. More explicitly, we choose c_1 and c_2 to vary between $-M$ and M (the choice of M will be explained later), so the factor base should also contain the integers $H + c$ for $-M \leqslant c \leqslant M$. Finally, note that for certain values of c_1 and c_2, we have negative values for $T(c_1, c_2)$, that is, -1 should also be included in the factor base. Therefore, we take

$$B = \{-1\} \cup \{p_1, p_2, \ldots, p_t\} \cup \{H + c \mid -M \leqslant c \leqslant M\}.$$

The size of this factor base is $n = 2M + t + 2$. By letting c_1 and c_2 vary in the range $-M \leqslant c_1 \leqslant c_2 \leqslant M$, we generate $m - 1$ relations of the form mentioned above. We assume that the base g of discrete logarithms is itself a small prime p_i in the factor base. This gives us a *free* relation: $\operatorname{ind}_g(p_i) \equiv 1 \pmod{p-1}$.

The resulting system of congruences has a coefficient matrix of size $m \times n$. We adjust M and t, so that $m \gg n$ (say, $m \approx 2n$). We hope that the coefficient matrix is of full column rank, and the system furnishes a unique solution for the indices of the elements of the factor base (Exercises 7.11 and 8.12).

Example 7.6 As an illustration of the linear sieve method, we take $p = 719$ and $g = 11$. Thus, $H = \lceil \sqrt{p} \rceil = 27$, and $J = H^2 - p = 10$. We take $t = 5$ and $M = 7$, that is, the factor base B consists of -1, the first five primes $2, 3, 5, 7, 11$, and the integers $H + c$ for $-M \leqslant c \leqslant M$, that is, $20, 21, 22, \ldots, 34$. The size of the factor base is $n = 2M + t + 2 = 21$. By considering all pairs (c_1, c_2) with $-M \leqslant c_1 \leqslant c_2 \leqslant M$, we obtain the following 30 relations.

c_1	c_2	$T(c_1, c_2)$	
-7	4	-99	$= (-1) \times 3^2 \times 11$
-6	2	-110	$= (-1) \times 2 \times 5 \times 11$
-6	7	-5	$= (-1) \times 5$
-5	-1	-147	$= (-1) \times 3 \times 7^2$
-5	0	-125	$= (-1) \times 5^3$
-5	2	-81	$= (-1) \times 3^4$
-5	5	-15	$= (-1) \times 3 \times 5$
-5	6	7	$= 7$
-4	-2	-144	$= (-1) \times 2^4 \times 3^2$
-4	-1	-121	$= (-1) \times 11^2$
-4	0	-98	$= (-1) \times 2 \times 7^2$
-4	1	-75	$= (-1) \times 3 \times 5^2$
-4	4	-6	$= (-1) \times 2 \times 3$
-4	6	40	$= 2^3 \times 5$
-4	7	63	$= 3^2 \times 7$

c_1	c_2	$T(c_1, c_2)$	
-3	3	1	$= 1$
-3	4	25	$= 5^2$
-3	5	49	$= 7^2$
-2	0	-44	$= (-1) \times 2^2 \times 11$
-2	2	6	$= 2 \times 3$
-2	4	56	$= 2^3 \times 7$
-2	5	81	$= 3^4$
-1	1	9	$= 3^2$
-1	2	35	$= 5 \times 7$
-1	7	165	$= 3 \times 5 \times 11$
0	0	10	$= 2 \times 5$
0	2	64	$= 2^6$
1	3	121	$= 11^2$
2	4	180	$= 2^2 \times 3^2 \times 5$
4	4	242	$= 2^1 \times 11^2$

Moreover, we have a free relation $\operatorname{ind}_g(11) = 1$. Let C be the 31×21 matrix

$$
C = \begin{pmatrix}
1 & 0 & 2 & 0 & 0 & 1 & -1 & 0 & 0 & 0 & 0 & 0 & 0 & 0 & 0 & 0 & 0 & -1 & 0 & 0 & 0 \\
1 & 1 & 0 & 1 & 0 & 1 & 0 & -1 & 0 & 0 & 0 & 0 & 0 & 0 & 0 & -1 & 0 & 0 & 0 & 0 & 0 \\
1 & 0 & 0 & 1 & 0 & 0 & 0 & -1 & 0 & 0 & 0 & 0 & 0 & 0 & 0 & 0 & 0 & 0 & 0 & 0 & -1 \\
1 & 0 & 1 & 0 & 2 & 0 & 0 & 0 & -1 & 0 & 0 & 0 & -1 & 0 & 0 & 0 & 0 & 0 & 0 & 0 & 0 \\
1 & 0 & 0 & 3 & 0 & 0 & 0 & 0 & -1 & 0 & 0 & 0 & 0 & -1 & 0 & 0 & 0 & 0 & 0 & 0 & 0 \\
1 & 0 & 4 & 0 & 0 & 0 & 0 & 0 & -1 & 0 & 0 & 0 & 0 & 0 & -1 & 0 & 0 & 0 & 0 & 0 & 0 \\
1 & 0 & 1 & 1 & 0 & 0 & 0 & 0 & -1 & 0 & 0 & 0 & 0 & 0 & 0 & 0 & 0 & -1 & 0 & 0 & 0 \\
0 & 0 & 0 & 0 & 1 & 0 & 0 & 0 & -1 & 0 & 0 & 0 & 0 & 0 & 0 & 0 & 0 & 0 & -1 & 0 & 0 \\
1 & 4 & 2 & 0 & 0 & 0 & 0 & 0 & 0 & -1 & 0 & -1 & 0 & 0 & 0 & 0 & 0 & 0 & 0 & 0 & 0 \\
1 & 0 & 0 & 0 & 0 & 2 & 0 & 0 & 0 & -1 & 0 & 0 & -1 & 0 & 0 & 0 & 0 & 0 & 0 & 0 & 0 \\
1 & 1 & 0 & 0 & 2 & 0 & 0 & 0 & 0 & -1 & 0 & 0 & -1 & 0 & 0 & 0 & 0 & 0 & 0 & 0 & 0 \\
1 & 0 & 1 & 2 & 0 & 0 & 0 & 0 & 0 & -1 & 0 & 0 & 0 & -1 & 0 & 0 & 0 & 0 & 0 & 0 & 0 \\
1 & 1 & 1 & 0 & 0 & 0 & 0 & 0 & 0 & -1 & 0 & 0 & 0 & 0 & 0 & 0 & -1 & 0 & 0 & 0 & 0 \\
0 & 3 & 0 & 1 & 0 & 0 & 0 & 0 & 0 & -1 & 0 & 0 & 0 & 0 & 0 & 0 & 0 & 0 & 0 & -1 & 0 \\
0 & 0 & 2 & 0 & 1 & 0 & 0 & 0 & 0 & -1 & 0 & 0 & 0 & 0 & 0 & 0 & 0 & 0 & 0 & 0 & -1 \\
0 & 0 & 0 & 0 & 0 & 0 & 0 & 0 & 0 & 0 & -1 & 0 & 0 & 0 & 0 & 0 & -1 & 0 & 0 & 0 & 0 \\
0 & 0 & 0 & 2 & 0 & 0 & 0 & 0 & 0 & 0 & -1 & 0 & 0 & 0 & 0 & 0 & -1 & 0 & 0 & 0 & 0 \\
0 & 0 & 0 & 0 & 2 & 0 & 0 & 0 & 0 & 0 & -1 & 0 & 0 & 0 & 0 & 0 & 0 & -1 & 0 & 0 & 0 \\
1 & 2 & 0 & 0 & 0 & 1 & 0 & 0 & 0 & 0 & -1 & 0 & -1 & 0 & 0 & 0 & 0 & 0 & 0 & 0 & 0 \\
0 & 1 & 1 & 0 & 0 & 0 & 0 & 0 & 0 & 0 & -1 & 0 & 0 & 0 & -1 & 0 & 0 & 0 & 0 & 0 & 0 \\
0 & 3 & 0 & 0 & 1 & 0 & 0 & 0 & 0 & 0 & -1 & 0 & 0 & 0 & 0 & -1 & 0 & 0 & 0 & 0 & 0 \\
0 & 0 & 4 & 0 & 0 & 0 & 0 & 0 & 0 & 0 & -1 & 0 & 0 & 0 & 0 & 0 & -1 & 0 & 0 & 0 & 0 \\
0 & 0 & 2 & 0 & 0 & 0 & 0 & 0 & 0 & 0 & 0 & -1 & 0 & -1 & 0 & 0 & 0 & 0 & 0 & 0 & 0 \\
0 & 0 & 0 & 1 & 1 & 0 & 0 & 0 & 0 & 0 & 0 & -1 & 0 & 0 & -1 & 0 & 0 & 0 & 0 & 0 & 0 \\
0 & 0 & 1 & 1 & 0 & 1 & 0 & 0 & 0 & 0 & 0 & -1 & 0 & 0 & 0 & 0 & 0 & 0 & 0 & 0 & -1 \\
0 & 1 & 0 & 1 & 0 & 0 & 0 & 0 & 0 & 0 & 0 & 0 & -2 & 0 & 0 & 0 & 0 & 0 & 0 & 0 & 0 \\
0 & 6 & 0 & 0 & 0 & 0 & 0 & 0 & 0 & 0 & 0 & 0 & -1 & 0 & -1 & 0 & 0 & 0 & 0 & 0 & 0 \\
0 & 0 & 0 & 0 & 0 & 2 & 0 & 0 & 0 & 0 & 0 & 0 & 0 & -1 & 0 & -1 & 0 & 0 & 0 & 0 & 0 \\
0 & 2 & 2 & 1 & 0 & 0 & 0 & 0 & 0 & 0 & 0 & 0 & 0 & -1 & 0 & -1 & 0 & 0 & 0 & 0 & 0 \\
0 & 1 & 0 & 0 & 0 & 2 & 0 & 0 & 0 & 0 & 0 & 0 & 0 & 0 & 0 & -2 & 0 & 0 & 0 & 0 & 0 \\
0 & 0 & 0 & 0 & 0 & 1 & 0 & 0 & 0 & 0 & 0 & 0 & 0 & 0 & 0 & 0 & 0 & 0 & 0 & 0 & 0
\end{pmatrix}.
$$

We have generated the following linear system (where $x_i = \operatorname{ind}_{11}(i)$):

$$
C \left(x_{-1} \quad x_2 \quad x_3 \quad x_5 \quad x_7 \quad x_{11} \quad x_{20} \quad x_{21} \quad \cdots \quad x_{34} \right)^{\mathrm{t}}
$$
$$
\equiv \left(0 \quad 0 \quad \cdots \quad 0 \quad 1 \right)^{\mathrm{t}} \pmod{718}.
$$

We have the prime factorization $p - 1 = 718 = 2 \times 359$, that is, we need to solve the system modulo 2 and modulo 359. The coefficient matrix C has full column rank (that is, 21) modulo 359, whereas it has a column rank of 20 modulo 2. Thus, there are two solutions modulo 2 and one solution modulo 359. Combining by CRT gives two solutions modulo 718. We can easily verify the correctness of a solution by computing powers of g.

The correct solution modulo 2 turns out to be

$$(1 \; 0 \; 0 \; 0 \; 0 \; 1 \; 0 \; 0 \; 1 \; 1 \; 0 \; 0 \; 0 \; 0 \; 0 \; 0 \; 0 \; 0 \; 1 \; 1)^t,$$

whereas the (unique) solution modulo 359 is

$$(0 \; 247 \; 42 \; 5 \; 291 \; 1 \; 140 \; 333 \; 248 \; 344 \; 65 \; 10 \; 17 \; 126 \; 67 \; 279 \; 294 \; 304 \; 158 \; 43 \; 31)^t.$$

Combining by the CRT gives the final solution:

$\mathrm{ind}_{11}(-1) = 359,$	$\mathrm{ind}_{11}(21) = 692,$	$\mathrm{ind}_{11}(28) = 426,$
$\mathrm{ind}_{11}(2) = 606,$	$\mathrm{ind}_{11}(22) = 607,$	$\mathrm{ind}_{11}(29) = 638,$
$\mathrm{ind}_{11}(3) = 42,$	$\mathrm{ind}_{11}(23) = 703,$	$\mathrm{ind}_{11}(30) = 294,$
$\mathrm{ind}_{11}(5) = 364,$	$\mathrm{ind}_{11}(24) = 424,$	$\mathrm{ind}_{11}(31) = 304,$
$\mathrm{ind}_{11}(7) = 650,$	$\mathrm{ind}_{11}(25) = 10,$	$\mathrm{ind}_{11}(32) = 158,$
$\mathrm{ind}_{11}(11) = 1,$	$\mathrm{ind}_{11}(26) = 376,$	$\mathrm{ind}_{11}(33) = 43,$
$\mathrm{ind}_{11}(20) = 140,$	$\mathrm{ind}_{11}(27) = 126,$	$\mathrm{ind}_{11}(34) = 31.$ □

7.2.2.2 Sieving

Let me now explain how sieving can be carried out to generate the relations of the linear sieve method. First, fix a value of c_1 in the range $-M \leqslant c_1 \leqslant M$, and allow c_2 to vary in the interval $[c_1, M]$. Let q be a small prime in the factor base ($q = p_i$ for $1 \leqslant i \leqslant t$), and h a small positive exponent. We need to determine all the values of c_2 for which $q^h | T(c_1, c_2)$ (for the fixed choice of c_1). The condition $T(c_1, c_2) \equiv 0 \pmod{q^h}$ gives the linear congruence in c_2:

$$(H + c_1)c_2 \equiv -(J + c_1 H) \pmod{q^h},$$

which can be solved by standard techniques (see Section 1.4).

We initialize an array A indexed by c_2 to $A_{c_2} = \log |T(c_1, c_2)|$. For each appropriate choice of q and h, we subtract $\log q$ from all of the array locations A_{c_2}, where $c_2 \in [c_1, M]$ satisfies the above congruence. After all the choices for q and h are considered, we find those array locations A_{c_2} which store values close to zero. These correspond to precisely all the smooth values of $T(c_1, c_2)$ for the fixed c_1. We make trial divisions of these $T(c_1, c_2)$ values by the small primes $p_1, p_2, \ldots, p_t$ in the factor base.

The details of this sieving process are similar to those discussed in connection with the quadratic sieve method for factoring integers (Section 6.6), and are omitted here. Many variants that are applicable to the QSM (like large prime variation and incomplete sieving) apply to the linear sieve method too.

7.2.2.3 Running Time

Let me now prescribe values for t and M so that the linear sieve method runs in $L[1]$ time. We include all primes $\leqslant L[1/2]$ in the factor base, that is, t is again of the form $L[1/2]$. The probability that an integer of absolute value $\tilde{O}(\sqrt{p})$ is smooth with respect to these primes is then $L[-\frac{1}{2}]$ (by Corollary 6.14). We choose $M = L[1/2]$ also. That is, the size of the factor base is again $L[1/2]$. The total number of pairs (c_1, c_2), for which $T(c_1, c_2)$ is tested for smoothness, is $\Theta(M^2)$ which is of the order of $L[1]$. Since each of these values of $T(c_1, c_2)$ is $\tilde{O}(\sqrt{p})$, the expected number of relations is $L[1]L[-\frac{1}{2}] = L[1/2]$, the same as the size of the factor base.

It is easy to argue that for these choices of t and M, the entire sieving process (for all values of c_1) takes a time of $L[1]$. Straightforward Gaussian elimination involving $L[1/2]$ equations in $L[1/2]$ variables takes a running time of $L[3/2]$. However, the system of congruences generated by the linear sieve method is necessarily sparse. Employing some efficient sparse system solver reduces the running time to $L[1]$. To sum up, the first stage of the linear sieve method can be arranged to run in $L[1]$ time. The space requirement is $L[1/2]$.

7.2.2.4 Second Stage

The first stage gives us the discrete logarithms of primes $\leqslant L[1/2]$. If we adopt a strategy similar to the second stage of the basic method (search for a smooth value of $ag^\alpha \pmod{p}$ for randomly chosen α), we spend $L[3/2]$ time to compute each individual discrete logarithm. That is too much. Moreover, we do not make use of the indices of $H + c$, available from the first stage.

Using a somewhat trickier technique, we can compute individual logarithms in only $L[1/2]$ time. This improved method has three steps.

- We choose α randomly, and try to express $ag^\alpha \pmod{p}$ as a product of primes $\leqslant L[2]$. We do not sieve using all primes $\leqslant L[2]$, because that list of primes is rather huge, and sieving would call for an unacceptable running time of $L[2]$. We instead use the elliptic curve method (Section 6.8) to detect the $L[2]$-smoothness of $ag^\alpha \pmod{p}$. This factoring algorithm is sensitive to the smoothness of the integer being factored. More concretely, it can detect the $L[2]$-smoothness of an integer of value $O(p)$ (and completely factor such an $L[2]$-smooth integer) in $L[1/4]$ time. Suppose that for some α, we obtain the following factorization:

$$ag^\alpha \equiv \left(\prod_{i=1}^{t} p_i^{\alpha_i}\right)\left(\prod_{j=1}^{k} q_j^{\beta_j}\right) \pmod{p},$$

where p_i are the primes in the factor base B, and q_j are primes $\leqslant L[2]$ not in the factor base. We then have

$$\operatorname{ind}_g a \equiv -\alpha + \sum_{i=1}^{t} \alpha_i \operatorname{ind}_g(p_i) + \sum_{j=1}^{k} \beta_j \operatorname{ind}_g(q_j) \pmod{p-1}.$$

By Corollary 6.14, an integer of value $O(p)$ is $L[2]$-smooth with probability $L[-\frac{1}{4}]$, that is, $L[1/4]$ choices of α are expected to suffice to find an $L[2]$-smooth value of $ag^\alpha \pmod{p}$. The expected running time for arriving at the last congruence is, therefore, $L[1/4] \times L[1/4] = L[1/2]$. Since $\mathrm{ind}_g(p_i)$ are available from the first stage, we need to compute the indices of the *medium-sized* primes $q_1, q_2, \ldots, q_k$. Since $k \leqslant \log n$, we achieve a running time of $L[1/2]$ for the second stage of the linear sieve method, provided that we can compute each $\mathrm{ind}_g(q_j)$ in $L[1/2]$ time.

- Let q be a medium-sized prime. We find an integer y close to $\sqrt{p}/q$ such that y is $L[1/2]$-smooth. Since the first stage gives us the indices of all primes $\leqslant L[1/2]$, the index of y can be computed. We use sieving to obtain such an integer y. Indeed, the value of y is $O^\sim(\sqrt{p})$, and so the probability that y is $L[1/2]$-smooth is $L[-\frac{1}{2}]$. This means that sieving around $\sqrt{p}/q$ over an interval of size $L[1/2]$ suffices.

- Let q be a medium-sized prime, and y an $L[1/2]$-smooth integer close to $\sqrt{p}/q$, as computed above. Consider the integer $T'(c) = (H+c)qy - p$ for a small value of c. We have $T'(c) = O(\sqrt{p})$, and so $T'(c)$ is $L[1/2]$-smooth with probability $L[-\frac{1}{2}]$, that is, sieving over the $L[1/2]$ values of c in the range $-M \leqslant c \leqslant M$ is expected to give an $L[1/2]$-smooth value

$$T'(c) = (H+c)qy - p = p_1^{\gamma_1} p_2^{\gamma_2} \cdots p_t^{\gamma_t}, \text{ that is,}$$

$$\mathrm{ind}_g q \equiv -\mathrm{ind}_g(H+c) - \mathrm{ind}_g y + \sum_{i=1}^{t} \gamma_i \, \mathrm{ind}_g(p_i) \pmod{p-1}.$$

Example 7.7 Let us continue with the database of indices available from the first stage described in Example 7.6. Suppose that we want to compute the index of $a = 123$ modulo $p = 719$ with respect to $g = 11$. We first obtain the relation $ag^{161} \equiv 182 \equiv 2 \times 7 \times 13 \pmod{719}$, that is, $\mathrm{ind}_g a \equiv -161 + \mathrm{ind}_g(2) + \mathrm{ind}_g(7) + \mathrm{ind}_g(13) \equiv -161 + 606 + 650 + \mathrm{ind}_g(13) \equiv 377 + \mathrm{ind}_g(13) \pmod{718}$. What remains is to compute the index of $q = 13$.

We look for an 11-smooth integer y close to $\sqrt{p}/13 \approx 2.0626$. We take $y = 3$. Example 7.6 gives $\mathrm{ind}_g(y) = 42$. (Since we are working here with an artificially small p, the value of y turns out to be abnormally small.)

Finally, we find an 11-smooth value of $(H+c)qy - p$ for $-7 \leqslant c \leqslant 7$. For $c = 4$, we have $(H+4)qy - p = 490 = 2 \times 5 \times 7^2$, that is, $\mathrm{ind}_g q \equiv -\mathrm{ind}_g(H+4) - \mathrm{ind}_g(y) + \mathrm{ind}_g(2) + \mathrm{ind}_g(5) + 2\mathrm{ind}_g(7) \equiv -304 - 42 + 606 + 364 + 2 \times 650 \equiv 488 \pmod{718}$. This gives the desired discrete logarithm $\mathrm{ind}_g(123) \equiv 377 + 488 \equiv 147 \pmod{718}$. $\qquad\square$

7.2.3 Residue-List Sieve Method (RLSM)

Like the linear sieve method, the residue list sieve method looks at integers near $\sqrt{p}$. However, instead of taking the product of two such integers, the residue-list sieve method first locates smooth integers near $\sqrt{p}$. After that, pairs of located smooth integers are multiplied.

The factor base now consists of -1, and the primes $p_1, p_2, \ldots, p_t \leqslant L[\frac{1}{2}]$. Let $H = \lceil \sqrt{p} \rceil$, $J = H^2 - p$, and $M = L[1]$. We first locate $L[\frac{1}{2}]$-smooth integers of the form $H + c$ with $-M \leqslant c \leqslant M$. Since each such candidate is $O(\sqrt{p})$, the smoothness probability is $L[-\frac{1}{2}]$, that is, among the $L[1]$ candidates of the form $H + c$, we expect to obtain $L[\frac{1}{2}]$ smooth values. Let $H + c_1$ and $H + c_2$ be two such smooth integers. Consider $T(c_1, c_2) = (H + c_1)(H + c_2) - p = J + (c_1 + c_2)H + c_1 c_2$. We have $T(c_1, c_2) = O(\sqrt{p})$, that is, the smoothness probability of $T(c_1, c_2)$ with respect to the factor base B is again $L[-\frac{1}{2}]$. Since there are $\Theta(L[\frac{1}{2}]^2) = L[1]$ pairs (c_1, c_2) with both $H + c_1$ and $H + c_2$ smooth, the expected number of smooth values of $T(c_1, c_2)$ is $L[\frac{1}{2}]$. These relations are solved to compute the indices of the $L[\frac{1}{2}]$ elements of the factor base B.

Example 7.8 Let us compute discrete logarithms modulo the prime $p = 863$ to the primitive base $g = 5$. We have $H = \lceil \sqrt{863} \rceil = 30$, and $J = H^2 - p = 37$. Take $B = \{-1, 2, 3, 5, 7, 11\}$, and $M = 7$. The smooth values of $H + c$ with $-M \leqslant c \leqslant M$ are shown in the left table below. Combination of these smooth values yields smooth $T(c_1, c_2)$ values as shown in the right table below.

c	$H + c$		$H + c_1$	$H + c_2$	$T(c_1, c_2)$
-6	$24 = 2^3 \times 3$		24	36	1
-5	$25 = 5^2$		25	32	$-63 = (-1) \times 3^2 \times 7$
-3	$27 = 3^3$		25	35	$12 = 2^2 \times 3$
-2	$28 = 2^2 \times 7$		27	32	1
0	$30 = 2 \times 3 \times 5$		27	33	$28 = 2^2 \times 7$
2	$32 = 2^5$		28	32	$33 = 3 \times 11$
3	$33 = 3 \times 11$				
5	$35 = 5 \times 7$				
6	$36 = 2^2 \times 3^2$				

The following relations are thus obtained. We use the notation x_a to stand for $\mathrm{ind}_g a$ (where $a \in B$).

$$(3x_2 + x_3) + (2x_2 + 2x_3) \equiv 0 \pmod{862},$$
$$(2x_5) + (5x_2) \equiv x_{-1} + 2x_3 + x_7 \pmod{862},$$
$$(2x_5) + (x_5 + x_7) \equiv 2x_2 + x_3 \pmod{862},$$
$$(3x_3) + (5x_2) \equiv 0 \pmod{862},$$
$$(3x_3) + (x_3 + x_{11}) \equiv 2x_2 + x_7 \pmod{862},$$
$$(2x_2 + x_7) + (5x_2) \equiv x_3 + x_{11} \pmod{862}.$$

Moreover, we have the free relation:

$$x_5 \equiv 1 \pmod{862}.$$

We also use the fact that $x_{-1} \equiv (p - 1)/2 \equiv 431 \pmod{862}$, since $g = 5$ is a primitive root of p. This gives us two solutions of the above congruences:

$$(x_{-1}, x_2, x_3, x_5, x_7, x_{11}) \; = \; (431, 161, 19, 1, 338, 584), \quad \text{and}$$
$$(x_{-1}, x_2, x_3, x_5, x_7, x_{11}) \; = \; (431, 592, 450, 1, 769, 153).$$

But $5^{161} \equiv -2 \pmod{863}$, whereas $5^{592} \equiv 2 \pmod{863}$. That is, the second solution gives the correct values of the indices of the factor-base elements. $\square$

The first stage of the residue-list sieve method uses two sieves. The first one is used to locate all the smooth values of $H + c$. Since c ranges over $L[1]$ values between $-M$ and M, this sieve takes a running time of the form $L[1]$.

In the second sieve, one combines pairs of smooth values of $H + c$ obtained from the first sieve, and identifies the smooth values of $T(c_1, c_2)$. But $H + c$ itself ranges over $L[1]$ values (although there are only $L[\frac{1}{2}]$ smooth values among them). In order that the second sieve too can be completed in $L[1]$ time, we, therefore, need to adopt some special tricks. For each small prime power q^h, we maintain a list of smooth $H + c$ values obtained from the first sieve. This list should be kept sorted with respect to the residues $(H + c) \operatorname{rem} q^h$. The name *residue-list sieve method* is attributed to these lists. Since there are $L[\frac{1}{2}]$ prime powers q^h, and there are $L[\frac{1}{2}]$ smooth values of $H + c$, the total storage requirement for all the residue lists is $L[1]$.

For determining the smoothness of $T(c_1, c_2) = (H + c_1)(H + c_2) - p$, one fixes c_1, and lets c_2 vary in the interval $c_1 \leqslant c_2 \leqslant M$. For each small prime power q^h, one calculates $(H + c) \operatorname{rem} q^h$ and $p \operatorname{rem} q^h$. One then solves for the value(s) of $(H + c_2) \pmod{q^h}$ from the congruence $T(c_1, c_2) \equiv 0 \pmod{q^h}$. For each solution χ, one consults the residue list for q^h to locate all the values of c_2 for which $(H + c_2) \operatorname{rem} q^h = \chi$. Since the residue list is kept sorted with respect to the residue values modulo q^h, binary search can quickly identify the desired values of c_2, leading to a running time of $L[1]$ for the second sieve.

The resulting sparse system with $L[\frac{1}{2}]$ congruences in $L[\frac{1}{2}]$ variables can be solved in $L[1]$ time. The second stage of the residue-list sieve method is identical to the second stage of the linear sieve method, and can be performed in $L[\frac{1}{2}]$ time for each individual logarithm. The second stage involves a sieve in the third step, which calls for the residue lists available from the first stage.

Let us now make a comparative study between the performances of the linear sieve method and the residue-list sieve method. The residue-list sieve method does not include any $H + c$ value in the factor base. As a result, the size of the factor base is smaller than that in the linear sieve method. However, maintaining the residue lists calls for a storage of size $L[1]$. This storage is *permanent* in the sense that the second stage (individual logarithm calculation) requires these lists. For the linear sieve method, on the other hand, the permanent storage requirement is only $L[\frac{1}{2}]$. Moreover, the (hidden) o(1) term in the exponent of the running time is higher in the residue-list sieve method than in the linear sieve method. In view of these difficulties, the residue-list sieve method turns out to be less practical than the linear sieve method.

7.2.4 Gaussian Integer Method (GIM)

An adaptation of ElGamal's method[10] for computing discrete logarithms in $\mathbb{F}_{p^2}$, the Gaussian integer method is an attractive alternative to the linear sieve method. It has a running time of $L[1]$ and a space requirement of $L[\frac{1}{2}]$.

Assume that (at least) one of the integers $-1, -2, -3, -7, -11, -19, -43$, -67 and -163 is a quadratic residue modulo p. Let $-r$ be such an integer. (For these values of r, $\mathbb{Z}[\sqrt{-r}]$ is a UFD. The algorithm can be made to work even if the assumption does not hold.) Let s be a square root of $-r$ modulo p. Consider the ring homomorphism

$$\Phi : \mathbb{Z}[\sqrt{-r}] \to \mathbb{F}_p, \text{ that maps } a + b\sqrt{-r} \mapsto (a + bs) \text{ rem } p.$$

By Cornacchia's algorithm (Algorithm 1.10), we compute integers u, v such that $p = u^2 + rv^2$. Since $s^2 \equiv -r \pmod{p}$, either $u + vs \equiv 0 \pmod{p}$ or $u - vs \equiv 0 \pmod{p}$. Suppose that $u + vs \equiv 0 \pmod{p}$. (One may first compute u, v and subsequently take $s \equiv -v^{-1}u \pmod{p}$.) The condition $u^2 + rv^2 = p$ implies that u, v are $O(\sqrt{p})$. For small integers c_1, c_2, the expression

$$T(c_1, c_2) = c_1 u + c_2 v$$

is $O(\sqrt{p})$. If we treat $c_1 u + c_2 v$ as an element of $\mathbb{Z}[\sqrt{-r}]$, we can write

$$c_1 u + c_2 v = c_1(u + v\sqrt{-r}) + v(c_2 - c_1\sqrt{-r}).$$

Application of the ring homomorphism Φ gives

$$T(c_1, c_2) \equiv c_1 u + c_2 v \equiv c_1(u + vs) + v\Phi(c_2 - c_1\sqrt{-r}) \pmod{p}.$$

But $u + vs \equiv 0 \pmod{p}$, so

$$T(c_1, c_2) \equiv v\Phi(c_2 - c_1\sqrt{-r}) \pmod{p}.$$

Let $B_1 = \{p_1, p_2, \ldots, p_t\}$ be the set of all (rational) primes $\leqslant L[\frac{1}{2}]$, and $B_2 = \{q_1, q_2, \ldots, q_{t'}\}$ the set of all (complex) primes $a + b\sqrt{-r}$ of $\mathbb{Z}[\sqrt{-r}]$ with $a^2 + b^2 r \leqslant L[\frac{1}{2}]$. Suppose that $T(c_1, c_2)$ factors completely over B_1, and $c_2 - c_1\sqrt{-r}$ factors completely over B_2, that is, we have

$$p_1^{\alpha_1} p_2^{\alpha_2} \cdots p_t^{\alpha_t} \equiv v\Phi(q_1)^{\beta_1}\Phi(q_2)^{\beta_2} \cdots \Phi(q_{t'})^{\beta_{t'}} \pmod{p}, \text{ that is,}$$

$$\alpha_1 \operatorname{ind}_g(p_1) + \alpha_2 \operatorname{ind}_g(p_2) + \cdots + \alpha_t \operatorname{ind}_g(p_t)$$
$$\equiv \operatorname{ind}_g(v) + \beta_1 \operatorname{ind}_g \Phi(q_1) + \beta_2 \operatorname{ind}_g \Phi(q_2) + \cdots + \beta_{t'} \operatorname{ind}_g \Phi(q_{t'}) \pmod{p-1}.$$

This is a relation in the Gaussian integer method.

The factor base B now consists of -1, the $t = L[\frac{1}{2}]$ rational primes p_1, p_2, $\ldots, p_t$, the images $\Phi(q_1), \Phi(q_2), \ldots, \Phi(q_{t'})$ of the $t' = L[\frac{1}{2}]$ complex primes, and the integer v. The size of B is $t + t' + 2$ which is $L[\frac{1}{2}]$. We let c_1, c_2 vary in

---

[10]Taher ElGamal, A subexponential-time algorithm for computing discrete logarithms over $GF(p^2)$, *IEEE Transactions on Information Theory*, 31, 473–481, 1985.

the interval $[-M, M]$ with $M = L[\frac{1}{2}]$. In order to avoid duplicate relations, we consider only those pairs with $\gcd(c_1, c_2) = 1$. There are $L[1]$ such pairs. Since each $T(c_1, c_2)$ is $O(\sqrt{p})$, the probability that it is smooth with respect to the $L[\frac{1}{2}]$ rational primes is $L[-\frac{1}{2}]$, that is, we expect to get $L[\frac{1}{2}]$ smooth values of $T(c_1, c_2)$. On the other hand, the complex number $c_2 - c_1\sqrt{-r}$ is smooth over B_2 with high (constant) probability ($c_2^2 + rc_1^2$ is $L[1]$, whereas B_2 contains all complex primes $a + b\sqrt{-r}$ with $a^2 + rb^2 \leqslant L[\frac{1}{2}]$). Thus, a significant fraction of (c_1, c_2) pairs, for which $T(c_1, c_2)$ is smooth, leads to relations for the Gaussian integer method, that is, we get $L[\frac{1}{2}]$ relations in $L[\frac{1}{2}]$ variables, as desired.

Example 7.9 Let us compute discrete logarithms modulo the prime $p = 997$ to the primitive base $g = 7$. We have $p \equiv 1 \pmod 4$, so -1 is a quadratic residue modulo p, and we may take $r = 1$. But then, we use the ring $\mathbb{Z}[i]$ of Gaussian integers (this justifies the name of this algorithm). We express $p = u^2 + v^2$ with $u = 31$ and $v = 6$, and take $s \equiv -v^{-1}u \equiv 161 \pmod p$ as the modular square root of -1, satisfying $u + vs \equiv 0 \pmod p$.

We take $t = 6$ small rational primes, that is, $B_1 = \{2, 3, 5, 7, 11, 13\}$. A set of pairwise non-associate complex primes $a + bi$ with $a^2 + b^2 \leqslant 13$ is $B_2 = \{1 + i, 2 + i, 2 - i, 2 + 3i, 2 - 3i\}$. The factor base is, therefore, given by

$$B = \{-1, 2, 3, 5, 7, 11, 13, v, \Phi(1 + i), \Phi(2 + i), \Phi(2 - i), \Phi(2 + 3i), \Phi(2 - 3i)\}$$
$$= \{-1, 2, 3, 5, 7, 11, 13, 6, 1 + s, 2 + s, 2 - s, 2 + 3s, 2 - 3s\}$$
$$= \{-1, 2, 3, 5, 7, 11, 13, 6, 162, 163, -159, 485, -481\}$$
$$= \{-1, 2, 3, 5, 7, 11, 13, 6, 162, 163, 838, 485, 516\}.$$

The prime integers 3, 7 and 11 remain prime in $\mathbb{Z}[i]$. We take (c_1, c_2) pairs with $\gcd(c_1, c_2) = 1$, so these primes do not occur in the factorization of $c_2 - c_1 i$.

The units in $\mathbb{Z}[i]$ are $\pm 1, \pm i$. We have

$$\text{ind}_g(\Phi(1)) = \text{ind}_g(1) = 0, \text{ and}$$
$$\text{ind}_g(\Phi(-1)) = \text{ind}_g(-1) = (p-1)/2 = 498.$$

Moreover, $\Phi(i) = s$ and $\Phi(-i) = -s$. One of $\pm s$ has index $(p-1)/4$, and the other has index $3(p-1)/4$. In this case, we have

$$\text{ind}_g(\Phi(i)) = \text{ind}_g(s) = (p-1)/4 = 249, \text{ and}$$
$$\text{ind}_g(\Phi(-i)) = \text{ind}_g(-s) = 3(p-1)/4 = 747.$$

Let us take $M = 5$, that is, we check all c_1, c_2 values between -5 and 5 with $\gcd(c_1, c_2) = 1$. There are 78 such pairs. For 37 of these pairs, the integer $T(c_1, c_2) = c_1 u + c_2 v$ is smooth with respect to B_1. Among these, 23 yield smooth values of $c_2 - c_1 i$ with respect to B_2 (see the table on the next page).

Let us now see how such a factorization leads to a relation. Consider $c_1 = -3$ and $c_2 = -4$. In this case, $T(c_1, c_2) = (-1) \times 3^2 \times 13$, and $c_2 - c_1 i = i(2 + i)^2$. This gives $\text{ind}_g(-1) + 2\text{ind}_g(3) + \text{ind}_g(13) \equiv \text{ind}_g(v) + \text{ind}_g(\Phi(i)) + 2\text{ind}_g(\Phi(2 + i)) \pmod{p-1}$, that is, $498 + 2\text{ind}_g(3) + \text{ind}_g(13) \equiv \text{ind}_g(6) + 249 + 2\text{ind}_g(163) \pmod{996}$. The reader is urged to convert the other 22 relations and solve the resulting system of congruences. $\square$

c_1	c_2	$T(c_1, c_2) = c_1 u + c_2 v$	$c_2 - c_1 i$
-3	-4	$-117 = (-1) \times 3^2 \times 13$	$-4 + 3i = (i) \times (2 + i)^2$
-3	-2	$-105 = (-1) \times 3 \times 5 \times 7$	$-2 + 3i = (-1) \times (2 - 3i)$
-3	-1	$-99 = (-1) \times 3^2 \times 11$	$-1 + 3i = (i) \times (1 + i) \times (2 - i)$
-3	2	$-81 = (-1) \times 3^4$	$2 + 3i = 2 + 3i$
-2	-3	$-80 = (-1) \times 2^4 \times 5$	$-3 + 2i = (i) \times (2 + 3i)$
-2	1	$-56 = (-1) \times 2^3 \times 7$	$1 + 2i = (i) \times (2 - i)$
-2	3	$-44 = (-1) \times 2^2 \times 11$	$3 + 2i = (i) \times (2 - 3i)$
-1	-3	$-49 = (-1) \times 7^2$	$-3 + i = (i) \times (1 + i) \times (2 + i)$
-1	1	$-25 = (-1) \times 5^2$	$1 + i = 1 + i$
-1	3	$-13 = (-1) \times 13$	$3 + i = (1 + i) \times (2 - i)$
-1	5	$-1 = (-1)$	$5 + i = (-i) \times (1 + i) \times (2 + 3i)$
0	1	$6 = 2 \times 3$	$1 = 1$
1	-5	$1 = 1$	$-5 - i = (i) \times (1 + i) \times (2 + 3i)$
1	-3	$13 = 13$	$-3 - i = (-1) \times (1 + i) \times (2 - i)$
1	-1	$25 = 5^2$	$-1 - i = (-1) \times (1 + i)$
1	3	$49 = 7^2$	$3 - i = (-i) \times (1 + i) \times (2 + i)$
2	-3	$44 = 2^2 \times 11$	$-3 - 2i = (-i) \times (2 - 3i)$
2	-1	$56 = 2^3 \times 7$	$-1 - 2i = (-i) \times (2 - i)$
2	3	$80 = 2^4 \times 5$	$3 - 2i = (-i) \times (2 + 3i)$
3	-2	$81 = 3^4$	$-2 - 3i = (-1) \times (2 + 3i)$
3	1	$99 = 3^2 \times 11$	$1 - 3i = (-i) \times (1 + i) \times (2 - i)$
3	2	$105 = 3 \times 5 \times 7$	$2 - 3i = 2 - 3i$
3	4	$117 = 3^2 \times 13$	$4 - 3i = (-i) \times (2 + i)^2$

Sieving to locate smooth $T(c_1, c_2)$ values is easy. After the sieve terminates, we throw away smooth values with $\gcd(c_1, c_2) > 1$. If $\gcd(c_1, c_2) = 1$ and $T(c_1, c_2)$ is smooth, we make trial division of $c_2 - c_1 \sqrt{-r}$ by the complex primes in B_2. The entire sieving process can be completed in $L[1]$ time.

The resulting sparse system of $L[\frac{1}{2}]$ equations in $L[\frac{1}{2}]$ variables can be solved in $L[1]$ time. The second stage of the Gaussian integer method is identical to the second stage of the linear sieve method, and can be performed in $L[\frac{1}{2}]$ time for each individual logarithm.

The size of the factor base in the Gaussian integer method is $t + t' + 2$, whereas the size of the factor base in the linear sieve method is $t + 2M + 2$. Since $t' \ll M$ (indeed, t' is roughly proportional to $M/\ln M$), the Gaussian integer method gives significantly smaller systems of linear congruences than the linear sieve method. In addition, the values of $T(c_1, c_2)$ are somewhat smaller in the Gaussian integer method than in the linear sieve method (we have $|c_1 u + c_2 v| \leqslant \sqrt{2} M \sqrt{p}$, whereas $J + (c_1 + c_2)H + c_1 c_2 \leqslant 2M \sqrt{p}$, approximately). Finally, unlike the residue-list sieve method, the Gaussian integer method is not crippled by the necessity of $L[1]$ permanent storage. In view of these advantages, the Gaussian integer method is practically the most preferred $L[1]$-time algorithm for computing discrete logarithms in prime fields.

7.2.5   Cubic Sieve Method (CSM)

Reyneri's cubic sieve method is a faster alternative to the $L[1]$-time algorithms discussed so far. Unfortunately, it is not clear how we can apply this method to a general prime p. In some special cases, this method is naturally applicable, and has a best running time of $L[\sqrt{2/3}]$ (nearly $L[0.816]$).

Suppose that we know a solution of the congruence

$$x^3 \equiv y^2 z \pmod{p}$$

with $x^3 \neq y^2 z$ (as integers), and with $x, y, z = O(p^{\xi})$ for $1/3 \leqslant \xi < 1/2$. For small integers c_1, c_2, c_3 with $c_1 + c_2 + c_3 = 0$, we have

$$
\begin{aligned}
&(x + c_1 y)(x + c_2 y)(x + c_3 y) \\
\equiv\ &x^3 + (c_1 + c_2 + c_3)x^2 y + (c_1 c_2 + c_1 c_3 + c_2 c_3)xy^2 + (c_1 c_2 c_3)y^3 \\
\equiv\ &y^2 z + (c_1 c_2 + c_1 c_3 + c_2 c_3)xy^2 + (c_1 c_2 c_3)y^3 \\
\equiv\ &y^2 [z + (c_1 c_2 + c_1 c_3 + c_2 c_3)x + (c_1 c_2 c_3)y] \pmod{p}.
\end{aligned}
$$

Let us denote

$$
\begin{aligned}
T(c_1, c_2, c_3) &= z + (c_1 c_2 + c_1 c_3 + c_2 c_3)x + (c_1 c_2 c_3)y \\
&= z - (c_1^2 + c_1 c_2 + c_2^2)x - c_1 c_2(c_1 + c_2)y.
\end{aligned}
$$

For small values of c_1, c_2, c_3, we have $T(c_1, c_2, c_3) = O(p^{\xi})$. We attempt to factor $T(c_1, c_2, c_3)$ over the first t primes. If the factorization attempt is successful, we get a relation of the form

$$(x + c_1 y)(x + c_2 y)(x + c_3 y) \equiv y^2 p_1^{\alpha_1} p_2^{\alpha_2} \cdots p_t^{\alpha_t} \pmod{p}, \text{ that is,}$$

$$
\begin{aligned}
&\mathrm{ind}_g(x + c_1 y) + \mathrm{ind}_g(x + c_2 y) + \mathrm{ind}_g(x + c_3 y) \\
\equiv\ &\mathrm{ind}_g(y^2) + \alpha_1 \mathrm{ind}_g(p_1) + \alpha_2 \mathrm{ind}_g(p_2) + \cdots + \alpha_t \mathrm{ind}_g(p_t) \pmod{p-1}.
\end{aligned}
$$

Therefore, we take a factor base of the following form (the choice of t and M to be explained shortly):

$$B = \{-1\} \cup \{p_1, p_2, \ldots, p_t\} \cup \{y^2\} \cup \{x + cy \mid -M \leqslant c \leqslant M\}.$$

Example 7.10 Let us compute discrete logarithms modulo the prime $p = 895189$ to the primitive base $g = 17$. A solution of the congruence $x^3 \equiv y^2 z \pmod{p}$ is $x = 139$, $y = 2$ and $z = 13$. This solution satisfies $x^3 = y^2 z + 3p$. We take $t = 25$ (all primes < 100), and $M = 50$. The size of the factor base is $2M + t + 3 = 128$. Letting c_1, c_2, c_3 vary in the range $-M \leqslant c_1 \leqslant c_2 \leqslant c_3 \leqslant M$ with $c_1 + c_2 + c_3 = 0$, we get 196 relations, some of which are shown in the table on the next page. In addition, we have a free relation

$$g^1 \equiv 17 \pmod{p}.$$

This results in a system of 197 linear congruences in 128 variables modulo $p - 1 = 895188$.   □

c_1	c_2	c_3	$T(c_1, c_2, c_3)$
-50	1	49	$-345576 = (-1) \times 2^3 \times 3 \times 7 \times 11^2 \times 17$
-50	21	29	$-323736 = (-1) \times 2^3 \times 3 \times 7 \times 41 \times 47$
-50	23	27	$-323268 = (-1) \times 2^2 \times 3 \times 11 \times 31 \times 79$
-49	15	34	$-312816 = (-1) \times 2^4 \times 3 \times 7^3 \times 19$
-48	1	47	$-318222 = (-1) \times 2 \times 3^3 \times 71 \times 83$
-48	22	26	$-295647 = (-1) \times 3 \times 11 \times 17^2 \times 31$
-47	9	38	$-291648 = (-1) \times 2^6 \times 3 \times 7^2 \times 31$
-47	11	36	$-289218 = (-1) \times 2 \times 3 \times 19 \times 43 \times 59$
-47	17	30	$-284088 = (-1) \times 2^3 \times 3 \times 7 \times 19 \times 89$
	$\cdots$		
-5	-5	10	$-9912 = (-1) \times 2^3 \times 3 \times 7 \times 59$
-5	2	3	$-2688 = (-1) \times 2^7 \times 3 \times 7$
-4	-2	6	$-3783 = (-1) \times 3 \times 13 \times 97$
-4	0	4	$-2211 = (-1) \times 3 \times 11 \times 67$
-3	-1	4	$-1770 = (-1) \times 2 \times 3 \times 5 \times 59$
-3	1	2	$-972 = (-1) \times 2^2 \times 3^5$
-2	-1	3	$-948 = (-1) \times 2^2 \times 3 \times 79$
-2	1	1	$-408 = (-1) \times 2^3 \times 3 \times 17$
-1	-1	2	$-400 = (-1) \times 2^4 \times 5^2$
-1	0	1	$-126 = (-1) \times 2 \times 3^2 \times 7$
0	0	0	$13 = 13$

Let me now specify the parameters t and M. Suppose that x, y, z are each $O(p^\xi)$. We take t as the number of primes $\leqslant L[\sqrt{\xi/2}]$. By the prime number theorem, t is again of the form $L[\sqrt{\xi/2}]$. We also take $M = L[\sqrt{\xi/2}]$. There are $\Theta(M^2) = L[\sqrt{2\xi}]$ triples (c_1, c_2, c_3) with $-M \leqslant c_1 \leqslant c_2 \leqslant c_3 \leqslant M$ and $c_1 + c_2 + c_3 = 0$. Each $T(c_1, c_2, c_3)$ is of value $O(p^\xi)$ and has a probability $L[-\frac{\xi}{2\sqrt{\xi/2}}] = L[-\sqrt{\xi/2}]$ of being smooth with respect to the t primes in the factor base. Thus, the expected number of relations for all choices of (c_1, c_2, c_3) is $L[\sqrt{\xi/2}]$. The size of the factor base is $t + 2M + 3$ which is also $L[\sqrt{\xi/2}]$.

In order to locate the smooth values of $T(c_1, c_2, c_3)$, we express $T(c_1, c_2, c_3)$ as a function of c_1 and c_2 alone. The conditions $-M \leqslant c_1 \leqslant c_2 \leqslant c_3 \leqslant M$ and $c_1 + c_2 + c_3 = 0$ imply that c_1 varies between $-M$ and 0, and c_2 varies between $\max(c_1, -(M + c_1))$ and $-c_1/2$ for a fixed c_1. We fix c_1, and let c_2 vary in this allowed range. For a small prime power q^h, we solve $T(c_1, c_2, c_3) \equiv 0 \pmod{q^h}$ for c_2, taking c_1 as constant and $c_3 = -(c_1 + c_2)$. This calls for solving a quadratic congruence in c_2. The details are left to the reader (Exercise 7.17).

The sieving process can be completed in $L[2\sqrt{\xi/2}] = L[\sqrt{2\xi}]$ time. Solving the resulting sparse system of $L[\sqrt{\xi/2}]$ linear congruences in $L[\sqrt{\xi/2}]$ variables also takes the same time. To sum up, the first stage of the cubic sieve method can be so implemented as to run in $L[\sqrt{2\xi}]$ time. The space requirement is $L[\sqrt{\xi/2}]$. If $\xi = 1/3$, the running time is $L[\sqrt{2/3}] \approx L[0.816]$, and the space requirement is $L[\sqrt{1/6}] \approx L[0.408]$.

The second stage of the cubic sieve method is costlier than those for the $L[1]$ methods discussed earlier. The trouble now is that the first stage supplies a smaller database of discrete logarithms ($L[\sqrt{\xi/2}]$ compared to $L[1/2]$). This indicates that we need to perform more than $L[1/2]$ work in order to compute individual logarithms. Here is a strategy that runs in $L[\sqrt{2\xi}]$ time.

- Express $ag^\alpha \pmod{p}$ as a product of primes $\leqslant L[2]$. This can be done in $L[1/2]$ time by the elliptic curve method. It remains to compute the index of each medium-sized prime q that appears in this factorization.

- Find values of $c_3 \geqslant 0$ such that $x + c_3 y$ is divisible by q and the cofactor $(x+c_3 y)/q$ is $L[\sqrt{\xi/2}]$-smooth. This can be done by running a sieve over values of $c_3 \geqslant 0$. We need $L[\sqrt{\xi/2}]$ such values of c_3 for the second stage to terminate with high probability. As a result, we may have to work with values of $c_3 > M$, that is, the indices of $x + c_3 y$ are not necessarily available in the database computed in the first stage. If $c_3 \leqslant M$, we obtain the value of $\mathrm{ind}_g q$, otherwise we proceed to the next step.

- For each $c_3 > M$ obtained above, we let c_2 vary in the range $-c_3 - M \leqslant c_2 \leqslant -c_3 + M$, and set $c_1 = -(c_2 + c_3)$. (We may have $c_2 < -M$, that is, $\mathrm{ind}_g(x + c_2 y)$ need not be available from the first stage.) We run a sieve over this range to locate a value of c_2 for which both $x + c_2 y$ and $T(c_1, c_2, c_3) = z + (c_1 c_2 + c_1 c_3 + c_2 c_3)x + (c_1 c_2 c_3)y$ are $L[\sqrt{\xi/2}]$-smooth. Since these values are $O(p^\xi)$, each of these is smooth with probability $L[-\sqrt{\xi/2}]$, that is, both are smooth with probability $L[-\sqrt{2\xi}]$. Since $2M + 1$ (that is, $L[\sqrt{\xi/2}]$) values of c_2 are tried, the probability that a suitable c_2 can be found for a fixed c_3 is only $L[-\sqrt{\xi/2}]$. If no suitable c_2 is found, we repeat this step with another value of c_3. After $L[\sqrt{\xi/2}]$ trials, a suitable c_2 is found with high probability. We then have

$$(x + c_1 y)(x + c_2 y)(x + c_3 y) \equiv y^2 T(c_1, c_2, c_3) \pmod{p}.$$

By our choice of c_3, the integer $x + c_3 y$ is q times an $L[\sqrt{\xi/2}]$-smooth value. The quantities $x + c_2 y$ and $T(c_1, c_2, c_3)$ are both $L[\sqrt{\xi/2}]$-smooth (by the choice of c_2). We forced $-M \leqslant c_1 \leqslant M$, that is, the index of $x + c_1 y$ is already available from the first stage (so also is $\mathrm{ind}_g(y^2)$). Therefore, we obtain the desired index of the medium-sized prime q.

Asymptotically, the cubic sieve method is faster than the $L[1]$-time algorithms discussed earlier. However, the quadratic and cubic terms in the subexponential values c_1, c_2, c_3 make the values of $T(c_1, c_2, c_3)$ rather large compared to p^ξ. As a result, the theoretically better performance of the cubic sieve method does not show up unless the bit size of p is rather large. Practical experiences suggest that for bit sizes $\geqslant 200$, the cubic sieve method achieves some speedup over the $L[1]$ algorithms. On the other hand, the number-field sieve method takes over for bit sizes $\geqslant 300$. To sum up, the cubic sieve method seems to be a good choice only for a narrow band (200–300) of bit sizes. Another problem with the cubic sieve method is that its second stage is as slow as its first stage, and much slower than the second stages of the $L[1]$ algorithms.

The biggest trouble attached to the cubic sieve method is that we do not know how to compute x, y, z as small as possible satisfying $x^3 \equiv y^2 z \pmod{p}$ and $x^3 \neq y^2 z$. A solution is naturally available only for some special primes p. General applicability of the cubic sieve method thus remains unclear.

7.2.6 Number-Field Sieve Method (NFSM)

The number-field sieve method[11] is the fastest known algorithm for computing discrete logarithms in prime fields. It is an adaptation of the method with the same name for factoring integers. It can also be viewed as a generalization of the Gaussian integer method.

We start with an irreducible (over $\mathbb{Q}$) polynomial $f(x) \in \mathbb{Z}[x]$ and an integer m satisfying $f(m) \equiv 0 \pmod{p}$. We extend $\mathbb{Q}$ by adjoining a root θ of $f(x)$: $K = \mathbb{Q}(\theta)$. The ring $\mathfrak{O}_K$ of integers of K is the number ring to be used in the algorithm. For simplicity, we assume that $\mathfrak{O}_K$ supports unique factorization of elements, and that elements of $\mathfrak{O}_K$ can be written as polynomials in θ with rational coefficients. The map Φ taking $\theta \mapsto m \pmod{p}$ extends to a ring homomorphism $\mathfrak{O}_K \to \mathbb{F}_p$.

For small coprime integers c_1, c_2, we consider the two elements $T_1(c_1, c_2) = c_1 + c_2\theta \in \mathfrak{O}_K$ and $T_2(c_1, c_2) = \Phi(T_1(c_1, c_2)) = c_1 + c_2 m \in \mathbb{Z}$. Suppose that $T_1(c_1, c_2)$ is smooth with respect to small primes q_i of $\mathfrak{O}_K$, and $T_2(c_1, c_2)$ is smooth with respect to small rational primes p_j, that is, $T_1(c_1, c_2) = \prod_i q_i^{\alpha_i}$, and $T_2(c_1, c_2) = \prod_j p_j^{\beta_j}$. Application of the homomorphism Φ gives

$$\Phi(T_1(c_1, c_2)) \equiv \Phi\left(\prod_i q_i^{\alpha_i}\right) \equiv \prod_i \Phi(q_i)^{\alpha_i} \equiv T_2(c_1, c_2) \equiv \prod_j p_j^{\beta_j} \pmod{p}.$$

This leads to the relation

$$\sum_i \alpha_i \operatorname{ind}_g(\Phi(q_i)) \equiv \sum_j \beta_j \operatorname{ind}_g(p_j) \pmod{p-1}.$$

In addition to the rational primes p_j, we should, therefore, include $\Phi(q_i)$ in the factor base B for small primes q_i of $\mathfrak{O}_K$. We also consider a set of generators of the group of units of $\mathfrak{O}_K$, and include $\Phi(u)$ in B for each such generator u. I do not go into further detail on the NFSM, but mention only that this method (under the assumptions mentioned above) runs in approximately $\exp\left((1.526 + o(1))(\ln n)^{1/3}(\ln \ln n)^{2/3}\right)$ time.

Element-wise unique factorization may fail in the ring $\mathfrak{O}_K$. If so, we need to look at unique factorization at the level of ideals. This variant of the NFSM runs in approximately $\exp\left((1.923 + o(1))(\ln n)^{1/3}(\ln \ln n)^{2/3}\right)$ time.

---

[11]Daniel M. Gordon, Discrete logarithms in GF(p) using the number field sieve, *SIAM Journal of Discrete Mathematics*, 6, 124–138, 1993.

7.3   Algorithms for Fields of Characteristic Two

Like prime fields, the finite fields $\mathbb{F}_{2^n}$ of characteristic two (also called binary fields) find diverse applications. Indeed, the arithmetic in binary fields can be implemented efficiently (compared to the general extension fields $\mathbb{F}_{p^n}$). In this section, we assume that $\mathbb{F}_{2^n}$ has the polynomial-basis representation $\mathbb{F}_2(\theta)$ with $f(\theta) = 0$, where $f(x) \in \mathbb{F}_2[x]$ is an irreducible polynomial of degree n. Moreover, we assume that a primitive element $g(\theta) \in \mathbb{F}_{2^n}^*$ is provided to us. We calculate logarithms of elements of $\mathbb{F}_{2^n}^*$ to the base $g(\theta)$.

All index calculus methods for $\mathbb{F}_{2^n}$ essentially try to factor non-zero polynomials of $\mathbb{F}_2[x]$ into irreducible polynomials of small degrees. The concept of *smoothness* of polynomials plays an important role in this context. Before proceeding to a discussion of the algorithms, we need to specify density estimates for smooth polynomials (a counterpart of Theorem 6.13).

Theorem 7.11 *A non-zero polynomial in $\mathbb{F}_2[x]$ of degree k has all its irreducible factors with degrees $\leqslant m$ with probability*

$$p(k, m) = \exp\left((-1 + o(1))\frac{k}{m}\ln\frac{k}{m}\right).$$

This estimate is valid for $k \to \infty$, $m \to \infty$, and $k^{1/100} \leqslant m \leqslant k^{99/100}$.

A polynomial $h(x)$ chosen uniformly randomly from the set of non-zero polynomials in $\mathbb{F}_2[x]$ of degrees $< l$ has a probability $2^k/(2^l - 1) \approx 2^{-(l-k)}$ to have degree exactly k. Therefore, such an $h(x)$ has all irreducible factors of degrees $\leqslant m$ with approximate probability

$$\sum_{k=0}^{l-1} 2^{-(l-k)} p(k, m) \approx p(l, m)\left[\frac{(le/m)^{1/m}}{2 - (le/m)^{1/m}}\right].$$

As $l \to \infty$, $m \to \infty$, and $l^{1/100} \leqslant m \leqslant l^{99/100}$, this probability is approximately equal to $p(l, m)$. ◁

A special case of this estimate is highlighted now.

Corollary 7.12 *Let $l = \alpha n$ and $m = \beta\sqrt{n \ln n}$ for some positive real constants α, β with $\alpha \leqslant 1$. Then,*

$$p(l, m) \approx \exp\left(-\frac{\alpha}{2\beta}\sqrt{n \ln n}\right) = L\left[-\frac{\alpha}{2\beta}\right],$$

where the (revised) notation $L[\gamma]$ stands for $\exp\left((\gamma + o(1))\sqrt{n \ln n}\right)$. ◁

7.3.1 Basic Index Calculus Method

We start with a factor base B consisting of non-constant irreducible polynomials $w(x) \in \mathbb{F}_2[x]$ of degrees $\leqslant m$ (the choice of m is specified later). In the first stage, we compute the discrete logarithms of all $w(\theta)$ to the base $g(\theta)$. To this end, we raise $g(\theta)$ to random exponents α, and try to express the canonical representatives of $g(\theta)^\alpha \in \mathbb{F}_{2^n}$ as products of polynomials $w(\theta)$ with $w(x) \in B$. If a factoring attempt is successful, we obtain a *relation*:

$$g(x)^\alpha \equiv \prod_{w(x) \in B} w(x)^{\gamma_w} \pmod{f(x)},$$

or equivalently,

$$g(\theta)^\alpha = \prod_{w(x) \in B} w(\theta)^{\gamma_w} \in \mathbb{F}_{2^n} = \mathbb{F}_2(\theta).$$

Taking the discrete logarithm of both sides, we get

$$\alpha \equiv \sum_{w \in B} \gamma_w \operatorname{ind}_{g(\theta)} w(\theta) \pmod{2^n - 1}.$$

This is a linear congruence in the variables $\operatorname{ind}_{g(\theta)} w(\theta)$. Let $|B| = t$. We vary α, and generate s relations with $t \leqslant s \leqslant 2t$. The resulting $s \times t$ system is solved modulo $2^n - 1$ to obtain the indices of the elements of the factor base.

The second stage computes individual logarithms using the database obtained from the first stage. Suppose that we want to compute the index of $a(\theta) \in \mathbb{F}_{2^n}^*$. We pick random α, and attempt to decompose $a(\theta)g(\theta)^\alpha$ into irreducible factors of degrees $\leqslant m$. A successful factoring attempt gives

$$a(\theta)g(\theta)^\alpha = \prod_{w \in B} w(\theta)^{\delta_w}, \text{ that is,}$$

$$\operatorname{ind}_{g(\theta)} a(\theta) \equiv -\alpha + \sum_{w \in B} \delta_w \operatorname{ind}_{g(\theta)} w(\theta) \pmod{2^n - 1}.$$

Example 7.13 Let us take $n = 17$, and represent $\mathbb{F}_{2^{17}} = \mathbb{F}_2(\theta)$, where $\theta^{17} + \theta^3 + 1 = 0$. The size of $\mathbb{F}_{2^n}^*$ is $2^n - 1 = 131071$ which is a prime. Therefore, every element of $\mathbb{F}_{2^n}$ other than $0, 1$ is a generator of $\mathbb{F}_{2^n}^*$. Let us compute discrete logarithms to the base $g(\theta) = \theta^7 + \theta^5 + \theta^3 + \theta$. We choose $m = 4$, that is, the factor base is $B = \{w_1, w_2, w_3, w_4, w_5, w_6, w_7, w_8\}$, where

$$
\begin{aligned}
w_1 &= \theta, \\
w_2 &= \theta + 1, \\
w_3 &= \theta^2 + \theta + 1, \\
w_4 &= \theta^3 + \theta + 1, \\
w_5 &= \theta^3 + \theta^2 + 1, \\
w_6 &= \theta^4 + \theta + 1, \\
w_7 &= \theta^4 + \theta^3 + \theta^2 + \theta + 1, \\
w_8 &= \theta^4 + \theta^3 + 1.
\end{aligned}
$$

In the first stage, we compute $g(\theta)^\alpha$ for random $\alpha \in \{1, 2, \ldots, 2^n - 2\}$. Some values of α, leading to B-smooth values of $g(\theta)^\alpha$, are shown below.

α	$g(\theta)^\alpha$
73162	$\theta^{16} + \theta^{15} + \theta^{12} + \theta^{11} + \theta^{10} + \theta^8 + \theta^6 + \theta^5 + \theta^4 + \theta^3 + 1$
	$= (\theta^2 + \theta + 1)^3 (\theta^3 + \theta + 1)^2 (\theta^4 + \theta + 1)$
87648	$\theta^{16} + \theta^{12} + \theta^{11} + \theta^8 + \theta^7 + \theta^5 + \theta^4 + \theta^2 + \theta + 1$
	$= (\theta + 1)^2 (\theta^2 + \theta + 1)^2 (\theta^3 + \theta + 1)(\theta^3 + \theta^2 + 1)(\theta^4 + \theta^3 + 1)$
18107	$\theta^{15} + \theta^{14} + \theta^{13} + \theta^{12} + \theta^8 + \theta^7 + \theta^6 + \theta^5$
	$= \theta^5 (\theta + 1)^4 (\theta^3 + \theta + 1)(\theta^3 + \theta^2 + 1)$
31589	$\theta^{16} + \theta^{14} + \theta^{12} + \theta^{11} + \theta^{10} + \theta^8 + \theta^7 + \theta^5 + \theta + 1$
	$= (\theta + 1)^7 (\theta^2 + \theta + 1)^3 (\theta^3 + \theta + 1)$
26426	$\theta^{14} + \theta^{13} + \theta^{11} + \theta^9 + \theta^8 + \theta^7 + \theta^6 + \theta^5 + \theta^4 + \theta^3 + \theta$
	$= \theta(\theta^3 + \theta + 1)^3 (\theta^4 + \theta^3 + \theta^2 + \theta + 1)$
74443	$\theta^{15} + \theta^{14} + \theta^{13} + \theta^{11} + \theta^{10} + \theta^6 + \theta^4 + \theta^3 + \theta + 1$
	$= (\theta + 1)(\theta^2 + \theta + 1)(\theta^3 + \theta + 1)^3 (\theta^3 + \theta^2 + 1)$
29190	$\theta^{16} + \theta^{14} + \theta^{13} + \theta^{10} + \theta^7 + \theta^3 + \theta$
	$= \theta(\theta^2 + \theta + 1)(\theta^3 + \theta^2 + 1)^3 (\theta^4 + \theta + 1)$
109185	$\theta^{16} + \theta^{15} + \theta^{14} + \theta^{13} + \theta^{11} + \theta^9 + \theta^8 + \theta^7 + \theta^6 + \theta^2 + \theta + 1$
	$= (\theta + 1)^3 (\theta^2 + \theta + 1)(\theta^3 + \theta^2 + 1)(\theta^4 + \theta^3 + 1)(\theta^4 + \theta^3 + \theta^2 + \theta + 1)$

Taking logarithms of the above relations leads to the following linear system. We use the notation $d_i = \text{ind}_g(w_i)$.

$$
\begin{pmatrix}
0 & 0 & 3 & 2 & 0 & 1 & 0 & 0 \\
0 & 2 & 2 & 1 & 1 & 0 & 0 & 1 \\
5 & 4 & 0 & 1 & 1 & 0 & 0 & 0 \\
0 & 7 & 3 & 1 & 0 & 0 & 0 & 0 \\
1 & 0 & 0 & 3 & 0 & 0 & 1 & 0 \\
0 & 1 & 1 & 3 & 1 & 0 & 0 & 0 \\
1 & 0 & 1 & 0 & 3 & 1 & 0 & 0 \\
0 & 3 & 1 & 0 & 1 & 0 & 1 & 1
\end{pmatrix}
\begin{pmatrix}
d_1 \\ d_2 \\ d_3 \\ d_4 \\ d_5 \\ d_6 \\ d_7 \\ d_8
\end{pmatrix}
\equiv
\begin{pmatrix}
73162 \\ 87648 \\ 18107 \\ 31589 \\ 26426 \\ 74443 \\ 29190 \\ 109185
\end{pmatrix}
\pmod{131071}.
$$

Solving the system gives the indices of the elements of the factor base:

$$
\left.
\begin{aligned}
d_1 &\equiv \text{ind}_g(w_1) \equiv & \text{ind}_g(\theta) & \equiv 71571 \\
d_2 &\equiv \text{ind}_g(w_2) \equiv & \text{ind}_g(\theta + 1) & \equiv 31762 \\
d_3 &\equiv \text{ind}_g(w_3) \equiv & \text{ind}_g(\theta^2 + \theta + 1) & \equiv 5306 \\
d_4 &\equiv \text{ind}_g(w_4) \equiv & \text{ind}_g(\theta^3 + \theta + 1) & \equiv 55479 \\
d_5 &\equiv \text{ind}_g(w_5) \equiv & \text{ind}_g(\theta^3 + \theta^2 + 1) & \equiv 2009 \\
d_6 &\equiv \text{ind}_g(w_6) \equiv & \text{ind}_g(\theta^4 + \theta + 1) & \equiv 77357 \\
d_7 &\equiv \text{ind}_g(w_7) \equiv & \text{ind}_g(\theta^4 + \theta^3 + \theta^2 + \theta + 1) & \equiv 50560 \\
d_8 &\equiv \text{ind}_g(w_8) \equiv & \text{ind}_g(\theta^4 + \theta^3 + 1) & \equiv 87095
\end{aligned}
\right\} \pmod{131071}.
$$

In the second stage, we compute the index of $a(\theta) = \theta^{15} + \theta^7 + 1$. For the choice $\alpha = 3316$, the element $a(\theta)g(\theta)^\alpha$ factors completely over B.

$$
\begin{aligned}
a(\theta)g(\theta)^{3316} &= \theta^{16} + \theta^{14} + \theta^9 + \theta^6 + \theta^4 + \theta^3 + \theta^2 + 1 \\
&= (\theta + 1)^3 (\theta^2 + \theta + 1)^3 (\theta^3 + \theta^2 + 1)(\theta^4 + \theta^3 + 1).
\end{aligned}
$$

This gives

$$\begin{aligned}
\operatorname{ind}_g a &\equiv -3316 + 3d_2 + 3d_3 + d_5 + d_8 \\
&\equiv -3316 + 3 \times 31762 + 3 \times 5306 + 2009 + 87095 \\
&\equiv 65921 \ (\mathrm{mod}\ 131071).
\end{aligned}$$

One can verify that $(\theta^7 + \theta^5 + \theta^3 + \theta)^{65921} = \theta^{15} + \theta^7 + 1$. $\qquad \square$

I now deduce the optimal running time of this basic index calculus method. In the first stage, α is chosen randomly from $\{1, 2, \ldots, 2^n - 2\}$, and accordingly, g^α is a random element of $\mathbb{F}_{2^n}^*$, that is, a polynomial of degree $O(n)$. We take $m = c\sqrt{n \ln n}$ for some positive real constant c. By Corollary 7.12, the probability that all irreducible factors of g^α have degrees $\leqslant m$ is $L\left[-\frac{1}{2c}\right]$, that is, $L\left[\frac{1}{2c}\right]$ random values of α need to be tried for obtaining a single relation.

By Corollary 3.7, the total number of irreducible polynomials in $\mathbb{F}_2[x]$ of degree k is nearly $2^k/k$. Therefore, the size of the factor base is $t = |B| \approx \sum_{k=1}^m 2^k/k$. Evidently, $2^m/m \leqslant t \leqslant m 2^m$, that is, $t = \exp\left((\ln 2 + o(1))m\right)$. Putting $m = c\sqrt{n \ln n}$ gives $t = \exp\left((c \ln 2 + o(1))\sqrt{n \ln n}\right) = L[c \ln 2]$. Since s relations (with $t \leqslant s \leqslant 2t$) need to be generated, an expected number of $L\left[\frac{1}{2c} + c \ln 2\right]$ random values of α need to be tried. Each such trial involves factoring $g(\theta)^\alpha$. We have polynomial-time (randomized) algorithms for factoring polynomials over finite fields (trial division by $L[c \ln 2]$ elements of B is rather costly), so the relation-collection stage runs in $L\left[\frac{1}{2c} + c \ln 2\right]$ time. The quantity $\frac{1}{2c} + c \ln 2$ is minimized for $c = 1/\sqrt{2 \ln 2}$, which leads to a running time of $L[\sqrt{2 \ln 2}] = L[1.1774\ldots]$ for the relation-collection phase.

The size of the factor base is $t = L\left[\sqrt{\frac{\ln 2}{2}}\right] = L[0.5887\ldots]$. Each relation contains at most $O(m)$ irreducible polynomials, so the resulting system of congruences is sparse and can be solved in (essentially) quadratic time, that is, in time $L[\sqrt{2 \ln 2}]$—the same as taken by the relation-collection phase.

The second stage involves obtaining a single relation (one smooth value of ag^α), and can be accomplished in expected time $L\left[\frac{1}{2c}\right] = L\left[\sqrt{\frac{\ln 2}{2}}\right] = L[0.5887\ldots]$, that is, much faster than the first stage.

7.3.1.1 A Faster Relation-Collection Strategy

Blake et al.[12] propose a heuristic trick to speed up the relation-collection stage. Although their trick does not improve the asymptotic running time, it significantly reduces the number of iterations for obtaining each relation. Let me explain this trick in connection with the first stage. An analogous improvement applies to the second stage too.

---

[12]Ian F. Blake, Ryoh Fuji-Hara, Ronald C. Mullin and Scott A. Vanstone, Computing logarithms in finite fields of characteristic two, *SIAM Journal of Algebraic and Discrete Methods*, 5, 276–285, 1984.

Let us denote $h(\theta) = g(\theta)^\alpha$ for a randomly chosen α. Instead of making an attempt to factor $h(\theta)$, we first run an extended gcd calculation on $f(x)$ and $h(x)$ in $\mathbb{F}_2[x]$. The gcd loop maintains an invariance $u_i(x)f(x) + v_i(x)h(x) = r_i(x)$ for polynomials $u_i(x), v_i(x), r_i(x) \in \mathbb{F}_2[x]$. Initially, $v_i(x) = 1$ is of low degree, and $r_i(x) = h(x)$ is of degree $O(n)$. Eventually, the remainder polynomial $r_i(x)$ becomes 1 (low degree), and $v_i(x)$ attains a degree $O(n)$. Somewhere in the middle, we expect to have $\deg v_i(x) \approx \deg r_i(x) \approx n/2$. When this happens, we stop the extended gcd loop. Since $f(\theta) = 0$, we get $v_i(\theta)h(\theta) = r_i(\theta)$, that is, $h(\theta) = v_i(\theta)^{-1}r_i(\theta)$ with both v_i and r_i of degrees nearly $n/2$.

We replace the attempt to factor $h(\theta)$ by an attempt to factor both $v_i(\theta)$ and $r_i(\theta)$. The probability that $h(\theta)$ is smooth with respect to B is about $p(n, m)$, whereas the probability that both $v_i(\theta)$ and $r_i(\theta)$ are smooth with respect to B is about $p(n/2, m)^2$. From Theorem 7.11, we have

$$\frac{p(n/2, m)^2}{p(n, m)} \approx \exp\left((1 + o(1))\frac{n}{m}\ln 2\right) \approx 2^{n/m}.$$

This means that we expect to obtain smooth values of $r_i(\theta)/v_i(\theta)$ about $2^{n/m}$ times more often than we expect to find smooth values of $h(\theta)$. Although this factor is absorbed in the $o(1)$ term in the running time $L[\sqrt{2\ln 2}]$ of the first stage, the practical benefit of this trick is clearly noticeable.

Example 7.14 Let us continue to use the parameters of Example 7.13. I demonstrate a situation where $h(\theta)$ fails to be smooth, whereas $r_i(\theta)/v_i(\theta)$ is smooth and leads to a relation.

Suppose we choose $\alpha = 39864$. We obtain $h(\theta) = g(\theta)^\alpha = \theta^{16} + \theta^{12} + \theta^8 + \theta^7 + \theta^6 + \theta + 1 = (\theta^2 + \theta + 1)^2(\theta^{12} + \theta^{10} + \theta^8 + \theta^3 + \theta^2 + \theta + 1)$, that is, $h(\theta)$ fails to factor completely over B (the degree-twelve polynomial being irreducible).

We now compute the extended gcd of $f(x)$ and $h(x)$. Since we are interested only in the v sequence, there is no need to keep track of the u sequence. The following table summarizes the iterations of the extended gcd loop, with the first two rows ($i = 0, 1$) corresponding to the initialization step.

i	q_i	r_i	v_i
0	—	$f(x) = x^{17} + x^3 + 1$	0
1	—	$h(x) = x^{16} + x^{12} + x^8 + x^7 + x^6 + x + 1$	1
2	x	$x^{13} + x^9 + x^8 + x^7 + x^3 + x^2 + x + 1$	x
3	x^3	$x^{11} + x^{10} + x^8 + x^7 + x^5 + x^4 + x^3 + x + 1$	$x^4 + 1$
4	$x^2 + x + 1$	$x^9 + x^8 + x^7 + x^5 + x^3 + x^2 + x$	$x^6 + x^5 + x^4 + x^2 + 1$
5	$x^2 + 1$	$x^7 + x^5 + x^3 + x^2 + 1$	$x^8 + x^7 + x^5 + x^4$

We stop the extended gcd loop as soon as $\deg r_i(x) \leqslant n/2$. In this example, this happens for $i = 5$. We have the factorizations:

$$\begin{aligned} r_5(\theta) &= \theta^7 + \theta^5 + \theta^3 + \theta^2 + 1 = (\theta^3 + \theta + 1)(\theta^4 + \theta + 1), \\ v_5(\theta) &= \theta^8 + \theta^7 + \theta^5 + \theta^4 = \theta^4(\theta + 1)^2(\theta^2 + \theta + 1). \end{aligned}$$

This implies that

$$g(\theta)^{39864} = h(\theta) = r_5(\theta)/v_5(\theta)$$
$$= \theta^{-4}(\theta+1)^{-2}(\theta^2+\theta+1)^{-1}(\theta^3+\theta+1)(\theta^4+\theta+1).$$

Taking logarithm to the base $g(\theta)$ yields the relation:

$$-4d_1 - 2d_2 - d_3 + d_4 + d_6 \equiv 39864 \pmod{131071}.$$

The above extended-gcd table shows that $\deg v_i(x) + \deg r_i(x) \approx n$ for all values of i. Therefore, when $\deg r_i(x) \approx n/2$, we have $\deg v_i(x) \approx n/2$ too. This is indeed the expected behavior. However, there is no theoretical guarantee (or proof) that this behavior is exhibited in all (or most) cases. As a result, the modification of Blake et al. is only heuristic. □

7.3.2 Linear Sieve Method (LSM)

The basic index calculus method checks the smoothness of polynomials of degrees nearly n. Blake et al.'s modification speeds this up by checking pairs of polynomials for smoothness, each having degree about $n/2$. The linear sieve method, an adaptation of the quadratic sieve method for factoring integers or the linear sieve method for prime fields, checks single polynomials of degrees about $n/2$ for smoothness. This gives the linear sieve method a running time of $L[0.8325\ldots]$, better than $L[1.1774\ldots]$ time taken by the basic method.

We assume that the defining polynomial $f(x)$ is of the special form $x^n + f_1(x)$, where $f_1(x)$ is a polynomial of low degree. For the linear sieve method, it suffices to have $f_1(x)$ of degree no more than $n/2$. Density estimates for irreducible polynomials in $\mathbb{F}_2[x]$ guarantee that we expect to find irreducible polynomials of the form $x^n + f_1(x)$ with $\deg f_1(x)$ as low as $O(\log n)$.

Let $\nu = \lceil n/2 \rceil$. We consider a factor base B with two parts: B_1 containing non-constant irreducible polynomials of $\mathbb{F}_2[x]$ having degrees $\leqslant m$ (the choice of m to be made precise later), and B_2 containing polynomials of the form $x^\nu + c(x)$ with polynomials $c(x) \in \mathbb{F}_2[x]$ of degrees $< m$ (or $\leqslant m$). The size of B_1 is between $2^m/m$ and $m2^m$, as deduced earlier. The size of B_2 is 2^m (or 2^{m+1} if we allow $\deg c(x) = m$). To sum up, $|B| = O^{\sim}(2^m)$.

Let us multiply two polynomials $x^\nu + c_1(x)$ and $x^\nu + c_2(x)$ from B_2:

$$(x^\nu + c_1(x))(x^\nu + c_2(x))$$
$$\equiv x^{2\nu} + (c_1(x)+c_2(x))x^\nu + c_1(x)c_2(x)$$
$$\equiv x^\epsilon f_1(x) + (c_1(x)+c_2(x))x^\nu + c_1(x)c_2(x) \pmod{f(x)}.$$

Here, $\epsilon = 2\lceil n/2 \rceil - n$ is 0 or 1 (depending on whether n is even or odd). Let us call $T(c_1, c_2) = x^\epsilon f_1(x) + (c_1(x)+c_2(x))x^\nu + c_1(x)c_2(x)$. If $m \ll n$ (in fact, $m = O(\sqrt{n \ln n})$, as we see later), $T(c_1, c_2)$ is of degree slightly larger than

$n/2$. We try to factor $T(c_1, c_2)$ completely over the irreducible polynomials of B_1. If the factoring attempt is successful, we get a *relation*:

$$(x^\nu + c_1(x))(x^\nu + c_2(x))$$
$$\equiv x^\epsilon f_1(x) + (c_1(x) + c_2(x))x^\nu + c_1(x)c_2(x)$$
$$\equiv \prod_{w \in B_1} w(x)^{\gamma_w} \pmod{f(x)},$$

or, equivalently,

$$(\theta^\nu + c_1(\theta))(\theta^\nu + c_2(\theta)) = \theta^\epsilon f_1(\theta) + (c_1(\theta) + c_2(\theta))\theta^\nu + c_1(\theta)c_2(\theta) = \prod_{w \in B_1} w(\theta)^{\gamma_w}.$$

Taking logarithm, we get

$$\mathrm{ind}_{g(\theta)}(\theta^\nu + c_1(\theta)) + \mathrm{ind}_{g(\theta)}(\theta^\nu + c_2(\theta)) \equiv \sum_{w \in B_1} \gamma_w \, \mathrm{ind}_{g(\theta)}(w(\theta)) \pmod{2^n - 1}.$$

This is a linear congruence in the indices of the elements of the factor base $B = B_1 \cup B_2$. We assume that $g(x)$ itself is an irreducible polynomial of small degree, that is, $g(x) = w_k(x) \in B_1$ for some k. This gives us a *free* relation:

$$\mathrm{ind}_{g(\theta)}(w_k(\theta)) \equiv 1 \pmod{2^n - 1}.$$

As c_1 and c_2 range over all polynomials in B_2, many relations are generated. Let $t = |B|$. The parameter m should be so chosen that all (c_1, c_2) pairs lead to an expected number s of relations, satisfying $t \leqslant s \leqslant 2t$. The resulting system of linear congruences is then solved modulo $2^n - 1$. This completes the first stage of the linear sieve method.

Example 7.15 Let us take $n = 17$, and represent $\mathbb{F}_{2^{17}} = \mathbb{F}_2(\theta)$, where $\theta^{17} + \theta^3 + 1 = 0$. Here, $f_1(x) = x^3 + 1$ is of degree much smaller than $n/2$. Since $|\mathbb{F}_{2^{17}}^*| = 131071$ is a prime, every element of $\mathbb{F}_{2^{17}}^*$, other than 1, is a generator of the multiplicative group $\mathbb{F}_{2^{17}}^*$.

We have $\nu = \lceil n/2 \rceil = 9$, and $\epsilon = 2\nu - n = 1$. We take $m = 4$, that is, B_1 consists of the eight irreducible polynomials $w_1, w_2, \ldots, w_8$ of Example 7.13, whereas B_2 consists of the $2^4 = 16$ polynomials $x^9 + a_3 x^3 + a_2 x^2 + a_1 x + a_0$ with each $a_i \in \{0, 1\}$. Let us name these polynomials as $w_{9 + (a_3 a_2 a_1 a_0)_2}$, that is, $B_2 = \{w_9, w_{10}, \ldots, w_{24}\}$. It follows that $|B| = |B_1| + |B_2| = 8 + 16 = 24$.

We vary c_1 and c_2 over all polynomials of B_2. In order to avoid repetitions, we take $c_1 = w_i$ for $i = 9, 10, \ldots, 24$, and, for each i, we take $c_2 = w_j$ for $j = i, i+1, \ldots, 24$. Exactly 24 smooth polynomials $T(c_1, c_2)$ are obtained.

Let us now see how these smooth values of $T(c_1, c_2)$ lead to linear congruences. As an example, consider the relation $(x^9 + x^2 + 1)(x^9 + x^2 + x) \equiv x^2(x+1)^2(x^3 + x + 1)(x^3 + x^2 + 1) \pmod{f(x)}$. Substituting $x = \theta$ gives $(\theta^9 + \theta^2 + 1)(\theta^9 + \theta^2 + \theta) = \theta^2(\theta + 1)^2(\theta^3 + \theta + 1)(\theta^3 + \theta^2 + 1)$, that is, $w_{14}w_{15} = w_1^2 w_2^2 w_4 w_5$, that is, $2d_1 + 2d_2 + d_4 + d_5 - d_{14} - d_{15} \equiv 0 \pmod{2^{17} - 1}$, where $d_i = \mathrm{ind}_{g(\theta)}(w_i(\theta))$.

c_1	c_2	$T(c_1, c_2)$
0	0	$x^4 + x = x(x+1)(x^2+x+1)$
0	$x^3 + 1$	$x^{12} + x^9 + x^4 + x = x(x+1)^9(x^2+x+1)$
1	1	$x^4 + x + 1$
x	x	$x^4 + x^2 + x = x(x^3 + x + 1)$
x	x^2	$x^{11} + x^{10} + x^4 + x^3 + x = x(x^2+x+1)^3(x^4+x+1)$
x	$x^3 + 1$	$x^{12} + x^{10} + x^9 = x^9(x^3+x+1)$
x	$x^3 + x^2 + 1$	$x^{12} + x^{11} + x^{10} + x^9 + x^3$
		$= x^3(x^2+x+1)(x^3+x^2+1)(x^4+x^3+x^2+x+1)$
$x+1$	$x+1$	$x^4 + x^2 + x + 1 = (x+1)(x^3+x^2+1)$
$x+1$	$x^3 + x^2$	$x^{12} + x^{11} + x^{10} + x^9 + x^2 + x$
		$= x(x+1)(x^2+x+1)^2(x^3+x+1)^2$
$x+1$	$x^3 + x^2 + x$	$x^{12} + x^{11} + x^9 = x^9(x^3+x^2+1)$
x^2	x^2	x
$x^2 + 1$	$x^2 + 1$	$x+1$
$x^2 + 1$	$x^2 + x$	$x^{10} + x^9 + x^3 + x^2$
		$= x^2(x+1)^2(x^3+x+1)(x^3+x^2+1)$
$x^2 + x$	$x^2 + x$	$x^2 + x = x(x+1)$
$x^2 + x$	$x^2 + x + 1$	x^9
$x^2 + x$	$x^3 + x^2 + x + 1$	$x^{12} + x^9 + x^5 + x^4 = x^4(x+1)^4(x^4+x+1)$
$x^2 + x + 1$	$x^2 + x + 1$	$x^2 + x + 1$
$x^2 + x + 1$	x^3	$x^{12} + x^{11} + x^{10} + x^9 + x^5 + x^3 + x$
		$= x(x^3+x+1)(x^4+x+1)(x^4+x^3+1)$
$x^3 + 1$	$x^3 + 1$	$x^6 + x^4 + x + 1 = (x+1)(x^2+x+1)(x^3+x+1)$
$x^3 + x$	$x^3 + x$	$x^6 + x^4 + x^2 + x = x(x+1)(x^4+x^3+1)$
$x^3 + x$	$x^3 + x + 1$	$x^9 + x^6 + x^4 + x^3 + x^2 = x^2(x^2+x+1)^2(x^3+x+1)$
$x^3 + x$	$x^3 + x^2$	$x^{11} + x^{10} + x^6 + x^5 + x^3 + x$
		$= x(x+1)^5(x^2+x+1)(x^3+x^2+1)$
$x^3 + x^2$	$x^3 + x^2$	$x^6 + x = x(x+1)(x^4+x^3+x^2+x+1)$
$x^3 + x^2 + x$	$x^3 + x^2 + x$	$x^6 + x^2 + x = x(x^2+x+1)(x^3+x^2+1)$

All the 24 equations generated above are homogeneous, and correspond to a solution $d_i = 0$ for all i. In order to make the system non-homogeneous, we take $g(\theta) = w_k(\theta)$ for some k, $1 \leqslant k \leqslant 8$. For example, if we take $g(\theta) = \theta^2 + \theta + 1$, we obtain the free relation $d_3 \equiv 1 \pmod{2^{17} - 1}$.

The resulting 25×24 system is now solved to compute all the unknown indices d_i for $1 \leqslant i \leqslant 24$. This step is not shown here. □

For deducing the running time of the LSM, we take $m = c\sqrt{n \ln n}$. The size of B is then $L[c \ln 2]$. The degree of each $T(c_1, c_2)$ is nearly $n/2$, so $T(c_1, c_2)$ is smooth with respect to B_1 with an approximate probability of $L[-\frac{1}{4c}]$ (by Corollary 7.12). The total number of $T(c_1, c_2)$ values that are checked for smoothness is $2^m(2^m + 1)/2 \approx 2^{2m-1} = L[2c \ln 2]$. We, therefore, expect to obtain $L[-\frac{1}{4c} + 2c \ln 2]$ relations. In order that the number of relations is between t and $2t$, we then require $-\frac{1}{4c} + 2c \ln 2 \approx c \ln 2$, that is, $c \approx \frac{1}{2\sqrt{\ln 2}}$.

Each $T(c_1, c_2)$ can be factored in (probabilistic) polynomial time. There are about $2^{2m-1} = L[2c \ln 2]$ values of the pair (c_1, c_2), so the relation-collection stage can be completed in $L[2c \ln 2]$ time. Solving $L[c \ln 2]$ sparse congruences in $L[c \ln 2]$ variables requires the same time. To sum up, the first stage of the linear sieve method for $\mathbb{F}_{2^n}$ runs in $L[2c \ln 2] = L[\sqrt{\ln 2}] = L[0.83255\ldots]$ time.

For the fields $\mathbb{F}_p$, we analogously need to factor all $T(c_1, c_2)$ values (see Section 7.2.2). In that case, $T(c_1, c_2)$ are integers, and we do not know any easy way to factor them. Using trial division by the primes in the factor base leads to subexponential time for each $T(c_1, c_2)$. That is why we used sieving in order to reduce the amortized effort of smoothness checking.

For $\mathbb{F}_{2^n}$, on the other hand, we know good (polynomial-time) algorithms (although probabilistic, but that does not matter) for factoring each polynomial $T(c_1, c_2)$. Trial division is a significantly slower strategy, since the factor base contains a subexponential number of irreducible polynomials of small degrees. Moreover, sieving is not required to achieve the running time $L[\sqrt{\ln 2}]$, and the name linear *sieve* method for $\mathbb{F}_{2^n}$ sounds like a bad choice.

However, a kind of polynomial sieving (Exercise 7.26) can be applied to all the *sieve* algorithms for $\mathbb{F}_{2^n}$. These sieves usually do not improve upon the running time in the $L[\ ]$ notation, because they affect only the $o(1)$ terms in the running times. But, in practice, these sieves do possess the potential of significantly speeding up the relation-collection stages.

In the second stage, we need to compute individual logarithms. If we use a strategy similar to the second stage of the basic method, we spend $L[\frac{1}{2c}] = L[\sqrt{\ln 2}] = L[0.83255\ldots]$ time for each individual logarithm. This is exactly the same as the running time of the first stage. The problem with this strategy is that we now have a smaller database of indices of small irreducible polynomials, compared to the basic method ($L[\frac{1}{2\sqrt{\ln 2}}]$ instead of $L[\frac{1}{\sqrt{2 \ln 2}}]$). Moreover, we fail to exploit the database of indices of the elements of B_2.

A strategy similar to the second stage of the linear sieve method for prime fields (Section 7.2.2) can be adapted to $\mathbb{F}_{2^n}$ (solve Exercise 7.19).

7.3.3   Cubic Sieve Method (CSM)

An adaptation of the cubic sieve method for factoring integers or computing discrete logarithms in prime fields, the cubic sieve method for $\mathbb{F}_{2^n}$ generates a set of smoothness candidates $T(c_1, c_2)$, each of degree about $n/3$. The reduction in the degrees of these $T(c_1, c_2)$ values leads to an improved running time of $L[0.67977\ldots]$. The cubic sieve method for factoring integers or for index calculations in prime fields encounters the problem of solving the congruence $x^3 \equiv y^2 z \pmod{n}$ or $x^3 \equiv y^2 z \pmod{p}$. The cubic sieve method for computing indices in $\mathbb{F}_{2^n}$ does not suffer from this drawback, and is straightaway applicable for any extension degree n. The only requirement now is that the defining polynomial $f(x)$ should be of the form $f(x) = x^n + f_1(x)$ with $\deg f_1(x) \leqslant n/3$. As argued in connection with the linear sieve method for $\mathbb{F}_{2^n}$, such an $f(x)$ exists with high probability, and can be easily detected.

Let us take $\nu = \lceil n/3 \rceil$, and $\epsilon = 3 \lceil n/3 \rceil - n \in \{0, 1, 2\}$. The factor base B now consists of two parts: B_1 containing irreducible polynomials in $\mathbb{F}_2[x]$ of degrees $\leqslant m$, and B_2 containing polynomials of the form $x^\nu + c(x)$ with

$c(x) \in \mathbb{F}_2[x]$, and $\deg c(x) < m$. Consider the product:

$$(x^{\nu} + c_1(x))(x^{\nu} + c_2(x))(x^{\nu} + c_1(x) + c_2(x))$$
$$\equiv x^{3\nu} + (c_1(x)^2 + c_1(x)c_2(x) + c_2(x)^2)x^{\nu} + c_1(x)c_2(x)(c_1(x) + c_2(x))$$
$$\equiv x^{\epsilon}f_1(x) + (c_1(x)^2 + c_1(x)c_2(x) + c_2(x)^2)x^{\nu} +$$
$$c_1(x)c_2(x)(c_1(x) + c_2(x)) \pmod{f(x)}.$$

Let us denote the last polynomial expression by $T(c_1, c_2)$ (or as $T(c_1, c_2, c_3)$ with $c_3 = c_1 + c_2$). But $\deg f_1(x) \leqslant n/3$ and $m \ll n$ (indeed, $m = \mathrm{O}(\sqrt{n \ln n})$), so the degree of $T(c_1, c_2)$ is slightly larger than $n/3$. We check whether $T(c_1, c_2)$ factors completely over irreducible polynomials in $\mathbb{F}_2[x]$ of degrees $\leqslant m$ (that is, over B_1). If so, we get a *relation*:

$$(x^{\nu} + c_1(x))(x^{\nu} + c_2(x))(x^{\nu} + c_1(x) + c_2(x)) \equiv \prod_{w \in B_1} w(x)^{\gamma_w} \pmod{f(x)},$$

or equivalently,

$$(\theta^{\nu} + c_1(\theta))(\theta^{\nu} + c_2(\theta))(\theta^{\nu} + c_1(\theta) + c_2(\theta)) = \prod_{w \in B_1} w(\theta)^{\gamma_w},$$

which leads to the linear congruence

$$\mathrm{ind}_{g(\theta)}(\theta^{\nu} + c_1(\theta)) + \mathrm{ind}_{g(\theta)}(\theta^{\nu} + c_2(\theta)) + \mathrm{ind}_{g(\theta)}(\theta^{\nu} + c_1(\theta) + c_2(\theta))$$
$$\equiv \sum_{w \in B_1} \gamma_w \, \mathrm{ind}_{g(\theta)}(w(\theta)) \pmod{2^n - 1}.$$

By varying c_1 and c_2, we collect many such relations. We also assume that $g(\theta) = w_k(\theta) \in B_1$ for some k. This gives us a *free* relation

$$\mathrm{ind}_{g(\theta)}(w_k(\theta)) \equiv 1 \pmod{2^n - 1}.$$

The resulting system is solved to compute the indices of the elements of B.

Example 7.16 It is difficult to illustrate the true essence of the cubic sieve method for small values of n, as we have done in Examples 7.13 and 7.15. This is because if n is too small, the expressions $c_1^2 + c_1 c_2 + c_2^2$ and $c_1 c_2(c_1 + c_2)$ lead to $T(c_1, c_2)$ values having degrees comparable to n.

We take $n = 31$, so $\nu = \lceil n/3 \rceil = 11$, and $\epsilon = 3\nu - n = 2$. The size of $\mathbb{F}_{2^{31}}^*$ is $2^{31} - 1 = 2147483647$ which is a prime. Thus, every element of $\mathbb{F}_{2^{31}}$, other than 0 and 1, is a generator of the group $\mathbb{F}_{2^{31}}^*$. We represent $\mathbb{F}_{2^{31}}$ as $\mathbb{F}_2(\theta)$, where $\theta^{31} + \theta^3 + 1 = 0$, that is, $f_1(x) = x^3 + 1$ is of suitably small degree.

Let us choose $m = 8$. B_1 contains all non-constant irreducible polynomials of $\mathbb{F}_2[x]$ of degrees $\leqslant 8$. There are exactly 71 such polynomials. B_2 consists of the $2^8 = 256$ polynomials of the form $x^{11} + c(x)$ with $\deg c(x) < 8$. The size of the factor base is, therefore, $|B| = |B_1| + |B_2| = 71 + 256 = 327$.

We generate the polynomials $T(c_1, c_2, c_3)$ (with $c_3 = c_1 + c_2$), and check the smoothness of these polynomials over B_1. The polynomial $T(c_1, c_2, c_3)$ remains the same for any of the six permutations of the arguments c_1, c_2, c_3. To avoid repetitions, we force the condition $c_1(2) \leqslant c_2(2) \leqslant c_3(2)$ before considering $T(c_1, c_2, c_3)$. There are exactly 11,051 such tuples. Out of them, only 487 tuples lead to smooth values of $T(c_1, c_2, c_3)$. I am not going to list all these values. Let us instead look at a single demonstration. Take $c_1(x) = x^5 + x^3 + x^2 + x + 1$, and $c_2(x) = x^7 + x^6 + x^4 + x^3 + x^2$, so $c_3(x) = c_1(x) + c_2(x) = x^7 + x^6 + x^5 + x^4 + x + 1$. Note that $c_1(2) = 47$, $c_2(2) = 220$, and $c_3(2) = 243$, so, for this (c_1, c_2, c_3), the polynomial $T(c_1, c_2, c_3)$ is computed and factored.

$$
\begin{aligned}
& (x^{11} + x^5 + x^3 + x^2 + x + 1) \times (x^{11} + x^7 + x^6 + x^4 + x^3 + x^2) \times \\
& (x^{11} + x^7 + x^6 + x^5 + x^4 + x + 1) \\
\equiv\ & x^2(x^3 + 1) + (c_1^2 + c_1 c_2 + c_2^2)x^{11} + c_1 c_2(c_1 + c_2) \\
\equiv\ & x^{25} + x^{22} + x^{20} + x^{19} + x^{16} + x^{15} + x^{14} + x^{12} + x^{10} + x^9 + x^8 + \\
& x^6 + x^5 + x^4 + x^3 \\
\equiv\ & x^3(x^3 + x^2 + 1)(x^5 + x^2 + 1)(x^6 + x + 1) \times \\
& (x^8 + x^7 + x^6 + x^4 + x^3 + x^2 + 1) \ (\mathrm{mod}\ f(x)).
\end{aligned}
$$

Putting $x = \theta$ gives

$$
\begin{aligned}
& (\theta^{11} + \theta^5 + \theta^3 + \theta^2 + \theta + 1) \times (\theta^{11} + \theta^7 + \theta^6 + \theta^4 + \theta^3 + \theta^2) \times \\
& (\theta^{11} + \theta^7 + \theta^6 + \theta^5 + \theta^4 + \theta + 1) \\
=\ & \theta^3(\theta^3 + \theta^2 + 1)(\theta^5 + \theta^2 + 1)(\theta^6 + \theta + 1) \times \\
& (\theta^8 + \theta^7 + \theta^6 + \theta^4 + \theta^3 + \theta^2 + 1).
\end{aligned}
$$

If we number the polynomials in B in an analogous fashion as in Example 7.15, this relation can be rewritten as

$$
w_{119} w_{292} w_{315} = w_1^3 w_5 w_9 w_{15} w_{67}.
$$

With the notation $d_i = \mathrm{ind}_{g(\theta)}(w_i(\theta))$, we have the linear congruence

$$
3d_1 + d_5 + d_9 + d_{15} + d_{67} - d_{119} - d_{292} - d_{315} \equiv 0 \ (\mathrm{mod}\ 2^{31} - 1).
$$

All these relations are homogeneous, so we need to include a free relation. For instance, taking $g(\theta) = \theta^7 + \theta + 1 = w_{24}$ gives $d_{24} \equiv 1 \ (\mathrm{mod}\ 2^{31} - 1)$. □

Let us now deduce the running time of the cubic sieve method for $\mathbb{F}_{2^n}$. We choose $m = c\sqrt{n \ln n}$ for some constant c to be determined. The size of B is $t = L[c \ln 2]$. Since each $T(c_1, c_2)$ is a polynomial of degree $\approx n/3$, its probability of being smooth is $L[-\frac{1}{6c}]$. The total number of triples (c_1, c_2, c_3) with $c_3 = c_1 + c_2$, distinct with respect to permutations of the elements, is exactly $\frac{2}{3} \times 4^{m-1} + 2^{m-1} + \frac{1}{3} = \Theta(2^{2m}) = L[2c \ln 2]$. Thus, we expect to get $L[-\frac{1}{6c} + 2c \ln 2]$ relations. In order to obtain a solvable system (more equations than variables), we require $-\frac{1}{6c} + 2c \ln 2 \approx c \ln 2$, that is, $c \approx 1/\sqrt{6 \ln 2}$. This choice leads to a running time of $L[2c \ln 2] = L[\sqrt{2(\ln 2)/3}\,]$ for relation collection. Solving a sparse system of $L[c \ln 2]$ congruences in as many variables takes the same time. Thus, the first stage of the cubic sieve method runs in $L[\sqrt{2(\ln 2)/3}] = L[0.67977\ldots]$ time.

The second stage of the cubic sieve method for prime fields can be adapted to work for $\mathbb{F}_{2^n}$. The details are left to the reader (solve Exercise 7.20).

7.3.4   Coppersmith's Method (CM)

Coppersmith's method[13] runs in $\exp\left((c + o(1))n^{1/3}(\ln n)^{2/3}\right)$ time, and is asymptotically faster than all $L[c]$ algorithms for computing discrete logarithms in $\mathbb{F}_{2^n}$, discussed so far. These earlier algorithms generate smoothness candidates of degrees proportional to n. Speedup (reduction of c in the running time $L[c]$) was obtained by decreasing the constant of proportionality. Coppersmith's method, on the other hand, generates smoothness candidates of degrees about $n^{2/3}$. This results in an asymptotic improvement in the running time. Note, however, that although Coppersmith's method has a running time similar to that achieved by the number-field sieve method for factoring integers or for computing indices in prime fields, Coppersmith's method is not an adaptation of the number-field sieve method, nor conversely.

We assume that the defining polynomial $f(x)$ is of the special form $f(x) = x^n + f_1(x)$ with $f_1(x)$ having degree $O(\log n)$. We argued earlier that such an irreducible polynomial $f(x)$ can be found with very high probability.

Coppersmith's method requires selecting some initial parameters. First, a positive integer k is chosen such that 2^k is $O((n/\ln n)^{1/3})$. We set $h = \lfloor \frac{n}{2^k} \rfloor + 1$, so $h = \tilde{O}(n^{2/3})$. Two bounds, m and b, are also chosen. The factor base B consists of all non-constant irreducible polynomials in $\mathbb{F}_2[x]$ of degrees $\leqslant m$, whereas b stands for a bound on the degrees of $c_1(x)$ and $c_2(x)$ (see below).

A relation in the first stage of Coppersmith's method is generated as follows. We choose polynomials $c_1(x), c_2(x)$ of degrees less than b. In order to avoid duplicate relations, we require $\gcd(c_1(x), c_2(x)) = 1$. Let

$$
\begin{aligned}
T_1(x) &= c_1(x)x^h + c_2(x), \text{ and} \\
T_2(x) &\equiv T_1(x)^{2^k} \equiv T_1(x^{2^k}) \equiv c_1(x^{2^k})x^{h2^k} + c_2(x^{2^k}) \\
&\equiv c_1(x^{2^k})x^{h2^k - n}f_1(x) + c_2(x^{2^k}) \pmod{f(x)}.
\end{aligned}
$$

The degree of $T_1(x)$ is less than $b + h$, whereas the degree of $T_2(x)$ is less than $(b+1)2^k$. Since $2^k = \tilde{O}(n^{1/3})$, choosing $b = \tilde{O}(n^{1/3})$ implies that both T_1 and T_2 are of degrees about $n^{2/3}$ (recall that $h = \tilde{O}(n^{2/3})$).

If both T_1 and T_2 factor completely over the factor base B, that is, if

$$
T_1(x) = \prod_{w \in B} w(x)^{\gamma_w}, \quad \text{and} \quad T_2(x) = \prod_{w \in B} w(x)^{\delta_w},
$$

---

[13]Don Coppersmith, Fast evaluation of logarithms in fields of characteristic two, *IEEE Transactions on Information Theory*, 30, 587–594, 1984.

the condition $T_2(x) \equiv T_1(x)^{2^k} \pmod{f(x)}$ leads to the congruence

$$\sum_{w \in B} \delta_w \operatorname{ind}_{g(\theta)}(w(\theta)) \equiv 2^k \sum_{w \in B} \gamma_w \operatorname{ind}_{g(\theta)}(w(\theta)) \pmod{2^n - 1}.$$

By varying c_1 and c_2, we generate more relations than variables (the size of the factor base). Moreover, we assume that $g(\theta)$ is itself an irreducible polynomial $w_j(\theta)$ of small degree, so that we have the *free* relation $\operatorname{ind}_{g(\theta)}(w_j(\theta)) \equiv 1 \pmod{2^n - 1}$. The resulting system of linear congruences is solved to obtain the indices of the elements of the factor base.

Example 7.17 As in Example 7.16, let us choose $n = 31$ and $f(x) = x^{31} + x^3 + 1$, so that $f_1(x) = x^3 + 1$ has a suitably small degree. We take $k = 2$, so $h = \lfloor n/2^k \rfloor + 1 = 8$. We choose $m = 7$, that is, the factor base B consists of all non-constant irreducible polynomials in $\mathbb{F}_2[x]$ of degrees $\leqslant 7$. There are exactly 41 of them, that is, $|B| = 41$. Finally, we take the bound $b = 5$.

I now demonstrate how a relation can be obtained. Take $c_1(x) = x^2 + 1$, and $c_2(x) = x^4 + x^2 + x$. We compute the two smoothness candidates as

$$
\begin{aligned}
T_1(x) &= c_1(x)x^h + c_2(x) &= x^{10} + x^8 + x^4 + x^2 + x, \\
T_2(x) &= c_1(x^{2^k})x^{h2^k - n} f_1(x) + c_2(x^{2^k}) &= x^{16} + x^{12} + x^9 + x^8 + x,
\end{aligned}
$$

both of which factor completely into irreducible polynomials of degrees $\leqslant 7$:

$$
\begin{aligned}
T_1(x) &= x(x^4 + x^3 + x^2 + x + 1)(x^5 + x^4 + x^3 + x^2 + 1), \\
T_2(x) &= x(x^4 + x + 1)(x^5 + x^2 + 1)(x^6 + x + 1).
\end{aligned}
$$

Putting $x = \theta$ gives the congruence

$$d_1 + d_8 + d_9 + d_{15} \equiv 4(d_1 + d_7 + d_{14}) \pmod{2^{31} - 1},$$

where $d_i = \operatorname{ind}_{g(\theta)} w_i(\theta)$, and w_i are numbered by a scheme as in Example 7.13.

There are 512 pairs $(c_1(x), c_2(x))$ with $c_1(x)$ and $c_2(x)$ coprime and of degrees $< b = 5$. We obtain 55 relations involving $d_1, d_2, \ldots, d_{41}$. If we assume $g(\theta) = w_{28}(\theta)$, then we have the free relation $d_{28} \equiv 1 \pmod{2^{31} - 1}$. □

Let us now look at the running time of Coppersmith's method. The probability that both T_1 and T_2 are smooth is about $p(b+h, m)p((b+1)2^k, m)$. There are exactly 2^{2b-1} pairs (c_1, c_2) for which the polynomials T_1 and T_2 are computed. So the expected number of relations is $2^{2b-1}p(b+h, m)p((b+1)2^k, m)$. The factor base contains $O^{\sim}(2^m)$ irreducible polynomials. In order to obtain a system of congruences with slightly more equations than variables, we require

$$2^{2b-1}p(b+h, m)p((b+1)2^k, m) \approx 2^m.$$

We take

$$2^k = \alpha n^{1/3}(\ln n)^{-1/3}, \quad m = \beta n^{1/3}(\ln n)^{2/3}, \quad b = \gamma n^{1/3}(\ln n)^{2/3}.$$

The optimal choices $\beta = \gamma = 2^{2/3}(3 \ln 2)^{-2/3} = 0.97436\ldots$ and $\alpha = \gamma^{-1/2} = 1.01306\ldots$ lead to the running time for the relation-collection phase as:

$$\exp\left((1.35075\ldots + o(1))n^{1/3}(\ln n)^{2/3}\right).$$

A sparse system of $\tilde{O}(2^m)$ congruences can be solved in the same time.

The second stage of Coppersmith's method is somewhat involved. Suppose that we want to compute the index of $a(\theta) \in \mathbb{F}_{2^n}^*$ to the base $g(\theta)$. As in the basic method, we choose random integers α and compute $a(\theta)g(\theta)^\alpha$. Now, we have a significantly smaller factor base than in the basic method. So each $a(\theta)g(\theta)^\alpha$ is smooth with a very small probability, and finding a suitable value of α would take too much time. We instead demand $a(\theta)g(\theta)^\alpha$ to have all irreducible factors with degrees $\leqslant n^{2/3}(\ln n)^{1/3}$. The probability of success per trial is $p(n, n^{2/3}(\ln n)^{1/3}) = \exp\left((1.0986\ldots + o(1))n^{1/3}(\ln n)^{2/3}\right)$. Since each trial involves factoring a polynomial of degree $< n$ (a task that can be finished in probabilistic polynomial time), we find a desired factorization in the same subexponential time. Suppose that some α gives

$$a(\theta)g(\theta)^\alpha = \prod_{i=1}^{r} u_i(\theta)$$

with the degree of each $u_i(\theta)$ no more than $n^{2/3}(\ln n)^{1/3}$. In order to determine $\mathrm{ind}_{g(\theta)}(a(\theta))$, it suffices to compute $\mathrm{ind}_{g(\theta)}(u_i(\theta))$ for all i.

We reduce the computation of each $\mathrm{ind}_{g(\theta)}(u_i(\theta))$ to the computation of $\mathrm{ind}_{g(\theta)}(u_{i,j}(\theta))$ for some polynomials $u_{i,j}$ of degrees smaller than that of u_i. This recursive process of replacing the index of a polynomial by the indices of multiple polynomials of smaller degrees is repeated until we eventually arrive at polynomials with degrees so reduced that they belong to the factor base.

In order to explain the reduction process, suppose that we want to compute the index of a polynomial $u(\theta)$ of degree $d \leqslant n^{2/3}(\ln n)^{1/3}$. The procedure is similar to the first stage of Coppersmith's method. We first choose a positive integer k satisfying $2^k \approx \sqrt{n/d}$, and take $h = \lfloor \frac{n}{2^k} \rfloor + 1$. For relatively prime polynomials $c_1(x), c_2(x)$ of small degrees, we consider the two polynomials

$$
\begin{aligned}
T_1(x) &= c_1(x)x^h + c_2(x), \text{ and} \\
T_2(x) &\equiv T_1(x)^{2^k} \equiv c_1(x^{2^k})x^{h2^k-n}f_1(x) + c_2(x^{2^k}) \pmod{f(x)}.
\end{aligned}
$$

We choose $c_1(x)$ and $c_2(x)$ in such a manner that $T_1(x)$ is a multiple of $u(x)$. We choose a degree d', and want both $T_1(x)/u(x)$ and $T_2(x)$ to split into irreducible factors of degrees $\leqslant d'$. For a successful factorization, we have

$$\prod_i v_i(x) \equiv 2^k \left[u(x) \prod_j w_j(x) \right] \pmod{f(x)},$$

where $v_i(x)$ and $w_j(x)$ are polynomials of degrees $\leqslant d'$. But then, $\mathrm{ind}_{g(\theta)}(u(\theta))$ can be computed if the indices of all $v_i(\theta)$ and $w_j(\theta)$ are computed.

It can be shown that we can take $d' \leqslant d/1.1$. In that case, the depth of recursion becomes $O(\log n)$ (recursion continues until the polynomials of B are arrived at). Moreover, each reduction gives a branching of no more than n new index calculations. That is, the total number of intermediate polynomials created in the process is only $n^{O(\log n)} = \exp\left[c(\ln n)^2\right]$ (for a positive constant c), and so the running time of the second stage of Coppersmith's method is dominated by the initial search for a suitable $a(\theta)g(\theta)^\alpha$. We argued earlier that this step runs in $\exp\left((1.0986\ldots + o(1))n^{1/3}(\ln n)^{2/3}\right)$ time.

7.4 Algorithms for General Extension Fields

Determination of indices in arbitrary fields $\mathbb{F}_{q^n}$ (where q may be any prime or already a power of a prime, and $n > 1$) is a relevant computational question. Not many algorithms are studied in this area. It appears that many subexponential-time algorithms used for prime and characteristic-two fields can be adapted to the case of $\mathbb{F}_{q^n}$ too. However, the applicability and analysis of these adaptations have not been rigorously studied. On the contrary, a specific adaptation of the number-field sieve method has been proposed and both theoretically and experimentally studied.

7.4.1 A Basic Index Calculus Method

We represent $\mathbb{F}_{q^n}$ as an extension of $\mathbb{F}_q$ by a monic irreducible polynomial $f(x) \in \mathbb{F}_q[x]$ of degree n. Suppose that q is not too large, so that it is feasible to work with factor bases B consisting of all monic irreducible polynomials in $\mathbb{F}_q[x]$ of degrees $\leqslant D$ for some bound $D \geqslant 1$. Let us compute indices with respect to a generator g of $\mathbb{F}_{q^n}^*$. For randomly chosen $\alpha \in \{1, 2, \ldots, q-2\}$, we compute g^α as a polynomial in $\mathbb{F}_q[x]$ of degree $< n$. We attempt to factor this polynomial completely over B. A successful factoring attempt gives a *relation*:

$$g^\alpha \equiv \kappa \prod_i w_i^{\beta_i} \pmod{f(x)},$$

where $\kappa \in \mathbb{F}_q^*$, and w_i are the (monic) irreducible polynomials in the factor base B. Taking logarithm to the base g gives

$$\alpha \equiv \text{ind}_g \kappa + \sum_i \beta_i \, \text{ind}_g w_i \pmod{q^n - 1}.$$

After collecting many relations, we solve the system to compute the unknown indices $\text{ind}_g w_i$ of the factor-base elements. See below to know how to handle the *unknown* quantities $\text{ind}_g \kappa$ for $\kappa \in \mathbb{F}_q^*$.

In the second stage, the individual logarithm of $h \in \mathbb{F}_{q^n}^*$ is computed by locating one B-smooth value of hg^α.

The problem with this method is that for large values of q, it may even be infeasible to work with factor bases B containing only all the linear polynomials of $\mathbb{F}_q[x]$ (there are q of them)—a situation corresponding to $D = 1$. A suitable subset of linear polynomials may then act as the factor base.

Another problem associated with the basic method is that g^α, even if smooth over B, need not be monic. If its leading coefficient is κ, we need to include $\text{ind}_g \kappa$ too in the relation. However, this is not a serious problem. If q is not very large, it is feasible to compute indices in $\mathbb{F}_q^*$. More precisely, $g' = g^{(q^n-1)/(q-1)}$ is an element of $\mathbb{F}_q^*$, and $\text{ind}_g \kappa = \left(\frac{q^n-1}{q-1}\right) \text{ind}_{g'} \kappa$.

7.4.2 Function-Field Sieve Method (FFSM)

Proposed by Adleman[14] and improved by Adleman and Huang[15] and by Joux and Lercier[16], the function-field sieve method adapts the number-field sieve method to computing indices in $\mathbb{F}_{q^n}$. A brief intuitive overview of Adleman and Huang's variant is presented here.

We represent $\mathbb{F}_{q^n}$ by a monic irreducible polynomial $f(x) \in \mathbb{F}_q[x]$ of degree n. A bivariate polynomial

$$H(x,y) = \sum_{i=0}^{d} \sum_{j=0}^{d'-1} h_{i,j} y^i x^j \in \mathbb{F}_q[x,y]$$

is then chosen. This polynomial should satisfy the following eight conditions.

1. $H(x,y)$ is irreducible in $\bar{\mathbb{F}}_q[x,y]$, where $\bar{\mathbb{F}}_q$ is the algebraic closure of $\mathbb{F}_q$.
2. $H(x,m(x))$ is divisible by the defining polynomial $f(x)$ for some (known) univariate polynomial $m(x) \in \mathbb{F}_q[x]$.
3. $h_{d,d'-1} = 1$.
4. $h_{d,0} \neq 0$.
5. $h_{0,d'-1} \neq 0$.
6. $\sum_{i=0}^{d} h_{i,d'-1} y^i \in \mathbb{F}_q[y]$ is square-free.
7. $\sum_{j=0}^{d'-1} h_{d,j} x^j \in \mathbb{F}_q[x]$ is square-free.
8. The size of the Jacobian $\mathbb{J}_{\mathbb{F}_{q^n}}(H)$ is coprime to $(q^n - 1)/(q - 1)$.

The curve $H(x,y)$ plays an important role here. The set $\mathbb{F}_q(H)$ of all rational functions on H is called the *function field* of H (Section 4.4.2). The set of all integers in $\mathbb{F}_q(H)$ is the set $\mathbb{F}_q[H]$ of all polynomial functions on H. Because of the Condition 2 above, the function taking $y \mapsto m(x) \pmod{f(x)}$ naturally extends to a ring homomorphism $\Phi : \mathbb{F}_q[H] \to \mathbb{F}_{q^n}[x]$.

---

[14] Leonard M. Adleman, The function field sieve, *ANTS*, 108–121, 1994.

[15] Leonard M. Adleman, Ming-Deh A. Huang, Function field sieve method for discrete logarithms over finite fields, *Information and Computation*, 151(1-2), 5–16, 1999.

[16] Antoine Joux and Reynald Lercier, The function field sieve in the medium prime case, *EuroCrypt*, 254–270, 2006.

Adleman and Huang propose the following method to determine the polynomial $H(x, y)$. The defining polynomial $f(x) \in \mathbb{F}_q[x]$ is monic of degree n. The y-degree d of H is chosen as a value about $n^{1/3}$. Let $d' = \lceil n/d \rceil$, and $\delta = dd' - n < d$. For any monic polynomial $m(x) \in \mathbb{F}_q[x]$ of degree $d' \approx n^{2/3}$, we then have $x^\delta f(x) = m(x)^d + H_{d-1}(x)m(x)^{d-1} + H_{d-2}(x)m(x)^{d-2} + \cdots + H_1(x)m(x) + H_0(x)$, where each $H_i(x) \in \mathbb{F}_q[x]$ is of degree $\leqslant d - 1$. Let

$$H(x, y) = y^d + H_{d-1}(x)y^{d-1} + H_{d-2}(x)y^{d-2} + \cdots + H_1(x)y + H_0(x).$$

By construction, $H(x, m(x)) \equiv 0 \pmod{f(x)}$. More concretely, we can take $f(x) = x^n + f_1(x)$ with $\deg f_1(x) < n^{2/3}$, and $m(x) = x^{d'}$. But then, $H(x, y) = y^d + x^\delta f_1(x)$. We vary $f_1(x)$ until $H(x, y)$ satisfies the above eight conditions.

Once a suitable polynomial $H(x, y)$ is chosen, we need to choose a factor base. Let S consist of all monic irreducible polynomials of $\mathbb{F}_q[x]$ with degrees no more than about $n^{1/3}$. For $r(x), s(x) \in \mathbb{F}_q[x]$ with degrees no more than about $n^{1/3}$ and with $\gcd(r(x), s(x)) = 1$, we consider two quantities: the polynomial $r(x)m(x) + s(x)$, and the polynomial function $r(x)y + s(x) \in \mathbb{F}_q(H)$. The polynomial $rm + s$ is attempted to factor completely over S, whereas the function $ry + s$ is attempted to factor completely in $\mathbb{F}_q(H)$ over a set of *primes* of $\mathbb{F}_q(H)$ of small norms. If both the factoring attempts are successful, we get a *relation* (also called a *doubly smooth pair*). Factorization in $\mathbb{F}_q(H)$ is too difficult a topic to be elaborated in this book. It suffices for the time being to note that factoring $r(x)y + s(x)$ essentially boils down to factoring its norm which is the univariate polynomial $r(x)^d H(x, -s(x)/r(x)) \in \mathbb{F}_q[x]$. Both $r(x)m(x) + s(x)$ and this norm are polynomials of degrees no more than about $n^{2/3}$. Sieving is carried out to identify the doubly smooth pairs.

Each trial for a relation in the FFSM, therefore, checks the smoothness of two polynomials of degrees about $n^{2/3}$. This is asymptotically better than trying to find one smooth polynomial of degree proportional to n (as in LSM or CSM). The FFSM achieves a running time of $L_{q^n}[1/3, (32/9)^{1/3}]$.

7.5   Algorithms for Elliptic Curves (ECDLP)

Conventionally, elliptic-curve groups are treated as additive groups. Let E be an elliptic curve defined over the finite field $\mathbb{F}_q$. The group $E_q = E_{\mathbb{F}_q}$ of rational points on the curve E is introduced in Chapter 4. The group E_q need not be cyclic. Let $P \in E_q$. The multiples λP of P constitute a cyclic subgroup G of E_q. Given $Q \in E_q$, our task is to determine a λ for which $Q = \lambda P$, provided that $Q \in G$. We call λ the discrete logarithm or the index of Q with respect to the base P, and denote this as $\mathrm{ind}_P Q$.

No subexponential algorithms are known to compute indices in elliptic curves. Ideas inherent in the index calculus methods for finite fields are not straightaway applicable to elliptic curves. Even when they are, they are likely to be ineffective and of no practical significance. The square-root methods of Section 7.1 are the only general-purpose algorithms known to solve the elliptic-curve discrete logarithm problem.

For some special types of elliptic curves, *good* algorithms are known. For anomalous elliptic curves, the SmartASS method is a linear-time probabilistic algorithm proposed by Smart,[17] Satoh and Araki,[18] and Semaev.[19] The MOV algorithm reduces the problem of computing discrete logarithms in E_q to the problem of computing indices in the finite field $\mathbb{F}_{q^k}$ for some k. For supersingular curves, the value of k is small ($k \leqslant 6$). The SmartASS method uses p-adic logarithms, whereas the MOV reduction uses pairing.

Silverman proposes the *xedni calculus method*[20] for computing indices in elliptic curves. Its essential idea is to lift elliptic curves over prime fields $\mathbb{F}_p$ to elliptic curve over rationals $\mathbb{Q}$. The xedni calculus method has, however, been proved to be inefficient, both theoretically and practically.

In what follows, I explain the MOV reduction algorithm. The SmartASS method and the xedni calculus method are beyond the scope of this book.

7.5.1 MOV/Frey–Rück Reduction

The MOV/Frey–Rück reduction is proposed independently by Menezes, Okamoto and Vanstone,[21] and by Frey and Rück.[22] Suppose that E is an elliptic curve defined over $\mathbb{F}_q$, and we want to compute the index of $Q \in E_q$ with respect to a base $P \in E_q$ of prime order m. Let e be a pairing (like Weil or reduced Tate pairing) of points in E_{q^k}, where k is the embedding degree. We take the first argument of e from the subgroup G of E_q, generated by P. In order that e is non-degenerate, we need to take the second argument of e from a subgroup $G' \neq G$ of E_{q^k}. Since m is prime, any non-zero element of G' is a generator of G'. We choose a random generator R of G', and reduce the computation of $\mathrm{ind}_P Q$ to the computation of a discrete logarithm in the finite field $\mathbb{F}_{q^k}$. This reduction is explained in Algorithm 7.1.

[17] Nigel P. Smart, The discrete logarithm problem on elliptic curves of trace one, *Journal of Cryptology*, 12, 193–196, 1999.

[18] T. Satoh and K. Araki, Fermat quotients and the polynomial time discrete log algorithm for anomalous elliptic curves, *Commentarii Mathematici Universitatis Sancti Pauli*, 47, 81–92, 1998.

[19] Igor A. Semaev, Evaluation of discrete logarithms on some elliptic curves, *Mathematics of Computation*, 67, 353–356, 1998.

[20] Joseph H. Silverman, The xedni calculus and the elliptic curve discrete logarithm problem, *Design, Codes and Cryptography*, 20, 5–40, 2000.

[21] Alfred J. Menezes, T. Okamoto and Scott A. Vanstone, Reducing elliptic curve logarithms to a finite field, *IEEE Transactions on Information Theory*, 39, 1639–1646, 1993.

[22] Gerhard Frey and Hans-Georg Rück, A remark concerning m-divisibility and the discrete logarithm problem in the divisor class group of curves, *Mathematics of Computation*, 62, 865–874, 1994.

Algorithm 7.1: THE MOV/FREY–RÜCK REDUCTION

```
Let k be the embedding degree for E, q and m.
Let G' be a subgroup of E_{q^k} of order m, not containing P.
Choose a random non-zero element R from G'.
Let α = e(P, R).
Let β = e(Q, R).
Compute λ = ind_α β in F_{q^k}.
Return λ.
```

If $k > 1$, then R can be chosen as follows. There is no need to construct the group G' explicitly. We keep on picking random non-zero points $R' \in E_{q^k} \setminus E_q$ and setting $R = (|E_{q^k}|/m)R'$ until we find an $R \neq \mathcal{O}$. If m is close to q, this random search succeeds within only a few iterations. Finding random elements R' in E_{q^k} and computing the multiple R of R' can be done in (probabilistic) polynomial time in $k \log q$.

Let $\lambda = \text{ind}_P Q$. By the choice of G' and R, the pairing values $\alpha, \beta \in \mathbb{F}_{q^k}$ are non-degenerate. By the bilinearity of e, we have $\alpha^\lambda = e(P, R)^\lambda = e(\lambda P, R) = e(Q, R) = \beta$. This establishes the correctness of Algorithm 7.1.

For computing the discrete logarithm $\text{ind}_\alpha \beta$ in the field $\mathbb{F}_{q^k}$, one may use a subexponential algorithm like those discussed earlier in this chapter. In order that this computation is feasible, k should be small. Pairing-friendly curves (like supersingular curves), therefore, admit subexponential solutions to ECDLP. For a general curve E, on the other hand, k is rather large (of the order of q), and Algorithm 7.1 does not yield a subexponential solution.

Example 7.18 Consider the curve $E : Y^2 = X^3 + 3X$ defined over $\mathbb{F}_{43}$ (See Example 4.68). We take the base point $P = (1, 2)$ which has order 11, and want to compute $\text{ind}_P Q$, where $Q = (36, 25)$. The embedding degree in this case is $k = 2$. We represent $\mathbb{F}_{43^2}$ as $\mathbb{F}_{43}(\theta)$, where $\theta^2 + 1 = 0$. A point of order 11 outside $E_{\mathbb{F}_{43}}$ is $R = (7 + 2\theta, 37 + 33\theta)$. Miller's algorithm for Weil pairing (Algorithm 4.2) gives $\alpha = e_m(P, R) = 7 + 34\theta$, and $\beta = e_m(Q, R) = 26 + 20\theta$. In the field $\mathbb{F}_{43^2}$, we have $\text{ind}_\alpha \beta = 7$. This gives $Q = 7P$. One can easily verify that this is the correct relation between P and Q. □

Exercises

1. Let $h \in \mathbb{F}_q^*$ have order m (a divisor of $q-1$). Prove that for $a \in \mathbb{F}_q^*$, the discrete logarithm $\text{ind}_h\, a$ exists if and only if $a^m = 1$.

2. Let E be an elliptic curve defined over $\mathbb{F}_q$, and $P \in E_q$ a point of order m. Prove or disprove: For a point $Q \in E_q$, the discrete logarithm $\text{ind}_P\, Q$ exists if and only if $mQ = \mathcal{O}$.

3. Suppose that g and g' are two primitive elements of $\mathbb{F}_q^*$. Show that if one can compute discrete logarithms to the base g in $O(f(\log q))$ time, then one can also compute discrete logarithms to the base g' in $O(f(\log q))$ time. (Assume that $f(\log q)$ is a super-polynomial expression in $\log q$.)

4. Suppose that g is a primitive element of a finite field $\mathbb{F}_q$, where q is a power of 2. Prove that computing $\text{ind}_g\, a$ is polynomial-time equivalent to the computation of the parity of $\text{ind}_g\, a$.

5. Explain how the baby-step-giant-step method can be used to compute the order of an element in a finite group of size n. (Assume that the prime factorization of n is unknown.)

6. Let G be a finite group, and $g \in G$. Suppose that an $a = g^x$ is given together with the knowledge that $i \leqslant x \leqslant j$ for some known i, j. Let $k = j - i + 1$. Describe how the baby-step-giant-step method to determine x can be modified so as to use a storage for only $O(\sqrt{k})$ group elements and a time of only $O\tilde{}(\sqrt{k})$ group operations.

7. Let $n = pq$ be the product of two distinct odd primes p, q of the same bit size.
   (a) Let $g \in \mathbb{Z}_n^*$. Prove that $\text{ord}_n\, g$ divides $\phi(n)/2$.
   (b) Conclude that $g^{(n+1)/2} \equiv g^x \pmod{n}$, where $x = (p+q)/2$.
   (c) Use Exercise 7.6 to determine x. Demonstrate how you can factor n from the knowledge of x.
   (d) Prove that this factoring algorithm runs in $O\tilde{}(n^{1/4})$ time.

8. Let $g_1, g_2, \ldots, g_t, a$ belong to a finite group G. The *multi-dimensional discrete-logarithm problem* is to find integers $x_1, x_2, \ldots, x_t$ such that $a = g_1^{x_1} g_2^{x_2} \cdots g_t^{x_t}$ (if such integers exist). Some r is given such that a has the above representation with $0 \leqslant x_i < r$ for all i. Devise a baby-step-giant-step method to compute $x_1, x_2, \ldots, x_t$ using only $O\tilde{}(r^{t/2})$ group operations.

9. Discuss how the Pollard rho and lambda methods (illustrated in Examples 7.2 and 7.3 for prime fields) can be modified to work for extension fields $\mathbb{F}_{p^n}$.

10. Let γ_{ij} be the exponent of the small prime p_j in the i-th relation in the basic index calculus method for a prime field $\mathbb{F}_p$. Assume that a random integer $g^\alpha \pmod{p}$ has probability $1/p_j^h$ for being divisible by p_j^h, where p_j is a small prime, and h is a small exponent. Determine the probability that $\gamma_{ij} \neq 0$.

11. Let C be the coefficient matrix obtained in the first stage of the LSM for the prime field $\mathbb{F}_p$. Count the expected number of non-zero entries in the j-th

column of C for the following cases:

    (1) $j = 0$ (the first column corresponding to $\mathrm{ind}_g(-1)$).

    (2) $1 \leqslant j \leqslant t$ (the next t columns corresponding to the variables $\mathrm{ind}_g p_j$).

    (3) $t+1 \leqslant j \leqslant 2M+t+1$ (the last $2M+1$ columns corresponding to the variables $\mathrm{ind}_g(H+c)$).

12. Suppose that in the linear sieve method for computing discrete logarithms in $\mathbb{F}_p$, we obtain an $m \times n$ system of congruences, where $n = t + 2M + 2$, and $m = 2n$. Assume that the $T(c_1, c_2)$ values behave as random integers (within a bound). Calculate the expected number of non-zero entries in the $m \times n$ coefficient matrix. You may use the asymptotic formula that, for a positive real number x, the sum of the reciprocals of the primes $\leqslant x$ is approximately $\ln \ln x + \mathcal{B}_1$, where $\mathcal{B}_1 = 0.2614972128\ldots$ is the *Mertens constant*. (**Remark:** The expected number of non-zero entries is significantly smaller than the obvious upper bound $\mathcal{O}(m \log p)$. However, the assumption that the probability that a small prime q divides a candidate $T(c_1, c_2)$ is the same as the probability that q divides a *smooth* value of $T(c_1, c_2)$ is heuristic.)

13. Suppose that we want to adapt the dual sieve of Exercise 6.18 to the linear sieve method over prime fields. We use two sieves each on a sieving interval of length half of the sieving length in the original LSM.

    (a) Count the total number of candidates checked for smoothness by the dual sieve. Identify what problem the dual sieve faces.

    (b) Explain how you can increase the sieving length of each sieve to avoid the problem of Part (a). What new problem does this increase introduce?

14. Describe the second stage for the residue-list sieve method for prime fields $\mathbb{F}_p$. Discuss how the residue lists available from the first stage continue to be used in the second stage. Note that the second stage must run in $L[1/2]$ time.

15. Describe the second stage for the Gaussian integer method for prime fields $\mathbb{F}_p$. The running time of the second stage should be $L[1/2]$.

16. Count the number of triples (c_1, c_2, c_3) (with $c_1 + c_2 + c_3 = 0$ and $-M \leqslant c_1 \leqslant c_2 \leqslant c_3 \leqslant M$) in the cubic sieve method for prime fields $\mathbb{F}_p$.

17. Describe how sieving is done in the cubic sieve method for prime fields, in order to find all the solutions of the congruence $T(c_1, c_2, c_3) \equiv 0 \pmod{q^h}$, where q is a small prime in the factor base, and h is a small positive integer.

18. Let $N(k, m)$ denote the number of polynomials in $\mathbb{F}_2[x]$ with degree k and with all irreducible factors of degree $\leqslant m$. Also, let $I(i)$ be the number of irreducible polynomials in $\mathbb{F}_2[x]$ of degree i. Prove that

$$N(k, m) = \sum_{i=1}^{m} \sum_{r \geqslant 1} N(k - ri, i - 1) \binom{r + I(i) - 1}{r}.$$

19. Modify the second stage of the linear sieve method for prime fields to work for the fields $\mathbb{F}_{2^n}$. What is the running time of this modified second stage?

20. Modify the second stage of the cubic sieve method for prime fields to work for the fields $\mathbb{F}_{2^n}$. What is the running time of this modified second stage?

21. Prove that the number of triples (c_1, c_2, c_3) in the cubic sieve method for the field $\mathbb{F}_{2^n}$ is $\frac{2}{3} \times 4^{m-1} + 2^{m-1} + \frac{1}{3}$.

22. Propose an adaptation of the residue-list sieve method for the fields $\mathbb{F}_{2^n}$.

23. Extend the concept of large prime variation to the case of the fields $\mathbb{F}_{2^n}$.

24. Prove that there are exactly 2^{2b-1} pairs $(c_1(x), c_2(x))$ in Coppersmith's method with $\deg c_1 < b$, $\deg c_2 < b$, and $\gcd(c_1, c_2) = 1$. Exclude the pair $(0, 1)$ in your count.

25. A *gray code*[23] $G_0^{(d)}, G_1^{(d)}, \ldots, G_{2^d-1}^{(d)}$ of dimension d is the enumeration of all d-bit strings, defined recursively as follows. For $d = 1$, we have $G_0^{(1)} = 0$ and $G_1^{(1)} = 1$, whereas for $d \geqslant 2$, we have $G_k^{(d)} = \begin{cases} 0 G_k^{(d-1)} & \text{if } 0 \leqslant k < 2^{d-1}, \\ 1 G_{2^d-k-1}^{(d-1)} & \text{if } 2^{d-1} \leqslant k < 2^d. \end{cases}$
Prove that for $1 \leqslant k < 2^d$, the bit strings $G_{k-1}^{(d)}$ and $G_k^{(d)}$ differ in exactly one bit position given by $v_2(k)$ (the multiplicity of 2 in k).

26. In this exercise, we explore Gordon and McCurley's polynomial sieving procedure[24] in connection with the linear sieve method for $\mathbb{F}_{2^n}$. Let $w(x)$ be an irreducible polynomial in B of low degree, h a small positive integer, and $\delta = m - 1 - h \deg w(x)$. We find polynomials c_1, c_2 of degrees $< m$ satisfying $T(c_1, c_2) \equiv x^\epsilon f_1(x) + (c_1(x) + c_2(x))x^\nu + c_1(x)c_2(x) \equiv 0 \pmod{w(x)^h}$. Suppose that, for a fixed c_1, a solution of this congruence is $\bar{c}_2(x)$. Then, all the solutions for $c_2(x)$ are $c_2(x) = \bar{c}_2(x) + u(x)w(x)^h$ for all polynomials $u(x)$ of degrees $< \delta$. Describe how the δ-dimensional gray code can be used to efficiently step through all these values of $c_2(x)$. The idea is to replace the product $u(x)w(x)^h$ by more efficient operations. Complete the description of the sieve, based upon this strategy. Deduce the running time for this sieve.

27. Argue that the basic index calculus method for $\mathbb{F}_{q^n}$ (with small q) can be designed to run in $L(q^n, 1/2, c)$ time.

28. Extend the linear sieve method for characteristic-two fields to compute indices in $\mathbb{F}_{q^n}$. Assume that q is *small*.

29. How can sieving be done in the extended linear sieve method of Exercise 7.28?

30. How can you modify Coppersmith's method in order to compute indices in non-binary fields of small characteristics (like three or five)?

Programming Exercises

Using the GP/PARI calculator, implement the following.

31. The baby-step-giant-step method for prime finite fields.

32. The Pollard rho method for binary finite fields.

33. The Pollard lambda method for elliptic curves over finite fields.

---

[23] The Gray code is named after the American physicist Frank Gray (1887–1969).
[24] Daniel M. Gordon and Kevin S. McCurley, Massively parallel computation of discrete logarithms, *CRYPTO*, 312–323, 1992.

34. The sieving step for the linear sieve method for prime fields.

35. The sieving step for the linear sieve method for binary finite fields.

36. Relation generation in the basic index calculus method in the field $\mathbb{F}_{3^n}$.

Chapter 8

Large Sparse Linear Systems

Algorithms for solving large sparse linear systems over the finite rings $\mathbb{Z}_M$ constitute the basic theme for this chapter. Arguably, this is not part of number theory. However, given the importance of our ability to *quickly* solve such systems in connection with factoring and discrete-logarithm algorithms, a book on *computational* number theory has only little option to ignore this topic. The sieving phases of factoring and discrete-log algorithms are massively parallelizable. On the contrary, the linear-algebra phases resist parallelization efforts, and may turn out to be the bottleneck in practical implementations.

Throughout this chapter, we plan to solve the linear system of congruences:

$$A\mathbf{x} \equiv \mathbf{b} \ (\mathrm{mod} \ M), \tag{8.1}$$

where A is an $m \times n$ matrix with elements form $\mathbb{Z}_M$, $\mathbf{x}$ an $n \times 1$ vector, $\mathbf{b}$ an $m \times 1$ vector, and $M \geqslant 2$ an integer modulus. In earlier chapters, the letter n indicates other quantities. In Chapter 6, n was the integer to be factored, whereas for computing discrete logarithms over $\mathbb{F}_{2^n}$, the extension degree was n. In this chapter, n (and also m) are subexponential expressions in the input sizes ($\log n$ for factoring, and n for discrete logarithms in $\mathbb{F}_{2^n}$). I hope that this notational inconsistency across chapters will not be a source of confusion.

In several ways, systems generated by integer-factoring algorithms differ from those generated by discrete-log algorithms. For factoring, the system is homogeneous (that is, $\mathbf{b} = \mathbf{0}$), and $n > m$. Our goal is to obtain a non-zero solution $\mathbf{x}$ to the system. Such a solution belongs to the null space of A, which has dimension $\geqslant n - m$. If $n \gg m$ (like $n = 2m$), we expect to find many such solutions $\mathbf{x}$ that lead to non-trivial Fermat congruences capable of splitting the input integer. Finally, the modulus is $M = 2$ for factoring algorithms.

On the contrary, discrete-log algorithms yield non-homogeneous systems (that is, $\mathbf{b} \neq \mathbf{0}$). We also desire the systems to be of full (column) rank, so we arrange to have $m \gg n$ (like $m = 2n$). Finally, the modulus M is usually a large integer (like $q - 1$ for computing discrete logarithms in $\mathbb{F}_q$).

In many senses, these basic differences do not matter, both theoretically and practically. For example, suppose that we like to compute a solution $\mathbf{x}$ of the homogeneous system $A\mathbf{x} = \mathbf{0}$. For a random $\mathbf{x}_1$, we compute $\mathbf{b} = A\mathbf{x}_1$. If $\mathbf{b} = \mathbf{0}$, we are done. So assume that $\mathbf{b} \neq \mathbf{0}$. Since $A(\mathbf{x} + \mathbf{x}_1) = A\mathbf{x} + A\mathbf{x}_1 = \mathbf{b}$, this means that if we can solve the non-homogeneous system $A\mathbf{y} = \mathbf{b}$, a solution for the corresponding homogeneous system can be obtained as $\mathbf{x} = \mathbf{y} - \mathbf{x}_1$. So, without loss of generality, we assume that $\mathbf{b} \neq \mathbf{0}$ in Eqn (8.1).

Many algorithms we are going to discuss generate a square system

$$B\mathbf{x} \equiv \mathbf{c} \pmod{M} \tag{8.2}$$

by multiplying Eqn (8.1) by A^{t} (transpose of A). Here, $B = A^{\mathrm{t}}A$ is an $n \times n$ matrix, and $\mathbf{c} = A^{\mathrm{t}}\mathbf{b}$ is an $n \times 1$ vector. Clearly, a solution of Eqn (8.1) is a solution of Eqn (8.2) too. The converse is also true with high probability.

The modulus M has some bearing on the choice of the algorithm for solving the system. The so-called block methods are specifically suited to the modulus $M = 2$. If $M \geqslant 3$, we usually employ the non-block variants. It is worthwhile to note here that M need not be a prime for every system we obtain from a discrete-log algorithm. This, in turn, implies that $\mathbb{Z}_M$ is not necessarily a field. Let us see how serious the consequences of this are.

First, suppose that a complete prime factorization of the modulus M is known. We solve the system modulo the prime divisors of M, and then lift the solutions to appropriate powers of these primes. The lifting procedure is similar to Hensel's lifting described in Section 1.5, and is studied in Exercise 8.1. In some cases, a partial factorization of M is only available. For discrete logarithms in $\mathbb{F}_q$ with q odd, the modulus $q - 1$ can be written as $q - 1 = 2^s t$ with t odd. It is a good idea to also remove factors of small primes (other than 2) from M, before applying a system solver for this modulus.

If M is a modulus with unknown factorization, an effort to factor M and then use the procedure described in the last paragraph is usually too much of an investment of time. We instead *pretend* that M is a prime (that is, $\mathbb{Z}_M$ is a field), and apply a system-solver algorithm with the modulus. If, during the running of this algorithm, an attempt is made to invert a non-zero non-invertible element, the algorithm fails. But then, we also discover a non-trivial factorization $M = M_1 M_2$. We continue solving the system modulo both M_1 and M_2, pretending again that M_1 and M_2 are primes.

A standard algorithm to solve linear systems is *Gaussian elimination* which converts the coefficient matrix A to a (reduced) row echelon form, and then obtains the values of the variables by backward substitution. For an $m \times n$ system with $m = \Theta(n)$, this method takes $O(m^3)$ (or $O(n^3)$) time which is unacceptably high for most factoring and discrete-log algorithms.

A commonness of all the sieving algorithms is that the linear systems produced by them are necessarily sparse, that is, there are only a few non-zero entries in each row (or column, not necessarily both). This sparsity needs to be effectively exploited to arrive at $O^{\sim}(m^2)$ (or $O^{\sim}(n^2)$) algorithms for completing the linear-algebra phase. This chapter is an introduction to these sparse system solvers. I do not make any effort here to explain the standard cubic algorithms like Gaussian elimination, but straightaway jump to the algorithms suitable for sparse systems. Structured Gaussian elimination applies to any modulus M. The standard Lanczos and Wiedemann methods are typically used for odd moduli M, although they are equally applicable to $M = 2$. The block versions of Lanczos and Wiedemann methods are significantly more efficient compared to their standard versions, but are meant for $M = 2$ only.

The structure of a typical sparse matrix A from the sieve algorithms merits a discussion in this context. Each row of A is sparse, but the columns in A have significant variations in their *weights* (that is, counts of non-zero elements). A randomly chosen integer is divisible by a small prime p with probability about $1/p$. The column in A corresponding to p is expected to contain about m/p non-zero entries. If p is small (like $p = 2$), the column corresponding to p is quite dense. On the contrary, if p is relatively large (like the millionth prime 15,485,863), the column corresponding to it is expected to be rather sparse. The columns in A corresponding to factor-base elements other than small primes (like $H + c$ in the linear sieve method) are expected to contain only a small constant number of non-zero elements. In view of these observations, we call some of the columns *heavy*, and the rest *light*. More concretely, we may choose a small positive real constant α (like $1/32$), and call a column of A heavy if its weight is more than αm, light otherwise.

Usually, each non-zero entry in A is a small positive or negative integer, even when the modulus M is large. Here, *small* means absolute values no larger than a few hundreds. It is preferable to represent a negative entry $-a$ by $-a$ itself, and not by the canonical representative $M - a$. This practice ensures that each entry in A can be represented by a single-precision signed integer. The matrix A is typically not stored in a dense format (except perhaps for $M = 2$, in which case multiple coefficients can be packed per word). We instead store only the non-zero entries in a row-major format.

In the situations where we solve Eqn (8.2) instead of Eqn (8.1), it is often not advisable to compute $B = A^t A$ explicitly, for B may be significantly denser than A. It is instead preferable to carry out a multiplication by B as two multiplications by the sparse matrices A and A^t. While multiplying by A^t, we often find it handy to have a row-major listing of the non-zero elements of A^t, or equivalently a column-major listing of A.

8.1 Structured Gaussian Elimination

Structured Gaussian elimination (SGE) is used to reduce the dimensions of a sparse matrix by eliminating some of its rows and columns. The reduction in the size of the matrix may be quite considerable. Indeed, SGE may reduce the size of a matrix to such an extent that applying standard Gaussian elimination on the reduced system becomes more practical than applying the quadratic sparse solvers we describe later in this chapter. This is particularly important for $M = 2$, since in this case multiple coefficients can be packed per word in a natural way, and standard Gaussian elimination can operate on words, thereby processing multiple coefficients in each operation.

There are some delicate differences in the SGE procedure between the cases of factorization and discrete-log matrices. Here, I supply a unified treatment for both $m \geqslant n$ and $m \leqslant n$. For dealing with matrices specifically from factoring algorithms, one may look at the paper by Bender and Canfield.[1] Matrices from discrete-log algorithms are studied by LaMacchia and Odlyzko.[2]

During the execution of SGE, we call certain columns of A heavy, and the rest light. An initial settlement of this discrimination may be based upon the weights (the *weight* of a row or column of A is the number of non-zero entries in that row or column) of the columns in comparison with αm for a predetermined small positive fraction α (like $1/32$). Later, when rows and columns are removed, this quantitative notion of heaviness or lightness may be violated. The steps of SGE attempt to keep the light columns light, perhaps at the cost of increasing the weights of the heavy columns.

Step 1: *Delete columns of weights zero and one.*
    (a) Remove all columns of weight zero. These columns correspond to variables which do not appear in the system at all, and can be discarded altogether.
    (b) Remove all columns of weight one and the rows containing these non-zero entries. Each such column refers to a variable that appears in exactly one equation. When the values of other variables are available, the value of this variable can be obtained from the equation (that is, row) being eliminated.

After the completion of Step 1, all columns have weights $\geqslant 2$. Since the matrix has potentially lost many light columns, it may be desirable to declare some light columns as heavy. The obvious choices are those having the highest weights among the light columns. This may be done so as to maintain the heavy-vs-light discrimination based upon the fraction α. Notice that the value of m (the number of rows) reduces after every column removal in Step 1(b).

Step 2: *Delete rows of weights zero and one.*
    (a) A row of weight zero stands for the equation $0 = 0$, and can be eliminated. Although such a row may be absent in the first round of the steps of SGE, they may appear in later rounds.
    (b) Let the row R_i contain only one non-zero entry. Suppose that this entry corresponds to the variable x_j. This row supplies the value of x_j (we assume that the non-zero entry is invertible modulo M). Substitute this value of x_j in all the equations (rows) where x_j occurs. Delete R_i and the column (light or heavy) corresponding to x_j. Repeat for all rows of weight one.

---

[1] Edward A. Bender and E. Rodney Canfield, An approximate probabilistic model for structured Gaussian elimination, *Journal of Algorithms*, 31(2), 271–290, 1999.

[2] Brian A. LaMacchia and Andrew M. Odlyzko, Solving large sparse linear systems over finite fields, *CRYPTO*, 109–133, 1991.

Step 3: *Delete rows of weight one in the light part.*

Consider each row R_i whose intersection with all the light columns contains exactly one non-zero entry. Let this entry correspond to the variable x_j. The row R_i allows x_j to be written in terms of some variables corresponding to heavy columns. In all the equations (rows) where x_j occurs, substitute this value of x_j to eliminate x_j from these other rows. Then, remove the row R_i and the column corresponding to the variable x_j.

After all the appropriate rows for Steps 2 and 3 (and the corresponding columns) are removed, it may be necessary to redefine heavy and light columns as done at the end of Step 1.

Step 4: *Delete redundant rows.*

If the matrix contains more rows than columns (as is typical in the case of discrete-log matrices), we may throw away some *redundant* rows. Rows with the largest number of non-zero entries in the light columns are good candidates for removal, because their removal makes the light part even lighter.

After Steps 2–4 are executed, some columns and/or rows may be exposed as having weights one and zero. It is, therefore, necessary to repeat the round of above steps until no further reduction is possible.

Example 8.1 Let me demonstrate the working of structured Gaussian elimination on a system of 25 equations in 25 variables. This system is not generated by any sieve method. The randomly generated entries ensure that the first few columns are heavy. We start with the following set of equations. All equations in this example are over $\mathbb{F}_2$, that is, congruences modulo $M = 2$.

$$
\begin{aligned}
x_{18} &= 1 \\
x_1 + x_8 + x_{17} &= 0 \\
x_1 + x_3 + x_6 + x_{11} + x_{23} &= 1 \\
x_2 + x_5 + x_{23} &= 1 \\
x_3 + x_6 + x_{21} + x_{22} + x_{23} &= 1 \\
x_2 + x_3 + x_{13} + x_{21} &= 0 \\
x_1 + x_2 + x_7 &= 1 \\
x_{22} &= 1 \\
x_1 + x_2 + x_5 + x_6 + x_9 + x_{10} + x_{21} &= 0 \\
x_1 + x_3 + x_{14} + x_{18} &= 0 \\
x_1 + x_2 &= 1 \\
x_1 + x_5 &= 1 \\
x_3 + x_4 + x_5 + x_{16} + x_{24} &= 0 \\
x_1 + x_3 + x_{13} + x_{20} &= 1 \\
x_1 + x_4 + x_6 + x_{13} + x_{14} + x_{24} &= 0 \\
x_2 &= 1
\end{aligned}
$$

$$x_3 + x_{23} = 0$$
$$x_4 + x_{10} + x_{16} + x_{20} = 0$$
$$x_1 + x_6 + x_{11} + x_{24} = 0$$
$$x_{10} + x_{11} + x_{16} + x_{21} = 1$$
$$x_1 + x_2 + x_3 + x_4 + x_{11} + x_{21} = 0$$
$$x_1 + x_2 + x_4 + x_{17} = 0$$
$$x_1 = 0$$
$$x_1 + x_4 + x_{18} = 0$$
$$x_1 + x_2 + x_{25} = 0$$

Below, the system is written in the matrix form. Only the reduced matrix A (initially the 25×25 coefficient matrix) and the reduced vector $\mathbf{b}$ are shown.

	1	2	3	4	5	6	7	8	9	10	11	12	13	14	15	16	17	18	19	20	21	22	23	24	25		b
	H	H	H	H	L	H	L	L	L	L	L	L	L	L	L	L	L	L	L	L	L	H	L	L	L		
1	0	0	0	0	0	0	0	0	0	0	0	0	0	0	0	0	0	1	0	0	0	0	0	0	0		1
2	1	0	0	0	0	0	0	1	0	0	0	0	0	0	0	0	1	0	0	0	0	0	0	0	0		0
3	1	0	1	0	0	1	0	0	0	0	1	0	0	0	0	0	0	0	0	0	0	0	1	0	0		1
4	0	1	0	0	1	0	0	0	0	0	0	0	0	0	0	0	0	0	0	0	0	0	1	0	0		1
5	0	0	1	0	0	1	0	0	0	0	0	0	0	0	0	0	0	0	0	0	1	1	1	0	0		1
6	0	1	1	0	0	0	0	0	0	0	0	0	1	0	0	0	0	0	0	0	1	0	0	0	0		0
7	1	1	0	0	0	0	1	0	0	0	0	0	0	0	0	0	0	0	0	0	0	0	0	0	0		1
8	0	0	0	0	0	0	0	0	0	0	0	0	0	0	0	0	0	0	0	0	1	0	0	0	0		1
9	1	1	0	0	1	1	0	0	1	1	0	0	0	0	0	0	0	0	0	0	1	0	0	0	0		0
10	1	0	1	0	0	0	0	0	0	0	0	0	0	1	0	0	0	1	0	0	0	0	0	0	0		0
11	1	1	0	0	0	0	0	0	0	0	0	0	0	0	0	0	0	0	0	0	0	0	0	0	0		1
12	1	0	0	0	1	0	0	0	0	0	0	0	0	0	0	0	0	0	0	0	0	0	0	0	0		1
13	0	0	1	1	1	0	0	0	0	0	0	0	0	0	0	1	0	0	0	0	0	0	0	1	0		0
14	1	0	1	0	0	0	0	0	0	0	0	0	1	0	0	0	0	0	1	0	0	0	0	0	0		1
15	1	0	0	1	0	1	0	0	0	0	0	0	1	1	0	0	0	0	0	0	0	0	0	1	0		0
16	0	1	0	0	0	0	0	0	0	0	0	0	0	0	0	0	0	0	0	0	0	0	0	0	0		1
17	0	0	1	0	0	0	0	0	0	0	0	0	0	0	0	0	0	0	0	0	0	1	0	0	0		0
18	0	0	0	1	0	0	0	0	0	1	0	0	0	0	0	1	0	0	0	1	0	0	0	0	0		0
19	1	0	0	0	0	1	0	0	0	0	1	0	0	0	0	0	0	0	0	0	0	0	0	1	0		0
20	0	0	0	0	0	0	0	0	0	1	1	0	0	0	0	1	0	0	0	0	1	0	0	0	0		1
21	1	1	1	1	1	0	0	0	0	0	0	1	0	0	0	0	0	0	0	0	0	1	0	0	0		0
22	1	1	0	1	0	0	0	0	0	0	0	0	0	0	0	0	1	0	0	0	0	0	0	0	0		0
23	1	0	0	0	0	0	0	0	0	0	0	0	0	0	0	0	0	0	0	0	0	0	0	0	0		0
24	1	0	0	1	0	0	0	0	0	0	0	0	0	0	0	0	1	0	0	0	0	0	0	0	0		0
25	1	1	0	0	0	0	0	0	0	0	0	0	0	0	0	0	0	0	0	0	0	0	0	0	1		0

The table is headed A (with column headers 1–25 and the heaviness row H/L) and $\mathbf{b}$.

The indices of the undeleted rows and columns are shown as row and column headers. Let us take the heaviness indicator fraction as $\alpha = 1/5$, that is, a column is called heavy if and only if contains at least $\alpha m = m/5$ non-zero entries, where m is the current count of rows in A. Heavy and light columns are marked by H and L respectively.

Round 1

Step 1(a): Columns indexed 12, 15 and 19 have no non-zero entries, that is, the variables x_{12}, x_{15} and x_{19} appear in no equations. We cannot solve for these variables form the given system. Eliminating these columns reduces the system to the following:

	A																						b
	1	2	3	4	5	6	7	8	9	10	11	13	14	16	17	18	20	21	22	23	24	25	
	H	H	H	H	L	H	L	L	L	L	L	L	L	L	L	L	L	H	L	L	L	L	
1	0	0	0	0	0	0	0	0	0	0	0	0	0	0	0	1	0	0	0	0	0	0	1
2	1	0	0	0	0	0	0	1	0	0	0	0	0	0	1	0	0	0	0	0	0	0	0
3	1	0	1	0	0	1	0	0	0	0	1	0	0	0	0	0	0	0	0	1	0	0	1
4	0	1	0	0	1	0	0	0	0	0	0	0	0	0	0	0	0	0	0	1	0	0	1
5	0	0	1	0	0	1	0	0	0	0	0	0	0	0	0	0	0	1	1	1	0	0	1
6	0	1	1	0	0	0	0	0	0	0	0	1	0	0	0	0	0	1	0	0	0	0	0
7	1	1	0	0	0	0	1	0	0	0	0	0	0	0	0	0	0	0	0	0	0	0	1
8	0	0	0	0	0	0	0	0	0	0	0	0	0	0	0	0	0	0	1	0	0	0	1
9	1	1	0	0	1	1	0	0	1	1	0	0	0	0	0	0	0	1	0	0	0	0	0
10	1	0	1	0	0	0	0	0	0	0	0	0	1	0	0	1	0	0	0	0	0	0	0
11	1	1	0	0	0	0	0	0	0	0	0	0	0	0	0	0	0	0	0	0	0	0	1
12	1	0	0	0	1	0	0	0	0	0	0	0	0	0	0	0	0	0	0	0	0	0	1
13	0	0	1	1	1	0	0	0	0	0	0	0	0	1	0	0	0	0	0	0	1	0	0
14	1	0	1	0	0	0	0	0	0	0	0	1	0	0	0	0	1	0	0	0	0	0	1
15	1	0	0	1	0	1	0	0	0	0	0	1	1	0	0	0	0	0	0	0	1	0	0
16	0	1	0	0	0	0	0	0	0	0	0	0	0	0	0	0	0	0	0	0	0	0	1
17	0	0	1	0	0	0	0	0	0	0	0	0	0	0	0	0	0	0	0	1	0	0	0
18	0	0	0	1	0	0	0	0	0	1	0	0	0	1	0	0	1	0	0	0	0	0	0
19	1	0	0	0	0	1	0	0	0	0	1	0	0	0	0	0	0	0	0	0	1	0	0
20	0	0	0	0	0	0	0	0	0	1	1	0	0	1	0	0	0	1	0	0	0	0	1
21	1	1	1	1	0	0	0	0	0	0	1	0	0	0	0	0	0	1	0	0	0	0	0
22	1	1	0	1	0	0	0	0	0	0	0	0	0	0	1	0	0	0	0	0	0	0	0
23	1	0	0	0	0	0	0	0	0	0	0	0	0	0	0	0	0	0	0	0	0	0	0
24	1	0	0	1	0	0	0	0	0	0	0	0	0	0	0	1	0	0	0	0	0	0	0
25	1	1	0	0	0	0	0	0	0	0	0	0	0	0	0	0	0	0	0	0	0	1	0

Step 1(b): The columns with single non-zero entries are indexed $7, 8, 9, 25$. We delete these columns and the rows containing these non-zero entries (rows $7, 2, 9, 25$). We will later make the following substitutions:

$$x_7 = 1 + x_1 + x_2, \tag{8.3}$$

$$x_8 = 0 + x_1 + x_{17}, \tag{8.4}$$

$$x_9 = 0 + x_1 + x_2 + x_5 + x_6 + x_{10} + x_{21}, \tag{8.5}$$

$$x_{25} = 0 + x_1 + x_2. \tag{8.6}$$

The system reduces as follows:

	1	2	3	4	5	6	10	11	13	14	16	17	18	20	21	22	23	24	b
	H	H	H	H	L	L	L	L	L	L	L	L	L	L	L	L	L	L	
1	0	0	0	0	0	0	0	0	0	0	0	0	1	0	0	0	0	0	1
3	1	0	1	0	0	1	0	1	0	0	0	0	0	0	0	0	1	0	1
4	0	1	0	0	1	0	0	0	0	0	0	0	0	0	0	0	1	0	1
5	0	0	1	0	0	1	0	0	0	0	0	0	0	0	1	1	1	0	1
6	0	1	1	0	0	0	0	0	1	0	0	0	0	0	1	0	0	0	0
8	0	0	0	0	0	0	0	0	0	0	0	0	0	0	0	1	0	0	1
10	1	0	1	0	0	0	0	0	0	1	0	0	1	0	0	0	0	0	0
11	1	1	0	0	0	0	0	0	0	0	0	0	0	0	0	0	0	0	1
12	1	0	0	0	1	0	0	0	0	0	0	0	0	0	0	0	0	0	1
13	0	0	1	1	1	0	0	0	0	0	1	0	0	0	0	0	0	1	0
14	1	0	1	0	0	0	0	0	1	0	0	0	0	1	0	0	0	0	1
15	1	0	0	1	0	1	0	0	1	1	0	0	0	0	0	0	0	1	0
16	0	1	0	0	0	0	0	0	0	0	0	0	0	0	0	0	0	0	1
17	0	0	1	0	0	0	0	0	0	0	0	0	0	0	0	1	0	0	0
18	0	0	0	1	0	0	1	0	0	0	1	0	0	1	0	0	0	0	0
19	1	0	0	0	0	1	0	1	0	0	0	0	0	0	0	0	0	1	0
20	0	0	0	0	0	0	1	1	0	0	1	0	0	0	1	0	0	0	1
21	1	1	1	1	0	0	0	1	0	0	0	0	0	0	1	0	0	0	0
22	1	1	0	1	0	0	0	0	0	0	0	1	0	0	0	0	0	0	0
23	1	0	0	0	0	0	0	0	0	0	0	0	0	0	0	0	0	0	0
24	1	0	0	1	0	0	0	0	0	0	0	0	1	0	0	0	0	0	0

The top of the matrix is labelled A (over the coefficient columns) and b (over the right-hand column).

Step 2(a): At this instant, there are no zero rows, so this step is not executed.

Step 2(b): We look at rows with single non-zero entries. Row 1 is the first case with the sole non-zero entry at Column 18. We get the immediate solution

$$x_{18} = 1 \tag{8.7}$$

The variable x_{18} occurs in Rows 10 and 24. We substitute the value of x_{18} in these equations, and delete Row 1 and Column 18. This step is subsequently repeated for the following rows of weight one.

Row index	Column index	Rows adjusted
8	22	5
16	2	$4, 6, 11, 21, 22$
23	1	$3, 10, 11, 12, 14, 15, 19, 21, 22, 24$
24	4	$13, 15, 18, 21, 22$

In the process, we obtain the solutions for the following variables:

$$x_{22} = 1 \tag{8.8}$$
$$x_2 = 1 \tag{8.9}$$

$$x_1 \;=\; 0 \tag{8.10}$$
$$x_4 \;=\; 1 \tag{8.11}$$

After all these five iterations of Step 2(b), the system reduces to:

	3	5	6	10	11	13	14	16	17	20	21	23	24		b
	H	L	H	L	H	L	L	L	L	L	H	H	L		
3	1	0	1	0	1	0	0	0	0	0	0	1	0		1
4	0	1	0	0	0	0	0	0	0	0	0	1	0		0
5	1	0	1	0	0	0	0	0	0	0	1	1	0		0
6	1	0	0	0	0	1	0	0	0	0	1	0	0		1
10	1	0	0	0	0	0	1	0	0	0	0	0	0		1
11	0	0	0	0	0	0	0	0	0	0	0	0	0		0
12	0	1	0	0	0	0	0	0	0	0	0	0	0		1
13	1	1	0	0	0	0	0	1	0	0	0	0	1		1
14	1	0	0	0	0	1	0	0	0	1	0	0	0		1
15	0	0	1	0	0	1	1	0	0	0	0	0	1		1
17	1	0	0	0	0	0	0	0	0	0	0	1	0		0
18	0	0	0	1	0	0	0	1	0	1	0	0	0		1
19	0	0	1	0	1	0	0	0	0	0	0	0	1		0
20	0	0	0	1	1	0	0	1	0	0	1	0	0		1
21	1	0	0	0	1	0	0	0	0	0	1	0	0		0
22	0	0	0	0	0	0	0	1	0	0	0	0	0		0

The columns are relabeled as heavy/light after all the iterations of Step 2(b). Now, there are 16 rows, so columns with at most three non-zero entries are light, and those with more than three entries are heavy.

Step 3: The first row to contain only one non-zero entry in the light columns is Row 4 with that entry being in Column 5. This gives us the following future possibility of back substitution:

$$x_5 = 0 + x_{23}. \tag{8.12}$$

We substitute this expression for x_5 in all other equations where x_5 appears. Here, these other equations are 12 and 13. We effect this substitution by a subtraction (same as addition modulo 2) of Row 4 from Rows 12 and 13. The modified system is shown below. Notice that the subtraction of (multiples of) the row being deleted from other rows may change the weights of some columns other than the one being deleted. Consequently, it is necessary to relabel each column (or at least the columns suffering changes) as heavy/light, and restart the search for rows with single non-zero entries in the new light columns.

	3	6	10	11	13	14	16	17	20	21	23	24	b
	H	*H*	*L*	*H*	*H*	*L*	*H*	*L*	*L*	*H*	*H*	*H*	
3	1	1	0	1	0	0	0	0	0	0	1	0	1
5	1	1	0	0	0	0	0	0	0	1	1	0	0
6	1	0	0	0	1	0	0	0	0	1	0	0	1
10	1	0	0	0	0	1	0	0	0	0	0	0	1
11	0	0	0	0	0	0	0	0	0	0	0	0	0
12	0	0	0	0	0	0	0	0	0	0	1	0	1
13	1	0	0	0	0	0	1	0	0	0	1	1	1
14	1	0	0	0	1	0	0	0	1	0	0	0	1
15	0	1	0	0	1	1	0	0	0	0	0	1	1
17	1	0	0	0	0	0	0	0	0	0	1	0	0
18	0	0	1	0	0	0	1	0	1	0	0	0	1
19	0	1	0	1	0	0	0	0	0	0	0	1	0
20	0	0	1	1	0	0	1	0	0	1	0	0	1
21	1	0	0	1	0	0	0	0	0	0	1	0	0
22	0	0	0	0	0	0	0	1	0	0	0	0	0

Subsequently, Step 3 is executed several times. I am not showing these reductions individually, but record the substitutions carried out in these steps.

$$x_{14} = 1 + x_3 \tag{8.13}$$
$$x_{20} = 1 + x_3 + x_{13} \tag{8.14}$$
$$x_{10} = 0 + x_3 + x_{13} + x_{16} \tag{8.15}$$
$$x_{16} = 1 + x_3 + x_{23} + x_{24} \tag{8.16}$$
$$x_{24} = 0 + x_3 + x_6 + x_{13} \tag{8.17}$$
$$x_{17} = 0 \tag{8.18}$$

After all these steps, the system reduces as follows:

	3	6	11	13	21	23	b
	H	*H*	*H*	*H*	*H*	*H*	
3	1	1	1	0	0	1	1
5	1	1	0	0	1	1	0
6	1	0	0	1	1	0	1
11	0	0	0	0	0	0	0
12	0	0	0	0	0	1	1
17	1	0	0	0	0	1	0
19	1	0	1	1	0	0	0
20	1	0	1	1	1	0	1
21	1	0	1	0	1	0	0

This completes the first round of Steps 1–3. We prefer to avoid Step 4 in this example, since we started with a square system, and the number of rows always remains close to the number of columns.

Round 2: Although all the columns are marked heavy now, there still remain reduction possibilities. So we go through another round of the above steps.

Step 1(a) and 1(b): There are no columns of weight zero or one. So these steps are not executed in this round.

Step 2(a): Row 11 contains only zero entries, and is deleted from the system.

Step 2(b): Row 12 contains a single non-zero entry giving

$$x_{23} = 1. \tag{8.19}$$

We also adjust Rows 3, 5 and 17 by substituting this value of x_{23}. This lets Row 17 have a single non-zero entry at Column 3, so we have another reduction

$$x_3 = 1, \tag{8.20}$$

which changes Rows 3, 5, 6, 19, 20 and 21. The reduced system now becomes

$$
\begin{array}{c}
\quad\quad A \quad\quad\quad \mathbf{b} \\
\begin{array}{cccc}
6 & 11 & 13 & 21 \\
H & H & H & H
\end{array} \\
\begin{array}{c}
3 \\
5 \\
6 \\
19 \\
20 \\
21
\end{array}
\left[
\begin{array}{cccc}
1 & 1 & 0 & 0 \\
1 & 0 & 0 & 1 \\
0 & 0 & 1 & 1 \\
0 & 1 & 1 & 0 \\
0 & 1 & 1 & 1 \\
0 & 1 & 0 & 1
\end{array}
\right]
\left[
\begin{array}{c}
1 \\
0 \\
0 \\
1 \\
0 \\
1
\end{array}
\right]
\end{array}
$$

Step 3: Since all columns are heavy at this point, this step is not executed.

Another round of SGE does not reduce the system further. So we stop at this point. The final reduced system consists of the following equations:

$$
\begin{aligned}
x_6 + x_{11} &= 1 \\
x_6 + x_{21} &= 0 \\
x_{13} + x_{21} &= 0 \\
x_{11} + x_{13} &= 1 \\
x_{11} + x_{13} + x_{21} &= 0 \\
x_{11} + x_{21} &= 1
\end{aligned}
$$

Standard Gaussian elimination on this system gives the unique solution:

$$
\begin{aligned}
x_6 &= 1 & (8.21) \\
x_{11} &= 0 & (8.22) \\
x_{13} &= 1 & (8.23) \\
x_{21} &= 1 & (8.24)
\end{aligned}
$$

Now, we work backwards to obtain the values of other variables using the substitution equations generated so far. The way SGE works ensures that when we use a substitution equation to determine the value of a variable, all variables appearing on the right side of the equation are already solved. For example, when we invoke Eqn (8.17) to calculate x_{24}, we know the values of $x_6, x_{11}, x_{13}, x_{21}, x_3, x_{23}, x_{17}$. The right side $(x_3 + x_6 + x_{13})$ contains a subset of only these variables. After x_{24} is computed, we use Eqn (8.16) to calculate x_{16} (which requires the solution for x_{24}). These back substitution steps are not elaborated here. Only the final solution is given below.

x_1	x_2	x_3	x_4	x_5	x_6	x_7	x_8	x_9	x_{10}	x_{11}	x_{12}	x_{13}	x_{14}	x_{15}	x_{16}	x_{17}	x_{18}	x_{19}	x_{20}	x_{21}	x_{22}	x_{23}	x_{24}	x_{25}
0	1	1	1	1	1	0	0	0	0	0	$-$	1	0	$-$	0	0	1	$-$	1	1	1	1	1	1

Here, I use $-$ to indicate that solutions for these variables cannot be (uniquely) obtained from the given system. $\square$

Structured Gaussian elimination is a heuristic algorithm, and lacks solid analytic foundation (Bender and Canfield provide some estimates though). In practice, structured Gaussian elimination can substantially reduce the size of a sparse system. For example, LaMacchia and Odlyzko report implementations of SGE achieving 50–90% size reduction.

8.2 Lanczos Method

The Lanczos[3] method is an iterative method for solving linear systems. Originally proposed for solving equations over $\mathbb{R}$, the Lanczos method can be adapted[4] to work for finite fields. For an $n \times n$ system, it takes $O\tilde{}(n^2)$ time.

To start with, consider the system defined over the field $\mathbb{R}$ of real numbers:

$$A\mathbf{x} = \mathbf{b},$$

where A is an $n \times n$ symmetric positive-definite[5] matrix, and $\mathbf{b} \neq \mathbf{0}$. The Lanczos method starts with an initial direction $\mathbf{w}_0$. Subsequently, it keeps on generating linearly independent directions $\mathbf{w}_1, \mathbf{w}_2, \ldots$ until a direction $\mathbf{w}_s$ is found which is linearly dependent upon $\mathbf{w}_0, \mathbf{w}_1, \ldots, \mathbf{w}_{s-1}$. Any $n + 1$ (or more) vectors in $\mathbb{R}^n$ are linearly dependent, so $s \leqslant n$, that is, the Lanczos loop terminates after at most n iterations. A solution to the original linear system is obtained as a linear combination of the direction vectors $\mathbf{w}_0, \mathbf{w}_1, \ldots, \mathbf{w}_{s-1}$.

Two vectors $\mathbf{y}_1$ and $\mathbf{y}_2$ are called A-orthogonal if $\mathbf{y}_1^{\mathsf{t}} A \mathbf{y}_2 = \mathbf{0}$. We take

$$\mathbf{w}_0 = \mathbf{b}. \tag{8.25}$$

Subsequently, for $i = 1, 2, 3, \ldots$, we generate

---

[3]Cornelius Lanczos (1893–1974) was a Hungarian mathematician and physicist.
[4]See the citation of Footnote 2.
[5]An $n \times n$ matrix A with real entries is called *positive-definite* if $\mathbf{y}^{\mathsf{t}} A \mathbf{y} > 0$ for all n-dimensional real vectors $\mathbf{y} \neq \mathbf{0}$.

$$\mathbf{w}_i = A\mathbf{w}_{i-1} - \sum_{j=0}^{i-1} c_{ij}\mathbf{w}_j, \text{ where } c_{ij} = \frac{\mathbf{w}_j^t A^2 \mathbf{w}_{i-1}}{\mathbf{w}_j^t A\mathbf{w}_j}. \tag{8.26}$$

One can prove by induction on i that $\mathbf{w}_i^t A\mathbf{w}_j = 0$ for $i > j$. Moreover, since A is symmetric, for $i < j$, we have $\mathbf{w}_i^t A\mathbf{w}_j = (\mathbf{w}_j^t A\mathbf{w}_i)^t = 0$. Therefore, the direction vectors $\mathbf{w}_0, \mathbf{w}_1, \mathbf{w}_2, \ldots$ are A-orthogonal to one another. Now, $\mathbf{w}_s$ is the first vector linearly dependent upon the previous direction vectors, that is, $\mathbf{w}_s = \sum_{j=0}^{s-1} a_j \mathbf{w}_j$. But then, $\mathbf{w}_s^t A\mathbf{w}_s = \sum_{j=0}^{s-1} a_j \mathbf{w}_s^t A\mathbf{w}_j = 0$ by A-orthogonality. The positive-definiteness of A implies that $\mathbf{w}_s = \mathbf{0}$, that is, the Lanczos iterations continue until we obtain a direction vector $\mathbf{w}_s = \mathbf{0}$. Take

$$\mathbf{x} = \sum_{j=0}^{s-1} b_j \mathbf{w}_j, \text{ where } b_j = \frac{\mathbf{w}_j^t \mathbf{b}}{\mathbf{w}_j^t A\mathbf{w}_j}. \tag{8.27}$$

Since $\mathbf{x}$ is a linear combination of $\mathbf{w}_0, \mathbf{w}_1, \ldots, \mathbf{w}_{s-1}$, it follows that $A\mathbf{x}$ is a linear combination of $A\mathbf{w}_0, A\mathbf{w}_1, \ldots, A\mathbf{w}_{s-1}$. By Eqn (8.26), $A\mathbf{w}_j$ is a linear combination of $\mathbf{w}_0, \mathbf{w}_1, \ldots, \mathbf{w}_{j+1}$. Therefore, $A\mathbf{x}$ can be expressed as a linear combination of $\mathbf{w}_0, \mathbf{w}_1, \ldots, \mathbf{w}_s$. But $\mathbf{w}_s = \mathbf{0}$, and $\mathbf{w}_0 = \mathbf{b}$, so we can write

$$A\mathbf{x} - \mathbf{b} = \sum_{j=0}^{s-1} d_j \mathbf{w}_j$$

for some real coefficients d_i. Also, by the A-orthogonality of the direction vectors and by Eqn (8.27), we have $\mathbf{w}_j^t A\mathbf{x} = b_j \mathbf{w}_j^t A\mathbf{w}_j = \mathbf{w}_j^t \mathbf{b}$ for all j, $0 \leqslant j \leqslant s-1$. But then, $(A\mathbf{x} - \mathbf{b})^t (A\mathbf{x} - \mathbf{b}) = \sum_{j=0}^{s-1} d_j \mathbf{w}_j^t (A\mathbf{x} - \mathbf{b}) = 0$. This implies that $A\mathbf{x} - \mathbf{b} = \mathbf{0}$, that is, $\mathbf{x}$ computed as above satisfies $A\mathbf{x} = \mathbf{b}$.

The A-orthogonality of the direction vectors implies that the coefficients $c_{i,j} = 0$ in Eqn (8.26) for $0 \leqslant j \leqslant i-3$, and simplifies this formula as

$$\mathbf{w}_i = A\mathbf{w}_{i-1} - \frac{(A\mathbf{w}_{i-1})^t (A\mathbf{w}_{i-1})}{\mathbf{w}_{i-1}^t (A\mathbf{w}_{i-1})} \mathbf{w}_{i-1} - \frac{(A\mathbf{w}_{i-2})^t (A\mathbf{w}_{i-1})}{\mathbf{w}_{i-2}^t (A\mathbf{w}_{i-2})} \mathbf{w}_{i-2} \quad (8.28)$$

for $i \geqslant 2$. We include the computation of $\mathbf{w}_1$ in the initialization step, and show the Lanczos procedure as Algorithm 8.1. The algorithm makes some optimizations for implementation, like storing the vector $\mathbf{v}_i = A\mathbf{w}_i$ and the two scalar quantities $\alpha_i = \mathbf{v}_i^t \mathbf{v}$ and $\beta_i = \mathbf{w}_{i-1}^t \mathbf{v}_i$. Although different values (vectors and scalars) are shown as suffixed, it suffices to remember these values only from (at most) two previous iterations.

Algorithm 8.1 works for real symmetric positive-definite matrices A with $\mathbf{b} \neq \mathbf{0}$ (for $\mathbf{b} = \mathbf{0}$, it gives the trivial solution $\mathbf{x} = \mathbf{0}$). If we want to adapt this method to work for a finite field, we encounter several problems. All these problems can be reasonably solved, and this method can be used to solve systems like (8.1) of our interest. We now discuss these adaptations in detail.

Algorithm 8.1: LANCZOS METHOD FOR SOLVING $A\mathbf{x} = \mathbf{b}$

```
/* Initialize for i = 0 */
```
Set $\mathbf{w}_0 = \mathbf{b}$.
```
/* Initialize for i = 1 */
```
Compute $\mathbf{v}_1 = A\mathbf{w}_0$, $\alpha_1 = \mathbf{v}_1^t\mathbf{v}_1$, and $\beta_1 = \mathbf{w}_0^t\mathbf{v}_1$.
Compute $\mathbf{w}_1 = \mathbf{v}_1 - \dfrac{\alpha_1}{\beta_1}\mathbf{w}_0$.

Compute $\mathbf{x} = \left(\dfrac{\mathbf{w}_0^t\mathbf{b}}{\beta_1}\right)\mathbf{w}_0$.

Set $i = 1$.
While ($\mathbf{w}_i \neq \mathbf{0}$) {
    Increment i by 1.
    Compute $\mathbf{v}_i = A\mathbf{w}_{i-1}$, $\alpha_i = \mathbf{v}_i^t\mathbf{v}_i$, and $\beta_i = \mathbf{w}_{i-1}^t\mathbf{v}_i$.

    Compute the new direction $\mathbf{w}_i = \mathbf{v}_i - \left(\dfrac{\alpha_i}{\beta_i}\right)\mathbf{w}_{i-1} - \left(\dfrac{\mathbf{v}_i^t\mathbf{v}_{i-1}}{\beta_{i-1}}\right)\mathbf{w}_{i-2}$.

    Update the solution $\mathbf{x} = \mathbf{x} + \left(\dfrac{\mathbf{w}_{i-1}^t\mathbf{b}}{\beta_i}\right)\mathbf{w}_{i-1}$.

}
Return $\mathbf{x}$.

In general, our coefficient matrix A is not symmetric. It is not even square. However, instead of solving $A\mathbf{x} = \mathbf{b}$, we can solve $(A^tA)\mathbf{x} = (A^t\mathbf{b})$. If A is of close-to-full rank (that is, of rank nearly $\min(m, n)$, where A is an $m \times n$ matrix), the two systems are equivalent with high probability. The matrix A^tA is not only square ($n \times n$), but symmetric too. As mentioned earlier, we do not compute the matrix A^tA explicitly, since it may be significantly less sparse than A. In Algorithm 8.1, A is needed during the computation of the vectors $\mathbf{v}_i$ only. We carry this out as $A^t(A\mathbf{w}_{i-1})$ (instead of as $(A^tA)\mathbf{w}_{i-1}$). The (modified) vector $A^t\mathbf{b}$ (needed during initialization, and updating of the solution $\mathbf{x}$) may be precomputed once before the loop.

The notion of positive-definiteness is, however, meaningless in a context of finite fields. For real matrices, this property of A implies that $\mathbf{y}^tA\mathbf{y} = 0$ if and only if $\mathbf{y} = \mathbf{0}$, that is, no non-zero vector can be A-orthogonal to itself, that is, the computations in Algorithm 8.1 proceed without error, since all β_i values are non-zero (these are the only quantities by which divisions are made). When A is a matrix over $\mathbb{F}_q$, a non-zero direction vector $\mathbf{w}_{i-1}$ may lead to the situation $\beta_i = \mathbf{w}_{i-1}^t\mathbf{v}_i = \mathbf{w}_{i-1}^tA\mathbf{w}_{i-1} = 0$. This problem is rather serious for small values of q, that is, for q smaller than or comparable to n (there are at most n iterations, and so at most n values of β_i are encountered in the entire algorithm). If $q \gg n$, the odds against encountering this bad situation are low, and Algorithm 8.1 is expected to terminate without troubles.

A way to get around this problem is by working in a suitable extension field $\mathbb{F}_{q^d}$ for which $q^d \gg n$. The original system $(A^tA)\mathbf{x} = (A^t\mathbf{b})$ leads to computations to proceed in $\mathbb{F}_q$ itself. So we choose a random invertible $m \times m$
```
```

diagonal matrix D with entries from $\mathbb{F}_{q^d}$, and solve the equivalent system $DAx = Db$, that is, feed the system $(DA)^t(DA)x = (DA)^t Db$ (that is, the system $(A^t D^2 A)x = (A^t D^2 b)$) to Algorithm 8.1. The introduction of D does not affect sparsity (A and DA have non-zero entries at exactly the same locations), but the storage of DA may take more space than storing D and A individually (since the elements of DA are from a larger field compared to those of A). So, we keep the modified coefficient matrix $A^t D^2 A$ in this factored form, and compute $\mathbf{v}_i = A^t(D^2(A\mathbf{w}_{i-1}))$. Although the final solution belongs to $(\mathbb{F}_q)^n$, the computation proceeds in $\mathbb{F}_{q^d}$, and is less likely to encounter the problem associated with the original matrix. Still, if the computation fails, we choose another random matrix D, and restart from the beginning.

Example 8.2 (1) Consider the 6×4 system modulo the prime $p = 97$:

$$
\begin{pmatrix}
0 & 0 & 49 & 56 \\
0 & 35 & 12 & 0 \\
77 & 21 & 0 & 3 \\
0 & 79 & 0 & 0 \\
26 & 0 & 68 & 24 \\
17 & 0 & 0 & 53
\end{pmatrix}
\begin{pmatrix} x_1 \\ x_2 \\ x_3 \\ x_4 \end{pmatrix}
=
\begin{pmatrix} 41 \\ 11 \\ 9 \\ 8 \\ 51 \\ 77 \end{pmatrix}.
\tag{8.29}
$$

The coefficient matrix is not square. Left multiplication by the transpose of this matrix gives the square system:

$$
\begin{pmatrix}
7 & 65 & 22 & 10 \\
65 & 50 & 32 & 63 \\
22 & 32 & 88 & 11 \\
10 & 63 & 11 & 31
\end{pmatrix}
\begin{pmatrix} x_1 \\ x_2 \\ x_3 \\ x_4 \end{pmatrix}
=
\begin{pmatrix} 30 \\ 42 \\ 80 \\ 62 \end{pmatrix}.
\tag{8.30}
$$

By an abuse of notation, we let A denote the coefficient matrix of this square system. In practice, we do not compute A explicitly, but keep it as the product $B^t B$, where B is the 6×4 coefficient matrix of the original system. We feed System (8.30) to Algorithm 8.1. The computations made by the algorithm are summarized in the following table.

i	$\mathbf{v}_i$	α_i	β_i	$\mathbf{w}_i$	$\mathbf{x}$
0	—	—	—	$\begin{pmatrix} 30 \\ 42 \\ 80 \\ 62 \end{pmatrix}$	—
1	$\begin{pmatrix} 82 \\ 40 \\ 26 \\ 25 \end{pmatrix}$	22	10	$\begin{pmatrix} 16 \\ 64 \\ 44 \\ 5 \end{pmatrix}$	$\begin{pmatrix} 21 \\ 10 \\ 56 \\ 24 \end{pmatrix}$
2	$\begin{pmatrix} 52 \\ 46 \\ 22 \\ 78 \end{pmatrix}$	39	90	$\begin{pmatrix} 79 \\ 80 \\ 46 \\ 19 \end{pmatrix}$	$\begin{pmatrix} 86 \\ 76 \\ 65 \\ 14 \end{pmatrix}$

i	$\mathbf{v}_i$	α_i	β_i	$\mathbf{w}_i$	$\mathbf{x}$
3	$\begin{pmatrix} 68 \\ 67 \\ 19 \\ 38 \end{pmatrix}$	54	9	$\begin{pmatrix} 58 \\ 85 \\ 49 \\ 69 \end{pmatrix}$	$\begin{pmatrix} 56 \\ 80 \\ 77 \\ 78 \end{pmatrix}$
4	$\begin{pmatrix} 36 \\ 64 \\ 46 \\ 77 \end{pmatrix}$	51	60	$\begin{pmatrix} 0 \\ 0 \\ 0 \\ 0 \end{pmatrix}$	$\begin{pmatrix} 64 \\ 75 \\ 57 \\ 34 \end{pmatrix}$

We get $\mathbf{w}_4 = \mathbf{0}$, and the Lanczos loop terminates. The value of $\mathbf{x}$ after this iteration gives the solution of System (8.30): $x_1 = 64, x_2 = 75, x_3 = 57, x_4 = 34$. This happens to be a solution of System (8.29) too.

(2) Let us now try to solve a 6×4 system modulo 7:

$$\begin{pmatrix} 0 & 0 & 4 & 5 \\ 0 & 3 & 1 & 0 \\ 5 & 2 & 0 & 3 \\ 0 & 3 & 0 & 0 \\ 2 & 0 & 4 & 6 \\ 1 & 0 & 0 & 5 \end{pmatrix} \begin{pmatrix} x_1 \\ x_2 \\ x_3 \\ x_4 \end{pmatrix} = \begin{pmatrix} 2 \\ 1 \\ 3 \\ 6 \\ 1 \\ 6 \end{pmatrix} \tag{8.31}$$

Multiplication by the transpose of the coefficient matrix gives the following square system which we feed to Algorithm 8.1.

$$\begin{pmatrix} 2 & 3 & 1 & 4 \\ 3 & 1 & 3 & 6 \\ 1 & 3 & 5 & 2 \\ 4 & 6 & 2 & 4 \end{pmatrix} \begin{pmatrix} x_1 \\ x_2 \\ x_3 \\ x_4 \end{pmatrix} = \begin{pmatrix} 2 \\ 6 \\ 6 \\ 6 \end{pmatrix} \tag{8.32}$$

Since the modulus 7 is small (comparable to $n = 4$), the Lanczos algorithm fails on this system, as shown below. We obtain $\beta_2 = 0$ even though $\mathbf{w}_1 \neq \mathbf{0}$.

i	$\mathbf{v}_i$	α_i	β_i	$\mathbf{w}_i$	$\mathbf{x}$
0	$-$	$-$	$-$	$\begin{pmatrix} 2 \\ 6 \\ 6 \\ 6 \end{pmatrix}$	$-$
1	$\begin{pmatrix} 3 \\ 3 \\ 6 \\ 3 \end{pmatrix}$	0	1	$\begin{pmatrix} 3 \\ 3 \\ 6 \\ 3 \end{pmatrix}$	$\begin{pmatrix} 0 \\ 0 \\ 0 \\ 0 \end{pmatrix}$

i	$\mathbf{v}_i$	α_i	β_i	$\mathbf{w}_i$	$\mathbf{x}$
2	$\begin{pmatrix} 5 \\ 6 \\ 6 \\ 5 \end{pmatrix}$	3	0	Failure	

(3) In order to make Algorithm 8.1 succeed on System (8.31), we use the quadratic extension $\mathbb{F}_{49} = \mathbb{F}_7(\theta)$ with $\theta^2 + 1 = 0$. Since 49 is somewhat large compared to $n = 4$, the computations are now expected to terminate without error. We multiply (8.31) by the 6×6 diagonal matrix

$$D = \begin{pmatrix} 6\theta & 0 & 0 & 0 & 0 & 0 \\ 0 & 4\theta+3 & 0 & 0 & 0 & 0 \\ 0 & 0 & 6\theta+1 & 0 & 0 & 0 \\ 0 & 0 & 0 & 3\theta+5 & 0 & 0 \\ 0 & 0 & 0 & 0 & \theta & 0 \\ 0 & 0 & 0 & 0 & 0 & 2\theta+6 \end{pmatrix}$$

over $\mathbb{F}_{49}$, and generate the equivalent system

$$
\begin{pmatrix}
0 & 0 & 3\theta & 2\theta \\
0 & 5\theta + 2 & 4\theta + 3 & 0 \\
2\theta + 5 & 5\theta + 2 & 0 & 4\theta + 3 \\
0 & 2\theta + 1 & 0 & 0 \\
2\theta & 0 & 4\theta & 6\theta \\
2\theta + 6 & 0 & 0 & 3\theta + 2
\end{pmatrix}
\begin{pmatrix}
x_1 \\
x_2 \\
x_3 \\
x_4
\end{pmatrix}
=
\begin{pmatrix}
5\theta \\
4\theta + 3 \\
4\theta + 3 \\
4\theta + 2 \\
\theta \\
5\theta + 1
\end{pmatrix},
\tag{8.33}
$$

which leads to the corresponding square system:

$$
\begin{pmatrix}
2\theta & \theta & 6 & 6\theta + 1 \\
\theta & 2\theta + 4 & 2\theta & 2\theta \\
6 & 2\theta & 3\theta + 3 & 5 \\
6\theta + 1 & 2\theta & 5 & \theta + 4
\end{pmatrix}
\begin{pmatrix}
x_1 \\
x_2 \\
x_3 \\
x_4
\end{pmatrix}
=
\begin{pmatrix}
2\theta + 1 \\
5\theta + 1 \\
3\theta + 2 \\
2\theta + 6
\end{pmatrix}.
\tag{8.34}
$$

Algorithm 8.1 works on System (8.34) as follows:

i	$\mathbf{v}_i$	α_i	β_i	$\mathbf{w}_i$	$\mathbf{x}$
0	–	–	–	$\begin{pmatrix} 2\theta + 1 \\ 5\theta + 1 \\ 3\theta + 2 \\ 2\theta + 6 \end{pmatrix}$	–
1	$\begin{pmatrix} 3\theta + 4 \\ 4\theta + 3 \\ 4\theta + 2 \\ 4\theta + 4 \end{pmatrix}$	$5\theta + 2$	$6\theta + 3$	$\begin{pmatrix} 6\theta + 1 \\ 6\theta + 3 \\ \theta + 2 \\ 2\theta + 4 \end{pmatrix}$	$\begin{pmatrix} 5\theta \\ 4\theta + 4 \\ \theta + 1 \\ 3\theta + 3 \end{pmatrix}$
2	$\begin{pmatrix} 2\theta \\ \theta + 2 \\ 5\theta + 3 \\ 5 \end{pmatrix}$	$6\theta + 1$	$5\theta + 2$	$\begin{pmatrix} 2\theta \\ 4 \\ 5\theta + 2 \\ 4\theta + 3 \end{pmatrix}$	$\begin{pmatrix} 2\theta + 2 \\ 4 \\ 6\theta + 3 \\ 6\theta \end{pmatrix}$
3	$\begin{pmatrix} 1 \\ 4\theta + 3 \\ 5\theta + 6 \\ 5\theta + 6 \end{pmatrix}$	$4\theta + 2$	$6\theta + 4$	$\begin{pmatrix} 6\theta + 4 \\ 6 \\ 4\theta + 3 \\ 3\theta + 5 \end{pmatrix}$	$\begin{pmatrix} 3\theta + 1 \\ 2\theta + 6 \\ 6\theta + 5 \\ 6\theta + 3 \end{pmatrix}$
4	$\begin{pmatrix} \theta \\ 4\theta + 4 \\ 4 \\ 2\theta \end{pmatrix}$	$4\theta + 4$	$5\theta + 3$	$\begin{pmatrix} 0 \\ 0 \\ 0 \\ 0 \end{pmatrix}$	$\begin{pmatrix} 5 \\ 2 \\ 2 \\ 3 \end{pmatrix}$

These computations do not encounter the situation that $\beta_i = 0$ for $\mathbf{w}_{i-1} \neq \mathbf{0}$. The final solution belongs to $\mathbb{F}_7$:

$$x_1 = 5, \quad x_2 = 2, \quad x_3 = 2, \quad x_4 = 3. \qquad \square$$

If A is an $m \times n$ matrix with $m = \Theta(n)$, each iteration of Algorithm 8.1 involves $O(nk)$ finite-field operations, where k is the maximum number of non-zero entries in a row of A. Moreover, there are at most n iterations of the Lanczos loop. Therefore, if $k = O(\log n)$, and n is a subexponential expression in the size of the underlying field (as typically hold in our cases of interest), the running time of Algorithm 8.1 is $\tilde{O}(n^2)$ (or $\tilde{O}(m^2)$).

8.3   Wiedemann Method

Wiedemann's method[6] for solving linear systems uses linear recurrent sequences. Let $a_0, a_1, a_2, \ldots$ be an infinite sequence of elements from a field K. The first d terms are supplied as initial conditions, and for all $k \geqslant d$, we have

$$a_k = c_{d-1}a_{k-1} + c_{d-2}a_{k-2} + \cdots + c_1 a_{k-d+1} + c_0 a_{k-d} \tag{8.35}$$

for some constant elements $c_0, c_1, \ldots, c_{d-1} \in K$. Given the order d of the sequence, and $2d$ terms $a_0, a_1, \ldots, a_{2d-1}$ in the sequence, one can determine $c_0, c_1, \ldots, c_{d-1}$ as follows. For every $k \geqslant d$, we can rewrite Eqn (8.35) as

$$\begin{pmatrix} a_{k-1} & a_{k-2} & \cdots & a_{k-d+1} & a_{k-d} \end{pmatrix} \begin{pmatrix} c_{d-1} \\ c_{d-2} \\ \vdots \\ c_1 \\ c_0 \end{pmatrix} = a_k.$$

Using this relation for $k = d, d+1, \ldots, 2d-1$, we write

$$\begin{pmatrix} a_{d-1} & a_{d-2} & \cdots & a_1 & a_0 \\ a_d & a_{d-1} & \cdots & a_2 & a_1 \\ \vdots & \vdots & \cdots & \vdots & \cdots \\ a_{2d-2} & a_{2d-3} & \cdots & a_d & a_{d-1} \end{pmatrix} \begin{pmatrix} c_{d-1} \\ c_{d-2} \\ \vdots \\ c_1 \\ c_0 \end{pmatrix} = \begin{pmatrix} a_d \\ a_{d+1} \\ \vdots \\ a_{2d-2} \\ a_{2d-1} \end{pmatrix}, \tag{8.36}$$

that is, the coefficients $c_0, c_1, \ldots, c_{d-1}$ in the recurrence (8.35) can be obtained by solving a linear system.

This method of computing $c_0, c_1, \ldots, c_{d-1}$ suffers from two problems. First, under standard matrix arithmetic, the procedure takes $O(d^3)$ time. Faster matrix arithmetic (like Strassen's matrix multiplication[7]) reduces the exponent below three, but the desired goal of $O(d^2)$ running time cannot be achieved. The second problem is that the coefficient matrix of Eqn (8.36) may be non-invertible. This may happen, for example, when a recurrence relation of order smaller than d also generates the sequence $a_0, a_1, a_2, \ldots$.

There are several ways to solve these problems. The coefficient matrix in Eqn (8.36) has a particular structure. Each of its rows (except the topmost) is obtained by right shifting the previous row (and introducing a new term at the leftmost position). Such a matrix is called a *Toeplitz matrix*. A Toeplitz system like (8.36) can be solved in $O(d^2)$ time using special algorithms. I will come back to this topic again while studying the block Wiedemann algorithm. Here,

---

[6]Douglas H. Wiedemann, Solving sparse linear equations over finite fields, *IEEE Transactions on Information Theory*, 32(1), 54–62, 1986.

[7]Volker Strassen, Gaussian elimination is not optimal, *Numerical Mathematics*, 13, 354–356, 1969.

I explain the *Berlekamp–Massey* algorithm[8] for computing $c_0, c_1, \ldots, c_{d-1}$. The generating function for the sequence $a_0, a_1, a_2, \ldots$ is

$$
\begin{aligned}
G(x) &= a_0 + a_1 x + a_2 x^2 + \cdots + a_{d-1} x^{d-1} + a_d x^d + a_{d+1} x^{d+1} + \cdots \\
&= (a_0 + a_1 x + a_2 x^2 + \cdots + a_{d-1} x^{d-1}) + \sum_{k \geqslant d} a_k x^k \\
&= (a_0 + a_1 x + a_2 x^2 + \cdots + a_{d-1} x^{d-1}) + \\
&\quad \sum_{k \geqslant d} (c_{d-1} a_{k-1} + c_{d-2} a_{k-2} + \cdots + c_1 a_{k-d+1} + c_0 a_{k-d}) x^k \\
&= R(x) + (c_{d-1} x + c_{d-2} x^2 + \cdots + c_0 x^d) G(x),
\end{aligned}
$$

that is,

$$
C(x) G(x) = R(x), \tag{8.37}
$$

where $R(x)$ is a polynomial of degree $\leqslant d - 1$, and

$$
C(x) = 1 - c_{d-1} x - c_{d-2} x^2 - \cdots - c_0 x^d. \tag{8.38}
$$

In order to compute $c_0, c_1, \ldots, c_{d-1}$, it suffices to compute $C(x)$. To that end, we use an extended gcd computation as follows. Let

$$
A(x) = a_0 + a_1 x + a_2 x^2 + \cdots + a_{2d-1} x^{2d-1}.
$$

Then, Eqn (8.37) can be rewritten as

$$
C(x) A(x) + B(x) x^{2d} = R(x)
$$

for some *polynomial* $B(x)$. These observations lead to Algorithm 8.2 for computing the coefficients $c_0, c_1, \ldots, c_{d-1}$ in Eqn (8.35), given the first $2d$ terms $a_0, a_1, \ldots, a_{2d-1}$ in the sequence. The extended gcd computation maintains an invariance of the form $B_1(x) x^{2d} + C_1(x) A(x) = R_1(x)$. Since the multiplier $B_1(x)$ is not needed, it is not explicitly computed.

Algorithm 8.2: BERLEKAMP–MASSEY ALGORITHM

Let $R_0(x) = x^{2d}$ and $R_1(x) = a_0 + a_1 x + a_2 x^2 + \cdots + a_{2d-1} x^{2d-1}$.
Initialize $C_0(x) = 0$ and $C_1(x) = 1$.
While $(\deg R_1(x) \geqslant d)$ {
    Let $Q(x) = R_0(x) \operatorname{quot} R_1(x)$ and $R(x) = R_0(x) \operatorname{rem} R_1(x)$.
    Update $C(x) = C_0(x) - Q(x) C_1(x)$.
    Prepare for next iteration:
        $R_0(x) = R_1(x), \; R_1(x) = R(x), \; C_0(x) = C_1(x)$ and $C_1(x) = C(x)$.
}
Divide $C_1(x)$ by its constant term.
Recover the coefficients $c_0, c_1, \ldots, c_{d-1}$ from $C_1(x)$.

---

[8] Originally, Elwyn R. Berlekamp proposed this algorithm for decoding BCH codes (see Berlekamp's 1968 book *Algebraic Coding Theory*). James Massey (Shift-register synthesis and BCH decoding, *IEEE Transactions on Information Theory*, 15(1), 122–127, 1969) simplified and modified Berlekamp's algorithm to the form presented in this book.

Example 8.3 (1) Consider the sequence from $\mathbb{F}_7$ defined as

$$a_0 = 0, \ a_1 = 1, \ a_2 = 2, \ a_k = 3a_{k-1} + a_{k-3} \text{ for } k \geqslant 3. \tag{8.39}$$

The first six terms in this sequence are $0, 1, 2, 6, 5, 3$. The extended gcd computation in Algorithm 8.2 is shown below:

i	$Q_i(x)$	$R_i(x)$	$C_i(x)$
0	$-$	x^6	0
1	$-$	$3x^5 + 5x^4 + 6x^3 + 2x^2 + x$	1
2	$5x + 1$	$5x^3 + 6x$	$2x + 6$
3	$2x^2 + x + 3$	$3x^2 + 4x$	$3x^3 + 2x + 4$

We obtain $C(x) = \frac{1}{4}(4 + 2x + 3x^3) = 1 + 4x + 6x^3 = 1 - 3x - x^3$ which is consistent with the recurrence (8.39).

(2) Now, consider the following sequence from $\mathbb{F}_7$:

$$a_0 = 0, \ a_1 = 1, \ a_2 = 5, \ a_k = 3a_{k-1} + a_{k-3} \text{ for } k \geqslant 3. \tag{8.40}$$

The first six terms in the sequence are $0, 1, 5, 1, 4, 3$, and Algorithm 8.2 proceeds as follows:

i	$Q_i(x)$	$R_i(x)$	$C_i(x)$
0	$-$	x^6	0
1	$-$	$3x^5 + 4x^4 + x^3 + 5x^2 + x$	1
2	$5x + 5$	$3x^4 + 5x^3 + 5x^2 + 2x$	$2x + 2$
3	$x + 2$	$4x$	$5x^2 + x + 4$

This gives $C(x) = \frac{1}{4}(4 + x + 5x^2) = 1 + 2x + 3x^2 = 1 - 5x - 4x^2$. This appears to contradict the recurrence (8.40). However, notice that (8.40) is not the recurrence of smallest order to generate this sequence. Indeed, the following is the recurrence of smallest order for this purpose.

$$a_0 = 0, \ a_1 = 1, \ a_k = 5a_{k-1} + 4a_{k-2} \text{ for } k \geqslant 2.$$

Note also that $1 - 5x - 4x^2$ divides $1 - 3x - x^3$ in $\mathbb{F}_7[x]$. □

It is easy to argue that the Berlekamp–Massey algorithm performs a total of only $O(d^2)$ basic operations in the field K.

Let us now come to Wiedemann's algorithm. Let $\mathbf{Ax} = \mathbf{b}$ be a sparse linear system (over a field K), where A is a *square* $n \times n$ matrix. A is not needed to be symmetric (or positive-definite). The *characteristic equation* of A is

$$\chi_A(x) = \det(xI - A),$$

where I is the $n \times n$ identity matrix. *Cayley–Hamilton theorem*[9] states that A satisfies $\chi_A(x)$, that is, $\chi_A(A) = 0$. The set of all polynomials in $K[x]$ satisfied

---

[9]This is named after the British mathematician Arthur Cayley (1821–1895) and the Irish mathematician and physicist William Rowan Hamilton (1805–1865).

by A is an ideal of $K[x]$. The monic generator of this ideal, that is, the monic non-zero polynomial of the smallest degree, which A satisfies, is called the *minimal polynomial* of A and denoted as $\mu_A(x)$. Clearly, $\mu_A(x)|\chi_A(x)$ in $K[x]$. Wiedemann's algorithm starts by probabilistically determining $\mu_A(x)$. Let

$$\mu_A(x) = x^d - c_{d-1}x^{d-1} - c_{d-2}x^{d-2} - \cdots - c_1 x - c_0 \in K[x] \tag{8.41}$$

with $d = \deg \mu_A(x) \leqslant n$. Since $\mu_A(A) = 0$, for any $n \times 1$ non-zero vector $\mathbf{v}$ and for any integer $k \geqslant d$, we have

$$A^k\mathbf{v} - c_{d-1}A^{k-1}\mathbf{v} - c_{d-2}A^{k-2}\mathbf{v} - \cdots - c_1 A^{k-d+1}\mathbf{v} - c_0 A^{k-d}\mathbf{v} = \mathbf{0}. \tag{8.42}$$

Let v_k be the element of $A^k\mathbf{v}$ at some particular position. The sequence v_k, $k \geqslant 0$, satisfies the recurrence relation

$$v_k = c_{d-1}v_{k-1} + c_{d-2}v_{k-2} + \cdots + c_1 v_{k-d+1} + c_0 v_{k-d}$$

for $k \geqslant d$. Using the Berlekamp–Massey algorithm, we compute the polynomial $C(x)$ of degree $d' \leqslant d$. But then, $x^{d'}C(1/x)$ (the opposite of $C(x)$—compare Eqns (8.38) and (8.41)) divides $\mu_A(x)$ in $K[x]$. Trying several such sequences (corresponding to different positions in $A^k\mathbf{v}$), we obtain many polynomials $x^{d'}C(1/x)$, whose lcm is expected to be the minimal polynomial $\mu_A(x)$.

In order that the Berlekamp–Massey algorithm works correctly in all these cases, we take the obvious upper bound n for d, and supply $2n$ vector elements $v_0, v_1, \ldots, v_{2n-1}$. This means that we need to compute the $2n$ matrix-vector products $A^i\mathbf{v}$ for $i = 0, 1, 2, \ldots, 2n-1$. Since A is a sparse matrix with $O^\sim(n)$ non-zero entries, the determination of $\mu_A(x)$ can be completed in $O^\sim(n^2)$ time.

For obtaining a solution of $A\mathbf{x} = \mathbf{b}$, we use $\mu_A(x)$ as follows. Putting $k = d$ and $\mathbf{v} = \mathbf{b}$ in Eqn (8.42) gives

$$A(A^{d-1}\mathbf{b} - c_{d-1}A^{d-2}\mathbf{b} - c_{d-2}A^{d-3}\mathbf{b} - \cdots - c_1 A\mathbf{b}) = c_0\mathbf{b},$$

that is, if $c_0 \neq 0$,

$$\mathbf{x} = c_0^{-1}(A^{d-1}\mathbf{b} - c_{d-1}A^{d-2}\mathbf{b} - c_{d-2}A^{d-3}\mathbf{b} - \cdots - c_1 A\mathbf{b}) \tag{8.43}$$

is a solution of $A\mathbf{x} = \mathbf{b}$. The basic time-consuming task here is the computation of the $d \leqslant n$ matrix-vector products $A^i\mathbf{b}$ for $i = 0, 1, 2, \ldots, d-1$—a task that can be completed in $O^\sim(n^2)$ time.

Example 8.4 We solve the following system in $\mathbb{F}_7$ by Wiedemann's method.

$$\begin{pmatrix} 2 & 6 & 0 & 4 \\ 0 & 4 & 6 & 4 \\ 6 & 0 & 0 & 5 \\ 4 & 0 & 1 & 0 \\ 0 & 0 & 4 & 6 \\ 0 & 2 & 0 & 2 \end{pmatrix} \begin{pmatrix} x_1 \\ x_2 \\ x_3 \\ x_4 \end{pmatrix} = \begin{pmatrix} 2 \\ 1 \\ 2 \\ 4 \\ 1 \\ 1 \end{pmatrix}. \tag{8.44}$$

This is not a square system. Multiplication by the transpose of the coefficient matrix yields a square system on which we apply Wiedemann's method.

$$\begin{pmatrix} 0 & 5 & 4 & 3 \\ 5 & 0 & 3 & 2 \\ 4 & 3 & 4 & 6 \\ 3 & 2 & 6 & 6 \end{pmatrix} \begin{pmatrix} x_1 \\ x_2 \\ x_3 \\ x_4 \end{pmatrix} = \begin{pmatrix} 4 \\ 4 \\ 0 \\ 2 \end{pmatrix}. \tag{8.45}$$

Let us choose the non-zero vector $\mathbf{v} = \begin{pmatrix} 0 \\ 5 \\ 2 \\ 0 \end{pmatrix}$, and compute $A^i\mathbf{v}$ for $i = 0, 1, 2, \ldots, 7$. For $i \geqslant 1$, we compute $A^i\mathbf{v}$ as $A(A^{i-1}\mathbf{v})$.

i	0	1	2	3	4	5	6	7
$A^i\mathbf{v}$	$\begin{pmatrix}0\\5\\2\\0\end{pmatrix}$	$\begin{pmatrix}5\\6\\2\\1\end{pmatrix}$	$\begin{pmatrix}6\\5\\3\\3\end{pmatrix}$	$\begin{pmatrix}4\\3\\6\\1\end{pmatrix}$	$\begin{pmatrix}0\\5\\6\\4\end{pmatrix}$	$\begin{pmatrix}5\\5\\0\\0\end{pmatrix}$	$\begin{pmatrix}4\\4\\0\\4\end{pmatrix}$	$\begin{pmatrix}4\\0\\3\\2\end{pmatrix}$

First, we apply the Berlekamp–Massey algorithm on the first (topmost) elements of $A^i\mathbf{v}$. We supply 8 terms $0, 5, 6, 4, 0, 5, 4, 4$ to Algorithm 8.2, and obtain the output $6x^3 + 2x^2 + 1$. But we are interested in the opposite of this polynomial, that is, $M_1(x) = x^3 + 2x + 6$. Substituting x by A in M_1 gives $M_1(A) = \begin{pmatrix} 0 & 0 & 0 & 0 \\ 0 & 3 & 1 & 3 \\ 0 & 1 & 5 & 1 \\ 0 & 3 & 1 & 3 \end{pmatrix}$, that is, the minimal polynomial of A is not computed yet. So, we apply the Berlekamp–Massey algorithm on the second position of $A^i\mathbf{v}$. Upon an input of the sequence $5, 6, 5, 3, 5, 5, 4, 0$, Algorithm 8.2 outputs $3x^4 + 2x^2 + 4x + 1$, the opposite of which is $M_2(x) = x^4 + 4x^3 + 2x^2 + 3$. Since $M_2(x)$ is of degree four, it must be the minimal polynomial of A (indeed, $M_1(x)$ divides $M_2(x)$ in $\mathbb{F}_7[x]$). That is, we have computed

$$\mu_A(x) = x^4 + 4x^3 + 2x^2 + 3 = x^4 - 3x^3 - 5x^2 - 4.$$

In order to obtain the solution $\mathbf{x}$, we now compute $A^i\mathbf{b}$ for $i = 0, 1, 2, 3$. Again, we use $A^i\mathbf{b} = A(A^{i-1}\mathbf{b})$ for $i \geqslant 1$.

i	0	1	2	3
$A^i\mathbf{b}$	$\begin{pmatrix}4\\4\\0\\2\end{pmatrix}$	$\begin{pmatrix}5\\3\\5\\4\end{pmatrix}$	$\begin{pmatrix}5\\6\\3\\5\end{pmatrix}$	$\begin{pmatrix}1\\2\\3\\5\end{pmatrix}$

This gives

$$\mathbf{x} = 4^{-1} \left[\begin{pmatrix} 1 \\ 2 \\ 3 \\ 5 \end{pmatrix} - 3 \begin{pmatrix} 5 \\ 6 \\ 3 \\ 5 \end{pmatrix} - 5 \begin{pmatrix} 5 \\ 3 \\ 5 \\ 4 \end{pmatrix} \right] = \begin{pmatrix} 6 \\ 1 \\ 1 \\ 3 \end{pmatrix}. \qquad \square$$

Although the Lanczos and the Wiedemann methods look different at the first sight, there is a commonness between them. For a non-zero vector $\mathbf{b}$, the i-th *Krylov vector* is defined as $\mathbf{u}_{i-1} = A^{i-1}\mathbf{b}$. Clearly, there exists $s \in \mathbb{N}$ such that $\mathbf{u}_0, \mathbf{u}_1, \ldots, \mathbf{u}_{s-1}$ are linearly independent, whereas $\mathbf{u}_0, \mathbf{u}_1, \ldots, \mathbf{u}_{s-1}, \mathbf{u}_s$ are linearly dependent. The span of $\mathbf{u}_0, \mathbf{u}_1, \ldots, \mathbf{u}_{s-1}$ is called the *Krylov space* for A and $\mathbf{b}$. Both Lanczos and Wiedemann methods express the solution of $A\mathbf{x} = \mathbf{b}$ as a linear combination of the Krylov vectors. In view of this, these methods are often called *Krylov space methods*.

8.4 Block Methods

We now specialize to systems $A\mathbf{x} = \mathbf{b}$ modulo 2. Each coefficient in A is now a bit (0 or 1), and one packs multiple coefficients (like 32 or 64) in a computer word. One operates in blocks, that is, does arithmetic at a word level so as to process all the coefficients in a word simultaneously. These block methods run more efficiently than the methods that handle coefficients individually. I shortly explain the block versions of the Lanczos and the Wiedemann methods. Among several block adaptations of the Lanczos solver, I concentrate on Montgomery's variant.[10] Another interesting variant is from Coppersmith.[11] The block Wiedemann algorithm is proposed by Coppersmith.[12] In this section, I assume that A is an $n \times n$ matrix over a field K. Our case of interest is $K = \mathbb{F}_2$, but I will present the equations in a form valid for any K.

8.4.1 Block Lanczos Method

Instead of working on pairwise A-orthogonal vectors $\mathbf{w}_0, \mathbf{w}_1, \ldots, \mathbf{w}_s$, we now work with pairwise A-orthogonal *subspaces* $\mathcal{W}_0, \mathcal{W}_1, \ldots, \mathcal{W}_s$ of K^n. For subspaces $\mathcal{V}, \mathcal{W}$ of K^n, we first define some standard operations.

$$
\begin{aligned}
\mathcal{V} + \mathcal{W} &= \{\mathbf{v} + \mathbf{w} \mid \mathbf{v} \in \mathcal{V}, \text{ and } \mathbf{w} \in \mathcal{W}\}, \\
\mathcal{V}^{\mathrm{t}}\mathcal{W} &= \{\mathbf{v}^{\mathrm{t}}\mathbf{w} \mid \mathbf{v} \in \mathcal{V}, \text{ and } \mathbf{w} \in \mathcal{W}\}, \\
A\mathcal{V} &= \{A\mathbf{v} \mid \mathbf{v} \in \mathcal{V}\}.
\end{aligned}
$$

$\mathcal{V}$ and $\mathcal{W}$ are called *A-orthogonal* if $\mathcal{V}^{\mathrm{t}}A\mathcal{W} = \{\mathbf{0}\}$, that is, if $\mathbf{v}^{\mathrm{t}}A\mathbf{w} = \mathbf{0}$ for all $\mathbf{v} \in \mathcal{V}$ and $\mathbf{w} \in \mathcal{W}$. Let the subspace $\mathcal{V} \subseteq K^n$ have dimension ν. Fix any basis $\mathbf{v}_0, \mathbf{v}_1, \ldots, \mathbf{v}_{\nu-1}$ of $\mathcal{V}$, and consider the $n \times \nu$ matrix $V = (\mathbf{v}_0 \ \mathbf{v}_1 \ \cdots \ \mathbf{v}_{\nu-1})$. $\mathcal{V}$ is called *A-invertible* if the $\nu \times \nu$ matrix $V^{\mathrm{t}}AV$ is invertible. Since different

[10] Peter L. Montgomery, A block Lanczos algorithm for finding dependencies over GF(2), *EuroCrypt*, 106–120, 1995.

[11] Don Coppersmith, Solving linear equations over GF(2): Block Lanczos algorithm, *Linear Algebra and its Applications*, 192, 33–60, 1993.

[12] Don Coppersmith, Solving homogeneous linear equations over GF(2) via block Wiedemann algorithm, *Mathematics of Computation*, 62(205), 333–350, 1994.

bases of the same subspace are related by invertible transformation matrices, the notion of A-invertibility does not depend on the choice of the basis of $\mathcal{V}$.

The block Lanczos method assumes that A is a symmetric matrix. The subspaces $\mathcal{W}_0, \mathcal{W}_1, \ldots, \mathcal{W}_{s-1}$ generated by the method satisfy the conditions:

$\mathcal{W}_i$ is A-invertible for all $i = 0, 1, 2, \ldots, s - 1$,
$\mathcal{W}_i$ and $\mathcal{W}_j$ are A-orthogonal for $i \neq j$, and
$A\mathcal{W} \subseteq \mathcal{W}$, where $\mathcal{W} = \mathcal{W}_0 + \mathcal{W}_1 + \cdots + \mathcal{W}_{s-1}$.

The iterations stop as soon as we obtain the zero space $\mathcal{W}_s$. For the subspace $\mathcal{W}_j$, we consider the matrix W_j whose columns constitute a basis of $\mathcal{W}_j$. Then, the solution of $A\mathbf{x} = \mathbf{b}$ is given by

$$\mathbf{x} = \sum_{j=0}^{s-1} W_j (W_j^t A W_j)^{-1} W_j^t \mathbf{b}.$$

In the original Lanczos method, $\mathcal{W}_i$ is the one-dimensional subspace of K^n, generated by $\mathbf{w}_i$. In the block Lanczos method, we generate higher-dimensional subspaces instead of individual vectors. This helps us in two ways. First, block operations on words process multiple dimensions simultaneously. Second, the number s of iterations reduces roughly by a factor of the word size.

It remains to explain how the subspaces $\mathcal{W}_i$ or equivalently the matrices W_i of basis vectors are generated. Let ν denote the word size (like 32 or 64). We iteratively generate $n \times \nu$ matrices $V_0, V_1, \ldots, V_s$, and select some columns from each V_i to obtain W_i. More concretely, we choose $\nu \times \nu_i$ selection matrices S_i, and obtain $W_i = V_i S_i$, where S_i dictates which of the ν_i columns of V_i are to be included in W_i. For instance, let $\nu = 4$, and $\nu_i = 3$. We take

$$S_i = \begin{pmatrix} 1 & 0 & 0 \\ 0 & 0 & 0 \\ 0 & 1 & 0 \\ 0 & 0 & 1 \end{pmatrix}$$

if we plan to select the first, third and fourth columns of V_i for inclusion in W_i. Each column of S_i contains exactly one 1 in that row indicating the column of V_i to be selected. Note that $S_i^t S_i$ is the $\nu_i \times \nu_i$ identity matrix I_{ν_i}.

We start with a randomly chosen $n \times \nu$ matrix V_0. We plan to keep as many rows of V_0 in W_0 as possible. The requirement is that W_0 has to be A-invertible. The selection matrix S_0 is chosen accordingly, and we set $W_0 = V_0 S_0$.

After this initialization, the Lanczos loop continues for $i = 1, 2, 3, \ldots$. In the i-th iteration, an $n \times \nu$ matrix V_i is first computed using the formula

$$V_i = A W_{i-1} S_{i-1}^t + V_{i-1} - \sum_{j=0}^{i-1} W_j C_{i,j}, \text{ where}$$

$$C_{i,j} = (W_j^t A W_j)^{-1} W_j^t A (A W_{i-1} S_{i-1}^t + V_{i-1}) \text{ for } j = 0, 1, \ldots, i - 1.$$

If $V_i^t A V_i = 0$, the iterations stop.

All W_j are needed to be A-invertible for the computation of V_i to succeed. The matrix V_i computed above is A-orthogonal to W_j for $j = 0, 1, 2, \ldots, i-1$. However, V_i need not be A-invertible, so we use a selection matrix S_i to choose as many columns from V_i as possible, and form an A-invertible matrix W_i:

$$W_i = V_i S_i. \tag{8.46}$$

Evidently, W_i remains A-orthogonal to all previous W_j. The A-invertibility of W_i is required in later iterations.

Like the original Lanczos algorithm, the formula for computing V_i can be significantly simplified. We can take $C_{i,j} = 0$ for $j < i - 3$, so each updating formula for V_i uses only three previous V_j matrices (for $j = i-1, i-2, i-3$). The simplified formula uses the following intermediate matrices:

$$
\begin{aligned}
W_{i-1}^{\text{inv}} &= S_{i-1}(W_{i-1}^{\text{t}} A W_{i-1})^{-1} S_{i-1}^{\text{t}} \\
&= S_{i-1}(S_{i-1}^{\text{t}} V_{i-1}^{\text{t}} A V_{i-1} S_{i-1})^{-1} S_{i-1}^{\text{t}}, \tag{8.47} \\
D_i &= W_{i-1}^{\text{inv}}(V_{i-1}^{\text{t}} A^2 V_{i-1} S_{i-1} S_{i-1}^{\text{t}} + V_{i-1}^{\text{t}} A V_{i-1}) - I_\nu, \tag{8.48} \\
E_i &= W_{i-2}^{\text{inv}} V_{i-1}^{\text{t}} A V_{i-1} S_{i-1} S_{i-1}^{\text{t}}, \tag{8.49} \\
F_i &= W_{i-3}^{\text{inv}}(I_\nu - V_{i-2}^{\text{t}} A V_{i-2} W_{i-2}^{\text{inv}}) \\
&\quad (V_{i-2}^{\text{t}} A^2 V_{i-2} S_{i-2} S_{i-2}^{\text{t}} + V_{i-2}^{\text{t}} A V_{i-2}) S_{i-1} S_{i-1}^{\text{t}}. \tag{8.50}
\end{aligned}
$$

The simplified formula is as follows:

$$V_i = A V_{i-1} S_{i-1} S_{i-1}^{\text{t}} - V_{i-1} D_i - V_{i-2} E_i - V_{i-3} F_i. \tag{8.51}$$

This formula is valid for $i \geqslant 3$. For $j < 0$, we take V_j and W_j^{inv} as the $n \times \nu$ and $\nu \times \nu$ zero matrices, and S_j as the identity matrix I_ν. If so, Eqn (8.51) holds for all $i \geqslant 1$. Algorithm 8.3 summarizes the block Lanczos method.

Algorithm 8.3: BLOCK LANCZOS METHOD FOR SOLVING $Ax = b$

```
/* Initialize for i < 0 */
```
$W_{-2}^{\text{inv}} = W_{-1}^{\text{inv}} = 0_{\nu \times \nu}$, $V_{-2} = V_{-1} = 0_{n \times \nu}$, and $S_{-2} = S_{-1} = I_\nu$.

```
/* Initialize for i = 0 */
```
Take V_0 as a random $n \times \nu$ matrix.
Select a maximal A-invertible set of columns of V_0 as $W_0 = V_0 S_0$.
Initialize the solution $x = 0$, and set $i = 0$.

```
/* Lanczos loop for i = 1, 2, 3, ..., s */
while (V_i^t A V_i ≠ 0) {
```
    Increment i by 1.
    Compute W_{i-1}^{inv}, D_i, E_i and F_i using Eqns (8.47)−(8.50).
    Compute V_i using Eqn (8.51).
    Select a maximal A-invertible set of columns of V_i as $W_i = V_i S_i$.
    Add $W_{i-1}(W_{i-1}^{\text{t}} A W_{i-1})^{-1} W_{i-1}^{\text{t}} b$ to the solution vector x.
```
}
Return x.
```

Example 8.5 Let me demonstrate the working of the block Lanczos method on artificially small parameters. We solve the following symmetric 10×10 system modulo two. Let us take the word size as $\nu = 4$.

$$\begin{pmatrix} 0 & 1 & 0 & 1 & 0 & 0 & 0 & 1 & 1 & 0 \\ 1 & 0 & 0 & 1 & 1 & 0 & 1 & 1 & 1 & 0 \\ 0 & 0 & 0 & 0 & 0 & 0 & 1 & 0 & 0 & 1 \\ 1 & 1 & 0 & 0 & 1 & 1 & 1 & 1 & 0 & 1 \\ 0 & 1 & 0 & 1 & 1 & 0 & 1 & 0 & 1 & 1 \\ 0 & 0 & 0 & 1 & 0 & 1 & 0 & 0 & 0 & 1 \\ 0 & 1 & 1 & 1 & 1 & 0 & 1 & 0 & 1 & 0 \\ 1 & 1 & 0 & 1 & 0 & 0 & 0 & 1 & 0 & 0 \\ 1 & 1 & 0 & 0 & 1 & 0 & 1 & 0 & 1 & 0 \\ 0 & 0 & 1 & 1 & 1 & 1 & 0 & 0 & 0 & 0 \end{pmatrix} \begin{pmatrix} x_1 \\ x_2 \\ x_3 \\ x_4 \\ x_5 \\ x_6 \\ x_7 \\ x_8 \\ x_9 \\ x_{10} \end{pmatrix} = \begin{pmatrix} 1 \\ 0 \\ 0 \\ 1 \\ 1 \\ 0 \\ 0 \\ 0 \\ 0 \\ 0 \end{pmatrix}.$$

Initialization for $i = -2, -1$: $W_{-2}^{\text{inv}} = W_{-1}^{\text{inv}} = \begin{pmatrix} 0 & 0 & 0 & 0 \\ 0 & 0 & 0 & 0 \\ 0 & 0 & 0 & 0 \\ 0 & 0 & 0 & 0 \end{pmatrix}$, $V_{-2} = V_{-1} =$

$$\begin{pmatrix} 0 & 0 & 0 & 0 \\ 0 & 0 & 0 & 0 \\ 0 & 0 & 0 & 0 \\ 0 & 0 & 0 & 0 \\ 0 & 0 & 0 & 0 \\ 0 & 0 & 0 & 0 \\ 0 & 0 & 0 & 0 \\ 0 & 0 & 0 & 0 \\ 0 & 0 & 0 & 0 \\ 0 & 0 & 0 & 0 \end{pmatrix}, \text{ and } S_{-2} = S_{-1} = \begin{pmatrix} 1 & 0 & 0 & 0 \\ 0 & 1 & 0 & 0 \\ 0 & 0 & 1 & 0 \\ 0 & 0 & 0 & 1 \end{pmatrix}.$$

Initialization for $i = 0$: We start with the following randomly generated

$$V_0 = \begin{pmatrix} 0 & 1 & 0 & 0 \\ 0 & 1 & 1 & 1 \\ 1 & 1 & 1 & 0 \\ 1 & 0 & 0 & 1 \\ 0 & 0 & 0 & 0 \\ 0 & 0 & 1 & 1 \\ 0 & 0 & 0 & 0 \\ 1 & 0 & 0 & 1 \\ 1 & 0 & 1 & 1 \\ 1 & 0 & 0 & 0 \end{pmatrix}.$$ A set of three A-invertible columns of V_0 is chosen by

the selection matrix $S_0 = \begin{pmatrix} 1 & 0 & 0 \\ 0 & 1 & 0 \\ 0 & 0 & 0 \\ 0 & 0 & 1 \end{pmatrix}$ giving $W_0 = V_0 S_0 = \begin{pmatrix} 0 & 1 & 0 \\ 0 & 1 & 1 \\ 1 & 1 & 0 \\ 1 & 0 & 1 \\ 0 & 0 & 0 \\ 0 & 0 & 1 \\ 0 & 0 & 0 \\ 1 & 0 & 1 \\ 1 & 0 & 1 \\ 1 & 0 & 0 \end{pmatrix}$ (the first, second and fourth columns of V_0). Finally, we initialize the solution vector: $\mathbf{x} = (0 \ \ 0 \ \ 0 \ \ 0 \ \ 0 \ \ 0 \ \ 0 \ \ 0 \ \ 0 \ \ 0)^t$.

Iteration for $i = 1$**:** First, we compute $V_0^t A V_0 = \begin{pmatrix} 0 & 1 & 1 & 0 \\ 1 & 0 & 1 & 1 \\ 1 & 1 & 0 & 1 \\ 0 & 1 & 1 & 1 \end{pmatrix}$ which is

non-zero. So we go inside the loop body, and compute the temporary matrices

$$W_0^{inv} = \begin{pmatrix} 1 & 1 & 0 & 1 \\ 1 & 0 & 0 & 0 \\ 0 & 0 & 0 & 0 \\ 1 & 0 & 0 & 1 \end{pmatrix}, D_1 = \begin{pmatrix} 0 & 1 & 1 & 0 \\ 0 & 1 & 1 & 1 \\ 0 & 0 & 1 & 0 \\ 1 & 1 & 0 & 0 \end{pmatrix}, E_1 = \begin{pmatrix} 0 & 0 & 0 & 0 \\ 0 & 0 & 0 & 0 \\ 0 & 0 & 0 & 0 \\ 0 & 0 & 0 & 0 \end{pmatrix}, \text{ and } F_1 =$$

$$\begin{pmatrix} 0 & 0 & 0 & 0 \\ 0 & 0 & 0 & 0 \\ 0 & 0 & 0 & 0 \\ 0 & 0 & 0 & 0 \end{pmatrix}. \text{ Next, we compute } V_1 = \begin{pmatrix} 1 & 0 & 1 & 1 \\ 0 & 1 & 0 & 0 \\ 1 & 0 & 1 & 1 \\ 1 & 0 & 1 & 1 \\ 1 & 1 & 0 & 1 \\ 1 & 1 & 1 & 0 \\ 1 & 0 & 0 & 1 \\ 1 & 0 & 1 & 1 \\ 0 & 0 & 0 & 0 \\ 0 & 0 & 1 & 0 \end{pmatrix}. \text{ A selection matrix}$$

for V_1 is $S_1 = \begin{pmatrix} 1 & 0 & 0 & 0 \\ 0 & 1 & 0 & 0 \\ 0 & 0 & 1 & 0 \\ 0 & 0 & 0 & 1 \end{pmatrix}$ yielding $W_1 = V_1$. Finally, we update the solution

vector to $\mathbf{x} = (1 \ \ 1 \ \ 0 \ \ 1 \ \ 0 \ \ 0 \ \ 0 \ \ 1 \ \ 1 \ \ 1)^t$.

Two more iterations are needed before the block Lanczos algorithm terminates. Computations in these iterations are shown in the table on the next page. The second iteration ($i = 2$) updates the solution vector to

$$\mathbf{x} = (1 \ \ 1 \ \ 0 \ \ 1 \ \ 1 \ \ 1 \ \ 1 \ \ 1 \ \ 1 \ \ 0)^t,$$

whereas the third iteration ($i = 3$) updates $\mathbf{x}$ to

$$\mathbf{x} = (0 \ \ 0 \ \ 1 \ \ 1 \ \ 1 \ \ 1 \ \ 0 \ \ 1 \ \ 1 \ \ 0).$$

The loop terminates after this iteration, so this is the final solution. □

Let us now investigate when Algorithm 8.3 may fail because of the lack of positive-definiteness of the matrix A. Example 8.5 illustrates the situation that V_3 has become 0, and so $V_3^t A V_3$ is 0 too. In the earlier iteration ($i = 2$), V_2 is non-zero, but $V_2^t A V_2$ is of rank 3 (less than its dimension 4). The selection matrix S_2 ensures that only three vectors are added to the Krylov space.

In general, any V_i offers $2^\nu - 1$ non-empty choices for the selection matrix S_i. If $V_i \neq 0$, any one of these choices, that yields an A-invertible W_i, suffices. Typically, $\nu \geqslant 32$, whereas we solve systems of size n no larger than a few hundreds of millions (larger systems are infeasible anyway). Moreover, during each iteration, multiple vectors are added to the Krylov space in general, that is, the number of iterations is expected to be substantially smaller than n.

	$i = 2$	$i = 3$	$i = 4$
$V_{i-1}^{t}AV_{i-1}$	$\begin{pmatrix} 0 & 0 & 1 & 0 \\ 0 & 0 & 0 & 1 \\ 1 & 0 & 0 & 0 \\ 0 & 1 & 0 & 1 \end{pmatrix}$	$\begin{pmatrix} 0 & 0 & 0 & 1 \\ 0 & 0 & 0 & 0 \\ 0 & 0 & 1 & 0 \\ 1 & 0 & 0 & 0 \end{pmatrix}$	$\begin{pmatrix} 0 & 0 & 0 & 0 \\ 0 & 0 & 0 & 0 \\ 0 & 0 & 0 & 0 \\ 0 & 0 & 0 & 0 \end{pmatrix}$
W_{i-1}^{inv}	$\begin{pmatrix} 0 & 0 & 1 & 0 \\ 0 & 1 & 0 & 1 \\ 1 & 0 & 0 & 0 \\ 0 & 1 & 0 & 0 \end{pmatrix}$	$\begin{pmatrix} 0 & 0 & 0 & 1 \\ 0 & 0 & 0 & 0 \\ 0 & 0 & 1 & 0 \\ 1 & 0 & 0 & 0 \end{pmatrix}$	
D_i	$\begin{pmatrix} 0 & 1 & 0 & 0 \\ 0 & 0 & 1 & 0 \\ 0 & 1 & 0 & 1 \\ 1 & 1 & 1 & 1 \end{pmatrix}$	$\begin{pmatrix} 0 & 0 & 1 & 0 \\ 0 & 1 & 0 & 0 \\ 0 & 0 & 1 & 1 \\ 1 & 0 & 0 & 0 \end{pmatrix}$	
E_i	$\begin{pmatrix} 0 & 1 & 1 & 0 \\ 0 & 0 & 1 & 0 \\ 0 & 0 & 0 & 0 \\ 0 & 1 & 1 & 1 \end{pmatrix}$	$\begin{pmatrix} 0 & 0 & 1 & 0 \\ 1 & 0 & 0 & 0 \\ 0 & 0 & 0 & 1 \\ 0 & 0 & 0 & 0 \end{pmatrix}$	
F_i	$\begin{pmatrix} 0 & 0 & 0 & 0 \\ 0 & 0 & 0 & 0 \\ 0 & 0 & 0 & 0 \\ 0 & 0 & 0 & 0 \end{pmatrix}$	$\begin{pmatrix} 0 & 0 & 0 & 0 \\ 0 & 0 & 0 & 0 \\ 0 & 0 & 0 & 0 \\ 0 & 0 & 0 & 0 \end{pmatrix}$	
V_i	$\begin{pmatrix} 1 & 0 & 0 & 0 \\ 1 & 0 & 0 & 0 \\ 0 & 0 & 0 & 1 \\ 0 & 0 & 1 & 1 \\ 0 & 0 & 0 & 0 \\ 0 & 0 & 1 & 1 \\ 1 & 0 & 1 & 1 \\ 0 & 0 & 0 & 0 \\ 1 & 0 & 1 & 0 \\ 0 & 0 & 0 & 0 \end{pmatrix}$	$\begin{pmatrix} 0 & 0 & 0 & 0 \\ 0 & 0 & 0 & 0 \\ 0 & 0 & 0 & 0 \\ 0 & 0 & 0 & 0 \\ 0 & 0 & 0 & 0 \\ 0 & 0 & 0 & 0 \\ 0 & 0 & 0 & 0 \\ 0 & 0 & 0 & 0 \\ 0 & 0 & 0 & 0 \\ 0 & 0 & 0 & 0 \end{pmatrix}$	
S_i	$\begin{pmatrix} 1 & 0 & 0 \\ 0 & 0 & 0 \\ 0 & 1 & 0 \\ 0 & 0 & 1 \end{pmatrix}$	$\begin{pmatrix} \\ \\ \end{pmatrix}$	
W_i	$\begin{pmatrix} 1 & 0 & 0 \\ 1 & 0 & 0 \\ 0 & 0 & 1 \\ 0 & 1 & 1 \\ 0 & 0 & 0 \\ 0 & 1 & 1 \\ 1 & 1 & 1 \\ 0 & 0 & 0 \\ 1 & 1 & 0 \\ 0 & 0 & 0 \end{pmatrix}$	$\begin{pmatrix} \\ \\ \\ \\ \end{pmatrix}$	

Thus, it is highly probable that we can always find a suitable subset of the columns of a non-zero V_i to form an A-invertible W_i. On the other hand, if $V_i = 0$, no selection matrix S_i can produce an A-invertible W_i. This means that although the modulus is small (only 2), there is not much of a need to work in extension fields for the algorithm to succeed.

8.4.2 Block Wiedemann Method

Suppose that we want to solve an $m \times n$ system $A\mathbf{x} = \mathbf{b}$ of linear equations over the field K. If $\mathbf{b} \neq \mathbf{0}$, we introduce a new variable x_{n+1}, and convert the original system to the homogeneous form

$$(A \quad -\mathbf{b}) \begin{pmatrix} \mathbf{x} \\ x_{n+1} \end{pmatrix} = \mathbf{0}_{n+1}.$$

Any solution of this $m \times (n+1)$ system with $x_{n+1} = 1$ gives a solution of the original system $A\mathbf{x} = \mathbf{b}$. Without loss in generality, we can, therefore, take $\mathbf{b} = \mathbf{0}$. Moreover, there is no harm in taking $m = n$ (premultiply by A^t), that is, we concentrate upon a *square* $(n \times n)$ *homogeneous* system $A\mathbf{x} = \mathbf{0}$.

In the original Wiedemann algorithm, we compute the minimal polynomial of the scalar sequence $\mathbf{u}^t A^i \mathbf{v}$ for a randomly chosen vector $\mathbf{v}$ and for a projection vector $\mathbf{u}$. In the block Wiedemann method, we take a block of μ vectors as U and a block of ν vectors as V, that is, U is an $n \times \mu$ matrix, whereas V is an $n \times \nu$ matrix. We typically have $\mu = \nu =$ the size of a computer word (like 32 or 64). Take $W = AV$. Consider the sequence

$$M_i = U^t A^i W \quad \text{for } i \geqslant 0$$

of $\mu \times \nu$ matrices. Let $d = \lceil n/\nu \rceil$. There exist coefficient vectors $\mathbf{c}_0, \mathbf{c}_1, \ldots, \mathbf{c}_d$ of dimension $\nu \times 1$ such that the sequence M_i is linearly generated as

$$M_k \mathbf{c}_d + M_{k-1} \mathbf{c}_{d-1} + M_{k-2} \mathbf{c}_{d-2} + \cdots + M_{k-d+1} \mathbf{c}_1 + M_{k-d} \mathbf{c}_0 = \mathbf{0}_\mu$$

for all $k \geqslant d$. Let $e = \left\lceil \frac{\nu(d+1)}{\mu} \right\rceil$. Applying the above recurrence for $k = d, d+1, \ldots, d+e-1$ yields the system

$$\begin{pmatrix} M_d & M_{d-1} & M_{d-2} & \cdots & M_1 & M_0 \\ M_{d+1} & M_d & M_{d-1} & \cdots & M_2 & M_1 \\ \cdots & \cdots & \cdots & \vdots & \cdots & \cdots \\ M_{d+e-1} & M_{d+e-2} & M_{d+e-3} & \cdots & M_e & M_{e-1} \end{pmatrix} \begin{pmatrix} \mathbf{c}_d \\ \mathbf{c}_{d-1} \\ \mathbf{c}_{d-2} \\ \vdots \\ \mathbf{c}_1 \\ \mathbf{c}_0 \end{pmatrix} = \mathbf{0}_{\mu e}. \quad (8.52)$$

This $(\mu e) \times \nu(d+1)$ system is the block analog of Eqn (8.36). For a moment, assume that a solution for $\mathbf{c}_0, \mathbf{c}_1, \ldots, \mathbf{c}_d$ is provided to us. With high probability, the sequence of $n \times \nu$ matrices $A^i W$, $i \geqslant 0$, also satisfies the recurrence as done by their $\mu \times \nu$ projections M_i. For all $k \geqslant d$, we then have

$$A^k W \mathbf{c}_d + A^{k-1} W \mathbf{c}_{d-1} + A^{k-2} W \mathbf{c}_{d-2} + \cdots + A^{k-d+1} W \mathbf{c}_1 + A^{k-d} W \mathbf{c}_0 = \mathbf{0}_n.$$

Putting $k = d$ and using the fact that $W = AV$, we get

$$A(A^d V \mathbf{c}_d + A^{d-1} V \mathbf{c}_{d-1} + A^{d-2} V \mathbf{c}_{d-2} + \cdots + AV \mathbf{c}_1 + V \mathbf{c}_0) = \mathbf{0}_n,$$

that is, a solution of $A\mathbf{x} = \mathbf{0}$ is given by

$$\mathbf{x} = A^d V \mathbf{c}_d + A^{d-1} V \mathbf{c}_{d-1} + A^{d-2} V \mathbf{c}_{d-2} + \cdots + AV \mathbf{c}_1 + V \mathbf{c}_0.$$

Let us now see how we can solve System (8.52). Coppersmith (Footnote 12) proposes a generalization of the Berlekamp–Massey algorithm for linear sequences generated by vectors. The procedure runs in the desired $O(n^2)$ time, but is somewhat complicated. Here, I explain a conceptually simpler algorithm from Kaltofen[13], which achieves the same running time. Kaltofen's algorithm exploits the fact that the coefficient matrix in (8.52) is in the Toeplitz form with scalars replaced by $\mu \times \nu$ matrix blocks. In practice, System (8.52) need not be square (in terms of both elements and blocks). Kaltofen carefully handles square subsystems which may slightly lose the Toeplitz property. Although these *almost Toeplitz systems* can be solved in $O(n^2)$ time, I present a simplified version of the algorithm assuming that $\mu = \nu$, so that we have both $e = d + 1$ and $\mu e = \nu(d + 1)$. This iterative algorithm is proposed by Levinson, and modified by Durbin, Trench and Zohar.[14]

To start with, let us solve a Toeplitz system of *scalars*, that is, let

$$T\mathbf{x} = \mathbf{b} \tag{8.53}$$

be an $n \times n$ system with the i, j-th entry of T given by $t_{i-j} \in K$ (a function of $i - j$). This could be an attempt to solve (8.36) in the original Wiedemann method.[15] Let $T^{(i)}$ be the $i \times i$ submatrix sitting at the top left corner of T:

$$T^{(i)} = \begin{pmatrix} t_0 & t_{-1} & t_{-2} & \cdots & t_{-i+1} \\ t_1 & t_0 & t_{-1} & \cdots & t_{-i+2} \\ \vdots & \vdots & \vdots & \cdots & \vdots \\ t_{i-1} & t_{i-2} & t_{i-3} & \cdots & t_0 \end{pmatrix}.$$

Clearly, $T = T^{(n)}$. Likewise, let $\mathbf{b}^{(i)}$ denote the $i \times 1$ vector obtained by the top i elements of $\mathbf{b}$ (we have $\mathbf{b} = \mathbf{b}^{(n)}$). We iteratively solve the system

$$T^{(i)} \mathbf{x}^{(i)} = \mathbf{b}^{(i)}$$

for $i = 1, 2, 3, \ldots, n$. We also keep on computing and using two auxiliary vectors $\mathbf{y}^{(i)}$ and $\mathbf{z}^{(i)}$ satisfying

$$T^{(i)} \mathbf{y}^{(i)} = \begin{pmatrix} \epsilon^{(i)} \\ \mathbf{0}_{i-1} \end{pmatrix} \quad \text{and} \quad T^{(i)} \mathbf{z}^{(i)} = \begin{pmatrix} \mathbf{0}_{i-1} \\ \epsilon^{(i)} \end{pmatrix} \tag{8.54}$$

---

[13]Erich Kaltofen, Analysis of Coppersmith's block Wiedemann algorithm for the parallel solution of sparse linear systems, *Mathematics of Computation*, 64(210), 777–806, 1995.

[14]For a nice survey, look at: Bruce R. Musicus, Levinson and fast Choleski algorithms for Toeplitz and almost Toeplitz matrices, *Technical Report* 538, Research Laboratory of Electronics, Massachusetts Institute of Technology, December 1988.

[15]We intend to solve System (8.36) or (8.52) where the variables are denoted by c_i or $\mathbf{c}_i$. Here, we use x_i (or X_i in Example 8.7) for variables. Likewise, for $\mathbf{b}$ (or B). This notational inconsistency is motivated by that solving Toeplitz systems is itself of independent interest.

for a suitable choice of the scalar $\epsilon^{(i)}$ to be specified later. In this section, I use the parenthesized superscript $^{(i)}$ to indicate quantities in the i-th iteration. In an actual implementation, it suffices to remember the quantities from only the previous iteration. Superscripts are used for logical clarity. Subscripts are used for matrix and vector elements (like t_i) and dimensions (like $\mathbf{0}_{i-1}$).

For $i = 1$, we have $t_0 x_1 = b_1$ which immediately gives $\mathbf{x}^{(1)} = (t_0^{-1} b_1)$, provided that $t_0 \neq 0$. We also take $\mathbf{y}^{(1)} = \mathbf{z}^{(1)} = (1)$, that is, $\epsilon^{(1)} = t_0$.

Suppose that $T^{(i)}\mathbf{x}^{(i)} = \mathbf{b}^{(i)}$ is solved, and we plan to compute a solution of $T^{(i+1)}\mathbf{x}^{(i+1)} = \mathbf{b}^{(i+1)}$. At this stage, the vectors $\mathbf{y}^{(i)}$ and $\mathbf{z}^{(i)}$, and the scalar $\epsilon^{(i)}$ are known. We write

$$T^{(i+1)} = \begin{pmatrix} & & t_{-i} \\ T^{(i)} & & \vdots \\ t_i \ t_{i-1} \ \cdots & t_0 \end{pmatrix} = \begin{pmatrix} t_0 \ t_{-1} \ \cdots \ t_{-i} \\ \vdots & T^{(i)} \\ t_i \end{pmatrix}.$$

This implies that the following equalities hold:

$$T^{(i+1)} \begin{pmatrix} \mathbf{y}^{(i)} \\ 0 \end{pmatrix} = \begin{pmatrix} \epsilon^{(i)} \\ \mathbf{0}_{i-1} \\ -\epsilon^{(i)}\xi^{(i+1)} \end{pmatrix} \quad \text{and} \quad T^{(i+1)} \begin{pmatrix} 0 \\ \mathbf{z}^{(i)} \end{pmatrix} = \begin{pmatrix} -\epsilon^{(i)}\zeta^{(i+1)} \\ \mathbf{0}_{i-1} \\ \epsilon^{(i)} \end{pmatrix},$$

where

$$\xi^{(i+1)} = -\frac{1}{\epsilon^{(i)}} \begin{pmatrix} t_i & t_{i-1} & \cdots & t_1 \end{pmatrix} \mathbf{y}^{(i)}, \quad \text{and} \tag{8.55}$$

$$\zeta^{(i+1)} = -\frac{1}{\epsilon^{(i)}} \begin{pmatrix} t_{-1} & t_{-2} & \cdots & t_{-i} \end{pmatrix} \mathbf{z}^{(i)}. \tag{8.56}$$

We compute $\mathbf{y}^{(i+1)}$ and $\mathbf{z}^{(i+1)}$ as linear combinations of $\begin{pmatrix} \mathbf{y}^{(i)} \\ 0 \end{pmatrix}$ and $\begin{pmatrix} 0 \\ \mathbf{z}^{(i)} \end{pmatrix}$:

$$\mathbf{y}^{(i+1)} = \begin{pmatrix} \mathbf{y}^{(i)} \\ 0 \end{pmatrix} + \xi^{(i+1)} \begin{pmatrix} 0 \\ \mathbf{z}^{(i)} \end{pmatrix}, \quad \text{and} \tag{8.57}$$

$$\mathbf{z}^{(i+1)} = \begin{pmatrix} 0 \\ \mathbf{z}^{(i)} \end{pmatrix} + \zeta^{(i+1)} \begin{pmatrix} \mathbf{y}^{(i)} \\ 0 \end{pmatrix}. \tag{8.58}$$

This requires us to take

$$\epsilon^{(i+1)} = \epsilon^{(i)}(1 - \xi^{(i+1)}\zeta^{(i+1)}). \tag{8.59}$$

Finally, we update the solution $\mathbf{x}^{(i)}$ to $\mathbf{x}^{(i+1)}$ by noting that

$$T^{(i+1)} \begin{pmatrix} \mathbf{x}^{(i)} \\ 0 \end{pmatrix} = \begin{pmatrix} \mathbf{b}^{(i)} \\ \eta^{(i+1)} \end{pmatrix} = \mathbf{b}^{(i+1)} + \left(\frac{\eta^{(i+1)} - b_{i+1}}{\epsilon^{(i+1)}} \right) T^{(i+1)}\mathbf{z}^{(i+1)},$$

where

$$\eta^{(i+1)} = \begin{pmatrix} t_i & t_{i-1} & \cdots & t_1 \end{pmatrix} \mathbf{x}^{(i)}, \tag{8.60}$$

that is,

$$\mathbf{x}^{(i+1)} = \begin{pmatrix} \mathbf{x}^{(i)} \\ 0 \end{pmatrix} + \left(\frac{b_{i+1} - \eta^{(i+1)}}{\epsilon^{(i+1)}} \right) \mathbf{z}^{(i+1)}. \tag{8.61}$$

Example 8.6 Let us solve the following 6×6 non-homogeneous Toeplitz system modulo the prime $p = 97$:

$$\begin{pmatrix} 85 & 92 & 79 & 6 & 15 & 72 \\ 39 & 85 & 92 & 79 & 6 & 15 \\ 87 & 39 & 85 & 92 & 79 & 6 \\ 42 & 87 & 39 & 85 & 92 & 79 \\ 82 & 42 & 87 & 39 & 85 & 92 \\ 30 & 82 & 42 & 87 & 39 & 85 \end{pmatrix} \begin{pmatrix} x_1 \\ x_2 \\ x_3 \\ x_4 \\ x_5 \\ x_6 \end{pmatrix} \equiv \begin{pmatrix} 82 \\ 53 \\ 31 \\ 12 \\ 14 \\ 48 \end{pmatrix} \pmod{97} \qquad (8.62)$$

We start with the initialization

$$\epsilon^{(1)} = 85, \ \mathbf{y}^{(1)} = \mathbf{z}^{(1)} = (\,1\,), \ \mathbf{x}^{(1)} = (\,85^{-1} \times 82\,) = (\,74\,).$$

Subsequently, for $i = 2, 3, 4, 5, 6$, we run the Levinson iteration. In each iteration, $\xi^{(i+1)}$ and $\zeta^{(i+1)}$ are first computed using Eqns (8.55) and (8.56), and then $\epsilon^{(i+1)}$ is computed using Eqn (8.59). The scalars $\xi^{(i+1)}$ and $\zeta^{(i+1)}$ allow us to compute $\mathbf{y}^{(i+1)}$ and $\mathbf{z}^{(i+1)}$ from Eqns (8.57) and (8.58). Finally, $\eta^{(i+1)}$ is computed by Eqn (8.60), and the updated solution vector $\mathbf{x}^{(i+1)}$ is obtained using Eqn (8.61). The following table illustrates the iterations.

i	$\xi^{(i)}$	$\zeta^{(i)}$	$\epsilon^{(i)}$	$\mathbf{y}^{(i)}$	$\mathbf{z}^{(i)}$	$\eta^{(i)}$	$\mathbf{x}^{(i)}$
2	76	40	93	$\begin{pmatrix} 1 \\ 76 \end{pmatrix}$	$\begin{pmatrix} 40 \\ 1 \end{pmatrix}$	73	$\begin{pmatrix} 80 \\ 5 \end{pmatrix}$
3	11	91	23	$\begin{pmatrix} 1 \\ 31 \\ 11 \end{pmatrix}$	$\begin{pmatrix} 91 \\ 69 \\ 1 \end{pmatrix}$	74	$\begin{pmatrix} 87 \\ 70 \\ 15 \end{pmatrix}$
4	90	44	26	$\begin{pmatrix} 1 \\ 73 \\ 13 \\ 90 \end{pmatrix}$	$\begin{pmatrix} 44 \\ 0 \\ 68 \\ 1 \end{pmatrix}$	47	$\begin{pmatrix} 80 \\ 70 \\ 13 \\ 77 \end{pmatrix}$
5	25	78	57	$\begin{pmatrix} 1 \\ 9 \\ 13 \\ 44 \\ 25 \end{pmatrix}$	$\begin{pmatrix} 78 \\ 15 \\ 44 \\ 7 \\ 1 \end{pmatrix}$	54	$\begin{pmatrix} 61 \\ 85 \\ 57 \\ 84 \\ 1 \end{pmatrix}$
6	5	60	29	$\begin{pmatrix} 1 \\ 11 \\ 88 \\ 70 \\ 60 \\ 5 \end{pmatrix}$	$\begin{pmatrix} 60 \\ 36 \\ 19 \\ 65 \\ 52 \\ 1 \end{pmatrix}$	14	$\begin{pmatrix} 31 \\ 67 \\ 96 \\ 3 \\ 72 \\ 48 \end{pmatrix}$

These computations lead to the solution $x_1 = 31$, $x_2 = 67$, $x_3 = 96$, $x_4 = 3$, $x_5 = 72$, and $x_6 = 48$ (where all equalities are modulo 97). □

Clearly, Levinson's algorithm terminates after only $\Theta(n^2)$ field operations (Exercise 8.7). The algorithm fails if $\epsilon^{(i)} = 0$ for some i. By Exercise 8.8, a successful termination of the algorithm demands each $T^{(i)}$ to be non-singular.

This is a very restrictive condition, even when T is of full rank, particularly since our interests focus on systems over small fields K (like $\mathbb{F}_2$). If K is large compared to n, this problem is not so serious, probabilistically.

Kaltofen suggests a way to get rid of this problem. For an $n \times n$ Toeplitz matrix T over a field K, Kaltofen considers the matrix $\hat{T} = UTV$, where U is an upper triangular Toeplitz matrix, and V is a lower triangular Toeplitz matrix. The elements on the main diagonals of U and V are 1. The elements of U above the main diagonal and the elements of V below the main diagonal are chosen randomly from a suitably large extension K^s. If T is of rank n, then all $\hat{T}^{(i)}$, $i = 1, 2, \ldots, n$, are invertible with probability at least $1 - n(n-1)/|K|^s$. Exercise 8.9 deals with the case of non-invertible T.

Levinson's algorithm continues to work even if the individual coefficients t_i of T are replaced by matrix blocks. In that case, other scalar variables in the algorithm must also be replaced by blocks or vectors of suitable dimensions. The details are left to the reader (Exercise 8.10). I demonstrate the block Levinson algorithm by an example.

Example 8.7 Let us again solve System (8.62) of Example 8.6. We take 2×2 blocks, and rewrite the system as

$$\begin{pmatrix} T_0 & T_{-1} & T_{-2} \\ T_1 & T_0 & T_{-1} \\ T_2 & T_1 & T_0 \end{pmatrix} \begin{pmatrix} X_1 \\ X_2 \\ X_3 \end{pmatrix} \equiv \begin{pmatrix} B_1 \\ B_2 \\ B_3 \end{pmatrix} \pmod{97},$$

where $T_0 = \begin{pmatrix} 85 & 92 \\ 39 & 85 \end{pmatrix}$, $T_{-1} = \begin{pmatrix} 79 & 6 \\ 92 & 79 \end{pmatrix}$, $T_{-2} = \begin{pmatrix} 15 & 72 \\ 6 & 15 \end{pmatrix}$, $T_1 = \begin{pmatrix} 87 & 39 \\ 42 & 87 \end{pmatrix}$, $T_2 = \begin{pmatrix} 82 & 42 \\ 30 & 82 \end{pmatrix}$, $X_1 = \begin{pmatrix} x_1 \\ x_2 \end{pmatrix}$, $X_2 = \begin{pmatrix} x_3 \\ x_4 \end{pmatrix}$, $X_3 = \begin{pmatrix} x_5 \\ x_6 \end{pmatrix}$, $B_1 = \begin{pmatrix} 82 \\ 53 \end{pmatrix}$, $B_2 = \begin{pmatrix} 31 \\ 12 \end{pmatrix}$, and $B_3 = \begin{pmatrix} 14 \\ 48 \end{pmatrix}$.

Since matrix multiplication is not commutative in general, we need to use two potentially different $\epsilon^{(i)}$ and $\epsilon'^{(i)}$ satisfying the block version of Eqn (8.54):

$$T^{(i)} Y^{(i)} = \begin{pmatrix} \epsilon^{(i)} \\ 0_{(i-1) \times \nu} \end{pmatrix} \quad \text{and} \quad T^{(i)} Z^{(i)} = \begin{pmatrix} 0_{(i-1) \times \nu} \\ \epsilon'^{(i)} \end{pmatrix}.$$

The corresponding updating equations for $\xi^{(i+1)}$, $\zeta^{(i+1)}$, $\epsilon^{(i+1)}$, $\epsilon'^{(i+1)}$ and $\mathbf{x}^{(i+1)}$ should be adjusted accordingly (solve Exercise 8.10).

Initialization $(i = 1)$: The initial solution is

$$X^{(1)} = T_0^{-1} B_1 = \begin{pmatrix} 80 \\ 5 \end{pmatrix}.$$

Let us plan to take

$$Y^{(1)} = Z^{(1)} = I_2 = \begin{pmatrix} 1 & 0 \\ 0 & 1 \end{pmatrix}, \quad \text{so that } \epsilon^{(1)} = \epsilon'^{(1)} = T_0 = \begin{pmatrix} 85 & 92 \\ 39 & 85 \end{pmatrix}.$$

Iteration for $i = 2$: The sequence of computations goes as follows.

$$\xi^{(2)} = -\left(\epsilon'^{(1)}\right)^{-1} T_1 Y^{(1)} = \begin{pmatrix} 78 & 31 \\ 63 & 11 \end{pmatrix},$$

$$\zeta^{(2)} = -\left(\epsilon^{(1)}\right)^{-1} T_{-1} Z^{(1)} = \begin{pmatrix} 91 & 64 \\ 69 & 61 \end{pmatrix},$$

$$\epsilon^{(2)} = \epsilon^{(1)}(I_2 - \zeta^{(2)}\xi^{(2)}) = \begin{pmatrix} 29 & 85 \\ 67 & 23 \end{pmatrix},$$

$$\epsilon'^{(2)} = \epsilon'^{(1)}(I_2 - \xi^{(2)}\zeta^{(2)}) = \begin{pmatrix} 23 & 85 \\ 67 & 29 \end{pmatrix},$$

$$Y^{(2)} = \begin{pmatrix} Y^{(1)} \\ 0_{2\times 2} \end{pmatrix} + \begin{pmatrix} 0_{2\times 2} \\ Z^{(1)} \end{pmatrix} \xi^{(2)} = \begin{pmatrix} 1 & 0 \\ 0 & 1 \\ 78 & 31 \\ 63 & 11 \end{pmatrix},$$

$$Z^{(2)} = \begin{pmatrix} 0_{2\times 2} \\ Z^{(1)} \end{pmatrix} + \begin{pmatrix} Y^{(1)} \\ 0_{2\times 2} \end{pmatrix} \zeta^{(2)} = \begin{pmatrix} 91 & 64 \\ 69 & 61 \\ 1 & 0 \\ 0 & 1 \end{pmatrix},$$

$$\eta^{(2)} = T_1 X^{(1)} = \begin{pmatrix} 74 \\ 12 \end{pmatrix},$$

$$X^{(2)} = \begin{pmatrix} X^{(1)} \\ 0_2 \end{pmatrix} + Z^{(2)} \left(\epsilon'^{(2)}\right)^{-1} (B_2 - \eta^{(2)}) = \begin{pmatrix} 80 \\ 70 \\ 13 \\ 77 \end{pmatrix}.$$

Iteration for $i = 3$: In this iteration, we have the following computations.

$$\xi^{(3)} = -\left(\epsilon'^{(2)}\right)^{-1} (T_2 \quad T_1) Y^{(2)} = \begin{pmatrix} 61 & 44 \\ 21 & 25 \end{pmatrix},$$

$$\zeta^{(3)} = -\left(\epsilon^{(2)}\right)^{-1} (T_{-1} \quad T_{-2}) Z^{(2)} = \begin{pmatrix} 78 & 78 \\ 15 & 32 \end{pmatrix},$$

$$\epsilon^{(3)} = \epsilon^{(2)}(I_2 - \zeta^{(3)}\xi^{(3)}) = \begin{pmatrix} 41 & 43 \\ 52 & 57 \end{pmatrix},$$

$$\epsilon'^{(3)} = \epsilon'^{(2)}(I_2 - \xi^{(3)}\zeta^{(3)}) = \begin{pmatrix} 57 & 43 \\ 52 & 41 \end{pmatrix},$$

$$Y^{(3)} = \begin{pmatrix} Y^{(2)} \\ 0_{2\times 2} \end{pmatrix} + \begin{pmatrix} 0_{2\times 2} \\ Z^{(2)} \end{pmatrix} \xi^{(3)} = \begin{pmatrix} 1 & 0 \\ 0 & 1 \\ 86 & 9 \\ 24 & 13 \\ 61 & 44 \\ 21 & 25 \end{pmatrix},$$

$$Z^{(3)} = \begin{pmatrix} 0_{2\times 2} \\ Z^{(2)} \end{pmatrix} + \begin{pmatrix} Y^{(2)} \\ 0_{2\times 2} \end{pmatrix} \zeta^{(3)} = \begin{pmatrix} 78 & 78 \\ 15 & 32 \\ 44 & 59 \\ 7 & 89 \\ 1 & 0 \\ 0 & 1 \end{pmatrix},$$

$$\eta^{(3)} = (\, T_2 \quad T_1 \,) X^{(2)} = \begin{pmatrix} 54 \\ 59 \end{pmatrix},$$

$$X^{(3)} = \begin{pmatrix} X^{(2)} \\ 0_2 \end{pmatrix} + Z^{(3)} \left(\epsilon'^{(3)} \right)^{-1} (B_3 - \eta^{(3)}) = \begin{pmatrix} 31 \\ 67 \\ 96 \\ 3 \\ 72 \\ 48 \end{pmatrix}.$$

Note that the solutions $X^{(1)}$, $X^{(2)}$ and $X^{(3)}$ in the block version are the same as the solutions $\mathbf{x}^{(2)}$, $\mathbf{x}^{(4)}$ and $\mathbf{x}^{(6)}$ in the original version (Example 8.6). $\quad\square$

Exercises

1. [*Lifting*] Let $p \in \mathbb{P}$ and $e \in \mathbb{N}$. Describe how you can lift solutions of $Ax \equiv b \pmod{p^e}$ to solutions of $Ax \equiv b \pmod{p^{e+1}}$.

2. For linear systems arising out of factorization algorithms, we typically need multiple solutions of homogeneous systems. Describe how the standard Lanczos method can be tuned to meet this requirement.

3. Repeat Exercise 8.2 for the block Lanczos algorithm. More precisely, show how a block of solutions to the homogeneous system can be obtained.

4. In order to address the problem of self-orthogonality in the Lanczos algorithm, we modified the system $Ax = b$ to $A^t D^2 Ax = A^t D^2 b$ for a randomly chosen invertible diagonal matrix D with entries from a suitable extension field. What is the problem if we plan to solve the system $DA^t Ax = DA^t b$ instead?

5. For the block Lanczos method, describe a method to identify the selection matrix S_i of Eqn (8.46)

6. The Wiedemann algorithm chooses elements of $A^k v$ at particular positions. This amounts to multiplying Eqn (8.42) from left by suitable *projection vectors* u. Generalize this concept to work for any non-zero vectors u.

7. Prove that Levinson's algorithm for solving Eqn (8.53), as presented in the text, performs a total of about $3n^2$ multiplications and about $3n^2$ additions.

8. Prove that Eqn (8.53) is solvable by Levinson's iterative algorithm if and only if the matrices $T^{(i)}$ are invertible for all $i = 1, 2, \ldots, n$.

9. Let T be an $n \times n$ Toeplitz matrix with rank $r \leqslant n$. We call T to be of *generic rank profile* if $T^{(i)}$ is invertible for all $i = 1, 2, \ldots, r$. Describe a strategy to let Levinson's algorithm generate random solutions of a solvable system $Tx = b$, where T is a Toeplitz matrix of generic rank profile with rank $r < n$.

10. Write the steps of the block version of Levinson's algorithm. For simplicity, assume that each block of T is a square matrix of size $\nu \times \nu$, and the entire coefficient matrix T is also square of size $n \times n$ with $\nu \mid n$.

11. Block algorithms are intended to speed up solving linear systems over $\mathbb{F}_2$. They are also suitable for parallelization. Explain how.

12. Let $A = (a_{ij})$ be an $m \times n$ matrix with entries from $\mathbb{F}_q$. Suppose that $m \geqslant n$. Let r denote the rank of A, and $d = n - r$ the *rank deficit* (also called *defect*) of A. Denote the j-th column of A by A_j. A non-zero n-tuple $(c_1, c_2, \ldots, c_n) \in (\mathbb{F}_q)^n$, for which $\sum_{j=1}^{n} c_j A_j = 0$, is called a *linear dependency* of the columns of A. Let l denote the number of linear dependencies of the columns of A.
   (a) Prove that $l + 1 = q^d$.
   (b) Let the entries of A be randomly chosen. Prove that $E(r) \geqslant n - \log_q(E(l) + 1)$, where $E(X)$ is the expected value of the random variable X.
   (c) How can you compute $E(l)$, given a probability distribution for each a_{ij}?

Chapter 9

Public-Key Cryptography

In this chapter, we study some engineering applications of number-theoretic algorithms. The American mathematician Leonard Eugene Dickson (1874–1954) commented: *Thank God that number theory is unsullied by any application.* This assertion is no longer true. The development of error-correcting codes in the 1950s/60s involved first serious engineering applications of finite fields. The advent of public-key cryptography in late nineteen seventies opens yet another avenue of application. Almost everything that appears in earlier chapters has profound implications in public-key cryptography. This chapter provides an introductory exposure to public-key algorithms. This is not a book on cryptography. Nonetheless, a treatment of some practical applications of the otherwise theoretical study may be motivating to the readers.

I start with classical algorithms of public-key cryptography, and then discuss pairing-based protocols, a new branch of public-key cryptography. The main problems that cryptography deals with are listed below. A study of breaking cryptographic protocols is referred to as *cryptanalysis*. *Cryptology* refers to the combined study of cryptography and cryptanalysis.

- **Message confidentiality:** Alice wants to send a private message M to Bob. Use of a public channel for transferring the message M allows any eavesdropper to access M, resulting in a loss in the privacy and confidentiality of the message. In order to avoid this problem, Alice first transforms M to a ciphertext $C = E(M)$ by applying an *encryption* function E, and sends C (instead of M) through the public channel. Bob, upon receiving C, uses a *decryption* function D to recover $M = D(C)$.

  Eve, the eavesdropper, must not be able to generate M from C in feasible time. This requirement can be met in two ways. First, the functions E and D may be secretly chosen by Alice and Bob. But then, they need to set up these functions before any transmission. Every pair of communicating parties requires a secret algorithm. Moreover, this strategy is known to be weak from the angle of information theory. A better approach is to use the same algorithm for *every* pair of communicating parties. Confidentiality is achieved by *keys*. Alice encrypts as $C = E_K(M)$, and Bob decrypts as $M = D_{K'}(C)$. If the decryption key K' is not disclosed to Eve, it should be infeasible for her to generate M from C.

- **Key agreement:** Although encryption helps Alice and Bob to exchange private messages over a public channel, they must have a mechanism to agree upon the keys K and K'. *Symmetric* or *secret-key* cryptography deals with the situation that $K = K'$. This common key may be set up by a private communication between Alice and Bob. Another alternative is to run a *key-agreement* or a *key-exchange* protocol, where Alice and Bob generate random secrets, and exchange masked versions of these secrets over a public channel. Combining the personal secret and the masked secret from the other party, each of Alice and Bob computes a common value which they later use as the symmetric key $K = K'$. Eve, from the knowledge of only the masked secrets, cannot compute this common value in feasible time. In this way, the necessity of a private communication between Alice and Bob is eliminated.

  In an *asymmetric* or *public-key* cryptographic system, we have $K' \neq K$. The encryption key (also called the *public key*) K is made public (even to eavesdroppers). Anybody having access to K can encrypt messages for Bob. However, only Bob can decrypt these messages. The knowledge of K' (called the *private key*) is vital in the decryption process. Although K and K' are sort of matching keys, it should be infeasible to compute K' from K. This is how an asymmetric cryptosystem derives its security. Asymmetric cryptosystems alleviate the need of any key agreement between Alice and Bob, secretly or publicly.

- **Digital signatures:** Like hand-written signatures, a digital signature binds a digital message (or document) M to an entity (say, Bob). Digital signature schemes are based upon asymmetric keys. The *signing key K'* is known only to Bob. The *verification key K* is made public. In order to sign M, Bob computes a short representative m of M, and uses the signing key K' to generate $s = S_{K'}(m)$. The signed message is the pair (M, s). Anybody (say, Alice) having access to the verification key K can verify the authenticity of the signed message (M, s) as follows. Alice first computes the short representative m from M, and applies a verification function to generate $m' = V_K(s)$. The signature is accepted as authentic if and only if $m' = m$. A digital signature scheme is like a public-key encryption scheme with the sequence of using the keys K, K' reversed. A signature scheme requires that no party without the knowledge of the signing key K' can generate a verifiable signature s on a message M.

- **Entity authentication:** The authenticity of an entity Alice is realized by Alice's knowledge of a secret piece of information Σ. In order to prove her identity to a verifier, Bob, Alice demonstrates her knowledge of Σ to Bob. Any party not having access to Σ cannot impersonate Alice.

  Entity authentication can be achieved by *passwords*. Bob stores $f(\Sigma)$, where f is a one-way function (a function that cannot be easily inverted). During an authentication session, Alice reveals her password Σ to Bob. Bob accepts Alice if and only if $f(\Sigma)$ equals Bob's stored value. The use of f is necessitated as a safeguard against impersonation attempts by parties having access to Bob's storage. Password-based authentication is *weak*, since Alice has to disclose her secret Σ to the verifier Bob.

  In a *strong* authentication scheme, Alice does not reveal Σ directly to Bob. Instead they run a protocol which allows Alice to succeed (with high probability) if and only if she knows Σ. Strong authentication schemes are based upon public-key cryptosystems.

- **Certification:** Public-key protocols use key pairs (K, K'). The encryption (or verification) key K is made public, whereas the decryption (or signing) key K' is kept secret. While using Bob's public key K, one must be certain that K really belongs to Bob. This binding is achieved by *digital certificates*. A *trusted third party*, also called the *certification authority* (CA), embeds K along with other identifying information (like name, address, e-mail ID of Bob) in a certificate Γ. The CA digitally signs Γ by its signing key. Anybody willing to use Bob's public key verifies CA's signature on the certificate. If the signature is verified, the identifying information in the certificate is scrutinized. In case all these pieces of information ensure that they correspond to Bob, the public key of Bob, embedded in the certificate, is used.

In order to solve the above cryptographic problems, number theory is used. There are several computational problems (like factoring large composite integers, and computing discrete logarithms in certain groups) which cannot be solved efficiently by best existing algorithms (subexponential algorithms). Public-key cryptography is based upon the assumption that these problems are not solvable by polynomial-time algorithms. Although this assumption is not exactly justified, this is how this technology is developed. Interestingly enough, this is how this technology must be developed. Computational problems that have provably high lower bounds appear to be unsuitable in realizing cryptographic protocols. Intuitively, these problems are so difficult that a decryption function with an instrumental role of the private key cannot be designed. Public-key cryptography turns out to be an intriguing application that exploits our inability to solve certain computational problems efficiently.

Symmetric ciphers occasionally use number-theoretic tools. For example, stream ciphers are sometimes implemented using linear feedback shift registers (LFSRs). The theory of LFSRs is dependent upon properties of polynomials over the finite field $\mathbb{F}_2$. The Berlekamp–Massey algorithm can be used to cryptanalyze such stream ciphers. Some nonlinearity is introduced in the output of LFSRs in order that this attack cannot be mounted. Another sample application of finite fields is the use of the field $\mathbb{F}_{256} = \mathbb{F}_{2^8}$ in the Rijndael cipher currently adopted as a standard (AES—the advanced encryption standard). This block cipher exploits the high nonlinearity in the multiplicative inverse operation of $\mathbb{F}_{256}$. Despite these sporadic uses of number theory, the study of symmetric ciphers is not classified as applied number theory. On the contrary, number theory is omnipresent in all facets of public-key cryptography. In view of this, I concentrate only on public-key technology in the rest of this chapter.

Symmetric and asymmetric ciphers are, however, not competitors of one another. Asymmetric technology makes certain things possible (like realization of digital signatures) that cannot be achieved by symmetric techniques. On the contrary, public-key cryptographic functions are orders of magnitude slower than symmetric cryptographic functions. In practice, a combination of symmetric and asymmetric algorithms is used. For example, long messages are encrypted by symmetric ciphers (block ciphers, typically). The symmetric key between the communicating parties is established by either a key-agreement protocol or a public-key encryption of the symmetric key. In cryptography, both symmetric and asymmetric ciphers are useful and important. Here, our focus is on number theory, and so on public-key techniques only.

A particular type of functions is often used in public-key cryptosystems, most notably in signature schemes and pairing-based schemes. These are called *hash functions*. A hash function H maps bit strings of any length to bit strings of a fixed length n. A hash function H suitable for cryptography is required to have the three following properties.

First preimage resistance Given an n-bit string y, it should, in general, be difficult to find a string x (of any length) with $H(x) = y$. Of course, one may choose certain values of x, and store $(x, H(x))$ pairs for these values of x. It is evidently easy to invert H on these values of $H(x)$. However, since the storage of this table is restricted by availability of memory (for example, if $n = 160$, it is infeasible to store a table of size 2^{160}), the general complexity of inverting H should be high.

Second preimage resistance Given a string x, it should be difficult to find a different string y such that $H(x) = H(y)$. Such a pair (x, y) is called a *collision* for H. Since H maps an infinite domain to a finite range, collisions must exist for any hash function. However, it should be difficult to find a y colliding with any given x.

Collision resistance It should be difficult to find any two strings x, y with $H(x) = H(y)$. Unlike second preimage resistance, the choice of both x and y are free here.

The relations between these properties of hash functions are explored in Exercise 9.1. An obviously needed property for a cryptographic hash function is that it should be easy to compute. A function provably possessing all these properties is not known. There are practical constructions of hash functions. An example is provided by the secure hash family of functions like SHA-1.[1] For some cryptosystems, we often modify the range of a hash function to some set A other than $\{0,1\}^n$. The set A could be $\mathbb{Z}_r$ or even an elliptic-curve group over a finite field. A practical hash function (like SHA-1) can often be used as the basic building block for constructing such modified hash functions. However, such constructions are not necessarily trivial. In this book, I will make no attempt to construct hash functions. At present, these constructions are not really applications of number theory. For us, it suffices to know that hash functions, secure for practical cryptographic applications, do exist.

9.1 Public-Key Encryption

Among a host of public-key encryption algorithms available in the literature, I pick only two in this section. Historically, these turn out to be the first two public-key encryption algorithms to appear in the literature.

9.1.1 RSA Encryption

The first public-key encryption and signature algorithm RSA[2] happens to be the most popular public-key algorithm. This popularity stems from its simplicity and resistance against cryptanalytic studies for over three decades. The RSA algorithm is dependent upon the difficulty of factoring large composite integers. It involves the following steps.

---

[1] Federal Information Processing Standard http://csrc.nist.gov/publications/fips/fips180-3/fips180-3_final.pdf.

[2] Ronald Linn Rivest, Adi Shamir and Leonard Max Adleman, A method for obtaining digital signatures and public-key cryptosystems, *Communications of the ACM*, 21(2), 120–126, 1978.

Key generation: The entity willing to receive RSA-encrypted messages generates two random large primes p and q of nearly the same bit length. In order to achieve a decent amount of security, this bit length should be at least 512. The product $n = pq$ and the Euler function $\phi(n) = (p-1)(q-1)$ are then computed. Finally, an integer e coprime to $\phi(n)$ is chosen, and its inverse d modulo $\phi(n)$ is computed by the extended gcd algorithm. The pair (n, e) is published as the public key, whereas d is kept secret as the private key.

The public key e need not be a random element of $\mathbb{Z}^*_{\phi(n)}$. In order to speed up RSA encryption, it is a good idea to take e as small as possible (but greater than 1). A possibility is to take as e the smallest prime not dividing $\phi(n)$.

I now argue that the public knowledge of n and e does not allow an adversary to compute the secret d. (This problem is often referred to as the *RSA key-inversion problem* (RSAKIP).) If the adversary can factor n, she can compute $\phi(n) = (p-1)(q-1)$ and then $d \equiv e^{-1} \pmod{\phi(n)}$. This does not imply that an adversary *has* to factor n to obtain d from the knowledge of n and e only. However, it turns out that RSAKIP is computationally equivalent to factoring n, that is, if one can compute d from the knowledge of n and e only, one can factor n too. A probabilistic algorithm for this is supplied below.

Write $ed - 1 = 2^s t$ with t odd and $s \geqslant 2$ ($ed - 1$ is a multiple of $\phi(n) = (p-1)(q-1)$). Let $a \in \mathbb{Z}^*_n$. Since $\operatorname{ord}_n a$ divides $\phi(n)$, it divides $ed - 1$ too, so $\operatorname{ord}_n a = 2^{s'} t'$ with $0 \leqslant s' \leqslant s$ and $t'|t$. But then, $\operatorname{ord}_n(a^t) = \frac{2^{s'} t'}{\gcd(2^{s'} t', t)} = 2^{s'}$. Let us look at the multiplicative orders of a^t modulo p and q individually.

Let g be a primitive root modulo p, and $a \equiv g^k \pmod{p}$. We have $\operatorname{ord}_p g = p - 1 = 2^v r$ with $v \geqslant 1$, and r odd. If k is odd, then $\operatorname{ord}_p(a) = 2^v r'$ for some $r'|r$. If k is even, then $\operatorname{ord}_p(a) = 2^{v'} r'$ for some $v' < v$ and $r'|r$. Consequently, $\operatorname{ord}_p(a^t)$ equals 2^v if k is odd, or $2^{v'}$ for some $v' < v$ if k is even. Likewise, $\operatorname{ord}_q(a^t)$ is $2^{w'}$ for some $w' \leqslant w$, where w is the multiplicity of 2 in $q - 1$. Moreover, $w' = w$ if and only if $\operatorname{ind}_h a$ is odd for some primitive root h of q.

We randomly choose $a \in \{2, 3, \ldots, n - 1\}$. If $\gcd(a, n) \neq 1$, this gcd is a non-trivial factor of n. But this has a very low probability. So assume that $a \in \mathbb{Z}^*_n$. We compute $b \equiv a^t \pmod{n}$. We have argued that $\operatorname{ord}_p b = 2^{v'}$ and $\operatorname{ord}_q b = 2^{w'}$ for some $v', w' \in \{0, 1, 2, \ldots, s\}$. If $v' < w'$, then $b^{2^{v'}} \equiv 1 \pmod{p}$, whereas $b^{2^{v'}} \not\equiv 1 \pmod{q}$, so $\gcd(b^{2^{v'}} - 1, n) = p$. Likewise, if $v' > w'$, then $\gcd(b^{2^{w'}} - 1, n) = q$. In short, if $v' \neq w'$, there exists an $s' \in \{0, 1, 2, \ldots, s - 1\}$ for which $\gcd(b^{2^{s'}} - 1, n)$ is a non-trivial factor of n. We keep on computing $b, b^2, b^4, b^8, \ldots, b^{2^{s-1}}$ modulo n by successive squaring. For each $b^{2^{s'}}$ so computed, we compute $\gcd(b^{2^{s'}} - 1, n)$. If some s' gives a non-trivial factor of n, we are done. Otherwise, we choose another random a, and repeat.

Let us now investigate how likely the occurrence of the useful case $v' \neq w'$ is. Exactly half of the elements of $\mathbb{Z}^*_p$ have $v' = v$, and exactly half of the elements of $\mathbb{Z}^*_q$ have $w' = w$ (the quadratic non-residues). If $v = w$, then a is useful if it is a quadratic residue modulo p but a non-residue modulo q, or if it is a quadratic non-residue modulo p but a residue modulo q. The count of such useful values of a is, therefore, $2 \times \left(\frac{p-1}{2}\right)\left(\frac{q-1}{2}\right) = \phi(n)/2$. If $v < w$, then a is useful if it is a quadratic non-residue modulo q. If $v > w$, then a is useful if it is a quadratic non-residue modulo p. Therefore, at least half of the elements of $\mathbb{Z}_n^*$ are capable of factoring n, so a randomly chosen a splits n non-trivially with probability $\geqslant 1/2$, that is, only a few choices of a suffice.

The RSA key-inversion problem is, therefore, probabilistic polynomial-time equivalent to factoring n. Coron and May[3] prove that the RSAKIP is *deterministic* polynomial-time equivalent to factoring.

Example 9.1 Suppose that Bob publishes the public RSA key:

$$n = 35394171409,$$
$$e = 7.$$

Somehow, it is leaked to Eve that Bob's private key is

$$d = 15168759223.$$

Let us see how this knowledge helps Eve to factor n. Since $ed - 1 = 2^{11} \times 51846345$, we have $s = 11$ and $t = 51846345$. Eve chooses $a = 5283679203$, and computes $b \equiv a^t \equiv 90953423 \pmod{n}$. Subsequently, for $s' = 0, 1, 2, \ldots, 10$, Eve computes $\gcd(b^{2^{s'}} - 1, n)$. It turns out that for $s' = 0, 1, 2, 3$, this gcd is 1, and for $s' = 4, 5, \ldots, 10$, this gcd is n itself. This indicates that b has the same order (namely, 2^4) modulo both the prime factors p and q of n.

Eve then tries $a = 985439671$ for which $b \equiv a^t \equiv 12661598494 \pmod{n}$. The gcd of $b^{2^{s'}} - 1$ with n is now 1 for $s' = 0, 1, 2, 3$, it is 132241 for $s' = 4, 5$, and n itself for $s' = 6, 7, 8, 9, 10$. As soon as the non-trivial gcd $p = 132241$ is obtained (for $s' = 4$), Eve stops and computes the cofactor $q = n/p = 267649$. The complete gcd trail is shown here to illustrate that Eve is successful in this attempt, because $\mathrm{ord}_p\, b = 2^4$, whereas $\mathrm{ord}_q\, b = 2^6$, in this case. $\qquad\square$

Encryption: In order to encrypt a message M for Bob, Alice obtains Bob's public key (n, e). The message M is converted to an element $m \in \mathbb{Z}_n$ (this procedure is called *encoding*, and is not needed to be done securely). Alice computes and sends $c \equiv m^e \pmod{n}$ to Bob.

Decryption: Upon receiving c, Bob recovers m as $m \equiv c^d \pmod{n}$ using his knowledge of the private key d.

To show that this decryption correctly recovers m, write $ed - 1 = k\phi(n) = k(p-1)(q-1)$ for some positive integer k. In view of the CRT, it suffices to show that m^{ed} is congruent to m modulo both p and q. If $p|m$, then $m^{ed} \equiv 0 \equiv m \pmod{p}$. On the other hand, if $\gcd(m, p) = 1$, we have $m^{p-1} \equiv 1 \pmod{p}$ by Fermat's little theorem, so $m^{ed} \equiv m^{ed-1} \times m \equiv \left(m^{p-1}\right)^{k(q-1)} \times m \equiv m \pmod{p}$. In both the cases, $m^{ed} \equiv m \pmod{p}$. Likewise, $m^{ed} \equiv m \pmod{q}$.

---

[3] Jean-Sebastien Coron and Alexander May, Deterministic polynomial-time equivalence of computing the RSA secret key and factoring, *Journal of Cryptology*, 20(1), 39–50, 2007.

Let us now investigate the connection between factoring n and RSA decryption. Since the ciphertext c is sent through a public channel, an eavesdropper can intercept c. If she can factor n, she computes $d \equiv e^{-1} \pmod{\phi(n)}$, and subsequently decrypts c as Bob does. However, the converse capability of the eavesdropper is not mathematically established. This means that it is not known whether factoring n is necessary to decrypt c, or, in other words, if Eve can decrypt c from the knowledge of n and e only, she can also factor n. At present, no algorithms other than factoring n is known (except for some pathological parameter values) to decrypt RSA-encrypted messages (without the knowledge of the decryption exponent d). In view of this, we often say that the security of RSA is based upon the intractability of the integer factoring problem. It, however, remains an open question whether RSA decryption (knowing n, e only) is equivalent to or easier than factoring n.

Example 9.2 Let us work with the modulus of Example 9.1.[4] Bob chooses the two primes $p = 132241$ and $q = 267649$, and computes $n = pq = 35394171409$ and $\phi(n) = (p-1)(q-1) = 35393771520$. The smallest prime that does not divide $\phi(n)$ is 7, so Bob takes $e = 7$. Extended gcd of e and $\phi(n)$ gives $d \equiv e^{-1} \equiv 15168759223 \pmod{\phi(n)}$. Bob publishes $(n, e) = (35394171409, 7)$ as his public key, and keeps the private key $d = 15168759223$ secret.

For encrypting the message $M = $ "Love", Alice converts M to an element m of $\mathbb{Z}_n$. Standard 7-bit ASCII encoding gives $m = 76 \times 128^3 + 111 \times 128^2 + 118 \times 128 + 101 = 161217381$. Alice obtains Bob's public key (n, e), and encrypts m as $c \equiv m^e \equiv 26448592996 \pmod{n}$. This value of c (in some encoding suitable for transmission) is sent to Bob via a public channel.

Bob decrypts as $m \equiv c^d \equiv 161217381 \pmod{n}$, so M is decoded from m correctly as "Love." If any key other than Bob's decryption exponent d is used to decrypt c, the output ciphertext will be different from the message sent. For example, if Eve uses $d' = 4238432571$, she gets $m' \equiv c^{d'} \equiv 21292182600 \pmod{n}$, whose 7-bit ASCII decoding gives the string "O(sXH." □

9.1.2 ElGamal Encryption

ElGamal's encryption algorithm[5] is based on the Diffie–Hellman problem (DHP). Let G be a cyclic group with a generator g. The DHP is the problem of computing g^{ab} from the knowledge of g^a and g^b. If one can compute discrete logarithms in G, one can solve the DHP in G. In view of this, it is necessary to take the group G as one in which it is infeasible to compute discrete logarithms.

Key generation: ElGamal's scheme uses two types of keys. For each entity, a *permanent key pair* is used for encrypting all messages, whereas a

---

[4]The moduli used in the examples of this chapter are artificially small. They are meant only for illustrating the working of the algorithms. In order to achieve a decent level of security, practical implementations have to use much larger parameter values.

[5]Taher ElGamal, *A public-key cryptosystem and a signature scheme based on discrete logarithms*, *IEEE Transactions on Information Theory*, 31(4), 469–472, 1985.

temporary or a *session key pair* is used for each encryption. The session key must be different in different runs of the encryption algorithm. Both these key pairs are of the form (d, g^d). Here, d is a random integer between 2 and $|G| - 1$, and is used as the private key, whereas g^d is to be used as the public key. Obtaining an ElGamal private key from the corresponding public key is the same as computing a discrete logarithm in G, that is, the ElGamal key-inversion problem is computationally equivalent to the DLP in G.

ElGamal encryption: In order to send a message $m \in G$ (M is encoded as an element in G) to Bob, Alice obtains Bob's public key g^d. Alice generates a random integer $d' \in \{2, 3, \ldots, |G| - 1\}$ (session private key), and computes $s = g^{d'}$ (session public key). Alice then masks the message m by the quantity $g^{dd'}$ as $t = m \left(g^d\right)^{d'}$. The encrypted message (s, t) is sent to Bob.

ElGamal decryption: Upon receiving (s, t) from Alice, Bob recovers m as $m = ts^{-d}$ using his permanent private key d. The correctness of this decryption is based on that $ts^{-d} = mg^{dd'}(g^{d'})^{-d} = m$.

An eavesdropper possesses knowledge of G, g, g^d and $s = g^{d'}$. If she can compute $g^{dd'}$, she decrypts the message as $m = t(g^{dd'})^{-1}$. Conversely, if Eve can decrypt (s, t), she computes $g^{dd'} = tm^{-1}$. ElGamal decryption (using public information only) is, therefore, as difficult as solving the DHP in G.

Example 9.3 The prime $p = 35394171431$ is chosen by Bob, along with the generator $g = 31$ of $G = \mathbb{F}_p^*$. These parameters G and g may be used by multiple (in fact, all) entities in a network. This is in contrast with RSA where each entity should use a different n. Bob takes $d = 4958743298$ as his private key, and computes and publishes the public key $y \equiv 31^d \equiv 628863325 \pmod{p}$.

In order to encrypt $m = 161217381$ (the 7-bit ASCII encoding of "Love"), Alice first generates a session key $d' = 19254627018$, and computes $s \equiv g^{d'} \equiv 33303060050 \pmod{p}$ and $t \equiv my^d \equiv 3056015643 \pmod{p}$. Alice sends the pair $(33303060050, 3056015643)$ to Bob over a public channel.

Bob decrypts the ciphertext (s, t) as $m \equiv ts^{-d} \equiv 161217381 \pmod{p}$. If any private key other than d is used, one expects to get a different recovered message. If Eve uses $d = 21375157906$, she decrypts (s, t) as $m' \equiv ts^{-d} \equiv 24041362599 \pmod{p}$, whose 7-bit ASCII decoding is "YGh)'." $\qquad\square$

9.2 Key Agreement

The Diffie–Hellman key-agreement protocol[6] is the first published public-key algorithm. It is based on the Diffie–Hellman problem. Indeed, the ElGamal encryption scheme is an adaptation of the Diffie–Hellman protocol.

---

[6] Whitfield Diffie and Martin Edward Hellman, New directions in cryptography, *IEEE Transactions on Information Theory*, 22, 644–654, 1976.

In order to share a secret, Alice and Bob publicly agree upon a suitable finite cyclic group G with a generator g. Alice chooses a random integer $d \in \{2, 3, \ldots, |G|-1\}$, and sends g^d to Bob. Bob, in turn, chooses a random integer $d' \in \{2, 3, \ldots, |G| - 1\}$, and sends $g^{d'}$ to Alice. Alice computes $g^{dd'} = (g^{d'})^d$, whereas Bob computes $g^{dd'} = (g^d)^{d'}$. The element $g^{dd'} \in G$ is the common secret exchanged publicly by Alice and Bob. An eavesdropper knows g^d and $g^{d'}$ only, and cannot compute $g^{dd'}$ if the DHP is infeasible in the group G.

Example 9.4 Alice and Bob publicly decide to use the group $G = \mathbb{F}_p^*$, where $p = 35394171431$. They also decide the generator $g = 31$ publicly. Alice chooses $d = 5294364$, and sends $y \equiv g^d \equiv 21709635652 \pmod{p}$ to Bob. Likewise, Bob chooses $d' = 92215703$, and sends $y' \equiv g^{d'} \equiv 31439131289 \pmod{p}$ to Alice. Alice computes $\alpha \equiv y'^d \equiv 10078655355 \pmod{p}$, whereas Bob computes $\beta \equiv y^{d'} \equiv 10078655355 \pmod{p}$. The shared secret $\alpha = \beta$ may be used for future use. For example, Alice and Bob may use the 7-bit ASCII decoding "%Ep&{" of $\alpha = \beta$ as the key of a block cipher. The task of an eavesdropper is to compute this shared secret knowing G, g, y and y' only. This is equivalent to solving an instance of a Diffie–Hellman problem in G. $\qquad\square$

9.3   Digital Signatures

I now introduce four digital signature algorithms. RSA signatures are adapted from the RSA encryption algorithm, and ElGamal signatures are adapted from the ElGamal encryption algorithm. DSA (the *digital signature standard*) and its elliptic-curve variant ECDSA are ElGamal-like signatures, that are accepted as federal information processing standards.

In the rest of this section, I assume that M is the message to be signed by Bob. In what is called a *signature scheme with appendix*, Bob first generates a small representative m of M. A *hash function* is typically used to generate $m = H(M)$. The signature generation primitive is applied on the representative m.

9.3.1   RSA Signature

Bob selects RSA keys (n, e) and d as in the case of RSA encryption. Bob generates the appendix as $s = m^d \pmod{n}$. The signed message is (M, s). For verifying Bob's signature, Alice obtains Bob's public key (n, e), and computes $m' \equiv s^e \pmod{n}$. Alice accepts (M, s) if and only if $m' = H(M)$. RSA signature is like RSA encryption with the application of the keys reversed. The correctness and security of RSA signatures are as for RSA encryption.

Example 9.5 Bob generates the primes $p = 241537$ and $q = 382069$, and computes $n = pq = 92283800053$ and $\phi(n) = (p - 1)(q - 1) = 92283176448$.

The smallest prime not dividing $\phi(n)$ is chosen as the public exponent $e = 5$. The corresponding private exponent is $d \equiv e^{-1} \equiv 55369905869 \pmod{\phi(n)}$.

Suppose that Bob plans to sign the message $m = 1234567890$ (may be, because he wants to donate these many dollars in a charity fund). He generates the appendix as $s \equiv m^d \equiv 85505674365 \pmod{n}$.

To verify Bob's signature, Alice obtains his public key $e = 5$, and computes $m' \equiv s^e \equiv 1234567890 \pmod{n}$. Since $m' = m$, the signature is verified.

A forger uses $d' = 2137532490$ to generate the signature $s' \equiv m^{d'} \equiv 84756771448 \pmod{n}$ on m. Verification by Bob's public key gives $m' \equiv s'^e \equiv 23986755072 \pmod{n}$. Since $m' \neq m$, the forged signature is not verified. $\qquad \Box$

9.3.2 ElGamal Signature

ElGamal's original signature scheme was proposed for the group $G = \mathbb{F}_p^*$, where p is a suitably large prime. Although it is possible to generalize this construction to any (finite) cyclic group G, let us concentrate on the case $G = \mathbb{F}_p^*$ only. We assume that a primitive root g modulo p is provided to us.

After fixing p and g, Bob chooses a random $d \in \{2, 3, \ldots, p-2\}$ as his private key, and publishes $y \equiv g^d \pmod{p}$ as his public key. For signing (a message representative) m, Bob chooses a random session key $d' \in \{2, 3, \ldots, p-2\}$ coprime to $p-1$. Subsequently, the quantities $s \equiv g^{d'} \pmod{p}$ and $t \equiv (m - ds)d'^{-1} \pmod{p-1}$ are computed. Bob's signature on m is the pair (s, t). The computation of t requires $\gcd(d', p-1) = 1$.

We have $m \equiv td' + sd \pmod{p-1}$, so that $g^m \equiv (g^{d'})^t (g^d)^s \equiv s^t y^s \pmod{p}$. Therefore, to verify Bob's signature (s, t), Alice needs to check whether the congruence $g^m \equiv s^t y^s \pmod{p}$ holds. This requires using Bob's public key y.

Let us investigate the security of ElGamal's signature scheme. The knowledge of the permanent public key y and the session public key s does not reveal the corresponding private keys d and d', under the assumption that it is infeasible to solve the DLP modulo p. Clearly, if d is known to an adversary, she can generate a valid signature on m in exactly the same way as Bob does (the session secret d' can be *chosen* by the adversary). Conversely, suppose that Eve can produce a valid signature (s, t) of Bob on m. The equation $m \equiv td' + sd \pmod{p-1}$ has $\Theta(p)$ solutions in the unknown quantities d, d' which cannot be uniquely identified unless one of d, d' is provided to Eve. But then, d' can be chosen by the adversary, implying that she can solve for d, that is, can compute the discrete logarithm of y. This argument *intuitively* suggests that the security of the ElGamal signature scheme is based upon the difficulty of computing discrete logarithms in $\mathbb{F}_p$. (This is not a formal security proof. Indeed, a silly signature scheme may reveal d to the adversary, and the effort spent by the adversary in the process may be less than necessary to solve the DLP in $\mathbb{F}_p$. Hopefully, ElGamal signatures do not leak such silly information. To the best of my knowledge, the equivalence of forgery of ElGamal signatures with solving the DLP is not established yet.)

Example 9.6 Let us choose the ElGamal parameters $p = 92283800099$ and $g = 19$. Bob chooses the private key $d = 23499347910$, and the corresponding public key is $y \equiv g^d \equiv 66075503407 \pmod{p}$. Let $m = 1234567890$ be the message (representative) to be signed by Bob.

Bob chooses the random session secret $d' = 9213753243$, and generates $s \equiv g^{d'} \equiv 85536409136 \pmod{p}$ and $t \equiv (m - ds)d'^{-1} \equiv 22134180366 \pmod{p-1}$. Bob's signature on m is the pair $(85536409136, 22134180366)$.

A verifier computes $g^m \equiv 44505409554 \pmod{p}$ and $s^t y^s \equiv 44505409554 \pmod{p}$. These two quantities being equal, the signature is verified.

A forger generates s by choosing d' like Bob. Computing t, however, uses d which the forger guesses as $d = 14762527324$. This gives $t' \equiv (m - ds)d'^{-1} \equiv 43362818978 \pmod{p-1}$. For the forged signature (s, t'), we have $g^m \equiv 44505409554 \pmod{p}$ as before, whereas $s^{t'} y^s \equiv 52275374670 \pmod{p}$. These two quantities are not equal, so the forged signature is not verified. □

9.3.3 DSA

The digital signature algorithm (DSA)[7] is an efficient variant of ElGamal's signature scheme. The reduction in the running time arises from that DSA works in a subgroup of $\mathbb{F}_p^*$. Let q be a prime divisor of $p - 1$, having bit length 160 or more. To compute a generator g of the (unique) subgroup G of $\mathbb{F}_p^*$ of size q, we choose random $h \in \mathbb{F}_p^*$ and compute $g \equiv h^{(p-1)/q} \pmod{p}$ until we have $g \not\equiv 1 \pmod{p}$. The parameters for DSA are p, q and g.

Bob sets up a permanent key pair by choosing a random $d \in \{2, 3, \ldots, q-1\}$ (the private key) and then computing $y \equiv g^d \pmod{p}$ (the public key).

For signing the message (representative) m, Bob chooses a random session secret $d' \in \{2, 3, \ldots, q - 1\}$, and computes $s = (g^{d'} \pmod{p}) \pmod{q}$ and $t \equiv (m + ds)d'^{-1} \pmod{q}$. Bob's signature on m is the pair (s, t).

For verifying this signature, one computes $w \equiv t^{-1} \pmod{q}$, $u_1 \equiv mw \pmod{q}$, $u_2 \equiv sw \pmod{q}$, and $v \equiv (g^{u_1} y^{u_2} \pmod{p}) \pmod{q}$, and accepts if and only if $v = s$. The correctness of this procedure is easy to establish.

The basic difference between ElGamal's scheme and DSA is that all exponents in ElGamal's scheme are needed modulo $p - 1$, whereas all exponents in DSA are needed modulo q. In order that the DLP in $\mathbb{F}_p$ is difficult, one needs to take the bit size of p at least 1024. On the contrary, q may be as small as only 160 bits long. Thus, exponentiation time in DSA decreases by a factor of (at least) six. The signature size also decreases by the same factor. The DSA standard recommends a particular way of generating the primes p and q.

Example 9.7 Choose the prime $p = 92283800153$, for which we have $p - 1 = 2^3 \times 21529 \times 535811$. We take $q = 21529$. To locate an element g of order q in $\mathbb{F}_p^*$, we take $h = 3284762809$ which gives $g \equiv h^{(p-1)/q} \equiv 34370710159 \pmod{p}$.

---

[7]Federal Information Processing Standard http://csrc.nist.gov/publications/fips/fips186-3/fips_186-3.pdf.

Bob chooses the permanent key pair as: $d = 14723$ (the private key) and $y \equiv g^d \equiv 62425452257 \pmod{p}$ (the public key).

Let $m = 1234567890$ be the message (representative) to be signed by Bob. Bob chooses the session secret $d' = 9372$, for which $g^{d'} \equiv 58447941827 \pmod{p}$. Reduction modulo q of this value gives $s = 764$. Bob computes $t \equiv (m + ds)d'^{-1} \equiv 17681 \pmod{q}$. Bob's signature on m is the pair $(764, 17681)$.

To verify this signature, Alice computes $w \equiv t^{-1} \equiv 19789 \pmod{q}$, $u_1 \equiv mw \equiv 12049 \pmod{q}$, $u_2 \equiv sw \equiv 5438 \pmod{q}$, and $v = (g^{u_1} y^{u_2} \pmod{p}) \pmod{q} = 58447941827 \pmod{q} = 764$. Since $v = s$, the signature is verified.

Let us see how verification fails on the forged signature $(764, 8179)$. We now have $w \equiv t^{-1} \equiv 8739 \pmod{q}$, $u_1 \equiv mw \equiv 7524 \pmod{q}$, $u_2 \equiv sw \equiv 2606 \pmod{q}$, and $v = (g^{u_1} y^{u_2} \pmod{p}) \pmod{q} = 9836368153 \pmod{q} = 4872$. Since $v \neq s$, the forged signature is not verified. $\qquad\square$

9.3.4 ECDSA

The elliptic-curve digital signature algorithm is essentially DSA adapted to elliptic-curve groups. The same FIPS document (Footnote 7) that accepts DSA as a standard includes ECDSA too.

Setting up the domain parameters for ECDSA involves some work. First, a finite field $\mathbb{F}_q$ is chosen, where q is either a prime or a power of 2. We choose two random elements $a, b \in \mathbb{F}_q$, and consider the curve $E : y^2 = x^3 + ax + b$ if q is prime, or the curve $y^2 + xy = x^3 + ax^2 + b$ if q is a power of 2. In order to avoid the MOV/Frey–Rück attack, it is necessary to take the curve as non-supersingular. Let n be a prime divisor of $|E_q|$, having bit length $\geqslant 160$, and let $h = |E_q|/n$ denote the corresponding cofactor. A random point G in E_q of order n is chosen by first selecting a random P on the curve and then computing $G = hP$ until one $G \neq \mathcal{O}$ is found. In order to determine n and to check that the curve E is not cryptographically weak (like supersingular or anomalous), it is necessary to compute the order $|E_q|$ by a point-counting algorithm. This is doable in reasonable time, since q is typically restricted to be no more than about 512 bits in length. The field size q (along with a representation of $\mathbb{F}_q$), the elements a, b of $\mathbb{F}_q$ defining the curve E, the integers n, h, and the point G constitute the domain parameters for ECDSA.

The signer (Bob) chooses a random integer $d \in \{2, 3, \dots, n-1\}$ (the private key), and publishes the elliptic-curve point $Y = dG$ (the public key).

To sign a message M, Bob maps M to a representative $m \in \{0, 1, 2, \dots, n-1\}$. A random session key $d' \in \{2, 3, \dots, n-1\}$ is generated, and the point $S = d'G$ is computed. The x-coordinate $x(S)$ of S is reduced modulo n to generate the first part of the signature: $s \equiv x(S) \pmod{n}$. In order to generate the second part, Bob computes $t \equiv (m + ds)d'^{-1} \pmod{n}$. Bob's signature on m (or M) is the pair (s, t).

In order to verify this signature, Alice obtains Bob's permanent public key Y, and computes the following: $w \equiv t^{-1} \pmod{n}$, $u_1 \equiv mw \pmod{n}$,

$u_2 \equiv sw \pmod{n}$, $V = u_1 G + u_2 Y$, and $v \equiv x(V) \pmod{n}$. The signature is accepted if and only if $v = s$. The correctness of this verification procedure is analogous to that for ElGamal or DSA signatures.

Example 9.8 Take the curve $E : Y^2 = X^3 + 3X + 6$ defined over the prime field $\mathbb{F}_{997}$. By Example 4.79, $E_{\mathbb{F}_{997}}$ has order $1043 = 7 \times 149$, so we take $n = 149$ and $h = 7$. A random point of order n on E is $G = h \times (14, 625) = (246, 540)$.

Bob generates the key pair (d, Y), where $d = 73$ and $Y = dG = (698, 240)$.

Suppose that the message representative to be signed by Bob is $m = 123$. Choosing the session secret as $d' = 107$, Bob computes $S = d'G = (543, 20)$, $s = 543 \operatorname{rem} 149 = 96$, and $t \equiv (m + ds)d'^{-1} \equiv 75 \pmod{n}$.

To verify the signature $(96, 75)$, Alice computes $w \equiv t^{-1} \equiv 2 \pmod{n}$, $u_1 \equiv mw \equiv 97 \pmod{n}$, $u_2 \equiv sw \equiv 43 \pmod{n}$, $V = u_1 G + u_2 Y = (543, 20)$, and $v = 543 \operatorname{rem} 149 = 96$. Since $v = s$, the signature is verified.

The forged signature $(s, t') = (96, 112)$ leads to the computations: $w \equiv t'^{-1} \equiv 4 \pmod{n}$, $u_1 \equiv mw \equiv 45 \pmod{n}$, $u_2 \equiv sw \equiv 86 \pmod{n}$, $V = u_1 G + u_2 Y = (504, 759)$, and $v = 504 \operatorname{rem} 149 = 57 \neq s$. □

9.4 Entity Authentication

Strong authentication schemes are often referred to as *challenge-response authentication schemes* for the following reason. Suppose that Alice wants to prove to Bob her identity, that is, her knowledge of a secret σ, without disclosing σ directly to Bob. Bob generates a random *challenge* for Alice. Alice can successfully *respond* to this challenge if she knows σ. Lack of the knowledge of σ lets an impersonation attempt fail with high probability.

9.4.1 Simple Challenge-Response Schemes

A challenge-response scheme may be based on public-key encryption or digital signatures. Alice's knowledge of her private key enables her to successfully decrypt an encrypted message sent as a challenge by Bob, or to successfully generate a verifiable signature. These two approaches are elaborated now.

Algorithm 9.1 explains an authentication protocol based on Alice's capability to decrypt ciphertext messages. Bob generates a random plaintext message r, and sends a witness w of r to Alice. The witness ensures that Bob really possesses the knowledge of r. In order that third parties cannot make a successful replay of this protocol, it is required that Bob change the string r in different sessions. The function f should be one-way (not easily invertible) and collision-resistant (it should be difficult to find two different r, r' with $f(r) = f(r')$). A cryptographic hash function may be used as f.

Algorithm 9.1: CHALLENGE-RESPONSE SCHEME BASED ON ENCRYPTION

Bob generates a random string r, and computes the witness $w = f(r)$.
Bob encrypts r using Alice's public key e: $c = E_e(r)$.
Bob sends the witness w and the encrypted message c to Alice.
Alice decrypts c using her private key d: $r' = D_d(c)$.
If $w \neq f(r')$, Alice quits the protocol.
Alice sends the response r' to Bob.
Bob accepts Alice if and only if $r' = r$.

Bob then encrypts r, and sends the challenge $c = E_e(r)$ to Alice. If Alice knows the corresponding private key, she can decrypt c to recover r. If she sends r back to Bob, Bob becomes sure of Alice's capability to decrypt his challenge. A third party having no knowledge of d can only guess a value of r, and succeeds with a very little probability.

Before sending the decrypted r' back to Bob, Alice must check that Bob is participating honestly in the protocol. If $w \neq f(r')$, Alice concludes that the witness w does not establish Bob's knowledge of r. The implication could be that Bob is trying to make Alice decrypt some ciphertext message in order to obtain the corresponding unknown plaintext message. In such a situation, Alice should not proceed further with the protocol.

Example 9.9 Let me illustrate the working of Algorithm 9.1 in tandem with RSA encryption. Suppose that Alice sets up her keys as follows: $p = 132241$, $q = 267649$, $n = pq = 35394171409$, $\phi(n) = (p-1)(q-1) = 35393771520$, $e = 7$, $d \equiv e^{-1} \equiv 15168759223 \pmod{\phi(n)}$. Alice's knowledge of her private key d is to be established in an authentication interaction with Bob.

Bob chooses the random element $r = 2319486374$, and sends its sum of digits $w = 2+3+1+9+4+8+6+3+7+4 = 47$ to Alice as a witness of his knowledge of r. Cryptographically, this is a very bad realization of f, since it is neither one-way nor collision-resistant. But then, this is only a demonstrating example, not a real-life implementation.

Bob encrypts r as $c \equiv r^e \equiv 22927769204 \pmod{n}$. Upon receiving the challenge c, Alice first decrypts it: $r' \equiv c^d \equiv 2319486374 \pmod{n}$. Since the sum of digits in r' is $w = 47$, she gains the confidence that Bob really knows r'. She sends r' back to Bob. Finally, since $r' = r$, Bob accepts Alice. $\square$

Algorithm 9.2 describes a challenge-response scheme that uses Alice's ability to generate verifiable signatures. The message to be signed is generated partially by both the parties Alice and Bob. Bob's contribution c is necessary to prevent replay attacks by eavesdroppers. If c were absent in m, Eve can always send c' and $s = S_d(c')$ to Bob and impersonate as Alice, after a real interaction between Alice and Bob is intercepted. On the other hand, Alice's contribution c' in m is necessary to preclude possible attempts by Bob to let Alice use her private key on messages selected by Bob. The strings c and c' can be combined in several ways (like concatenation or modular addition).

Algorithm 9.2: CHALLENGE-RESPONSE SCHEME BASED ON SIGNATURES

```
Bob sends a random challenge c to Alice.
Alice generates a random string c'.
Alice combines c and c' to get a message m = combine(c, c').
Alice generates the signature s = S_d(m) on m.
Alice sends c' and s simultaneously to Bob.
Bob uses Alice's verification key to generate m' = V_e(s).
Bob accepts Alice if and only if m' = combine(c, c').
```

Example 9.10 Let us use the same RSA parameters and key pair of Alice as in Example 9.9. Bob sends the random string $c = 21321368768$ to Alice. Alice, in turn, generates the random string $c' = 30687013256$, and combines c and c' as $m \equiv c + c' \equiv 16614210615 \pmod{n}$. Alice then generates her signature $s \equiv m^d \equiv 26379460389 \pmod{n}$ on m, and sends c' and s together to Bob. Bob uses the RSA verification primitive to get $m' \equiv s^e \equiv 16614210615 \pmod{n}$. Since $m' \equiv c + c' \pmod{n}$, Bob accepts Alice as authentic. $\square$

9.4.2 Zero-Knowledge Protocols

A challenge-response authentication protocol does not reveal Alice's secrets straightaway to Bob or any eavesdropper, but may leak some partial information on this secret. Upon repeated use, Bob may increasingly acquire the capability of choosing strategic challenges. A *zero-knowledge protocol* is a form of challenge-response authentication scheme that comes with a mathematical *proof* that no partial information is leaked to Bob (or any eavesdropper). This means that all the parties provably continue to remain as ignorant of Alice's secret as they were before the protocol started. As a result, the security of the protocol does not degrade with continued use.

A zero-knowledge protocol typically consists of three stages. Suppose that Alice (the *claimant* or the *prover*) wants to prove her identity to Bob (the *verifier*). First, Alice chooses a random *commitment* and sends a *witness* of this commitment to Bob. Bob, in turn, sends a random *challenge* to Alice. Finally, Alice sends her *response* to the challenge back to Bob. If Alice knows the secret, she can send a valid response during every interaction, whereas an eavesdropper without the knowledge of Alice's secret can succeed with only some limited probability. If this probability is not too small, the protocol may be repeated a requisite number of times in order to reduce an eavesdropper's chance in succeeding in *all* these interactions to below a very small value.

In what follows, I explain Fiat and Shamir's zero-knowledge protocol.[8] Algorithm 9.3 details the steps in this protocol.

---

[8] Amos Fiat and Adi Shamir, How to prove yourself: Practical solutions to identification and signature problems, *Crypto*, 186–194, 1986.

Algorithm 9.3: THE FIAT–SHAMIR ZERO-KNOWLEDGE PROTOCOL

Setting up of domain parameters:
A trusted third party (TTP) selects two large primes p and q.
The TTP publishes the product $n = pq$ and a small integer k.

Setting up of Alice's secret:
The TTP selects k secret integers $s_1, s_2, \ldots, s_k \in \mathbb{Z}_n^*$.
The TTP computes the squares $v_i \equiv s_i^2 \pmod{n}$ for $i = 1, 2, \ldots, k$.
The TTP transfers $s_1, s_2, \ldots, s_k$ to Alice securely.
The TTP makes $v_1, v_2, \ldots, v_k$ public.

Authentication of Alice (claimant) by Bob (verifier):
Alice selects a random commitment $c \in \mathbb{Z}_n$.
Alice sends the witness $w \equiv c^2 \pmod{n}$ to Bob.
Bob sends k random bits (challenges) $e_1, e_2, \ldots, e_k$ to Alice.
Alice sends the response $r \equiv c \displaystyle\prod_{\substack{i=1 \\ e_i=1}}^{k} s_i \pmod{n}$ to Bob.

Bob accepts Alice if and only if $r^2 \equiv w \displaystyle\prod_{\substack{i=1 \\ e_i=1}}^{k} v_i \pmod{n}$.

In order to see how the Fiat–Shamir protocol works, first take the simple case $k = 1$, and drop all suffixes (so v_i becomes v, for example). Alice proves her identity by demonstrating her knowledge of s, the square v of which is known publicly. Under the assumption that computing square roots modulo a large integer n with unknown factorization is infeasible, no entity in the network (except the TTP) can compute s from the published value v.

In an authentication interaction, Alice first chooses a random commitment c which she later uses as a random mask for her response. Her commitment to c is reflected by the witness $w \equiv c^2 \pmod{n}$. Again because of the intractability of the modular square-root problem, neither Bob nor any eavesdropper can recover c from w. Now, Bob chooses a random challenge bit e. If $e = 0$, Alice sends the commitment c to Bob. If $e = 1$, Alice sends cs to Bob. In both cases, Bob squares this response, and checks whether this square equals w (for $e = 0$) or wv (for $e = 1$). In order that the secret s is not revealed to Bob, Alice must choose different commitments in different authentication interactions.

Let us now see how an adversary, Eve, can impersonate Alice. When Eve selects a commitment, she is unaware of the challenge that Bob is going to throw in future. In that case, her ability to supply valid responses to both the challenge bits is equivalent to her knowledge of s. If Eve does not know s, she can succeed with a probability of $1/2$ as follows. Suppose that Eve sends $w \equiv c^2 \pmod{n}$ to Bob for some c chosen by her. If Bob challenges with $e = 0$, she can send the correct response c to Bob. On the other hand, if Bob's challenge is $e = 1$, she cannot send the correct response cs, because

s is unknown to her. Eve may also prepare for sending the correct response for $e = 1$ as follows. She chooses a commitment c but sends the improper commitment $c^2/v \pmod{n}$ to Bob. If Bob challenges with $e = 1$, she sends the verifiable response c. On the other hand, if Bob sends $e = 0$, the verifiable response would be c/s which is unknown to Eve.

If a value of $k > 1$ is used, Eve succeeds in an interaction with Bob with probability 2^{-k}. This is because Eve's commitment can successfully handle exactly one of the 2^k different challenges $(e_1, e_2, \ldots, e_k)$ from Bob. Example 9.11 illustrates the case $k = 2$. If 2^{-k} is not small enough, the protocol is repeated t times, and the chance that Eve succeeds in *all* these interactions is 2^{-kt}. By choosing k and t suitably, this probability can be made as low as one desires.

A modification of the Fiat–Shamir protocol and a proof of the zero-knowledge property of the modified protocol are from Feige, Fiat and Shamir.[9] I will not deal with the Feige–Fiat–Shamir (FFS) protocol in this book.

Example 9.11 Suppose that the TTP chooses the composite integer $n = 148198401661$ to be used in the Fiat–Shamir protocol. It turns out that n is the product of two primes, but I am not disclosing these primes, since the readers, not being the TTP, are not supposed to know them.

Take $k = 2$. The TTP chooses the secrets $s_1 = 18368213879$ and $s_2 = 94357932436$ for Alice, and publishes their squares $v_1 \equiv s_1^2 \equiv 119051447029 \pmod{n}$ and $v_2 \equiv s_2^2 \equiv 100695453825 \pmod{n}$.

In an authentication protocol with Bob, Alice chooses the commitment $c = 32764862846 \pmod{n}$, and sends the witness $w \equiv c^2 \equiv 87868748231 \pmod{n}$ to Bob. The following table shows the responses of Alice for different challenges from Bob. The square r^2 and the product $w v_1^{e_1} v_2^{e_2}$ match in all the cases.

e_1	e_2	$r \pmod{n}$		$r^2 \pmod{n}$	$w v_1^{e_1} v_2^{e_2} \pmod{n}$
0	0	$c =$	32764862846	87868748231	87868748231
1	0	$c s_1 =$	50965101270	49026257157	49026257157
0	1	$c s_2 =$	102354808490	32027759547	32027759547
1	1	$c s_1 s_2 =$	65287381609	57409938890	57409938890

Let us now look at an impersonation attempt by Eve having no knowledge of Alice's secrets s_1, s_2 and Bob's challenges e_1, e_2. She can prepare for exactly one challenge of Bob, say, $(0, 1)$. She randomly chooses $c = 32764862846$, and sends the (improper) witness $w \equiv c^2/v_2 \equiv 72251816136 \pmod{n}$ to Bob. If Bob sends the challenge $(0, 1)$, Eve responds by sending $r = c = 32764862846$ for which $r^2 \equiv 87868748231 \pmod{n}$, and $w v_2 \equiv 87868748231$ too, that is, Eve succeeds. However, if Eve has to succeed for the other challenges $(0, 0)$, $(1, 0)$ and $(1, 1)$, she needs to send the responses c/s_2, $c s_1/s_2$ and $c s_1$, respectively, which she cannot do since she knows neither s_1 nor s_2. This illustrates that if all of the four possible challenges of Bob are equally likely, Eve succeeds with a probability of $1/4$ only. $\square$

---

[9]Uriel Feige, Amos Fiat and Adi Shamir, Zero knowledge proofs of identity, *Journal of Cryptology*, 1, 77–94, 1988.

9.5 Pairing-Based Cryptography

Using bilinear pairings in cryptographic protocols is a relatively new area in public-key cryptography. This development began with the work of Sakai, Ohgishi and Kasahara[10], but the paper was written in Japanese and did not immediately attract worldwide attention. Joux's three-party key-agreement protocol[11] and Boneh and Franklin's identity-based encryption scheme[12], published shortly afterwards, opened a floodgate in pairing-based cryptography.

Section 4.5 of this book already gives a reasonably detailed mathematical exposure to certain realizable pairing functions. Here, it suffices to review only the basic notion of a bilinear map. Let G_1 and G_2 be additive groups and G_3 a multiplicative group, each of prime order r. A bilinear pairing $e : G_1 \times G_2 \to G_3$ is a map satisfying $e(aP, bQ) = e(P, Q)^{ab}$ for all $P \in G_1$, $Q \in G_2$, and $a, b \in \mathbb{Z}$. For such a map to be useful, we require two conditions. First, e should be easily computable, and second, e should not be degenerate (that is, there must exist $P \in G_1$ and $Q \in G_2$ for which $e(P, Q)$ is not the identity element of G_3). Recall that Weil and (reduced) Tate pairings satisfy these conditions for appropriately chosen elliptic-curve groups G_1 and G_2 and for an appropriate subgroup G_3 of the multiplicative group of a finite field.

A special attention is given to the case $G_1 = G_2$ (call this group G). A bilinear map $e : G \times G \to G_3$ can, for example, be obtained from Weil or Tate pairing in conjunction with distortion maps on supersingular elliptic curves.

Before we jump to computationally difficult problems associated with bilinear pairing maps, let us review the Diffie–Hellman problem (introduced near the beginning of Chapter 7). Let P be a generator of the additive group G of prime order r. The *computational Diffie–Hellman problem* (CDHP) in G is the problem of computing abP from a knowledge of P, aP and bP only. The *decisional Diffie–Hellman problem* (DDHP) in G, on the other hand, refers to deciding, given P, aP, bP, zP, whether $z \equiv ab \pmod{r}$. In certain groups (like suitably large elliptic-curve groups), both these problems are computationally infeasible, at least to the extent we know about the discrete-logarithm problem in these groups. Of course, this assertion should not be taken to mean that one has to solve the discrete-logarithm problem to solve these Diffie–Hellman problems. It is only that at present, no better methods are known.

Difficulties arise in the presence of bilinear maps associated with G. First, consider the special (but practically important) case that $e : G \times G \to G_3$ is a bilinear map. Solving the DDHP in G becomes easy, since $z \equiv ab \pmod{r}$ if and only if $e(aP, bP) = e(P, zP)$. The CDHP in G is, however, not known

---

[10] R. Sakai, K. Ohgishi and M. Kasahara, Cryptosystems based on pairing, *SCIS*, 2000.

[11] Antoine Joux, A one-round protocol for tripartite Diffie–Hellman, *ANTS-4*, 385–394, 2004.

[12] Dan Boneh and Matthew K. Franklin, Identity based encryption from the Weil pairing, *Crypto*, 213–229, 2001.

to be easily solvable in the presence of bilinear maps $e : G \times G \to G_3$. Indeed, the CDHP in the presence of bilinear maps is believed to remain as difficult as it was without such maps. The group G is called a *gap Diffie–Hellman* (GDH) group if the DDHP is easy in G, but the CDHP is difficult in G. Note, however, that the presence of bilinear maps is not necessary (but only sufficient) for creating GDH groups. The notion of GDH groups readily extends to the general case of two groups G_1, G_2 admitting a bilinear map $e : G_1 \times G_2 \to G_3$.

The above discussion highlights the need for modifying the notion of Diffie–Hellman problems in the presence of bilinear maps. The (computational) *bilinear Diffie–Hellman problem* (BDHP) in the context of a bilinear map $e : G \times G \to G_3$ is to compute $e(P, P)^{abc}$ given the elements $P, aP, bP, cP \in G$ only. The *decisional bilinear Diffie–Hellman problem* (DBDHP), on the other hand, refers to the problem of deciding, from a knowledge of $P, aP, bP, cP, zP \in G$ only, whether $z \equiv abc \pmod{r}$ (or equivalently, whether $e(P, P)^z = e(P, P)^{abc}$). These two new problems are believed not to be assisted by the existence of the bilinear map e, that is, if the discrete-logarithm problem is difficult in G, it is assumed that the BDHP and DBDHP are difficult too, even in the presence of (efficiently computable) bilinear maps e.

Let us now turn our attention to the more general setting that $G_1 \neq G_2$, and we have a bilinear map $e : G_1 \times G_2 \to G_3$. We continue to assume that G_1, G_2 are additive groups and G_3 is a multiplicative group, each of prime order r. The presence of e is assumed not to make the DDHP in G_1 or G_2 easier. This is referred to as the *external Diffie–Hellman* (XDH) assumption.

In the context of bilinear maps $e : G_1 \times G_2 \to G_3$ (with $G_1 \neq G_2$), we talk about some other related computational problems which are again believed to be difficult (given that the discrete-logarithm problem is difficult in G_1 and also in G_2). Let P and Q be generators of G_1 and G_2, respectively. Since e is non-degenerate, $e(P, Q)$ is not the identity element of G_3. The Co-CDHP is the problem of computing $abQ \in G_2$, given $P, aP \in G_1$ and $Q, bQ \in G_2$ only. The special case $b = 1$ (computing aQ from P, aP, Q) is also often referred to by the same name. Indeed, these two variants are computationally equivalent, since the general case can be solved from the special case by replacing Q by bQ. Here, it is assumed that $b \not\equiv 0 \pmod{r}$ (otherwise the problem is trivial). Note also that the variant with $a = 1$ (computing bP from the knowledge of P, Q, bQ only) is analogously a problem equivalent to the general Co-CDHP. The Co-DDHP is to decide, given $P, aP \in G_1$ and $Q, bQ, zQ \in G_2$ only, whether $z \equiv ab \pmod{r}$. Again, the special variants corresponding to $b = 1$ or $a = 1$ are computationally equivalent to the general problem.

The two bilinear Diffie–Hellman problems in the context of the bilinear map $e : G_1 \times G_2 \to G_3$ are the following. The Co-BDHP is the problem of computing $e(P, Q)^{ab}$ from the knowledge of $P, aP, bP \in G_1$ and $Q \in G_2$ only. Finally, the co-DBDHP is the problem of deciding whether $z \equiv ab \pmod{r}$ from the knowledge of $P, aP, bP \in G_1$, $Q \in G_2$ and $e(P, Q)^z \in G_3$ only.

I am now going to describe a few cryptographic protocols that exploit the presence of bilinear maps and that derive their securities from the assumption that the different versions of the Diffie–Hellman problem discussed above are difficult to solve for suitable choices of the groups. If the assumption $G_1 = G_2$ ($= G$) simplifies the exposition (but without a loss of security), I will not hesitate to make this assumption. The reader should, however, keep in mind that these protocols may often be generalized to the case $G_1 \neq G_2$.

9.5.1   Identity-Based Encryption

Long before pairings have been employed in cryptography, Shamir[13] introduced the notion of identity-based encryption and signature schemes. Although Shamir proposed an identity-based signature scheme in this paper, he could not provide a concrete realization of an identity-based encryption scheme. Boneh and Franklin's work (2001) provided the first such realization, thereby bringing pairing to the forefront of cryptology research.

In a traditional encryption or signature scheme, the authenticity of an entity's public key is established by certificates. This requires a *certification authority* (CA) to sign every public key (and other identifying information). When an entity verifies the CA's signature on a public key (and the associated identifying information), (s)he gains the confidence in using the key.

An identity-based scheme also requires a *trusted authority* (TA) responsible for generating keys for each entity. In view of this, the TA is also often called the *key-generation center* (KGC) or the *private-key generator* (PKG). As in the case of certificates, every entity needs to meet the TA privately for obtaining its keys. However, there is no signature-verification process during the use of a public key. Instead, Alice can herself *generate* Bob's (authentic) public key from the identity of Bob (like Bob's e-mail address). No involvement of the TA is necessary at this stage. This facility makes identity-based schemes much more attractive than certification-based schemes.

9.5.1.1   Boneh–Franklin Identity-Based Encryption

The Boneh–Franklin IBE scheme has four stages discussed one by one below. We consider the simplified case of pairing given by $G_1 = G_2 = G$.

Setup: The TA (or PKG or KGC) first identifies suitable groups G, G_3 of prime order r, a generator P of G, and also a pairing map $e : G \times G \to G_3$. The TA chooses a random *master secret key* $s \in \mathbb{Z}_r^*$, and computes $P_{pub} = sP$. Two (cryptographic) hash functions $H_1 : \{0,1\}^* \to G$ and $H_2 : G_3 \to \{0,1\}^n$ for some suitable n are also chosen. The master secret key s is kept secret by the TA. The parameters $r, G, G_3, e, P, P_{pub}, n, H_1, H_2$ are publicly disclosed.

Key generation: Suppose that Bob wants to register to the TA. Let the identity of Bob be given by his e-mail address `bob@p.b.cr`. The TA first hashes the identity of Bob, and then multiplies the hash by the master secret s in

---

[13] Adi Shamir, Identity based cryptosystems and signature schemes, *Crypto'84*, 47–53, 1985.

order to generate Bob's decryption key:

$$P_{Bob} = H_1(\texttt{bob@p.b.cr}),$$
$$D_{Bob} = sP_{Bob}.$$

Here, P_{Bob} is the hashed identity of Bob, computable by anybody who knows Bob's e-mail address. The decryption key D_{Bob} is handed over to Bob (by the TA) securely, and is to be kept secret.

Encryption: In order to send an n-bit message M to Bob, Alice performs the following steps.

1. Alice computes Bob's hashed identity $P_{Bob} = H_1(\texttt{bob@p.b.cr}) \in G$.
2. Alice computes $g = e(P_{Bob}, P_{pub}) \in G_3$.
3. Alice chooses a random element $a \in \mathbb{Z}_r^*$.
4. Alice computes the ciphertext $C = (aP, M \oplus H_2(g^a)) \in G \times \{0,1\}^n$.

Here, $H_2(g^a)$ is used as a mask to hide the message M.

Decryption: Bob decrypts a ciphertext $C = (U, V) \in G \times \{0,1\}^n$ as

$$M = V \oplus H_2(e(D_{Bob}, U)).$$

This process involves use of Bob's private key D_{Bob}.

Let us first establish that this decryption procedure works, that is, we start with $U = aP$ and $V = M \oplus H_2(g^a)$. It suffices to show $g^a = e(D_{Bob}, aP)$. By bilinearity, it follows that $e(D_{Bob}, aP) = e(sP_{Bob}, aP) = e(P_{Bob}, P)^{sa} = e(P_{Bob}, sP)^a = e(P_{Bob}, P_{pub})^a = g^a$.

Let us now scrutinize the security of this scheme. Let the discrete logarithm of P_{Bob} to the base P be b. This is unknown to all the parties, but that does not matter. From public (or intercepted) information, an eavesdropper, Eve, knows P (a system parameter), $U = aP$ (part of the ciphertext), $P_{Bob} = bP$ (computable by anybody), and $P_{pub} = sP$ (the public key of the TA). The mask (before hashing) is $g^a = e(P_{Bob}, P_{pub})^a = e(bP, sP)^a = e(P, P)^{abs}$ (by bilinearity). Eve's ability to decrypt only from public knowledge is equivalent to computing the mask $H_2(e(P, P)^{abs})$ (under cryptographic assumptions about the hash function H_2). That is, Eve needs to solve the computational bilinear Diffie–Hellman problem (BDHP). Under the assumption that the BDHP is infeasible for the chosen parameters, the Boneh–Franklin scheme is secure.

Alice knows the session secret a (in addition to P, aP, bP, sP), so she can compute the mask $e(P, P)^{abs} = e(bP, sP)^a$. Bob knows neither of a, b, s, but can still compute the mask as $e(P, P)^{abs} = e(s(bP), aP)$ using his knowledge of $D_{Bob} = sbP = sP_{Bob}$.

Example 9.12 In all the examples of pairing-based protocols, I use the supersingular curve $E : y^2 = x^3 + x$ defined over the prime field $\mathbb{F}_p$, $p = 744283$. We have $|E_p| = p+1 = 4r$, where $r = 186071$ is a prime. The point $P = (6, 106202)$ of order r generates the subgroup $G = G_1$ of E_p. In order to fix the second group for defining a bilinear map, we take $\mathbb{F}_{p^2} = \mathbb{F}_p(\theta)$ with $\theta^2 + 1 = 0$. An element of $\mathbb{F}_{p^2}$ is represented as $a\theta + b$ with $a, b \in \mathbb{F}_p$. These elements follow the same arithmetic as do the complex numbers in the standard representation. The group E_{p^2} contains the subgroup G_2 of order r, generated by the point $Q = (-6, 106202\theta) = (744277, 106202\theta)$. Since P and Q are linearly independent, the Weil pairing $e_r : G_1 \times G_2 \to G_3$ is a degenerate bilinear map, where G_3 is the multiplicative group of r-th roots of unity in $\mathbb{F}_{p^2}^*$.

We often use a simplified pairing $e : G \times G \to G_3$. Since E is supersingular, it admits a distortion map $\varphi : G_1 \to G_2$ given by $\varphi(a, b) = (-a, b\theta)$. (In fact, $\varphi(P) = Q$.) In this case, we take the distorted Weil pairing $e(P_1, P_2) = e_r(P_1, \varphi(P_2))$ as the relevant bilinear map e (with both $P_1, P_2 \in G_1 = G$).

In the setup phase of the Boneh–Franklin scheme, the TA fixes these parameters r, G, G_3, e, P. Suppose that the TA chooses the master secret key $s = 314095$, for which the public key is $P_{pub} = sP = (246588, 425427)$. The TA should also specify n, H_1 and H_2. A hash function from $\{0, 1\}^*$ to the group G involves some involved constructions. I will not go into the details of such a construction, but assume that the hashed public identities of entities (like P_{Bob}) are provided to us. I, however, make a concrete proposal for $H_2 : G_3 \to \{0, 1\}^n$. Each element of G_3 is of the form $a\theta + b$ with $a, b \in \mathbb{F}_p$. Since p is a 20-bit prime, we will concatenate 20-bit representations of a and b, and call it $H_2(a\theta + b)$. This concatenated value equals $2^{20}a + b$, so $n = 40$. This is certainly not a good hash function at all, but is supplied here just as a placeholder to demonstrate the working of the Boneh–Franklin scheme.

In the registration phase, Bob's hashed public identity is obtained as $P_{Bob} = (267934, 76182)$ (we assume that this is computable by anybody knowing Bob's identity). Bob's secret key is then $D_{Bob} = sP_{Bob} = (505855, 273372)$.

Suppose that Alice wants to encrypt the 40-bit message $M = 2^{39} + 2^{29} + 2^{19} + 2^9 = 550293209600$ whose bit-wise representation is

$$M = 1000000000010000000000100000000001000000000.$$

Alice computes Bob's hashed public identity as $P_{Bob} = (267934, 76182)$ for which $g = e(P_{Bob}, P_{pub}) = 239214\theta + 737818$. Now, Alice chooses the random value $a = 60294$, and obtains

$$U = aP = (58577, 21875)$$

and $g^a = 609914\theta + 551077$. Applying H_2 on g^a gives

$$H_2(g^a) = 609914 \times 2^{20} + 551077 = 639541733541$$
$$= 1001010011100111101010000110100010100101,$$

where the last expression for $H_2(g^a)$ is its 40-bit binary representation. Using bit-wise XOR with M gives the second part of the ciphertext as

$$V = M \oplus H_2(g^a) = 0001010011000111101000000110101010100101.$$

Let us now see how the ciphertext (U, V) is decrypted by Bob. Bob first computes $e(D_{Bob}, U) = 609914\theta + 551077$. This is same as g^a computed by

Alice. Therefore, this quantity after hashing by H_2 and bit-wise xor-ing with V recovers the message M that Alice encrypted.

An eavesdropper Eve intercepts (U, V), and uses $D'_{Bob} = (215899, 48408) \neq D_{Bob}$ for decryption. This gives $e(D'_{Bob}, U) = 291901\theta + 498758$. Application of H_2 on this gives the bit string 01000111010000111101011110011110001000110. When this is xor-ed with V, Eve recovers the message

$$M' = \texttt{01010011100001000111011111111011011100011} \neq M.$$ □

9.5.2 Key Agreement Based on Pairing

The two pairing-based key-agreement protocols I am going to discuss now are based on the BDHP. Indeed, the Boneh–Franklin encryption scheme is a direct adaptation of these key-agreement protocols. This is quite similar to the fact that the ElGamal encryption scheme is a direct adaptation of the Diffie–Hellman key-agreement protocol.

9.5.2.1 Sakai–Ohgishi–Kasahara Two-Party Key Agreement

As in the Boneh–Franklin scheme, the TA generates public parameters: a prime r, groups G, G_3 of order r, a pairing map $e : G \times G \to G_3$, a generator P of G, TA's public key P_{pub} $(= sP$, where s is the master secret key), and a hash function $H_1 : \{0,1\}^* \to G$ for hashing identities of parties. The bit length n and the second hash function H_2 used by the Boneh–Franklin scheme are not needed in the Sakai–Ohgishi–Kasahara (SOK) protocol.

The TA hashes an individual party's identity, and then gives the product of s with this hashed identity to that party. For instance, Alice (with e-mail address `alice@p.b.cr`) has the public hashed identity $P_{Alice} = H_1(\texttt{alice@p.b.cr})$, and receives the private key $D_{Alice} = sP_{Alice}$ from the TA.

Suppose that two registered parties Alice and Bob plan to establish a shared secret. Alice computes Bob's hashed identity $P_{Bob} = H_1(\texttt{bob@p.b.cr})$, and generates the secret $S_{Alice} = e(D_{Alice}, P_{Bob})$. Likewise, Bob computes Alice's hashed identity $P_{Alice} = H_1(\texttt{alice@p.b.cr})$ and subsequently the secret $S_{Bob} = e(P_{Alice}, D_{Bob})$. We have $S_{Alice} = e(D_{Alice}, P_{Bob}) = e(sP_{Alice}, P_{Bob}) = e(P_{Alice}, P_{Bob})^s = e(P_{Alice}, sP_{Bob}) = e(P_{Alice}, D_{Bob}) = S_{Bob}$. This key-agreement protocol is *non-interactive* in the sense that no message transmission takes place between the communicating parties.

In order to see how this protocol is related to the BDHP, let a and b be the (unknown) discrete logarithms of the hashed identities P_{Alice} and P_{Bob} to the base P. All parties can compute aP, bP. Moreover, P and $sP = P_{pub}$ are public knowledge. The shared secret between Alice and Bob is $e(D_{Alice}, P_{Bob}) = e(P_{Alice}, P_{Bob})^s = e(aP, bP)^s = e(P, P)^{abs}$. This means that a derivation of this shared secret from the knowledge of P, aP, bP, sP only is equivalent to solving an instance of the BDHP. On the other hand, Alice knows $D_{Alice} = sP_{Alice} = saP$, so she can compute $e(P, P)^{abs} = e(saP, bP)$, whereas Bob knows $D_{Bob} = sbP$, so he too can compute $e(P, P)^{abs} = e(aP, sbP)$.

Example 9.13 The TA uses the elliptic curve and the distorted Weil pairing $e : G \times G \to G_3$ as in Example 9.12. Let the master secret key be $s = 592103$, for which the TA's public key is $P_{pub} = sP = (199703, 717555)$.

Suppose that Alice's hashed public identity is $P_{Alice} = (523280, 234529)$, so Alice's secret key is $D_{Alice} = (360234, 27008)$. Likewise, if Bob's hashed identity is $P_{Bob} = (267934, 76182)$, Bob's secret key is $D_{Bob} = (621010, 360227)$.

In the key-agreement phase, Alice and Bob compute each other's hashed identity. Subsequently, Alice computes $e(D_{Alice}, P_{Bob}) = 238010\theta + 137679$, and Bob computes $e(P_{Alice}, D_{Bob}) = 238010\theta + 137679$. Thus, the secret shared by Alice and Bob is the group element $238010\theta + 137679$. Since the distorted Weil pairing is symmetric about its two arguments, we have $e(D_{Alice}, P_{Bob}) = e(P_{Bob}, D_{Alice})$ and $e(P_{Alice}, D_{Bob}) = e(D_{Bob}, P_{Alice})$, so it is not necessary to decide which party's keys go in the first argument. □

9.5.2.2 Joux Three-Party Key Agreement

Joux's three-party protocol is conceptually similar to the SOK protocol with a few differences. First, there is no role played by the TA. The three communicating parties decide upon the parameters r, G, G_3, e, P publicly without involving any trusted party. Joux's protocol is not identity-based, that is, the master secret key and the private keys of entities are absent from the protocol. Finally, the protocol is interactive. The communicating parties need to carry out one round of message broadcasting before they establish a common secret.

Alice, Bob and Carol individually and secretly generate random elements $a, b, c \in \mathbb{Z}_r^*$, respectively. Alice broadcasts aP to Bob and Carol, Bob broadcasts bP to Alice and Carol, and Carol broadcasts cP to Alice and Bob. After this transmission, the parties proceed as follows.

1. Alice computes $e(bP, cP)^a = e(P, P)^{abc}$,
2. Bob computes $e(aP, cP)^b = e(P, P)^{abc}$, and
3. Carol computes $e(aP, bP)^c = e(P, P)^{abc}$.

So the shared secret is $e(P, P)^{abc}$. A passive eavesdropper Eve, listening to the communication, gains the knowledge of P, aP, bP, cP only. Her ability to compute $e(P, P)^{abc}$ amounts to solving an instance of the BDHP.

Example 9.14 Alice, Bob and Carol plan to use the elliptic curve, the point P, and the distorted Weil pairing $e : G \times G \to G_3$ as in Example 9.12. The computations done by Alice, Bob and Carol are summarized below.

Entity	x	xP	Shared value
Alice	$a = 328764$	$aP = (676324, 250820)$	$e(bP, cP)^a = 140130\theta + 718087$
Bob	$b = 76532$	$bP = (182560, 387188)$	$e(aP, cP)^b = 140130\theta + 718087$
Carol	$c = 127654$	$cP = (377194, 304569)$	$e(aP, bP)^c = 140130\theta + 718087$

Alice broadcasts aP after she computes this point in G. Likewise, Bob broadcasts bP, and Carol broadcasts cP. At the end of the protocol, the secret shared by the three parties is the element $140130\theta + 718087 \in \mathbb{F}_{p^2}$. □

9.5.3 Identity-Based Signature

An identity-based signature scheme is similar to an identity-based encryption scheme in the sense that a verifier derives the signer's public key directly from the public identity of the signer. The authenticity of this public key is not to be established by means of certificates.

9.5.3.1 Shamir Scheme

Adi Shamir happens to be the first to propose a concrete realization of an identity-based signature scheme (Footnote 13). Shamir's scheme is not based on pairing, but, being a pioneer, is worth mentioning in this context.

Setup: The TA generates two large primes p, q, and computes $n = pq$ and $\phi(n) = (p-1)(q-1)$. In addition, an integer $e \in \{2, 3, \ldots, \phi(n) - 2\}$, coprime to $\phi(n)$, is chosen by the TA. Although Shamir remarks that e should also be a large prime, it is not clear whether this is a real necessity. Any $e \geqslant 2$ (could be a prime) appears to make the scheme sufficiently secure. The integer $d \equiv e^{-1} \pmod{\phi(n)}$ is also computed by the TA.

The TA publishes (n, e) as public parameters to be used by all the entities across the network. Furthermore, a public hash function $H : \{0, 1\}^* \to \mathbb{Z}_n$ is fixed *a priori*. The factorization of n (knowledge of p or q or d or $\phi(n)$) is kept secret. The TA needs to use this secret knowledge to generate keys.

Key generation: For registering Bob, the TA generates the hashed identity $I_{Bob} = H(\texttt{bob@s.i.b.cr})$ (Bob is now not in the *domain* of pairing-based cryptography!). Bob's secret key $D_{Bob} \equiv I_{Bob}^d \pmod{n}$ is secretly computed and securely handed over to Bob by the TA. Note that $I_{Bob} \equiv D_{Bob}^e \pmod{n}$.

Signing: In order to sign a message $m \in \mathbb{Z}_n$, Bob first generates a random non-zero $x \in \mathbb{Z}_n$, and computes

$$s \equiv x^e \pmod{n}.$$

Subsequently, Bob uses his private key to generate

$$t \equiv D_{Bob} \times x^{H(s,m)} \pmod{n}.$$

Bob's signature on m is the pair (s, t).

Verification: Raising the last congruence (for t) to the e-th power gives

$$t^e \equiv D_{Bob}^e \times (x^e)^{H(s,m)} \equiv I_{Bob} \times s^{H(s,m)} \pmod{n}.$$

That is, the verifier checks whether the congruence $t^e \equiv I_{Bob} \times s^{H(s,m)} \pmod{n}$ holds. Here, Bob's *public key* I_{Bob} is obtained by hashing his identity.

A forger can start with a random non-zero x of her choice, and computes x^e and $x^{H(s,m)}$ modulo n. But then, the ability of the forger to generate the correct t is equivalent to her knowledge of D_{Bob}. Obtaining D_{Bob} from I_{Bob} amounts to RSA decryption without knowing the TA's decryption key d.

Example 9.15 The TA chooses the primes $p = 142871$ and $q = 289031$, and computes $n = pq = 41294148001$ and $\phi(n) = (p-1)(q-1) = 41293716100$. The prime $e = 103319$ (not a factor of $\phi(n)$) is chosen, and its inverse $d \equiv e^{-1} \equiv 35665134679$ (mod $\phi(n)$) is computed. The values of e and n are published.

In 7-bit ASCII encoding, the string Bob evaluates to $I_{Bob} = 2^{14} \times 66 + 2^7 \times 111 + 98 = 1095650$. Let us take this as Bob's public identity. His private key is generated by the TA as $D_{Bob} \equiv I_{Bob}^d \equiv 32533552181$ (mod n).

Bob wants to sign the message $m = 1627384950 \in \mathbb{Z}_n$. He first chooses $x = 32465921980$, and computes $s \equiv x^e \equiv 30699940025$ (mod n). The hash of s and m is computed as $H(s,m) \equiv sm \equiv 22566668067$ (mod n) (this is not a good hash function, but is used only for illustration). Finally, the second part of the signature is computed as $t \equiv D_{Bob} \times x^{H(s,m)} \equiv 7434728537$ (mod n).

For verifying the signature (s,t), one computes $t^e \equiv 22911772376$ (mod n) and $I_{Bob} \times s^{H(s,m)} \equiv 22911772376$ (mod n) (I_{Bob} is derived from the string Bob as the TA did). These two quantities are equal, so the signature is verified.

Let (s,t') be a forged signature, where $s = 30699940025$ as before (for the choice $x = 32465921980$), but $t' = 21352176809 \neq t$. The quantity $I_{Bob} \times s^{H(s,m)} \equiv 22911772376$ (mod n) remains the same as in the genuine signature, but $t'^e \equiv 9116775652$ (mod n) changes, thereby invalidating the signature. $\square$

9.5.3.2 Paterson Scheme

Shortly after the first identity-based signature scheme using pairing was proposed by Sakai, Ohgishi and Kasahara in 2000, many other similar schemes appeared in the literature. These schemes differ somewhat with respect to formal security guarantee and/or efficiency. For pedagogical reasons (close resemblance with the ElGamal signature scheme), I pick the proposal of Paterson,[14] although this scheme is neither the most secure nor the most efficient among the lot. The four phases of this scheme are now described.

Setup: The TA generates the parameters: a prime r, suitable groups G, G_3 of order r, a bilinear map $e : G \times G \to G_3$, a generator P of G, a master secret key s, and the point $P_{pub} = sP \in G$. Three hash functions $H_1 : \{0,1\}^* \to G$, $H_2 : \{0,1\}^* \to \mathbb{Z}_r$ and $H_3 : G \to \mathbb{Z}_r$ are also predetermined. The TA publishes $r, G, G_3, e, P, P_{pub}, H_1, H_2, H_3$. The key s is kept secret.

Key generation: For registering Bob, the TA derives the hashed identity $P_{Bob} = H_1(\text{bob@p.b.cr})$, and gives Bob the private key $D_{Bob} = sP_{Bob}$ securely.

Signing: Let M be the message to be signed by Bob, and $m = H_2(M)$. Bob generates a random $d' \in \mathbb{Z}_r$, and generates the signature (S,T) as follows. Notice that signature generation involves no pairing computation.

$$S = d'P, \quad \text{and}$$
$$T = d'^{-1}(mP - H_3(S)D_{Bob}).$$

---

[14]Kenny G. Paterson, ID-based signatures from pairings on elliptic curves, *Electronics Letters*, 38(18), 1025–1026, 2002.

Verification: Bob's signature (S, T) on M is verified if and only if the following condition is satisfied (where $m = H_2(M)$):

$$e(P, P)^m = e(S, T)e(P_{pub}, P_{Bob})^{H_3(S)}$$

If (S, T) is a valid signature of Bob on M (with $m = H_2(M)$), we have $mP = d'T + H_3(S)D_{Bob} = d'T + H_3(S)sP_{Bob}$. By bilinearity, it follows that

$$\begin{aligned}
e(P, P)^m &= e(P, mP) = e(P, d'T + H_3(S)sP_{Bob}) \\
&= e(P, d'T)e(P, H_3(S)sP_{Bob}) = e(d'P, T)e(sP, P_{Bob})^{H_3(S)} \\
&= e(S, T)e(P_{pub}, P_{Bob})^{H_3(S)}.
\end{aligned}$$

This establishes the correctness of the verification procedure.

The security of this scheme resembles that of the ElGamal signature scheme. Signatures in the Paterson scheme are generated much in the same way as in the ElGamal scheme, and this process involves working in G only. Verification proceeds in G_3 after the application of the bilinear map e.

Example 9.16 Let us continue to use the supersingular curve E, the generator P of the group G, and the distorted Weil pairing $e : G \times G \to G_3$ as in Example 9.12. Suppose the the TA's master secret key is $s = 219430$, for which the public key is $P_{pub} = sP = (138113, 152726)$. Suppose also that Bob's hashed identity is $P_{Bob} = (267934, 76182)$ which corresponds to the private key $D_{Bob} = sP_{Bob} = (334919, 466375)$.

Bob wants to sign the message M which hashes to $m = 12345 \in \mathbb{Z}_r$. Bob chooses the session secret $d' = 87413$, and gets $S = d'P = (513155, 447898)$. For computing the second part T of the signature, Bob needs to use a hash function $H_3 : G \to \mathbb{Z}_r$. Let us take $H_3(a, b) = ab \pmod{r}$. This is again not a good hash function but is used here as a placeholder. For this H_3, Bob gets $H_3(S) = 553526$, so $T = d'^{-1}(mP - H_3(S)D_{Bob}) = (487883, 187017)$.

For signature verification, Alice computes Bob's hashed public identity $P_{Bob} = (267934, 76182)$. Then, Alice computes $W_1 = e(P, P)^m = 459680\theta + 325199$, $W_2 = e(S, T) = 139295\theta + 53887$, and $W_3 = e(P_{pub}, P_{Bob})^{H_3(S)} = 610330\theta + 645472$. Since $W_1 = W_2 W_3$, the signature is verified.

Now, let us see how a forged signature is not verified. A forger can generate $S = d'P = (513155, 447898)$ for the choice $d' = 87413$, as Bob did. However, since D_{Bob} is unknown to the forger, she uses a random $T' = (446698, 456705)$, and claims that (S, T') is the signature of Bob on the message M (or m). For verifying this forged signature, one computes $W_1 = e(P, P)^m = 459680\theta + 325199$ and $W_3 = e(P_{pub}, P_{Bob})^{H_3(S)} = 610330\theta + 645472$ as in the case of the genuine signature. However, we now have $W_2' = e(S, T') = 638462\theta + 253684 \neq W_2$. This gives $W_2' W_3 = 367570\theta + 366935 \neq W_1$. □

An advantage of identity-based schemes using pairing over Shamir's scheme is a reduction in signature sizes. For standard security, n in Shamir's scheme should be a 1024-bit composite integer, and two elements modulo n (a total of about 2048 bits) constitute the signature. On the other hand,

Paterson's scheme generates two elliptic-curve points as the signature. With suitable choices of supersingular curves (like those of embedding degree six), the total size of a Paterson signature can be smaller than 2048 bits.

9.5.4 Boneh–Lynn–Shacham (BLS) Short Signature Scheme

As an application of pairing in cryptography, I finally present the short signature scheme proposed by Boneh et al.[15] This is a conventional signature scheme (that is, not identity-based). However, its attractiveness stems from the fact that compared to other such schemes (like DSA or ECDSA), the BLS scheme can produce significantly shorter signatures at the same security level. However, in order to ensure both the shortness of signatures and comparable security with DSA-like schemes, it is preferable to work on a general pairing function $e : G_1 \times G_2 \to G_3$ with $G_1 \neq G_2$. Supersingular curves supporting $G_1 = G_2$ may fail to give short signatures at the same security level of DSA.

Three groups G_1, G_2, G_3 of prime order r, and a bilinear map $e : G_1 \times G_2 \to G_3$ are chosen as parameters, along with a generator Q of G_2. Bob selects a random non-zero $d \in \mathbb{Z}_r$ as his private key. His public key is $Y = dQ \in G_2$.

In order to sign a message M, Bob computes a short representative $R = H(M) \in G_1$. Bob generates the signature as $\sigma = dR$, that is, the signature now consists of only one element of G_1. With a suitable choice of the underlying elliptic curve on which the pairing e is defined, this may lead to signatures as small as about 160 bits (only the x-coordinate of σ and a disambiguating bit to identify its correct y-coordinate suffice). On the contrary, a DSA or ECDSA signature requires about 320 bits.

It is interesting to note that the BLS scheme does not require a session secret (unlike other conventional schemes described earlier). We now have two groups G_1, G_2. The elements Q and $Y = dQ$ in G_2 are related in the same way as do R and $\sigma = dR$ in G_1. This implies that the verification of σ on M is the same as checking whether the equality $e(\sigma, Q) = e(R, Y)$ holds (both sides are equal to $e(R, Q)^d$), where R is computed as $H(M)$.

A forger, on the other hand, needs to solve an instance of Co-CDHP in order to generate a valid signature. Signature verification is easy, since the Co-DDHP is easy for G_1, G_2, whereas forging is difficult, since the Co-CDHP is difficult. It follows that any pair of gap Diffie–Hellman (GDH) groups G_1, G_2 forms a perfect setting for the BLS scheme.

Example 9.17 We use the supersingular elliptic curve E of Example 9.12. For BLS signatures, we consider the original Weil pairing

$$e = e_r : G_1 \times G_2 \to G_3,$$

where G_1 is the subgroup of E_p of order r generated by the point $P = (6, 106202)$, and G_2 is the subgroup of E_{p^2} of order r generated by the point

---

[15]Dan Boneh, Ben Lynn and Hovav Shacham, Short signatures from the Weil pairing, *Journal of Cryptology*, 17, 297–319, 2004.

$Q = \varphi(P) = (-6, 106202\theta) = (744277, 106202\theta)$. Let Bob's private key be $d = 73302$. This gives his public key as $Y = dQ = (40987, 640815\theta)$.

Let $R = (128039, 742463)$ be the hashed message that Bob wants to sign. Bob only computes $\sigma = dR = (626836, 439558)$.

In order to verify Bob's signature σ on R, one computes $W_1 = e(\sigma, Q) = 521603\theta + 230328$, and $W_2 = e(R, Y) = 521603\theta + 230328$. Since $W_1 = W_2$, the signature is verified.

Let $\sigma' = (221920, 287578)$ be a forged signature on the same R. Now, we get $W_1' = e(\sigma', Q) = 226963\theta + 361018 \neq W_1$, whereas $W_2 = e(R, Y) = 521603\theta + 230328$ remains the same as for the genuine signature. Since $W_1' \neq W_2$, the forged signature is not verified.   □

Exercises

1. Let H be a hash function.

   (a) Prove that if H is collision resistant, then H is second preimage resistant.

   (b) Give an example to corroborate that H may be second preimage resistant, but not collision resistant.

   (c) Corroborate by an example that H may be first preimage resistant, but not second preimage resistant.

   (d) Corroborate by an example that H may be second preimage resistant, but not first preimage resistant.

2. Let M be the message to be signed by a digital signature scheme. Using a hash function H, one obtains a representative $m = H(M)$, and the signature is computed as a function of m and the signer's private key. Describe how the three desirable properties of H are required for securing the signature scheme.

3. Prove that for an n-bit hash function H, collisions can be found with high probability after making about $2^{n/2}$ evaluations of H on random input strings.

4. For the RSA encryption scheme, different entities are required to use different primes p, q. Argue why. Given that the RSA modulus is of length t bits (like $t = 1024$), estimate the probability that two entities in a network of N entities accidentally use a common prime p or q. You may assume that each entity chooses p and q independently and randomly from the set of $t/2$-bit primes.

5. Let, in the RSA encryption scheme, the ciphertexts corresponding to messages $m_1, m_2 \in \mathbb{Z}_n$ be c_1, c_2 for the same recipient (Bob). Argue that the ciphertext corresponding to $m_1 m_2 \pmod{n}$ is $c_1 c_2 \pmod{n}$. What problem does this relation create? How can you remedy this problem?

6. Let Alice encrypt, using RSA, the same message for e identities sharing the same public key e (under pairwise coprime moduli). How can Eve identify the (common) plaintext message by intercepting the e ciphertext messages? Notice that this situation is possible in practice, since the RSA encryption exponent is often chosen as a small prime. How can this problem be remedied?

7. Let $m \in \mathbb{Z}_n$ be a message to be encrypted by RSA. Count how many messages m satisfy the identity $m^e \equiv m \pmod{n}$. These are precisely the messages which do not change after encryption.

8. Prove that the RSA decryption algorithm may fail to work for $p = q$, even if one correctly takes $\phi(p^2) = p^2 - p$. (If $p = q$, factoring $n = p^2$ is trivial, and the RSA scheme forfeits its security. Worse still, it does not work at all.)

9. To speed up RSA decryption, let Bob store the primes p, q (in addition to d), compute c^d modulo both p and q, and combine these two residues by the CRT. Complete the details of this decryption procedure. What speedup is produced by this modified decryption procedure (over directly computing $c^d \pmod{n}$)?

10. Establish why a new session secret d' is required for every invocation of the ElGamal encryption algorithm.

11. Suppose that in the Diffie–Hellman key-agreement protocol, the group size $|G|$ has a small prime divisor. Establish how an active adversary (an adversary who, in addition to intercepting a message, can modify the message and send the modified message to the recipient) can learn a shared secret between Alice and Bob. How can you remedy this attack?

12. Suppose that the group G in the Diffie–Hellman key-agreement protocol is cyclic of size m, whereas $g \in G$ has order n with $n|m$. Let $f = m/n$ be the cofactor. Suppose that f has a small prime divisor u, and Bob sends $g^{d'}h$ or h to Alice (but Alice sends g^d to Bob), where $h \in G$ is an element of order u. Suppose that Alice later uses a symmetric cipher to encrypt some message for Bob using the (shared) secret key computed by Alice. Explain how Bob can easily obtain d modulo u upon receiving the ciphertext. Explain that using $g^{fdd'}$ as the shared secret, this problem can be remedied.

13. Let G be a cyclic multiplicative group (like a subgroup of $\mathbb{F}_q^*$) with a generator g. Assume that the DLP is computationally infeasible in G. Suppose that Alice, Bob and Carol plan to agree upon a common shared secret by the Burmester–Desmedt protocol[16] which works as follows.

    1. Alice generates random a, and broadcasts $Z_a = g^a$.

    2. Bob generates random b, and broadcasts $Z_b = g^b$.

    3. Carol generates random c, and broadcasts $Z_c = g^c$.

    4. Alice broadcasts $X_a = (Z_b/Z_c)^a$.

    5. Bob broadcasts $X_b = (Z_c/Z_a)^b$.

    6. Carol broadcasts $X_c = (Z_a/Z_b)^c$.

    7. Alice computes $K_a = Z_c^{3a} X_a^2 X_b$.

    8. Bob computes $K_b = Z_a^{3b} X_b^2 X_c$.

    9. Carol computes $K_c = Z_b^{3c} X_c^2 X_a$.

Prove that $K_a = K_b = K_c = g^{ab+bc+ca}$.

14. Assume that Bob uses the same RSA key pair for both encryption and signature. Suppose also that Alice sends a ciphertext c to Bob, and the corresponding plaintext is m. Finally, assume that Bob is willing to sign a message in $\mathbb{Z}_n$ supplied by Eve. Bob only ensures that he does not sign a message (like c) which has been sent by him as a ciphertext. Describe how Eve can still arrange a message $\mu \in \mathbb{Z}_n$ such that Bob's signature on μ reveals m to Eve.

15. Show that a situation as described in Exercise 9.5 can happen for RSA signatures too. This is often termed as *existential forgery* of signatures. Explain, in this context, the role of the hash function H used for computing $m = H(M)$.

16. Describe how existential forgery is possible for ElGamal signatures.

---

[16]Mike Burmester and Yvo Desmedt, A secure and scalable group key exchange system, *Information Processing Letters*, 94(3), 137–143, 2005.

17. Explain why a new session key d' is required during each invocation of the ElGamal signature-generation procedure.

18. Suppose that for a particular choice of m and d' in the ElGamal signature generation procedure, one obtains $t = 0$. Argue why this situation must be avoided. (If one gets $t = 0$, one should choose another random d', and repeat the signing procedure until $t \neq 0$. For simplicity, this intricacy is not mentioned in the ElGamal signature generation procedure given in the text.)

19. Show that in the DSA signature generation procedure, it is possible to have $s = 0$ or $t = 0$. Argue why each of these cases must be avoided.

20. Show that for each message $m \in \mathbb{Z}_n$, there are at least two valid ECDSA signatures (s, t_1) and (s, t_2) of Bob with the same s. Identify a situation where there are more than two valid ECDSA signatures (s, t) with the same s.

21. In a *blind signature scheme*, Bob signs a message m without knowing the message m itself. Bob is presented a masked version μ of m. For example, Bob may be a bank, and the original message m may pertain to an electronic coin belonging to Alice. Since money spending is usually desired to be anonymous, Bob should not be able to identify Alice's identity from the coin. However, Bob's active involvement (signature on μ) is necessary to generate his signature on the original message m. Assume that the RSA signature scheme is used. Describe a method of masking m to generate μ such that Bob's signature on m can be easily recovered from his signature on μ.[17]

22. A *batch-verification* algorithm for signatures $s_1, s_2, \ldots, s_k$ on k messages (or message representatives) $m_1, m_2, \ldots, m_k$ returns "*signature verified*" if each s_i is a valid signature on m_i for $i = 1, 2, \ldots, k$. If one or more s_i is/are not valid signature(s) on the corresponding message(s) m_i, the algorithm should, in general, return "*signature not verified*." A batch-verification algorithm is useful when its running time on a batch of k signatures is significantly smaller than the total time needed for k individual verifications.
    (a) Suppose that $s_1, s_2, \ldots, s_k$ are RSA signatures of the *same* entity on messages $m_1, m_2, \ldots, m_k$. Describe a batch-verification procedure for these k signatures. Establish the speedup produced by your batch-verification algorithm. Also explain how the algorithm declares a batch of signatures as *verified* even when one or more signatures are not individually verifiable.
    (b) Repeat Part (a) on k DSA signatures $(s_1, t_1), (s_2, t_2), \ldots, (s_k, t_k)$ from the same entity on messages $m_1, m_2, \ldots, m_k$.
    (c) Repeat Part (b) when the k signatures come from k different entities.
    (d) What is the problem in adapting the algorithm of Part (b) or (c) to the batch verification of ECDSA signatures?

23. Describe how a zero-knowledge authentication scheme can be converted to a signature scheme.

24. Let $n = pq$ be a product of two primes p, q each congruent to 3 modulo 4.

---

[17]David Chaum, Blind signatures for untraceable payments, *Crypto*, 199–202, 1982.

(a) Prove that every quadratic residue modulo this n has four square roots, of which exactly one is again a quadratic residue modulo n. Argue that from the knowledge of p and q, this unique square root can be easily determined.

(b) Suppose that Alice wants to prove her knowledge of the factorization of n to Bob using the following authentication protocol. Bob generates a random $x \in \mathbb{Z}_n$, and sends the challenge $c \equiv x^4 \pmod{n}$ to Alice. Alice computes the unique square root r of c which is a square in $\mathbb{Z}_n$. Alice sends r to Bob. Bob accepts Alice's identity if and only if $r \equiv x^2 \pmod{n}$. Show that Bob can send a *malicious* challenge c (not a fourth power) to Alice such that Alice's response r reveals the factorization of n to Bob.

25. Let $e : G \times G \rightarrow G_3$ be a bilinear map (easily computable). Prove that the DLP in G is no more difficult than the DLP in G_3.

26. In this exercise, we deal with Boneh and Boyen's identity-based encryption scheme.[18] Let G, G_3 be groups of prime order r, P a generator of G, and $e : G \times G \rightarrow G_3$ a bilinear map. The master secret key of the TA consists of two elements $s_1, s_2 \in \mathbb{Z}_r^*$, and the public keys are $Y_1 = s_1 P$ and $Y_2 = s_2 P$. In the registration phase for Bob, the TA generates a random $t \in \mathbb{Z}_r^*$, and computes $K = (P_{Bob} + s_1 + s_2 t)^{-1} P$, where $P_{Bob} \in \mathbb{Z}_r^*$ is the hashed public identity of Bob, and where the inverse is computed modulo r. Bob's private key is the pair $D_{Bob} = (t, K)$.

In order to encrypt a message $m \in G_3$ for Bob, Alice generates a random $k \in \mathbb{Z}_r^*$, and computes the ciphertext $(U, V, W) \in G \times G \times G_3$, where $U = k P_{Bob} P + k Y_1$, $V = k Y_2$, and $W = M \times e(P, P)^k$.

Describe how the ciphertext (U, V, W) is decrypted by Bob.

27. Generalize the Sakai–Ohgishi–Kasahara key-agreement protocol to the general setting of a bilinear map $e : G_1 \times G_2 \rightarrow G_3$ (with $G_1 \neq G_2$, in general).[19]

28. Okamoto and Okamoto propose a three-party non-interactive key-agreement scheme.[20] The TA sets up a bilinear map $e : G \times G \rightarrow G_3$ with $r = |G| = |G_3|$. The TA also chooses a secret polynomial of a suitable degree k:

$$f(x) = d_0 + d_1 x + \cdots + d_k x^k \in \mathbb{Z}_r[x].$$

For a generator P of G, the TA computes $V_i = d_i P$ for $i = 0, 1, \ldots, k$. The TA publishes $r, G, G_3, e, P, V_0, V_1, \ldots, V_k$.

For an entity A with hashed identity $P_A \in \mathbb{Z}_r^*$, the private key is computed by the TA as $D_A \equiv f(P_A) \pmod{r}$.

Describe how three parties Alice, Bob and Carol with hashed public identities P_A, P_B, P_C (respectively) can come up with the shared secret $e(P, P)^{D_A D_B D_C}$. Comment on the choice of the degree k of f.

---

[18]Dan Boneh and Xavier Boyen, Efficient selective-ID secure identity based encryption without random oracles, *EuroCrypt*, 223–238, 2004.

[19]Régis Dupont and Andreas Enge, Provably secure non-interactive key distribution based on pairings, *Discrete Applied Mathematics*, 154(2), 270–276, 2006.

[20]Eiji Okamoto and Takeshi Okamoto, Cryptosystems based on elliptic curve pairing, *Modeling Decisions for Artificial Intelligence—MDAI*, 13–23, 2005.

29. Sakai, Ohgishi and Kasahara propose an identity-based signature scheme (in the same paper where they introduced their key-agreement scheme). The public parameters r, G, G_3, e, P, P_{pub}, and the master secret key s are as in the SOK key-agreement scheme. Also, let $H : \{0,1\}^* \to G$ be a hash function. Bob has the hashed public identity P_{Bob} and the private key $D_{Bob} = sP_{Bob}$. In order to sign a message M, Bob chooses a random $d \in \mathbb{Z}_r$, and computes $U = dP \in G$. For $h = H(P_{Bob}, M, U)$, Bob also computes $V = D_{Bob} + dh \in G$. Bob's signature on M is (U, V). Describe how this signature can be verified.

30. Cha and Cheon's identity-based signature scheme[21] uses a bilinear map $e : G \times G \to G_3$, and hash functions $H_1 : \{0,1\}^* \to G$ and $H_2 : \{0,1\}^* \times G \to \mathbb{Z}_r$, where $r = |G| = |G_3|$. For the master secret key s, the TA's public identity is $P_{pub} = sP$, where P is a generator of G. Bob's hashed (by H_1) public identity is P_{Bob}, and Bob's private key is $D_{Bob} = sP_{Bob}$. Bob's signature on a message M is (U, V), where, for a randomly chosen $t \in \mathbb{Z}_r$, Bob computes $U = tP_{Bob}$, and $V = (t + H_2(M, U))D_{Bob}$. Explain how verification of (U, V) is done in the Cha–Cheon scheme. Discuss how the security of the Cha–Cheon scheme is related to the bilinear Diffie–Hellman problem. Compare the efficiency of the Cha–Cheon scheme with the Paterson scheme.

31. Boneh and Boyen propose short signature schemes (not identity-based).[22] These schemes do not use hash functions. In this exercise, we deal with one such scheme. Let $e : G_1 \times G_2 \to G_3$ be a bilinear map with G_1, G_2, G_3 having prime order r. Let P be a generator of G_1, Q a generator of G_2, and $g = e(P, Q)$. The public parameters are $r, G_1, G_2, G_3, e, P, Q, g$. Bob selects a random $d \in \mathbb{Z}_r^*$ (private key), and makes $Y = dQ$ public. Bob's signature on the message $m \in \mathbb{Z}_r$ is $\sigma = (d + m)^{-1}P \in G_1$. Here, $(d + m)^{-1}$ is computed modulo r, and is taken to be 0 if $r|(d + m)$. Describe a verification procedure for this scheme. Argue that this verification procedure can be implemented to be somewhat faster than verification in the BLS scheme.

32. Boneh and Boyen's scheme presented in Exercise 9.31 is weakly secure in some sense. In order to make the scheme strongly secure, they propose a modification which does not yield very short signatures. Indeed, the signature size is now comparable to that in DSA or ECDSA. For this modified scheme, the parameters $r, G_1, G_2, G_3, e, P, Q, g$ are chosen as in Exercise 9.31. Two random elements $d_1, d_2 \in \mathbb{Z}_r^*$ are chosen by Bob as his private key, and the elements $Y_1 = d_1Q \in G_2$ and $Y_2 = d_2Q \in G_2$ are made public. In order to sign a message $m \in \mathbb{Z}_r$, Bob selects a random $t \in \mathbb{Z}_r$ with $t \not\equiv -(d_1 + m)d_2^{-1} \pmod{r}$, and computes $\sigma = (d_1 + d_2t + m)^{-1}P \in G_1$, where the inverse is computed modulo r. Bob's signature on m is the pair (t, σ). Describe a verification procedure for this scheme. Argue that this verification procedure can be implemented to be somewhat faster than BLS verification.

---

[21] Jae Choon Cha and Jung Hee Cheon, An identity-based signature from gap Diffie–Hellman groups, *PKC*, 18–30, 2003.

[22] Dan Boneh and Xavier Boyen, Short signatures without random oracles, *Journal of Cryptology*, 21(2), 149–177, 2008.

Programming Exercises

Using GP/PARI, implement the following functions.

33. RSA encryption and decryption.

34. ElGamal encryption and decryption.

35. RSA signature generation and verification.

36. ElGamal signature generation and verification.

37. DSA signature generation and verification.

38. ECDSA signature generation and verification.

Assuming that GP/PARI functions are available for Weil and distorted Weil pairing on supersingular elliptic curves, implement the following functions.

39. Boneh–Franklin encryption and decryption.

40. Paterson signature generation and verification.

41. BLS short signature generation and verification.

Appendices

Appendix A

Background

In this chapter, I review some background material needed as a prerequisite for understanding the contents of the earlier chapters. Readers who are already conversant with these topics may quickly browse through this appendix to become familiar with my notations and conventions.

A.1 Algorithms and Their Complexity

A basic notion associated with the performance of an algorithm is its running time. The actual running time of an algorithm depends upon its implementation (in a high-level language like C or Java, or in hardware). But the implementation alone cannot characterize the running time in any precise unit, like seconds, machine instructions or CPU cycles, because these measurements depend heavily on the architecture, the compiler, the version of the compiler, the run-time conditions (such as the current load of the machine, availability

of cache), and so on. It is, therefore, customary to express the running time of
an algorithm (or an implementation of it) in more abstract (yet meaningful)
terms. An algorithm can be viewed as a transformation that converts its input
I to an output O. The size $n = |I|$ of I is usually the parameter, in terms of
which the running time of the algorithm is specified. This specification is good
if it can be expressed as a simple function of n. This leads to the following
order notations. These notations are not invented by computer scientists, but
have been used by mathematicians for ages. Computer scientists have only
adopted them in the context of analyzing algorithms.

A.1.1   Order Notations

Order notations compare the rates of growth of functions. The most basic
definition in this context is given now. While analyzing algorithms, we typically restrict our attention to positive (or non-negative) real-valued functions
of positive (or non-negative) integers. The argument of such a function is the
input size which must be a non-negative integer. The value of this function is
typically the running time (or space requirement) of an algorithm, and cannot
be negative. In the context of analyzing algorithms, an unqualified use of the
term *function* indicates a function of this type.

Definition A.1 [*Big-O notation*]   We say that a function $f(n)$ is of the order
of $g(n)$, denoted $f(n) = O(g(n))$, if there exist a positive real constant c and
a non-negative integer n_0 such that $f(n) \leqslant cg(n)$ for all $n \geqslant n_0$.[1]   ◁

Intuitively, $f(n) = O(g(n))$ implies that f does not grow faster than g
up to multiplication by a positive constant value. Moreover, initial patterns
exhibited by f and g (that is, their values for $n < n_0$) are not of concern to
us. The inequality $f(n) \leqslant cg(n)$ must hold *for all sufficiently large n*.

Example A.2 (1)   Let $f(n) = 4n^3 - 16n^2 + 4n + 25$, and $g(n) = n^3$. We
see that $f(n) = 4(n+1)(n-2)(n-3) + 1 > 0$ for all integers $n \geqslant 0$. That
$f(2.5) = -2.5 < 0$ does not matter, since we are not interested in evaluating
f at fractional values of n. We have $f(n) \leqslant 4n^3 + 4n + 25$. But $n \leqslant n^3$ and
$1 \leqslant n^3$ for all $n \geqslant 1$, so $f(n) \leqslant 4n^3 + 4n^3 + 25n^3 = 33n^3$ for all $n \geqslant 1$, that is,
$f(n) = O(g(n))$. Conversely, $g(n) = n^3 = \frac{1}{3}(4n^3 - n^3) \leqslant \frac{1}{3}(4n^3 - n^3 + 4n + 25) \leqslant
\frac{1}{3}(4n^3 - 16n^2 + 4n + 25)$ for all $n \geqslant 16$, so $g(n) = O(f(n))$ too.

(2)   The example of Part (1) can be generalized. Let $f(n) = a_d n^d +
a_{d-1}n^{d-1} + \cdots + a_1 n + a_0$ be a polynomial with $a_d > 0$. For all sufficiently
large n, the term $a_d n^d$ dominates over the other non-zero terms of $f(n)$, and
consequently, $f(n) = O(n^d)$, and $n^d = O(f(n))$.

(3)   Let $f(n) = 4n^3 - 16n^2 + 4n + 25$ as in Part (1), but $g(n) = n^4$. For
all $n \geqslant 1$, we have $f(n) \leqslant 10n^3 \leqslant 10n^4$, so that $f(n) = O(n^4)$. We prove by

---

[1]In this context, some authors prefer to say that f is big-O of g. For them, f is of the
order of g if and only if $f(n) = \Theta(g(n))$ (Definition A.3).

contradiction that $g(n) = n^4$ is *not* $O(f(n))$. Suppose that $g(n) = O(f(n))$. This implies that there exist constants $c > 0$ (real) and $n_0 \geqslant 0$ (integer) such that $n^4 \leqslant c(4n^3 - 16n^2 + 4n + 25)$, that is, $n^4 - c(4n^3 - 16n^2 + 4n + 25) \leqslant 0$ for $n \geqslant n_0$. But $\frac{1}{4}n^4 > 4cn^3$ for $n > 16c = r_1$ (say), $\frac{1}{4}n^4 > 4cn$ for $n > \sqrt[3]{16c} = r_2$, and $\frac{1}{4}n^4 > 25c$ for $n > \sqrt[4]{100c} = r_3$. For any integer $n > \max(n_0, r_1, r_2, r_3)$, we have $n^4 - c(4n^3 - 16n^2 + 4n + 25) = \frac{1}{4}n^4 + \left(\frac{1}{4}n^4 - 4cn^3\right) + 16cn^2 + \left(\frac{1}{4}n^4 - 4cn\right) + \left(\frac{1}{4}n^4 - 25c\right) > \frac{1}{4}n^4 + 16cn^2 > 0$, a contradiction.

More generally, if $f(n)$ is a polynomial of degree d, and $g(n)$ a polynomial of degree e with $0 \leqslant d \leqslant e$, then $f(n)$ is $O(g(n))$. If $e = d$, then $g(n)$ is $O(f(n))$ too. But if $e > d$, then $g(n)$ is not $O(f(n))$. Thus, the degree of a polynomial function determines its rate of growth under the $O(\ )$ notation.

(4) Let $f(n) = 2 + \sin n$, and $g(n) = 2 + \cos n$. Because of the bias 2, the functions f and g are positive-valued. Evidently, $f(n) > g(n)$ infinitely often, and also $g(n) > f(n)$ infinitely often. But $1 \leqslant f(n) \leqslant 3$ and $1 \leqslant g(n) \leqslant 3$ for all $n \geqslant 0$, that is, $f(n) \leqslant 3g(n)$ and $g(n) \leqslant 3f(n)$ for all $n \geqslant 0$. Thus, $f(n)$ is $O(g(n))$ and $g(n)$ is $O(f(n))$. Indeed, both these functions are of the order of the constant function 1, and conversely. This is intuitively clear, since $f(n)$ and $g(n)$ remain confined in the band $[1, 3]$, and do not *grow* at all.

(5) Let $f(n) = \begin{cases} n^2 & \text{if } n \text{ is even} \\ n^3 & \text{if } n \text{ is odd} \end{cases}$, and $g(n) = \begin{cases} n^2 & \text{if } n \text{ is odd} \\ n^3 & \text{if } n \text{ is even} \end{cases}$. We see that f grows strictly faster than g for odd values of n, whereas g grows strictly faster than f for even values of n. This implies that neither $f(n)$ is $O(g(n))$ nor $g(n)$ is $O(f(n))$. $\square$

The big-O notation of Definition A.1 leads to some other related notations.

Definition A.3 Let $f(n)$ and $g(n)$ be functions.
  (1) [*Big-Omega notation*] If $f(n) = O(g(n))$, we write $g(n) = \Omega(f(n))$. (Big-O indicates upper bound, whereas big-Omega indicates lower bound.)
  (2) [*Big-Theta notation*] If $f(n) = O(g(n))$ and $f(n) = \Omega(g(n))$, we write $f(n) = \Theta(g(n))$. (In this case, f and g exhibit the same rate of growth up to multiplication by positive constants.)
  (3) [*Small-o notation*] We say that $f(n) = o(g(n))$ if for every positive constant c (however small), there exists $n_0 \in \mathbb{N}_0$ such that $f(n) \leqslant cg(n)$ for all $n \geqslant n_0$. (Here, g is an upper bound on f, which is not tight.)
  (4) [*Small-omega notation*] If $f(n) = o(g(n))$, we write $g(n) = \omega(f(n))$. (This means that f is a loose lower bound on g.)
All these order notations are called *asymptotic*, since they compare the growths of functions for all sufficiently large n. $\triangleleft$

Example A.4 (1) For any non-negative integer d and real constant $a > 1$, we have $n^d = o(a^n)$. In words, any exponential function asymptotically grows faster than any polynomial function. Likewise, $\log^k n = o(n^d)$ for any positive k and d, that is, any logarithmic (or poly-logarithmic) function asymptotically grows more slowly than any polynomial function.

(2) Consider the subexponential function $L(n) = \exp\left(\sqrt{n \ln n}\right)$. We have $L(n) = o(a^n)$ for any real constant $a > 1$. Also, $L(n) = \omega(n^d)$ for any integer constant $d \geqslant 0$. This is why $L(n)$ is called a subexponential function of n.

(3) It may be tempting to conclude that $f(n) = o(g(n))$ if and only if $f(n) = O(g(n))$ but $g(n) \neq O(f(n))$. For most functions we encounter during analysis of algorithms, these two notions of loose upper bounds turn out to be the same. However, there exists a subtle difference between them. As an illustration, take $f(n) = n$ for all $n \geqslant 0$, whereas $g(n) = \begin{cases} n & \text{if } n \text{ is odd} \\ n^2 & \text{if } n \text{ is even} \end{cases}$. We have $f(n) = O(g(n))$ and $g(n) \neq O(f(n))$. But $f(n) \neq o(g(n))$, since for the choice $c = 1/2$, we cannot find an n_0 such that $f(n) \leqslant cg(n)$ for *all* $n \geqslant n_0$ (look at the odd values of n). $\qquad\square$

Some comments about the input size n are in order now. There are several units in which n can be expressed. For example, n could be the number of bits in (some reasonable encoding of) the input I. When we deal with an array of n integers each fitting into a standard 32-bit (or 64-bit) machine word, the bit size of the array is $32n$ (or $64n$). Since we are interested in asymptotic formulas with constant factors neglected, it is often convenient to take the size of the array as n. Physically too, this makes sense, since now the input size is measured in units of words (rather than bits).

It may be the case that the input size is specified by two or more independent parameters. An $m \times n$ matrix of integers has the input size mn (in terms of words, or $32mn$ in terms of bits). However, m and n can be independent of one another, and we often express the running time of a matrix-manipulation algorithm as a function of two arguments m and n (instead of a function of one argument mn). If $m = n$, that is, for square ($n \times n$) matrices, the input size is n^2, but running times are expressed in terms of n, rather than of n^2. Here, n is not the input size, but a parameter that dictates the input size.

In computational number theory, we deal with large integers which do not fit in individual machine words. Treating each such integer as having a constant size is not justifiable. An integer k fits in $\lceil \log_{2^{32}} k \rceil$ 32-bit machine words. A change in the base of logarithms affects this size by a constant factor, so we take the size of k as $\log k$. For a polynomial input of degree d with coefficients modulo a large integer m, the input size is $\leqslant (d+1) \log m$, since the polynomial contains at most $d+1$ non-zero coefficients, and $\lceil \log m \rceil$ is an upper bound on the size of each coefficient. We may treat the size of the polynomial as consisting of two independent parameters d and $\log m$.

The order notations introduced in Definition A.1 and A.3 neglect constant factors. Sometimes, it is useful to neglect logarithmic factors too.

Definition A.5 [*Soft-O notation*] Let $f(n)$ and $g(n)$ be functions. We say that $f(n) = \tilde{O}(g(n))$ if $f(n) = O(g(n) \log^t g(n))$ for some constant $t \geqslant 0$. $\qquad\triangleleft$

Example A.6 Some exponential algorithms for factoring m (like the Pollard rho method) take running times of the form $O(\sqrt[4]{m} \log^t m)$. We say that the running time of such an algorithm is $O\tilde{}(\sqrt[4]{m})$. The idea is that $\sqrt[4]{m}$ grows so fast (exponentially) with $\log m$ that it is not of a great importance to look closely at the polynomial factor $\log^t m$, despite that this factor is non-constant. By using $O\tilde{}(\ )$ instead of $O(\ )$, we make it clear that we have neglected the factor $\log^t m$. The only information lost in the soft-O notation is the correct value of t, but that is insignificant, since irrespective of the value of the constant t, we have $\sqrt[4]{m} \log^t m = O(m^{\frac{1}{4}+\epsilon})$ for any $\epsilon > 0$ (however small). $\qquad\square$

A.1.2 Recursive Algorithms

A recursive algorithm invokes itself while solving a problem. Suppose that we want to solve a problem on an input of size n. We generate subproblems on input sizes $n_1, n_2, \ldots, n_k$, recursively call the algorithm on each of these k subproblems, and finally combine the solutions of these subproblems to obtain the solution of the original problem. Typically, the input size n_i of each subproblem is smaller than the input size n of the original problem. Recursion stops when the input size of a problem becomes so small that we can solve this instance of the problem directly, that is, without further recursive calls on even smaller instances. Such a recursive algorithm is often called a *divide-and-conquer* algorithm. If the number k of subproblems and the sizes $n_1, n_2, \ldots, n_k$ are appropriately small, this leads to good algorithms. However, if the number of subproblems is too large and/or the sizes n_i are not much reduced compared to n, the behavior of the divide-and-conquer algorithm may be quite bad (so this now may become a divide-and-get-conquered algorithm).

Let $T(n)$ be the running time of the recursive algorithm described in the last paragraph. Suppose that n_0 is the size such that we make recursive calls if and only if $n \geqslant n_0$. Then, $T(0), T(1), \ldots, T(n_0 - 1)$ are some constant values (the times required to directly solve the problem for these small instances). On the other hand, for $n \geqslant n_0$, we have

$$T(n) = T(n_1) + T(n_2) + \cdots + T(n_k) + c(n),$$

where $n_1, n_2, \ldots, n_k$ (and perhaps also k) depend upon n, and $c(n)$ is the cost associated with the generation of the subproblems and with the combination of the solutions of the k subproblems. Such an expression of a function in terms of itself (but with different argument values) is called a *recurrence relation*. Solving a recurrence relation to obtain a closed-form expression of the function is an important topic in the theory of algorithms.

Example A.7 The Fibonacci numbers are defined recursively as:

$$
\begin{aligned}
F_0 &= 0, \\
F_1 &= 1, \\
F_n &= F_{n-1} + F_{n-2} \text{ for } n \geqslant 2.
\end{aligned}
$$

Consider a recursive algorithm for computing F_n, given the input n. If $n = 0$ or $n = 1$, the algorithm immediately returns 0 and 1, respectively. If $n \geqslant 2$, it recursively calls the algorithm on $n-1$ and $n-2$, adds the values returned by these recursive calls, and returns this sum. This is a bad divide-and-conquer algorithm. In order to see why, let me count the number $s(n)$ of additions performed by this recursive algorithm.

$$
\begin{aligned}
s(0) &= 0, \\
s(1) &= 0, \\
s(n) &= s(n-1) + s(n-2) + 1 \quad \text{for } n \geqslant 2.
\end{aligned}
$$

We have $s(1) = F_1 - 1$ and $s(2) = s(0) + s(1) + 1 = 1 = F_2 - 1$. For $n \geqslant 3$, we inductively get $s(n) = s(n-1) + s(n-2) + 1 = F_{n-1} - 1 + F_{n-2} - 1 + 1 = F_n - 1$. Therefore, $s(n) = F_n - 1$ for all $n \geqslant 1$. We know that $F_n \approx \frac{1}{\sqrt{5}} \rho^n$ for all large n, that is, $s(n) = \Theta(\rho^n)$, where $\rho = \frac{1+\sqrt{5}}{2}$ is the *golden ratio*. This analysis implies that this recursive algorithm runs in time exponential in n.

The following iterative (that is, non-recursive) algorithm performs much better. We start with $F_0 = 0$ and $F_1 = 1$. From these two values, we compute $F_2 = F_1 + F_0$. We then compute $F_3 = F_2 + F_1$, $F_4 = F_3 + F_2$, and so on, until F_n for the given n is computed. This requires only $n - 1$ additions.

Since the size of F_n grows linearly with n (because $F_n \approx \frac{1}{\sqrt{5}} \rho^n$), we cannot compute F_n in $o(n)$ time. This is precisely what the iterative algorithm does (not exactly, it takes $\Theta(n^2)$ time, since we should consider adding integers of size proportional to n). However, if F_n needs to be computed modulo $m \approx n$ (see Algorithm 5.5), we can finish the computation in $\log^3 n$ time. This is now truly polynomial in the input size, since the size of n is $\log n$. $\quad\square$

The above example demonstrates a bad use of recursion. In many situations, however, recursion can prove to be rather useful.

Example A.8 Let A be an array of n elements (say, single-precision integers). Our task is to rearrange the elements in the increasing order (or non-decreasing order if A contains repetitions of elements). *Merge sort* is a divide-and-conquer algorithm to solve this problem. The array is broken in two halves, each of size about $n/2$. The two halves are recursively sorted. Finally, these sorted arrays of sizes (about) $n/2$ each are merged to a single sorted array of size n.

The splitting of A (the *divide* step) is trivial. We let the two parts contain the first $n/2$ and the last $n/2$ elements of A, respectively. If n is odd, then the sizes of the subarrays may be taken as $\lceil n/2 \rceil$ and $\lfloor n/2 \rfloor$.

The merging procedure (that is, the *combine* step) is non-trivial. Let A_1 and A_2 be two sorted arrays of sizes n_1 and n_2, respectively. We plan to merge the two arrays into a sorted array B of size $n = n_1 + n_2$. We maintain two indices i_1 and i_2 for reading from A_1 and A_2, respectively. These indices are initialized to point to the first locations (that is, the smallest elements) in the respective arrays. We also maintain an index j for writing to B. The index j is initialized to 1 (assuming that array indices start from 1).

So long as either $i_1 \leqslant n_1$ or $i_2 \leqslant n_2$, we carry out the following iteration. If $i_1 > n_1$, the array A_1 is exhausted, so we copy the i_2-th element of A_2 to the j-th location in B, and increment i_2 (by one). The symmetric case $i_2 > n_2$ is analogously handled. If $i_1 \leqslant n_1$ and $i_2 \leqslant n_2$, we compare the element a_1 of A_1 at index i_1 with the element a_2 of A_2 at index i_2. If $a_1 \leqslant a_2$, we copy a_1 to the j-th location in B, and increment i_1. If $a_1 > a_2$, we copy a_2 to the j-th location in B, and increment i_2. After each copy of an element of A_1 or A_2 to B, we increment j so that it points to the next available location for writing.

In each iteration of the merging loop, we make a few comparisons, a copy of an element from A_1 or A_2 to B, and two index increments, that is, each iteration takes $\Theta(1)$ time. There are exactly n iterations, since each iteration involves a single writing in B. Therefore, the merging step takes $\Theta(n)$ time.

The basis case for merge sort is $n = 1$ (arrays with only single elements). In this case, the array is already sorted, and the call simply returns.

Let $T(n)$ denote the time for merge soring an array of size n. We have:

$$
\begin{aligned}
T(1) &= 1, \\
T(n) &= T(\lceil n/2 \rceil) + T(\lfloor n/2 \rfloor) + n \text{ for } n \geqslant 2.
\end{aligned}
$$

We like to determine $T(n)$ as a closed-form expression in n. But the recurrence relation involves floor and ceiling functions, making the analysis difficult. For us, it suffices if we can provide a big-Θ estimate on $T(n)$.

We can get rid of all floor and ceiling expressions if and only if n is a power of 2, that is, $n = 2^t$ for some $t \in \mathbb{N}_0$. In this case, we have:

$$
\begin{aligned}
T(n) = T(2^t) &= 2T(2^{t-1}) + 2^t \\
&= 2(2T(2^{t-2}) + 2^{t-1}) + 2^t = 2^2 T(2^{t-2}) + 2 \times 2^t \\
&= 2^2(2T(2^{t-3}) + 2^{t-2}) + 2 \times 2^t = 2^3 T(2^{t-3}) + 3 \times 2^t \\
&= \cdots \\
&= 2^t T(1) + t2^t = 2^t + t2^t = (1+t)2^t = (1 + \log_2 n)n.
\end{aligned}
$$

We plan to generalize this result to the claim that $T(n) = \Theta(n \log n)$ (for all values of n). Proving this claim rigorously involves some careful considerations. For any $n \geqslant 1$, we can find a t such that $2^t \leqslant n < 2^{t+1}$. Since we already know $T(2^t)$ and $T(2^{t+1})$, we are tempted to write $T(2^t) \leqslant T(n) \leqslant T(2^{t+1})$ in order to obtain a lower bound and an upper bound on $T(n)$. But the catch is that we are permitted to write these inequalities provided that $T(n)$ is an increasing function of n, that is, $T(n) \leqslant T(n+1)$ for all $n \geqslant 1$. I now prove this property of $T(n)$ by induction on n.

Since $T(1) = 1$ and $T(2) = 2T(1) + 2 = 4$, we have $T(1) \leqslant T(2)$. For the inductive step, take $n \geqslant 2$, and assume that $T(m) \leqslant T(m+1)$ for all $m < n$. If $n = 2m$ (even), then $T(n) = 2T(m) + 2m$, whereas $T(n+1) = T(m+1) + T(m) + (2m+1)$, that is, $T(n+1) - T(n) = T(m+1) - T(m) + 1 \geqslant 0$ (by the induction hypothesis), that is, $T(n) \leqslant T(n+1)$. If $n = 2m+1$ (odd), then $T(n) = T(m+1) + T(m) + (2m+1)$, whereas $T(n+1) = 2T(m+1) + (2m+2)$, that is, $T(n+1) - T(n) = T(m+1) - T(m) + 1 \geqslant 0$, again by the induction hypothesis. Thus, $T(n)$ is an increasing function of n, as required.

We now manipulate the inequalities $2^t \leqslant n < 2^{t+1}$ and $T(2^t) \leqslant T(n) \leqslant T(2^{t+1})$. First, $T(n) \leqslant T(2^{t+1}) = (t+2)2^{t+1}$. Also, $n \geqslant 2^t$, that is, $T(n) \leqslant 2(\log_2 n + 2)n$. On the other hand, $T(n) \geqslant T(2^t) = (t+1)2^t$. Since $n < 2^{t+1}$, we have $T(n) \geqslant (\log_2 n)n/2$. It therefore follows that $T(n) = \Theta(n \log n)$. $\square$

This analysis for merge sort can be easily adapted to a generalized setting of certain types of divide-and-conquer algorithms. Suppose that an algorithm, upon an input of size n, creates $a \geqslant 1$ sub-instances, each of size (about) n/b. Suppose that the total effort associated with the divide and combine steps is $\Theta(n^d)$ for some constant $d \geqslant 0$. The following theorem establishes the (tight) order of the running time of this algorithm.

Theorem A.9 [*Master theorem for divide-and-conquer recurrences*] *Let $T(n)$ be an increasing function of n, which satisfies*

$$T(n) = aT(n/b) + \Theta(n^d),$$

whenever $n > 1$ is a power of b. Let $\tau = \log_b a$. Then, we have:
- *If $\tau > d$, then $T(n) = \Theta(n^\tau)$.*
- *If $\tau < d$, then $T(n) = \Theta(n^d)$.*
- *If $\tau = d$, then $T(n) = \Theta(n^d \log n)$.* ◁

Informally, τ is the parameter indicating the cost of *conquer*, whereas d is the parameter standing for the cost of *divide* and *combine*. If $\tau > d$, the cost of conquer dominates over the divide and combine cost. If $\tau < d$, the divide and combine cost dominates over the cost of conquer. If $\tau = d$, these two costs are the same at each level of the recursion, so we multiply this cost by the (maximum) depth of recursion.

Example A.10 (1) The running time of merge sort corresponds to $a = 2$, $b = 2$ and $d = 1$, so $\tau = \log_2 2 = 1 = d$. By the master theorem, this running time is $\Theta(n \log n)$.

(2) Let $A = (a_i, a_{i+1}, \ldots, a_j)$ be a *sorted* array of n elements. We want to determine whether a key x belongs to A, and if so, at which index. The *binary search* algorithm for accomplishing this task proceeds as follows. If $n = 1$ (that is, $i = j$), we check whether $x = a_i$, and return i or an invalid index depending upon the outcome of the comparison (equality or otherwise). If $n > 1$, we compute the central index $k = \lfloor (i+j)/2 \rfloor$ in the array, and compare x with a_k. If $x \leqslant a_k$, we recursively search for x in the subarray $(a_i, a_{i+1}, \ldots, a_k)$, otherwise we recursively search for x in the subarray $(a_{k+1}, a_{k+2}, \ldots, a_j)$. The correctness of this strategy follows from the fact that A is sorted. Since k is the central index in A, the recursive call is made on a subarray of size (about) $n/2$. Therefore, the running time of binary search can be expressed as

$$T(n) = T(n/2) + \Theta(1).$$

In the notation of the master theorem, we have $a = 1$, $b = 2$ and $d = 0$, so $\tau = \log_2 1 = 0 = d$, and it follows that $T(n) = \Theta(\log n)$.

Since recursive function calls incur some overhead, it is preferable to implement binary search as an iterative algorithm using a simple loop. The running-time analysis continues to hold for the iterative version too. For merge sort, however, replacing recursion by an iterative program is quite non-trivial. □

A.1.3 Worst-Case and Average Complexity

So far, we have characterized the running time of an algorithm as a function of only the input size n. However, the exact running time may change with what input is fed to the algorithm, even when the size of the input remains the same. So long as this variation is bounded by some constant factors, the asymptotic order notations are good enough to capture the running time as a function of n. However, there exist cases where the exact function of n changes with the choice of the input (of size n). In that case, it is necessary to give a more precise connotation to the notion of running time.

The *worst-case* running time of an algorithm on an input of size n is the maximum of the running times over all inputs of size n. The *average* running time of an algorithm on an input of size n is the average of the running times, taken over all inputs of size n. Likewise, we may define the *best-case* running time of an algorithm. For merge sort, all three running times happen to be $\Theta(n \log n)$. Depending upon the input, the constant hidden in the order notation may vary, but the functional relationship with n is always $\Theta(n \log n)$. I shortly describe another popular sorting algorithm for which the worst-case and the average running times differ in the functional expression.

Clearly, the worst-case running time gives an upper bound on the running time of an algorithm. On the other hand, the average running time indicates the expected performance of the algorithm on an average (or random) input, but provides no guarantee against *bad* inputs. In view of this, an unqualified use of the phrase *running time* stands for the worst-case running time. Average or expected running times are specifically mentioned to be so.

Example A.11 *Quick sort* is a recursive algorithm for sorting an array of elements (say, single-precision integers). Let the input array A have size n. The algorithm uses an element a in A as a *pivot*. It makes a partition of the array to a form LEG, where the subarray L consists of elements of A smaller than the pivot, E is the subarray consisting of the elements equal to the pivot, and the subarray G consists of the elements of A greater than the pivot. In the decomposition LEG, the block E is already in place for the sorting task. The blocks L and G are recursively quick sorted. When the recursive calls return, the entire array is sorted. Recursion terminates when an array of size 0 or 1 is passed as argument. Such an array is already sorted.

What remains to complete the description of quick sort is the way to partition A in the form LEG. If we are allowed to use an additional array, there is little difficulty in partitioning. However, it is not necessary to use an additional array. By suitable exchanges of elements of A, the LEG decomposition can be computed. I explain a variant of this *in-place* partitioning. The description also illustrates the useful concept of *loop invariance*.

Let us use the first element of A as the pivot a. We run a loop which maintains the invariance that A is always represented in the form $LEUG$, where L, E and G are as described earlier. The block U stands for the unprocessed elements, that is, the elements that have not yet been classified for inclusion in L, E or G. We maintain three indices i, j, k pointing to the L-E, E-U and U-G boundaries. The following figure describes this decomposition.

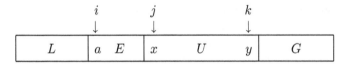

Before entering the partitioning loop, the first element of A belongs to the block E, whereas all the remaining $n-1$ elements belong to the block U. If we imagine that L is the empty block sitting before E and that G is the empty block sitting after U, we start with an $LEUG$ decomposition of A.

Inside the partitioning loop, the first element of U (that is, the element x pointed to by j) is considered for processing. Depending upon the result of comparison of x with the pivot a, one of the following three actions are made.

- If $x = a$, the region E grows by including x. This is effected by incrementing the index j (by one).

- If $x > a$, then x should go to the block G, that is, x may go to the location indexed by k. But this location already contains an unprocessed element y. The elements x and y are, therefore, swapped, and k is decremented to mark the growth of the block G. But j is not altered, since it now points to the unprocessed element y to be considered in the next iteration.

- If $x < a$, the element x should join the region L. Since L grows by one cell, the entire block E should shift by one cell. However, since E contains only elements with values equal to a, this shift of E can be more easily implemented by exchanging x with the first element of E, that is, by exchanging the elements pointed to by i and j. Both i and j should then be incremented by 1 in order to indicate the advance of the region E. The other end of U (that is, the index k) is not altered.

Each iteration of this loop processes one element from U. After exactly $n - 1$ iterations, U shrinks to a block of size 0, that is, the desired LEG decomposition of A is achieved. The partitioning procedure takes $\Theta(n)$ time, since each of the $n - 1$ iterations involves only a constant amount of work.

Let us now investigate the running time of the quick-sort algorithm. The original array A contains n elements. Suppose that, after the LEG decomposition, the block L contains n_1 elements, whereas the block G contains n_2 elements. Since the pivot a was chosen as an element of the array A, the block

E contains at least one element, that is, $n_1 + n_2 \leqslant n - 1$. If $T(n)$ denotes the running time to quick sort an array of n elements, we have:

$$
\begin{aligned}
T(0) &= 1, \\
T(1) &= 1, \\
T(n) &= T(n_1) + T(n_2) + n \text{ for } n \geqslant 2.
\end{aligned}
$$

A marked difference of this recurrence with the recurrence of merge sort is that now n_1 and n_2 depend heavily on the input.

In order to simplify our analysis, let us assume that the elements of A are distinct from one another, that is, $n_1 + n_2 = n - 1$. Intuitively, the algorithm exhibits the worst-case performance, when the partitioning is very skew, that is, either $n_1 = 0$ or $n_2 = 0$. In that case, a recursive call is made on an array of size $n - 1$, that is, the reduction in the size of the instance is as bad as possible. If every recursion experiences this situation, we have:

$$
\begin{aligned}
T(n) &= T(n-1) + T(0) + n = T(n-1) + (n+1) \\
&= T(n-2) + n + (n+1) \\
&= T(n-3) + (n-1) + n + (n+1) \\
&= \cdots \\
&= T(0) + 2 + 3 + \cdots + (n-1) + n + (n+1) \\
&= 1 + 2 + 3 + \cdots + (n-1) + n + (n+1) \\
&= (n+1)(n+2)/2,
\end{aligned}
$$

that is, $T(n) = \Theta(n^2)$.

Quick sort exhibits the best performance, when the splitting of A is rather balanced, that is, $n_1 \approx n_2 \approx n/2$. If such a situation happens in every recursive call, the recurrence is quite similar to the recurrence for merge sort, and we obtain $T(n) = \Theta(n \log n)$. Therefore, the functional behavior of the running time of quick sort in the worst case differs from that in the best case.

But then, what is the average behavior of quick sort? There are $n!$ permutations of n elements. The probability that the pivot a is the i-th smallest element of A is $(n-1)!/n! = 1/n$ for each $i = 1, 2, \ldots, n$. Therefore, the expected running time of quick sort satisfies the recurrence relation

$$
\begin{aligned}
T(n) &= \frac{1}{n}\Big[T(0) + T(n-1) + n\Big] + \frac{1}{n}\Big[T(1) + T(n-2) + n\Big] + \\
&\quad \frac{1}{n}\Big[T(2) + T(n-3) + n\Big] + \cdots + \frac{1}{n}\Big[T(n-1) + T(0) + n\Big] \\
&= \frac{2}{n}\Big[T(n-1) + T(n-2) + \cdots + T(1) + T(0)\Big] + n \text{ for } n \geqslant 2,
\end{aligned}
$$

that is,

$$
nT(n) = 2[T(n-1) + T(n-2) + \cdots + T(1) + T(0)] + n^2.
$$

For $n \geqslant 3$, we have

$$(n-1)T(n-1) = 2[T(n-2) + T(n-3) + \cdots + T(1) + T(0)] + (n-1)^2.$$

Therefore, $nT(n) - (n-1)T(n-1) = 2T(n-1) + 2n - 1$, that is,

$$T(n) = \left(\frac{n+1}{n}\right)T(n-1) + \frac{2n-1}{n} \quad \text{for } n \geqslant 3.$$

Unfolding this recurrence gives

$$
\begin{aligned}
T(n) &= \left(\frac{n+1}{n}\right)T(n-1) + \frac{2n-1}{n} \\
&= \left(\frac{n+1}{n}\right)\left[\left(\frac{n}{n-1}\right)T(n-2) + \frac{2n-3}{n-1}\right] + \frac{2n-1}{n} \\
&= \left(\frac{n+1}{n-1}\right)T(n-2) + (n+1)\left[\frac{2n-1}{n(n+1)} + \frac{2n-3}{(n-1)n}\right] \\
&= \left(\frac{n+1}{n-1}\right)\left[\left(\frac{n-1}{n-2}\right)T(n-3) + \frac{2n-5}{n-2}\right] + \\
&\quad (n+1)\left[\frac{2n-1}{n(n+1)} + \frac{2n-3}{(n-1)n}\right] \\
&= \left(\frac{n+1}{n-2}\right)T(n-3) + \\
&\quad (n+1)\left[\frac{2n-1}{n(n+1)} + \frac{2n-3}{(n-1)n} + \frac{2n-5}{(n-2)(n-1)}\right] \\
&= \cdots \\
&= \left(\frac{n+1}{3}\right)T(2) + (n+1)\left[\frac{2n-1}{n(n+1)} + \frac{2n-3}{(n-1)n} + \right. \\
&\quad \left. \frac{2n-5}{(n-2)(n-1)} + \cdots + \frac{5}{3\times 4}\right] \\
&= \frac{4}{3}(n+1) + (n+1)\left[\left(\frac{3}{n+1} - \frac{1}{n}\right) + \left(\frac{3}{n} - \frac{1}{n-1}\right) + \right. \\
&\quad \left. \left(\frac{3}{n-1} - \frac{1}{n-2}\right) + \cdots + \left(\frac{3}{4} - \frac{1}{3}\right)\right] \\
&= \frac{4}{3}(n+1) + 2(n+1)\left[\frac{1}{n} + \frac{1}{n-1} + \cdots + \frac{1}{4}\right] + 3 - \frac{1}{3}(n+1) \\
&= (n+1) + 3 + 2(n+1)\left[H_n - \frac{1}{1} - \frac{1}{2} - \frac{1}{3}\right] \\
&= 2(n+1)H_n - \frac{8}{3}(n+1) + 3.
\end{aligned}
$$

Here, $H_n = \frac{1}{1} + \frac{1}{2} + \cdots + \frac{1}{n}$ is the n-th harmonic number. We have $\ln(n+1) \leqslant H_n \leqslant \ln n + 1$ for $n \geqslant 1$. It therefore follows that the average running time of quick sort is $\Theta(n \log n)$. $\square$

A.1.4   Complexity Classes P and NP

So far, we have considered the complexity of algorithms. It is interesting to talk also about the complexity of computational problems, rather than of specific algorithms for solving them. Let P be a solvable problem, that is, there exists at least one algorithm to solve P. If we consider all algorithms to solve P, and take the minimum of the running times of these algorithms (in an order notation), we obtain the complexity of the problem P.

There are difficulties associated with this notion of complexity of problems. There may be infinitely many algorithms to solve a particular problem. Enumerating all these algorithms and taking the minimum of their running times is not feasible, in general. This limitation can be overcome by providing lower bounds on the complexity of problems. We prove analytically that any algorithm to solve some problem P must run in $\Omega(f(n))$ time. If we come up with an algorithm A to solve P in the worst-case running time $O(f(n))$, then A is called an *optimal algorithm* for P. A classical example is the sorting problem. It can be proved that under reasonable assumptions, an algorithm to sort an array of n elements must take $\Omega(n \log n)$ running time. Since merge sort takes $O(n \log n)$ time, it is an optimal sorting algorithm. We conclude that the complexity of sorting is $n \log n$ (in the sense of the big-Θ notation).

A bigger difficulty attached to the determination of the complexity of problems is that not all algorithms for solving a problem are known to us. Although we can supply a lower bound on the complexity of a problem, we may fail to come up with an optimal algorithm for that problem. For example, the complexity of multiplying two $n \times n$ matrices is certainly $\Omega(n^2)$, since any algorithm for multiplying two such matrices must read the two input matrices, and must output the product. But then, can there be an $O(n^2)$ algorithm (that is, an optimal algorithm) for matrix multiplication? The answer is not known to us. The obvious algorithm based on the formula $c_{ij} = \sum_{k=1}^{n} a_{ik} b_{kj}$ takes $\Theta(n^3)$ running time. Strassen discovers an $O(n^{\log_2 7})$-time, that is, an $O(n^{2.81})$-time, divide-and-conquer algorithm for matrix multiplication. The best known matrix-multiplication algorithm is from Coppersmith and Winograd, and runs in about $O(n^{2.376})$ time. There exists no proof that this is an optimal algorithm, that is, no algorithm can multiply two $n \times n$ matrices faster. The complexity of matrix multiplication remains unknown to this date.

In computational number theory, our current knowledge often prohibits us from concluding whether there at all exist polynomial-time algorithms for solving certain problems. Until the disclosure of the AKS test (August 2002), it was unknown whether the primality-testing problem can at all be solved in polynomial time. Today, we do not know whether the integer-factoring or the discrete-logarithm problem can be solved in polynomial time. There is no solid evidence that they cannot be. We cannot, as well, say that they can be.

The class of all problems that can be solved in polynomial time is called the complexity class P. All problems that have *known* polynomial-time algorithms belong to this class. But that is not all. The class P may contain problems for which polynomial-time algorithms are still not known. For example, primality testing has always been in the class P. It is only in August 2002 when we *know* that this problem is indeed in P. This indicates that our understanding of the boundary of the class P can never be clear. For certain problems, we can prove superpolynomial lower bounds, so these problems are naturally outside P. But problems like integer factorization would continue to bother us.

Intuitively, the class P contains precisely those problems that are *easily* solvable. Of course, an $O(n^{100})$-time algorithm would be practically as worthless as an $O(2^n)$-time algorithm. Nonetheless, treating *easy* synonymously as *polynomial-time* is a common perception in computer science.

An introduction to the class NP requires some abstraction. The basic idea is to imagine *algorithms that can guess*. Suppose that we want to sort an array A of n integers. Let there be an algorithm which, upon the input of A and n, guesses the index i at which the maximum of A resides. The algorithm then swaps the last element of A with the element at index i. Now that the maximum of A is in place, we reduce the original problem to that of sorting an array of $n - 1$ elements. By repeatedly guessing the maximum, we sort A in n iterations, provided that each guess made in this process is correct.

There are two intricacies involved here. First, what is the running time of this algorithm? We assume that each guess can be done in unit time. If so, the algorithm runs in $\Theta(n)$ time. But who will guarantee that the guesses the algorithm makes are correct? Nobody! One possibility to view this guessing procedure is to create parallel threads, each handling a guess. In this case, we talk about the *parallel* running time of this algorithm. Since the parallel algorithm must behave gracefully for all input sizes n, there must be an infinite number of computing elements to allow perfect parallelism among all guesses. A second way to realize a guess is to make all possible guesses one after another in a sequential manner. For the sorting example, the first guess involves n possible indices, the second $n-1$ indices, the third $n-2$, and so on. Thus, there is a total of $n!$ guesses, among which exactly one gives the correct result. Since $n! = \omega(2^n)$, we end up with an exponential-time (sequential) simulation of the guessing algorithm. A third way of making correct guesses is an availability of the guesses before the algorithm runs. Such a sequence of correct guesses is called a *certificate*. But then, who will supply a certificate? Nobody. We can only say that if a certificate is provided, we can sort in linear time.

For the sorting problem, there is actually not a huge necessity to guess. We can compute the index i of the maximum in A in only $O(n)$ time. When guessing is replaced by this computation, we come up with an $O(n^2)$-time sorting algorithm, popularly known as *selection sort*. For some other computational problems, it is possible to design efficient guessing algorithms, for which there is no straightforward way to replace guessing by an easy computation.

As an example, let us compute the discrete logarithm of $a \in \mathbb{F}_p^*$ to a primitive element g of $\mathbb{F}_p^*$ (where $p \in \mathbb{P}$). We seek for an integer x such that $a \equiv g^x \pmod{p}$. Suppose that a bound s on the bit size of x is given. Initially, we take $0 \leqslant x \leqslant p - 2$, so the bit size of p supplies a bound on s. Let us write

$x = 2^{s-1}\epsilon + x'$, where ϵ is the most significant bit of x with respect to the size constraint s. Let us imagine an algorithm that can guess the bit ϵ. Since $a \equiv g^x \equiv (g^{2^{s-1}})^\epsilon g^{x'} \pmod{p}$, we can compute $b \equiv g^{x'} \equiv a(g^{2^{s-1}})^{-\epsilon} \pmod{p}$ in time polynomial in $\log p$, and the problem reduces to computing the discrete logarithm x' of b to the base g. Since the bit size of x' is $s - 1$, this process of successively guessing the most significant bits of indices leads to a polynomial-time algorithm for the discrete-logarithm problem, provided that each guess can be done in unit time, and that each guess made in this process is correct. Like the sorting example, this guessing can be realized in three ways: a parallel implementation on an imaginary machine with infinitely many processing elements to handle inputs of any size, an exponential-time sequential simulation (the exhaustive search) of the bit-guessing procedure, and an availability of the correct guesses (a certificate) from an imaginary source. Evidently, if there is a polynomial-time algorithm to compute the most significant bit ϵ (from p, g, s), the discrete-logarithm problem can be solved in polynomial time too. To this date, no such polynomial-time algorithm for computing ϵ is known.

A k-way guess can be replaced by $\log_2 k$ guesses of bits. It is often desirable to assume that each bit can be guessed in unit time. Therefore, the above guessing algorithm for sorting takes $O(n \log n)$ time, since each guess involves choosing one of at most n possibilities. Guessing helps us to reduce the running time from $O(n^2)$ to $O(n \log n)$. But sorting is a problem which already has $O(n \log n)$-time algorithms (like merge sort) which do not require guessing.

Let A be a guessing algorithm for a problem P. Suppose that the number of bit guesses made by A on an input of size n is bounded from above by a polynomial in n. Suppose also that if each bit is correctly guessed, the correct solution is supplied by A in time polynomial in n. If guessing each bit takes unit time, A is a polynomial-time guessing algorithm for P. We say that A is a *non-deterministic polynomial-time algorithm* for solving P. Such an algorithm can be equivalently characterized in terms of certificates. A certificate is a specification of the sequence of correct guesses. Given a certificate, we can solve the problem P using A in polynomial time. Moreover, since there are only a polynomial number of guesses made by A, each certificate must be *succinct* (that is, of size no larger than a polynomial in n). A non-trivial example of certificates is provided in Chapter 5 (Pratt certificates for primality).

The class of problems having non-deterministic polynomial-time algorithms is called NP. A non-deterministic algorithm is allowed to make zero guesses, so P $\subseteq$ NP. Whether NP $\subseteq$ P too is not known. The P $\overset{?}{=}$ NP problem is the deepest open problem of computer science. The Clay Mathematical Foundation has declared an award of one million US dollars for solving it.

A.1.5   Randomized Algorithms

For certain computational problems in the complexity class NP, there is an interesting way to deal with non-determinism. Suppose that a significant fraction (like $1/2$) of all possible guesses for any given input leads to a solution of the problem. We replace guesses by random choices, and run the polynomial-time verifier on these choices. If this procedure is repeated for a few number of times, we hope to arrive at the solution in at least one of these runs.

First, consider *decision problems* (problems with *Yes/No* answers). For instance, the complement of the primality-testing problem, that is, checking the compositeness of $n \in \mathbb{N}$, is in NP, since a non-trivial divisor d of n (a divisor in the range $2 \leqslant d \leqslant n - 1$) is a succinct certificate for the compositeness of n. This certificate can be verified easily by carrying out a division of n by d.

But then, how easy is it to *guess* such a non-trivial divisor of a composite n? If n is the square of a prime, there exists only one such non-trivial divisor. A random guess reveals this divisor with a probability of about $1/n$, which is exponentially small in the input size $\log n$. Even when n is not of this particular form, it has only a few non-trivial divisors, and trying to find one by chance is like searching for a needle in a haystack.

An idea based upon Fermat's little theorem leads us to more significant developments. The theorem states that if p is prime, and a is not a multiple of p, then $a^{p-1} \equiv 1 \pmod{p}$. Any a (coprime to n) satisfying $a^{n-1} \not\equiv 1 \pmod{n}$ is a witness (certificate) to the compositeness of n. We know that if a composite n has at least one witness, then at least half of $\mathbb{Z}_n^*$ are witnesses too. Therefore, it makes sense that we randomly choose an element of $\mathbb{Z}_n$, and verify whether our choice is really a witness. If n is composite (and has a witness), then after only a few random choices, we hope to locate one witness, and become certain about the fact that n is composite. However, if no witness is located in several iterations, there are two possibilities: n does not have a witness at all, or despite n having witnesses, we have been so unlucky that we missed them in all these iterations. In this case, we declare n as prime with the understanding that this decision may be wrong, albeit with a small probability.

The Fermat test exemplifies how randomization helps us to arrive at practical solutions to computational problems. A *Monte Carlo algorithm* is a randomized algorithm which always runs fast but may supply a wrong answer (with low probability). The Fermat test is *No-biased*, since the answer *No* comes with zero probability of error. *Yes-biased* Monte Carlo algorithms, and Monte Carlo algorithms with *two-sided errors* may be conceived of.

The problem with the Fermat test is that it deterministically fails to find witnesses for Carmichael numbers. The Solovay–Strassen and the Miller–Rabin tests are designed to get around this problem. There are deterministic polynomial-time primality tests (like the AKS test), but the randomized tests are much more efficient and practical than the deterministic tests.

Another type of randomized algorithms needs mention in this context. A *Las Vegas algorithm* is a randomized algorithm that always produces the correct answer, but has a fast *expected* running time. This means that almost always we expect a Las Vegas algorithm to terminate fast, but on rare occasions, we may be so unlucky about the random guesses made in the algorithm that the algorithm fails to supply the (correct) answer for a very long time.

Root-finding algorithms for polynomials over large finite fields (Section 3.2) are examples of Las Vegas algorithms. Let $f(x) \in \mathbb{F}_q[x]$ be a polynomial with q odd. For a random $\alpha \in \mathbb{F}_q$, the polynomial $\gcd((x+\alpha)^{(q-1)/2} - 1, f(x))$ is a non-trivial factor of $f(x)$ with probability $\geqslant 1/2$. Therefore, computing this gcd for a few random values of α is expected to produce a non-trivial split of $f(x)$. The algorithm is repeated recursively on the two factors of $f(x)$ thus revealed. If f is of degree d, we need $d-1$ splits to obtain all the roots of f. However, we may be so unlucky that a very huge number of choices for α fails to produce a non-trivial split of a polynomial. Certainly, such a situation is rather unlikely, and that is the reason why Las Vegas algorithms are useful in practice. It is important to note that no deterministic algorithm that runs in time polynomial in $\log q$ is known to solve this root-finding problem.

As another example, let us compute a random prime of a given bit length l. By the prime number theorem, the number of primes $< 2^l$ is about $\frac{2^l}{0.693l}$. Therefore, if we randomly try $O(l)$ l-bit integers, we expect with high probability that at least one of these candidates is a prime. We subject each candidate to a polynomial-time (in l) primality test (deterministic or randomized) until a prime is located. The expected running time of this algorithm is polynomial in l. But chances remain, however small, that even after trying a large number of candidates, we fail to encounter a prime.

Randomized algorithms (also called *probabilistic algorithms*) are quite useful in number-theoretic computations. They often are the most practical among all known algorithms, and sometimes the only known polynomial-time algorithms. However, there are number-theoretic problems for which even randomization does not help much. For example, the best known algorithms for factoring integers and for computing discrete logarithms in finite fields have randomized flavors, and are better than the best deterministic algorithms known for solving these problems, but the improvement in the running time is from exponential to subexponential only.

A.2   Discrete Algebraic Structures

Computational number theory often deploys algebraic tools. It is, therefore, useful to have an understanding of discrete mathematical structures.

A.2.1   Functions and Operations

A *function* or *map* $f : A \to B$ is an association of each element $a \in A$ with an element $b \in B$. We say that a maps to b, or b is the image of a under f, or $f(a) = b$. A function $f : A \to B$ is called B-valued. The set A is called the *domain* of f, and the set B the *range* of f.

For functions $f : A \to B$ and $g : B \to C$, the *composition* $g \circ f$ is the function $A \to C$ defined as $(g \circ f)(a) = g(f(a))$ for all $a \in A$.

A function $f : A \to B$ is called *injective* or *one-one* if for any two different elements $a_1, a_2 \in A$, the images $f(a_1)$ and $f(a_2)$ are different.

A function $f : A \to B$ is called *surjective* or *onto* if for every $b \in B$, there exists (at least) one $a \in A$ with $b = f(a)$.

A function $f : A \to B$ which is both injective and surjective is called *bijective* or a *bijection* or a *one-to-one correspondence* between A and B. For a bijective function $f : A \to B$, the inverse function $f^{-1} : B \to A$ is defined as: $f^{-1}(b) = a$ if and only if $f(a) = b$. Since f is bijective, $f^{-1}(b)$ is a unique element of A for every $b \in B$. In this case, we have $f \circ f^{-1} = \mathrm{id}_B$ and $f^{-1} \circ f = \mathrm{id}_A$, where id_X is the *identity function* on the set X, that is, the function $X \to X$ that maps every element $x \in X$ to itself. Conversely, if $f : A \to B$ and $g : B \to A$ satisfy $f \circ g = \mathrm{id}_B$ and $g \circ f = \mathrm{id}_A$, then both f and g are bijections, and $g = f^{-1}$ and $f = g^{-1}$.

A *unary operation* on a set A is a function $A \to A$. Negation of integers is, for instance, a unary operator on the set $\mathbb{Z}$ of all integers.

A *binary operation* on a set A is a function $A \times A \to A$ (where $\times$ stands for the Cartesian product of two sets). If $* : A \times A \to A$ is a binary operation, the value $*(a_1, a_2)$ is usually specified in the infix notation, that is, as $a_1 * a_2$. Addition is, for instance, a binary operation on $\mathbb{Z}$, and we customarily write $a_1 + a_2$ instead of $+(a_1, a_2)$.

An *n-ary operation* on a set A is a function $\underbrace{A \times A \times \cdots \times A}_{n \text{ times}} \to A$.

A.2.2 Groups

We now study sets with operations.

Definition A.12 A *group* is a set G together with a binary operation $\diamond$ satisfying the following three properties.

(1) $a \diamond (b \diamond c) = (a \diamond b) \diamond c$ for all $a, b, c \in G$ (Associativity).

(2) There exists an element $e \in G$ such that $e \diamond a = a \diamond e = a$ for all $a \in G$. The element e is called the *identity element* of G. The identity element of a group is unique. If $\diamond$ is an addition operation, we often write the identity as 0. If $\diamond$ is a multiplicative operation, we often write the identity as 1.

(3) For every element $a \in G$, there exists an element $b \in G$ such that $a \diamond b = b \diamond a = e$. The element b is called the *inverse* of a, and is unique for every element a. If $\diamond$ is an addition operation, we often write $b = -a$, whereas if $\diamond$ is a multiplication operation, we write $b = a^{-1}$.

If, in addition, G satisfies the following property, we call G a *commutative group* or an *Abelian group*.

(4) $a \diamond b = b \diamond a$ for all $a, b \in G$. ◁

Example A.13 (1) The set $\mathbb{Z}$ of integers is an Abelian group under addition. The identity in this group is 0, and the inverse of a is $-a$. Multiplication is an associative and commutative operation on $\mathbb{Z}$, and 1 is the multiplicative identity, but $\mathbb{Z}$ is not a group under multiplication, since multiplicative inverses exist in $\mathbb{Z}$ only for the elements ± 1.

(2) The set $\mathbb{Q}$ of rational numbers is an Abelian group under addition. The set $\mathbb{Q}^* = \mathbb{Q} \setminus \{0\}$ of non-zero rational numbers is a group under multiplication. Likewise, $\mathbb{R}$ (the set of real numbers) and $\mathbb{C}$ (complex numbers) are additive groups, and their multiplicative groups are $\mathbb{R}^* = \mathbb{R} \setminus \{0\}$ and $\mathbb{C}^* = \mathbb{C} \setminus \{0\}$.

(3) The set $A[x]$ of polynomials over A in one variable x (where A is $\mathbb{Z}$, $\mathbb{Q}$, $\mathbb{R}$ or $\mathbb{C}$) is a group under polynomial addition. Non-zero polynomials do not form a group under polynomial multiplication, since inverses do not exist for all elements of $A[x]$ (like x). These results can be generalized to the set $A[x_1, x_2, \ldots, x_n]$ of multivariate polynomials over A.

(4) The set of all $m \times n$ matrices (with integer, rational, real or complex entries) is a group under matrix addition. The set of all invertible $n \times n$ matrices with rational, real or complex entries is a group under matrix multiplication. $\square$

Existence of inverses in groups leads to the following *cancellation laws*.

Proposition A.14 *Let a, b, c be elements of a group G (with operation $\diamond$). If $a \diamond b = a \diamond c$, then $b = c$. Moreover, if $a \diamond c = b \diamond c$, then $a = b$.* ◁

Definition A.15 Let G be a group under $\diamond$. A subset H of G is called a subgroup of G if H is also a group under the operation $\diamond$ inherited from G. In order that a subset H of G is a subgroup, it suffices that H is closed under the operation $\diamond$ and also under taking inverses, or equivalently if $a \diamond b^{-1} \in H$ for all $a, b \in H$, where b^{-1} is the inverse of b in G. ◁

Example A.16 (1) $\mathbb{Z}$ is a subgroup of $\mathbb{R}$ (under addition).

(2) $\mathbb{Z}$ is a subgroup of $\mathbb{Z}[x]$ (under addition). The set of all polynomials in $\mathbb{Z}[x]$ with even constant terms is another subgroup of $\mathbb{Z}[x]$.

(3) The set of $n \times n$ matrices with determinant 1 is a subgroup of all $n \times n$ invertible matrices (under matrix multiplication). $\square$

For the time being, let us concentrate on multiplicative groups, that is, groups under some multiplication operations. This is done only for notational convenience. The theory is applicable to groups under any operations.

Definition A.17 Let G be a group, and H a subgroup. For $a \in G$, the set $aH = \{ah \mid h \in H\}$ is called a *left coset* of H in G. Likewise, for $a \in G$, the set $Ha = \{ha \mid h \in H\}$ is called a *right coset* of H in G. ◁

Proposition A.18 *Let G be group (multiplicatively written), H a subgroup, and $a, b \in G$. The left cosets aH and bH are in bijection with one another. For every $a, b \in G$, we have either $aH = bH$ or $aH \cap bH = \emptyset$. We have $aH = bH$ if and only if $a^{-1}b \in H$.* ◁

A similar result holds for right cosets too. The last statement should be modified as: $Ha = Hb$ if and only if $ab^{-1} \in H$.

Definition A.19 Let H be a subgroup of G. The count of cosets (left or right, not both) of H in G is called the *index* of H in G, denoted $[G : H]$. ◁

Proposition A.18 tells that the left cosets (also the right cosets) form a partition of G. It therefore follows that:

Corollary A.20 *[Lagrange's theorem] Let G be a finite group, and H a subgroup. Then, the size of G is an integral multiple of the size of H. Indeed, we have $|G| = [G : H]|H|$.* ◁

A particular type of subgroups is of important concern to us.

Proposition A.21 *For a subgroup H of G, the following conditions are equivalent:*
* (a) *$aH = Ha$ for all $a \in G$.*
* (b) *$aHa^{-1} = H$ for all $a \in G$ (where $aHa^{-1} = \{aha^{-1} \mid h \in H\}$).*
* (c) *$aha^{-1} \in H$ for all $a \in G$ and for all $h \in H$.* ◁

Definition A.22 If H satisfies these equivalent conditions, it is called a *normal subgroup* of G. Every subgroup of an Abelian group is normal. ◁

Definition A.23 Let H be a normal subgroup of G, and G/H denote the set of cosets (left or right) of H in G. Define an operation on G/H as

$$(aH)(bH) = abH.$$

It is easy to verify that this is a well-defined binary operation on G/H, and that G/H is again a group under this operation. $H = eH$ (where e is the identity in G) is the identity in G/H, and the inverse of aH is $a^{-1}H$. We call G/H the *quotient* of G with respect to H. ◁

Example A.24 (1) Take the group $\mathbb{Z}$ under addition, $n \in \mathbb{N}$, and $H = n\mathbb{Z} = \{na \mid a \in \mathbb{Z}\}$. Then, H is a subgroup of $\mathbb{Z}$. Since $\mathbb{Z}$ is Abelian, H is normal. We have $[\mathbb{Z} : H] = n$. Indeed, all the cosets of H in $\mathbb{Z}$ are $a + n\mathbb{Z}$ for $a = 0, 1, 2, \ldots, n - 1$. We denote the set of these cosets as $\mathbb{Z}_n = \mathbb{Z}/n\mathbb{Z}$.

(2) Let $G = \mathbb{Z}[x]$ (additive group), and H the set of polynomials in G with even constant terms. H is normal in G. In this case, $[G : H] = 2$. The quotient group G/H contains only two elements: H and $1 + H$. □

In many situations, we deal with quotient groups. In general, the elements of a quotient group are sets. It is convenient to identify some particular element of each coset as the *representative* of that coset. When we define the group operation on the quotient group, we compute the representative of the result from the representatives standing for the operands. For example, $\mathbb{Z}_n$ is often identified as the set $\{0, 1, 2, \ldots, n - 1\}$, and the addition of $\mathbb{Z}_n$ is rephrased in terms of modular addition. Algebraically, an element $a \in \mathbb{Z}_n$ actually stands for the coset $a + n\mathbb{Z} = \{a + kn \mid k \in \mathbb{Z}\}$. The addition of cosets $(a + n\mathbb{Z}) + (b + n\mathbb{Z}) = (a + b) + n\mathbb{Z}$ is consistent with addition modulo n.

Definition A.25 Let G_1, G_2 be (multiplicative) groups. A function $f : G_1 \to G_2$ is called a (group) *homomorphism* if f commutes with the group operations, that is, $f(ab) = f(a)f(b)$ for all $a, b \in G_1$. A bijective homomorphism is called an *isomorphism*. For an isomorphism $f : G_1 \to G_2$, the inverse $f^{-1} : G_2 \to G_1$ is again a homomorphism (indeed, an isomorphism too). ◁

Theorem A.26 [*Isomorphism theorem*] *Let* $f : G_1 \to G_2$ *be a group homomorphism. Define the kernel of* f *as* $\ker f = \{a \in G_1 \mid f(a) = e_2\}$ *(where* e_2 *is the identity of* G_2*). Define the image of* f *as* $\operatorname{Im} f = \{b \in G_2 \mid b = f(a)$ *for some* $a \in G_1\}$*. Then,* $\ker f$ *is a normal subgroup of* G_1*,* $\operatorname{Im} f$ *is a subgroup of* G_2*, and* $G_1/\ker f$ *is isomorphic to* $\operatorname{Im} f$ *under the map* $a \ker f \mapsto f(a)$*.* ◁

Definition A.27 Let S be a subset of a group G (multiplicative). The set of all finite products of elements of S and their inverses is a subgroup of G, called the subgroup *generated* by S. If G is generated by a single element g, we call G a *cyclic group* generated by g, and g a *generator* of G. ◁

Example A.28 (1) $\mathbb{Z}$ (under addition) is a cyclic group generated by 1.

(2) For every $n \in \mathbb{N}$, the additive group $\mathbb{Z}_n$ is cyclic (and generated by 1).

(3) The subset of $\mathbb{Z}_n$, consisting of elements invertible modulo n, is a group under multiplication modulo n. We denote this group by $\mathbb{Z}_n^*$. $\mathbb{Z}_n^*$ is cyclic if and only if $n = 1, 2, 4, p^r, 2p^r$, where p is any odd prime, and $r \in \mathbb{N}$. □

Theorem A.29 *Let* G *be a cyclic group. If* G *is infinite, then* G *is isomorphic to the additive group* $\mathbb{Z}$*. If* G *is finite, it is isomorphic to the additive group* $\mathbb{Z}_n$ *for some* $n \in \mathbb{N}$*.* ◁

Proposition A.30 *Let* G *be a finite cyclic group of size* n*. The number of generators of* G *is* $\phi(n)$*.* ◁

Example A.31 The size of $\mathbb{Z}_n^*$ is $\phi(n)$. If $\mathbb{Z}_n^*$ is cyclic, then the number of generators of $\mathbb{Z}_n^*$ is $\phi(\phi(n))$. In particular, for a prime p, the number of generators of $\mathbb{Z}_p^*$ is $\phi(p-1)$. □

Proposition A.32 *Let* G *be a finite cyclic group (multiplicative) with identity* e*, and* H *a subgroup of* G *of size* m*. Then,* H *is cyclic too, and consists precisely of those elements* a *of* G *which satisfy* $a^m = e$*.* ◁

Definition A.33 Let G be a group (multiplicative), and $a \in G$. If the elements $a, a^2, a^3, \ldots$ are all distinct from one another, we say that a is of *infinite order*. In that case, a generates a subgroup of G, isomorphic to $\mathbb{Z}$. If a is not of infinite order, then the smallest positive integer r for which $a^r = \underbrace{a \times a \times \cdots \times a}_{r \text{ times}} = e$ (the identity of G) is called the *order* of a, denoted $\operatorname{ord} a$. If G is finite, $\operatorname{ord} a$ divides the size of G (by Corollary A.20). ◁

Definition A.34 A group G generated by a finite subset S of G is called *finitely generated*. In particular, cyclic groups are finitely generated. ◁

Example A.35 (1) The Cartesian product $\mathbb{Z} \times \mathbb{Z}$ (under component-wise addition) is generated by the two elements $(0,1)$ and $(1,0)$, but is not cyclic. More generally, the r-fold Cartesian product $\mathbb{Z}^r$ (under addition) is generated by r elements $(1,0,0,\ldots,0)$, $(0,1,0,\ldots,0)$, $\ldots$, $(0,0,\ldots,0,1)$.

(2) The group $\mathbb{Z}_8^*$ (under multiplication) is not cyclic, but generated by -1 and 5. Indeed, for any $t \geqslant 3$, the group $\mathbb{Z}_{2^t}^*$ is generated by -1 and 5. □

Theorem A.36 [*Structure theorem for finitely generated Abelian groups*] *Any finitely generated Abelian group G is isomorphic to the additive group $\mathbb{Z}^r \times \mathbb{Z}_{n_1} \times \mathbb{Z}_{n_2} \times \cdots \times \mathbb{Z}_{n_s}$ for some integers $r, s \geqslant 0$ and $n_1, n_2, \ldots, n_s > 0$ with the properties that $n_i | n_{i-1}$ for all $i = 2, 3, \ldots, s$.* ◁

Definition A.37 The integer r in Theorem A.36 is called the *rank* of the Abelian group G. (For finite Abelian groups, $r = 0$.) All elements of G with finite orders form a subgroup of G, isomorphic to $\mathbb{Z}_{n_1} \times \mathbb{Z}_{n_2} \times \cdots \times \mathbb{Z}_{n_s}$, called the *torsion subgroup* of G. ◁

A.2.3 Rings and Fields

Rings are sets with two binary operations.

Definition A.38 A *ring* is a set R with two binary operations $+$ (addition) and $\cdot$ (multiplication) satisfying the following conditions.

(1) R is an Abelian group under $+$.

(2) For every $a, b, c \in R$, we have $a(bc) = (ab)c$ (Associative).

(3) There exists an element $1 \in R$ such that $1 \cdot a = a \cdot 1 = a$ for all $a \in R$. The element 1 is called the *multiplicative identity* of A. (In some texts, the existence of the multiplicative identity is not part of the definition of a ring. We, however, assume that rings always come with multiplicative identities.)

(4) The operation $\cdot$ distributes over $+$, that is, for all $a, b, c \in R$, we have $a(b + c) = ab + ac$ and $(a + b)c = ac + bc$.

In addition to these requirements, we assume that:

(5) Multiplication in R is commutative, that is, $ab = ba$ for all $a, b \in R$. (This is again not part of the standard definition of a ring.) ◁

Definition A.39 A ring in which $0 = 1$ is the ring $R = \{0\}$. This is called the *zero ring*. ◁

Definition A.40 Let R be a non-zero ring. An element $a \in R$ is called a *zero divisor* if $ab = 0$ for some non-zero $b \in R$. Clearly, 0 is a zero divisor in R. If R contains no non-zero zero divisors, we call R an *integral domain*. ◁

In an integral domain, cancellation holds with respect to multiplication, that is, $ab = ac$ with $a \neq 0$ implies $b = c$.

Definition A.41 Let R be a non-zero ring. An element $a \in R$ is called a *unit* if $ab = 1$ for some $b \in R$. The set of all units in R is a group under the ring multiplication. We denote this group by R^*. If $R^* = R \setminus \{0\}$ (that is, if every non-zero element of R is a unit), we call R a field. ◁

Example A.42 (1) $\mathbb{Z}$ is an integral domain, but not a field. The only units of $\mathbb{Z}$ are ± 1.

(2) $\mathbb{Z}[i] = \{a + ib \mid a, b \in \mathbb{Z}\} \subseteq \mathbb{C}$ is an integral domain called the *ring of Gaussian integers*. The only units of $\mathbb{Z}[i]$ are $\pm 1, \pm i$.

(3) $\mathbb{Q}, \mathbb{R}, \mathbb{C}$ are fields.

(4) $\mathbb{Z}_n$ under addition and multiplication modulo n is a ring. The units of $\mathbb{Z}_n$ constitute the group $\mathbb{Z}_n^*$. $\mathbb{Z}_n$ is a field if and only if n is prime. For a prime p, the field $\mathbb{Z}_p$ is also denoted as $\mathbb{F}_p$.

(5) If R is a ring, the set $R[x]$ of all univariate polynomials with coefficients from R is a ring. If R is an integral domain, so too is $R[x]$. $R[x]$ is never a field. We likewise have the ring $R[x_1, x_2, \ldots, x_n]$ of multivariate polynomials.

(6) Let R be a ring. The set $R[[x]]$ of (infinite) power series over R is a ring. If R is an integral domain, so also is $R[[x]]$.

(7) Let R be a field. The set $R(x) = \{f(x)/g(x) \mid f(x), g(x) \in R[x], g(x) \neq 0\}$ of rational functions over R is again a field.

(8) The set of all $n \times n$ matrices (with elements from a field) is a *non-commutative* ring. It contains non-zero zero divisors (for $n > 1$).

(9) The Cartesian product $R_1 \times R_2 \times \cdots \times R_n$ of rings $R_1, R_2, \ldots, R_n$ is again a ring under element-wise addition and multiplication operations. □

Proposition A.43 *A field is an integral domain. A finite integral domain is a field.* ◁

Definition A.44 Let R be a (non-zero) ring. If the elements $1, 1+1, 1+1+1, \ldots$ are all distinct from one another, we say that the *characteristic* of R is zero. Otherwise, the smallest positive integer r for which $\underbrace{1 + 1 + \cdots + 1}_{r \text{ times}} = 0$

is called the characteristic of R, denoted char R. ◁

Example A.45 (1) $\mathbb{Z}, \mathbb{Q}, \mathbb{R}, \mathbb{C}$ are fields of characteristic zero.

(2) The rings $R, R[x], R[x_1, x_2, \ldots, x_n]$ are of the same characteristic.

(3) The characteristic of $\mathbb{Z}_n$ is n.

(4) If an integral domain has a positive characteristic c, then c is prime. □

Like normal subgroups, a particular type of subsets of a ring takes part in the formation of quotient rings.

Definition A.46 Let R be a ring. A subset $I \subseteq R$ is called an *ideal* of R if I is a subgroup of R under addition, and if $ra \in I$ for all $r \in R$ and $a \in I$.   ◁

Example A.47 (1) All ideals of $\mathbb{Z}$ are $n\mathbb{Z} = \{na \mid a \in \mathbb{Z}\}$ for $n \in \mathbb{N}_0$.

(2) Let S be any subset of a ring R. The set of all finite sums of the form ra with $r \in R$ and $a \in S$ is an ideal of R. We say that this ideal is *generated* by S. This ideal is called *finitely generated* if S is a finite set. The ideal $n\mathbb{Z}$ of $\mathbb{Z}$ is generated by the single integer n.

(3) The only ideals of a field K are $\{0\}$ and K.   □

Definition A.48 An ideal in a ring, generated by a single element, is called a *principal ideal*. An integral domain in which every ideal is principal is called a *principal ideal domain* or a *PID*.   ◁

Example A.49 (1) $\mathbb{Z}$ is a principal ideal domain.

(2) For a field K, the polynomial ring $K[x]$ is a principal ideal domain.

(3) The polynomial ring $K[x, y]$ over a field K is not a PID. Indeed, the ideal of $K[x, y]$, generated by $\{x, y\}$, is not principal.   □

Definition A.50 Let R be an integral domain.

(1) Let $a, b \in R$. We say that a divides b (in R) if $b = ac$ for some $c \in R$. We denote this as $a|b$.

(2) A non-zero non-unit a of R is called a *prime* in R if $a|(bc)$ implies $a|b$ or $a|c$ for all $b, c \in R$.

(3) A non-zero non-unit a of R is called *irreducible* if any factorization $a = bc$ of a in R implies that either b or c is a unit.   ◁

Theorem A.51 *A prime in an integral domain is irreducible. An irreducible element in a PID is prime.*   ◁

Definition A.52 An integral domain in which every non-zero element can be uniquely expressed as a product of primes is called a *unique factorization domain* or a *UFD*. Here, uniqueness is up to rearrangement of the prime factors and up to multiplications of the factors by units.   ◁

Theorem A.53 *Every PID is a UFD. In particular, $\mathbb{Z}$ and $K[x]$ (where K is a field) are UFDs.*   ◁

Example A.54 Not every integral domain supports unique factorization. Consider $\mathbb{Z}[\sqrt{-5}] = \{a + b\sqrt{-5} \mid a, b \in \mathbb{Z}\}$. It is easy to see that $\mathbb{Z}[\sqrt{-5}]$ is a ring. Being a subset of $\mathbb{C}$, it is an integral domain too. We have $6 = 2 \times 3 = (1 + \sqrt{-5})(1 - \sqrt{-5})$. Here, $2, 3, 1 \pm \sqrt{-5}$ are irreducible but not prime.   □

Let us now look at the formation of quotient rings.

Definition A.55 Let I be an ideal in a ring R. Then, I is a (normal) subgroup of the additive Abelian group R, and so R/I is a group under the addition law $(a + I) + (b + I) = (a + b) + I$. We can also define multiplication on R/I as $(a + I)(b + I) = (ab) + I$. These two operations make R/I a ring called the *quotient ring* of R with respect to I. ◁

Example A.56 For the ideal $n\mathbb{Z} = \{an \mid a \in \mathbb{Z}\}$ of $\mathbb{Z}$ generated by $n \in \mathbb{N}$, the quotient ring $\mathbb{Z}/n\mathbb{Z}$ is denoted by $\mathbb{Z}_n$. If the cosets of $n\mathbb{Z}$ are represented by $0, 1, 2, \ldots, n - 1$, this is same as $\mathbb{Z}_n$ defined in terms of modular operations. □

Definition A.57 Let R_1, R_2 be rings. A function $f : R_1 \to R_2$ is called a *(ring) homomorphism* if f commutes with the addition and multiplication operations of the rings, that is, if $f(a + b) = f(a) + f(b)$ and $f(ab) = f(a)f(b)$ for all $a, b \in R_1$. We additionally require f to map the multiplicative identity of R_1 to the multiplicative identity of R_2.

A bijective homomorphism is called an *isomorphism*. For an isomorphism $f : R_1 \to R_2$, the inverse $f^{-1} : R_2 \to R_1$ is again an isomorphism. ◁

Theorem A.58 [*Isomorphism theorem*] *Let $f : R_1 \to R_2$ be a ring homomorphism. The set* $\ker f = \{a \in R_1 \mid f(a) = 0\}$ *is called the kernel of f, and the set* $\mathrm{Im}\, f = \{b \in R_2 \mid b = f(a)$ *for some $a \in R_1\}$ is called the image of f. Then, $\ker f$ is an ideal of R_1, $\mathrm{Im}\, f$ is a ring (under the operations inherited from R_2), and $R_1/\ker f$ is isomorphic to $\mathrm{Im}\, f$ under the map $a + \ker f \mapsto f(a)$.* ◁

Definition A.59 Let I, J be ideals of a ring R. By $I + J$, we denote the set of all elements $a + b$ of R with $a \in I$ and $b \in J$. $I + J$ is again an ideal of R. The ideals I and J are called *coprime* (or *relatively prime*) if $I + J = R$.

The set $I \cap J$ is also an ideal of R, called the *intersection* of I and J.

The set of all finite sums of the form $\sum_{i=1}^{n} a_i b_i$ with $n \geqslant 0$, $a_i \in I$ and $b_i \in J$ is again an ideal of R, called the *product* IJ of the ideals I and J. ◁

Theorem A.60 [*Chinese remainder theorem*] *Let $I_1, I_2, \ldots, I_n$ be ideals in a ring R with $I_i + I_j = R$ for all i, j with $i \neq j$. The ring $R/(I_1 \cap I_2 \cap \cdots \cap I_n)$ is isomorphic to the Cartesian product $(R/I_1) \times (R/I_2) \times \cdots \times (R/I_n)$ under the map $a + (I_1 \cap I_2 \cap \cdots \cap I_n) \mapsto (a + I_1, a + I_2, \ldots, a + I_n)$.* ◁

Definition A.61 If an ideal I of R contains a unit u, then by definition $1 = u^{-1}u \in I$ too, so every $a \in R$ is in I too ($a = a \times 1$), that is, $I = R$. This is why R is called the *unit ideal* of R. Every ideal which is a proper subset of R does not contain any unit of R, and is called a *proper ideal* of R. The singleton subset $\{0\}$ in any ring is an ideal called the *zero ideal*. ◁

Definition A.62 A proper ideal I of a ring R is called *prime* if the following condition holds: Whenever a product ab of elements a, b of R is a member of I, we have either $a \in I$ or $b \in I$ (or both). ◁

Definition A.63 A proper ideal I of a ring R is called *maximal* if for any ideal J of R satisfying $I \subseteq J \subseteq R$, we have either $J = I$ or $J = R$. ◁

Theorem A.64 *An ideal I in a ring R is prime if and only if the quotient ring R/I is an integral domain. An ideal I in a ring R is maximal if and only if the quotient ring R/I is a field. Since every field is an integral domain, every maximal ideal in a ring is prime.* ◁

Example A.65 (1) Let $n \in \mathbb{N}_0$. The set $n\mathbb{Z}$ of all multiples of n is an ideal of $\mathbb{Z}$. This ideal is prime if and only if n is either zero or a prime.

(2) Let K be a field, and $f(x) \in K[x]$. All the polynomial multiples of $f(x)$ form an ideal in $K[x]$, which is prime if and only if $f(x)$ is either zero or an irreducible polynomial in $K[x]$.

(3) In $\mathbb{Z}$ and $K[x]$ (where K is a field), a non-zero (proper) ideal is prime if and only if it is maximal.

(4) A ring R is an integral domain if and only if the zero ideal is prime.

(5) The converse of Theorem A.64 is not true. In the ring $R = \mathbb{Z}[x]$ of univariate polynomials with integer coefficients, the ideals $I = \{xf(x) \mid f(x) \in \mathbb{Z}[x]\}$ and $J = \{xf(x) + 2g(x) \mid f(x), g(x) \in \mathbb{Z}[x]\}$ are both prime. Indeed, I consists of all polynomials with zero constant terms, whereas J consists of all polynomials with even constant terms. I is properly contained in J and cannot be maximal. J is a maximal ideal in $R = \mathbb{Z}[x]$. The quotient ring R/J is isomorphic to the field $\mathbb{Z}_2$. On the other hand, the quotient ring R/I is isomorphic to $\mathbb{Z}$ which is an integral domain but not a field. □

A.2.4 Vector Spaces

Vector spaces play very important roles in algebra and linear algebra.

Definition A.66 Let K be a field. An additive Abelian group V is called a *vector space* over K (or a K-vector space) if there is a scalar multiplication map $\cdot : K \times V \to V$ satisfying the following properties:

(1) $1 \cdot x = x$ for all $x \in V$.

(2) $(a + b)x = ax + bx$ for all $a, b \in K$ and $x \in V$.

(3) $a(x + y) = ax + ay$ for all $a \in K$ and $x, y \in V$.

(4) $a(bx) = (ab)x$ for all $a, b \in K$ and $x \in V$. ◁

Example A.67 (1) For any $n \in \mathbb{N}$, the Cartesian product K^n is a K-vector space under the scalar multiplication $a(x_1, x_2, \ldots, x_n) = (ax_1, ax_2, \ldots, ax_n)$. In particular, K itself is a K-vector space.

(2) More generally, if $V_1, V_2, \ldots, V_n$ are K-vector spaces, then so also is their Cartesian product $V_1 \times V_2 \times \cdots \times V_n$.

(3) Let K, L be fields with $K \subseteq L$. Then, L is a K-vector space with scalar multiplication defined as the multiplication of L. For example, $\mathbb{C}$ is a vector space over $\mathbb{R}$, and $\mathbb{R}$ is a vector space over $\mathbb{Q}$.

(4) The polynomial rings $K[x]$ and $K[x_1, x_2, \ldots, x_n]$ for any $n \in \mathbb{N}$ are K-vector spaces. □

Definition A.68 Let V be a vector space over K.

A subset $S \subseteq V$ is (or the elements of S are) called *linearly independent* over K if any finite sum of the form $a_1 x_1 + a_2 x_2 + \cdots + a_n x_n$ with $n \in \mathbb{N}_0$, $a_i \in K$, and $x_i \in S$ is zero only if $a_1 = a_2 = \cdots = a_n = 0$.

A subset $S \subseteq V$ is (or the elements of S are) said to *generate* V if every element of V can be written in the form $a_1 x_1 + a_2 x_2 + \cdots + a_n x_n$ with $n \in \mathbb{N}_0$, $a_i \in K$, and $x_i \in S$. If S is finite, V is called *finitely generated* over K. ◁

Theorem A.69 *For a subset S of a K-vector space V, the following two conditions are equivalent.*

(a) S is a maximal linearly independent (over K) subset of V (that is, $S \cup \{x\}$ is linearly dependent over K for any $x \in V \setminus S$).

(b) S is a minimal generating subset of V over K (that is, no proper subset of S generates V as a K-vector space). ◁

Definition A.70 A subset S of a K-vector space V, satisfying the two equivalent conditions of Theorem A.69, is called a *K-basis* of V. ◁

Theorem A.71 *Any two K-bases S, T of a K-vector space V are in bijection with one another. In particular, if S, T are finite, they are of the same size.* ◁

Definition A.72 The size of any K-basis of a K-vector space V is called the *dimension* of V over K, denoted by $\dim_K V$. If $\dim_K V$ is finite, V is called a *finite-dimensional* vector space over K. ◁

Example A.73 (1) The dimension of K^n over K is n. More generally, the dimension of $V_1 \times V_2 \times \cdots \times V_n$ is $d_1 + d_2 + \cdots + d_n$, where d_i is the dimension of the K-vector space V_i.

(2) The dimension of $\mathbb{C}$ over $\mathbb{R}$ is 2, since $1, i$ constitute an $\mathbb{R}$-basis of $\mathbb{C}$. $\mathbb{R}$ is not a finite-dimensional vector space over $\mathbb{Q}$.

(3) For a field K, the polynomial ring $K[x]$ has a basis $\{1, x, x^2, \ldots\}$. In particular, $K[x]$ is not finite-dimensional as a K-vector space.

(4) Let $x_1, x_2, \ldots, x_n$ be Boolean (that is, $\{0, 1\}$-valued) variables. The set V of all Boolean functions of $x_1, x_2, \ldots, x_n$ is an $\mathbb{F}_2$-vector space. The size of V is 2^{2^n}, so $\dim_{\mathbb{F}_2} V = 2^n$. The product (logical AND) $y_1 y_2 \cdots y_n$ with each $y_i \in \{x_i, \bar{x}_i\}$ is called a *minterm*. The 2^n minterms constitute an $\mathbb{F}_2$-basis of V, that is, each Boolean function can be written as a unique $\mathbb{F}_2$-linear combination (XOR, same as OR in this context) of the minterms. □

Definition A.74 Let V be a K-vector space. A subset $U \subseteq V$ is called a *vector subspace* of V if U is a subgroup of V, and if U is closed under the scalar multiplication map. If U is a subspace of V, then $\dim_K U \leqslant \dim_K V$. ◁

Definition A.75 Let V_1, V_2 be K-vector spaces. A function $f : V_1 \to V_2$ is called a vector-space *homomorphism* or a *linear transformation* if $f(ax + by) = af(x) + bf(y)$ for all $a, b \in K$ and $x, y \in V$. A bijective homomorphism is called an *isomorphism*. If $f : V_1 \to V_2$ is a vector-space isomorphism, then the inverse $f^{-1} : V_2 \to V_1$ is again a vector-space isomorphism. ◁

Theorem A.76 [*Isomorphism theorem*] *Let $f : V_1 \rightarrow V_2$ be a K-linear transformation. The set $\ker f = \{x \in V_1 \mid f(x) = 0\}$ is called the kernel of f. The set $\operatorname{Im} f = \{y \in V_2 \mid y = f(x) \text{ for some } x \in V_1\}$ is called the image of f. Then, $\ker f$ is a subspace of V_1, $\operatorname{Im} f$ is a subspace of V_2, and $V_1/\ker f$ is isomorphic to $\operatorname{Im} f$ under the map $x + \ker f \mapsto f(x)$.* ◁

Definition A.77 For a K-linear transformation $f : V_1 \rightarrow V_2$, the dimension $\dim_K(\ker f)$ is called the *nullity* of f, and $\dim_K(\operatorname{Im} f)$ the *rank* of f. ◁

Theorem A.78 [*Rank-nullity theorem*] *For a K-linear map $f : V_1 \rightarrow V_2$, the sum of the rank and the nullity of f is equal to $\dim_K V_1$.* ◁

Theorem A.79 *Let V, W be K-vector spaces. The set of all K-linear maps $V \rightarrow W$, denoted $\operatorname{Hom}_K(V, W)$, is a K-vector space with addition defined as $(f+g)(x) = f(x)+g(x)$ for all $x \in V$, and with scalar multiplication defined as $(af)(x) = af(x)$ for all $a \in K$ and $x \in V$. If $m = \dim_K V$ and $n = \dim_K W$ are finite, then the dimension of $\operatorname{Hom}_K(V, W)$ as a K-vector space is mn.* ◁

Definition A.80 For a K-vector space V, the K-vector space $\operatorname{Hom}_K(V, K)$ is called the *dual space* of V. The K-vector spaces V and $\operatorname{Hom}_K(V, K)$ are isomorphic (and have the same dimension over K). ◁

A.2.5   Polynomials

Polynomials play a crucial role in the algebra of fields. Let K be a field. Since the polynomial ring $K[x]$ is a PID, irreducible polynomials are same as prime polynomials in $K[x]$, and are widely used for defining field extensions.

Let $f(x) \in K[x]$ be an irreducible polynomial of degree n. The ideal $I = f(x)K[x] = \{f(x)a(x) \mid a(x) \in K[x]\}$ generated by $f(x) \in K[x]$ plays a role similar to that played by ideals generated by integer primes. The quotient ring $L = K[x]/I$ is a field. We have $K \subseteq L$, so L is a K-vector space. The dimension of L over K is $n = \deg f$. We call n the *degree* of the field extension $K \subseteq L$, denoted $[L : K]$. L contains a root $\alpha = x + I$ of $f(x)$. We say that L is obtained by *adjoining* the root α of f to K. Other roots of $f(x)$ may or may not belong to L. Elements of L can be written as polynomials $a_0 + a_1\alpha + a_2\alpha^2 + \cdots + a_{n-1}\alpha^{n-1}$ with unique $a_i \in K$. This representation of L also indicates that L has the dimension n as a vector space over K. Indeed, $1, \alpha, \alpha^2, \ldots, \alpha^{n-1}$ constitute a K-basis of L. We write $L = K(\alpha)$.

Example A.81 (1) The polynomial $x^2 + 1$ is irreducible in $\mathbb{R}[x]$. Adjoining a root i of this polynomial to $\mathbb{R}$ gives the field $\mathbb{C}$. The other root $-$i of $x^2 + 1$ is also included in $\mathbb{C}$. We have $[\mathbb{C} : \mathbb{R}] = 2$.

(2) The polynomial $x^3 - 2$ is irreducible in $\mathbb{Q}[x]$. Let α be a root of $x^3 - 2$ in $\mathbb{C}$. Adjoining α to $\mathbb{Q}$ gives the field $\mathbb{Q}(\alpha)$ of extension degree three over $\mathbb{Q}$. Every element of $\mathbb{Q}(\alpha)$ is of the form $a_0 + a_1\alpha + a_2\alpha^2$ with $a_0, a_1, a_2 \in \mathbb{Q}$.

There are three possibilities for α, namely, $2^{1/3}$, $2^{1/3}\omega$ and $2^{1/3}\omega^2$, where $2^{1/3}$ is the real cube root of 2, and $\omega = \frac{-1+i\sqrt{3}}{2}$ is a complex cube root of unity. Adjoining any of these roots gives essentially the same field, that is, there exist isomorphisms among the fields $\mathbb{Q}(2^{1/3})$, $\mathbb{Q}(2^{1/3}\omega)$ and $\mathbb{Q}(2^{1/3}\omega^2)$. Indeed, each of these fields is isomorphic to the quotient field $\mathbb{Q}[x]/(x^3 - 2)\mathbb{Q}[x]$.

Adjoining a root of $x^3 - 2$ to $\mathbb{Q}$ does not add the other two roots of $x^3 - 2$. For $\alpha = 2^{1/3}$, the extension $\mathbb{Q}(2^{1/3})$ lies inside $\mathbb{R}$, and cannot contain the two properly complex roots. The polynomial $(x - 2^{1/3}\omega)(x - 2^{1/3}\omega^2) = x^2 + 2^{1/3}x + 2^{2/3}$ is irreducible in $\mathbb{Q}(2^{1/3})[x]$. Adjoining a root of this polynomial to $\mathbb{Q}(2^{1/3})$ gives a field of extension degree two over $\mathbb{Q}(2^{1/3})$ and six over $\mathbb{Q}$.

(3) If K is a finite field, adjoining a root of an irreducible polynomial to K also adds the other roots of the polynomial to the extension. □

Theorem A.82 *Let $f(x) \in K[x]$ be an irreducible polynomial of degree n. There exists a field extension L of K with $[L : K] \leqslant n!$ such that all the roots of f lie in L.* ◁

Theorem A.83 *Let K be a field. Then, there exists an extension L of K such that every irreducible polynomial has all its roots in L.* ◁

Definition A.84 A smallest (with respect to inclusion) field L satisfying Theorem A.83 is called an *algebraic closure* of K. Algebraic closures of K are unique up to field isomorphisms that fix K element-wise. We denote *the* algebraic closure of K as $\bar{K}$. K is called *algebraically closed* if $\bar{K} = K$. ◁

Example A.85 (1) The field $\mathbb{C}$ of complex numbers is algebraically closed (*fundamental theorem of algebra*).

(2) The algebraic closure of $\mathbb{R}$ is $\mathbb{C}$. However, $\mathbb{C}$ is not the algebraic closure of $\mathbb{Q}$, since $\mathbb{C}$ contains transcendental numbers like e and π.

(3) Let $K = \{a_1, a_2, \ldots, a_n\}$ be a finite field of size n. The polynomial $(x - a_1)(x - a_2) \cdots (x - a_n) + 1$ does not have a root in K, that is, no finite field is algebraically closed. In particular, the algebraic closure of a finite field is an infinite field. □

A.3 Linear Algebra

Vector spaces and linear transformations are the basic objects of study in linear algebra. We have already studied some properties of these objects. Here, I concentrate mostly on some computational problems of linear algebra.

A.3.1   Linear Transformations and Matrices

Matrices are compact and handy representations of linear transformations between finite-dimensional vector spaces. Let V and W be K-vector spaces of respective dimensions n and m. Choose any K-basis $\alpha_1, \alpha_2, \ldots, \alpha_n$ of V, and any K-basis $\beta_1, \beta_2, \ldots, \beta_m$ of W. Let $f : V \to W$ be a linear transformation. In view of linearity, f is fully specified if only the elements $f(\alpha_j)$ for $j = 1, 2, \ldots, n$ are provided. For each $j = 1, 2, \ldots, n$, write

$$f(\alpha_j) = c_{1,j}\beta_1 + c_{2,j}\beta_2 + \cdots + c_{m,j}\beta_m$$

with $c_{i,j} \in K$. The $m \times n$ matrix M whose i,j-th entry is $c_{i,j}$ is a compact representation of the linear map f. For any element

$$\alpha = a_1\alpha_1 + a_2\alpha_2 + \cdots + a_n\alpha_n = (\alpha_1 \quad \alpha_2 \quad \cdots \quad \alpha_n) \begin{pmatrix} a_1 \\ a_2 \\ \vdots \\ a_n \end{pmatrix},$$

we have

$$
\begin{aligned}
f(\alpha) &= a_1 f(\alpha_1) + a_2 f(\alpha_2) + \cdots + a_n f(\alpha_n) \\[2mm]
&= (f(\alpha_1) \quad f(\alpha_2) \quad \cdots \quad f(\alpha_n)) \begin{pmatrix} a_1 \\ a_2 \\ \vdots \\ a_n \end{pmatrix} \\[2mm]
&= (\beta_1 \quad \beta_2 \quad \cdots \quad \beta_m) \begin{pmatrix} c_{1,1} & c_{1,2} & \cdots & c_{1,n} \\ c_{2,1} & c_{2,2} & \cdots & c_{2,n} \\ \vdots & \vdots & \cdots & \vdots \\ c_{m,1} & c_{m,2} & \cdots & c_{m,n} \end{pmatrix} \begin{pmatrix} a_1 \\ a_2 \\ \vdots \\ a_n \end{pmatrix} \\[2mm]
&= (\beta_1 \quad \beta_2 \quad \cdots \quad \beta_m) M \begin{pmatrix} a_1 \\ a_2 \\ \vdots \\ a_n \end{pmatrix},
\end{aligned}
$$

where $M = (c_{i,j})$ is the $m \times n$ transformation matrix for f. The argument α of f is specified by the scalars $a_1, a_2, \ldots, a_n$, whereas the image $f(\alpha) = b_1\beta_1 + b_2\beta_2 + \cdots + b_m\beta_m$ is specified by the scalars $b_1, b_2, \ldots, b_m$. We have

$$\begin{pmatrix} b_1 \\ b_2 \\ \vdots \\ b_m \end{pmatrix} = M \begin{pmatrix} a_1 \\ a_2 \\ \vdots \\ a_n \end{pmatrix},$$

that is, application of f is equivalent to premultiplication by the matrix M.

Now, take two linear maps $f, g \in \mathrm{Hom}_K(V, W)$ (see Theorem A.79) with transformation matrices M and N respectively. The linear map $f + g$ has the transformation matrix $M + N$, and the scalar product af has the transformation matrix aM. This means that the above description of linear maps in terms of matrices respects the vector-space operations of $\mathrm{Hom}_K(V, W)$.

In order to see how matrix multiplication is related to linear maps, take three K-vector spaces U, V, W of respective dimensions n, m, l. Choose bases $\alpha_1, \alpha_2, \ldots, \alpha_n$ of U, $\beta_1, \beta_2, \ldots, \beta_m$ of V, and $\gamma_1, \gamma_2, \ldots, \gamma_l$ of W. Finally, let $f : U \rightarrow V$ and $g : V \rightarrow W$ be linear maps given by

$$
\begin{aligned}
f(\alpha_k) &= c_{1,k}\beta_1 + c_{2,k}\beta_2 + \cdots + c_{m,k}\beta_m, \\
g(\beta_j) &= d_{1,j}\gamma_1 + d_{2,j}\gamma_2 + \cdots + d_{l,j}\gamma_l,
\end{aligned}
$$

for $k = 1, 2, \ldots, n$ and $j = 1, 2, \ldots, m$. The transformation matrices for f and g are respectively $M = (c_{j,k})_{m \times n}$ and $N = (d_{i,j})_{l \times m}$. But then, we have

$$
\begin{aligned}
(g \circ f)(\alpha_k) = g(f(\alpha_k)) &= g(c_{1,k}\beta_1 + c_{2,k}\beta_2 + \cdots + c_{m,k}\beta_m) \\
&= c_{1,k}g(\beta_1) + c_{2,k}g(\beta_2) + \cdots + c_{m,k}g(\beta_m) \\
&= c_{1,k}(d_{1,1}\gamma_1 + d_{2,1}\gamma_2 + \cdots + d_{l,1}\gamma_l) + \\
&\quad c_{2,k}(d_{1,2}\gamma_1 + d_{2,2}\gamma_2 + \cdots + d_{l,2}\gamma_l) + \\
&\quad \cdots + \\
&\quad c_{m,k}(d_{1,m}\gamma_1 + d_{2,m}\gamma_2 + \cdots + d_{l,m}\gamma_l) \\
&= (d_{1,1}c_{1,k} + d_{1,2}c_{2,k} + \cdots + d_{1,m}c_{m,k})\gamma_1 + \\
&\quad (d_{2,1}c_{1,k} + d_{2,2}c_{2,k} + \cdots + d_{2,m}c_{m,k})\gamma_2 + \\
&\quad \cdots + \\
&\quad (d_{l,1}c_{1,k} + d_{l,2}c_{2,k} + \cdots + d_{l,m}c_{m,k})\gamma_l \\
&= e_{1,k}\gamma_1 + e_{2,k}\gamma_2 + \cdots + e_{l,k}\gamma_l,
\end{aligned}
$$

where $e_{i,k}$ is the (i, k)-th entry of the matrix product NM (an $l \times n$ matrix). This indicates that the transformation matrix for $g \circ f : U \rightarrow W$ is the matrix product NM, that is, matrix products correspond to compositions of linear maps. More correctly, matrix multiplication was defined in such a weird way in order that it conforms with compositions of linear maps.

A.3.2   Gaussian Elimination

Gaussian elimination is a fundamentally useful computational tool needed for solving a variety of problems in linear algebra. Suppose that we have m linear equations in n variables $x_1, x_2, \ldots, x_n$:

$$
\begin{aligned}
a_{1,1}x_1 + a_{1,2}x_2 + \cdots + a_{1,n}x_n &= b_1, \\
a_{2,1}x_1 + a_{2,2}x_2 + \cdots + a_{2,n}x_n &= b_2, \\
&\cdots \\
a_{m,1}x_1 + a_{m,2}x_2 + \cdots + a_{m,n}x_n &= b_m.
\end{aligned}
\tag{A.1}
$$

In terms of matrices, we can rewrite this system as

$$
A\mathbf{x} = \mathbf{b}, \tag{A.2}
$$

where A is the $m \times n$ coefficient matrix $(a_{i,j})$, $\mathbf{x} = (x_1 \; x_2 \; \cdots \; x_n)^t$ is the vector of variables, and $\mathbf{b} = (b_1 \; b_2 \; \cdots \; b_m)^t$. Here, the superscript ^{t} stands for matrix transpose. If $\mathbf{b} = \mathbf{0}$, the system is called *homogeneous*.

In order to solve a system like (A.1), we convert the system (or the matrix A) to a *row echelon form* (REF). A matrix M is said to be in the row echelon form if the following three conditions are satisfied.

1. All zero rows of M are below the non-zero rows of M.

2. The first non-zero entry (called *pivot*) in a non-zero row is 1.[2]

3. The pivot of a non-zero row stays in a column to the left of the pivots of all the following non-zero rows.

For example,

$$\begin{pmatrix} 1 & 2 & 5 & 3 & 7 \\ 0 & 0 & 1 & 0 & 5 \\ 0 & 0 & 0 & 1 & 3 \\ 0 & 0 & 0 & 0 & 0 \end{pmatrix}$$

is a matrix in the row echelon form. A matrix M in REF is said to be in the *reduced row echelon form* (RREF) if the only non-zero entry in each column containing a pivot of some row is that pivot itself. For example, the RREF of the matrix in the above example is

$$\begin{pmatrix} 1 & 2 & 0 & 0 & -27 \\ 0 & 0 & 1 & 0 & 5 \\ 0 & 0 & 0 & 1 & 3 \\ 0 & 0 & 0 & 0 & 0 \end{pmatrix}$$

We can convert any matrix A to REF or RREF using Gaussian elimination. The REF of A is not unique, but the RREF of A is unique. Let me first explain how a system (or matrix) can be converted to an REF. Gaussian elimination involves a sequence of the following *elementary row operations*:

1. Exchange two rows.

2. Multiply a row by a non-zero element.

3. Subtract a non-zero multiple of a row from another row.

To start with, we process the first row of A. We find the leftmost column containing at least one non-zero element. If no such column exists, the REF conversion procedure is over. So let the l-th column be the leftmost column to contain a non-zero element. We bring such a non-zero element to the $(1, l)$-th position in A by a row exchange (unless that position already contained a non-zero element). By multiplying the first row with the inverse of the element at position $(1, l)$, we convert the pivot to 1. Subsequently, from each of the following rows containing a non-zero element at the l-th column, we subtract a suitable multiple of the first row in order to reduce the element at the

---

[2]Some authors do not impose this condition. For them, any non-zero entry is allowed as a pivot. This is not a serious issue anyway.

l-th column to zero. This completes the processing of the first row. We then recursively reduce the submatrix of A, obtained by removing the first row and the first l columns. When processing a system of equations (instead of its coefficient matrix), we apply the same elementary transformations on the right sides of the equations.

Example A.86 Consider the following system over $\mathbb{F}_7$:

$$
\begin{aligned}
5x_2 + 3x_3 + x_4 &= 6, \\
6x_1 + 5x_2 + 2x_3 + 5x_4 + 6x_5 &= 2, \\
2x_1 + 5x_2 + 5x_3 + 2x_4 + 6x_5 &= 6, \\
2x_1 + 2x_2 + 6x_3 + 2x_4 + 3x_5 &= 0.
\end{aligned}
$$

The matrix of coefficients (including the right sides) is:

$$
\left(\begin{array}{ccccc|c}
0 & 5 & 3 & 1 & 0 & 6 \\
6 & 5 & 2 & 5 & 6 & 2 \\
2 & 5 & 5 & 2 & 6 & 6 \\
2 & 2 & 6 & 2 & 3 & 0
\end{array} \right)
$$

We convert this system to an REF using the following steps.

$$
\left(\begin{array}{ccccc|c}
6 & 5 & 2 & 5 & 6 & 2 \\
0 & 5 & 3 & 1 & 0 & 6 \\
2 & 5 & 5 & 2 & 6 & 6 \\
2 & 2 & 6 & 2 & 3 & 0
\end{array} \right)
$$
[Exchanging Row 1 with Row 2]

$$
\left(\begin{array}{ccccc|c}
1 & 2 & 5 & 2 & 1 & 5 \\
0 & 5 & 3 & 1 & 0 & 6 \\
2 & 5 & 5 & 2 & 6 & 6 \\
2 & 2 & 6 & 2 & 3 & 0
\end{array} \right)
$$
[Multiplying Row 1 by $6^{-1} \equiv 6 \pmod 7$]

$$
\left(\begin{array}{ccccc|c}
1 & 2 & 5 & 2 & 1 & 5 \\
0 & 5 & 3 & 1 & 0 & 6 \\
0 & 1 & 2 & 5 & 4 & 3 \\
2 & 2 & 6 & 2 & 3 & 0
\end{array} \right)
$$
[Subtracting 2 times Row 1 from Row 3]

$$
\left(\begin{array}{ccccc|c}
1 & 2 & 5 & 2 & 1 & 5 \\
0 & 5 & 3 & 1 & 0 & 6 \\
0 & 1 & 2 & 5 & 4 & 3 \\
0 & 5 & 3 & 5 & 1 & 4
\end{array} \right)
$$
[Subtracting 2 times Row 1 from Row 4]
(Processing of Row 1 over)

$$
\left(\begin{array}{ccccc|c}
1 & 2 & 5 & 2 & 1 & 5 \\
0 & 1 & 2 & 3 & 0 & 4 \\
0 & 1 & 2 & 5 & 4 & 3 \\
0 & 5 & 3 & 5 & 1 & 4
\end{array} \right)
$$
[Multiplying Row 2 by $5^{-1} \equiv 3 \pmod 7$]

$$
\left(\begin{array}{ccccc|c}
1 & 2 & 5 & 2 & 1 & 5 \\
0 & 1 & 2 & 3 & 0 & 4 \\
0 & 0 & 0 & 2 & 4 & 6 \\
0 & 5 & 3 & 5 & 1 & 4
\end{array} \right)
$$
[Subtracting 1 times Row 2 from Row 3]

$$\begin{pmatrix} 1 & 2 & 5 & 2 & 1 & 5 \\ 0 & 1 & 2 & 3 & 0 & 4 \\ 0 & 0 & 0 & 2 & 4 & 6 \\ 0 & 0 & 0 & 4 & 1 & 5 \end{pmatrix}$$ [Subtracting 5 times Row 2 from Row 4]
(Processing of Row 2 over)

$$\begin{pmatrix} 1 & 2 & 5 & 2 & 1 & 5 \\ 0 & 1 & 2 & 3 & 0 & 4 \\ 0 & 0 & 0 & 1 & 2 & 3 \\ 0 & 0 & 0 & 4 & 1 & 5 \end{pmatrix}$$ (Column 3 contains no non-zero element,
so we proceed to Column 4)

[Multiplying Row 3 by $2^{-1} \equiv 4 \pmod 7$)]

$$\begin{pmatrix} 1 & 2 & 5 & 2 & 1 & 5 \\ 0 & 1 & 2 & 3 & 0 & 4 \\ 0 & 0 & 0 & 1 & 2 & 3 \\ 0 & 0 & 0 & 0 & 0 & 0 \end{pmatrix}$$ [Subtracting 4 times Row 3 from Row 4]
(Processing of Row 3 over)

(Row 4 is zero and needs no processing)

This last matrix is an REF of the original matrix.   □

We can convert a matrix in REF to a matrix in RREF. In the REF-conversion procedure above, we have subtracted suitable multiples of the current row from the rows below the current row. If the same procedure is applied to rows above the current row, the RREF is obtained. Notice that this additional task may be done after the REF conversion (as demonstrated in Example A.87 below), or during the REF-conversion procedure itself. When the original matrix is sparse, the second alternative is preferable, since an REF of the original sparse matrix may be quite dense, so zeroing all non-pivot column elements while handling a pivot element preserves sparsity.

Example A.87 The following steps convert the last matrix of Example A.86 to the RREF. This matrix is already in REF.

$$\begin{pmatrix} 1 & 0 & 1 & 3 & 1 & 4 \\ 0 & 1 & 2 & 3 & 0 & 4 \\ 0 & 0 & 0 & 1 & 2 & 3 \\ 0 & 0 & 0 & 0 & 0 & 0 \end{pmatrix}$$ [Subtracting 2 times Row 2 from Row 1]
(Handling Column 2)

$$\begin{pmatrix} 1 & 0 & 1 & 0 & 2 & 2 \\ 0 & 1 & 2 & 3 & 0 & 4 \\ 0 & 0 & 0 & 1 & 2 & 3 \\ 0 & 0 & 0 & 0 & 0 & 0 \end{pmatrix}$$ [Subtracting 3 times Row 3 from Row 1]
(Handling Column 4)

$$\begin{pmatrix} 1 & 0 & 1 & 0 & 2 & 2 \\ 0 & 1 & 2 & 0 & 1 & 2 \\ 0 & 0 & 0 & 1 & 2 & 3 \\ 0 & 0 & 0 & 0 & 0 & 0 \end{pmatrix}$$ [Subtracting 3 times Row 3 from Row 2]
(Handling Column 4)

In the last matrix, Columns 1, 2 and 4 contain pivot elements, so all non-pivot entries in these columns are reduced to zero. Columns 3 and 5 do not contain pivot elements, and are allowed to contain non-zero entries.   □

Let us now come back to our original problem of solving a linear system $A\mathbf{x} = \mathbf{b}$. Using the above procedure, we convert $(A \mid \mathbf{b})$ to an REF (or RREF)

B. If there is a zero row in the A part of B, containing a non-zero element in the **b** part, the given system is inconsistent, and is not solvable. So we assume that each row of B is either entirely zero or contains a non-zero element (a pivot) in the A part. The columns that do not contain a pivot element (in the A part) correspond to *free* or *independent* variables, whereas the columns that contain pivot elements correspond to *dependent* variables. Let $x_{i_1}, x_{i_2}, \ldots, x_{i_k}$ be the free variables, and $x_{j_1}, x_{j_2}, \ldots, x_{j_l}$ the dependent variables. Suppose that $i_1 < i_2 < \cdots < i_k$ and $j_1 < j_2 < \cdots < j_l$. From the non-zero rows of the REF, we express each dependent variable as a linear combination of the other variables. If we plug in values for the free variables $x_{i_1}, x_{i_2}, \ldots, x_{i_k}$, we can solve for $x_{j_l}, x_{j_{l-1}}, \ldots, x_{j_1}$ in that sequence. This process of obtaining values for the dependent variables is called *back* (or *backward*) *substitution*. For each tuple of values for $x_{i_1}, x_{i_2}, \ldots, x_{i_k}$, we obtain a different solution. If there are no free variables (that is, if $k = 0$), the solution is unique.

If the system is reduced to the RREF, then writing $x_{i_t} = x_{i_t}$ for $t = 1, 2, \ldots, k$ allows us to express the solutions as

$$\mathbf{x} = \mathbf{u} + x_{i_1}\mathbf{v}_1 + x_{i_2}\mathbf{v}_2 + \cdots + x_{i_k}\mathbf{v}_k \tag{A.3}$$

for some constant $n \times 1$ vectors $\mathbf{u}, \mathbf{v}_1, \mathbf{v}_2, \ldots, \mathbf{v}_k$. For a system reduced to an REF (but not the RREF), we cannot write the solutions in this form.

Example A.88 (1) From the REF of the system in Example A.86, we see that x_3 and x_5 are independent variables, whereas x_1, x_2 and x_4 are dependent variables. The REF gives

$$
\begin{aligned}
x_4 &= 3 - 2x_5 &&= 3 + 5x_5, \\
x_2 &= 4 - (2x_3 + 3x_4) &&= 4 + 5x_3 + 4x_4, \\
x_1 &= 5 - (2x_2 + 5x_3 + 2x_4 + x_5) &&= 5 + 5x_2 + 2x_3 + 5x_4 + 6x_5.
\end{aligned}
$$

For each pair of values for x_3, x_5, we solve for x_4, x_2, x_1 by backward substitution. Since the system is over $\mathbb{F}_7$ and has two free variables, there are exactly $7^2 = 49$ solutions.

(2) For the RREF of Example A.87, we express the solutions as

$$
\begin{aligned}
x_1 &= 2 - (x_3 + 2x_5) &&= 2 + 6x_3 + 5x_5, \\
x_2 &= 2 - (2x_3 + x_5) &&= 2 + 5x_3 + 6x_5, \\
x_3 &= x_3, \\
x_4 &= 3 - 2x_5 &&= 3 + 5x_5, \\
x_5 &= x_5.
\end{aligned}
$$

These solutions can be written also as

$$
\mathbf{x} = \begin{pmatrix} x_1 \\ x_2 \\ x_3 \\ x_4 \\ x_5 \end{pmatrix} = \begin{pmatrix} 2 \\ 2 \\ 0 \\ 3 \\ 0 \end{pmatrix} + x_3 \begin{pmatrix} 6 \\ 5 \\ 1 \\ 0 \\ 0 \end{pmatrix} + x_5 \begin{pmatrix} 5 \\ 6 \\ 0 \\ 5 \\ 1 \end{pmatrix}.
$$

The two vectors $(6 \quad 5 \quad 1 \quad 0 \quad 0)^t$ and $(5 \quad 6 \quad 0 \quad 5 \quad 1)^t$ are clearly linearly independent (look at the positions of the free variables). □

The Gaussian-elimination procedure (that is, conversion to REF or RREF or solving a system in the form (A.3)) makes $O(\max(m, n)^3)$ elementary operations in the field K.

A.3.3 Inverse and Determinant

Gaussian elimination is used for a variety of purposes other than solving linear systems. Now, we discuss some such applications of Gaussian elimination. Let A be a square matrix (of size $n \times n$), that is, a linear transformation f from an n-dimensional vector space V to an n-dimensional vector space W (we may have $V = W$). If f is an isomorphism (equivalently, if f is invertible as a function), the inverse transformation $f^{-1} : W \to V$ is represented by a matrix B satisfying $AB = BA = I_n$, where I_n is the $n \times n$ identity matrix. The matrix B is called the inverse of A, and is denoted as A^{-1}.

If A (that is, f) is invertible, any (consistent) system of the form $A\mathbf{x} = \mathbf{b}$ has a unique solution, that is, there are no free variables, that is, an REF of A contains pivot elements (1, to be precise) at every position of the main diagonal, that is, the RREF of A is the identity matrix. In order to compute A^{-1}, we convert A to I_n by elementary row operations. If we apply the same sequence of row operations on I_n, we get A^{-1}. If, during the RREF-conversion procedure, we ever encounter a situation where the i-th row does not contain a pivot at the i-th column, A is not invertible, that is, A^{-1} does not exist.

Example A.89 Let us compute the inverse of $A = \begin{pmatrix} 0 & 5 & 3 & 6 & 4 \\ 4 & 3 & 1 & 4 & 5 \\ 3 & 6 & 2 & 4 & 3 \\ 4 & 1 & 0 & 3 & 6 \\ 4 & 6 & 6 & 3 & 6 \end{pmatrix}$

defined over the field $\mathbb{F}_7$. To this end, we start with $(A \mid I_5)$, reduce A (the left part) to the identity matrix I_5. We also carry out the same sequence of row operations on the right part (initially I_5), and let it get converted to A^{-1}. The steps are given below. Unlike in Examples A.86 and A.87, we straightaway reduce A to RREF (instead of first to REF and then to RREF).

(The initial matrix $(A \mid I_5)$) $\quad \left(\begin{array}{ccccc|ccccc} 0 & 5 & 3 & 6 & 4 & 1 & 0 & 0 & 0 & 0 \\ 4 & 3 & 1 & 4 & 5 & 0 & 1 & 0 & 0 & 0 \\ 3 & 6 & 2 & 4 & 3 & 0 & 0 & 1 & 0 & 0 \\ 4 & 1 & 0 & 3 & 6 & 0 & 0 & 0 & 1 & 0 \\ 4 & 6 & 6 & 3 & 6 & 0 & 0 & 0 & 0 & 1 \end{array} \right)$

[Exchange Row 1 with Row 2] $\quad \left(\begin{array}{ccccc|ccccc} 4 & 3 & 1 & 4 & 5 & 0 & 1 & 0 & 0 & 0 \\ 0 & 5 & 3 & 6 & 4 & 1 & 0 & 0 & 0 & 0 \\ 3 & 6 & 2 & 4 & 3 & 0 & 0 & 1 & 0 & 0 \\ 4 & 1 & 0 & 3 & 6 & 0 & 0 & 0 & 1 & 0 \\ 4 & 6 & 6 & 3 & 6 & 0 & 0 & 0 & 0 & 1 \end{array} \right)$

[Multiply Row 1 by $4^{-1} \equiv 2 \pmod 7$]

$$\left(\begin{array}{ccccc|ccccc} 1 & 6 & 2 & 1 & 3 & 0 & 2 & 0 & 0 & 0 \\ 0 & 5 & 3 & 6 & 4 & 1 & 0 & 0 & 0 & 0 \\ 3 & 6 & 2 & 4 & 3 & 0 & 0 & 1 & 0 & 0 \\ 4 & 1 & 0 & 3 & 6 & 0 & 0 & 0 & 1 & 0 \\ 4 & 6 & 6 & 3 & 6 & 0 & 0 & 0 & 0 & 1 \end{array}\right)$$

[Subtract $3 \times$ Row 1 from Row 3]

$$\left(\begin{array}{ccccc|ccccc} 1 & 6 & 2 & 1 & 3 & 0 & 2 & 0 & 0 & 0 \\ 0 & 5 & 3 & 6 & 4 & 1 & 0 & 0 & 0 & 0 \\ 0 & 2 & 3 & 1 & 1 & 0 & 1 & 1 & 0 & 0 \\ 4 & 1 & 0 & 3 & 6 & 0 & 0 & 0 & 1 & 0 \\ 4 & 6 & 6 & 3 & 6 & 0 & 0 & 0 & 0 & 1 \end{array}\right)$$

[Subtract $4 \times$ Row 1 from Row 4]

$$\left(\begin{array}{ccccc|ccccc} 1 & 6 & 2 & 1 & 3 & 0 & 2 & 0 & 0 & 0 \\ 0 & 5 & 3 & 6 & 4 & 1 & 0 & 0 & 0 & 0 \\ 0 & 2 & 3 & 1 & 1 & 0 & 1 & 1 & 0 & 0 \\ 0 & 5 & 6 & 6 & 1 & 0 & 6 & 0 & 1 & 0 \\ 4 & 6 & 6 & 3 & 6 & 0 & 0 & 0 & 0 & 1 \end{array}\right)$$

[Subtract $4 \times$ Row 1 from Row 5]

$$\left(\begin{array}{ccccc|ccccc} 1 & 6 & 2 & 1 & 3 & 0 & 2 & 0 & 0 & 0 \\ 0 & 5 & 3 & 6 & 4 & 1 & 0 & 0 & 0 & 0 \\ 0 & 2 & 3 & 1 & 1 & 0 & 1 & 1 & 0 & 0 \\ 0 & 5 & 6 & 6 & 1 & 0 & 6 & 0 & 1 & 0 \\ 0 & 3 & 5 & 6 & 1 & 0 & 6 & 0 & 0 & 1 \end{array}\right)$$

[Multiply Row 2 by $5^{-1} \equiv 3 \pmod 7$]

$$\left(\begin{array}{ccccc|ccccc} 1 & 6 & 2 & 1 & 3 & 0 & 2 & 0 & 0 & 0 \\ 0 & 1 & 2 & 4 & 5 & 3 & 0 & 0 & 0 & 0 \\ 0 & 2 & 3 & 1 & 1 & 0 & 1 & 1 & 0 & 0 \\ 0 & 5 & 6 & 6 & 1 & 0 & 6 & 0 & 1 & 0 \\ 0 & 3 & 5 & 6 & 1 & 0 & 6 & 0 & 0 & 1 \end{array}\right)$$

[Subtract $6 \times$ Row 2 from Row 1]

$$\left(\begin{array}{ccccc|ccccc} 1 & 0 & 4 & 5 & 1 & 3 & 2 & 0 & 0 & 0 \\ 0 & 1 & 2 & 4 & 5 & 3 & 0 & 0 & 0 & 0 \\ 0 & 2 & 3 & 1 & 1 & 0 & 1 & 1 & 0 & 0 \\ 0 & 5 & 6 & 6 & 1 & 0 & 6 & 0 & 1 & 0 \\ 0 & 3 & 5 & 6 & 1 & 0 & 6 & 0 & 0 & 1 \end{array}\right)$$

[Subtract $2 \times$ Row 2 from Row 3]

$$\left(\begin{array}{ccccc|ccccc} 1 & 0 & 4 & 5 & 1 & 3 & 2 & 0 & 0 & 0 \\ 0 & 1 & 2 & 4 & 5 & 3 & 0 & 0 & 0 & 0 \\ 0 & 0 & 6 & 0 & 5 & 1 & 1 & 1 & 0 & 0 \\ 0 & 5 & 6 & 6 & 1 & 0 & 6 & 0 & 1 & 0 \\ 0 & 3 & 5 & 6 & 1 & 0 & 6 & 0 & 0 & 1 \end{array}\right)$$

[Subtract $5 \times$ Row 2 from Row 4]

$$\left(\begin{array}{ccccc|ccccc} 1 & 0 & 4 & 5 & 1 & 3 & 2 & 0 & 0 & 0 \\ 0 & 1 & 2 & 4 & 5 & 3 & 0 & 0 & 0 & 0 \\ 0 & 0 & 6 & 0 & 5 & 1 & 1 & 1 & 0 & 0 \\ 0 & 0 & 3 & 0 & 4 & 6 & 6 & 0 & 1 & 0 \\ 0 & 3 & 5 & 6 & 1 & 0 & 6 & 0 & 0 & 1 \end{array}\right)$$

[Subtract 3 × Row 2 from Row 5]
$$\begin{pmatrix} 1 & 0 & 4 & 5 & 1 & 3 & 2 & 0 & 0 & 0 \\ 0 & 1 & 2 & 4 & 5 & 3 & 0 & 0 & 0 & 0 \\ 0 & 0 & 6 & 0 & 5 & 1 & 1 & 1 & 0 & 0 \\ 0 & 0 & 3 & 0 & 4 & 6 & 6 & 0 & 1 & 0 \\ 0 & 0 & 6 & 1 & 0 & 5 & 6 & 0 & 0 & 1 \end{pmatrix}$$

[Multiply Row 3 by $6^{-1} \equiv 6 \pmod 7$]
$$\begin{pmatrix} 1 & 0 & 4 & 5 & 1 & 3 & 2 & 0 & 0 & 0 \\ 0 & 1 & 2 & 4 & 5 & 3 & 0 & 0 & 0 & 0 \\ 0 & 0 & 1 & 0 & 2 & 6 & 6 & 6 & 0 & 0 \\ 0 & 0 & 3 & 0 & 4 & 6 & 6 & 0 & 1 & 0 \\ 0 & 0 & 6 & 1 & 0 & 5 & 6 & 0 & 0 & 1 \end{pmatrix}$$

[Subtract 4 × Row 3 from Row 1]
$$\begin{pmatrix} 1 & 0 & 0 & 5 & 0 & 0 & 6 & 4 & 0 & 0 \\ 0 & 1 & 2 & 4 & 5 & 3 & 0 & 0 & 0 & 0 \\ 0 & 0 & 1 & 0 & 2 & 6 & 6 & 6 & 0 & 0 \\ 0 & 0 & 3 & 0 & 4 & 6 & 6 & 0 & 1 & 0 \\ 0 & 0 & 6 & 1 & 0 & 5 & 6 & 0 & 0 & 1 \end{pmatrix}$$

[Subtract 2 × Row 3 from Row 2]
$$\begin{pmatrix} 1 & 0 & 0 & 5 & 0 & 0 & 6 & 4 & 0 & 0 \\ 0 & 1 & 0 & 4 & 1 & 5 & 2 & 2 & 0 & 0 \\ 0 & 0 & 1 & 0 & 2 & 6 & 6 & 6 & 0 & 0 \\ 0 & 0 & 3 & 0 & 4 & 6 & 6 & 0 & 1 & 0 \\ 0 & 0 & 6 & 1 & 0 & 5 & 6 & 0 & 0 & 1 \end{pmatrix}$$

[Subtract 3 × Row 3 from Row 4]
$$\begin{pmatrix} 1 & 0 & 0 & 5 & 0 & 0 & 6 & 4 & 0 & 0 \\ 0 & 1 & 0 & 4 & 1 & 5 & 2 & 2 & 0 & 0 \\ 0 & 0 & 1 & 0 & 2 & 6 & 6 & 6 & 0 & 0 \\ 0 & 0 & 0 & 0 & 5 & 2 & 2 & 3 & 1 & 0 \\ 0 & 0 & 6 & 1 & 0 & 5 & 6 & 0 & 0 & 1 \end{pmatrix}$$

[Subtract 6 × Row 3 from Row 5]
$$\begin{pmatrix} 1 & 0 & 0 & 5 & 0 & 0 & 6 & 4 & 0 & 0 \\ 0 & 1 & 0 & 4 & 1 & 5 & 2 & 2 & 0 & 0 \\ 0 & 0 & 1 & 0 & 2 & 6 & 6 & 6 & 0 & 0 \\ 0 & 0 & 0 & 0 & 5 & 2 & 2 & 3 & 1 & 0 \\ 0 & 0 & 0 & 1 & 2 & 4 & 5 & 6 & 0 & 1 \end{pmatrix}$$

[Exchange Row 4 with Row 5]
$$\begin{pmatrix} 1 & 0 & 0 & 5 & 0 & 0 & 6 & 4 & 0 & 0 \\ 0 & 1 & 0 & 4 & 1 & 5 & 2 & 2 & 0 & 0 \\ 0 & 0 & 1 & 0 & 2 & 6 & 6 & 6 & 0 & 0 \\ 0 & 0 & 0 & 1 & 2 & 4 & 5 & 6 & 0 & 1 \\ 0 & 0 & 0 & 0 & 5 & 2 & 2 & 3 & 1 & 0 \end{pmatrix}$$

[Subtract 5 × Row 4 from Row 1]
$$\begin{pmatrix} 1 & 0 & 0 & 0 & 4 & 1 & 2 & 2 & 0 & 2 \\ 0 & 1 & 0 & 4 & 1 & 5 & 2 & 2 & 0 & 0 \\ 0 & 0 & 1 & 0 & 2 & 6 & 6 & 6 & 0 & 0 \\ 0 & 0 & 0 & 1 & 2 & 4 & 5 & 6 & 0 & 1 \\ 0 & 0 & 0 & 0 & 5 & 2 & 2 & 3 & 1 & 0 \end{pmatrix}$$

$$
\text{[Subtract } 4 \times \text{Row 4 from Row 2]} \quad
\left(\begin{array}{ccccc|ccccc}
1 & 0 & 0 & 0 & 4 & 1 & 2 & 2 & 0 & 2 \\
0 & 1 & 0 & 0 & 0 & 3 & 3 & 6 & 0 & 3 \\
0 & 0 & 1 & 0 & 2 & 6 & 6 & 6 & 0 & 0 \\
0 & 0 & 0 & 1 & 2 & 4 & 5 & 6 & 0 & 1 \\
0 & 0 & 0 & 0 & 5 & 2 & 2 & 3 & 1 & 0
\end{array}\right)
$$

$$
\text{[Multiply Row 5 by } 5^{-1} \equiv 3 \text{ (mod 7)]} \quad
\left(\begin{array}{ccccc|ccccc}
1 & 0 & 0 & 0 & 4 & 1 & 2 & 2 & 0 & 2 \\
0 & 1 & 0 & 0 & 0 & 3 & 3 & 6 & 0 & 3 \\
0 & 0 & 1 & 0 & 2 & 6 & 6 & 6 & 0 & 0 \\
0 & 0 & 0 & 1 & 2 & 4 & 5 & 6 & 0 & 1 \\
0 & 0 & 0 & 0 & 1 & 6 & 6 & 2 & 3 & 0
\end{array}\right)
$$

$$
\text{[Subtract } 4 \times \text{Row 5 from Row 1]} \quad
\left(\begin{array}{ccccc|ccccc}
1 & 0 & 0 & 0 & 0 & 5 & 6 & 1 & 2 & 2 \\
0 & 1 & 0 & 0 & 0 & 3 & 3 & 6 & 0 & 3 \\
0 & 0 & 1 & 0 & 2 & 6 & 6 & 6 & 0 & 0 \\
0 & 0 & 0 & 1 & 2 & 4 & 5 & 6 & 0 & 1 \\
0 & 0 & 0 & 0 & 1 & 6 & 6 & 2 & 3 & 0
\end{array}\right)
$$

$$
\text{[Subtract } 2 \times \text{Row 5 from Row 3]} \quad
\left(\begin{array}{ccccc|ccccc}
1 & 0 & 0 & 0 & 0 & 5 & 6 & 1 & 2 & 2 \\
0 & 1 & 0 & 0 & 0 & 3 & 3 & 6 & 0 & 3 \\
0 & 0 & 1 & 0 & 0 & 1 & 1 & 2 & 1 & 0 \\
0 & 0 & 0 & 1 & 2 & 4 & 5 & 6 & 0 & 1 \\
0 & 0 & 0 & 0 & 1 & 6 & 6 & 2 & 3 & 0
\end{array}\right)
$$

$$
\text{[Subtract } 2 \times \text{Row 5 from Row 4]} \quad
\left(\begin{array}{ccccc|ccccc}
1 & 0 & 0 & 0 & 0 & 5 & 6 & 1 & 2 & 2 \\
0 & 1 & 0 & 0 & 0 & 3 & 3 & 6 & 0 & 3 \\
0 & 0 & 1 & 0 & 0 & 1 & 1 & 2 & 1 & 0 \\
0 & 0 & 0 & 1 & 0 & 6 & 0 & 2 & 1 & 1 \\
0 & 0 & 0 & 0 & 1 & 6 & 6 & 2 & 3 & 0
\end{array}\right)
$$

The last matrix is of the form $(I_5 \mid A^{-1})$, that is, $A^{-1} =$

$$
\left(\begin{array}{ccccc}
5 & 6 & 1 & 2 & 2 \\
3 & 3 & 6 & 0 & 3 \\
1 & 1 & 2 & 1 & 0 \\
6 & 0 & 2 & 1 & 1 \\
6 & 6 & 2 & 3 & 0
\end{array}\right).
$$

One can verify that $AA^{-1} = A^{-1}A = I_5$ (modulo 7). $\qquad\square$

Definition A.90 The *determinant* of an $n \times n$ matrix $A = (a_{i,j})$ is defined as

$$
\det A = \sum_{\pi \in S_n} \left[\text{sign}(\pi) \prod_{i=1}^{n} a_{i,\pi(i)} \right],
$$

where the sum is over all permutations π of $1, 2, \ldots, n$, and the sign of π is $+1$ for even permutations π and -1 for odd permutations π.[3] $\qquad\triangleleft$

---

[3]The set of all permutations of $1, 2, \ldots, n$ is denoted by S_n. Every permutation in S_n can be obtained from $1, 2, \ldots, n$ by a sequence of transpositions (that is, swapping pairs of elements). Given a permutation $\pi \in S_n$, there are many (in fact, infinitely many) transposition sequences to obtain π from $1, 2, \ldots, n$, but the parity of the count of transpositions

The determinant of A is an important mathematical property of A. For example, A is invertible if and only if $\det A \neq 0$.

The determinant of a square matrix A can be computed by Gaussian elimination. We reduce the matrix A to an REF (not necessarily RREF). If this procedure reveals that some column does not contain a pivot element, A is not invertible, that is, $\det A = 0$. Otherwise, the REF-conversion procedure places a non-zero s_i at $A[i][i]$ for every $i = 1, 2, \ldots, n$. We force $A[i][i] = 1$ by dividing the entire i-th row by s_i (or multiplying by s_i^{-1}). Finally, let t be the number of row exchanges done during the REF-conversion procedure. Then,

$$\det A = (-1)^t \prod_{i=1}^{n} s_i.$$

Example A.91 Let us compute $\det A$ for the matrix A of Example A.89. If we are interested in computing only the determinant of A, it is not necessary to convert I_5 to A^{-1}, that is, the REF conversion may be restricted only to the first five columns. We have $s_1 = 4$, $s_2 = 5$, $s_3 = 6$, $s_4 = 1$, $s_5 = 5$, and $t = 2$, that is, $\det A \equiv (-1)^2 \times 4 \times 5 \times 6 \times 1 \times 5 \equiv 5 \pmod{7}$. □

The inverse or the determinant of an $n \times n$ matrix can be computed using $O(n^3)$ elementary operations in the underlying field K.

A.3.4 Rank and Nullspace

Definition A.92 Let A be an $m \times n$ matrix. Now, we may have $m \neq n$. The rows of A can be treated as n-tuples over K, that is, members of the K-vector space K^n. A maximum number of linearly independent (over K) rows of A is called the *row rank* of A. Likewise, we define the *column rank* of A as the maximum number of linearly independent columns of A. ◁

Theorem A.93 *For every matrix A, its row rank is the same as its column rank. We refer to this common value as the rank of A or rank(A). If A is the matrix of a K-linear map $f : V \to W$ (with $\dim_K V = n$ and $\dim_K W = m$), the rank of A is the same as the rank of f.* ◁

Definition A.94 The *nullspace* of A is the set of all solutions of the homogeneous system $A\mathbf{x} = \mathbf{0}$. These solutions, treated as n-tuples over K, form a subspace of K^n. The dimension of the nullspace of A is called the *nullity* of A, denoted as nullity(A). ◁

---

in each such sequence is the same for a given π. We call π an even or an odd permutation according as whether this parity is even or odd, respectively. It turns out that (for $n \geqslant 2$) exactly half of the $n!$ permutations in S_n are even, and the rest odd. S_n is a group under composition. The set A_n of even permutations in S_n is a subgroup of S_n (of index 2).

For example, for $n = 5$, consider the following sequence of transpositions: $1, 2, 3, 4, 5 \to 1, 5, 3, 4, 2 \to 1, 4, 3, 5, 2 \to 3, 4, 1, 5, 2 \to 2, 4, 1, 5, 3$. It follows that $2, 4, 1, 5, 3$ is an even permutation, whereas $3, 4, 1, 5, 2$ is an odd permutation of $1, 2, 3, 4, 5$.

Theorem A.95 *If A is the matrix of a linear map $f : V \to W$ (with $\dim_K V = n$ and $\dim_K W = m$), the nullity of A is the same as the nullity of f.* ◁

The rank-nullity theorem for linear maps implies the following:

Corollary A.96 *For an $m \times n$ matrix A, we have $\operatorname{rank}(A) + \operatorname{nullity}(A) = n$.* ◁

Both the rank and a basis of the nullspace of a matrix A can be computed by Gaussian elimination. We reduce A to an REF. Let k denote the number of free variables, and l the number of dependent variables. Then, k is the nullity of A, and l is the rank of A. If A is converted to the RREF, Eqn (A.3) allows us to write the solutions of $A\mathbf{x} = \mathbf{0}$ as follows (since we are dealing with a homogeneous system, we have $\mathbf{u} = \mathbf{0}$ in Eqn (A.3)):

$$\mathbf{x} = x_{i_1}\mathbf{v}_1 + x_{i_2}\mathbf{v}_2 + \cdots + x_{i_k}\mathbf{v}_k,$$

where $x_{i_1}, x_{i_2}, \ldots, x_{i_k}$ are the free variables, and $\mathbf{v}_1, \mathbf{v}_2, \ldots, \mathbf{v}_k$ are linearly independent vectors constituting a basis of the nullspace of A.

Example A.97 Let us compute the rank of $A = \begin{pmatrix} 0 & 5 & 3 & 1 & 0 \\ 6 & 5 & 2 & 5 & 6 \\ 2 & 5 & 5 & 2 & 6 \\ 2 & 2 & 6 & 2 & 3 \end{pmatrix}$ of Example A.86 over $\mathbb{F}_7$. We have seen that x_3, x_5 are free variables, and x_1, x_2, x_4 are dependent variables. Therefore, $\operatorname{rank}(A) = 3$ and $\operatorname{nullity}(A) = 2$.

The RREF of A (Example A.87 and Example A.88(2)) indicates that all solutions of $A\mathbf{x} = \mathbf{0}$ can be written as

$$\mathbf{x} = x_3 \begin{pmatrix} 6 \\ 5 \\ 1 \\ 0 \\ 0 \end{pmatrix} + x_5 \begin{pmatrix} 5 \\ 6 \\ 0 \\ 5 \\ 1 \end{pmatrix}.$$

So $(6\ 5\ 1\ 0\ 0)^{\mathrm{t}}$ and $(5\ 6\ 0\ 5\ 1)^{\mathrm{t}}$ form a basis of the nullspace of A. □

Example A.98 As an application, let us find linear dependencies among the Boolean functions $f_0, f_1, \ldots, f_{n-1}$ in s Boolean variables $x_0, x_1, \ldots, x_{s-1}$. Let $m = 2^s$, and $T_0, T_1, \ldots, T_{m-1}$ the minterms in the variables $x_0, x_1, \ldots, x_{s-1}$. We want to determine all n-tuples $(x_0, x_1, \ldots, x_{n-1}) \in \mathbb{F}_2^n$ such that

$$x_0 f_0 + x_1 f_1 + \cdots + x_{n-1} f_{n-1} = 0, \tag{A.4}$$

where 0 indicates the zero function. Each f_j can be written uniquely as

$$f_j = a_{0,j} T_0 + a_{1,j} T_1 + \cdots + a_{m-1,j} T_{m-1}, \tag{A.5}$$

where $a_{i,j}$ is the value of f_j for the values of $x_0, x_1, \ldots, x_{s-1}$ given by the minterm T_i. Substituting Eqn (A.5) in Eqn (A.4) for all $j = 0, 1, \ldots, n-1$ gives $(a_{0,0}x_0 + a_{0,1}x_1 + \cdots + a_{0,n-1}x_{n-1})T_0 + (a_{1,0}x_0 + a_{1,1}x_1 + \cdots +$

$a_{1,n-1}x_{n-1})T_1 + \cdots + (a_{m-1,0}x_0 + a_{m-1,1}x_1 + \cdots + a_{m-1,n-1}x_{n-1})T_{m-1} = 0.$
Since $T_0, T_1, \ldots, T_{m-1}$ are linearly independent over $\mathbb{F}_2$, we have:

$$
\begin{aligned}
a_{0,0}x_0 + a_{0,1}x_1 + \cdots + a_{0,n-1}x_{n-1} &= 0, \\
a_{1,0}x_0 + a_{1,1}x_1 + \cdots + a_{1,n-1}x_{n-1} &= 0, \\
&\cdots \\
a_{m-1,0}x_0 + a_{m-1,1}x_1 + \cdots + a_{m-1,n-1}x_{n-1} &= 0.
\end{aligned}
$$

Let $A = (a_{i,j})_{\substack{0 \leqslant i \leqslant m-1 \\ 0 \leqslant j \leqslant n-1}}$. We need to solve the homogeneous system $A\mathbf{x} = \mathbf{0}$ over $\mathbb{F}_2$. The solutions of Eqn (A.4) are precisely the vectors in the nullspace of A.

As a numeric example, consider the following ten functions $f_0, f_1, \ldots, f_9$ in three Boolean variable x_0, x_1, x_2:

minterm	x_0	x_1	x_2	f_0	f_1	f_2	f_3	f_4	f_5	f_6	f_7	f_8	f_9
$T_0 = \bar{x}_0\bar{x}_1\bar{x}_2$	0	0	0	1	1	1	1	1	0	1	0	0	0
$T_1 = \bar{x}_0\bar{x}_1 x_2$	0	0	1	0	1	0	1	1	0	1	1	0	1
$T_2 = \bar{x}_0 x_1\bar{x}_2$	0	1	0	0	1	0	0	0	1	0	1	0	0
$T_3 = \bar{x}_0 x_1 x_2$	0	1	1	1	0	1	0	1	0	0	0	0	0
$T_4 = x_0\bar{x}_1\bar{x}_2$	1	0	0	1	0	1	1	1	1	0	0	0	0
$T_5 = x_0\bar{x}_1 x_2$	1	0	1	1	0	0	0	1	0	1	1	0	0
$T_6 = x_0 x_1\bar{x}_2$	1	1	0	1	1	0	0	1	1	0	1	0	0
$T_7 = x_0 x_1 x_2$	1	1	1	1	1	1	1	0	0	0	0	1	0

We denote by A the 8×10 matrix formed by the values of the functions (the last ten columns in the table above):

$$
A = \begin{pmatrix}
1 & 1 & 1 & 1 & 1 & 0 & 1 & 0 & 0 & 0 \\
0 & 1 & 0 & 1 & 1 & 0 & 1 & 1 & 0 & 1 \\
0 & 1 & 0 & 0 & 0 & 1 & 0 & 1 & 0 & 0 \\
1 & 0 & 1 & 0 & 1 & 0 & 0 & 0 & 0 & 0 \\
1 & 0 & 1 & 1 & 1 & 1 & 0 & 0 & 0 & 0 \\
1 & 0 & 0 & 0 & 1 & 0 & 1 & 1 & 0 & 0 \\
1 & 1 & 0 & 0 & 1 & 1 & 0 & 1 & 0 & 0 \\
1 & 1 & 1 & 1 & 0 & 0 & 0 & 0 & 1 & 0
\end{pmatrix}.
$$

The RREF of A is (the steps for calculating this are not shown here):

$$
\begin{pmatrix}
1 & 0 & 0 & 0 & 0 & 0 & 0 & 1 & 0 & 1 \\
0 & 1 & 0 & 0 & 0 & 1 & 0 & 1 & 0 & 0 \\
0 & 0 & 1 & 0 & 0 & 0 & 0 & 0 & 0 & 0 \\
0 & 0 & 0 & 1 & 0 & 1 & 0 & 0 & 0 & 0 \\
0 & 0 & 0 & 0 & 1 & 0 & 0 & 1 & 0 & 1 \\
0 & 0 & 0 & 0 & 0 & 0 & 1 & 1 & 0 & 0 \\
0 & 0 & 0 & 0 & 0 & 0 & 0 & 0 & 1 & 1 \\
0 & 0 & 0 & 0 & 0 & 0 & 0 & 0 & 0 & 0
\end{pmatrix}.
$$

The free variables are x_5, x_7, x_9, and all the solutions of $A\mathbf{x} = \mathbf{0}$ are

$$\mathbf{x} = \begin{pmatrix} x_0 \\ x_1 \\ x_2 \\ x_3 \\ x_4 \\ x_5 \\ x_6 \\ x_7 \\ x_8 \\ x_9 \end{pmatrix} = x_5 \begin{pmatrix} 0 \\ 1 \\ 0 \\ 1 \\ 0 \\ 1 \\ 0 \\ 0 \\ 0 \\ 0 \end{pmatrix} + x_7 \begin{pmatrix} 1 \\ 1 \\ 0 \\ 0 \\ 1 \\ 0 \\ 1 \\ 1 \\ 0 \\ 0 \end{pmatrix} + x_9 \begin{pmatrix} 1 \\ 0 \\ 0 \\ 0 \\ 1 \\ 0 \\ 0 \\ 0 \\ 1 \\ 1 \end{pmatrix} \quad \text{with } x_5, x_7, x_9 \in \{0, 1\}.$$

Therefore, three linearly independent linear equations involving $f_0, f_1, \ldots, f_9$ are $f_1 + f_3 + f_5 = 0$, $f_0 + f_1 + f_4 + f_6 + f_7 = 0$, and $f_0 + f_4 + f_8 + f_9 = 0$. All dependencies among these functions can be obtained by the $\mathbb{F}_2$-linear combinations of these independent equations. There are $2^3 - 1 = 7$ non-zero linear equations in the given ten functions. □

A.3.5 Characteristic and Minimal Polynomials

Let A be an $n \times n$ (square) matrix with elements from a field K. Choose any n-dimensional vector $\mathbf{v}$. The $n + 1$ vectors $\mathbf{v}, A\mathbf{v}, A^2\mathbf{v}, \ldots, A^n\mathbf{v}$ must be linearly dependent, that is, there exist $c_0, c_1, \ldots, c_n$, not all zero, such that

$$c_0\mathbf{v} + c_1 A\mathbf{v} + c_2 A^2\mathbf{v} + \cdots + c_n A^n\mathbf{v} = \mathbf{0},$$

that is, $p(A)\mathbf{v} = \mathbf{0}$, where $p(x) = c_0 + c_1 x + c_2 x^2 + \cdots + c_n x^n \in K[x]$ is a non-zero polynomial. If $\mathbf{v}_1, \mathbf{v}_2, \ldots, \mathbf{v}_n$ constitute a basis of K^n, and if $p_i(A)\mathbf{v}_i = \mathbf{0}$ for non-zero polynomials $p_i(x) \in K[x]$, $i = 1, 2, \ldots, n$, then $q(A)\mathbf{v} = \mathbf{0}$ for all $\mathbf{v} \in K^n$, where $q(x) = \operatorname{lcm}(p_1(x), p_2(x), \ldots, p_n(x))$. But then, $q(A) = 0_{n \times n}$, that is, the matrix A satisfies non-zero polynomial equations over K.

Definition A.99 The monic polynomial of smallest positive degree, satisfied by A is called the *minimal polynomial* $\mu_A(x) \in K[x]$ of A. ◁

It turns out that $p(A) = 0$ for some $p(x) \in K[x]$ if and only if $\mu_A(x)|p(x)$. One particular polynomial is always satisfied by A:

Definition A.100 The monic polynomial $\det(xI_n - A)$ of degree n is called the *characteristic polynomial* $\chi_A(x)$ of A. ◁

Theorem A.101 [*Cayley–Hamilton theorem*] $\chi_A(A) = 0$. ◁

Computation of $\chi_A(x)$ is easy, since this amounts to computing the determinant of the matrix $xI_n - A$. Suppose that $\chi_A(x)$ factors over a suitable extension of K as

$$\chi_A(x) = (x - \epsilon_1)^{r_1}(x - \epsilon_2)^{r_2} \cdots (x - \epsilon_k)^{r_k}$$

with $\epsilon_i \in \bar{K}$ and $r_i \in \mathbb{N}$. The minimal polynomial of A must be of the form

$$\mu_A(x) = (x - \epsilon_1)^{s_1}(x - \epsilon_2)^{s_2} \cdots (x - \epsilon_k)^{s_k}$$

for some s_i satisfying $0 \leqslant s_i \leqslant r_i$ for all i. It turns out that the value $s_i = 0$ is not possible, that is, the roots of $\mu_A(x)$ are precisely the roots of $\chi_A(x)$.

Example A.102 For the matrix $A = \begin{pmatrix} 5 & 6 & 2 \\ 2 & 1 & 3 \\ 1 & 1 & 4 \end{pmatrix}$ defined over $\mathbb{F}_7$, we have

$$xI_3 - A = \begin{pmatrix} x+2 & 1 & 5 \\ 5 & x+6 & 4 \\ 6 & 6 & x+3 \end{pmatrix}. \text{ Therefore,}$$

$$\chi_A(x) = \det(xI_3 - A) = x^3 + 4x^2 + 5x + 2 = (x+1)^2(x+2).$$

Thus, the minimal polynomial of A is either $\chi_A(x)$ itself or $(x+1)(x+2)$. Since $(A + I_3)(A + 2I_3) = 0$, we have $\mu_A(x) = (x+1)(x+2) = x^2 + 3x + 2$. $\square$

The roots $\epsilon_1, \epsilon_2, \ldots, \epsilon_k$ of $\chi_A(x)$ or $\mu_A(x)$ are called the *eigenvalues* of A. These are important quantities associated with A. However, keeping in mind the current content of this book, I will not discuss eigenvalues further.

A.4   Probability

In a *random experiment*, there are several possibilities for the output. For example, if a coin is tossed, the outcome may be either a head (H) or a tail (T). If a die is thrown, the outcome is an integer in the range 1–6. The life of an electric bulb is a positive real number. To every possible outcome of a random experiment, we associate a quantity called the *probability* of the outcome, that quantifies the likelihood of the occurrence of that event.

In general, there is no good way to determine the probabilities of events in a random experiment. In order to find the probability of obtaining H in the toss of a coin, we may toss the coin n times, and count the number h of occurrences of H. Define $p_n = h/n$. The probability of H is taken as the limit $p = \lim_{n \to \infty} p_n$ (provided that the limit exists). The probability of T is then $q = 1 - p$. Likewise, we determine the probability of each integer $1, 2, \ldots, 6$ by throwing a die an infinite number of times. Computing the probability of the life of an electric bulb is more complicated. First, there are infinitely many (even uncountable) possibilities. Second, we need to measure the life of many bulbs in order to identify a pattern. As the number of bulbs tends to infinity, the pattern tends to the probability distribution of the life of a bulb.

Evidently, these methods for determining the probabilities of events in a random experiment are impractical (usually infeasible). To get around this difficulty, we make certain simplifying assumptions. For example, we assume that a coin is unbiased. This means that both H and T are equally likely to occur, that is, each has a probability of $1/2$. If we toss an unbiased coin 100 times, we cannot certainly say that we obtain exactly 50 heads and exactly 50 tails, but this would be the most likely event. The probability of each integer $1, 2, \ldots, 6$ in the throw of an unbiased die is likewise taken as $1/6$, since each of the six possibilities are equally likely to occur. The life of an electric bulb is modeled as a non-negative real number with exponentially decreasing likelihood of survival as the lifetime increases.

These abstract models of probability turn out to be practically useful in a variety of contexts. In what follows, I assume that the probabilities of events in a random experiment are provided to us. I focus on how we can play with these probabilities in order to perform things useful to us. I also expect the readers to have some acquaintance with the basic notion of probability.

A.4.1 Random Variables and Probability Distributions

A *random variable* x assumes values from the set X of all outcomes of a random experiment. The set X is called the *sample space* for x. To the members of X, we associate probabilities (idealized, computed or approximated). If X is a countable set (not necessarily finite), we call the random variable x *discrete*. If X is uncountable, we call x a *continuous* random variable.

Let x be a discrete random variable with sample space $X = \{x_1, x_2, x_3, \ldots\}$ (the set X may be infinite). The *probability distribution* of X is the association of a non-negative real number $p_i = \Pr(x = x_i)$ to each i. We must have $\sum_i p_i = 1$, indicating that each p_i lies in the real interval $[0, 1]$.[4]

Defining the probability distribution of a continuous random variable x involves more effort. For simplicity, assume that the sample space X is a subset of $\mathbb{R}$. We do not assign probabilities to individual elements of X. We instead supply a non-negative real-valued function f defined on X. The probability associated with the real interval $[a, b]$ (the interval may be open at one or both ends) is given by the integral $\int_{x=a}^b f(x)\,dx$. We may set $f(x) = 0$ for $x \in \mathbb{R} \setminus X$ so that f is defined on the entire real line. The probability distribution function f must satisfy $\int_{x=-\infty}^\infty f(x)\,dx = \int_{x \in X} f(x)\,dx = 1$.

Example A.103 (1) The random variable C that stands for the outcome of the toss of a coin assumes two values H and T. If the coin is assumed to be unbiased, we have $\Pr(C = H) = \Pr(C = T) = 1/2$.

(2) The random variable D that stands for the outcome of the throw of a die has the sample space $\{1, 2, 3, 4, 5, 6\}$. If the die is unbiased, the probability distribution for D is $\Pr(D = 1) = \Pr(D = 2) = \Pr(D = 3) = \Pr(D = 4) = \Pr(D = 5) = \Pr(D = 6) = 1/6$.

---

[4]Let a, b be real numbers with $a < b$. The *open interval* (a, b) is the set $\{x \in \mathbb{R} \mid a < x < b\}$, and the *closed interval* $[a, b]$ is $\{x \in \mathbb{R} \mid a \leqslant x \leqslant b\}$. We also talk about intervals $[a, b)$ and $(a, b]$, closed at one end and open at the other.

(3) Consider the random experiment of choosing an element from $\mathbb{Z}_n$. The random variable U standing for this event has sample space $\mathbb{Z}_n$. If all elements are equally likely to occur in the choice, then $\Pr(x = i) = 1/n$ for each $i \in \mathbb{Z}_n$.

(4) [*Exponential distribution*] The random variable V representing the life of a bulb is continuous. Its probability distribution function is often modeled as $f(x) = \lambda e^{-\lambda x}$ for $x \geqslant 0$, where λ is a positive real constant. The probability that the life of a bulb is between a and b (with $a < b$) is $\int_{x=a}^{b} \lambda e^{-\lambda x}\, dx = e^{-\lambda a} - e^{-\lambda b}$. For the interval $[0, \infty)$, this integral evaluates to 1. $\qquad\square$

We can think about more complicated events, as exemplified below.

Example A.104 (1) [*Binomial distribution*] Let the random variable A stand for the number of heads in n throws of an unbiased coin. The sample space for A is $\{0, 1, 2, \ldots, n\}$. For an integer k in this set, the probability is $p_k = \Pr(A = k) = \binom{n}{k}/2^n$. This is because n tosses have 2^n possible outcomes, and k heads can occur in $\binom{n}{k}$ ways. Since the coin is unbiased, all of the 2^n possible outcomes are equally likely.

In general, if the probability of head is p and that for tail is q (so $p+q = 1$), the probability that A takes the value k is $\binom{n}{k}p^k q^{n-k}$.

(2) [*Geometric distribution*] Let B denote the number of tosses of a coin until we obtain a head. The sample space for B is $\mathbb{N}$, which is infinite but countable. If p and q are the probabilities of head and tail in one toss of the coin, then $\Pr(B = k) = q^{k-1}p$. For an unbiased coin, this probability is $1/2^k$.

(3) [*Hypergeometric distribution*] An urn contains n_1 red and n_2 blue balls, and m balls are randomly taken out of the urn together. The probability that exactly k red balls are taken out is $\binom{n_1}{k}\binom{n_2}{m-k}/\binom{n_1+n_2}{m}$.

(4) [*Uniform distribution*] Let $X = \{x_1, x_2, \ldots, x_n\}$ be a finite set. An element is drawn from X. If all the elements have equal likelihood of being drawn, the probability that x_i is drawn is $1/n$.

For a real interval $[a, b]$ of finite length $l = b - a$, we define the continuous uniform distribution as $f(x) = 1/(b - a)$. The probability associated with a subinterval $[c, d]$ of $[a, b]$ is $(d - c)/(b - a)$.

We can carry the concept of uniform distribution to higher dimensions too. For example, let us define the uniform distribution on a unit square by the function $f(x) = 1$. For a sub-region A of this square, the probability is given by the double integral $\int \int_A dx\, dy$ which evaluates to the area of A.

(5) [*Poisson distribution*] Suppose that we expect λ occurrences of some event in one unit of time. The probability that exactly k of this type of events happen in one unit of time is given by $\frac{\lambda^k e^{-\lambda}}{k!}$ for all integers $k \geqslant 0$.

(6) [*Normal (or Gaussian) distribution*] Many natural quantities (like weight or height of people in a group, velocity of a particle) flock around a mean value μ with some deviation characterized by σ. In these cases, the probability distribution is modeled as $f_{\mu,\sigma}(x) = \frac{1}{\sigma\sqrt{2\pi}}e^{-\frac{(x-\mu)^2}{2\sigma^2}}$ for $-\infty < x < \infty$.

The probability that x lies within $\pm\sigma$ of μ is $\int_{x=\mu-\sigma}^{\mu+\sigma} f_{\mu,\sigma}(x)\,dx \approx 0.6827$. The probability for $x \in [\mu - 2\sigma, \mu + 2\sigma]$ is about 0.9545, for $x \in [\mu - 3\sigma, \mu + 3\sigma]$ is about 0.9973, and for $x \in [\mu - 6\sigma, \mu + 6\sigma]$ is about $1 - 1.973 \times 10^{-9}$.

The normal distribution has theoretical importance too. In 1810, Laplace proved the *central limit theorem*, which states that the mean of a real-valued sample of size n tends to follow a normal distribution as $n \to \infty$. More importantly, this result holds irrespective of the original probability distribution followed by the objects of the sample. □

If U and V are two discrete random variables with sample spaces X and Y (both subsets of $\mathbb{R}$), the random variable $U + V$ has sample space $X + Y = \{x + y \mid x \in X, y \in Y\}$. For $z \in X + Y$, the probability that the random variable $U + V$ assumes the value z is $\sum_{\substack{x,y \\ x+y=z}} \Pr(U = x, V = y)$, where the probability $\Pr(U = x, V = y)$ is the *joint probability* that $U = x$ and $V = y$. The random variables U and V are called *independent* if $\Pr(U = x, V = y) = \Pr(U = x) \times \Pr(V = y)$ for all $x \in X$ and $y \in Y$.

The product UV is again a random variable with sample space $XY = \{xy \mid x \in X, y \in Y\}$. For $z \in XY$, the probability that the random variable UV assumes the value z is $\sum_{\substack{x,y \\ xy=z}} \Pr(U = x, V = y)$.

The sum and product of continuous random variables can also be defined (although I do not do it here).

The *expectation* of a real-valued discrete random variable U with sample space X is defined as

$$E(U) = \sum_{x \in X} x \Pr(U = x).$$

If U is a real-valued continuous random variable with probability distribution function $f(x)$, the *expectation* of U is defined as

$$E(U) = \int_{-\infty}^{\infty} x f(x)\,dx.$$

For two random variables U and V, we always have $E(U + V) = E(U) + E(V)$. If U and V are *independent*, we have $E(UV) = E(U)E(V)$.

A.4.2 Birthday Paradox

A very important result from probability theory, used on multiple occasions in this book, is the birthday paradox. Let S be a finite set of size n. We keep on choosing elements from S uniformly randomly. That is, in each draw, each of the n elements of S is equally likely (has probability $1/n$) to be chosen. The question is: how many elements must be chosen from S in order to obtain a *collision* with high probability? More precisely, let x_i be the element chosen from S in the i-th draw. If k elements are drawn (with replacement) from S, what is the chance (as a function of n and k) of having $x_i = x_j$ for at least one pair (i, j) satisfying $1 \leqslant i < j \leqslant k$?

In the worst case, $n+1$ elements should be chosen to ensure that there is at least one collision. However, we expect collisions to occur much earlier, that is, for much smaller values of k. More precisely, for $k \approx 1.18\sqrt{n}$, the probability of collision is more than $1/2$, whereas for $k \approx 3.06\sqrt{n}$, the probability is more than 0.99. In short, $\Theta(\sqrt{n})$ choices suffice to obtain collisions with very high probability. Assuming that each of the 365 days in the year is equally likely to be the date of birth of a human, only 23 randomly chosen people have a chance of at least half to contain a pair with the same birthday. For a group of 58 randomly chosen people, this probability is at least 0.99.

It is not difficult to derive the probability of collision. The probability that there is no collision in k draws is $n(n-1)(n-2)\cdots(n-k+1)/n^k$, that is, the probability of at least one collision is

$$
\begin{aligned}
p_{\text{collision}}(n,k) &= 1 - \frac{n(n-1)(n-2)\cdots(n-k+1)}{n^k} \\
&= 1 - \left(1 - \frac{1}{n}\right)\left(1 - \frac{2}{n}\right)\cdots\left(1 - \frac{k-1}{n}\right).
\end{aligned}
$$

For a positive real value x much smaller than 1, we have $1 - x \approx e^{-x}$. So long as k is much smaller than n, we then have

$$
p_{\text{collision}}(n,k) \approx 1 - e^{-\frac{1+2+\cdots+(k-1)}{n}} = 1 - e^{\frac{-k(k-1)}{2n}} \approx 1 - e^{\frac{-k^2}{2n}}.
$$

Stated differently, we have

$$
k \approx \sqrt{-2\ln(1 - p_{\text{collision}}(n,k))} \times \sqrt{n}.
$$

Plugging in $p_{\text{collision}}(n,k) = 1/2$ gives $k \approx 1.1774\sqrt{n}$, whereas $p_{\text{collision}}(n,k) = 0.99$ gives $k \approx 3.0349\sqrt{n}$.

A.4.3 Random-Number Generators

Many number-theoretic algorithms require a source of random number generators (RNGs). For example, the Miller–Rabin primality test needs random bases in $\mathbb{Z}_n^*$, Dixon's integer-factoring method requires randomly generated elements $x_i \in \mathbb{Z}_n$ in a search for smooth values $x_i^2 \pmod{n}$, it is often needed to locate random points on an elliptic curve, and so on. In general, the problem of random-number generation can be simplistically stated as the generation of sequences of bits $b_0 b_1 b_2 \ldots$ such that each bit is of probability $1/2$ of being zero and of probability $1/2$ of being one, and the knowledge of the bits $b_0, b_1, \ldots, b_i$ cannot let one predict the next bit b_{i+1}. If one clubs multiple bits together, one can talk about random sequences of integers or even floating-point numbers.

RNGs can be classified into two broad categories. *Hardware generators* sample random properties of physical objects to generate random bit streams. Some example sources are noise in electronic circuits or nature, and quantum-mechanical phenomena like spin of electrons or polarization of photons. One may also consider the load of a computing processor and even user inputs, as sources of random bits. Hardware RNGs are called *true* RNGs, because they possess good statistical properties, at least theoretically. But hardware RNGs are costly and difficult to control, and it is usually impossible to use them to generate the same random sequence at two different times and/or locations.

Software RNGs offer practical solutions to the random-number generation problem. They are called *pseudorandom number generators* (PRNG), since they use known algorithms to generate sequences of random bits (or integers or floating-point numbers). A PRNG operates on a *seed* initialized to a value s_0. For $i = 1, 2, 3, \ldots$, two functions f and g are used. The next element in the pseudorandom sequence is generated as $x_i = f(s_i)$, and the seed is updated to $s_{i+1} = g(s_i)$. In many cases, the seed itself is used as the random number, that is, $x_i = s_i$. PRNGs are easy to implement (in both hardware and software), and are practically usable if their outputs *look* random.

Most commonly, PRNGs are realized using *linear congruential generators* in which the pseudorandom sequence is generated as $x_{i+1} \equiv ax_i + b \pmod{m}$ for some suitable modulus m, multiplier a, and increment b. The parameters a, b, m should be carefully chosen so as to avoid pseudorandom sequences of poor statistical properties. An example of linear congruential generator is the ANSI-C generator defined by $m = 2^{31}$, $a = 1103515245$, and $b = 12345$. The seed is initialized to $x_0 = 12345$. This PRNG is known to have many flaws but can be used in many practical situations.

PRNGs with provably good statistical properties are also known. In 1986, Lenore Blum, Manuel Blum and Michael Shub propose a PRNG called the *Blum-Blum-Shub* or *BBS generator*. This PRNG uses a modulus $m = pq$, where both p and q are suitably large primes congruent to 3 modulo 4. The sequence generation involves a modular squaring: $x_{i+1} \equiv x_i^2 \pmod{m}$. It is not practical, since modular operations on multiple-precision integers are not very efficient. Moreover, a statistical drawback of the BBS generator is that each x_i (except perhaps for $i = 0$) is a quadratic residue modulo m, that is, all elements in $\mathbb{Z}_m^*$ are not generated by the BBS generator. It, however, is proved that if only $O(\log \log m)$ least significant bits of x_i are used in the output stream, then the output bit sequence is indistinguishable from random, so long as factoring the modulus m is infeasible. In view of this, the BBS generator is often used in cryptographic applications.

So far, we have concentrated on generating uniformly random samples (bits or elements of $\mathbb{Z}_m$). Samples following other probability distributions can also be generated using a uniform PRNG. The concept of cumulative probability helps us in this context.

Let U be a random variable with sample space X. For simplicity, suppose that $X = \{x_1, x_2, \ldots, x_n\}$ is finite. Let $p_i = \Pr(U = x_i)$. We break the real interval $[0, 1)$ in n disjoint sub-intervals: $I_1 = [0, p_1)$, $I_2 = [p_1, p_1 + p_2)$, $I_3 = [p_1 + p_2, p_1 + p_2 + p_3), \ldots, I_n = [p_1 + p_2 + \cdots + p_{n-1}, 1)$. We then generate a uniformly random floating-point number $x \in [0, 1)$. There is a unique k such that I_k contains x. We find out this k (for example, using binary search), and output x_k. It is easy to argue that the output follows the distribution of U.

If the elements of X are ordered as $x_1 < x_2 < x_3 < \cdots < x_n$, the key role is played here by the *cumulative probabilities* $\Pr(U \leqslant x_k) = p_1 + p_2 + \cdots + p_k$.

In the case of a continuous random variable U, the summation is to be replaced by integration. As an example, consider the exponential distribution standing for the lifetime of an electric bulb: $f(x) = e^{-x}$ for all $x \geqslant 0$ (see Example A.103(4), we have taken $\lambda = 1$ for simplicity). In order to generate a bulb sample with a lifetime following this distribution, we obtain the *cumulative probability distribution*: $F(x) = \Pr(U \leqslant x) = \int_{x=0}^{x} e^{-x} \, dx = 1 - e^{-x}$. We generate a uniformly random floating-point value $y \in [0, 1)$, and output x satisfying $y = F(x) = 1 - e^{-x}$, that is, $x = \ln\left(\frac{1}{1-y}\right)$.

Appendix B

Solutions to Selected Exercises

Chapter 1 Arithmetic of Integers

3. Let $a = (a_{s-1}a_{s-2} \ldots a_1 a_0)_B$ be the multiple-precision integer whose square needs to be computed.

In the schoolbook multiplication method, we run a doubly nested loop on i, j. The outer loop variable i runs in the range $0, 1, 2, \ldots, s - 1$, whereas for each i, the inner loop variable j runs in the range $0, 1, 2, \ldots, i$. If $i = j$, a_i^2 is computed as the double-precision word $(hl)_B$. The digits h and l are added (with carry adjustments) to the $(2i + 1)$-st and the $2i$-th words of the product, respectively. If $i \neq j$, only one product $a_i a_j$ is computed as a double-precision word (h, l), and the digits h and l are added twice each to the $(i + j + 1)$-st and the $(i + j)$-th words of the product, respectively. Here, the products $a_i a_j$ and $a_j a_i$ are computed only once (but added twice). This saves a significant number of word multiplications compared to the general-purpose multiplication routine. In practice, one typically achieves a speedup by 20–30% using this special trick for squaring.

Instead of adding the two words of $a_i a_j$ twice to the product, one may compute $2a_i a_j$ by left-shifting the product $a_i a_j$ by one bit. The computation of $2a_i a_j$ may lead to an overflow (in a double-precision space). If so, 1 is added to the $(i + j + 2)$-nd word of the product. The other two words of $2a_i a_j$ are added once each to the $(i + j + 1)$-st and the $(i + j)$-th words of the product.

For Karatsuba squaring, note that $2A_1 A_0 = A_1^2 + A_0^2 - (A_1 - A_0)^2$. This means that three squares of integers of half sizes are computed recursively, and combined using one addition and one subtraction.

5. Let us divide $a = (a_{s-1}a_{s-2} \ldots a_1a_0)_B$ by $b = B^l + m$. If $l \geqslant s$, then $a < b$, and we are done. Otherwise, we multiply m by a_{s-1} to obtain the double-precision word $(hl)_B$. Notice that for $l \geqslant 2$, the $(s-1)$-st word of $a_{s-1}B^{s-1-l}b$ is a_{s-1}, so we set the $(s-1-l)$-th word of the quotient to a_{s-1}, set a_{s-1} to zero, and subtract h and l from the $(s-l)$-th and the $(s-1-l)$-th words of a with borrow adjustments. If the borrow is not finally absorbed, we add $B^{s-1-l}b$ to the negative difference, and subtract 1 from the $(s-1-l)$-th word of the quotient. Now, a is reduced to an integer of the form $a' = (a'_{s'-1}a'_{s'-2} \ldots a'_1a'_0)_B$ with $s' = s-1$. We then repeat the above steps to reduce $a'_{s'-1}$ to zero. This process is continued until a reduces to an integer smaller than b. This reduced value is the desired remainder.

For dividing a by $b = B^l - m$, the words h and l are added (instead of subtracted) to the $(s-l)$-th and the $(s-1-l)$-st words of a with carry adjustments. In this case, a can never become negative, that is, the overhead of adding b to an intermediate negative a is absent.

8. Write the operands as

$$\begin{aligned} a &= A_2 R^2 + A_1 R + A_0, \\ b &= B_1 R + B_0, \end{aligned}$$

where R is a suitable integral power of the base B. The product is a polynomial of degree three:

$$c = C_3 R^3 + C_2 R^2 + C_1 R + C_0,$$

where

$$\begin{aligned} C_3 &= A_2 B_1, \\ C_2 &= A_2 B_0 + A_1 B_1, \\ C_1 &= A_2 B_0 + A_0 B_1, \\ C_0 &= A_0 B_0. \end{aligned}$$

Instead of computing all the six products $A_i B_j$, we evaluate $c = ab$ at $R = \infty, 0, \pm 1$ to obtain

$$\begin{aligned} C_3 &= c(\infty) &= A_2 B_1, \\ C_0 &= c(0) &= A_0 B_0, \\ C_3 + C_2 + C_1 + C_0 &= c(1) &= (A_2 + A_1 + A_0)(B_1 + B_0), \\ -C_3 + C_2 - C_1 + C_0 &= c(-1) &= (A_2 - A_1 + A_0)(-B_1 + B_0). \end{aligned}$$

The four products on the right side are computed by a good multiplication algorithm (these products have balanced operands). Since

$$\begin{pmatrix} 1 & 0 & 0 & 0 \\ 0 & 0 & 0 & 1 \\ 1 & 1 & 1 & 1 \\ -1 & 1 & -1 & 1 \end{pmatrix} \begin{pmatrix} C_3 \\ C_2 \\ C_1 \\ C_0 \end{pmatrix} = \begin{pmatrix} c(\infty) \\ c(0) \\ c(1) \\ c(-1) \end{pmatrix},$$

the coefficients C_3, C_2, C_1, C_0 can be expressed in terms of the subproducts as

$$\begin{pmatrix} C_3 \\ C_2 \\ C_1 \\ C_0 \end{pmatrix} = \begin{pmatrix} 1 & 0 & 0 & 0 \\ 0 & -1 & \frac{1}{2} & \frac{1}{2} \\ -1 & 0 & \frac{1}{2} & -\frac{1}{2} \\ 0 & 1 & 0 & 0 \end{pmatrix} \begin{pmatrix} c(\infty) \\ c(0) \\ c(1) \\ c(-1) \end{pmatrix}.$$

12. (a) For each i in the range $2 \leqslant i \leqslant k + 1$, we have $r_{i-2} = q_i r_{i-1} + r_i$. Since $r_{i-2} > r_{i-1}$, we have $q_i \geqslant 1$, so $r_{i-2} \geqslant r_{i-1} + r_i$. Moreover, $r_k \neq 0$, that is, $r_k \geqslant 1 = F_2$, and $r_{k-1} > r_k$, that is, $r_{k-1} \geqslant 2 = F_3$. But then, $r_{k-2} \geqslant r_{k-1} + r_k \geqslant F_3 + F_2 = F_4$, $r_{k-3} \geqslant r_{k-2} + r_{k-1} \geqslant F_4 + F_3 = F_5$, and so on. Proceeding in this way, we can show that $r_1 \geqslant F_{k+1}$, and $r_0 \geqslant F_{k+2}$.
(b) Compute $r_i = r_{i-2} \text{ rem } r_{i-1}$, where $0 \leqslant r_i \leqslant r_{i-1} - 1$. If $r_i > \frac{1}{2}r_{i-1}$, replace r_i by $r_{i-1} - r_i$. The correctness of this variant is based on the fact that $\gcd(r_{i-1}, r_i) = \gcd(r_{i-1}, -r_i) = \gcd(r_{i-1}, r_{i-1} - r_i)$.
(c) In the original Euclidean algorithm, we have $r_1 \geqslant F_{k+1} \approx \frac{1}{\sqrt{5}}\rho^{k+1}$, that is, $k \leqslant -1 + (\log(\sqrt{5}r_1))/\log \rho$. For the modified algorithm, let k' denote the number of iterations. We have $r_1 \geqslant 2r_2 \geqslant 2^2 r_3 \geqslant \cdots \geqslant 2^{k'-1}r_{k'} \geqslant 2^{k'-1}$, that is, $k' \leqslant 1 + \log(r_1)/\log(2)$. Since $2 > \rho$, the modified algorithm has the potential of reducing the number of iterations of the Euclidean loop by a factor of $\log 2/\log \rho \approx 1.440$.

13. Let a, b be the two integers of which the gcd is to be computed. For the time being, assume that a, b are both odd. In the extended binary gcd algorithm, we keep track of three sequences r_i, u_i, v_i satisfying the invariance

$$u_i a + v_i b = r_i.$$

We initialize $u_0 = 1, v_0 = 0, r_0 = a$, and $u_1 = 0, v_1 = 1, r_1 = b$, so the invariance is satisfied for $i = 0, 1$.

In the binary gcd loop, the smaller of r_{i-2}, r_{i-1} is subtracted from the other to obtain an even integer r_i. Suppose that $r_{i-2} > r_{i-1}$, so we compute $r_i = r_{i-2} - r_{i-1}$ (the other case can be symmetrically handled). By the invariance for $i - 1, i - 2$, we already have

$$\begin{aligned} u_{i-2}a + v_{i-2}b &= r_{i-2}, \\ u_{i-1}a + v_{i-1}b &= r_{i-1}. \end{aligned}$$

After subtraction, we should continue to have $u_i a + v_i b = r_i$, so we set

$$\begin{aligned} u_i &= u_{i-2} - u_{i-1}, \\ v_i &= v_{i-2} - v_{i-1}. \end{aligned}$$

The next step is more complicated. Now, r_i is even, so an appropriate number of 2's should be factored out of it so as make it odd. Here, I describe the removal of a single factor of 2. Since $u_i a + v_i b = r_i$, dividing r_i by 2 requires dividing both u_i and v_i by 2 so that the invariance is maintained. But there is no guarantee that u_i, v_i are even. However, since a, b are odd, exactly one of u_i, v_i cannot be odd. If both u_i, v_i are even, we extract a factor of 2 from both. If both u_i, v_i are odd, we rewrite the invariance as

$$(u_i + b)a + (v_i - a)b = r_i.$$

Now, $u_i + b$ and $v_i - a$ are even, so a factor of 2 can be extracted from all of $u_i + b, v_i - a, r_i$.

Notice that each iteration of the binary gcd loop uses values from only two previous iterations. So it is not necessary to store the entire sequences u_i, v_i, r_i. Moreover, it is not necessary to explicitly maintain (and update) both the u and v sequences. If only the u sequence is maintained, one can compute $v = (r - ua)/b$. This is done after the gcd loop terminates.

Let us now remove the restriction that a, b are odd, and write $a = 2^s a'$ and $b = 2^t b'$ with a', b' odd. By the above binary gcd loop, we compute the extended gcd of a', b' giving

$$u'a' + v'b' = d' = \gcd(a', b').$$

Let $r = \min(s, t)$. Then, the gcd d of a, b is

$$2^r u'a' + 2^r v'b' = u'(2^r a') + v'(2^r b') = 2^r d' = d.$$

If $s = t = r$, we are done. So suppose that $s > t$ (the case $s < t$ can be symmetrically handled). In that case, we have $d = u'2^t a + v'b = d$. It is not guaranteed that u' itself would supply the remaining $s - t$ factors of 2 at this stage. For example, this initial u' may be odd. Still, we proceed as follows. Suppose that at some point of time, we have arrived at a readjusted Bézout relation of the form

$$u'2^\tau a + v'b = d$$

with $t \leqslant \tau < s$ and with u' odd. We rewrite this invariance as

$$(u' + b')2^\tau a + (v' - 2^{\tau-t}a)b = d.$$

Now, $u' + b'$ is even, so factor(s) of 2 can be extracted from it and absorbed in 2^τ so long as necessary and/or possible.

Like the main binary gcd loop, it suffices to update only one of the u and v sequences in this final adjustment loop. Moreover, only the value from the previous iteration is needed for the current iteration to proceed.

16. (a) We have $u_k = u_{k-2} - q_k u_{k-1}$ with $q_k \geqslant 1$, for all $k \geqslant 2$. It is easy to establish that the sequence $u_2, u_3, u_4, \ldots$ alternate in sign, and $|u_k| > q_k |u_{k-1}| \geqslant |u_{k-1}|$ (unless $u_{k-2} = 0$ which is true only for $k = 3$). The given inequalities for the v sequence can be analogously established.

(b) We start with $u_0 = 1$, $u_1 = 0$, and obtain $u_2 = u_0 - q_2 u_1 = 1 = K_0()$, $u_3 = u_1 - q_3 u_2 = -q_3$, that is, $|u_3| = K_1(q_3)$. In general, suppose that $|u_i| = K_{i-2}(q_3, \ldots, q_i)$ and $|u_{i+1}| = K_{i-1}(q_3, \ldots, q_{i+1})$. But then,

$|u_{i+2}| = |u_i| + q_{i+2}|u_{i+1}| = q_{i+2}K_{i-1}(q_3,\ldots,q_{i+1}) + K_{i-2}(q_3,\ldots,q_i) = K_i(q_3,\ldots,q_{i+2})$. Thus, we have:

$$|u_i| = K_{i-2}(q_3,\ldots,q_i) \text{ for all } i \geqslant 2.$$

Analogously, we have:

$$|v_i| = K_{i-1}(q_2,\ldots,q_i) \text{ for all } i \geqslant 1.$$

(c) Take $n = i - 2$, and substitute $x_1 = q_2$, $x_2 = q_3$, $\ldots$, $x_{n+1} = q_i$ in Exercise 1.15(e) to get $|u_i|K_{i-2}(q_3,\ldots,q_i) - |v_i|K_{i-3}(q_3,\ldots,q_{i-1}) = (-1)^{i-2}$ for all $i \geqslant 3$. Moreover, since $u_0 = v_1 = u_2 = 1$, we have:

$$\gcd(u_i, v_i) = 1 \text{ for all } i \geqslant 0.$$

(d) If $r_j = 0$, that is, $r_{j-1} = \gcd(a,b) = d$, we have $u_j a + v_j b = 0$, that is, $u_j\left(\frac{a}{d}\right) = -v_j\left(\frac{b}{d}\right)$. By Part (c), $\gcd(u_j, v_j) = 1$. Moreover, $\gcd\left(\left(\frac{a}{d}\right), \left(\frac{b}{d}\right)\right) = 1$. It therefore follows that $u_j = \pm\left(\frac{b}{d}\right)$ and $v_j = \mp\left(\frac{a}{d}\right)$.

(e) The case $b|a$ is easy to handle. So assume that $a > b > 0$ and $b \nmid a$. Then, the gcd loop ends with $j \geqslant 4$. In this case, we have $|u_3| < |u_4| < \cdots < |u_{j-1}| < |u_j| = \frac{b}{d}$ and $|v_2| < |v_3| < \cdots < |v_{j-1}| < |v_j| = \frac{a}{d}$, that is,

$$|u_{j-1}| < \frac{b}{d} \text{ and } |v_{j-1}| < \frac{a}{d}.$$

21. Write $a = (a_{s-1}a_{s-2}\ldots a_1 a_0)_B$ and $b = (b_{t-1}b_{t-2}\ldots b_1 b_0)_B$, where each a_i and each b_j are base-B words. We assume that B is a power of 2, and that a and b are odd. If so, b_0 is an odd integer, and $b_0^{-1} \pmod{B}$ exists. We compute the multiplier $\mu = b_0^{-1}a_0 \pmod{B}$. The least significant word of $a - \mu b$ is zero. If $B = 2^r$, we can remove at least r factors of 2 from $a - \mu b$.

Computing the inverse $b_0^{-1} \pmod{B}$ can be finished by a single-precision extended gcd computation. Moreover, the multiplication $b_0^{-1}a_0 \pmod{B}$ is again of single-precision integers. In most CPUs, this is the value returned by the single-precision multiplication function (the more significant word is ignored). Therefore, μ can be computed efficiently. The multiplication μb and the subtraction $a - \mu b$ are also efficient (taking time proportional to s).

But then, we require $(a - \mu b)/B$ to have word length (at least) one smaller than that of a. This is ensured if $t < s$, that is, if the word length of b is at least one smaller than that of a. If $s = t$ (we cannot have $s < t$ since $a > b$), then μb may be an $(s+1)$-word integer, so $a - \mu b$ is again an $(s+1)$-bit negative integer. Ignoring the sign and removing the least significant word of $a - \mu b$, we continue to have an s-word integer.

25. Consider the product $x = n(n-1)(n-2)\cdots(n-r+1)$ of r consecutive integers. If any of the factors $n - i$ of x is zero, we have $x = 0$ which is a multiple of $r!$. If all factors $n - i$ of x are negative, we can write x as $(-1)^r$ times a product of r consecutive positive integers. Therefore, we can assume without loss of generality that $1 \leqslant r \leqslant n$. But then, $x = n!/(n-r)!$. For any prime p and any $k \in \mathbb{N}$, we have $\lfloor n/p^k \rfloor \geqslant \lfloor (n-r)/p^k \rfloor + \lfloor r/p^k \rfloor$. By Exercise 1.24, we conclude that $v_p(x) \geqslant v_p(r!)$.

28. (a) The modified square-and-multiply algorithm is elaborated below.

Let $r = (r_{l-1} r_{l-2} \ldots r_1 r_0)_2$, and $s = (s_{l-1} s_{l-2} \ldots s_1 s_0)_2$.
Precompute $xy \pmod{n}$.
Initialize $prod = 1$.
For $i = l - 1, l - 2, \ldots, 1, 0$ {
    Set $prod = prod^2 \pmod{n}$.
    If $(r_i = 1)$ and $(s_i = 1)$, set $prod = prod \times (xy) \pmod{n}$,
    else if $(r_i = 1)$ and $(s_i = 0)$, set $prod = prod \times x \pmod{n}$,
    else if $(r_i = 0)$ and $(s_i = 1)$, set $prod = prod \times y \pmod{n}$.
}

The modified algorithm reduces the number of square operations to half of that performed by two independent calls of the repeated square-and-multiply algorithm. The number of products depends on the bit patterns of the exponents r and s. For random exponents, about half of the bits are one, that is, two exponentiations make about l modular multiplications. The modified algorithm skips the multiplication only when both the bits are 0—an event having a probability of $1/4$ for random exponents. Thus, the expected number of multiplications done by the modified algorithm is $0.75l$. The precomputation involves only one modular multiplication, and has negligible overhead.

41. Let p be a prime divisor of the modulus m, and e the multiplicity of p in m. If $p|a$, both a^m and $a^{m-\phi(m)}$ (and so their difference too) are divisible by p^e, since p^{e-1} (and so p^e too) are $\geqslant e$ for all $p \geqslant 2$ and $e \geqslant 1$. On the other hand, if $p \nmid a$, we have $a^{\phi(p^e)} \equiv 1 \pmod{p^e}$ by the original theorem of Euler. But then, $a^{\phi(m)} \equiv 1 \pmod{p^e}$ (since $\phi(p^e)|\phi(m)$), that is, $a^m - a^{m-\phi(m)} = a^{m-\phi(m)}(a^{\phi(m)} - 1)$ is again divisible by p^e.

45. [Only if] Let $a \in \mathbb{Z}$ be a solution to the t congruences $x \equiv a_i \pmod{m_i}$. But then for all i, j with $i \neq j$, we have $a = a_i + k_i m_i = a_j + k_j m_j$ for some integers k_i, k_j, that is, $a_i - a_j = k_j m_j - k_i m_i$ is a multiple of $\gcd(m_i, m_j)$.

[If] We proceed by induction on t. As the base case, we take $t = 2$, that is, we look at the two congruences $x \equiv a_1 \pmod{m_1}$ and $x \equiv a_2 \pmod{m_2}$ with $d = \gcd(m_1, m_2)$ dividing $a_1 - a_2$. There exist $u, v \in \mathbb{Z}$ such that $um_1 + vm_2 = d$. Consider $x = a_1 + \left(\frac{a_2 - a_1}{d}\right) um_1$. Since $d|(a_2 - a_1)$ by hypothesis, $\left(\frac{a_2 - a_1}{d}\right)$ is an integer, so $x \equiv a_1 \pmod{m_1}$. Moreover, $um_1 = d - vm_2$, so that $a_1 + \left(\frac{a_2 - a_1}{d}\right) um_1 \equiv a_1 + (a_2 - a_1) - \left(\frac{a_2 - a_1}{d}\right) vm_2 \equiv a_2 \pmod{m_2}$. Thus, $x = a_1 + \left(\frac{a_2 - a_1}{d}\right) um_1$ is a simultaneous solution of the two given congruences.

Now, take $t \geqslant 3$, and assume that the result holds for $t - 1$ congruences. As in the base case, the first two of the t given congruences are simultaneously solvable. Let a_0 be any particular solution of the first two congruences, and $m_0 = \text{lcm}(m_1, m_2)$. Any solution of the first two congruences satisfy $x \equiv a_0 \pmod{m_0}$ (see the uniqueness proof below). We now look at the $t - 1$ congruences $x \equiv a_i \pmod{m_i}$ for $i = 0, 3, 4, \ldots, t$. Take any $i \in \{3, 4, \ldots, t\}$, and

write $\gcd(m_0, m_i) = \gcd(\text{lcm}(m_1, m_2), m_i) = \text{lcm}(\gcd(m_1, m_i), \gcd(m_2, m_i))$.
Now, $\gcd(m_1, m_i) | (a_1 - a_i)$ (by hypothesis), $a_1 - a_i = (a_1 - a_0) + (a_0 - a_i)$,
$a_0 \equiv a_1 \pmod{m_1}$ (since a_0 is a solution of the first of the initial congruences), and $\gcd(m_1, m_i) | m_1$. It follows that $\gcd(m_1, m_i) | (a_0 - a_i)$. Likewise,
$\gcd(m_2, m_i) | (a_0 - a_i)$, and so $\gcd(m_0, m_i) = \text{lcm}(\gcd(m_1, m_i), \gcd(m_2, m_i))$
divides $a_0 - a_i$. By induction hypothesis, the $t-1$ congruences $x \equiv a_i \pmod{m_i}$
for $i = 0, 3, 4, \ldots, t$ are simultaneously solvable.

For proving the uniqueness, let a and b be two solutions of the given congruences. For all i, we then have $a \equiv a_i \pmod{m_i}$ and $b \equiv a_i \pmod{m_i}$, that
is, $a \equiv b \pmod{m_i}$, that is, $m_i | (a - b)$. Therefore, $\text{lcm}(m_1, m_2, \ldots, m_t) | (a - b)$.

47. [If] Since $\gcd(e, \phi(m)) = 1$, an extended gcd calculation gives d satisfying
$ed \equiv 1 \pmod{\phi(m)}$. It suffices to show that $a^{ed} \equiv a \pmod{m}$ for all $a \in \mathbb{Z}_m$ (so exponentiation to the d-th power is the inverse of exponentiation
to the e-th power). Take any prime divisor p_i of the modulus m. We show
that $a^{ed} \equiv a \pmod{p_i}$. By CRT, it then follows that $a^{ed} \equiv a \pmod{m}$.
If $p_i | a$, both a^{ed} and a are congruent to zero modulo p_i. So assume that
$p_i \nmid a$. Write $ed = t\phi(m) + 1$ for some integer t. By Fermat's little theorem,
$a^{p_i-1} \equiv 1 \pmod{p_i}$. Since $\phi(m)$ and so $t\phi(m)$ too are multiples of $p_i - 1$, we
have $a^{t\phi(m)} \equiv 1 \pmod{p_i}$, that is, $a^{ed} \equiv a \pmod{p_i}$.

[Only if] Suppose that $s = \gcd(e, p_i - 1) > 1$ for some $i \in \{1, 2, \ldots, k\}$. Let
g be a primitive root of p_i. Then, $g^{(p_i-1)/s} \not\equiv 1 \pmod{p_i}$, but $(g^{(p_i-1)/s})^e \equiv$
$(g^{p_i-1})^{e/s} \equiv 1 \pmod{p_i}$. Consider two elements $a, b \in \mathbb{Z}_m$ satisfying $a \equiv b \equiv$
$1 \pmod{p_j}$ for $j \neq i$, but $a \equiv g^{(p_i-1)/s} \pmod{p_i}$ and $b \equiv 1 \pmod{p_i}$. By
CRT, $a \not\equiv b \pmod{m}$, but $a^e \equiv b^e \equiv 1 \pmod{m}$, that is, the e-th-power-
exponentiation map is not bijective. Therefore, $\gcd(e, \phi(p_i)) = 1$ for all i, that
is, $\gcd(e, \phi(m)) = 1$.

48. (a) We have $b_p \equiv b \pmod{p}$ and $b_q \equiv b \pmod{q}$, so we have to combine
these two values by the CRT. Let $\beta = b_q + tq$. Then, $\beta \equiv b_q \pmod{q}$.
Also, $tq \equiv b_p - b_q \pmod{p}$, so $\beta \equiv b_q + (b_p - b_q) \equiv b_p \pmod{p}$. Therefore,
$\beta \equiv b \pmod{pq}$.

(b) Let $s = |n|$ be the bit size of m. We then have the bit sizes $|p| \approx s/2$ and
$|q| \approx s/2$. Since modular exponentiation is done in cubic time, computing the
two modular exponentiations to obtain b_p and b_q takes a total time which is
about $1/4$-th of that for computing $b \equiv a^e \pmod{m}$ directly. The remaining
operations in the modified algorithm can be done in $O(s^2)$ time. Thus, we get
a speed-up of about four.

50. In view of the CRT, it suffices to count the solutions of $x^{m-1} \equiv 1 \pmod{p^e}$
for a prime divisor p of m with $e = v_p(m)$. Since the derivative of $x^{m-1} - 1$
is non-zero $(-x^{m-2})$ modulo p, each solution of $x^{m-1} \equiv 1 \pmod{p}$ lifts
uniquely to a solution of $x^{m-1} \equiv 1 \pmod{p^e}$. Therefore, it suffices to count
the solutions of $x^{m-1} \equiv 1 \pmod{p}$. Let g be a primitive root of p, and g^α a
solution of $x^{m-1} \equiv 1 \pmod{p}$. This implies that $(m-1)\alpha \equiv 0 \pmod{p-1}$
(since $\text{ord}_p g = p - 1$). This is a linear congruence in α, and has exactly
$\gcd(p-1, m-1)$ solutions.

53. **(a)** Let $f(x) = a_d x^d + a_{d-1} x^{d-1} + \cdots + a_1 x + a_0$. The binomial theorem with the substitution $x = \xi'$ gives

$$
\begin{aligned}
f(\xi') &= a_d(\xi + kp^e)^d + a_{d-1}(\xi + kp^e)^{d-1} + \cdots + a_1(\xi + kp^e) + a_0 \\
&= f(\xi) + kp^e f'(\xi) + p^{2e} \times t
\end{aligned}
$$

for some integer t. The condition $f(\xi') \equiv 0 \pmod{p^{2e}}$ implies that $f(\xi) + kp^e f'(\xi) \equiv 0 \pmod{p^{2e}}$, that is, $f'(\xi)k \equiv -\left(\frac{f(\xi)}{p^e}\right) \pmod{p^e}$. Each solution of this linear congruence modulo p^e gives a lifted root ξ' of $f(x)$ modulo p^{2e}.
(b) Here, $f(x) = 2x^3 + 4x^2 + 3$, so $f'(x) = 6x^2 + 8x$. For $p = 5$, $e = 2$ and $\xi = 14$, we have $f(\xi) = 2 \times 14^3 + 4 \times 14^2 + 3 = 6275$, that is, $f(\xi)/25 \equiv 251 \equiv 1 \pmod{25}$. Also, $f'(\xi) \equiv 6 \times 14^2 + 8 \times 14 \equiv 1288 \equiv 13 \pmod{25}$. Thus, we need to solve $13k \equiv -1 \pmod{25}$. Since $13^{-1} \equiv 2 \pmod{25}$, we have $k \equiv -2 \equiv 23 \pmod{25}$. It follows that the only solution of $2x^3 + 4x^2 + 3 \equiv 0 \pmod{625}$ is $14 + 23 \times 25 \equiv 589 \pmod{625}$.

56. Since we can extract powers of two easily from a, we assume that a is odd. For the Jacobi symbol, b is odd too. If $a = b$, then $\left(\frac{a}{b}\right) = 0$. If $a < b$, we use the quadratic reciprocity law to write $\left(\frac{a}{b}\right)$ in terms of $\left(\frac{b}{a}\right)$. So it remains only to analyze the case of $\left(\frac{a}{b}\right)$ with a, b odd and $a > b$. Let $\alpha = a - b$. We write $\alpha = 2^r a'$ with $r \in \mathbb{N}$ and a' odd. If r is even, then $\left(\frac{a}{b}\right) = \left(\frac{a'}{b}\right)$, whereas if r is odd, then $\left(\frac{a}{b}\right) = \left(\frac{2}{b}\right)\left(\frac{a'}{b}\right) = (-1)^{(b^2-1)/8}\left(\frac{a'}{b}\right)$. So, the problem reduces to computing $\left(\frac{a'}{b}\right)$ with both a', b odd.

59. **(a)** By Euler's criterion, we have $b^{(p-1)/2} \equiv -1 \pmod{p}$. By Fermat's little theorem, we also have $b^{p-1} \equiv 1 \pmod{p}$. Therefore, the order of $g \equiv b^q \pmod{p}$ is 2^v. If $\left(\frac{a}{p}\right) = 1$, we have $\left(\frac{a^q}{p}\right) = \left(\frac{a}{p}\right)^q = 1$. Moreover, the order of a^q modulo p is a divisor of 2^{v-1}, that is, $a^q \equiv g^s \pmod{p}$ for some even integer s. We can rewrite this as $a^q g^t \equiv 1 \pmod{p}$, where $t = 2^v - s$ (if $s = 0$, we take $t = 0$). But then, $(a^{(q+1)/2} g^{t/2})^2 \equiv a \pmod{p}$, that is, $a^{(q+1)/2} g^{t/2}$ is a square root of a modulo p.

   The loop in Tonelli–Shanks algorithm determines the one bits of $t/2$. Since $t/2 < 2^{v-1}$, we can write $t/2 = t_{v-2} 2^{v-2} + t_{v-3} 2^{v-3} + \cdots + t_1 2 + t_0$ with each $t_i \in \{0, 1\}$. Assume that at the beginning of some iteration of the loop, bits $t_0, t_1, \ldots, t_{v-i-2}$ are determined, and $t_{v-i-1} = 1$. At that point, x stores the value $a^{(q+1)/2} g^{(t_{v-i-2} \cdots t_1 t_0)_2} \pmod{p}$. (Initially, $x \equiv a^{(q+1)/2} \pmod{p}$, and no bits t_i are determined.) Since $(a^{(q+1)/2} g^{t/2})^2 \equiv a \pmod{p}$, we have

$$
x^2 a^{-1} \equiv \left(g^{-2^{v-i-1}(t_{v-2} t_{v-3} \ldots t_{v-i-1})_2}\right)^2 \equiv g^{-2^{v-i}(t_{v-2} t_{v-3} \ldots t_{v-i-1})_2} \pmod{p}.
$$

The order of g modulo p is 2^v, and the integer $(t_{v-2} t_{v-3} \ldots t_{v-i-1})_2$ is odd. Therefore, i is smallest among all integers j for which $(x^2 a^{-1})^{2^j} \equiv 1 \pmod{p}$. Once this i is detected, we know $t_{v-i-1} = 1$, and multiply x by $g^{2^{v-i-1}}$, so x now stores the value $a^{(q+1)/2} g^{(t_{v-i-1} t_{v-i-2} \cdots t_1 t_0)_2} \pmod{p}$. When all the bits

$t_0, t_1, \ldots, t_{v-2}$ are determined, x stores the value $a^{(q+1)/2} g^{t/2} \pmod{p}$. At this point, we have $x^2 a^{-1} \equiv (x^2 a^{-1})^{2^0} \equiv 1 \pmod{p}$.

(b) There are $(p-1)/2$ quadratic residues and $(p-1)/2$ quadratic non-residues in $\mathbb{Z}_p^*$. Therefore, a randomly chosen element in $\mathbb{Z}_p^*$ is a quadratic non-residue with probability $1/2$. That is, trying a constant number of random candidates gives us a non-residue, and locating b is expected to involve only a constant number of Legendre-symbol calculations. The remaining part of the algorithm involves two modular exponentiations to compute g and the initial value of x. Moreover, $a^{-1} \pmod{p}$ can be precomputed outside the loop. Each iteration of the loop involves at most $v - 1$ modular squaring operations to detect i. This is followed by one modular multiplication. It therefore follows that Algorithm 1.9 runs in probabilistic polynomial time.

(c) By Conjecture 1.74, the smallest quadratic non-residue modulo p is less than $2 \ln^2 p$. Therefore, we may search for b in Algorithm 1.9 deterministically in the sequence $1, 2, 3, \ldots$ until a non-residue is found. The search succeeds in less than $2 \ln^2 p$ iterations.

61. (a) Let $r = \lfloor \sqrt{p} \rfloor$. Since p is not a perfect square, we have $r^2 < p < (r+1)^2$. Consider the $(r+1)^2$ integers $(u + vx)$ rem p with $u, v \in \{0, 1, 2, \ldots, r\}$. By the pigeon-hole principle, there exist unequal pairs (u_1, v_1) and (u_2, v_2) such that $u_1 + xv_1 \equiv u_2 + xv_2 \pmod{p}$. Let $a = u_1 - u_2$ and $b = v_2 - v_1$. Then, $a \equiv bx \pmod{p}$. By the choice of u_1, u_2, v_1, v_2, we have $-r \leqslant a \leqslant r$ and $-r \leqslant b \leqslant r$ with either $a \neq 0$ or $b \neq 0$. Furthermore, since $x \neq 0$, both a and b must be non-zero.

(b) For $p = 2$, take $a = b = 1$. So, we assume that p is an odd prime.

If $p \equiv 1 \pmod{4}$, the congruence $x^2 \equiv -1 \pmod{p}$ is solvable by Exercise 1.60. Using this value of x in Part (a) gives us a non-zero pair (a, b) satisfying $a \equiv bx \pmod{p}$, that is, $a^2 \equiv b^2 x^2 \equiv -b^2 \pmod{p}$, that is, $a^2 + b^2 \equiv 0 \pmod{p}$. By Part (a), $0 < a^2 + b^2 \leqslant 2r^2 < 2p$, that is, $a^2 + b^2 = p$.

Finally, take a prime $p \equiv 3 \pmod{4}$. If $p = a^2 + b^2$ for some a, b (both must be non-zero), we have $(ab^{-1})^2 \equiv -1 \pmod{p}$. But by Exercise 1.60, the congruence $x^2 \equiv -1 \pmod{p}$ does not have a solution.

(c) [If] Let $m = p_1 p_2 \cdots p_s q_1^{2e_1} q_2^{2e_2} \cdots q_t^{2e_t}$, where $p_1, p_2, \ldots, p_t$ are all the prime divisors (not necessarily distinct from one another) of m, that are not of the form $4k + 3$, and where $q_1, q_2, \ldots, q_t$ are all the prime divisors (distinct from one another) of m, that are of the form $4k + 3$. Since $(a^2 + b^2)(c^2 + d^2) = (ac + bd)^2 + (ad - bc)^2$, Part (b) establishes that $p_1 p_2 \cdots p_s$ can be expressed as $\alpha^2 + \beta^2$. Now, take $a = \alpha q_1^{e_1} q_2^{e_2} \cdots q_t^{e_t}$ and $b = \beta q_1^{e_1} q_2^{e_2} \cdots q_t^{e_t}$.

[Only if] Let $m = a^2 + b^2$ for some integers a, b, and let $q \equiv 3 \pmod{4}$ be a prime divisor of m. Since $q | (a^2 + b^2)$, and the congruence $x^2 \equiv -1 \pmod{q}$ is not solvable by Exercise 1.60, we must have $q | a$ and $q | b$. Let $e = \min(v_q(a), v_q(b))$. Since $m = a^2 + b^2$, we have $m/(q^e)^2 = (a/q^e)^2 + (b/q^e)^2$, that is, $m/(q^e)^2$ is again a sum of two squares. If q divides $m/(q^e)^2$, then q divides both a/q^e and b/q^e as before, a contradiction to the choice of e.

70. The result is obvious for $e = 1$, so take $e \geqslant 2$.

Lemma: For every $e \geqslant 2$, we have $(1+ap)^{p^{e-2}} \equiv 1 + ap^{e-1} \pmod{p^e}$.

Proof We proceed by induction on e. For $e = 2$, both sides of the congruence are equal to the integer $1 + ap$. So assume that the given congruence holds for some $e \geqslant 2$. We investigate the value of $(1+ap)^{p^{e-1}}$ modulo p^{e+1}. By the induction hypothesis, $(1+ap)^{p^{e-2}} = 1 + ap^{e-1} + up^e$ for some integer u. Raising both sides of this equality to the p-th power gives

$$
\begin{aligned}
(1+ap)^{p^{e-1}} &= (1 + ap^{e-1} + up^e)^p \\
&= 1 + \binom{p}{1}(ap^{e-1} + up^e) + \binom{p}{2}(ap^{e-1} + up^e)^2 + \cdots + \\
&\quad \binom{p}{p-1}(ap^{e-1} + up^e)^{p-1} + (ap^{e-1} + up^e)^p \\
&= 1 + ap^e + p^{e+1} \times v
\end{aligned}
$$

for some integer v (since p is prime and so $p \mid \binom{p}{k}$ for $1 \leqslant k \leqslant p-1$, and since the last term in the binomial expansion is divisible by $p^{p(e-1)}$, in which the exponent $p(e-1) \geqslant e+1$ for all $p \geqslant 3$ and $e \geqslant 2$). $\bullet$

Let us now derive the order of $1+ap$ modulo p^e. Using the lemma for $e+1$ indicates $(1+ap)^{p^{e-1}} \equiv 1 + ap^e \pmod{p^{e+1}}$ and, in particular, $(1+ap)^{p^{e-1}} \equiv 1 \pmod{p^e}$. Therefore, $\text{ord}_{p^e}(1+ap) \mid p^{e-1}$. The lemma also implies that $(1+ap)^{p^{e-2}} \not\equiv 1 \pmod{p^e}$ (for a is coprime to p), that is, $\text{ord}_{p^e}(1+ap) \nmid p^{e-2}$. We, therefore, have $\text{ord}_{p^e}(1+ap) = p^{e-1}$.

76. The infinite simple continued fraction expansion of $\sqrt{2}$ is $\langle a_0, a_1, a_2, \ldots \rangle = \langle 1, \overline{2} \rangle$. Let $h_n/k_n = \langle a_0, a_1, \ldots, a_n \rangle$ be the n-th convergent to $\sqrt{2}$. But then, for every $n \in \mathbb{N}$, we have $\left| \sqrt{2} - \frac{h_n}{k_n} \right| < \frac{1}{k_n k_{n+1}} = \frac{1}{k_n(2k_n + k_{n-1})} \leqslant \frac{1}{2k_n^2}$, that is, $\sqrt{2}k_n - \frac{1}{2k_n} < h_n < \sqrt{2}k_n + \frac{1}{2k_n}$, that is, $-\sqrt{2} + \frac{1}{4k_n^2} < h_n^2 - 2k_n^2 < \sqrt{2} + \frac{1}{4k_n^2}$. Since $k_n \geqslant 1$, it follows that $h_n^2 - 2k_n^2 \in \{0, 1, -1\}$ for all $n \in \mathbb{N}$. But $\sqrt{2}$ is irrational, so we cannot have $h_n^2 - 2k_n^2 = 0$. Furthermore, $\frac{h_n}{k_n} < \sqrt{2}$ for even n, whereas $\frac{h_n}{k_n} > \sqrt{2}$ for odd n. Consequently, $h_n^2 - 2k_n^2 = \begin{cases} -1 & \text{if } n \text{ is even,} \\ 1 & \text{if } n \text{ is odd.} \end{cases}$

78. (a) We compute $\sqrt{5} = \langle 2, 4, 4, 4, \ldots \rangle = \langle 2, \overline{4} \rangle$ as follows:

$$\xi_0 = \sqrt{5} = 2.236\ldots, \qquad a_0 = \lfloor \xi_0 \rfloor = 2$$

$$\xi_1 = \frac{1}{\xi_0 - a_0} = \frac{1}{\sqrt{5} - 2} = \sqrt{5} + 2 = 4.236\ldots, \qquad a_1 = \lfloor \xi_1 \rfloor = 4$$

$$\xi_2 = \frac{1}{\xi_1 - a_1} = \frac{1}{\sqrt{5} - 2} = \sqrt{5} + 2 = 4.236\ldots, \qquad a_2 = \lfloor \xi_2 \rfloor = 4$$

$$\cdots$$

(b) The first convergent is $r_0 = \frac{h_0}{k_0} = \langle 2 \rangle = 2/1$, that is, $h_0 = 2$ and $k_0 = 1$. But $h_0^2 - 5k_0^2 = -1$. Then, we have $r = \frac{h_1}{k_1} = \langle 2, 4 \rangle = 2 + \frac{1}{4} = \frac{9}{4}$, that is, $h_1 = 9$ and $k_1 = 4$. We have $h_1^2 - 5k_1^2 = 1$. Since $k_0 \leqslant k_1 < k_2 < k_3 < \cdots$, the smallest solution is $(9, 4)$.

(c) We proceed by induction on n. For $n = 0$, $(x_0, y_0) = (a, b) = (9, 4)$ is a solution of $x^2 - 5y^2 = 1$ by Part (b). So assume that $n \geqslant 1$, and that $x_{n-1}^2 - 5y_{n-1}^2 = 1$. But then

$$
\begin{aligned}
x_n^2 - 5y_n^2 &= (ax_{n-1} + 5by_{n-1})^2 - 5(bx_{n-1} + ay_{n-1})^2 \\
&= a^2(x_{n-1}^2 - 5y_{n-1}^2) - 5b^2(x_{n-1}^2 - 5y_{n-1}^2) \\
&= a^2 - 5b^2 = 1.
\end{aligned}
$$

80. Let $r_0, r_1, r_2, \ldots, r_j$ be the remainder sequence computed by the Euclidean gcd algorithm with $r_0 = a$, $r_1 = b$, and $r_j = 0$. We assume that $a \geqslant b$ so that $r_0 \geqslant r_1 > r_2 > \cdots > r_{j-1} > r_j$. Let d_i be the bit length of r_i. We then have $d_0 \geqslant d_1 \geqslant d_2 \geqslant \cdots \geqslant d_{j-1} \geqslant d_j$. The crucial observation here is that if some d_i is slightly larger than d_{i+1}, then Euclidean division of r_i by r_{i+1} is quite efficient, but we do not expect to have a huge size reduction in r_{i+2} compared to r_i. On the other hand, if d_i is much larger than d_{i+1}, then Euclidean division of r_i by r_{i+1} takes quite some time, but the size reduction of r_{i+2} compared to r_i is also substantial.

In order to make this observation more precise, we note that Euclidean division of an s-bit integer by a t-bit integer (with $s \geqslant t$) takes time roughly proportional to $t(s-t)$. (In practice, we do word-level operations on multiple-precision integers, and so the running time of a Euclidean division is roughly proportional to $t'(s' - t')$ where s', t' are the word lengths of the operands. But since bit lengths are roughly proportional to word lengths, we may talk about bit lengths only.) Therefore, the running time of the Euclidean gcd loop is roughly proportional to $d_1(d_0 - d_1) + d_2(d_1 - d_2) + d_3(d_2 - d_3) + \cdots + d_{j-1}(d_{j-2} - d_{j-1}) \leqslant d_1[(d_0 - d_1) + (d_1 - d_2) + (d_2 - d_3) + \cdots + (d_{j-2} - d_{j-1})] = d_1(d_0 - d_{j-1}) \leqslant d_0 d_1 \leqslant d_0^2 = O(\lg^2 a)$.

81. A GP/PARI code implementing this search is given below.

```
nprime = 0; nsol = 0;
for (p=2, 10^6, \
   if (isprime(p), \
      nprime++; \
      t = p; s = 0; \
      while (t > 0, s += t % 7; t = floor(t / 7)); \
      if ((s > 1) && (!isprime(s)), \
         nsol++; \
         print("p = ", p, ", S7(p) = ", s); \
      ) \
   ) \
)
print("Total number of primes less that 10^6 is ", nprime);
print("Total number of primes for which S7(p) is composite is ", nsol);
```

This code reveals that among 78498 primes $p < 10^6$, only 13596 lead to composite values of $S_7(p)$. Each of these composite values is either 25 or 35.

Let $p = a_{k-1}7^{k-1} + a_{k-2}7^{k-2} + \cdots + a_1 7 + a_0$ be the base-7 representation of p, so $S_7(p) = a_{k-1} + a_{k-2} + \cdots + a_1 + a_0$. One easily sees that $S_7(p) - p$ is a multiple of 6. It suffices to consider the primes $p > 7$, so p must be of the form $6s \pm 1$, that is, $S_7(p)$ too is of the form $6t \pm 1$. For $p < 10^7$, we have $k \leqslant 8$ (where k is the number of non-zero 7-ary digits of p), that is, $S_7(p) \leqslant 48$. All integers r in the range $1 < r \leqslant 48$ and of the form $6t \pm 1$ are $5, 7, 11, 13, 17, 19, 23, 25, 29, 31, 35, 37, 41, 43, 47$. All of these except 25 and 35 are prime.

82. The following GP/PARI function implements the search for integral values of $(a^2 + b^2)/(ab + 1)$. Since the function is symmetric about the two arguments, we restrict the search to $1 \leqslant a \leqslant b \leqslant B$.

```
getsol (B) = \
   for (a=1, B, \
      for (b=a, B, \
         c = a^2 + b^2; \
         d = a*b + 1; \
         if (c % d == 0, \
            print("a = ", a, ", b = ", b, ", (a^2+b^2)/(ab+1) = ", c / d); \
         ) \
      ) \
   )
```

The observation is that whenever $(a^2 + b^2)/(ab + 1)$ is an integer, that integer is a perfect square. For proving this, let a, b be non-negative integers with $a \geqslant b > 0$ such that $(a^2 + b^2)/(ab + 1) = n \in \mathbb{N}$. But then, we have

$$a^2 - (nb)a + (b^2 - n) = 0.$$

This is a quadratic equation in a (treating b as constant). Let the other solution of this equation be a'. We have $a + a' = nb$ and $aa' = b^2 - n$. The first of these equations implies that $a' = nb - a$ is an integer. Since $(a'^2 + b^2)/(a'b + 1)$ is positive, a' cannot be negative. Moreover, $a' = (b^2 - n)/a \leqslant (a^2 - n)/a < a^2/a = a$. Thus, we can replace the solution (a, b) of $(a^2 + b^2)/(ab + 1) = n$ by a strictly smaller solution (b, a') (with the same n). This process cannot continue indefinitely, that is, we must eventually encounter a pair $(\bar{a}, \bar{b})$ for which $(\bar{a}^2 + \bar{b}^2)/(\bar{a}\bar{b} + 1) = n$ with $\bar{b} = 0$. But then, $n = \bar{a}^2$.

Chapter 2   Arithmetic of Finite Fields

1. We adopt the convention that the degree of the zero polynomial is $-\infty$. For any two polynomials $f(x), g(x)$, we then have $\deg(f(x)g(x)) = \deg f(x) + \deg g(x)$. Moreover, we can include the case $r(x) = 0$ in the case $\deg r(x) < \deg g(x)$.
   (a) Let $m = \deg f(x)$ and $n = \deg g(x)$. Since the result is trivial for $n = 0$ (constant non-zero polynomials $g(x)$), we assume that $n \geqslant 1$, and proceed by

induction on m. If $m < n$, we take $q(x) = 0$ and $r(x) = f(x)$. So consider $m \geqslant n$, and assume that the result holds for all polynomials $f_1(x)$ of degrees $< m$. If a and b are the leading coefficients of f and g, we construct the polynomial $f_1(x) = f(x) - (a/b)x^{m-n}g(x)$. Clearly, $\deg f_1(x) < m$, and so by the induction hypothesis, $f_1(x) = q_1(x)g(x) + r_1(x)$ for some polynomials q_1 and r_1 with $\deg r_1 < \deg g$. But then, $f(x) = (q_1(x) + (a/b)x^{m-n})g(x) + r_1(x)$, that is, we take $q(x) = q_1(x) + (a/b)x^{m-n}$ and $r(x) = r_1(x)$.

In order to prove the uniqueness of the quotient and the remainder polynomials, suppose that $f(x) = q(x)g(x) + r(x) = \bar{q}(x)g(x) + \bar{r}(x)$ with both r and $\bar{r}$ having degrees less than $\deg g$. But then, $(q(x) - \bar{q}(x))g(x) = \bar{r}(x) - r(x)$. If $r \neq \bar{r}$, then the right side is a non-zero polynomial of degree less than n, whereas the left side, if non-zero, is a polynomial of degree $\geqslant n$. This contradiction indicates that we must have $q = \bar{q}$ and $r = \bar{r}$.

(b) Since $r(x) = f(x) - q(x)g(x)$, any common divisor of $f(x)$ and $g(x)$ divides $r(x)$ and so $\gcd(g(x), r(x))$ too. Likewise, $f(x) = q(x)g(x) + r(x)$ implies that any common divisor of $g(x)$ and $r(x)$ divides $f(x)$ and so $\gcd(f(x), g(x))$ too. In particular, $\gcd(f, g) \mid \gcd(g, r)$ and $\gcd(g, r) \mid \gcd(f, g)$. If both these gcds are taken as monic polynomials, they must be equal.

(c) We follow a procedure similar to the Euclidean gcd of integers. We generate three sequences $r_i(x), u_i(x), v_i(x)$ maintaining the invariance $u_i(x)f(x) + v_i(x)g(x) = r_i(x)$ for all $i \geqslant 0$. We initialize the sequences as $r_0(x) = f(x)$, $u_0(x) = 1$, $v_0(x) = 0$, $r_1(x) = g(x)$, $u_1(x) = 0$, $v_1(x) = 1$. Subsequently, for $i = 2, 3, 4, \ldots$, we compute the quotient $q_i(x)$ and $r_i(x)$ of Euclidean division of $r_{i-2}(x)$ by $r_{i-1}(x)$. We also update the u and v sequences as $u_i(x) = u_{i-2}(x) - q_i(x)u_{i-1}(x)$ and $v_i(x) = v_{i-2}(x) - q_i(x)v_{i-1}(x)$. The algorithm terminates, since the r sequence consists of polynomials with strictly decreasing degrees. If j is the smallest index for which $r_j(x) = 0$, then $\gcd(f(x), g(x)) = r_{j-1}(x) = u_{j-1}(x)f(x) + v_{j-1}(x)g(x)$.

(d) Let $d(x) = \gcd(f(x), g(x)) = u(x)f(x) + v(x)g(x)$ for some polynomials u, v. For any polynomial $q(x)$, we have $d(x) = (u(x) - q(x)g(x))f(x) + (v(x) + q(x)f(x))g(x)$. In particular, we can take $q(x) = u(x)$ quot $g(x)$, and assume that $\deg u < \deg g$ in the Bézout relation $d = uf + vg$. But then, $\deg vg = \deg v + \deg g = \deg(d - uf) \leqslant \max(\deg d, \deg uf) = \deg uf = \deg u + \deg f < \deg g + \deg f$, that is, $\deg v < \deg f$.

9. (c) We use a window of size t. For simplicity, t should divide the bit size w of a word. If $w = 32$ or 64, natural choices for t are $2, 4, 8$. For each t-bit pattern $(a_{t-1}a_{t-2} \ldots a_1a_0)$, the $2t$-bit pattern $(0a_{t-1}0a_{t-2} \ldots 0a_10a_0)$ is precomputed and stored in a table of size 2^t. In the squaring loop, t bits of the operand are processed simultaneously. For a t-bit chunk in the operand, the square is read from the precomputed table and XOR-ed with the output with an appropriate shift. Note that the precomputed table is an absolutely constant table, that is, independent of the operand.

10. We first extract the coefficients of x^{255} through x^{233} from γ_3:

$$\mu = \text{RIGHT-SHIFT}(\gamma_3, 41).$$

We then make these bits in γ_3 zero as follows:

γ_3 is AND-ed with the constant integer 0x1FFFFFFFFFFF.

What remains is to add $\mu f_1 = \mu(x^{74} + 1) = x^{64}(x^{10}\mu) + \mu$ to γ. Since μ is a 23-bit value, this is done as follows:

γ_1 is XOR-ed with LEFT-SHIFT$(\mu, 10)$,

γ_0 is XOR-ed with μ.

13. Initialize the u sequence as $u_0 = \beta$ and $u_1 = 0$. The rest of the extended gcd algorithm remains the same. Now, the extended gcd loop maintains the invariance $u_i\beta^{-1}\alpha + v_i f = r_i$ (where f is the defining polynomial). If $r_j = 1$, we have $u_j\beta^{-1}\alpha \equiv 1 \pmod{f}$, that is, $\beta\alpha^{-1} = u_j$.

15. (a) We have $\alpha^{2^{2k}-1} = \alpha^{(2^k-1)(2^k+1)} = (\alpha^{2^k}-1)^{2^k}\alpha^{2^k-1}$. Moreover, $\alpha^{2^{2k+1}-1} = \alpha^{2^{2k+1}-2+1} = (\alpha^{2^{2k}-1})^2\alpha$.

(b) The following algorithm resembles left-to-right exponentiation.

Let $n - 1 = (n_{s-1}n_{s-2}\ldots n_1 n_0)_2$ with $n_{s-1} = 1$.
Initialize $prod = \alpha$ and $k = 1$.
/* Loop for computing $\alpha^{2^{n-1}-1}$ */
For $i = s - 2, s - 3, \ldots, 2, 1, 0$, repeat: {
    /* Here, $k = (n_{s-1}n_{s-2}\ldots n_{i+1})_2$, and $prod = \alpha^{2^k-1}$ */
    Set $t = prod$.   /* Remember α^{2^k-1} */
    For $j = 1, 2, \ldots, k$, set $prod = prod^2$.   /* $prod = (\alpha^{2^k-1})^{2^k}$ */
    Set $prod = prod \times t$.   /* $prod = \alpha^{2^{2k}-1} = (\alpha^{2^k-1})^{2^k}\alpha^{2^k-1}$ */
    Set $k = 2k$.   /* $k = (n_{s-1}n_{s-2}\ldots n_{i+1}0)_2$ */
    If $(n_i = 1)$ {   /* $(n_{s-1}n_{s-2}\ldots n_{i+1}n_i)_2 = (n_{s-1}n_{s-2}\ldots n_{i+1}0)_2 + 1$ */
        Set $prod = prod^2 \times \alpha$ and $k = k + 1$.
    }
}
Return $prod^2$.

(c) Let $N_i = (n_{s-1}n_{s-2}\ldots n_i)_2$. The number of squares (in the field) performed by the loop is $\leqslant (N_{s-1}+N_{s-2}+\cdots+N_1)+(s-1) = \lfloor(n-1)/2^{s-1}\rfloor + \lfloor(n-1)/2^{s-2}\rfloor+\cdots+\lfloor n/2\rfloor+(s-1) \leqslant (n-1)(\frac{1}{2^{s-1}}+\frac{1}{2^{s-2}}+\cdots+\frac{1}{2})+(s-1) \leqslant (n-1)+(s-1) \leqslant n+s$. The number of field multiplications performed by the loop is $\leqslant 2s$. The algorithm of Exercise 2.14(b), on the other hand, performs about n square and n multiplication operations in the field. Since $s \approx \lg n$, the current algorithm is expected to be faster than the algorithm of Exercise 2.14(b) (unless n is too small).

24. Let us represent elements of $\mathbb{F}_{p^n}$ in a basis $\beta_0, \beta_1, \ldots, \beta_{n-1}$. Take an element $\alpha = a_0\beta_0 + a_1\beta_1 + a_2\beta_2 + \cdots + a_{n-1}\beta_{n-1}$ with each $a_i \in \mathbb{F}_p$. We then have $a_i^p = a_i$ for all i. Therefore, $\alpha^p = a_0\beta_0^p + a_1\beta_1^p + a_2\beta_2^p + \cdots + a_{n-1}\beta_{n-1}^p$. If we precompute and store each β_i^p as an $\mathbb{F}_p$-linear combination of $\beta_0, \beta_1, \ldots, \beta_{n-1}$, computing α^p can be finished in $O(n^2)$ time.

If $\beta_0, \beta_1, \beta_2, \ldots, \beta_{n-1}$ constitute a normal basis of $\mathbb{F}_{p^n}$ over $\mathbb{F}_p$ with $\beta_i = \beta^{p^i}$, then we have $\beta_i^p = \beta_{(i+1) \text{ rem } n}$. Therefore, the p-th power exponentiation of $(a_0, a_1, \ldots, a_{n-1})$ is the cyclic shift $(a_{n-1}, a_0, a_1, \ldots, a_{n-2})$. That is, p-th power exponentiation with respect to a normal basis is very efficient.

26. (b) We first write the input operands as $(a_0 + a_1\theta) + (a_2)\theta^2$ and $(b_0 + b_1\theta) + (b_2)\theta^2$. The first level of Karatsuba–Ofman multiplication involves computing the three products $(a_0 + a_1\theta)(b_0 + b_1\theta)$, a_2b_2 and $(a_0 + a_2 + a_1\theta)(b_0 + b_2 + b_1\theta)$, of which only one (a_2b_2) is an $\mathbb{F}_q$-multiplication. Applying a second level of Karatsuba–Ofman multiplication on $(a_0 + a_1\theta)(b_0 + b_1\theta)$ requires three $\mathbb{F}_q$-multiplications: a_0b_0, a_1b_1, and $(a_0 + a_1)(b_0 + b_1)$. Likewise, computing $(a_0 + a_2 + a_1\theta)(b_0 + b_2 + b_1\theta)$ involves three $\mathbb{F}_q$-multiplications: $(a_0 + a_2)(b_0 + b_2)$, a_1b_1, and $(a_0 + a_1 + a_2)(b_0 + b_1 + b_2)$. Finally, note that the product a_1b_1 appears in both the second-level Karatsuba–Ofman multiplications, and needs to be computed only once.

(d) Let us write the input operands as $(a_0 + a_1\theta + a_2\theta^2) + (a_3 + a_4\theta)\theta^3$ and $(b_0 + b_1\theta + b_2\theta^2) + (b_3 + b_4\theta)\theta^3$. In the first level of Karatsuba–Ofman multiplication, we need the three products $(a_0 + a_1\theta + a_2\theta^2)(b_0 + b_1\theta + b_2\theta^2)$ (requiring six $\mathbb{F}_q$-multiplications by Part (b)), $(a_3 + a_4\theta)(b_3 + b_4\theta)$ (requiring three $\mathbb{F}_q$-multiplications by Part (a)), and $((a_0 + a_3) + (a_1 + a_4)\theta + a_2\theta^2)((b_0 + b_3) + (b_1 + b_4)\theta + b_2\theta^2)$ (requiring six $\mathbb{F}_q$-multiplications again by Part (b)). However, the $\mathbb{F}_q$-product a_2b_2 is commonly required in the first and the third of these three first-level products, and needs to be computed only once.

36. (a) The monic linear irreducible polynomials over $\mathbb{F}_4$ are $x, x+1, x+\theta, x+\theta+1$. The products of any two (including repetition) of these polynomials are the reducible monic quadratic polynomials—there are ten of them: x^2, $x^2 + 1$, $x^2 + \theta + 1$, $x^2 + \theta$, $x^2 + x$, $x^2 + \theta x$, $x^2 + (\theta+1)x$, $x^2 + (\theta+1)x + \theta$, $x^2 + \theta x + (\theta+1)$, and $x^2 + x + 1$. The remaining six monic quadratic polynomials are irreducible: $x^2 + x + \theta$, $x^2 + x + (\theta + 1)$, $x^2 + \theta x + 1$, $x^2 + \theta x + \theta$, $x^2 + (\theta + 1)x + 1$, and $x^2 + (\theta + 1)x + (\theta + 1)$.

(b) Let us use the polynomial $x^2 + x + \theta$ to represent $\mathbb{F}_{16}$. That is, $\mathbb{F}_{16} = \mathbb{F}_4(\psi)$, where $\psi^2 + \psi + \theta = 0$. Let us take two elements

$$
\begin{aligned}
\alpha &= (a_3\theta + a_2)\psi + (a_1\theta + a_0), \\
\beta &= (b_3\theta + b_2)\psi + (b_1\theta + b_0)
\end{aligned}
$$

in $\mathbb{F}_{16}$. The formula for their sum is simple:

$$\alpha + \beta = [(a_3 + b_3)\theta + (a_2 + b_2)]\psi + [(a_1 + b_1)\theta + (a_0 + b_0)].$$

The product involves reduction with respect to both θ and ψ.

$$
\begin{aligned}
\alpha\beta &= [(a_3\theta + a_2)(b_3\theta + b_2)]\psi^2 + [(a_3\theta + a_2)(b_1\theta + b_0) + (a_1\theta + a_0)(b_3\theta + b_2)]\psi + \\
&\quad [(a_1\theta + a_0)(b_1\theta + b_0)] \\
&= [(a_3b_3 + a_3b_2 + a_2b_3)\theta + (a_3b_3 + a_2b_2)]\psi^2 + \\
&\quad [(a_3b_1 + a_3b_0 + a_2b_1 + a_1b_3 + a_1b_2 + a_0b_3)\theta + (a_3b_1 + a_2b_0 + a_1b_3 + a_0b_2)]\psi +
\end{aligned}
$$

$$[(a_1b_1+a_1b_0+a_0b_1)\theta+(a_1b_1+a_0b_0)]$$
$$= [(a_3b_3+a_3b_2+a_2b_3)\theta+(a_3b_3+a_2b_2)](\psi+\theta)+$$
$$[(a_3b_1+a_3b_0+a_2b_1+a_1b_3+a_1b_2+a_0b_3)\theta+(a_3b_1+a_2b_0+a_1b_3+a_0b_2)]\psi+$$
$$[(a_1b_1+a_1b_0+a_0b_1)\theta+(a_1b_1+a_0b_0)]$$
$$= \Big[(a_3b_3+a_3b_2+a_3b_1+a_3b_0+a_2b_3+a_2b_1+a_1b_3+a_1b_2+a_0b_3)\theta+$$
$$(a_3b_3+a_3b_1+a_2b_2+a_2b_0+a_1b_3+a_0b_2)\Big]\psi+$$
$$\Big[(a_3b_3+a_3b_2+a_2b_3)\theta^2+(a_3b_3+a_2b_2+a_1b_1+a_1b_0+a_0b_1)\theta+(a_1b_1+a_0b_0)\Big]$$
$$= \Big[(a_3b_3+a_3b_2+a_3b_1+a_3b_0+a_2b_3+a_2b_1+a_1b_3+a_1b_2+a_0b_3)\theta+$$
$$(a_3b_3+a_3b_1+a_2b_2+a_2b_0+a_1b_3+a_0b_2)\Big]\psi+$$
$$\Big[(a_3b_2+a_2b_3+a_2b_2+a_1b_1+a_2b_0+a_0b_1)\theta+(a_3b_3+a_3b_2+a_2b_3+a_1b_1+a_0b_0)\Big]$$

(c) We have $|\mathbb{F}_{16}^*| = 15 = 3 \times 5$, $\psi^3 = (\theta+1)\psi+\theta \neq 1$ and $\psi^5 = \theta \neq 1$, so ψ is a primitive element of $\mathbb{F}_{16}$.

(d) We have the following powers of $\gamma = (\theta+1)\psi+1$:

$$\gamma \;=\; (\theta+1)\psi+1,$$
$$\gamma^2 \;=\; (\theta)\psi+(\theta),$$
$$\gamma^4 \;=\; (\theta+1)\psi+(\theta),$$
$$\gamma^8 \;=\; (\theta)\psi.$$

Thus, the minimal polynomial of γ over $\mathbb{F}_2$ is $(x+\gamma)(x+\gamma^2)(x+\gamma^4)(x+\gamma^8) = x^4 + x^3 + x^2 + x + 1$.

(e) The minimal polynomial of γ over $\mathbb{F}_4$ is $(x+\gamma)(x+\gamma^4) = (x+(\theta+1)\psi+1)(x+(\theta+1)\psi+\theta) = x^2 + (\theta+1)x + 1$.

41. As in Exercise 2.32, we represent $\mathbb{F}_{16} = \mathbb{F}_2(\phi)$ with $\phi^4 + \phi + 1 = 0$. The representation of $\mathbb{F}_{16}$ in Exercise 2.36 is $\mathbb{F}_{16} = \mathbb{F}_2(\theta)(\psi)$, where $\theta^2 + \theta + 1 = 0$ and $\psi^2 + \psi + \theta = 0$. We need to compute the change-of-basis matrix from the polynomial basis $(1, \phi, \phi^2, \phi^3)$ to the composite basis $(1, \theta, \psi, \theta\psi)$. To that effect, we note that ϕ satisfies $x^4 + x + 1 = 0$, and obtain a root of this polynomial in the second representation. Squaring $\psi^2 + \psi + \theta = 0$ gives $\psi^4 + \psi^2 + \theta^2 = 0$, that is, $\psi^4 + (\psi^2 + \psi + \theta) + \psi + (\theta^2 + \theta) = 0$, that is, $\psi^4 + \psi + 1 = 0$. We consider the linear map μ taking ϕ to ψ, and obtain:

$$\mu(1) \;=\; 1,$$
$$\mu(\phi) \;=\; \psi,$$
$$\mu(\phi^2) \;=\; \psi^2 \;=\; \psi+\theta,$$
$$\mu(\phi^3) \;=\; \psi(\psi+\theta) \;=\; \psi^2+\psi\theta \;=\; \theta+\psi+\psi\theta.$$

Therefore, the change-of-basis matrix is

$$T = \begin{pmatrix} 1 & 0 & 0 & 0 \\ 0 & 0 & 1 & 0 \\ 0 & 1 & 1 & 0 \\ 0 & 1 & 1 & 1 \end{pmatrix}$$

42. We iteratively find elements $\beta_0, \beta_1, \ldots, \beta_{n-1}$ to form an $\mathbb{F}_p$-basis of $\mathbb{F}_{p^n}$. Initially, any non-zero element of $\mathbb{F}_{p^n}$ can be taken as β_0, so the number of choices is $p^n - 1$. Now, suppose that i linearly independent elements $\beta_0, \beta_1, \ldots, \beta_{i-1}$ are chosen. The number of all possible $\mathbb{F}_p$-linear combinations of these i elements is exactly p^i. We choose any β_i which is not a linear combination of $\beta_0, \beta_1, \ldots, \beta_{i-1}$, that is, the number of choices for β_i is exactly $p^n - p^i$.

44. Both the parts follow from the following result.

Claim: Let $d = \gcd(m, n)$. Then, g decomposes in $\mathbb{F}_{2^m}[x]$ into a product of d irreducible polynomials each of degree n/d.

Proof Take any root $\alpha \in \overline{\mathbb{F}}_p$ of g. The conjugates of α over $\mathbb{F}_{p^m}$ are α, α^{p^m}, $\alpha^{(p^m)^2}, \ldots, \alpha^{(p^m)^{t-1}}$, where t is the smallest integer for which $\alpha^{(p^m)^t} = \alpha$. On the other hand, $\deg g = n$, and g is irreducible over $\mathbb{F}_p$, implying that $\alpha^{p^k} = \alpha$ if and only if k is a multiple of n. Therefore, $mt \equiv 0 \pmod{n}$. The smallest positive integral solution for t is n/d. That is, the degree of α over $\mathbb{F}_{p^m}$ is exactly n/d. Since this is true for any root of g, the claim is established. ●

53. If α is a t-th power residue, then $\beta^t = \alpha$ for some $\beta \in \mathbb{F}_q^*$. But then, $\alpha^{(q-1)/d} = (\beta^t)^{(q-1)/d} = (\beta^{q-1})^{t/d} = 1$ by Fermat's little theorem.

Proving the converse requires more effort. Let γ be a primitive element in $\mathbb{F}_q^*$. Then, an element γ^i is a t-th power residue if and only if $\gamma^i = (\gamma^y)^t$ for some y, that is, the congruence $ty \equiv i \pmod{q-1}$ is solvable for y, that is, $\gcd(t, q-1) | i$. Thus, the values of $i \in \{0, 1, 2, \ldots, q-2\}$ for which γ^i is a t-th power residue are precisely $0, d, 2d, \ldots, (\frac{q-1}{d} - 1)d$, that is, there are exactly $(q-1)/d$ t-th power residues in $\mathbb{F}_q^*$. All these t-th power residues satisfy $x^{(q-1)/d} = 1$. But then, since $x^{(q-1)/d} - 1$ cannot have more than $(q-1)/d$ roots, no t-th power non-residue can satisfy $x^{(q-1)/d} = 1$.

54. If $q = 2^n$, take $x = 0$ and $y = a^{2^{n-1}}$. So assume that q is odd, and write the given equation as $x^2 = \alpha - y^2$. As y ranges over all values in $\mathbb{F}_q$, the quantity y^2 ranges over a total of $(q+1)/2$ values (zero, and all the quadratic residues), that is, $\alpha - y^2$ too assumes $(q+1)/2$ distinct values. Not all these values can be quadratic non-residues, since there are only $(q-1)/2$ non-residues in $\mathbb{F}_q$.

58. (d) If $\alpha = \gamma^p - \gamma$, then by additivity of the trace function, we have $\text{Tr}(\alpha) = \text{Tr}(\gamma^p) - \text{Tr}(\gamma) = (\gamma^p + \gamma^{p^2} + \gamma^{p^3} + \cdots + \gamma^{p^n}) - (\gamma + \gamma^p + \gamma^{p^2} + \cdots + \gamma^{p^{n-1}}) = 0$, since $\gamma^{p^n} = \gamma$ by Fermat's little theorem.

Conversely, suppose that $\text{Tr}(\alpha) = 0$. It suffices to show that the polynomial $x^p - x - \alpha$ has at least one root in $\mathbb{F}_{p^n}$. Since $x^{p^n} - x$ is the product of all monic linear polynomials in $\mathbb{F}_{p^n}[x]$, the number of roots of $x^p - x - \alpha$ is the degree of the gcd of $x^p - x - \alpha$ with $x^{p^n} - x$. In order to compute this gcd, we compute $x^{p^n} - x$ modulo $x^p - x - \alpha$. But $x^p \equiv x + \alpha \pmod{x^p - x - \alpha}$, so

$$
\begin{aligned}
x^{p^n} - x &\equiv (x+\alpha)^{p^{n-1}} - x \\
&\equiv x^{p^{n-1}} + \alpha^{p^{n-1}} - x \\
&\equiv (x+\alpha)^{p^{n-2}} + \alpha^{p^{n-1}} - x \\
&\equiv x^{p^{n-2}} + \alpha^{p^{n-2}} + \alpha^{p^{n-1}} - x \\
&\equiv \cdots \\
&\equiv x + \alpha + \alpha^p + \alpha^{p^2} + \cdots + \alpha^{p^{n-2}} + \alpha^{p^{n-1}} - x \\
&\equiv \mathrm{Tr}(\alpha) \\
&\equiv 0 \pmod{x^p - x - \alpha}.
\end{aligned}
$$

Therefore, $\gcd(x^p - x - \alpha, x^{p^n} - x) = x^p - x - \alpha$, that is, $\alpha = \gamma^p - \gamma$ for p distinct elements of $\mathbb{F}_{p^n}$.

61. (a) Let $\theta_0, \theta_1, \ldots, \theta_{n-1}$ constitute an $\mathbb{F}_p$-basis of $\mathbb{F}_{p^n}$. Let A_i denote the i-th column of A (for $i = 0, 1, 2, \ldots, n-1$). Suppose that $a_0 A_0 + a_1 A_1 + \cdots + a_{n-1} A_{n-1} = 0$. Let $\alpha = a_0 \theta_0 + a_1 \theta_1 + \cdots + a_{n-1} \theta_{n-1}$. Since $a_i^p = a_i$ for all i, we then have $a_0 \mathrm{Tr}(\theta_i \theta_0) + a_1 \mathrm{Tr}(\theta_i \theta_1) + \cdots + a_{n-1} \mathrm{Tr}(\theta_i \theta_{n-1}) = \mathrm{Tr}(\theta_i (a_0 \theta_0 + a_1 \theta_1 + \cdots + a_{n-1} \theta_{n-1})) = \mathrm{Tr}(\theta_i \alpha) = 0$ for all i. Since $\theta_0, \theta_1, \ldots, \theta_{n-1}$ constitute a basis of $\mathbb{F}_{p^n}$ over $\mathbb{F}_p$, it follows that $\mathrm{Tr}(\beta \alpha) = 0$ for all $\beta \in \mathbb{F}_{p^n}$. If $\alpha \neq 0$, this in turn implies that $\mathrm{Tr}(\gamma) = 0$ for all $\gamma \in \mathbb{F}_{p^n}$. But the polynomial $x + x^p + x^{p^2} + \cdots + x^{p^{n-1}}$ can have at most p^{n-1} roots. Therefore, we must have $\alpha = 0$. But then, by the linear independence of $\theta_0, \theta_1, \ldots, \theta_{n-1}$, we conclude that $a_0 = a_1 = \cdots = a_{n-1} = 0$, that is, the columns of A are linearly independent, that is, $\Delta(\theta_0, \theta_1, \ldots, \theta_{n-1}) \neq 0$.

Conversely, if $\theta_0, \theta_1, \ldots, \theta_{n-1}$ are linearly dependent, then $a_0 \theta_0 + a_1 \theta_1 + \cdots + a_{n-1} \theta_{n-1} = 0$ for some $a_0, a_1, \ldots, a_{n-1} \in \mathbb{F}_p$, not all zero. But then, for all $i \in \{0, 1, 2, \ldots, n-1\}$, we have $a_0 \theta_i \theta_0 + a_1 \theta_i \theta_1 + \cdots + a_{n-1} \theta_i \theta_{n-1} = 0$, that is, $a_0 \mathrm{Tr}(\theta_i \theta_0) + a_1 \mathrm{Tr}(\theta_i \theta_1) + \cdots + a_{n-1} \mathrm{Tr}(\theta_i \theta_{n-1}) = 0$, that is, the columns of A are linearly dependent, that is, $\Delta(\theta_0, \theta_1, \ldots, \theta_{n-1}) = 0$.

(c) Consider the van der Monde matrix

$$
V(\lambda_0, \lambda_1, \ldots, \lambda_{n-1}) =
\begin{pmatrix}
1 & 1 & 1 & \cdots & 1 \\
\lambda_0 & \lambda_1 & \lambda_2 & \cdots & \lambda_{n-1} \\
\lambda_0^2 & \lambda_1^2 & \lambda_2^2 & \cdots & \lambda_{n-1}^2 \\
\vdots & \vdots & \vdots & \cdots & \vdots \\
\lambda_0^{n-1} & \lambda_1^{n-1} & \lambda_2^{n-1} & \cdots & \lambda_{n-1}^{n-1}
\end{pmatrix}.
$$

If $\lambda_i = \lambda_j$, the determinant of this matrix is 0. It therefore follows that

$$
\det V(\lambda_0, \lambda_1, \ldots, \lambda_{n-1}) = \pm \prod_{0 \leqslant i < j \leqslant n-1} (\lambda_i - \lambda_j).
$$

If we take $\theta_i = \theta^i$ in Part (b), we see that $B^{\mathrm{t}} = V(\theta, \theta^p, \theta^{p^2}, \ldots, \theta^{p^{n-1}})$. Finally, $\det B = \det B^{\mathrm{t}}$, and $\det A = (\det B)^2$.

66. The following GP/PARI function accepts as input the element $a(x)$ that we want to invert, the characteristic p, and the defining polynomial $f(x)$.

```
BinInv(a,p,f) = \
    local(r1,r2,u1,u2); \
    r1 = Mod(1,p) * a; r2 = Mod(1,p) * f; \
    u1 = Mod(1,p); u2 = Mod(0,p); \
    while (1, \
        while(polcoeff(r1,0)==Mod(0,p), \
            r1 = r1 / (Mod(1,p) * x); \
            if (polcoeff(u1,0) != Mod(0,p), \
                u1 = u1 - (polcoeff(u1,0) / polcoeff(f,0)) * f \
            ); \
            u1 = u1 / (Mod(1,p) * x); \
            if (poldegree(r1) == 0, return(lift(u1/polcoeff(r1,0)))); \
        ); \
        while(polcoeff(r2,0)==Mod(0,p), \
            r2 = r2 / (Mod(1,p) * x); \
            if (polcoeff(u2,0) != Mod(0,p), \
                u2 = u2 - (polcoeff(u2,0) / polcoeff(f,0)) * f \
            ); \
            u2 = u2 / (Mod(1,p) * x); \
            if (poldegree(r2) == 0, return(lift(u2/polcoeff(r2,0)))); \
        ); \
        if (poldegree(r1) >= poldegree(r2), \
            c = polcoeff(r1,0)/polcoeff(r2,0); r1 = r1 - c*r2; u1 = u1 - c*u2, \
            c = polcoeff(r2,0)/polcoeff(r1,0); r2 = r2 - c*r1; u2 = u2 - c*u1 \
        ) \
    )
```

A couple of calls of this function follow.

```
BinInv(x^6+x^3+x^2+x, 2, x^7+x^3+1)
BinInv(9*x^4+7*x^3+5*x^2+3*x+2, 17, x^5+3*x^2+5)
```

69. First, we write two functions for computing the trace and the norm of $a \in \mathbb{F}_{p^n}$. The characteristic p and the defining polynomial f are also passed to these functions. The extension degree n is determined from f.

```
abstrace(p,f,a) = \
    local(n,s,u); \
    f = Mod(1,p) * f; \
    a = Mod(1,p) * a; \
    n = poldegree(f); \
    s = u = a; \
    for (i=1,n-1, \
        u = lift(Mod(u,f)^p); \
        s = s + u; \
    ); \
    return(lift(s));
```

```
absnorm(p,f,a) = \
    local(n,t,u); \
    f = Mod(1,p) * f; \
    a = Mod(1,p) * a; \
    n = poldegree(f); \
    t = u = a; \
    for (i=1,n-1, \
        u = lift(Mod(u,f)^p); \
        t = (t * u) % f; \
    ); \
    return(lift(t));
```

The following statements print the traces and norms of all elements of $\mathbb{F}_{64} = \mathbb{F}_2(\theta)$, where $\theta^6 + \theta + 1 = 0$.

```
f = x^6 + x + 1;
p = 2;
for (i=0,63, \
    a = 0; t = i; \
    for (j=0, 5, c = t % 2; a = a + c * x^j; t = floor(t/2)); \
    print("a = ", a, ", Tr(a) = ", abstrace(p,f,a), ", N(a) = ", absnorm(p,f,a)) \
)
```

Chapter 3 Arithmetic of Polynomials

2. **(a)** Let $I_{q,n}$ denote the product of all monic irreducible polynomials of degree n. We have $x^{q^n} - x = \prod_{d|n} I_{q,d}$. By the multiplicative form of the Möbius inversion formula, we have

$$I_{q,n} = \prod_{d|n}(x^{q^d} - x)^{\mu(n/d)} = \prod_{d|n}(x^{q^{n/d}} - x)^{\mu(d)}.$$

(b) For $q = 2$ and $n = 6$, we have

$$\begin{aligned}
I_{2,6} &= (x^{2^6} + x)^{\mu(1)}(x^{2^3} + x)^{\mu(2)}(x^{2^2} + x)^{\mu(3)}(x^{2^1} + x)^{\mu(6)} \\
&= \frac{(x^{64} + x)(x^2 + x)}{(x^8 + x)(x^4 + x)} \\
&= x^{54} + x^{53} + x^{51} + x^{50} + x^{48} + x^{46} + x^{45} + x^{43} + x^{42} + x^{33} + \\
&\quad x^{32} + x^{30} + x^{29} + x^{27} + x^{25} + x^{24} + x^{22} + x^{21} + x^{12} + x^{11} + \\
&\quad x^9 + x^8 + x^6 + x^4 + x^3 + x + 1.
\end{aligned}$$

(c) Now, take $q = 4$ and $n = 3$ to obtain:

$$I_{4,3} = (x^{4^3} + x)^{\mu(1)}(x^{4^1} + x)^{\mu(3)} = \frac{x^{64} + x}{x^4 + x}$$

$$= x^{60} + x^{57} + x^{54} + x^{51} + x^{48} + x^{45} + x^{42} + x^{39} + x^{36} + x^{33} +$$
$$x^{30} + x^{27} + x^{24} + x^{21} + x^{18} + x^{15} + x^{12} + x^9 + x^6 + x^3 + 1.$$

8. (a) This follows from Exercise 2.52.

(b) Let $R_\alpha = \{\beta - \alpha \mid \beta$ is a quadratic residue in $\mathbb{F}_q\}$, and $N_\alpha = \{\beta - \alpha \mid \beta$ is a quadratic non-residue in $\mathbb{F}_q\}$. Then, R_α consists of all the roots of $v_\alpha(x)$, and N_α of all the roots of $w_\alpha(x)$. Let γ be any root of f. For a random α, we conclude, under the given reasonable assumptions, that $\gamma = \alpha$ with probability $1/q$, $\gamma \in R_\alpha$ with probability about $1/2$, and $\gamma \in N_\alpha$ with probability about $1/2$ again. The first case ($\gamma = \alpha$) is very unlikely, and can be ignored. Therefore, all the d roots of f are in R_α with probability $1/2^d$, and all are in N_α with probability $1/2^d$. Therefore, the probability that $\gcd(v_\alpha(x), f(x))$ is a non-trivial factor of $f(x)$ is $1 - 2 \times \frac{1}{2^d} = 1 - \frac{1}{2^{d-1}} \geqslant 1/2$ for $d \geqslant 2$.

(c) By Part (b), we need to try a constant number of random elements $\alpha \in \mathbb{F}_q$ before we expect to split f non-trivially. Each trial involves an exponentiation modulo f and a gcd calculation of polynomials of degrees $\leqslant d$. Exactly $d - 1$ splits are necessary to obtain all the roots of f.

9. (a) Assume that $q \gg d/r$, and that the quadratic residues are randomly distributed in $\mathbb{F}_{q^r}^*$. Since q is odd, $\xi \in \mathbb{F}_{q^r}^*$ is a quadratic residue if and only if all its conjugates ξ^{q^i} are. Moreover, if $\alpha \in \mathbb{F}_q$, we have $(\xi - \alpha)^{q^i} = \xi^{q^i} - \alpha$, that is, the conjugates of $\xi - \alpha$ are $\xi^{q^i} - \alpha$.

For $\alpha \in \mathbb{F}_q$, define R_α as the set of all the roots of $(x + \alpha)^{(q^r-1)/2} - 1$ in $\mathbb{F}_{q^r}$, and N_α as the set of all the roots of $(x+\alpha)^{(q^r-1)/2}+1$ in $\mathbb{F}_{q^r}$. If $g(x)$ is an irreducible (over $\mathbb{F}_q$) factor of $f(x)$, then all the roots of $g(x)$ belong to $\mathbb{F}_{q^r}$. Moreover, all these roots are simultaneously present either in R_α or in N_α. Therefore, $g(x)$ is a factor of $\gcd((x + \alpha)^{(q^r-1)/2} - 1, f(x))$ with probability $1/2$, and so the probability of obtaining a non-trivial split of f (for a randomly chosen α) is $1 - 2 \times \frac{1}{2^{d/r}}$ which is $\geqslant 1/2$ for $d \geqslant 2r$.

10. (f) Let $\alpha_1, \alpha_2, \alpha_3, \ldots, \alpha_{q-1}$ be an ordering of the elements of $\mathbb{F}_q^*$. For example, if $\mathbb{F}_{2^n} = \mathbb{F}_2(\theta)$, we can take $\alpha_i = a_{n-1}\theta^{n-1} + a_{n-2}\theta^{n-2} + \cdots + a_1\theta + a_0$, where $(a_{n-1}a_{n-2}\ldots a_1a_0)_2$ is the binary representation of i. We can choose $u(x)$ in Algorithm 3.8 as $\alpha_i x^j$, where j increases in the sequence $1, 3, 5, 7, \ldots$, and for each j, the index i runs from 1 to $q - 1$. (That is, we choose $u(x)$ sequentially as $\alpha_1 x, \alpha_2 x, \ldots, \alpha_{q-1} x, \alpha_1 x^3, \alpha_2 x^3, \ldots, \alpha_{q-1} x^3, \alpha_1 x^5, \alpha_2 x^5, \ldots, \alpha_{q-1} x^5, \alpha_1 x^7, \ldots$ until a non-trivial split is obtained.) Moreover, in a recursive call, the search for $u(x)$ should start from the element in this sequence, that is next to the polynomial that yielded a non-trivial split of $f(x)$. This method is effective for small q. For large values of q, one may select $u(x) = \alpha x^s$ for random $\alpha \in \mathbb{F}_q^*$ and for small odd degrees s.

30. (a) In view of the CRT for polynomials (Exercise 3.28), it suffices to show that $h(x)^q \equiv h(x) \pmod{g(x)}$ has exactly q solutions for $h(x)$ of degrees $< \delta$, where $g(x)$ is an irreducible factor of $f(x)$ of degree δ. Let us represent $\mathbb{F}_{q^\delta}$ by adjoining a root of $g(x)$ to $\mathbb{F}_q$. But then, all solutions for $h(x)$ are those elements γ of $\mathbb{F}_{q^\delta}$ that satisfy $\gamma^q = \gamma$. These are precisely all the elements of $\mathbb{F}_q$. In other words, the only solutions of $h(x)^q \equiv h(x) \pmod{g(x)}$ are $h(x) \equiv \gamma \pmod{g(x)}$, where $\gamma \in \mathbb{F}_q$.

(b) For $r = 0, 1, 2, \ldots, d-1$, write

$$x^{rq} \equiv \beta_{r,0} + \beta_{r,1}x + \beta_{r,2}x^2 + \cdots + \beta_{r,d-1}x^{d-1} \pmod{f(x)}.$$

The condition $h(x)^q \equiv h(x) \pmod{f(x)}$ implies that

$$
\begin{aligned}
&\alpha_0 + \alpha_1 x^q + \alpha_2 x^{2q} + \cdots + \alpha_{d-1}x^{(d-1)q} \\
\equiv\ & \alpha_0(\beta_{0,0} + \beta_{0,1}x + \beta_{0,2}x^2 + \cdots + \beta_{0,d-1}x^{d-1}) + \\
& \alpha_1(\beta_{1,0} + \beta_{1,1}x + \beta_{1,2}x^2 + \cdots + \beta_{1,d-1}x^{d-1}) + \\
& \alpha_2(\beta_{2,0} + \beta_{2,1}x + \beta_{r,2}x^2 + \cdots + \beta_{2,d-1}x^{d-1}) + \\
& \cdots + \\
& \alpha_{d-1}(\beta_{d-1,0} + \beta_{d-1,1}x + \beta_{d-1,2}x^2 + \cdots + \beta_{d-1,d-1}x^{d-1}) \\
\equiv\ & (\alpha_0\beta_{0,0} + \alpha_1\beta_{1,0} + \alpha_2\beta_{2,0} + \cdots + \alpha_{d-1}\beta_{d-1,0}) + \\
& (\alpha_0\beta_{0,1} + \alpha_1\beta_{1,1} + \alpha_2\beta_{2,1} + \cdots + \alpha_{d-1}\beta_{d-1,1})x + \\
& (\alpha_0\beta_{0,2} + \alpha_1\beta_{1,2} + \alpha_2\beta_{2,2} + \cdots + \alpha_{d-1}\beta_{d-1,2})x^2 + \\
& \cdots + \\
& (\alpha_0\beta_{0,d-1} + \alpha_1\beta_{1,d-1} + \alpha_2\beta_{2,d-1} + \cdots + \alpha_{d-1}\beta_{d-1,d-1})x^{d-1} \\
\equiv\ & \alpha_0 + \alpha_1 x + \alpha_2 x^2 + \cdots + \alpha_{d-1}x^{d-1} \pmod{f(x)}.
\end{aligned}
$$

Equating coefficients of x^i for $i = 0, 1, 2, \ldots, d-1$ gives the following linear system over $\mathbb{F}_q$:

$$
\begin{aligned}
\alpha_0\beta_{0,0} + \alpha_1\beta_{1,0} + \alpha_2\beta_{2,0} + \cdots + \alpha_{d-1}\beta_{d-1,0} &= \alpha_0, \\
\alpha_0\beta_{0,1} + \alpha_1\beta_{1,1} + \alpha_2\beta_{2,1} + \cdots + \alpha_{d-1}\beta_{d-1,1} &= \alpha_1, \\
\alpha_0\beta_{0,2} + \alpha_1\beta_{1,2} + \alpha_2\beta_{2,2} + \cdots + \alpha_{d-1}\beta_{d-1,2} &= \alpha_2, \\
&\cdots \\
\alpha_0\beta_{0,d-1} + \alpha_1\beta_{1,d-1} + \alpha_2\beta_{2,d-1} + \cdots + \alpha_{d-1}\beta_{d-1,d-1} &= \alpha_{d-1}.
\end{aligned}
$$

Consider the $d \times d$ matrix Q whose r, s-th element is $\beta_{s,r} - \delta_{r,s}$ for $0 \leqslant r \leqslant d-1$ and $0 \leqslant s \leqslant d-1$, where $\delta_{r,s} = \begin{cases} 1 & \text{if } r = s \\ 0 & \text{otherwise} \end{cases}$ is the Kronecker delta. All the solutions for $h(x)$ in $h(x)^q \equiv h(x) \pmod{f(x)}$ can be obtained by solving

the homogeneous linear system $Q \begin{pmatrix} \alpha_0 \\ \alpha_1 \\ \alpha_2 \\ \vdots \\ \alpha_{d-1} \end{pmatrix} = \begin{pmatrix} 0 \\ 0 \\ 0 \\ \vdots \\ 0 \end{pmatrix}$.

(c) By Part (a), this system has exactly q^t solutions, that is, the nullity of Q is t, and so its rank is $d - t$.

(d) There exists a unique solution (modulo $f(x)$) to the set of congruences: $h(x) \equiv \gamma_1 \pmod{f_1(x)}$, $h(x) \equiv \gamma_2 \pmod{f_2(x)}$, and $h(x) \equiv 0 \pmod{g(x)}$ for any other irreducible factor g of f. Since $\gamma_1, \gamma_2, 0 \in \mathbb{F}_q$, this $h(x)$ belongs to the nullspace V of Q (by Part (a)).

Let $h_1, h_2, \ldots, h_t$ constitute an $\mathbb{F}_q$-basis of V (here h_i are polynomials identified with their coefficient vectors). If $h_i(x) \pmod{f_1(x)}$ and $h_i(x) \pmod{f_2(x)}$ are equal for all $i = 1, 2, \ldots, t$, then for any $\mathbb{F}_q$-linear combination $h(x)$ of $h_1, h_2, \ldots, h_t$, the elements $h(x) \pmod{f_1(x)}$ and $h(x) \pmod{f_2(x)}$ of $\mathbb{F}_q$ are equal. On the contrary, we have $h(x) \in V$ such that $h(x) \equiv 0 \pmod{f_1(x)}$ and $h(x) \equiv 1 \pmod{f_2(x)}$. Therefore, for at least one i, the elements $\gamma_1 = h_i(x) \pmod{f_1(x)}$ and $\gamma_2 = h_i(x) \pmod{f_2(x)}$ must be distinct.

(e) If we choose $\gamma = \gamma_1$ as in Part (d), then $f_1(x) | (h_i(x) - \gamma)$, but $f_2(x) \nmid (h_i(x) - \gamma)$. Therefore, $\gcd(f(x), h_i(x) - \gamma)$ is a non-trivial factor of $f(x)$. However, this γ is not known in advance. If q is small, we try all possible $\gamma \in \mathbb{F}_q$ and all $h_1, h_2, \ldots, h_t$ until all irreducible factors of f are discovered.

Generate the matrix Q as in Part (b).
Compute the nullity t and a basis $h_1, h_2, \ldots, h_t$ of the nullspace V of Q.
Repeat until t irreducible factors of f are discovered: {
    For $i = 1, 2, \ldots, t$ {
        For each $\gamma \in \mathbb{F}_q$ {
            Compute the gcds of reducible factors of $f(x)$ with $h_i(x) - \gamma$.
            Add all the non-trivial factors found to the list of factors of $f(x)$.
            Mark the new factors that are irreducible.

        }

    }
}

---

31. We first express x^{2r} modulo $f(x) = x^8 + x^5 + x^4 + x + 1$ for $r = 0, 1, 2, \ldots, d-1$:

$$
\left.
\begin{aligned}
x^0 &\equiv 1 \\
x^2 &\equiv x^2 \\
x^4 &\equiv x^4 \\
x^8 &\equiv x^5 + x^4 + x + 1 \\
x^{10} &\equiv x^7 + x^6 + x^3 + x^2 \\
x^{12} &\equiv x^6 + x^5 + x^2 + 1 \\
x^{14} &\equiv x^7 + x^5 + x^2 + x + 1
\end{aligned}
\right\} \pmod{x^8 + x^5 + x^4 + x + 1}.
$$

Therefore, the matrix Q is

$$
Q = \begin{pmatrix}
0 & 0 & 0 & 0 & 1 & 0 & 1 & 1 \\
0 & 1 & 0 & 0 & 1 & 0 & 0 & 1 \\
0 & 1 & 1 & 0 & 0 & 1 & 1 & 1 \\
0 & 0 & 0 & 1 & 0 & 1 & 0 & 0 \\
0 & 0 & 1 & 0 & 0 & 0 & 0 & 0 \\
0 & 0 & 0 & 0 & 1 & 1 & 1 & 1 \\
0 & 0 & 0 & 1 & 0 & 1 & 0 & 0 \\
0 & 0 & 0 & 0 & 0 & 1 & 0 & 0
\end{pmatrix}.
$$

The nullspace of Q is generated by the two vectors $\begin{pmatrix} 1 \\ 0 \\ 0 \\ 0 \\ 0 \\ 0 \\ 0 \\ 0 \end{pmatrix}$ and $\begin{pmatrix} 0 \\ 1 \\ 0 \\ 0 \\ 1 \\ 0 \\ 1 \\ 0 \end{pmatrix}$, so $t = 2$,

that is, $f(x)$ has two irreducible factors. These two vectors are identified with the polynomials $h_1(x) = 1$ and $h_2(x) = x^6 + x^4 + x$.

We then calculate $\gcd(h_i(x) - a, f(x))$ for $i = 1, 2$ and $a = 0, 1$. Since $h_1(x) = 1$, each $\gcd(h_1(x) - a, f(x))$ is either 1 or $f(x)$, that is, there is no need to compute these gcds. However, $h_2(x) \neq 1$, and we can successfully split $f(x)$ for $i = 2$ and $a = 0$ as follows.

$$\begin{aligned} f_1(x) &= \gcd(h_2(x), f(x)) = x^5 + x^3 + 1, \\ f_2(x) &= f(x)/f_1(x) = x^3 + x + 1. \end{aligned}$$

36. (i) Let us write $\Phi_n(x) = f(x)g(x)$ with f non-constant and irreducible in $\mathbb{Z}[x]$. Let $\xi \in \mathbb{C}$ be a root of $f(x)$. Choose any prime p that does not divide n. Assume that ξ^p is not a root of f. Since ξ^p is a primitive n-th root of unity, $\Phi_n(\xi^p) = f(\xi^p)g(\xi^p) = 0$. Therefore, $f(\xi^p) \neq 0$ implies that $g(\xi^p) = 0$. But $f(x)$ is the minimal polynomial of ξ, and so $f(x)|g(x^p)$ in $\mathbb{Z}[x]$.

Let $\bar{a}(x)$ denote the modulo-p reduction of a polynomial $a(x) \in \mathbb{Z}[x]$. Since $f(x)|g(x^p)$ in $\mathbb{Z}[x]$, we must have $\bar{f}(x)|\bar{g}(x^p)$ in $\mathbb{F}_p[x]$. We have $\bar{g}(x^p) = \bar{g}(x)^p$ in $\mathbb{F}_p[x]$. Therefore, $\bar{f}(x)|\bar{g}(x)^p$ implies that there exists a common irreducible factor $\bar{h}(x)$ of $\bar{f}(x)$ and $\bar{g}(x)$. This, in turn, implies that $\bar{h}(x)^2|\bar{\Phi}_n(x)$. Moreover, $\Phi_n(x)$ divides $x^n - 1$ in $\mathbb{Z}[x]$, so $\bar{\Phi}_n(x)$ divides $x^n - 1$ in $\mathbb{F}_p[x]$, that is, $\bar{h}(x)^2$ divides $x^n - 1$ in $\mathbb{F}_p[x]$. The formal derivative of $x^n - 1$ is $nx^{n-1} \neq 0$ in $\mathbb{F}_p[x]$ since $p \nmid n$. Therefore, $\gcd(x^n - 1, nx^{n-1}) = 1$, that is, $x^n - 1$ is square-free, a contradiction to the fact that $\bar{h}(x)^2|(x^n - 1)$ in $\mathbb{F}_p[x]$.

This contradiction proves that ξ^p must be a root of $f(x)$. Repeatedly applying this result proves that for all k with $\gcd(k, n) = 1$, ξ^k is again a root of $f(x)$, that is, all primitive n-th roots of unity are roots of $f(x)$.

38. (1) We can convert $\mathrm{Syl}(f, g)$ to $\mathrm{Syl}(g, f)$ by mn number of interchanges of adjacent rows.

(2) Let us write $f(x) = q(x)g(x) + r(x)$ with $\deg r < \deg g$. Write $q(x) = q_{m-n}x^{m-n} + q_{m-n-1}x^{m-n-1} + \cdots + q_1x + q_0$. Subtract q_{m-n} times the $(n+1)$-st row, q_{m-n-1} times the $(n+2)$-nd row, and so on from the first row in order to convert the first row to the coefficients of $r(x)$ treated as a polynomial of formal degree m. Likewise, from the second row subtract q_{m-n} times the $(n+2)$-nd row, q_{m-n-1} times the $(n+3)$-rd row, and so on. This is done for each of the first n rows. We then make mn interchanges of adjacent rows in order to bring the last m rows in the first m row positions. This gives us a matrix of the form $S = \begin{pmatrix} T & U \\ 0_{2n \times (m-n)} & \mathrm{Syl}(g, r) \end{pmatrix}$. Here, T is an $(m - n) \times (m - n)$

upper triangular matrix with each entry in the main diagonal being equal to b_n. Moreover, r is treated as a polynomial of formal degree n. Therefore,

$$\text{Res}(f, g) = (-1)^{mn} \det S = (-1)^{mn} (\det T) \text{Res}(g, r) = (-1)^{mn} b_n^{m-n} \text{Res}(g, r).$$

(3) Clearly, the last three of the given expressions are equal to each other, so it suffices to show that $\text{Res}(f, g)$ is equal to any of these expressions. We assume that $m \geqslant n$ (if not, use Part (1)). We proceed by induction on the actual degree of the second argument. If $\deg g = 0$ (that is, $g(x) = a_0$ is a constant), we have $\text{Res}(f, g) = a_0^m = a_m^0 \prod_{i=1}^m g(\alpha_i)$. Moreover, we also need to cover the case $g = 0$. In this case, we have $\text{Res}(f, g) = 0 = a_m^0 \prod_{i=1}^m g(\alpha_i)$. Now, suppose that $n > 0$. By Part (2), we have $\text{Res}(f, g) = (-1)^{mn} b_n^{m-n} \text{Res}(g, r)$ with r treated as a polynomial of formal degree n. But r is a polynomial of actual degree $\leqslant n - 1$, so the induction assumption is that $\text{Res}(g, r) = b_n^n \prod_{j=1}^n r(\beta_j)$. Since $f = qg + r$ and each β_j is a root of g, we have $r(\beta_j) = f(\beta_j)$. It follows that $\text{Res}(f, g) = (-1)^{mn} b_n^{m-n} b_n^n \prod_{j=1}^n f(\beta_j) = (-1)^{mn} b_n^m \prod_{j=1}^n f(\beta_j)$.

42. Consider the Sylvester matrix S of f and g. The first $n - 1$ rows contain the coefficients of $z^n - 1$, and the last n rows the coefficients of $g(z)$. We make the $n \times (n - 1)$ block at the bottom left corner of S zero by subtracting suitable multiples of the first $n-1$ rows from the last n rows. For example, from the n-th row, we subtract α times the first row, α^p times the second row, $\ldots$, $\alpha^{p^{n-2}}$ times the $(n-1)$-st row. These row operations do not change the determinant of S. The converted matrix is now of the form $T = \begin{pmatrix} I_{n-1} & C \\ 0_{n \times (n-1)} & D \end{pmatrix}$, where I_{n-1} is the $(n-1) \times (n-1)$ identity matrix, and

$$D = \begin{pmatrix} \alpha^{p^{n-1}} & \alpha & \alpha^p & \alpha^{p^2} & \cdots & \alpha^{p^{n-2}} \\ \alpha^{p^{n-2}} & \alpha^{p^{n-1}} & \alpha & \alpha^p & \cdots & \alpha^{p^{n-3}} \\ \alpha^{p^{n-3}} & \alpha^{p^{n-2}} & \alpha^{p^{n-1}} & \alpha & \cdots & \alpha^{p^{n-4}} \\ \vdots & \vdots & \vdots & \vdots & \cdots & \vdots \\ \alpha & \alpha^p & \alpha^{p^2} & \alpha^{p^3} & \cdots & \alpha^{p^{n-1}} \end{pmatrix}.$$

We have $\text{Res}(f, g) = \det S = \det D$. By making $n(n - 1)/2$ exchanges of adjacent rows, we convert D to the matrix B of Exercise 2.61 with $\theta_i = \alpha^{p^i}$. Now, α is a normal element in $\mathbb{F}_{p^n}$ if and only if $\det B \neq 0$, that is, $\det D \neq 0$, that is, $\det S \neq 0$, that is, $\gcd(f, g) = 1$.

43. Let $b_1^*, b_2^*, \ldots, b_n^*$ be the Gram–Schmidt orthogonalization of $b_1, b_2, \ldots, b_n$ (Algorithm 3.10). Let M^* denote the matrix whose columns are $b_1^*, b_2^*, \ldots, b_n^*$. Since $b_i = b_i^* + \sum_{j=1}^{i-1} \mu_{i,j} b_j^*$, we have $M = AM^*$, where

$$A = \begin{pmatrix} 1 & \mu_{2,1} & \mu_{3,1} & \mu_{4,1} & \cdots & \mu_{n,1} \\ 0 & 1 & \mu_{3,2} & \mu_{4,2} & \cdots & \mu_{n,2} \\ 0 & 0 & 1 & \mu_{4,3} & \cdots & \mu_{n,3} \\ \vdots & \vdots & \vdots & \vdots & \cdots & \vdots \\ 0 & 0 & 0 & 0 & \cdots & 1 \end{pmatrix}.$$

But then, $\det M = (\det A)(\det M^*) = \det M^*$. Since the vectors $\mathbf{b}_i^*$ are orthogonal to one another, we have $|\det M^*| = \prod_{i=1}^{n} |\mathbf{b}_i^*|$. Let $\langle \mathbf{x}, \mathbf{y} \rangle$ denote the inner product of the vectors $\mathbf{x}$ and $\mathbf{y}$. We have

$$
\begin{aligned}
|\mathbf{b}_i^*|^2 &= \langle \mathbf{b}_i^*, \mathbf{b}_i^* \rangle \\
&= \langle \mathbf{b}_i - \sum_{j=1}^{i-1} \mu_{i,j} \mathbf{b}_j^*, \mathbf{b}_i^* \rangle \\
&= \langle \mathbf{b}_i, \mathbf{b}_i^* \rangle - \sum_{j=1}^{i-1} \mu_{i,j} \langle \mathbf{b}_j^*, \mathbf{b}_i^* \rangle \\
&= \langle \mathbf{b}_i, \mathbf{b}_i^* \rangle \\
&= \langle \mathbf{b}_i, \mathbf{b}_i - \sum_{j=1}^{i-1} \mu_{i,j} \mathbf{b}_j^* \rangle \\
&= \langle \mathbf{b}_i, \mathbf{b}_i \rangle - \sum_{j=1}^{i-1} \mu_{i,j} \langle \mathbf{b}_i, \mathbf{b}_j^* \rangle \\
&= \langle \mathbf{b}_i, \mathbf{b}_i \rangle - \sum_{j=1}^{i-1} \frac{\langle \mathbf{b}_i, \mathbf{b}_j^* \rangle^2}{\langle \mathbf{b}_j^*, \mathbf{b}_j^* \rangle} \\
&= |\mathbf{b}_i|^2 - \sum_{j=1}^{i-1} \left(\frac{\langle \mathbf{b}_i, \mathbf{b}_j^* \rangle}{|\mathbf{b}_j^*|} \right)^2 \\
&\leqslant |\mathbf{b}_i|^2,
\end{aligned}
$$

that is, $|\mathbf{b}_i^*| \leqslant |\mathbf{b}_i|$ for all i, and so

$$
|\det M| = |\det M^*| = \prod_{i=1}^{n} |\mathbf{b}_i^*| \leqslant \sum_{i=1}^{n} |\mathbf{b}_i|.
$$

50. (a) By the definition of reduced bases (Eqns (3.1) and (3.2)), we have

$$
|\mathbf{b}_i^*|^2 \geqslant \left(\frac{3}{4} - \mu_{i,i-1}^2 \right) |\mathbf{b}_{i-1}^*|^2 \geqslant \frac{1}{2} |\mathbf{b}_{i-1}^*|^2.
$$

Applying this result $i - j$ times shows that for $1 \leqslant j \leqslant i \leqslant n$, we have

$$
|\mathbf{b}_j^*|^2 \leqslant 2^{i-j} |\mathbf{b}_i^*|^2.
$$

Gram–Schmidt orthogonalization gives $\mathbf{b}_j = \mathbf{b}_j^* + \sum_{k=1}^{j-1} \mu_{j,k} \mathbf{b}_k^*$. Since the vectors $\mathbf{b}_1^*, \mathbf{b}_2^*, \dots, \mathbf{b}_n^*$ are orthogonal to one another, we then have:

$$
\begin{aligned}
|\mathbf{b}_j|^2 &= |\mathbf{b}_j^*|^2 + \sum_{k=1}^{j-1} \mu_{j,k}^2 |\mathbf{b}_k^*|^2 \\
&\leqslant |\mathbf{b}_j^*|^2 + \frac{1}{4} \sum_{k=1}^{j-1} |\mathbf{b}_k^*|^2 \qquad \text{[by Eqn (3.1)]}
\end{aligned}
$$

$$\leqslant \ |\mathbf{b}_j^*|^2 + \frac{1}{4}\sum_{k=1}^{j-1}2^{j-k}|\mathbf{b}_j^*|^2 \qquad \text{[proved above]}$$

$$= \ \left(2^{j-2}+\frac{1}{2}\right)|\mathbf{b}_j^*|^2$$

$$\leqslant \ 2^{j-1}|\mathbf{b}_j^*|^2 \qquad\qquad\quad \text{[since } j \geqslant 1]$$

$$\leqslant \ 2^{j-1}2^{i-j}|\mathbf{b}_i^*|^2 \qquad\quad \text{[proved above]}$$

$$= \ 2^{i-1}|\mathbf{b}_i^*|^2.$$

(b) By Hadamard's inequality, $d(L) \leqslant \prod_{i=1}^n |\mathbf{b}_i|$, and by Part (a), $|\mathbf{b}_i| \leqslant 2^{(i-1)/2}|\mathbf{b}_i^*|$. But $\mathbf{b}_1^*, \mathbf{b}_2^*, \ldots, \mathbf{b}_n^*$ form an orthogonal basis of L, so $d(L) = \prod_{i=1}^n |\mathbf{b}_i^*|$. Thus, $\prod_{i=1}^n |\mathbf{b}_i| \leqslant 2^{[0+1+2+\cdots+(n-1)]/2}d(L) = 2^{n(n-1)/4}d(L)$.
(c) By Part (a), $|\mathbf{b}_1| \leqslant 2^{(i-1)}|\mathbf{b}_i^*|$ for all $i = 1, 2, \ldots, n$. Therefore, $|\mathbf{b}_1|^n \leqslant 2^{[0+1+2+\cdots+(n-1)]/2}\prod_{i=1}^n |\mathbf{b}_i^*| = 2^{n(n-1)/4}d(L)$.

51. We can write $\mathbf{x} = \sum_{i=1}^m u_i\mathbf{b}_i = \sum_{i=1}^m u_i^*\mathbf{b}_i^*$ with integers u_i and real numbers u_i^*, where $m \in \{1, 2, \ldots, n\}$ is the largest integer for which $u_m \neq 0$. By the Gram–Schmidt orthogonalization formula, we must have $u_m = u_m^*$, so

$$|\mathbf{x}|^2 \geqslant (u_m^*)^2|\mathbf{b}_m^*|^2 = u_m^2|\mathbf{b}_m^*|^2 \geqslant |\mathbf{b}_m^*|^2.$$

Exercise 3.50(a) gives $|\mathbf{b}_i|^2 \leqslant 2^{m-1}|\mathbf{b}_m^*|^2$ for all i, $1 \leqslant i \leqslant m$. In particular, for $i = 1$, we have

$$|\mathbf{b}_1|^2 \leqslant 2^{m-1}|\mathbf{b}_m^*|^2 \leqslant 2^{m-1}|\mathbf{x}|^2 \leqslant 2^{n-1}|\mathbf{x}|^2.$$

Now, take any n linearly independent vectors $\mathbf{x}_1, \mathbf{x}_2, \ldots, \mathbf{x}_n$ in L. Let m_j be the value of m for $\mathbf{x}_j$. As in the last paragraph, we can prove that

$$|\mathbf{b}_i|^2 \leqslant 2^{n-1}|\mathbf{x}_j|^2$$

for all i, $1 \leqslant i \leqslant m_j$. Since $\mathbf{x}_1, \mathbf{x}_2, \ldots, \mathbf{x}_n$ are linearly independent, we must have $m_j = n$ for at least one j. For this j, we have $|\mathbf{b}_i|^2 \leqslant 2^{n-1}|\mathbf{x}_j|^2$ for all i in the range $1 \leqslant i \leqslant n$.

57. The function `EDF()` takes four arguments: the polynomial $f(x)$, the prime modulus p, the degree r of each irreducible factor of f, and a bound B. This bound dictates the maximum degree of $u(x)$ used in Algorithm 3.7. The choice $B = 1$ corresponds to Algorithm 3.5.

```
EDF(f,p,r,B) = \
   local(u,d,i,g,h,e); \
   f = Mod(1,p) * f; \
   if (poldegree(f) == 0, return; ); \
   if (poldegree(f) == r, print("Factor found: ", lift(f)); return; ); \
   e = (p^r - 1) / 2; \
   while (1, \
      u = Mod(0,p); \
      d = 1 + random() % B; \
      for (i=0, d, u = u + Mod(random(),p) * x^i ); \
```

```
    u = Mod(u,f); g = u^e; \
    h = gcd(lift(g)-Mod(1,p),f); h = h / polcoeff(h,poldegree(h)); \
    if ((poldegree(h) > 0) && (poldegree(h) < poldegree(f)), \
        EDF(h,p,r,B); \
        EDF(f/h,p,r,B); \
        return; \
    ); \
);

EDF(x^8 + x^7 + 2*x^6 + 2*x^5 + x^4 + 2*x^3 + 2*x^2 + x + 1, 3, 4, 2)
EDF(x^6 + 16*x^5 + 3*x^4 + 16*x^3 + 8*x^2 + 8*x + 14, 17, 2, 1)
```

---

58. Following the recommendation of Exercise 3.8(f), we choose $u(x)$ for Algorithm 3.8 in the sequence $x, x^3, x^5, x^7, \ldots$. We pass the maximum degree $2d+1$ such that x^{2d+1} has already been tried as $u(x)$. The next recursive call starts with $u(x) = x^{2d+3}$. The outermost call should pass -1 as d.

```
EDF2(f,r,d) = \
    local(u,s,i,h); \
    f = Mod(1,2) * f; \
    if (poldegree(f) == 0, return); \
    if (poldegree(f) == r, print("Factor found: ", lift(f)); return); \
    while (1, \
        d = d + 2; u = Mod(1,2) * x^d; s = u; \
        for (i=1,r-1, u= u^2 % f; s = s + u; ); \
        h = gcd(f,s); \
        if ((poldegree(h) > 0) && (poldegree(h) < poldegree(f)), \
            EDF2(h,r,d); \
            EDF2(f/h,r,d); \
            return; \
        ); \
    );

EDF2(x^18+x^17+x^15+x^11+x^6+x^5+1, 6, -1)
EDF2(x^20+x^18+x^17+x^16+x^15+x^12+x^10+x^9+x^7+x^3+1, 5, -1)
```

---

59. The following GP/PARI functions implement Berlekamp's Q-matrix factorization over a prime field $\mathbb{F}_p$. The first function computes Q, its nullity t, and a basis of the nullspace of Q. It then calls the second function in order to compute $\gcd(h_i(x) - \gamma, f)$ for $i = 1, 2, \ldots, t$ and for $\gamma \in \mathbb{F}_p$. The functions assume that $f(x)$ is square-free. The computation of the d powers $x^{rp} \pmod{f(x)}$ can be optimized further. Two sample calls of BQfactor() follow.

```
BQfactor(f,p) = \
    f = Mod(1,p) * f; d = poldegree(f); \
    Q = matrix(d,d); for (i=0,d-1, Q[i+1,1] = Mod(0,p)); \
    for (r=1,d-1, \
        h = lift(lift(Mod(Mod(1,p)*x,f)^(r*p))); \
        for (i=0,d-1, Q[i+1,r+1] = Mod(polcoeff(h,i),p)); \
        Q[r+1,r+1] = Q[r+1,r+1] - Mod(1,p); \
    ); \
```

```
    V = matker(Q); t = matsize(V)[2]; \
    if (t==1, print(lift(f)); return); \
    decompose(V,t,f,p,d);

decompose(V,t,f,p,d) = \
    d = poldegree(f); \
    for (i=1,t, \
        h = 0; \
        for (j=0,d-1, h = h + Mod(V[j+1,i],p) * x^j); \
        for (a=0,p-1, \
            f1 = gcd(h-Mod(a,p), f); \
            d1 = poldegree(f1); \
            f1 = f1 / polcoeff(f1,d1); \
            if ((d1 > 0) && (d1 < d), \
                if (polisirreducible(f1), \
                    print(lift(f1)), \
                    decompose(V,t,f1,p,d); \
                ); \
                f2 = f / f1; \
                if (polisirreducible(f2), \
                    print(lift(f2)), \
                    decompose(V,t,f2,p,d); \
                ); \
                return; \
            ); \
        ); \
    );

BQfactor(x^8+x^5+x^4+x+1, 2)
BQfactor(x^8+x^5+x^4+2*x+1, 3)
```

60. We first write a function `polyliftonce()` to lift a factorization $f = gh$ modulo p^n to a factorization modulo p^{n+1}. This function is called multiple times by `polylift()` in order to lift the factorization modulo p to the factorization modulo p^n. Sample calls are also supplied to demonstrate the working of these functions on the data of Example 3.34.

```
polyliftonce(f,g,h,p,n) = \
    local(q,i,j,w,A,c,b,r,s,t); \
    q = p^n; \
    w = (f - g * h) / q; \
    r = poldegree(g); s = poldegree(h); t = poldegree(f); \
    b = matrix(t+1,1); A = matrix(t+1,t+1); \
    for(i=0, t, b[t-i+1,1] = Mod(polcoeff(w,i),p)); \
    for (j=0, s, \
        for (i=0, r, \
            A[i+j+1,j+1] = Mod(polcoeff(g,r-i),p); \
        ); \
    ); \
    for (j=0, r-1, \
        for (i=0, s, \
            A[i+j+2,s+j+2] = Mod(polcoeff(h,s-i),p); \
        ); \
    ); \
```

```
    c = A^(-1) * b; \
    u = 0; v = 0; \
    for (i=0, r-1, u = u + lift(c[s+i+2,1]) * x^(r-i-1)); \
    for (i=0, s, v = v + lift(c[i+1,1]) * x^(s-i)); \
    return([g+q*u,h+q*v]);

polylift(f,g,h,p,n) = \
    local(i,j,q,L); \
    q = p; \
    for (i=1, n-1, \
        L = polyliftonce(f,g,h,p,i); \
        q = q * p; \
        g = lift(L[1]); \
        for (j=0, poldegree(g), if (polcoeff(g,j) > q/2, g = g - q * x^j) ); \
        h = lift(L[2]); \
        for (j=0, poldegree(h), if (polcoeff(h,j) > q/2, h = h - q * x^j) ); \
    ); \
    return([g,h]);

polylift(35*x^5-22*x^3+10*x^2+3*x-2, x^2+2*x-2, -4*x^3-5*x^2+6*x+1, 13, 1)
polylift(35*x^5-22*x^3+10*x^2+3*x-2, x^2+2*x-2, -4*x^3-5*x^2+6*x+1, 13, 2)
polylift(35*x^5-22*x^3+10*x^2+3*x-2, x^2+2*x-2, -4*x^3-5*x^2+6*x+1, 13, 3)
```

Chapter 4   Arithmetic of Elliptic Curves

9. Let us use the equation $Y^2 = f(X)$ for the curve, where $f(X) = X^3 + aX^2 + bX + c$. The condition $3P = \mathcal{O}$ is equivalent to the condition $2P = -P$. Therefore, if $P = (h, k)$ is a (finite) point of order three, we have $x(2P) = x(-P) = x(P)$. By Exercise 4.7, we then have $h^4 - 2bh^2 - 8ch + (b^2 - 4ac) = 4h(h^3 + ah^2 + bh + c)$, that is,

$$\psi(h) = 3h^4 + 4ah^3 + 6bh^2 + 12ch + (4ac - b^2) = 0.$$

This is a quartic equation in h alone, and can have at most four roots for h. For each root h, we obtain at most two values of k satisfying $k^2 = h^3 + ah^2 + bh + c$. Thus, there are at most eight points of order three on the curve.

If the field K is algebraically closed, the above quartic equation has exactly four roots. For each such root, we cannot get $k = 0$, because a point of the form $(h, 0)$ is of order two. We also need to argue that the roots of the quartic equation cannot be repeated. To that effect, we compute the discriminants:

$$\mathrm{Discr}(f) = -4a^3c + a^2b^2 + 18abc - 4b^3 - 27c^2,$$

$$\mathrm{Discr}(\psi) = 2^8 \times 3^3 \Big(-16a^6c^2 + 8a^5b^2c - a^4b^4 + 144a^4bc^2 - 68a^3b^3c - 216a^3c^3 +$$

$$8a^2b^5 - 270a^2b^2c^2 + 144ab^4c + 972abc^3 - 16b^6 - 216b^3c^2 - 729c^4 \Big)$$

$$= -2^8 \times 3^3 \times \big(\mathrm{Discr}(f) \big)^2.$$

If char $K \neq 2, 3$, we conclude that ψ has multiple roots if and only if f has multiple roots. (Note that $\psi(h) = 2f(h)f''(h) - f'(h)^2$ and $\psi'(h) = 12f(h)$, so a common root of ψ and ψ' is also a common root of f and f', and conversely.)

15. If we substitute $X \equiv 0 \pmod{p}$, we get a unique solution for Y, namely, $Y \equiv 0 \pmod{p}$. Now, substitute a value of $X \not\equiv 0 \pmod{p}$. If $X^2 + a \equiv 0 \pmod{p}$, then each of the two values $\pm X$ gives the unique solution $Y \equiv 0 \pmod{p}$. So assume that $X^2 + a \not\equiv 0 \pmod{p}$. We have $\left(\frac{X^3 + aX}{p}\right) = \left(\frac{X}{p}\right)\left(\frac{X^2 + a}{p}\right)$.
Since $p \equiv 3 \pmod{4}$, $\left(\frac{-X}{p}\right) = -\left(\frac{X}{p}\right)$. Moreover, $\left(\frac{X^2 + a}{p}\right) = \left(\frac{(-X)^2 + a}{p}\right)$.
Therefore, one of the two values $\pm X$ yields two solutions for Y, and the other no solution for Y.

22. The only $\mathbb{F}_2$-rational points on $Y^2 + Y = X^3 + X$ are $\mathcal{O}$, $(0,0)$, $(0,1)$, $(1,0)$ and $(1,1)$, so the trace of Frobenius at 2 is -2. The roots of $W^2 + 2W + 2 = 0$ are $-1 \pm i$. Therefore, the size of the group of this curve over $\mathbb{F}_{2^n}$ is

$$
\begin{aligned}
S(n) &= 2^n + 1 - ((-1 + i)^n + (-1 - i)^n) \\
&= 2^n + 1 - (-1)^n(\sqrt{2})^n \left(e^{-\frac{in\pi}{4}} + e^{\frac{in\pi}{4}}\right) \\
&= 2^n + 1 - (-1)^n 2^{\frac{n}{2}+1} \cos\frac{n\pi}{4}.
\end{aligned}
$$

On the curve $Y^2 + Y = X^3 + X + 1$, the only $\mathbb{F}_2$-rational point is $\mathcal{O}$, that is, the trace of Frobenius at 2 is 2. The roots of $W^2 - 2W + 2 = 0$ are $1 \pm i$. As above, we deduce that the size of the elliptic-curve group over $\mathbb{F}_{2^n}$ is

$$
T(n) = 2^n + 1 - 2^{\frac{n}{2}+1} \cos\frac{n\pi}{4}.
$$

The following table lists the values of $S(n)$ and $T(n)$ for different values of n modulo 8.

$n \pmod 8$	$S(n)$	$T(n)$
0	$2^n + 1 - 2^{\frac{n}{2}+1}$	$2^n + 1 - 2^{\frac{n}{2}+1}$
1	$2^n + 1 + 2^{\frac{n+1}{2}}$	$2^n + 1 - 2^{\frac{n+1}{2}}$
2	$2^n + 1$	$2^n + 1$
3	$2^n + 1 - 2^{\frac{n+1}{2}}$	$2^n + 1 + 2^{\frac{n+1}{2}}$
4	$2^n + 1 + 2^{\frac{n}{2}+1}$	$2^n + 1 + 2^{\frac{n}{2}+1}$
5	$2^n + 1 - 2^{\frac{n+1}{2}}$	$2^n + 1 + 2^{\frac{n+1}{2}}$
6	$2^n + 1$	$2^n + 1$
7	$2^n + 1 + 2^{\frac{n+1}{2}}$	$2^n + 1 - 2^{\frac{n+1}{2}}$

For each $n \in \mathbb{N}$, the trace is 0, $\pm 2^{\frac{n}{2}+1}$ (only if n is even), or $\pm 2^{\frac{n+1}{2}}$ (only if n is odd). The trace is divisible by 2 in all these cases. So over all extensions $\mathbb{F}_{2^n}$, the curves $Y^2 + Y = X^3 + X$ and $Y^2 + Y = X^3 + X + 1$ are supersingular.

27. Let us use the simplest form of Weierstrass equation:

$$
E : Y^2 = X^3 + aX + b.
$$

Let all of $P = (h_1, k_1)$, $Q = (h_2, k_2)$, $P + Q = (h_3, k_3)$ and $2P + Q = (h_4, k_4)$ be finite points. If we compute $P + Q$ first, we need to compute the slope λ_1 of the line passing through P and Q. Next, we need to compute the slope λ_2 of the line passing through P and $P + Q$. We have:

$$\lambda_1 = \frac{k_2 - k_1}{h_2 - h_1}, \text{ and}$$

$$\lambda_2 = \frac{k_3 - k_1}{h_3 - h_1} = \frac{\lambda_1(h_1 - h_3) - k_1 - k_1}{h_3 - h_1} = -\lambda_1 - \frac{2k_1}{h_3 - h_1}.$$

For computing λ_2 (and subsequently h_4 and k_4), we, therefore, do not need the value of k_3. That is, we can avoid the computation of $k_3 = \lambda_1(h_1 - h_3) - k_1$.

Addition, subtraction and multiplication by two and three being efficient (linear-time) in the field size, let us concentrate on the times of multiplication (M), squaring (S) and inversion (I). Each addition of distinct finite points takes time (about) $1M + 1S + 1I$, and each doubling of a finite point takes time $1M + 2S + 1I$. Therefore, the computation of $(P + P) + Q$ takes time $2M + 3S + 2I$. On the contrary, the computation of $(P + Q) + P$ takes time $2M + 2S + 2I$. Moreover, if the intermediate y-coordinate is not computed, one multiplication is saved, and the computation of $(P + Q) + P$ takes time $1M + 2S + 2I$.

29. Let us write the equation of the curve as

$$C : f_d(X, Y) + f_{d-1}(X, Y) + \cdots + f_1(X, Y) + f_0(X, Y) = 0,$$

where $f_i(X, Y)$ the sum of all non-zero terms of degree i in $f(X, Y)$. The homogenization of C is then

$$C^{(h)} : f_d(X, Y) + Z f_{d-1}(X, Y) + \cdots + Z^{d-1} f_1(X, Y) + Z^d f_0(X, Y) = 0.$$

In order to find the points at infinity on $C^{(h)}$, we put $Z = 0$ and get $f_d(X, Y) = 0$. Let us write this equation as

$$c_d X^d + c_{d-1} X^{d-1} Y + c_{d-2} X^{d-2} Y^2 + \cdots + c_1 XY^{d-1} + c_0 Y^d = 0,$$

where $c_i \in K$ are not all zero. If c_0 is the only non-zero coefficient, we have $Y^d = 0$, and the only point at infinity on the curve is $[1, 0, 0]$. If $c_i \neq 0$ for some $i > 0$, we rewrite this equation as

$$c_d (X/Y)^d + c_{d-1} (X/Y)^{d-1} + c_{d-2} (X/Y)^{d-2} + \cdots + c_1 (X/Y) + c_0 = 0.$$

This is a univariate equation having roots $\alpha_1, \alpha_2, \ldots, \alpha_t$ with $t \leq d$. For each root α_i, we have $X/Y = \alpha_i$, that is, $X = \alpha_i Y$, that is, $[\alpha_i, 1, 0]$ is a point at infinity on the curve.

30. (a) We have

$$\lambda = \frac{\frac{k_2}{l_2} - \frac{k_1}{l_1}}{\frac{h_2}{l_2} - \frac{h_1}{l_1}} = \frac{k_2 l_1 - k_1 l_2}{h_2 l_1 - h_1 l_2}.$$

Therefore,

$$\frac{h}{l} = \lambda^2 - \frac{h_1}{l_1} - \frac{h_2}{l_2} = \frac{l_1 l_2 (k_2 l_1 - k_1 l_2)^2 - (h_2 l_1 - h_1 l_2)^2 (h_1 l_2 + h_2 l_1)}{l_1 l_2 (h_2 l_1 - h_1 l_2)^2},$$

and

$$\frac{k}{l} = \lambda \left(\frac{h_1}{l_1} - \frac{h}{l} \right) - \frac{k_1}{l_1}.$$

Substituting the values of λ and h/l gives an explicit expression for k/l. These expressions are too clumsy. Fortunately, there are many common subexpressions used in these formulas. Computing these intermediate subexpressions allows us to obtain h, k, l as follows:

$$
\begin{aligned}
T_1 &= k_2 l_1 - k_1 l_2, \\
T_2 &= h_2 l_1 - h_1 l_2, \\
T_3 &= T_2^2, \\
T_4 &= T_2 T_3, \\
T_5 &= l_1 l_2 T_1^2 - T_4 - 2 h_1 l_2 T_3, \\
h &= T_2 T_5, \\
k &= T_1 (h_1 l_2 T_3 - T_5) - k_1 l_2 T_4, \\
l &= l_1 l_2 T_4.
\end{aligned}
$$

These expressions can be further optimized by using temporary variables to store the values of $h_1 l_2$, $k_1 l_2$ and $l_1 l_2$.

(b) We proceed as in the case addition. Here, I present only the final formulas.

$$
\begin{aligned}
T_1 &= 3 h_1^2 + a l_1^2, \\
T_2 &= k_1 l_1, \\
T_3 &= h_1 k_1 T_2, \\
T_4 &= T_1^2 - 8 T_3, \\
T_5 &= T_2^2, \\
h' &= 2 T_2 T_4, \\
k' &= T_1 (4 T_3 - T_4) - 8 k_1^2 T_5, \\
l' &= 8 T_2 T_5.
\end{aligned}
$$

(c) Computing the affine coordinates requires a division in the field. If this division operation is much more expensive than multiplication and squaring in the field, avoiding this operation inside the loop (but doing it only once after the end of the loop) may speed up the point-multiplication algorithm. However, projective coordinates increase the number of multiplication (and squaring) operations substantially. Therefore, it is not clear whether avoiding one division in the loop can really provide practical benefits. Implementers report contradictory results in the literature. The practical performance heavily depends on the library used for the field arithmetic.

31. The second point can be treated as having projective coordinates $[h_2, k_2, 1]$, that is, $l_2 = 1$. The formulas in Exercise 4.30(a) can be used with $l_2 = 1$. This saves the three multiplications $h_1 l_2$, $k_1 l_2$ and $l_1 l_2$.

In the conditional addition part of the point-multiplication algorithm, the second summand is always the base point P which is available in affine coordinates. Inside the loop, we always keep the sum S in projective coordinates. However, the computation of $S + P$ benefits from using mixed coordinates.

32. (b) Let $[h, k, l]_{c,d}$ be a finite point on C. We have $l \neq 0$, and so $[h, k, l]_{c,d} \sim [h/l^c, k/l^d, 1]_{c,d}$. Therefore, we identify $[h, k, l]_{c,d}$ with the point $(h/l^c, k/l^d)$. Conversely, to a point (h, k) in affine coordinates, we associate the point $[h, k, 1]_{c,d}$ with projective coordinates. It is easy to verify that these associations produce a bijection of all finite points on C with all points on $C^{(c,d)}$ with non-zero Z-coordinates.

(c) We need to set $Z = 0$ in order to locate the points at infinity. This gives us the polynomial $g(X, Y) = f^{(c,d)}(X, Y, 0)$. In general, g is not a homogeneous polynomial in the standard sense. However, if we give a weight of c to X and a weight of d to Y, each non-zero term in g is of the same total weight. Let $X^i Y^j$ and $X^{i'} Y^{j'}$ have the same weight, that is, $ci + dj = ci' + dj'$, that is, $c(i - i') = d(j' - j)$. For the sake of simplicity, let us assume that $\gcd(c, d) = 1$ (the case $\gcd(c, d) > 1$ can be gracefully handled). But then, $i \equiv i' \pmod{d}$ and $j \equiv j' \pmod{c}$. In view of this, we proceed as follows.

If X divides g, we get the point $[0, 1, 0]_{c,d}$ at infinity. If Y divides g, we get the point $[1, 0, 0]_{c,d}$ at infinity. We remove all factors of X and Y from g, and assume that g is divisible by neither X nor Y. By the argument in the last paragraph, we can write $g(X, Y) = h(X^d, Y^c)$. Call $X' = X^d$ and $Y' = Y^c$. Then, $h(X', Y')$ is a homogeneous polynomial in X', Y'. We find all the roots for X'/Y' from this equation. Each root α corresponds to a point $\mathcal{O}_\alpha$ at infinity on the curve. Since $\gcd(c, d) = 1$, we have $uc + vd = 1$ for some integers u, v. The choices $X = \alpha^v$ and $Y = \alpha^{-u}$ are consistent with $X' = X^d$, $Y' = Y^c$ and $X'/Y' = \alpha$, so we take $\mathcal{O}_\alpha = [\alpha^v, \alpha^{-u}, 0]_{c,d}$. There is a small catch here, namely, the values u, v in Bézout's relation are not unique. Given any solution u, v of $uc + vd = 1$, all solutions are given by $(u + rd)c + (v - rc)d = 1$ for any $r \in \mathbb{Z}$. But then, $[\alpha^{v-rc}, \alpha^{-(u+rd)}, 0]_{c,d} = [\frac{\alpha^v}{(\alpha^r)^c}, \frac{\alpha^{-u}}{(\alpha^r)^d}, 0]_{c,d} = [\alpha^v, \alpha^{-u}, 0]_{c,d}$, that is, the point $\mathcal{O}_\alpha$ does not depend on the choice of the pair (u, v).

33. (a) Substituting X by X/Z^2 and Y by Y/Z^3 in the equation of the curve and multiplying by Z^6, we obtain

$$E^{(2,3)} : Y^2 = X^3 + aXZ^4 + bZ^6.$$

(b) Put $Z = 0$ to get $X^3 - Y^2 = 0$. Now, set $X' = X^3$ and $Y' = Y^2$ to get $X' - Y' = 0$. The only root of this equation is $X'/Y' = 1$. Therefore, the point at infinity on $E^{(2,3)}$ is $[1, 1, 0]_{2,3}$ (since 1 raised to any integral power is again 1, we do not need to compute Bézout's relation involving 2, 3).

(c) The point $[h, k, l]_{2,3}$ (with $l \neq 0$) has affine coordinates $(h/l^2, k/l^3)$. The opposite of this point is $(h/l^2, -k/l^3)$, that is, the point $[h/l^2, -k/l^3, 1]_{2,3} = [h, -k, l]_{2,3} = [h, k, -l]_{2,3}$.

(d) Let $P_1 = [h_1, k_1, l_1]_{2,3}$ and $P_2 = [h_2, k_2, l_2]_{2,3}$ be two finite points on E with $P_1 \neq \pm P_2$. In order to compute the sum $P_1 + P_2 = [h, k, l]_{2,3}$, we compute the sum of the points $(h_1/l_1^2, k_1/l_1^3)$ and $(h_2/l_2^2, k_2/l_2^3)$. We proceed as in Exercise 4.30(a), and obtain the following formulas.

$$
\begin{aligned}
T_1 &= h_1 l_2^2, \\
T_2 &= h_2 l_1^2, \\
T_3 &= k_1 l_2^3, \\
T_4 &= k_2 l_1^3, \\
T_5 &= T_2 - T_1, \\
T_6 &= T_4 - T_3, \\
h &= -T_5^3 - 2T_1 T_5^2 + T_6^2, \\
k &= -T_3 T_5^3 + T_6(T_1 T_5^2 - h), \\
l &= l_1 l_2 T_5.
\end{aligned}
$$

(e) The double of $[h, k, l]_{c,d}$ is the point $[h', k', l']_{c,d}$ computed as follows.

$$
\begin{aligned}
T_1 &= 4hk^2, \\
T_2 &= 3h^2 + al^4, \\
h' &= -2T_1 + T_2^2, \\
k' &= -8k^4 + T_2(T_1 - h'), \\
l' &= 2kl.
\end{aligned}
$$

41. Since $x^2 - y^2 = 1$ on the curve, we have

$$
\begin{aligned}
& 2y^4 - 2y^3 x - y^2 + 2yx - 1 \\
={} & (2y^4 - 2y^2) - (2y^3 x - 2yx) + (y^2 - 1) \\
={} & (y^2 - 1)(2y^2 - 2yx + 1) \\
={} & (y - 1)(y + 1)(y - x)^2,
\end{aligned}
$$

and

$$
\begin{aligned}
& y^2 + yx + y + x + 1 \\
={} & x^2 + yx + x + y \\
={} & (x + 1)(y + x).
\end{aligned}
$$

Therefore,

$$
R(x, y) = \frac{(y - 1)(y + 1)(y - x)^2}{(x + 1)(y + x)}.
$$

Zeros of $y - 1$: The line $y - 1 = 0$ cuts the hyperbola at $P_1 = (\sqrt{2}, 1)$ and $P_2 = (-\sqrt{2}, 1)$. We can take $y - 1$ as a uniformizer at each of these two points, and conclude that P_1 and P_2 are simple zeros of R.

Zeros of $y + 1$: The line $y + 1 = 0$ cuts the hyperbola at $P_3 = (\sqrt{2}, -1)$ and $P_4 = (-\sqrt{2}, -1)$. We can take $y + 1$ as a uniformizer at each of these two points, and conclude that P_3 and P_4 are simple zeros of R.

Zeros of $x + 1$: The line $x + 1 = 0$ touches the hyperbola at $P_5 = (-1, 0)$. At this point, the non-tangent line y can be taken as the uniformizer. We have

$$R(x, y) = y^{-2} \left[\frac{(y - 1)(y + 1)(y - x)^2 (x - 1)}{(y + x)} \right].$$

Thus, P_5 is a pole of R of multiplicity two.

Zeros of $(y - x)^2$: The line $y - x = 0$ does not meet the curve at any finite point. However, it touches the curve at $\mathcal{O}_1 = [1, 1, 0]$, one of its point at infinity. The vertical line $x = \infty$ (or $1/x = 0$) meets but is not tangential to the hyperbola at $\mathcal{O}_1$. Thus, $1/x$ can be taken as a uniformizer at $\mathcal{O}_1$. But

$$R(x, y) = \frac{(y - 1)(y + 1)(y^2 - x^2)^2}{(x + 1)(y + x)^3} = (1/x)^2 \left[\frac{(\frac{y}{x} - \frac{1}{x})(\frac{y}{x} + \frac{1}{x})(-1)^2}{(1 + \frac{1}{x})(\frac{y}{x} + 1)^3} \right].$$

At $\mathcal{O}_1$, we have $y/x = 1$ and $1/x = 0$. It follows that $\mathcal{O}_1$ is a zero of R of multiplicity two.

Zeros of $y + x$: The only intersection of the line $y + x$ with the hyperbola is at $\mathcal{O}_2$ (the second point at infinity on the curve). We can again take $1/x$ as the uniformizer at $\mathcal{O}_2 = [1, -1, 0]$. We write

$$R(x, y) = \frac{(y - 1)(y + 1)(y - x)^3}{(x + 1)(y^2 - x^2)} = (1/x)^{-4} \left[\frac{(\frac{y}{x} - \frac{1}{x})(\frac{y}{x} + \frac{1}{x})(\frac{y}{x} - 1)^3}{(1 + \frac{1}{x})(-1)} \right].$$

But $y/x = -1$ and $1/x = 0$ at $\mathcal{O}_2$, so $\mathcal{O}_2$ is a pole of R of multiplicity four.

43. For elliptic curves, we prefer to use the explicit formulas given as Eqns (4.13), (4.14) and (4.15). The given rational function is written as $G(x, y)/H(x, y)$ with $G(x, y) = x$ and $H(x, y) = y$.

    Finite zeros of $G(x, y)$: The only zero of x is the special point $P_1 = (0, 0)$. We have $e = 1$ and $l = 0$, so the multiplicity of P_1 is $2e = 2$.

    Finite zeros of $H(x, y)$: The zeros of y are the special points $P_1 = (0, 0)$, $P_2 = (1, 0)$ and $P_3 = (-1, 0)$. For each of these points, we have $e = 0$ and $l = 1$, so $\mathrm{ord}_{P_i}(y) = 1$ for $i = 1, 2, 3$.

    Zeros and poles of $R = G/H$: P_1 is a zero of R of multiplicity $2 - 1 = 1$. P_2 and P_3 are poles of multiplicity one. Moreover, $\mathrm{ord}_{\mathcal{O}}(G) = -2$ and $\mathrm{ord}_{\mathcal{O}}(H) = -3$, so that $\mathrm{ord}_{\mathcal{O}}(R) = -2 + 3 = 1$, that is, R has a simple zero at $\mathcal{O}$.

52. **Bilinearity:** We claim that $([P] - [\mathcal{O}]) + ([Q] - [\mathcal{O}]) \sim [P + Q] - [\mathcal{O}]$. If one (or both) of P, Q is/are $\mathcal{O}$, this is certainly true. If $P = -Q$, the right side is 0, and the left side is the divisor of the vertical line $L_{P,Q}$ (that is, a principal divisor, that is, equivalent to 0). If P, Q are finite points that are not opposites of one another, then let $U = P + Q$, and observe that

$[P] + [Q] + [-U] - 3[\mathcal{O}] = \text{Div}(L_{P,Q})$ and $[U] + [-U] - 2[\mathcal{O}] = \text{Div}(L_{U,-U})$. Therefore, $([P] - [\mathcal{O}]) + ([Q] - [\mathcal{O}]) = [U] - [\mathcal{O}] + \text{Div}(L_{P,Q}/L_{U,-U})$.

For $V = P, Q, P + Q, R$, let D_V be a divisor equivalent to $[V] - [\mathcal{O}]$, and let f_V be the rational function such that $\text{Div}(f_V) = mD_V$. By definition,

$$e_m(P + Q, R) = f_{P+Q}(D_R)/f_R(D_{P+Q}).$$

Since $([P] - [\mathcal{O}]) + ([Q] - [\mathcal{O}]) \sim [P + Q] - [\mathcal{O}]$, we have $D_P + D_Q \sim D_{P+Q}$, that is, $D_{P+Q} = D_P + D_Q + \text{Div}(h)$ for some rational function h. Moreover, $\text{Div}(f_{P+Q}) = mD_{P+Q} = mD_P + mD_Q + m\,\text{Div}(h) = \text{Div}(f_P) + \text{Div}(f_Q) + \text{Div}(h^m) = \text{Div}(f_P f_Q h^m)$, so we can take $f_{P+Q} = f_P f_Q h^m$. Therefore,

$$e_m(P + Q, R) = \frac{f_P(D_R)f_Q(D_R)h(D_R)^m}{f_R(D_P + D_Q + \text{Div}(h))} = \frac{f_P(D_R)f_Q(D_R)h(D_R)^m}{f_R(D_P)f_R(D_Q)f_R(\text{Div}(h))}.$$

But $h(D_R)^m = h(mD_R) = h(\text{Div}(f_R)) = f_R(\text{Div}(h))$ (by Weil's reciprocity law), and so $e_m(P + Q, R) = e_m(P, R)e_m(Q, R)$.

One can analogously check that $e_m(P, Q + R) = e_m(P, Q)e_m(P, R)$.

Alternation: Take two divisors D_1, D_2 with disjoint supports, each equivalent to $[P] - [\mathcal{O}]$. But then, $D_1 \sim D_2$, that is, $D_1 = D_2 + \text{Div}(h)$ for some rational function h. Also, choose functions f_1, f_2 such that $\text{Div}(f_1) = mD_1$ and $\text{Div}(f_2) = mD_2$. Since $\text{Div}(f_1) = mD_1 = mD_2 + m\,\text{Div}(h) = \text{Div}(f_2) + \text{Div}(h^m) = \text{Div}(f_2 h^m)$, we can take $f_1 = f_2 h^m$. Therefore,

$$e_m(P, P) = \frac{f_1(D_2)}{f_2(D_1)} = \frac{f_2 h^m(D_2)}{f_2(D_2 + \text{Div}(h))} = \frac{f_2(D_2)h(\text{Div}(f_2))}{f_2(D_2)f_2(\text{Div}(h))} = 1.$$

Skew symmetry: By the previous two properties, we have

$$\begin{aligned}
1 &= e_m(P + Q, P + Q) &= e_m(P, P)e_m(P, Q)e_m(Q, P)e_m(Q, Q) \\
&= e_m(P, Q)e_m(Q, P).
\end{aligned}$$

58. Let $\text{Div}(f) = m[P] - m[\mathcal{O}]$, $D \sim [Q] - [\mathcal{O}]$ and $D' \sim [Q'] - [\mathcal{O}]$, where $Q = mQ'$. Since $Q - \mathcal{O} - m(Q' - \mathcal{O}) = \mathcal{O}$, we have $D \sim mD'$, that is, $D = mD' + \text{Div}(h)$ for some rational function h. But then,

$$\begin{aligned}
\langle P, Q \rangle_m &= f(D) &= f(mD' + \text{Div}(h)) &= f(mD')f(\text{Div}(h)) \\
&= f(D')^m h(\text{Div}(f)) &= f(D')^m h(m[P] - m[\mathcal{O}]) \\
&= \left(f(D')h([P] - [\mathcal{O}])\right)^m.
\end{aligned}$$

59. Let us restrict our attention to the simple Weierstrass equation $Y^2 = X^3 + aX + b$. Let $L : y + \mu x + \lambda = 0$ be the equation of a non-vertical straight line. The conjugate of L is $\hat{L} = -y + \mu x + \lambda$. For a point Q, we have:

$$L(-Q) = \hat{L}(Q).$$

Moreover,

$$N(L) = L\hat{L} = -y^2 + (\mu x + \lambda)^2 = -(x^3 + ux + vx + w)$$

for some u, v, w. Suppose that L is the line through the points $P_1 = (h_1, k_1)$ and $P_2 = (h_2, k_2)$ on the curve (we may have $P_1 = P_2$). The third point of intersection of L with the curve is $(h_3, -k_3)$, where $P_1 + P_2 = (h_3, k_3)$. Clearly, h_1, h_2, h_3 are roots of $N(L)$, and it follows that

$$N(L) = -(x - h_1)(x - h_2)(x - h_3).$$

In particular, if $P_1 = P_2$, we have

$$N(L) = -(x - h_1)^2(x - h_3).$$

In the parts below, we take $Q = (h, k)$.

(a) We have

$$\frac{L_{U,U}(Q)}{L_{U,-U}^2(Q)L_{2U,-2U}(Q)} = \frac{L_{U,U}(Q)}{(h - x(U))^2(h - x(2U))} = -\frac{1}{\hat{L}_{U,U}(Q)}$$

$$= -\frac{1}{L_{U,U}(-Q)}.$$

(b) Let us denote $tU = (h_t, y_t)$. But then,

$$\frac{L_{(k+1)U,kU}(Q)}{L_{(k+1)U,-(k+1)U}(Q)L_{(2k+1)U,\ (2k+1)U}(Q)}$$

$$= \frac{L_{(k+1)U,kU}(Q)}{(h - h_{k+1})(h - h_{2k+1})}$$

$$= \frac{L_{(k+1)U,kU}(Q)\hat{L}_{(k+1)U,kU}(Q)}{(h - h_{k+1})(h - h_{2k+1})\hat{L}_{(k+1)U,kU}(Q)}$$

$$= \frac{L_{(k+1)U,kU}(Q)\hat{L}_{(k+1)U,kU}(Q)(h - h_k)}{(h - h_k)(h - h_{k+1})(h - h_{2k+1})\hat{L}_{(k+1)U,kU}(Q)}$$

$$= -\frac{h - h_k}{\hat{L}_{(k+1)U,kU}(Q)}$$

$$= -\frac{L_{kU,-kU}(Q)}{L_{(k+1)U,kU}(-Q)}.$$

(c) Put $k = 1$ in Part (b).

61. For $n \geqslant 2$, let us define the function

$$g_{n,P} = f_{n,P}L_{nP,-nP}.$$

If we can compute $g_{n,P}$, we can compute $f_{n,P}$, and conversely. Moreover, if $mP = \mathcal{O}$ (this is usually the case in Weil- and Tate-pairing computations), we have $L_{mP,-mP} = 1$, and so $g_{m,P} = f_{m,P}$, that is, no final adjustment is necessary to convert the value of g to the value of f. In view of this, we can modify Miller's algorithm so as to compute $g_{n,P}(Q)$. Write $n = (1n_{s-1}n_{s-2}\ldots n_1n_0)_2$. We first consume two leftmost bits from n. If $n_{s-1} = 0$, we compute

$$g_{2,P}(Q) = f_{2,P}L_{2P,-2P}(Q) = L_{P,P}(Q).$$

On the other hand, if $n_{s-1} = 1$, we compute

$$g_{3,P}(Q) = f_{3,P}(Q)L_{3P,-3P} = f_{2,P}(Q)L_{2P,P}(Q) = \frac{L_{P,P}(Q)L_{2P,P}(Q)}{L_{2P,-2P}(Q)}.$$

In these two cases, we initialize U to $2P$ and $3P$, respectively.

The Miller loop runs for $i = s - 2, s - 3, \ldots, 1, 0$. In the i-th iteration, we consume the bit n_i, and update the value of the function f (or g) and the point U appropriately. Suppose that at the beginning of the i-th iteration, we have computed $g_{k,P}$ and $U = kP$. If $n_i = 0$, we compute $g_{2k,P}$ and $2kP = 2U$ in the current iteration. If $n_i = 1$, we compute $g_{2k+1,P}$ and $(2k+1)P = 2U + P$. We consider these two cases separately.

If $n_i = 0$, we have

$$
\begin{aligned}
g_{2k,P}(Q) &= f_{2k,P}(Q)L_{2kP,-2kP}(Q) \\
&= f_{k,P}^2(Q)L_{kP,kP}(Q) \\
&= \left(\frac{g_{k,P}(Q)}{L_{kP,-kP}(Q)}\right)^2 L_{kP,kP}(Q) \\
&= g_{k,P}^2(Q)\left(\frac{L_{kP,kP}(Q)}{L_{kP,-kP}^2(Q)}\right) \\
&= -g_{k,P}^2(Q)\left(\frac{L_{2kP,-2kP}(Q)}{L_{kP,kP}(-Q)}\right) \quad \text{[by Exercise 4.59(a)]} \\
&= -g_{k,P}^2(Q)\left(\frac{L_{2U,-2U}(Q)}{L_{U,U}(-Q)}\right).
\end{aligned}
$$

If $n_i = 1$, we have

$$
\begin{aligned}
g_{2k+1,P}(Q) &= f_{2k+1,P}(Q)L_{(2k+1)P,-(2k+1)P}(Q) \\
&= f_{2k,P}L_{2kP,P}(Q) \\
&= g_{2k,P}(Q)\left(\frac{L_{2kP,P}(Q)}{L_{2kP,-2kP}(Q)}\right) \\
&= -g_{k,P}^2(Q)\left(\frac{L_{2U,-2U}(Q)}{L_{U,U}(-Q)}\right)\left(\frac{L_{2U,P}(Q)}{L_{2U,-2U}(Q)}\right) \\
&= -g_{k,P}^2(Q)\left(\frac{L_{2U,P}(Q)}{L_{U,U}(-Q)}\right).
\end{aligned}
$$

Here, each of the two cases $n_i = 0$ and $n_i = 1$ requires the computation of only two line functions. On the contrary, the original Miller loop requires two or four line-function computations for these two cases, respectively. Therefore, if many bits in n are 1, the modified loop is expected to be somewhat more efficient than the original Miller loop.

63. For any two points U, V on the curve, we have:

$$[U] + [V] - [U + V] - [\mathcal{O}] = \mathrm{Div}\left(\frac{L_{U,V}}{L_{U+V,-(U+V)}}\right).$$

(a) Since $\text{Div}(f_{0,P,S}) = 0$, we can take $f_{0,P,S} = 1$. Moreover, $\text{Div}(f_{1,P,S}) = [P+S] - [S] - [P] + [\mathcal{O}] = \text{Div}\left(\frac{L_{(P+S),-(P+S)}}{L_{P,S}}\right)$.

(b) We have $\text{Div}(f_{n+1,P,S}) - \text{Div}(f_{n,P,S}) = [P+S] - [S] - [(n+1)P] - [nP] = ([P+S] - [P] - [S] + [\mathcal{O}]) + ([nP] + [P] - [(n+1)P] - [\mathcal{O}]) = \text{Div}(f_{1,P,S}) + \text{Div}\left(\frac{L_{nP,P}}{L_{(n+1)P,-(n+1)P}}\right)$.

(c) We have $\text{Div}(f_{n+n',P,S}) = (n+n')[P+S] - (n+n')[S] - [(n+n')P] + [\mathcal{O}] = (n[P+S] - n[S] - [nP] + [\mathcal{O}]) + (n'[P+S] - n'[S] - [n'P] + [\mathcal{O}]) + ([nP] + [n'P] - [(n+n')P] - [\mathcal{O}]) = \text{Div}(f_{n,P,S}) + \text{Div}(f_{n',P,S}) + \text{Div}\left(\frac{L_{nP,n'P}}{L_{(n+n')P,-(n+n')P}}\right)$.

65. Let $Q = aP$ for some a, where $P, Q \in E_q$ are points of order m. For the distorted Weil pairing, we have $e_m(P, \phi(Q)) = e_m(P, \phi(aP)) = e_m(P, a\phi(P)) = e_m(P, \phi(P))^a = e_m(aP, \phi(P)) = e_m(Q, \phi(P))$.

The twisted pairing is defined on $G \times G'$ with two different groups G, G', so the question of symmetry is not legitimate in the context of twisted pairings.

75. For simplicity, we assume that the curve is defined over a field of characteristic $\neq 2, 3$ by the equation $E : Y^2 = X^3 + aX + b$. The following program translates the algorithm of Exercise 4.57 to GP/PARI.

```
L(U,V,Q) = \
   local(l,m); \
   if ((U == [0]) && (V == [0]), return(1)); \
   if (U == [0], return(Q[1] - V[1])); \
   if (V == [0], return(Q[1] - U[1])); \
   if ((U[1] == V[1]) && (U[2] == -V[2]), return(Q[1] - U[1])); \
   if ((U[1] == V[1]) && (U[2] == V[2]), \
      l = (3 * U[1]^2 + E[4]) / (2 * U[2]), \
      l = (V[2] - U[2]) / (V[1] - U[1]) \
   ); \
   m = l * U[1] - U[2]; \
   return(Q[2] - l * Q[1] + m);

p = 43; L([Mod(1,p),Mod(2,p)],[Mod(1,p),Mod(2,p)],[x,y])
```

76. We first fix the prime p, a supersingular curve E of the given form, and the parameters m and k. We also need a representation of the field $\mathbb{F}_{p^k}$. Since $p \equiv 3 \pmod 4$ and $k = 2$, we can represent this extension as $\mathbb{F}_p(\theta)$ with $\theta^2 + 1 = 0$. We pass two points for which the reduced Tate pairing needs to be computed. We invoke the line-function primitive L(U,V,Q) implemented in Exercise 4.75. We use the formula $\langle P, Q \rangle_m = f_{m,P}(Q)$. If the computation fails, we conclude that this is a case of degenerate output, and return 1.

```
p = 43; E = ellinit([0,0,0,Mod(3,p),0]);
m = (p + 1) / 4; k = 2; T = Mod(t, Mod(1,p)*t^2 + Mod(1,p))

Tate(P,Q) = \
   local(s,U,V,fnum,fden); \
   s = ceil(log(p)/log(2)); while (bittest(m,s)==0, s--); \
```

```
    fnum = fden = 1; U = P; s--; \
    while (s >= 0, \
        fnum = fnum^2 * L(U,U,Q); \
        U = elladd(E,U,U); \
        V = U; if (matsize(U)==[1,2], V[2] = -U[2]); \
        fden = fden^2 * L(U,V,Q); \
        if (bittest(m,s) == 1, \
            fnum = fnum * L(U,P,Q); \
            U = elladd(E,U,P); \
            V = U; if(matsize(U)==[1,2], V[2] = -U[2]); \
            fden = fden * L(U,V,Q); \
        ); \
        s--; \
    ); \
    if ((fnum == 0) && (fden == 0), return(1)); \
    return((fnum / fden) ^ ((p^k - 1) / m));

lift(lift(Tate([1, 2], [15 + 22*T, 5 + 14*T])))
```

Chapter 5 Primality Testing

2. The sieve of Eratosthenes marks many composite integers multiple times. For example, 30 is marked thrice—as multiples of 2, 3, and 5. To improve the running time of the sieve, we plan to mark a composite integer $x \leqslant n$ only once, namely, as a proper multiple of its least prime factor p. If $x = pf$ with the cofactor f, then $p \leqslant \sqrt{x} \leqslant \sqrt{n}$, and $p \leqslant r$ for any prime factor r of f. In the following algorithm, the outer loop runs over all possible values of f (between 2 and $\lfloor n/2 \rfloor$). Let q denote the least prime factor of one value of f in its range of variation. So long as $p \leqslant q$, the least prime factor of pf is p, so we mark pf as composite (provided that $pf \leqslant n$). But q is not known in advance, so we keep on dividing f by p (along with marking pf) until $p|f$. At this point of time, we terminate the iteration for the current cofactor f. The algorithm requires a (sorted) list $p_1, p_2, \ldots, p_t$ of all primes $\leqslant \sqrt{n}$.

Prepare the sorted list $p_1, p_2, \ldots, p_t$ of all primes $\leqslant \sqrt{n}$.
Initialize each element of the array P (indexed 1 through n) to *true*.
For $f = 2, 3, 4, \ldots, \lfloor n/2 \rfloor$, repeat: {
    For $i = 1, 2, \ldots, t$, repeat: {
        If $(p_i f > n)$, break the inner *for* loop.
        Mark $P[p_i f]$ as *false*.
        If $(p_i | f)$, break the inner *for* loop.
    }
}
For $i = 2, 3, \ldots, n$, repeat: { If $P[i]$ is *true*, output i. }

The primes $\leqslant \sqrt{n}$ may be obtained recursively. Even the original sieve of Eratosthenes can generate these primes in $O(\sqrt{n} \ln \ln n)$ (that is, $o(n)$) time.

3. (a) By the prime number theorem, the number of primes $\leqslant x$ is nearly $x/\ln x$, that is, a randomly chosen integer $\leqslant x$ is prime with probability about $1/\ln x$. Under the assumption that x and $2x+1$ behave as random integers, one $n+i$ is a Sophie Germain prime with probability about $1/[\ln(n+M)\ln(2(n+M)+1)]$ which is approximately $1/\ln^2 n$. Therefore, we should take $M = \ln^2 n$ (or a small multiple of $\ln^2 n$).

(b) We use an array A indexed by i in the range $0 \leqslant i \leqslant M$. It is not essential to know the exact factorizations of $n+i$. Detecting only that $n+i$ or $2(n+i)+1$ is divisible by any p_j suffices to throw away $n+i$.

In view of this, we initialize each array location A_i to 1. Now, take $q = p_j$ for some $j \in \{1,2,\ldots,t\}$. The condition $q \mid (n+i)$ implies $i \equiv -n \pmod{q}$, so we set $A_i = 0$ for all values of i satisfying this congruence. Moreover, for $q \neq 2$, the condition $q \mid [2(n+i)+1]$ implies $i \equiv -n - 2^{-1} \pmod{q}$, that is, we set $A_i = 0$ for all values of i satisfying this second congruence.

After all primes $p_1, p_2, \ldots, p_t$ are considered, we check the primality of $n+i$ and $2(n+i)+1$ only for those i, for which we still have $A_i = 1$.

(c) Let $P = p_1 p_2 \cdots p_t$ and $Q = p_2 p_3 \cdots p_t$. The probability that a random $n+i$ is not divisible by any p_j is about $\phi(P)/P$. Likewise, the probability that a random $2(n+i)+1$ is not divisible by any p_j is about $\phi(Q)/Q$. Let us assume that the two events *divisibility of $n+i$ by p_j* and *divisibility of $2(n+i)+1$ by p_j* are independent. But then, we check the primality of $n+i$ and $2(n+i)+1$ for about $(M+1)\frac{\phi(P)\phi(Q)}{PQ}$ values of i. Therefore, the speedup obtained is close to $\frac{PQ}{\phi(P)\phi(Q)}$. For $t = 10$, this speedup is about 20; for $t = 100$, it is about 64; and for $t = 1000$, it is about 128. Note that for a suitably chosen t, we may neglect the sieving time which is $O(t + M \log t)$, that is, $O(t + (\log^2 n)(\log t))$. In contrast, each primality test (like Miller–Rabin) takes $O(\log^3 n)$ time.

9. Since the three primes r, p, q must be distinct from one another, we take $p < q$. Since there are only finitely many pairs (p,q) satisfying $p < q < r$, we may assume, without loss of generality, that $r < q$. Moreover, if $p < r < q$, then by the result to be proved, there are only finitely many Carmichael numbers of the form prq with the smallest factor $p < r$. Therefore, we assume that $r < p < q$, that is, r is the smallest prime factor of $n = rpq$.

We have $n - 1 = pqr - 1 = pq(r-1) + (pq - 1)$. Since n is a Carmichael number, we have $(r-1)\mid(n-1)$, that is, $(r-1)\mid(pq-1)$. A similar result holds for p and q too, that is, for some positive integers u, v, w, we have:

$$
\begin{aligned}
pq - 1 &= u(r - 1), \\
qr - 1 &= v(p - 1), \\
pr - 1 &= w(q - 1).
\end{aligned}
$$

Since p, q, r are odd primes with $r < p < q$, we have $q - 1 > p$, that is, $pr - 1 = w(q - 1) > wp$, that is, $pr > wp$, that is, $w < r$, that is,

$$w \leqslant r - 1.$$

Now, $qr = 1 + v(p-1) = r\left(\frac{pr-1}{w} + 1\right)$, that is,

$$p - 1 = \frac{(r-1)(r+w)}{vw - r^2}.$$

Since $r - 1, r + w, p - 1$ are positive, $vw - r^2$ is positive too. Finally, since $vw - r^2$ is an integer, we have

$$p - 1 \leqslant (r-1)(r+w) \leqslant (r-1)(2r-1).$$

Given r, there are, therefore, only finitely many possibilities for p. For each such p, we get only finitely many primes q satisfying $pr - 1 = w(q-1)$.

11. **(a)** In view of Exercise 5.10(a), it suffices to concentrate only on Carmichael numbers n. We can write $n = p_1 p_2 \cdots p_r$ with pairwise distinct odd primes $p_1, p_2, \ldots, p_r, r \geqslant 3$, and with $(p_i - 1)|(n-1)$ for all $i = 1, 2, \ldots, r$. We now consider two cases.

Case 1: All $\frac{n-1}{p_i-1}$ are even.

We choose a base $a \in \mathbb{Z}_n^*$ such that $\left(\frac{a}{p_1}\right) = -1$, whereas $\left(\frac{a}{p_i}\right) = +1$ for $i = 2, 3, \ldots, r$. By the definition of the Jacobi symbol, we have $\left(\frac{a}{n}\right) = -1$. By Euler's criterion, $a^{(p_1-1)/2} \equiv -1 \pmod{p_1}$. Since $\frac{n-1}{p_1-1} = \frac{(n-1)/2}{(p_1-2)/2}$ is even by hypothesis, we have $a^{(n-1)/2} \equiv 1 \pmod{p_1}$. On the other hand, for $i = 2, 3, \ldots, r$, we have $a^{(p_i-1)/2} \equiv 1 \pmod{p_i}$, that is, $a^{(n-1)/2} \equiv 1 \pmod{p_i}$. By CRT, we then have $a^{(n-1)/2} \equiv 1 \pmod{n}$, that is, $a^{(n-1)/2} \not\equiv \left(\frac{a}{n}\right) \pmod{n}$, that is, n is not an Euler pseudoprime to base a.

Case 2: Some $\frac{n-1}{p_i-1}$ is odd.

Without loss of generality, assume that $\frac{n-1}{p_1-1}$ is odd. Again take $a \in \mathbb{Z}_n^*$ with $\left(\frac{a}{p_1}\right) = -1$ and $\left(\frac{a}{p_i}\right) = +1$ for $i = 2, 3, \ldots, r$. By the definition of the Jacobi symbol, we then have $\left(\frac{a}{n}\right) = -1$. On the other hand, by Euler's criterion, we have $a^{(n-1)/2} \equiv -1 \pmod{p_1}$ and $a^{(n-1)/2} \equiv 1 \pmod{p_i}$ for $i = 2, 3, \ldots, r$. By CRT, we conclude that $a^{(n-1)/2} \not\equiv \pm 1 \pmod{n}$, that is, $a^{(n-1)/2} \not\equiv \left(\frac{a}{n}\right) \pmod{n}$, that is, n is not an Euler pseudoprime to base a.

(b) Suppose that n is an Euler pseudoprime to the bases $a_1, a_2, \ldots, a_t \in \mathbb{Z}_n^*$ only. Let a be a base to which n is not an Euler pseudoprime. (Such a base exists by Part (a).) We have $a^{(n-1)/2} \not\equiv \left(\frac{a}{n}\right) \pmod{n}$. On the other hand, $a_i^{(n-1)/2} \equiv \left(\frac{a_i}{n}\right) \pmod{n}$ for $i = 1, 2, \ldots, t$. It follows that $(aa_i)^{(n-1)/2} \equiv a^{(n-1)/2} a_i^{(n-1)/2} \not\equiv \left(\frac{a}{n}\right)\left(\frac{a_i}{n}\right) \equiv \left(\frac{aa_i}{n}\right) \pmod{n}$, that is, n is not an Euler pseudoprime to each of the bases aa_i, that is, there are at least t bases to which n is not an Euler pseudoprime.

13. **(a)** We have $n - 1 = pq - 1 = p(2p-1) - 1 = (p-1)(2p+1)$. Let $\alpha \equiv a^{(n-1)/2} \pmod{p}$, and $\beta \equiv a^{(n-1)/2} \pmod{q}$. Modulo p, we have $\alpha \equiv (a^{(p-1)/2})^{2p+1} \equiv \left(\frac{a}{p}\right)^{2p+1} \equiv \left(\frac{a}{p}\right) \pmod{p}$. In particular, $\alpha = \pm 1$.

Since $\left(\frac{a}{n}\right) = \left(\frac{a}{p}\right)\left(\frac{a}{q}\right)$, we conclude that n is an Euler pseudoprime to base a only if $\beta = \pm 1$.

The determination of β is more involved. If a is a quadratic residue modulo n, then $a^{(q-1)/2} \equiv 1 \pmod{q}$. Exactly half of these residues satisfy $a^{(q-1)/4} \equiv 1 \pmod{q}$, and the remaining half $a^{(q-1)/4} \equiv -1 \pmod{q}$. Since $2p+1$ is odd, it follows that $\beta \equiv a^{(n-1)/2} \equiv 1 \pmod{q}$ for half of the quadratic residues modulo q and $\beta \equiv a^{(n-1)/2} \equiv -1 \pmod{q}$ for the remaining half of the quadratic residues modulo q. If a is a quadratic non-residue modulo q, then $a^{2p+1} \pmod{q}$ is again a quadratic non-residue modulo q (since $2p+1$ is odd). Therefore, $\beta \equiv a^{(n-1)/2} \equiv (a^{2p+1})^{(q-1)/4} \not\equiv \pm 1 \pmod{q}$, that is, it suffices to concentrate only on the case that a is a quadratic residue modulo q.

Let n be an Euler pseudoprime to base a. If $\alpha = \left(\frac{a}{p}\right) = 1$, we should have $\left(\frac{a}{q}\right) = 1$ and $\beta = 1$. There are exactly $\left(\frac{p-1}{2}\right)\left(\frac{q-1}{4}\right) = \frac{(p-1)(q-1)}{8}$ such bases in $\mathbb{Z}_n^*$. For each such base, $\left(\frac{a}{n}\right) = \left(\frac{a}{p}\right)\left(\frac{a}{q}\right) = 1$ and $a^{(n-1)/2} \equiv 1 \pmod{n}$. On the other hand, if $\alpha = \left(\frac{a}{p}\right) = -1$, we must have $\left(\frac{a}{q}\right) = 1$ and $\beta = -1$. There are again exactly $\left(\frac{p-1}{2}\right)\left(\frac{q-1}{4}\right) = \frac{(p-1)(q-1)}{8}$ such bases in $\mathbb{Z}_n^*$. For each such base, we have $\left(\frac{a}{n}\right) = \left(\frac{a}{p}\right)\left(\frac{a}{q}\right) = -1$ and $a^{(n-1)/2} \equiv -1 \pmod{n}$.

(b) If $p \equiv 3 \pmod 4$, we have $q \equiv 1 \pmod 4$, that is, $n \equiv 3 \pmod 4$, that is, $(n-1)/2$ is odd. Now, n is a strong pseudoprime to base a if and only if $a^{(n-1)/2} \equiv \pm 1 \pmod{n}$. As shown in Part (a), this condition is the same as the condition $a^{(n-1)/2} \equiv \left(\frac{a}{n}\right) \pmod{n}$.

18. (a) If $b = 1$, we have $\alpha\beta = 1$, that is, $\beta = \alpha^{-1}$. Therefore, $(\alpha/\beta)^t = 1$ implies $\alpha^{2t} = 1$, that is, $\alpha^t = \pm 1$ (since $\mathbb{F}_p$ and $\mathbb{F}_{p^2}$ are fields). But then, $\alpha^t = \beta^t = \pm 1$, and $V_t = \alpha^t + \beta^t = \pm 2$.

21. (b) We prove this by contradiction. Suppose that n is composite. Let p be the smallest prime divisor of n. Then $3 \leqslant p \leqslant \sqrt{n}$. By the given condition, $a^{(n-1)/2} \equiv -1 \not\equiv 1 \pmod{p}$, whereas $a^{n-1} \equiv (-1)^2 \equiv 1 \pmod{p}$, that is, $\mathrm{ord}_p a = t2^r$ for some odd $t \geqslant 1$. But $\mathrm{ord}_p a | (p-1)$, that is, $t2^r | (p-1)$, that is, $2^r | (p-1)$. But $p \neq 1$, so $p - 1 \geqslant 2^r$, that is, $p \geqslant 2^r + 1 > \sqrt{k2^r} + 1 = \sqrt{n-1} + 1 \geqslant \sqrt{n}$, a contradiction to the choice of p.

(c) Assume that the input integer n is already a Proth number.

Repeat the following steps for t times: {
    Choose a random base a in the range $1 \leqslant a \leqslant n - 1$.
    If $a^{(n-1)/2} \equiv -1 \pmod{n}$, return *yes*.
}
Return *no*.

The running time of this algorithm is dominated by (at most) t modular exponentiations. So long as t is a constant (or a polynomial in $\log n$), the running time of this algorithm is bounded by a polynomial in $\log n$.

22. (a) Since $a^{n-1} \equiv 1 \pmod{n}$, we have $a^{n-1} \equiv 1 \pmod{p}$. Moreover, since $p \mid n$, we have $\gcd(a^{(n-1)/q} - 1, p) = 1$, that is, $a^{(n-1)/q} \not\equiv 1 \pmod{p}$. It follows that $\operatorname{ord}_p(a) = ut$ for some $t \mid v$. But then, $b \equiv a^t \pmod{p}$ has order u modulo p. Since $\operatorname{ord}_p b$ divides $\phi(p) = p - 1$, we have $u \mid (p - 1)$, that is, $p \equiv 1 \pmod{u}$.
(b) Suppose that n is composite. Take any prime divisor p of n with $p \leqslant \sqrt{n}$. By Part (a), $p \geqslant u+1 \geqslant \sqrt{n} + 1$, a contradiction. Therefore, n must be prime.
(c) In order to convert the above observations to an efficient algorithm, we need to clarify two issues.

(1) The integer $n - 1$ can be written as uv with u, v as above and with $u \geqslant \sqrt{n}$. We can keep on making trial divisions of $n - 1$ by small primes $q_1 = 2$, $q_2 = 3$, $q_3 = 5$, ... until $n - 1$ reduces to a value $v \leqslant \sqrt{n}$. If $n - 1$ is not expressible in the above form, we terminate the procedure, and report *failure* after suitably many small primes are tried.

(2) We need an element a satisfying the two conditions $a^{n-1} \equiv 1 \pmod{n}$ and $\gcd(a^{(n-1)/q} - 1, n) = 1$ for all $q \mid u$. If n is prime, any of the $\phi(n-1)$ primitive elements modulo n satisfies these conditions. Thus, a suitable random base a is expected to be available within a few iterations.

27. The adaptation of the sieve of Eratosthenes is implemented by the following function. Here, k is the number of small primes to be used in the sieve. The sieve length is taken to be $M = 2t$. If no primes are found in the interval $[a, a + M - 1]$ for a randomly chosen a, the function returns -1.

```
randprime(t,k) = \
   local(M,a,A,i,p,r); \
   a = 0; \
   while (a <= 2^(t-1), a = random(2^t)); \
   M = 2 * t; A = vector(M); \
   for (i=0, M-1, A[i+1] = 1); \
   for (i=1, k, \
      p = prime(i); r = a % p; \
      if (r != 0, r = p - r); \
      while (r < M, A[r+1] = 0; r = r + p ); \
   ); \
   for (i=0, M-1, if (A[i+1] == 1, \
      if(isprime(a+i), return(a+i)); \
   )); \
   return(-1);
```

32. We write $n - 1 = 2^s t$ with t odd, where $l = \left(\frac{a^2 - 4b}{n} \right)$. The following function computes (and returns) V_t and V_{t+1} modulo n using the doubling formulas. The power $b^t \pmod{n}$ is also computed (and returned).

```
VMod(m,n,a,b) = \
   local(V,Vnext,B,i); \
   V = Mod(2,n); Vnext = Mod(a,n); B = Mod(1,n); \
   i = ceil(log(m)/log(2)) - 1; \
```

```
while (i >= 0, \
    if (bittest(m,i) == 0, \
        Vnext = V * Vnext - a * B; \
        V = V^2 - 2 * B; \
        B = B^2; \
        , \
        V =  V * Vnext - a * B; \
        Vnext = Vnext^2 - 2 * B * b; \
        B = B^2 * b; \
    ); \
    i--; \
); \
return([V,Vnext,B]);
```

Next, we compute U_t from V_t and V_{t+1}. If $U_t \equiv 0 \pmod{n}$, then n is a strong Lucas pseudoprime with parameters (a, b). If not, we check whether any of $V_{2^j t} \equiv 0 \pmod{n}$ for $j = 0, 1, 2, \ldots, s - 1$. If so, n is again a strong Lucas pseudoprime. If all these checks fail, then n is not a strong Lucas pseudoprime.

```
strongLucaspsp(n,a,b) = \
    local(D,l,U,s,t,V,Vnext,W,B); \
    D = a^2 - 4*b; l = kronecker(D,n); \
    if (l == 0, return(0)); \
    t = n - 1; s = 0; \
    while (bittest(t,s) == 0, s++); \
    t = t / (2^s); \
    W = VMod(t,n,a,b); \
    V = W[1]; Vnext = W[2]; B = W[3]; \
    U = (2*Vnext - a*V)/D; \
    if (U == Mod(0,n), return(1)); \
    while (s>0, \
        if (V == Mod(0,n), return(1)); \
        Vnext = V * Vnext - a * B; \
        V = V^2 - 2 * B; \
        B = B^2; \
        s--; \
    ); \
    return(0);
```

Chapter 6 Integer Factorization

4. To avoid the sequences $1, 1, 1, \ldots$ and $-1, -1, -1, \ldots$, respectively.

7. (a) Let $u = p_1^{e_1} p_2^{e_2} \cdots p_t^{e_t}$ for small primes $p_1, p_2, \ldots, p_t$ (distinct from the large prime v). The first stage computes $b \equiv a^E \pmod{n}$, where $E = p_1^{f_1} p_2^{f_2} \cdots p_t^{f_t}$ for some $f_i \geqslant e_i$. Now, $\mathrm{ord}_p\, a \,|\, (p - 1)$, and $\mathrm{ord}_p\, b = \mathrm{ord}_p\, a / \gcd(\mathrm{ord}_p\, a, E) \in \{1, v\}$. Therefore, if $b \not\equiv 1 \pmod{n}$, we have $\mathrm{ord}_p\, b = v$.

(b) We have $b^k \equiv 1 \pmod{p}$ if and only if $v|k$. Therefore, we may assume that $b_0, b_1, b_2, \ldots$ constitute a random sequence in the multiplicative group generated by v. By the birthday paradox, we expect $b_i \equiv b_j \pmod{p}$ with $i \neq j$ after $\Theta(\sqrt{v})$ iterations.

(c) If $b_i \equiv b_j \pmod{p}$, but $b_i \not\equiv b_j \pmod{n/p}$, then $\gcd(b_i - b_j, n)$ is a non-trivial factor of n.

(d) We store the sequence $b_0, b_1, b_2, \ldots$. Whenever a new b_i is generated, we compute $\gcd(b_i - b_j, n)$ for $j = 0, 1, 2, \ldots, i - 1$. Since $O(\sqrt{B'})$ terms in the sequence suffice to split n with high probability, the running time is essentially that of $O(\sqrt{B'})$ exponentiations and $O(B')$ gcd calculations. The storage requirement is that of $O(\sqrt{B'})$ elements of $\mathbb{Z}_n$.

The number of primes between B and B' is about $\frac{B'}{\ln B'} - \frac{B}{\ln B}$. If $B' \gg B$, this is about $\frac{B'}{\ln B'}$. The second stage described in Section 6.3.1 makes (at most) these many exponentiations and these many gcd computations, and calls for a storage of only a constant number of elements of $\mathbb{Z}_n$ (in addition to the absolutely constant list of primes between B and B').

It follows that the second stage of Section 6.3.1 has running time comparable with that of the variant discussed in this exercise (the new variant carries out fewer exponentiations but more gcd calculations than the original variant). The space requirements of the two variants are not directly comparable $(O(B'/\ln B')$ large primes against $O(\sqrt{B'})$ elements of $\mathbb{Z}_n$). If n is large, the original variant is better in terms of storage.

8. (d) Let $n = a^2 + b^2 = c^2 + d^2$, that is, $(a - c)(a + c) = (d - b)(d + b)$. Assume that $a > c$, so that $d > b$. Let $u = \gcd(a - c, b - d)$ and $v = \gcd(a + c, b + d)$. Write $a - c = u\alpha$, $d - b = u\beta$, $a + c = v\gamma$ and $b + d = v\delta$ for positive integers $\alpha, \beta, \gamma, \delta$. But then, $(a - c)(a + c) = (d - b)(d + b)$ implies $u\alpha v\gamma = u\beta v\delta$, that is, $\alpha\gamma = \beta\delta$. Since $\gcd(\alpha, \beta) = \gcd(\gamma, \delta) = 1$, this implies that $\alpha = \delta$ and $\beta = \gamma$, that is, we have

$$
\begin{aligned}
a - c &= u\alpha, \\
a + c &= v\beta, \\
d - b &= u\beta, \\
d + b &= v\alpha.
\end{aligned}
$$

This gives $a = (u\alpha + v\beta)/2$ and $b = (v\alpha - u\beta)/2$, that is, $n = a^2 + b^2 = (u^2 + v^2)(\alpha^2 + \beta^2)/4$. If $\alpha = \beta$, then $a = d$ and $b = c$, a contradiction. So at least one of α, β is $\geqslant 2$, that is, $\alpha^2 + \beta^2 \geqslant 5$.

Since n is odd, one of a, b is odd and the other even. Likewise, one of c, d is odd, and the other even. If a and c are of the same parity, then so also are b and d, that is, u and v are both even, that is, $(u^2 + v^2)/4$ is a non-trivial factor of n. On the other hand, if a, c are of opposite parities, then so also are b, d, and it follows that u, v, α, β are all odd. But then, $(u^2 + v^2)/2$ is a non-trivial factor of n.

12. (a) For a randomly chosen x, the integer $T(c) = (x + c)^2$ rem n is of value $O(n)$, and so has a probability of $L\left[\frac{-1}{2 \times \frac{1}{2}}\right] = L[-1]$ of being $L[1/2]$-smooth.

That is, $L[1]$ values of c need to be tried in order to obtain a single relation. Since we require about $2t$ (which is again $L[1/2]$) relations, the value of M should be $L[1] \times L[1/2] = L[3/2]$.

(b) Proceed as in the QSM. Let $x^2 = kn + J$ with $J \in \{0, 1, 2, \ldots, n-1\}$. We have $(x + c)^2 \equiv x^2 + 2xc + c^2 \equiv kn + J + 2xc + c^2 \equiv T(c) \pmod{n}$, where $T(c) = J + 2xc + c^2$. Use an array A indexed by c in the range $-M \leqslant c \leqslant M$. Initialize $A_c = \log |T(c)|$. For each small prime q and small exponent h, solve the congruence $(x + c)^2 \equiv kn \pmod{q^h}$. For all values of c in the range $-M \leqslant c \leqslant M$, that satisfy the above congruence, subtract $\log q$ from A_c. When all q and h values are considered, check which array locations A_c store values ≈ 0. Perform trial divisions on the corresponding $T(c)$ values.

(c) Follow the analysis of sieving in QSM. Initializing A takes $L[3/2]$ time. Solving all the congruences $(x + c)^2 \equiv kn \pmod{q^h}$ takes $L[1/2]$ time. Subtraction of all $\log q$ values takes $L[3/2]$ time. Trial division of $L[1/2]$ smooth values by $L[1/2]$ primes takes $L[1]$ time. Finally, the sparse system with $L[1/2]$ variables and $L[1/2]$ equations can be solved in $L[1]$ time.

18. (c) Let $H = \lceil \sqrt{n} \rceil$ (original QSM) and $H' = \lceil \sqrt{2n} \rceil$ (modified QSM). Let M be the optimal sieving limit when the original QSM runs alone. In the case of the dual QSM, we run both the original QSM and the modified QSM with a sieving limit of $M/2$. In the original QSM, $2M + 1$ candidates (for $-M \leqslant c \leqslant M$) are tried for smoothness. In the dual QSM, there are two sieves each handling $M + 1$ candidates (for $-M/2 \leqslant c \leqslant M/2$). The total number of candidates for the dual QSM is, therefore, $2M + 2$. In the original QSM, $2M+1$ candidates are expected to supply the requisite number of relations. So the dual QSM, too, is expected to supply nearly the same number of relations, provided that the candidates are not much larger in the dual QSM than in the original QSM. Indeed, we now show that the dual QSM actually reduces the absolute values of the candidates by a factor larger than 1.

The values of $|T(c)|$ for the original QSM are approximately proportional to H, whereas those for the modified QSM are roughly proportional to $H' \approx \sqrt{2}H$. In particular, the average value of $|T(c)|$ for the first sieve is nearly $2H \times (M/4) = MH/2$ (the sieving interval is $M/2$ now), and the average value of $|T(c)|$ for the second sieve is about $2H' \times (M/4) \approx \sqrt{2}\, MH/2$. The overall average is, therefore, $(1 + \sqrt{2})MH/4$. When the original QSM runs alone, this average is MH. Consequently, the smoothness candidates in the dual QSM are smaller than those for the original QSM by a factor of $4/(1 + \sqrt{2}) \approx 1.657$. As a result, the dual QSM is expected to supply more relations than the original QSM. Viewed from another angle, we can take slightly smaller values for M and/or t in the dual QSM than necessary for the original QSM, that is, the dual QSM is slightly more efficient than the original QSM.

The dual QSM does not consider the larger half of the $T(c)$ values (corresponding to $M/2 < |c| \leqslant M$) for smoothness tests. It instead runs another sieve. Although the smoothness candidates in the second sieve are about $\sqrt{2}$ times larger than the candidates in the original sieve, there is an overall reduction in the absolute values of $T(c)$ (averaged over the two sieves).

20. (a) Solve the following quadratic congruence for c_0:

$$c_0^2 + 2c_0 H + J \equiv 0 \pmod{q}.$$

If we run the original sieve in QSM, such a q and a corresponding value c_0 of c can be obtained from the sieving array itself (an entry holding a residual value between $\log p_t$ and $2 \log p_t$).
(b) We have

$$T_q(c) = \frac{T(c_0 + cq)}{q} = \frac{T(c_0)}{q} + 2c(H + c_0) + c^2 q.$$

This is an integer since $q | T(c_0)$. Since $T_q(c)$ is a quadratic polynomial in c, sieving can be done as in the original QSM. Moreover, c and c_0 are small, so the maximum value of $|T_q(c)|$ is $\approx 2MH$—the same as in the original QSM.
(c) A large prime q available from the (original) QSM is useful only when it appears in multiple relations. But $q > p_t$ is large, so the probability that it appears as the unfactored part in two (or more) relations among $2M + 1$ choices of c is small. The special-q variant allows us to systematically collect relations in which q is involved. If $T_q(c)$ is p_t-smooth for some c, we have $T(c_0 + cq)$ equal to q times this smooth value. We are interested in only a few such relations, and so we can run the special-q sieve only for a small sieving interval (of length $L[1/2]$ only, instead of $L[1]$ as in the original sieve).

25. (a) The point-multiplication algorithm is modified as follows. Two variables S and T are used in order to store $N_i P$ and $(N_i + 1)P$, respectively.

Initialize $S = \mathcal{O}$ and $T = P$.
For $i = s - 1, s - 2, \ldots, 1, 0$, repeat: {
    If $(n_i = 0)$ {   /* Update (S, T) to $(2S, 2S + P) = (2S, S + T)$ */
        Assign $T = S + T$ and $S = 2S$.
    } else {   /* Update (S, T) to $(2S + P, 2S + 2P) = (S + T, 2T)$ */
        Assign $S = S + T$ and $T = 2T$.
    }
}
Return S.

(b) Let $P_1 = (h_1, k_1)$, $P_2 = (h_2, k_2)$, $P_1 + P_2 = (h_3, k_3)$, and $P_1 - P_2 = (h_4, k_4)$. First, consider the case $P_1 \neq P_2$. The addition formula gives

$$\begin{aligned}
(h_1 - h_2)^2 h_3 &= (h_1 + h_2)(h_1 h_2 + a) + 2b - 2k_1 k_2, \\
(h_1 - h_2)^2 h_4 &= (h_1 + h_2)(h_1 h_2 + a) + 2b + 2k_1 k_2.
\end{aligned}$$

Multiplying these two formulas and substituting $k_1^2 = h_1^3 + ah_1 + b$ and $k_2^2 = h_2^3 + ah_2 + b$, we get

$$h_3 h_4 (h_1 - h_2)^2 = (h_1 h_2 - a)^2 - 4b(h_1 + h_2).$$

This implies that given h_1, h_2, h_4 alone, one can compute h_3. If $P_1 = P_2$ (the case of doubling), we have

$$4h_3(h_1^3 + ah_1 + b) = (h_1^2 - a)^2 - 8bh_1.$$

Given h_1 alone, we can compute h_3.

(c) The point-multiplication algorithm of Part (a) is a special case of Part (b), where $P_1 = N_i P$ and $P_2 = (N_i + 1)P$. Here, $P_1 - P_2 = -P$ has known X-coordinate. Therefore, for computing mP, we have the choice of never computing any Y-coordinate. This saves many arithmetic operations modulo n.

An iteration in the original point-multiplication algorithm computes only a doubling if $n_i = 0$, and a doubling and an addition if $n_i = 1$. An iteration in the algorithm of Part (a) makes one addition and one doubling irrespective of the bit value n_i. Despite that, ignoring all Y-coordinates is expected to bring down the running time. Since the X-coordinate alone suffices to detect failure in the computation of mP, the ECM continues to work.

28. (a) We assume that an integer of the form $a + bm$ is divisible by the medium prime q with probability $1/q$. Therefore, $N'/N \approx \sum_{q \in M} \frac{1}{q} = \sum_{p \in B} \frac{1}{p} - \sum_{p \in B'} \frac{1}{p} \approx \ln \ln B - \ln \ln B' = \ln \ln B - \ln \ln(kB) = \ln \ln B - \ln(\ln k + \ln B) = \ln \ln B - \ln(\ln B(1 + \ln k/\ln B)) = -\ln(1 + \ln k/\ln B) \approx -\ln k/\ln B = \ln(1/k)/\ln B$. For $B = 10^6$ and $k = 0.25$, this ratio is about 0.1.

(b) Here, we force a smooth value to be divisible by at least one medium prime. Therefore, the candidates having all prime factors $\leqslant B'$ (only the small primes) are missed in the lattice sieve.

Treat $a + bm \approx bm$ as random integers with absolute values no more than $C = b_{max}m$. Let $u = \ln C/\ln B$ and $u' = \ln C/\ln B'$. Then, the fraction of smooth integers that are missed by the lattice sieve is approximately $u'^{-u'}/u^{-u} = u^u/u'^{u'}$.

For the given numerical values, $C = 10^{36}$, $u \approx 6$ and $u' \approx 6.669$. Therefore, the fraction of smooth values missed by the lattice sieve is $\approx u^u/u'^{u'} \approx 0.149$. The implication is that we now check the smoothness of about 10% of the line-sieve candidates. Still, we obtain 85% of the smooth values.

(c) If (a, b) and (a', b') satisfy $a + bm \equiv 0 \pmod{q}$ and $a' + b'm \equiv 0 \pmod{q}$, and if $r \in \mathbb{Z}$, then $(a + a') + (b + b')m \equiv 0 \pmod{q}$ and $ra \equiv (rb)m \pmod{q}$.

(d) We need to locate all (a, b) pairs such that $a + bm$ is divisible by q and $(a + bm)/q$ is smooth over all primes $< q$. If $a + bm \equiv 0 \pmod{q}$, we can write $(a, b) = cV_1 + dV_2 = c(a_1, b_1) + d(a_2, b_2) = (ca_1 + da_2, cb_1 + db_2)$, that is, $a = ca_1 + da_2$ and $b = cb_1 + db_2$. Clearly, if a and b are coprime, then c and d are coprime too. So we concentrate on (c, d) pairs with $\gcd(c, d) = 1$.

We use a two-dimensional array A indexed by c, d in the ranges $-C \leqslant c \leqslant C$ and $1 \leqslant d \leqslant D$. We initialize $A[c, d]$ to $\log |c(a_1 + b_1 m) + d(a_2 + b_2 m)|$ if $\gcd(c, d) = 1$, or to $+\infty$ if $\gcd(c, d) > 1$. Let p be a prime $< q$, and h a small exponent. The condition $p^h | (a + bm)$ is equivalent to the condition that $cu_1 + du_2 \equiv 0 \pmod{p^h}$, where $u_1 = a_1 + b_1 m$ and $u_2 = a_2 + b_2 m$ correspond to the basis vectors V_1, V_2. Both u_1 and u_2 are divisible by q, but not by p. If

$p^h | u_1$ (but $p \nmid u_2$), then we subtract $\log p$ from all the entries in the d-th column if and only if $p^h | d$. Likewise, if $p^h | u_2$ (but $p \nmid u_1$), then $\log p$ is subtracted from the entire c-th row if and only if $p^h | c$. Finally, if $\gcd(u_1, p) = \gcd(u_2, p) = 1$, then for each row c, a solution for d is obtained from $cu_1 + du_2 \equiv 0 \pmod{p^h}$, and every p^h-th location is sieved in the c-th row.

Pollard suggests another sieving idea. The (c, d) pairs satisfying $cu_1 + du_2 \equiv 0 \pmod{p^h}$ form a sublattice. Let $W_1 = (c_1, d_1)$ and $W_2 = (c_2, d_2)$ constitute a reduced basis of this sublattice. The array locations $(c, d) = e(c_1, d_1) + f(c_2, d_2) = (ec_1 + fc_2, ed_1 + fd_2)$ are sieved for small coprime e and f.

35. Dixon's method (both the relation-collection and the linear-algebra phases) are implemented below. The decisions about t (the size of the factor base) and s (the number of relations to be collected) are not taken by this function.

```
Dixon(n,s,t) = \
   local(i,j,x,y,X,A,C,D,k,v,beta,d,e); \
   A = matrix(t,s); C = matrix(t,s); X = vector(s); i = 1; \
   while(i <= s, \
      X[i] = random(n); y = (X[i]^2) % n; \
      for (j=1,t, \
         A[j,i] = 0; \
         while (y % prime(j) == 0, A[j,i]++; y = y / prime(j)); \
      ); \
      if (y == 1, i++) \
   ); \
   print("Relations collected"); \
   for (i=1,t, for(j=1,s, C[i,j] = Mod(A[i,j],2))); \
   D = lift(matker(C)); k = matsize(D)[2]; \
   print("Nullspace computed: dimension = ", k); \
   v = vector(k); beta = vectorv(s); d = 1; \
   while ((d==1) || (d==n), \
      for (j=1,k, v[j] = random(2)); \
      for (i=1,s, \
         for(j=1,k, \
            beta[i] = beta[i] + v[j] * D[i,j] \
         ); \
         beta[i] = beta[i] % 2; \
      ); \
      e = mattranspose(A * beta); \
      x = Mod(1,n); for (i=1,s, if (beta[i], x = x * Mod(X[i],n))); \
      y = Mod(1,n); for (j=1,t, y = y * Mod(prime(j),n)^(e[j]/2)); \
      d = gcd(lift(x)-lift(y),n); \
      print("x = ", lift(x), ", y = ", lift(y), ", d = ", d); \
   ); \
   return(d);

Dixon(64349,20,15)
```

---

37. Given a bound b, we first compute the factor base B consisting of -1 and all primes $\leqslant b$, modulo which n is a quadratic residue. We also require the sieving range $[-M, M]$. In what follows, we implement the incomplete sieving strategy. The tolerance bound ξ should also be supplied.

```
getFactorBase(n,b) = \
   local(P,B,p,t,i); \
   P = vector(b); p = 2; t = 1; P[t] = -1; \
   while (p <= b, \
      if (kronecker(n,p) == 1, t++; P[t] = p); \
      p = nextprime(p+1); \
   ); \
   B = vector(t); for (i=1,t, B[i] = P[i]); \
   return([t,B]);

QSMsieve(n,b,M,xi) = \
   local(H,J,tB,t,B,logB,c,c1,c2,A,i,T,bnd); \
   H = 1 + sqrtint(n); J = H^2 - n; \
   tB = getFactorBase(n,b); t = tB[1]; B = tB[2]; tB = 0; \
   logB = vector(t); for (i=2,t, logB[i] = floor(1000 * log(B[i]))); \
   A = vector(2*M + 1); \
   for (c=-M,M, \
      T = J + 2*H*c + c^2; if (T < 0, T = -T); \
      A[c+M+1] = floor(1000 * log(T)); \
   ); \
   c = -M; if (J % 2 == 0, if (c % 2 == 1, c++), if (c % 2 == 0, c++)); \
   while (c < M, A[c+M+1] -= logB[2]; c += 2); \
   for (i=3,t, \
      c1 = sqrt(Mod(n,B[i])); c2 = B[i] - c1; \
      c1 = lift(c1 - Mod(H,B[i])); c2 = lift(c2 - Mod(H,B[i])); \
      c = c1; while (c >= -M, A[c+M+1] -= logB[i]; c -= B[i]); \
      c = c1 + B[i]; while (c <= M, A[c+M+1] -= logB[i]; c += B[i]); \
      c = c2; while (c >= -M, A[c+M+1] -= logB[i]; c -= B[i]); \
      c = c2 + B[i]; while (c <= M, A[c+M+1] -= logB[i]; c += B[i]); \
   ); \
   bnd = floor(xi * 1000 * log(nextprime(B[t]))); \
   for (c=-M, M, \
      if (A[c+M+1] <= bnd, \
         T = J + 2*H*c + c^2; \
         X = vector(t); if (T < 0, T = -T; X[1] = 1, X[1] = 0); \
         for (i=2,t, while(T % B[i] == 0, X[i]++; T = T / B[i])); \
         if (T==1, \
            print("Relation found for c = ", c, ", X = ", X), \
            print("False alarm for c = ", c)); \
         ); \
      ); \
   )

QSMsieve(713057,50,50,1.5);
```

Chapter 7   Discrete Logarithms

5. Let $h \in \{1, 2, \ldots, n\}$ be the order of an element a in a finite group G of size n, and $m = \lceil \sqrt{n} \rceil$. Write $h - 1 = jm + i$ for some $i, j \in \{0, 1, 2, \ldots, m - 1\}$. We have $e = a^h = a(a^m)^j a^i = (a^m)^j a^{i+1}$ (where e is the identity in G), that is, $a^{-mj} = a^{i+1}$. For $i = 1, 2, \ldots, m$, we compute and store the pairs (i, a^i) in a table sorted with respect to the second component. Then, we precompute a^{-m}, and find the smallest $j \in \{0, 1, 2, \ldots, m-1\}$ for which $a^{-mj} = (a^{-m})^j$ is the second element of some pair in the precomputed table. If multiple values of i correspond to the same value as a^{-mj}, we take the smallest such i. But then, $h = \mathrm{ord}(a) = jm + i$ for these j and i.

11. Number the columns of the coefficient matrix C by $0, 1, 2, \ldots, t + 2M + 1$. Column 0 corresponds to the *prime* -1, Columns 1 through t to the small primes $p_1, p_2, \ldots, p_t$, and Columns $t + 1$ through $t + (2M + 1)$ to the $H + c$ values for $-M \leqslant c \leqslant M$. Suppose also that the last row corresponds to the *free* relation $\mathrm{ind}_g(p_j) = 1$ for some j. This row has only one non-zero entry. We now count the number of non-zero entries in the first $m - 1$ rows.

(1) The expected number of non-zero entries in Column 0 is $(m - 1)/2$.

(2) For $1 \leqslant j \leqslant t$, the expected number of non-zero entries in Column j is $(m - 1)/p_j$, since a randomly chosen integer is divisible by the prime p_j with probability $1/p_j$.

(3) In the submatrix consisting of the first $m - 1$ rows and the last $2M + 1$ columns, each row has exactly two non-zero entries corresponding to the two values c_1, c_2 for a smooth $T(c_1, c_2)$. Of course, we allow the possibility $c_1 = c_2$ during sieving (in which case there is only one non-zero entry in a row), but this situation occurs with a low probability, and we expect to get at most only a small constant number of such rows. In view of this, we neglect the effects of these rows in our count, and conclude that the expected number of non-zero entries in each of the last $2M + 1$ columns is $2(m - 1)/(2M + 1) \approx m/M$.

15. We first express some ag^α as a product of small ($\leqslant L[1/2]$) and medium-sized ($\leqslant L[2]$) primes. In order to compute the index of a medium-sized prime q, we fix some c_1 and vary c_2 in order to locate a value of $c_1 u + c_2 v$ which is q times an $L[1/2]$-smooth value. We first locate one particular value γ of c_2 for which $c_1 u + c_2 v$ is a multiple of q, and then sieve over all $c_2 = \gamma + c_2' q$ with $|c_2'| \leqslant L[1/2]$. Since the smoothness candidates are $O(\sqrt{p})$, this sieve is expected to produce a few $L[1/2]$-smooth values of $(c_1 u + c_2 v)/q$. This sieve runs in $L[1/2]$ time, as required. Among these smooth values just obtained, we locate one for which $c_2 - c_1 \sqrt{-r} \in \mathbb{Z}[\sqrt{-r}]$ is smooth over the small complex primes of norms $\leqslant L[1/2]$. But then, we get a relation of the form

$$\mathrm{ind}_g q + \mathrm{ind}_g((c_1 u + c_2 v)/q) \equiv \mathrm{ind}_g v + \mathrm{ind}_g(\Phi(c_2 - c_1 \sqrt{-r})) \pmod{p - 1}.$$

Since $(c_1 u + c_2 v)/q$ is smooth over rational primes $\leqslant L[1/2]$, and $c_2 - c_1 \sqrt{-r}$ is smooth over complex primes of norms $\leqslant L[1/2]$, we obtain $\mathrm{ind}_g q$ using the precomputed database of indices of factor-base elements.

18. Take a polynomial $f(x)$ of degree k with all irreducible factors of degrees $\leqslant m$. Let $i \geqslant 1$ be the largest degree of an irreducible factor of $f(x)$. Write

$$f(x) = g(x)u_1(x)^{e_1} u_2(x)^{e_2} \cdots u_s(x)^{e_s},$$

where each $u_j(x)$ is an irreducible polynomial of degree i, and all irreducible factors of $g(x)$ are of degrees $\leqslant i - 1$. The degree of $g(x)$ is $k - ri$, where $r = e_1 + e_2 + \cdots + e_s \in \mathbb{N}$. The number of possibilities for g is $N(k - ri, i - 1)$. For a given sum r, the total number of choices of $s, e_1, e_2, \ldots, e_s$ is $\binom{r+I(i)-1}{r}$ (since there are $I(i)$ irreducible polynomials in $\mathbb{F}_2[x]$ of degree i, and the number of solutions in non-negative integers of $x_1 + x_2 + \cdots + x_{I(i)} = r$ is the binomial coefficient $\binom{r+I(i)-1}{r}$). If we vary both i and r, we get

$$N(k, m) = \sum_{i=1}^{m} \sum_{r \geqslant 1} N(k - ri, i - 1)\binom{r + I(i) - 1}{r}.$$

19. The second stage of the LSM for $\mathbb{F}_{2^n}$ has the following sub-stages.

- For random $\alpha \in \{0, 1, 2, \ldots, 2^n - 2\}$, we try to express ag^α as a product of irreducible polynomials of degrees $\leqslant \frac{\sqrt{n \ln n}}{\sqrt{\ln 2}}$. The polynomial ag^α is of degree $\approx n$, and is smooth with probability $L[-\frac{\sqrt{\ln 2}}{2}]$. Therefore, $L[\frac{\sqrt{\ln 2}}{2}] = L[0.416\ldots]$ random values of α are expected to produce one smooth ag^α. The irreducible factors of this ag^α with degrees $\leqslant \frac{\sqrt{n \ln n}}{2\sqrt{\ln 2}}$ are in the factor base. So it suffices to compute the indices of all irreducible factors of ag^α with degrees larger than $\frac{\sqrt{n \ln n}}{2\sqrt{\ln 2}}$ but not larger than $\frac{\sqrt{n \ln n}}{\sqrt{\ln 2}}$. Let $u(x)$ be such an irreducible polynomial. Since each ag^α can be factored in polynomial time, this sub-stage runs in $L[0.416\ldots]$ time.

- To compute the index of $u(x)$, we find a polynomial $v(x) = x^{\lceil n/2 \rceil} + d(x) = u(x)w(x)$ with all irreducible factors of $w(x)$ having degrees $\leqslant \frac{\sqrt{n \ln n}}{2\sqrt{\ln 2}}$. If one multiple of $u(x)$ of the form $x^{\lceil n/2 \rceil} + d_0(x)$ is located, we search among the candidates $x^{\lceil n/2 \rceil} + d_0(x) + e(x)u(x)$ with $\deg e$ as small as possible. We expect to get one such $v(x)$ after $L[\frac{\sqrt{\ln 2}}{2}] = L[0.416\ldots]$ trials. Indeed, $\deg e$ is expected to be as small as $O(\log n)$. Using polynomial-time factoring algorithms, we finish this sub-stage in $L[0.416\ldots]$ time. The index of $w(x)$ is computed from the database of indices available from the first stage. If $x^{\lceil n/2 \rceil} + d(x)$ is in the factor base, we get $\text{ind}_g u$. This is usually not the case, so we go to the next sub-stage.

- We look at the following polynomial $y(x)$ with $\deg c(x) \leqslant \frac{\sqrt{n \ln n}}{2\sqrt{\ln 2}}$.

$$
\begin{aligned}
y(x) &\equiv v(x)(x^{\lceil n/2 \rceil} + c(x)) \equiv (x^{\lceil n/2 \rceil} + d(x))(x^{\lceil n/2 \rceil} + c(x)) \\
&\equiv x^e f_1(x) + (c(x) + d(x))x^{\lceil n/2 \rceil} + c(x)d(x) \pmod{f(x)}.
\end{aligned}
$$

Since both $\deg c$ and $\deg d$ are much smaller compared to $\lceil n/2 \rceil$, we expect to find one $y(x)$ with all irreducible factors of degrees $\leqslant \frac{\sqrt{n \ln n}}{2\sqrt{\ln 2}}$, after trying all of the $L[\frac{\sqrt{\ln 2}}{2}]$ polynomials $x^{\lceil n/2 \rceil} + c(x)$ in the factor base. The index of this $x^{\lceil n/2 \rceil} + c(x)$ is available from the first stage, so we obtain $\text{ind}_g u$. This sub-stage too runs in $L[\frac{\sqrt{\ln 2}}{2}] = L[0.416\ldots]$ time.

21. To avoid duplicate counting of triples, force the condition $c_1(2) \geqslant c_2(2) \geqslant c_3(2)$. We have $c_3(2) = c_1(2)$ XOR $c_2(2)$. Let $i = \deg c_1$ and $j = \deg c_2$. If $i < j$, then $c_1(2) < c_2(2)$, whereas if $i > j$, then $c_2(2) < c_3(2)$. Therefore, we must have $i = j$. If $i = j = -\infty$, then $c_1 = c_2 = c_3 = 0$. If $i \geqslant 0$, there are $1 + 2 + 3 + \cdots + 2^i = 2^i(2^i + 1)/2$ possibilities for choosing c_1 and c_2 satisfying $c_1(2) \geqslant c_2(2)$. Each such choice gives c_3 satisfying $c_2(2) \geqslant c_3(2)$. Therefore, the total number of tuples (c_1, c_2, c_3) in the CSM for binary fields is

$$1 + \sum_{i=0}^{m-1} 2^i(2^i + 1)/2 = 1 + \frac{1}{2}\sum_{i=0}^{m-1} 4^i + \frac{1}{2}\sum_{i=0}^{m-1} 2^i$$

$$= 1 + \frac{1}{2}\left(\frac{4^m - 1}{3}\right) + \frac{1}{2}(2^m - 1) = \frac{2}{3} \times 4^{m-1} + 2^{m-1} + \frac{1}{3}.$$

24. If $c_1 = 0$, then $\gcd(c_1, c_2) = c_2$. This gcd is 1 if and only if $c_2 = 1$. Since the pair $(0, 1)$ is not to be counted, we may assume that c_1 is non-zero. We proceed by induction on b. If $b = 1$, the only possibilities for (c_1, c_2) are $(1, 0)$ and $(1, 1)$. On the other hand, $2^{2 \times 1 - 1} = 2$, that is, the induction basis holds.

Now, let $b \geqslant 2$. The number of non-zero values of c_1 is $2^b - 1$, and the number of values of c_2 is 2^b. Among these $(2^b - 1)2^b = 2^{2b} - 2^b$ pairs (c_1, c_2), those with non-constant gcds should be discarded. Let $d(x) = \gcd(c_1, c_2)$ have degree $\delta \geqslant 1$. Write $c_1(x) = d(x)u_1(x)$ and $c_2(x) = d(x)u_2(x)$ with degrees of u_1 and u_2 less than $b - \delta$ and with $\gcd(u_1, u_2) = 1$. But u_1 is non-zero, so by the induction hypothesis, the number of choices for (u_1, u_2) is $2^{2(b-\delta)-1}$. The gcd $d(x)$ can be chosen as any one of the 2^δ polynomials of degree δ. Therefore, the desired number of pairs (c_1, c_2) is

$$2^{2b} - 2^b - \sum_{\delta=1}^{b-1}\left(2^\delta \times 2^{2(b-\delta)-1}\right) = 2^{2b} - 2^b - 2^{2b-1}\sum_{\delta=1}^{b-1}\frac{1}{2^\delta}$$

$$= 2^{2b} - 2^b - 2^{2b-1}\left(1 - \frac{1}{2^{b-1}}\right) = 2^{2b-1}.$$

26. Fix c_1, and sieve over different values of c_2. We use an array A indexed by c_2. A reasonable strategy is to let $c_2(x)$ stand for the array index $c_2(2)$. To avoid duplicate generation of relations, we would not use the entire array (of size 2^m). For simplicity, however, we do not impose restrictions on c_2 here.

We initialize the sieving array A by setting $A_{c_2(2)} = \deg T(c_1, c_2)$. Then, we take $(w(x), h)$ pairs. The condition $w(x)^h | T(c_1, c_2)$ gives a linear congruence in c_2 (for the fixed c_1). If $\bar{c}_2(x)$ is a solution for c_2 of this congruence, all solutions for c_2 are $c_2(x) = \bar{c}_2(x) + u(x)w(x)^h$ for all polynomials $u(x)$ of degrees $< \delta = m - h \deg w$. For each such solution of $c_2(x)$, we subtract $\deg w$ from $A_{c_2(2)}$. At the end of the sieving, we identify all A_{c_2} locations storing zero. These correspond to all the smooth values of $T(c_1, c_2)$ for the fixed c_1.

Computing $c_2(x) = \bar{c}_2(x) + u(x)w(x)^h$ for all $u(x)$ involves many multiplications of the form $u(x)w(x)^h$. Using the δ-dimensional gray code, we can avoid these multiplications. We step through the possibilities of $u(x)$ using the

sequence $G_0, G_1, G_2, \ldots, G_{2^\delta-1}$. The i-th bit-string $G_i = a_{\delta-1}a_{\delta-2}\ldots a_1a_0$ is
identified with the polynomial $u_i(x) = a_{\delta-1}x^{\delta-1} + a_{\delta-2}x^{\delta-2} + \cdots + a_1x + a_0$.
We let $u(x)$ vary in the sequence $u_0, u_1, \ldots, u_{2^\delta-1}$. Initially, $u_0(x) = 0$, and
$c_2(x) = \bar{c}_2(x)$. Subsequently, we have $u_i(x) = u_{i-1}(x) + x^{v_2(i)}$. Therefore,
$\bar{c}_2(x) + u_i(x)w(x)^h = (\bar{c}_2(x) + u_{i-1}(x)w(x)^h) + x^{v_2(i)}w(x)^h$. Instead of multiplying $u_i(x)$ with $w(x)^h$, we can shift $w(x)^h$ (this was precomputed) by $v_2(i)$,
and add that shifted value to the previous value of $c_2(x)$. Since one multiplication is usually somewhat slower than the combination of one shift and one
addition, we gain some speedup in the sieving process.

   Theoretically, this polynomial sieve runs in the same time as the variant
with trial division. However, since sieving replaces trial divisions by simpler
operations, we expect to get some practical speedup with sieving.

32. In the following code, the σ_i values are stored in the array S, the τ_i values in T,
and the multipliers in M. The first two of these arrays are randomly generated.
A (pseudorandom) sequence of triples (w_i, s_i, t_i) is generated and stored in the
array W. We have $w_i = g^{s_i}a^{t_i}$ for all i. The array W stores $\Theta(\sqrt{2^n-1})$ triples.
It is with high likelihood that periodicity is detected within these many iterations. After the generation of the sequence, the array W is sorted with respect to
the first components of the triples. Finally, a pass is made through the sorted
W in order to detect repetitions. If we use an $O(m \log m)$-time algorithm to
sort an array of size m, this implementation runs in $\tilde{O}(\sqrt{2^n-1})$ time.

```
update(w,r,S,T,M,f) = \
   local(k,s,t,v); \
   k = 1 + substpol(lift(w[1]),x,2) % r; \
   s = (w[2] + S[k]) % (2^n-1); \
   t = (w[3] + T[k]) % (2^n-1); \
   v = (w[1] * M[k]) % f; \
   return([v,s,t]);

genseq(f,g,a,r) = \
   local(n,i,W,s,t,v,asz); \
   n = poldegree(f); \
   S = vector(r,x,random(2^n-1)); \
   T = vector(r,x,random(2^n-1)); \
   M = vector(r); \
   for(i=1,r, M[i] = lift((g^S[i])*(a^T[i]))); \
   asz = ceil(3*sqrt(2^n-1)); W = vector(asz); \
   s = random(2^n-1); t = random(2^n-1); v = lift((g^s)*(a^t)); \
   W[1] = [v,s,t]; \
   for (i=2, asz, W[i] = update(W[i-1],r,S,T,M,f)); \
   return(lift(W));

DLPrho(f,g,a,r) = \
   local(n,W,asz,i,s,t); \
   n = poldegree(f); \
   asz = ceil(3*sqrt(2^n-1)); \
   W = genseq(f,g,a,r); \
   W = sort(W,asz); \
   for (i=1,asz-1, \
```

```
    if (W[i][1] == W[i+1][1], \
        s = W[i+1][2] - W[i][2]; \
        t = W[i][3] - W[i+1][3]; \
        if (gcd(t,2^n-1)==1, return(lift(Mod(s,2^n-1)/Mod(t,2^n-1)))); \
    ); \
)

f = Mod(1, 2)*x^7 + Mod(1, 2)*x + Mod(1, 2);
g = Mod(Mod(1, 2)*x^3 + Mod(1,2)*x^2, f);
a = Mod(Mod(1, 2)*x^5 + Mod(1, 2)*x^2 + Mod(1, 2), f);
DLPrho(f,g,a,10)
```

---

35. A GP/PARI code for the LSM for $\mathbb{F}_{2^n}$ is somewhat involved. We first implement a function getFB1(m) to return the list of all (non-constant) irreducible polynomials of $\mathbb{F}_2[x]$ with degrees $\leqslant m$.

In the sieving stage, we index array elements by a polynomial $c_2(x)$. We associate the array location $c_2(2)$ with $c_2(x)$. We need two functions int2poly and poly2int in order convert $c_2(x)$ to $c_2(2)$, and conversely.

The function DLPLSM2n implements incomplete sieving by the irreducible polynomials in B_1 (degrees $\leqslant m$). We restrict the degrees of $c(x)$ in $x^\nu + c(x)$ to less than m. In order to avoid generation of duplicate relations, we let (c_1, c_2) pairs satisfy the constraints $0 \leqslant c_1(2) \leqslant c_2(2) \leqslant 2^m - 1$. For each fixed $c_1(x)$, we sieve an array indexed by $c_2(x)$ in the range $c_1(2) \leqslant c_2(2) \leqslant 2^m - 1$. The array is initialized by the degrees of $T(c_1, c_2)$. Subsequently, for each irreducible polynomial $u(x) \in B_1$, we solve $u|T(c_1, c_2)$. This is equivalent to solving the linear congruence $(x^\nu + c_1(x))c_2(x) \equiv x^\epsilon f_1(x) + x^\nu c_1(x) \pmod{u(x)}$. For a solution $c(x)$ of $c_2(x)$, all solutions are $c_2(x) = c(x) + u(x)v(x)$, where $v(x)$ is any polynomial of degree $\leqslant m - \deg(u) - 1$. After all $u \in B$ are considered, we look at the array indices where the leftover degrees are $\leqslant \xi m$. These are shortlisted candidates in the LSM smoothness test (for the fixed c_1). The actual smooth values are identified by polynomial factorizations. Gordon and McCurley's trick (use of gray codes) is not implemented.

```
getFB1(m) = \
    local(t,B,B1,F,i,j); \
    B = vector(2^(m+1)); t = 0; \
    for (i=1,m, \
        F = factor(Mod(1,2)*x^(2^i)+x); \
        for (j=1,matsize(F)[1], \
            if (poldegree(F[j,1]) == i, t++; B[t] = F[j,1]); \
        ); \
    ); \
    B1 = vector(t); for (i=1,t, B1[i] = B[i]); return(B1);

int2poly(d) = \
    local(c,i); \
    c = Mod(0,2); i = 0; \
    while(d>0, c += Mod(d,2)*x^i; d = floor(d/2); i++); \
    return(c);
```

```
poly2int(c) = \
   local(d,i); \
   d = 0; \
   for (i=0,poldegree(c), \
      if (polcoeff(c,i)==Mod(1,2), d += 2^i); \
   ); \
   return(d);

DLPLSM2n(f,m,xi) = \
   local(n,f1,nu,epsilon,B1,t,A,c1,c2,d1,d2,T,i,j,u,v,w); \
   n = poldegree(f); f1 = f + Mod(1,2) * x^n; \
   nu = ceil(n/2); epsilon = 2*nu - n; \
   B1 = getFB1(m); t = matsize(B1)[2]; \
   A = vector(2^m); \
   for (d1=0,2^m-1, \
      c1 = int2poly(d1); \
      for(d2=0,d1-1, A[d2+1]=-1); \
      for(d2=d1,2^m-1, \
         c2 = int2poly(d2); \
         T = Mod(1,2)*x^epsilon*f1 + (c1+c2)*x^nu + c1*c2; \
         A[d2+1] = poldegree(T); \
      ); \
      for (i=1,t, \
         u = B1[i]; \
         v = Mod(1,2)*x^nu+c1; w = Mod(1,2)*x^epsilon*f1+x^nu*c1; \
         if (poldegree(gcd(v,u)) > 0, \
            if (poldegree(gcd(w,u)) > 0, \
               for(j=d1,2^m-1, A[j+1] -= poldegree(u)); \
            ); \
            , \
            c = lift(Mod(w,u)/Mod(v,u)); \
            for (i=0, 2^(m-poldegree(u))-1, \
               j = poly2int(int2poly(i)*u+c); \
               if (j >= d1, A[j+1] -= poldegree(u)); \
            ); \
         ); \
      ); \
      for (d2=d1,2^m-1, \
         if(A[d2+1] <= xi*m, \
            c2 = int2poly(d2); \
            print1("c1 = ", lift(c1), ", c2 = ", lift(c2)); \
            T = Mod(1,2)*x^epsilon*f1 + (c1+c2)*x^nu + c1*c2; \
            if (poldegree(T) > 0, \
               T = factor(T); \
               if (poldegree(T[matsize(T)[1],1])>m, print1(" false alarm")); \
            ); \
            print(""); \
         ); \
      ); \
   );

f = Mod(1,2)*x^17 + Mod(1,2)*x^3 + Mod(1,2);
g = Mod(1,2)*x^7 + Mod(1,2)*x^5 + Mod(1,2)*x^3 + Mod(1,2);
DLPLSM2n(f,4,1.5);
```

Chapter 8 Large Sparse Linear Systems

3. Since we now handle blocks of size ν (like 32 or 64), we can find multiple null-space vectors for A using a single execution of the Lanczos method. Assume that A is a symmetric $n \times n$ matrix over $\mathbb{F}_2$. We find a random $n \times \nu$ matrix $\mathbf{Y}$ over $\mathbb{F}_2$, compute $\mathbf{B} = A\mathbf{Y}$, and solve for an $n \times \nu$ matrix $\mathbf{X}$ satisfying

$$A\mathbf{X} = \mathbf{B}.$$

This essentially means that we are solving ν systems $A\mathbf{x} = \mathbf{b}$ simultaneously. All the formulas for the block Lanczos method can be rephrased in terms of this block solution. In particular, the final solution is given by

$$\mathbf{X} = \sum_{j=0}^{s-1} W_j (W_j^t A W_j)^{-1} W_j^t \mathbf{B},$$

where W_j are determined exactly as in the block Lanczos method presented in Section 8.4.1. The ν columns of $\mathbf{X} - \mathbf{Y}$ are random vectors in the null-space of A. Any $\mathbb{F}_2$-linear combination of these column vectors gives us a solution of the homogeneous system $A\mathbf{x} = \mathbf{0}$.

5. For simplicity, let us remove the subscript i. Given an $n \times \nu$ matrix V, we plan to compute a selection matrix S for which $W = VS$ consists of a maximal set of A-invertible columns of V. The matrix A is assumed to be symmetric.

First, notice that $W^t A W = S^t Q S$, where $Q = V^t A V$ is a symmetric $\nu \times \nu$ matrix. Let $r = \text{Rank}(Q)$. By standard linear-algebra techniques (like conversion to row echelon form), we determine a set of r linearly independent rows of Q. The selection matrix S is so constructed as to take the r columns of Q with the same indices as the rows just chosen. Let Q_{11} denote the $r \times r$ symmetric submatrix of Q obtained by throwing away all the rows and columns other than those chosen above.

For proving that $\text{Rank}(Q) = \text{Rank}(Q_{11})$, we rearrange Q, if necessary, to assume that the first r rows and the first r columns of Q are chosen as Q_{11}. We write $Q = \begin{pmatrix} Q_{11} & Q_{12} \\ Q_{21} & Q_{22} \end{pmatrix}$, where Q_{22} is a symmetric $(\nu - r) \times (\nu - r)$ matrix, and $Q_{12}^t = Q_{21}$. Since the last $\nu - r$ columns of Q are linearly dependent upon the first r columns, we have $\begin{pmatrix} Q_{12} \\ Q_{22} \end{pmatrix} = \begin{pmatrix} Q_{11} \\ Q_{21} \end{pmatrix} T$ for an $r \times (\nu - r)$ matrix T.

Therefore, $Q_{12} = Q_{11}T$ and $Q_{22} = Q_{21}T = Q_{12}^t T = (Q_{11}T)^t T = T^t Q_{11}T$, so

$$Q = \begin{pmatrix} I_r & 0 \\ T^t & I_{\nu-r} \end{pmatrix} \begin{pmatrix} Q_{11} & 0 \\ 0 & 0 \end{pmatrix} \begin{pmatrix} I_r & T \\ 0 & I_{\nu-r} \end{pmatrix},$$

that is, $\text{Rank}(Q) = \text{Rank}(Q_{11})$. Finally, note that $W^t A W = S^t Q S$ is the same as Q_{11} chosen above.

8. Fix any $i \in \{1, 2, 3, \ldots, n\}$. Consider the following $i \times i$ matrix formed by the vectors $\mathbf{z}^{(j)}$ for $j = 1, 2, \ldots, i$:

$$Z^{(i)} = \begin{pmatrix} 1 & z_1^{(2)} & z_1^{(3)} & \cdots & z_1^{(i)} \\ 0 & 1 & z_2^{(3)} & \cdots & z_2^{(i)} \\ 0 & 0 & 1 & \cdots & z_3^{(i)} \\ \vdots & \vdots & \vdots & \cdots & \vdots \\ 0 & 0 & 0 & \cdots & 1 \end{pmatrix}.$$

By Eqn (8.54), we have

$$T^{(i)} Z^{(i)} = \begin{pmatrix} \epsilon_1 & 0 & 0 & \cdots & 0 \\ * & \epsilon_2 & 0 & \cdots & 0 \\ * & * & \epsilon_3 & \cdots & 0 \\ \vdots & \vdots & \vdots & \cdots & \vdots \\ * & * & * & \cdots & \epsilon_i \end{pmatrix},$$

and, therefore, $\det T^{(i)} \det Z^{(i)} = \prod_{j=1}^{i} \epsilon_j$. But $\det Z^{(i)} = 1$, so

$$\det T^{(i)} = \prod_{j=1}^{i} \epsilon_j.$$

Levinson's algorithm succeeds if and only if all $\epsilon^{(i)}$ are non-zero. This is equivalent to the condition that all $T^{(i)}$ are invertible.

9. First, consider the homogeneous system $T\mathbf{x} = \mathbf{0}$. Take a random n-dimensional vector $\mathbf{y}$, and compute $\mathbf{u} = T\mathbf{y}$. Denote the columns of T as $\mathbf{c}_i$ for $i = 1, 2, \ldots, n$. Also, let $\mathbf{y} = (\, y_1 \quad y_2 \quad \cdots \quad y_n \,)^{t}$. We can write $\mathbf{u} = y_1 \mathbf{c}_1 + y_2 \mathbf{c}_2 + \cdots + y_n \mathbf{c}_n$. By hypothesis, the columns $\mathbf{c}_{r+1}, \mathbf{c}_{r+2}, \ldots, \mathbf{c}_n$ can be written as linear combinations of the columns $\mathbf{c}_1, \mathbf{c}_2, \ldots, \mathbf{c}_r$. Therefore, we can rewrite

$$\mathbf{u} = T\mathbf{y} = z_1 \mathbf{c}_1 + z_2 \mathbf{c}_2 + \cdots + z_r \mathbf{c}_r = T\mathbf{z},$$

where $\mathbf{z} = (\, z_1 \quad z_2 \quad \cdots \quad z_r \quad 0 \quad 0 \quad \cdots \quad 0 \,)^{t}$ for some $z_1, z_2, \ldots, z_r$. But then, $T(\mathbf{z} - \mathbf{y}) = \mathbf{0}$, that is, $\mathbf{z} - \mathbf{y}$ is a solution of $T\mathbf{x} = \mathbf{0}$.

Now, consider a non-homogeneous system $T\mathbf{x} = \mathbf{b}$. For a randomly chosen $\mathbf{y}$, set $\mathbf{u} = T\mathbf{y} + \mathbf{b}$. If $T\mathbf{x} = \mathbf{b}$ is solvable, $\mathbf{u}$ is T times a vector. As in the last paragraph, we can then find $\mathbf{z} = (\, z_1 \quad z_2 \quad \cdots \quad z_r \quad 0 \quad 0 \quad \cdots \quad 0 \,)^{t}$ with the bottom $n - r$ entries zero, such that $\mathbf{u} = T\mathbf{z}$. But then, $T(\mathbf{z} - \mathbf{y}) = \mathbf{b}$, that is, $\mathbf{z} - \mathbf{y}$ is a solution of $T\mathbf{x} = \mathbf{b}$.

To determine $z_1, z_2, \ldots, z_r$, note that $T^{(r)} \mathbf{z}' = \mathbf{u}'$, where $\mathbf{u}'$ and $\mathbf{z}'$ are the r-dimensional vectors consisting of the top r entries of $\mathbf{u}$ and $\mathbf{z}$, respectively. But $T^{(r)}$ is a Toeplitz matrix. Moreover, each $T^{(i)}$ is given to be invertible for $i = 1, 2, \ldots, r$. Therefore, Levinson's algorithm can be used to compute $\mathbf{z}'$.

10. Let $N = n/\nu$, that is, we solve an $N \times N$ system of blocks. Denote the (i, j)-th block by T_{i-j}. Let $T^{(i)}$ denote the $i \times i$ block matrix (i is the number of blocks)

at the top left corner of the block Toeplitz matrix T. Also, let $B^{(i)}$ denote the subvector of $\mathbf{b}$ consisting of the top νi entries. We iteratively compute $X^{(i)}$, $Y^{(i)}$ and $Z^{(i)}$ satisfying

$$T^{(i)}X^{(i)} = B^{(i)}, \quad T^{(i)}Y^{(i)} = \begin{pmatrix} \epsilon^{(i)} \\ 0_{(i-1)\times\nu} \end{pmatrix} \quad \text{and} \quad T^{(i)}Z^{(i)} = \begin{pmatrix} 0_{(i-1)\times\nu} \\ \epsilon'^{(i)} \end{pmatrix}.$$

Here, $X^{(i)}$ is a vector with νi entries, whereas $Y^{(i)}$ and $Z^{(i)}$ are $\nu i \times \nu$ matrices, and $\epsilon^{(i)}$ and $\epsilon'^{(i)}$ are $\nu \times \nu$ matrices to be chosen later. Initialization goes as:

$$X^{(1)} = T_0^{-1}B_1, \quad Y^{(1)} = Z^{(1)} = I_\nu, \quad \epsilon^{(1)} = \epsilon'^{(1)} = T_0.$$

For $i = 1, 2, \ldots, N-1$, we iteratively proceed as follows. We have

$$T^{(i+1)}\begin{pmatrix} Y^{(i)} \\ 0_{\nu\times\nu} \end{pmatrix} = \begin{pmatrix} \epsilon^{(i)} \\ 0_{(i-1)\nu\times\nu} \\ -\epsilon'^{(i)}\xi^{(i+1)} \end{pmatrix} \quad \text{and} \quad T^{(i+1)}\begin{pmatrix} 0_{\nu\times\nu} \\ Z^{(i)} \end{pmatrix} = \begin{pmatrix} -\epsilon^{(i)}\zeta^{(i+1)} \\ 0_{(i-1)\nu\times\nu} \\ \epsilon'^{(i)} \end{pmatrix},$$

where

$$\xi^{(i+1)} = -\left(\epsilon'^{(i)}\right)^{-1}(T_i \quad T_{i-1} \quad \cdots \quad T_1)Y^{(i)} \quad \text{and}$$

$$\zeta^{(i+1)} = -\left(\epsilon^{(i)}\right)^{-1}(T_{-1} \quad T_{-2} \quad \cdots \quad T_{-i})Z^{(i)}.$$

In order to update Y and Z as

$$Y^{(i+1)} = \begin{pmatrix} Y^{(i)} \\ 0_{\nu\times\nu} \end{pmatrix} + \begin{pmatrix} 0_{\nu\times\nu} \\ Z^{(i)} \end{pmatrix}\xi^{(i+1)} \quad \text{and} \quad Z^{(i+1)} = \begin{pmatrix} 0_{\nu\times\nu} \\ Z^{(i)} \end{pmatrix} + \begin{pmatrix} Y^{(i)} \\ 0_{\nu\times\nu} \end{pmatrix}\zeta^{(i+1)},$$

we need to take

$$\epsilon^{(i+1)} = \epsilon^{(i)}(I_{\nu\times\nu} - \zeta^{(i+1)}\xi^{(i+1)}) \quad \text{and} \quad \epsilon'^{(i+1)} = \epsilon'^{(i)}(I_{\nu\times\nu} - \xi^{(i+1)}\zeta^{(i+1)}).$$

Finally, we update the solution as

$$X^{(i+1)} = \begin{pmatrix} X^{(i)} \\ \mathbf{0}_\nu \end{pmatrix} + Z^{(i+1)}\left(\epsilon'^{(i+1)}\right)^{-1}(B^{(i+1)} - \eta^{(i+1)}),$$

where

$$\eta^{(i+1)} = (T_i \quad T_{i-1} \quad \cdots \quad T_1)X^{(i)}.$$

12. **(a)** Without loss of generality, we may assume that $A_1, A_2, \ldots, A_r$ are linearly independent, and the remaining columns belong to the subspace V generated by the first r columns. Let $(c_1, c_2, \ldots, c_n)$ be a linear dependency of the columns. We have $\sum_{i=r+1}^n c_i A_i \in V$, that is, there is a unique way of expressing this vector as a linear combination of $A_1, A_2, \ldots, A_r$ (these vectors form a basis of V). This means that given any $c_{r+1}, c_{r+2}, \ldots, c_n$ in a linear dependency, the coefficients $c_1, c_2, \ldots, c_r$ are fixed. Moreover, all of $c_{r+1}, c_{r+2}, \ldots, c_n$ cannot be zero in a linear dependency, since that implies that $A_1, A_2, \ldots, A_r$ are not linearly independent.

(b) We have $\mathrm{E}(l) + 1 = \mathrm{E}(q^d) \geqslant q^{\mathrm{E}(d)}$ (since arithmetic mean of a non-negative-valued random variable is no less than its geometric mean). Therefore, $\mathrm{E}(d) \leqslant \log_q(\mathrm{E}(l) + 1)$. Finally, note that $d + r = n$ (a constant), so $\mathrm{E}(r) \geqslant n - \log_q(\mathrm{E}(l) + 1)$.

(c) Fix a non-zero tuple $(c_1, c_2, \ldots, c_n) \in \mathbb{F}_q^n$ with exactly k non-zero entries. We first determine the probability that this is a linear dependency of the columns of A. Let $c_{i_1}, c_{i_2}, \ldots, c_{i_k}$ be the non-zero entries in the tuple. Now, $c_{i_1} A_{i_1} + c_{i_2} A_{i_2} + \cdots + c_{i_k} A_{i_k} = \mathbf{0}$ gives n equations from the n rows, each of the form $c_{i_1} a_1 + c_{i_2} a_2 + \cdots + c_{i_k} a_k = 0$, where a_j are matrix entries following the given probability distribution. If P_k denotes the probability that such a linear combination of the scalars is zero, then the above linear combination of the columns is zero with probability P_k^n. Now, a tuple $(c_1, c_2, \ldots, c_n) \in \mathbb{F}_q^n$ with exactly k non-zero entries can be chosen in $\binom{n}{k}(q-1)^k$ ways. Finally, we also vary k to obtain

$$\mathrm{E}(l) = \sum_{k=1}^{n} \binom{n}{k}(q-1)^k P_k^n.$$

Determination of P_k depends upon the probability distribution of the matrix entries. For a case study, look at: Johannes Blömer, Richard Karp and Emo Welzl, The rank of sparse random matrices over finite fields, *Random Structures and Algorithms*, Vol. 10, 407–419, 1997.

Chapter 9 Public-Key Cryptography

1. (d) Let H' be an n-bit cryptographic hash function. Define an $(n+1)$-bit hash function as $H(x) = \begin{cases} 0x & \text{if } x \text{ is of bit length } n, \\ 1H'(x) & \text{otherwise.} \end{cases}$ For half of the $(n+1)$-bit strings (those starting with 0), it is easy to find a preimage. However, it is difficult to locate a second preimage of a hash value h. This is because if h starts with 0, it does not have a second preimage at all. On the other hand, if h starts with 1, it is difficult to find a second preimage under H'.

5. We have $c_1 \equiv m_1^e \pmod{n}$ and $c_2 \equiv m_2^e \pmod{n}$, so $c_1 c_2 \equiv (m_1 m_2)^e \pmod{n}$.

Suppose that an adversary Eve wants the plaintext m_1 corresponding to the ciphertext c_1. Bob is willing to decrypt ciphertext messages to Eve, provided that the ciphertext message is not c_1 (or a simple function of it, like $\pm c_1^{\pm 1} \pmod{n}$). Eve generates a random m_2, and encrypts it with Bob's public key to get c_2. Eve asks Bob to decrypt $c_1 c_2 \pmod{n}$ which conceals c_1 by the random mask c_2. When Bob presents the corresponding plaintext message $m_1 m_2$, Eve retrieves m_1 by multiplying this with $m_2^{-1} \pmod{n}$.

One way to prevent such an attack is to adopt a convention on the structure of plaintext messages. Since Eve has no control over $m_1 m_2$, it is with high probability that $m_1 m_2$ would not satisfy the structural constraints, and the decryption request is, therefore, rejected by Bob. A possible structural constraint is to use some parity-check bits in a message. If the number of such bits is $\geqslant 100$, then a random $m_1 m_2$ passes all these parity checks with a probability of only $1/2^{100}$.

6. We have $c_i \equiv m^e \pmod{n_i}$ for $i = 1, 2, \ldots, e$. By the CRT, one computes $c \pmod{n_1 n_2 \cdots n_e}$ such that $c \equiv c_i \pmod{n_i}$ for all $i = 1, 2, \ldots, e$. Since $m^e < n_1 n_2 \ldots n_e$, we have $c = m^e$ (as integers), that is, m can be obtained by taking the integer e-th root of c.

In order to avoid this (and many other) attacks, it is often advisable to append some random bits to every plaintext being encrypted. This process is called *salting*. Only a few random bits dramatically reduce the practicality of mounting the above attack, even when e is as small as three.

11. Let $n = \text{ord } g = uv$ with u small (may be prime). An active eavesdropper Eve mounts the following attack.

Alice selects $d \in \{2, 3, \ldots, n-1\}$ and sends g^d to Bob.
Eve intercepts g^d, computes and sends to Bob the element $(g^d)^v$.
Bob selects $d' \in \{2, 3, \ldots, n-1\}$ and sends $g^{d'}$ to Alice.
Eve intercepts $g^{d'}$, computes and sends to Bob the element $(g^{d'})^v$.
Alice and Bob compute the shared secret $g^{vdd'}$.

The order of the shared secret $g^{vdd'}$ divides u and is small. That is, there are only a few possibilities for the shared secret. If Alice and Bob make a future communication using the shared secret, the correct possibility can be worked out by Eve by an exhaustive search.

This attack can be avoided if $n = \text{ord } g$ does not have any small divisor. For example, we choose a suitably large finite field $\mathbb{F}_q$ with $|\mathbb{F}_q|$ having a prime factor n of bit size $\geqslant 160$. Although q may be so large that $q - 1$ cannot be (completely) factored in feasible time, a knowledge of n alone suffices. In that case, we work in the subgroup of $\mathbb{F}_q^*$ of order n.

16. Let (s, t) be the ElGamal signature on the message representative m, that is, $s \equiv g^{d'} \pmod{p}$ (for some d') and $t \equiv (m - ds)d'^{-1} \pmod{p-1}$. Take any k coprime to $p - 1$, and let $s' \equiv g^{kd'} \equiv s^k \pmod{p}$ and $t' \equiv (s^{k-1}k^{-1})t \equiv (ms^{k-1} - ds')(kd')^{-1} \pmod{p-1}$, that is, (s', t') is again a valid signature on the message representative ms^{k-1}.

20. The ECDSA signature on M with $m = H(M)$ is (s, t), where $S = d'G$ (for a random d'), $s \equiv x(S) \pmod{n}$, and $t \equiv (m + ds)d'^{-1} \pmod{n}$. If we choose $n - d'$ in place of d', we get $S' = -d'G = -S$, but $x(S') = x(S)$, whereas $(m + ds)(n - d')^{-1} \equiv -(m + ds)d'^{-1} \pmod{n}$, that is, $(s, n - t)$ is again a valid signature on M.

Now, assume that $n < q$ (this may happen even if the cofactor h is 1). This means that a given s may have multiple representatives in $\mathbb{F}_q$, and each representative produces two signatures with the same s.

22. (b) The verification of the i-th DSA signature (s_i, t_i) involves computing $w_i \equiv t_i^{-1} \pmod{q}$, $u_i \equiv m_i w_i \pmod{q}$ and $v_i \equiv s_i w_i \pmod{q}$, and checking whether $s_i \equiv (g^{u_i} y^{v_i} \pmod{p}) \pmod{q}$. Multiplying the k verification equations gives

$$\prod_{i=1}^{k} s_i \equiv \left(\left(g^{\sum_{i=1}^{k} u_i} \right) \left(y^{\sum_{i=1}^{k} v_i} \right) \pmod{p} \right) \pmod{q}.$$

This is the equation for the batch verification of the k DSA signatures. The sums of u_i and of v_i should be computed modulo q.

Modular exponentiations are the most expensive operations in DSA verification. The number of modular exponentiations drops from $2k$ (individual verification) to 2 (batch verification). So the speedup achieved is about k.

It is evident that individual signatures may be faulty, but their product qualifies for being verified as a batch. For randomly chosen messages and signatures, the probability of such an occurrence is, however, quite low.

(c) In Part (b), we assumed that y remains constant for all signatures. If not, we should modify the batch-verification criterion to

$$\prod_{i=1}^{k} s_i \equiv \left(\left(g^{\sum_{i=1}^{k} u_i} \right) y_1^{v_1} y_2^{v_2} \cdots y_k^{v_k} \pmod{p} \right) \pmod{q}.$$

This requires $k+1$ exponentiations, and the speedup is about $2k/(k+1) \leqslant 2$.

23. The following signature scheme is derived from a zero-knowledge authentication scheme. We suppose that Alice wants to sign a message M. We use the symbol $\|$ to represent concatenation of strings.

Alice generates a random commitment t and a witness w to t.

Alice uses a hash function H to generate the challenge $c = H(M\|w)$ herself. Alice's signature on M is the pair (w, r), where r is Alice's response on c.

A verifier checks whether r is consistent with the challenge $c = H(M\|w)$.

---

26. For a valid encrypted message (U, V, W), we have $U + tV = kP_{Bob}P + kY_1 + tkY_2 = kP_{Bob}P + ks_1P + tks_2P = k(P_{Bob} + s_1 + ts_2)P$, that is, $e(U+tV, K) = e(k(P_{Bob}+s_1+ts_2)P, (P_{Bob}+s_1+ts_2)^{-1}P) = e(kP, P) = e(P, P)^k$. Therefore, M is retrieved by Bob as $W/e(U + tV, K)$.

28. Alice computes $f_B = \sum_{i=0}^{k} P_B^i V_i$ and $f_C = \sum_{i=0}^{k} P_C^i V_i$. We have $f_B = \sum_{i=0}^{k} P_B^i d_i P = (\sum_{i=0}^{k} d_i P_B^i)P = f(P_B)P = D_B P$. Likewise, $f_C = D_C P$. Therefore, Alice obtains $e(f_B, f_C)^{D_A} = e(D_B P, D_C P)^{D_A} = e(P, P)^{D_A D_B D_C}$. Bob and Carol can analogously compute this shared value.

In order to reconstruct the secret polynomial $f(x)$ uniquely using interpolation, we need $f(P_i)$ for k known values of P_i. If n is the number of entities participating in the network, and if all of these entities disclose their private keys to one another, then too the polynomial cannot be uniquely determined, provided that $k > n$.

30. Let (U, V) be a valid signature of Bob on the message M. Then,

$$V = (t + H_2(M, U))D_{Bob} = s(t + H_2(M, U))P_{Bob}$$
$$= s(tP_{Bob} + H_2(M, U)P_{Bob}) = s(U + H_2(M, U)P_{Bob}).$$

Therefore,

$$e(P, V) = e(P, s(U + H_2(M, U)P_{Bob})) = e(sP, U + H_2(M, U)P_{Bob}),$$

that is, the signature is verified if and only if the following condition holds:

$$e(P, V) = e(P_{pub}, U + H_2(M, U)P_{Bob}).$$

If $U + H_2(M, U)P_{Bob} = aP$, the above condition is equivalent to checking whether $e(P, saP) = e(sP, aP)$. Under the gap Diffie–Hellman assumption, it is easy to verify a signature. However, generating a signature involves solving an instance of the (computational) bilinear Diffie–Hellman problem.

Each Cha–Cheon signature generation requires two point multiplications in G, whereas each Paterson signature generation requires three such point multiplications. Verification involves two pairing computations in the Cha–Cheon scheme, and three pairing computations and two field exponentiations in the Paterson scheme. It follows that the Cha–Cheon scheme is considerably more efficient than the Paterson scheme.

32. Since $g = e(P, Q) = e((d_1 + d_2t + m)^{-1}P, (d_1 + d_2t + m)Q) = e(\sigma, d_1Q + td_2Q + mQ)$, the signature is verified if and only if we have

$$e(\sigma, Y_1 + tY_2 + mQ) = g.$$

This verification procedure involves two point multiplications (and two additions) in G_2 and only one pairing computation. On the contrary, BLS verification involves two pairing computations, and would be slower than this Boneh–Boyen verification if one pairing computation takes more time than two point multiplications.

38. We assume that all ECDSA system parameters are fixed beforehand. The following functions sign and verify messages given these parameters and the appropriate keys as arguments.

```
ECDSAsgn(m,d,E,p,n,G) = \
   local(d1,S,s,t); \
   d1 = 2 + random(n-2); \
   S = ellpow(E,Mod(G,p),d1); \
   s = lift(S[1]) % n; \
   t = lift(Mod(m + d * s, n) / Mod(d1, n)); \
   return([s,lift(t)]);

ECDSAvrf(m,S,Y,E,p,n,G) = \
   local(w,u1,u2,V,v); \
   w = Mod(S[2],n)^(-1); u1 = Mod(m,n) * Mod(w,n); u2 = Mod(S[1],n) * w; \
   V = elladd(E,ellpow(E,Mod(G,p),lift(u1)),ellpow(E,Y,lift(u2))); \
   v = lift(V[1]) % n; \
   if (v == S[1], return(1), return(0));
```

Use of these functions is now demonstrated for the data in Example 9.8.

```
p = 997;
E = ellinit([Mod(0,p),Mod(0,p),Mod(0,p),Mod(3,p),Mod(6,p)]);
n = 149;
G = [246,540];
d = 73;
Y = ellpow(E,G,d); print("Y = ",lift(Y));
m = 123;
S = ECDSAsgn(m,d,E,p,n,G); print("S = ", S);
ECDSAvrf(m,S,Y,E,p,n,G)
ECDSAvrf(m,[96,75],Y,E,p,n,G)
ECDSAvrf(m,[96,112],Y,E,p,n,G)
```

---

39. In this exercise, we assume that the parameters E, q, k, r, P, Q and the relevant hash functions are defined globally. If symmetric pairing can be used, we invoke the distorted Weil pairing dWeil(), otherwise we invoke Weil(). The four stages of the Boneh–Franklin IBE scheme can be implemented as follows.

```
BF_TAkeys() = \
    local(s,Ppub); \
    s = 2 + random(r-2); \
    Ppub = ellpow(E,P,s); \
    return([s,lift(Ppub)]);

BFreg(s,PBob) = \
    local(DBob); \
    DBob = ellpow(E,PBob,s); \
    return(lift(DBob));

BFenc(M,Ppub,PBob) = \
    local(g,a,U,V); \
    g = dWeil(PBob,Ppub); \
    a = 2 + random(r-2);\
    U = ellpow(E,P,a);\
    h = lift(lift(g^a)); \
    V = H2(h); \
    V = bitxor(V,M); \
    return([U,V]);

BFdec(C,DBob) = \
    local(h,M); \
    h = H2(lift(lift(dWeil(DBob,C[1])))); \
    M = bitxor(C[2],h); \
    return(M);
```

Index

For Product Safety Concerns and Information please contact our EU
representative GPSR@taylorandfrancis.com
Taylor & Francis Verlag GmbH, Kaufingerstraße 24, 80331 München, Germany